CIRCUIT INTERRUPTION

ELECTRICAL ENGINEERING AND ELECTRONICS

A Series of Reference Books and Textbooks

Editors

Marlin O. Thurston
Department of Electrical
Engineering
The Ohio State University
Columbus, Ohio

William Middendorf
Department of Electrical
and Computer Engineering
University of Cincinnati
Cincinnati, Ohio

1. Rational Fault Analysis, *edited by Richard Saeks and S. R. Liberty*
2. Nonparametric Methods in Communications, *edited by P. Papantoni-Kazakos and Dimitri Kazakos*
3. Interactive Pattern Recognition, *Yi-tzuu Chien*
4. Solid-State Electronics, *Lawrence E. Murr*
5. Electronic, Magnetic, and Thermal Properties of Solid Materials, *Klaus Schröder*
6. Magnetic-Bubble Memory Technology, *Hsu Chang*
7. Transformer and Inductor Design Handbook, *Colonel Wm. T. McLyman*
8. Electromagnetics: Classical and Modern Theory and Applications, *Samuel Seely and Alexander D. Poularikas*
9. One-Dimensional Digital Signal Processing, *Chi-Tsong Chen*
10. Interconnected Dynamical Systems, *Raymond A. DeCarlo and Richard Saeks*
11. Modern Digital Control Systems, *Raymond G. Jacquot*
12. Hybrid Circuit Design and Manufacture, *Roydn D. Jones*
13. Magnetic Core Selection for Tranformers and Inductors: A User's Guide to Practice and Specification, *Colonel Wm. T. McLyman*
14. Static and Rotating Electromagnetic Devices, *Richard H. Engelmann*
15. Energy-Efficient Electric Motors: Selection and Application, *John C. Andreas*
16. Electromagnetic Compossibility, *Heinz M. Schlicke*
17. Electronics: Models, Analysis, and Systems, *James G. Gottling*

18. Digital Filter Design Handbook, *Fred J. Taylor*
19. Multivariable Control: An Introduction, *P. K. Sinha*
20. Flexible Circuits: Design and Applications, *Steve Gurley, with contributions by Carl A. Edstrom, Jr., Ray D. Greenway, and William P. Kelly*
21. Circuit Interruption: Theory and Techniques, *edited by Thomas E. Browne, Jr.*

Other Volumes in Preparation

CIRCUIT INTERRUPTION
Theory and Techniques

Edited by THOMAS E. BROWNE, JR.
*Westinghouse Research
and Development Center
Pittsburgh, Pennsylvania*

 MARCEL DEKKER, INC.   NEW YORK · BASEL

Library of Congress Cataloging in Publication Data
Main entry under title:

Circuit interruption.

(Electrical engineering and electronics; 21)
Includes index.
1. Electric circuit-breakers. 2. Electric arc.
3. Electric contacts. I. Browne, Thomas E. [date]
II. Series.
TK2842.C57 1984   621.31'7   84-12087
ISBN 0-8247-7177-X

Copyright © 1984 by Marcel Dekker, Inc.   All Rights Reserved

Neither this book nor any part may be reproduced or transmitted in any form or by any means, electronic or mechanical, including photocopying, microfilming, and recording, or by any information storage and retrieval system, without permission in writing from the publisher.

Marcel Dekker, Inc.
270 Madison Avenue, New York, New York 10016

Current printing (last digit):
10 9 8

Printed in the United States of America

DEDICATED

to the memory of

Joseph Slepian

Teacher, Inventor and Pioneer in Furthering
the Understanding of Circuit-Interrupting Devices

Preface

In the early growth of electric power systems, there was fairly complete understanding of mechanical and electromagnetic theory as applied to rotating machinery. The circuit-interrupting devices, however, were developed by "cut and try" methods and they all relied on the somewhat mysterious properties of the electric arc to switch the current. As short-circuit currents increased, some of the interrupting failures were quite spectacular. As an indication of the early state of the theory, while a young engineer I was told by an experienced breaker designer that oil circuit breaker ratings could be increased mainly by making the tanks stronger! Although patent records show much activity in circuit breaker development around the beginning of this century, it was only in the 1920s (see Chapter 1) that worldwide research began to solve some of the mysteries of high-power circuit interruption. Since then an imposing amount of published literature on arc interruption in circuit breakers has emerged, but relatively few books have been written on the subject (see Chapter 1, General References).

The aim of this book is to provide information to practicing engineers and students about the history and current state of circuit interruption theory and techniques by drawing on the knowledge and experience of twenty-two contributors who are or have recently been actively involved in research, development, or application of modern circuit breakers. The large number of contributors reflects the great diversity that characterizes the circuit interruption field today. Even in this book, oil and compressed-air circuit breakers are not covered because they are being superseded by the sulfur hexafluoride (SF_6) puffer breaker (Chapter 10). Power fuses are also not included since developments in the theory of these devices have not been extensive.

Because of the emphasis on physical principles, no attempt has been made to furnish complete descriptions of all presently manufactured circuit interrupters; only illustrative examples have been described. These examples have invariably been drawn from equipment with which these authors are most familiar, but products of other manufacturers are for the most part generally similar.

I wish to express my appreciation to my past and present associates, to many of whom I owe much of my own education in this field. My first mentor was the late Dr. Joseph Slepian, to whose memory this book is dedicated. Also appreciated is the encouragement and support of many others, including the large group of other contributors who have taken time from pressing schedules, and sometimes from retirement. Special mention is due Dr. Paul Slade and Dr. Clive Kimblin, who served together with the writer as an editorial committee. Dr. Peter Chantry gave valuable assistance in reviewing the Appendix, which I originally used as a teaching aid some years ago.

Essential to this undertaking has been the support from the Power Interruption and Plasma Systems Department of the Westinghouse R & D Center. I also appreciate the support received from other Westinghouse Electric Corporation divisions, especially the Power Circuit Breaker and Switchgear Divisions and the Transmission and Distribution Systems Engineering Department. Many illustrations have been taken from internal Westinghouse reports and descriptive bulletins.

I especially wish to thank Mrs. Toni McElhaney, who has been responsible for most of the typing, capably aided by Mrs. Bettyann Toth. Similar support has come from staffs of the University of Pittsburgh and Carnegie-Mellon University.

On the purely personal side, sincere thanks are due my wife, Edna, for her encouragement and forbearance during these months and years of delay in my entering a time of real retirement.

Thomas E. Browne, Jr.

Contents

Preface	v
Contributors	xiii

1 INTRODUCTION
Thomas E. Browne, Jr.

		1
1.1	History	1
1.2	Mechanics	5
1.3	Progress in Understanding	6
1.4	The Future	7
References		8

2 ELECTRICAL AND SYSTEM ASPECTS
R. Gerald Colclaser, Jr.

		11
2.1	Electrical Environment	11
2.2	Transient Switching Laws	12
2.3	Arc Characteristics	16
2.4	Direct-Current Interruption	21
2.5	Alternating-Current Interruption	25
References		37

3 CIRCUIT BREAKER APPLICATION
Charles L. Wagner

		39
3.1	Circuit Breaker Standards	40
3.2	Standard Rating Tables	42
3.3	Rated Voltage	46
3.4	Insulation Levels	47
3.5	Continuous Current Rating	62
3.6	Interrupting Current Ratings	65
3.7	Momentary or Closing and Latching Rating	82
3.8	Reclosing Duty Factors	83

3.9	Rated Interrupting Time	84
3.10	Transient Recovery Voltage	85
3.11	Capacitance Switching	105
3.12	Other Application Considerations	122
3.13	Summary	131
References		132

4 NATURE OF THE ELECTRIC ARC 135
Joachim V. R. Heberlein, Clive W. Kimblin, and Anthony Lee

4.1	Introduction	135
4.2	Physical Description	136
4.3	Description of the Recovery Phenomena	152
4.4	Summary	154
References		154

5 PHYSICAL THEORY OF THE ARC IN A GAS BLAST 157
John J. Lowke

5.1	Introduction	157
5.2	Arcs at High Current	159
5.3	Transient Arcs at Low Current	169
5.4	Summary	183
References		183

6 CALCULATION OF ARC-CIRCUIT INTERACTION 187
Leslie S. Frost and Thomas E. Browne, Jr.

Part 1	Computer Solution of Arc Models with Connected Circuit	188
6.1	Mathematical Models	188
6.2	Combined Model Solution	192
Part 2	Approximate Analytical Model Solutions	211
6.3	Impressed Current and Voltage	211
6.4	Interaction of the Arc with the Circuit	218
6.5	Relation of Model Parameters to Current Rate of Change	223
6.6	Use of the Cassie Criterion	227
6.7	Modeling of the Puffer Breaker	235
6.8	Nomenclature	236
References		239

7 POSTARC DIELECTRIC RECOVERY IN A BLAST ARC 241
David T. Tuma

7.1	Introduction	241
7.2	Experimental Investigations	243
7.3	Theoretical Investigations	257
7.4	Summary	271
References		271

Contents

8	DIELECTRIC PROPERTIES OF CIRCUIT BREAKERS Alan H. Cookson, Lyon Mandelcorn, Roy E. Wootton, and J. Franklin Roach	275
	8.1 Introduction	275
	8.2 Field Analysis	276
	8.3 Insulation Properties of Materials	278
	8.4 Bushings	337
	References	348
9	SF_6 BREAKER RESEARCH AND DEVELOPMENT Robert E. Friedrich	353
	9.1 Introduction	354
	9.2 Properties of SF_6	354
	9.3 Early Research	355
	9.4 Early SF_6 Applications	359
	9.5 Recent Research	367
	9.6 Materials Research	369
	9.7 High-Power Single-Pressure Breakers	369
	9.8 Conclusions	370
	References	373
10	SINGLE-PRESSURE SF_6 CIRCUIT BREAKERS Ben J. Calvino	377
	10.1 Introduction	377
	10.2 SF_6 Single-Pressure Interrupting Element	381
	10.3 Development of Single-Pressure SF_6 Interrupters	383
	10.4 Electrical Interruption Characteristics of a Puffer Circuit Breaker	397
	10.5 Mechanical and Electrical Reliability	404
	10.6 Industrial Circuit Breaker Configurations	408
	10.7 Conclusions	418
	10.8 Additional Considerations for Sec. 10.3.1	419
	References	423
11	MAGNETIC AIR CIRCUIT BREAKERS Thomas E. Browne, Jr. and James D. Finley	425
	11.1 Introduction	425
	11.2 Examples	426
	11.3 Theory	432
	References	452

12	INTERRUPTION IN VACUUM *Clive W. Kimblin, Paul G. Slade, and Roy E. Voshall*	455
	12.1 Historical Background	456
	12.2 Vacuum Interrupter Description	457
	12.3 Electrode Phenomena in Vacuum	459
	12.4 Arcing and Interruption in Vacuum Interrupters during an Ac Wave	462
	12.5 Range of Present-Day Applications	472
	References	480
13	VACUUM CIRCUIT BREAKER APPLICATION AND SWITCHING SURGE PROTECTION *John F. Perkins*	487
	13.1 Historical Perspective	488
	13.2 Characteristics of Vacuum Circuit Breakers	488
	13.3 Current Chopping	491
	13.4 Multiple Reignitions and Voltage Escalation	494
	13.5 Virtual Current Chopping	496
	13.6 Circuits Most Subject to Surge	498
	13.7 Determination of Switching Transient Voltages	498
	13.8 Vacuum Interrupter Surge Program Input Parameters	505
	13.9 Validation of the Computer Program by Switching Tests	510
	13.10 Typical Switching Applications	514
	13.11 Recommended Practice for System Protection	520
	13.12 Summary	524
	References	524
14	MOLDED-CASE LOW-VOLTAGE CIRCUIT BREAKERS *Anthony Lee and Paul G. Slade*	527
	14.1 Introduction	527
	14.2 Breaker Functions	532
	14.3 Breaker Components	535
	14.4 Interruption Process	549
	14.5 Trends in Low-Voltage Molded-Case Circuit Breakers	555
	References	561
15	ELECTRIC CONTACT PHENOMENA *Paul G. Slade*	565
	15.1 Introduction	565
	15.2 Contact Fundamentals	566
	15.3 Considerations in Contact Design	582
	15.4 Classes of Contact Materials	592
	References	603

Contents

16 THE MECHANICAL OPERATION OF POWER CIRCUIT BREAKERS RATED OVER 15kV   605
Roswell C. Van Sickle

16.1 Circuit Breaker Functions Affecting Mechanical Operations   606
16.2 Improving Opening Performance   606
16.3 Operating Mechanisms   607
16.4 Operating Linkage   610
16.5 Descriptions of Pneumatic Mechanisms   612
16.6 Resistor Switching Mechanism   617
16.7 Conclusions   619
References   620

17 INTERRUPTION TESTING   621
C. Donald Fahrnkopf and David Vallo

17.1 Introduction   621
17.2 Types of Interruption Tests   624
17.3 Interruption-Testing Techniques   627
References   639

APPENDIX: BASIC CONCEPTS OF GASEOUS CONDUCTION   641
Thomas E. Browne, Jr.

A.1 Nature of Gaseous Conductors: Kinetic Theory   642
A.2 Movement of Ions and Electrons   657
A.3 Ionization and Deionization   669
A.4 Self-Maintained Discharges   681
References   683

INDEX   685

Contributors

THOMAS E. BROWNE, JR.* Westinghouse Research and Development Center, Pittsburgh, Pennsylvania

BEN J. CALVINO Westinghouse Electric Corporation, Trafford, Pennsylvania

R. GERALD COLCLASER, JR. University of Pittsburgh, Pittsburgh, Pennsylvania

ALAN H. COOKSON Westinghouse Research and Development Center, Pittsburgh, Pennsylvania

C. DONALD FAHRNKOPF† Westinghouse Electric Corporation, East Pittsburgh, Pennsylvania

JAMES D. FINLEY Westinghouse Electric Corporation, East Pittsburgh, Pennsylvania

ROBERT E. FRIEDRICH† Westinghouse Electric Corporation, Trafford, Pennsylvania

LESLIE S. FROST* Westinghouse Research and Development Center, Pittsburgh, Pennsylvania

JOACHIM V. R. HEBERLEIN Westinghouse Research and Development Center, Pittsburgh, Pennsylvania

CLIVE W. KIMBLIN Westinghouse Research and Development Center, Pittsburgh, Pennsylvania

*Retired
†Currently Consulting Engineer, Pittsburgh, Pennsylvania

ANTHONY LEE Westinghouse Research and Development Center, Pittsburgh, Pennsylvania

JOHN J. LOWKE Commonwealth Scientific and Industrial Research Organization, Sydney, New South Wales, Australia

LYON MANDELCORN Westinghouse Research and Development Center, Pittsburgh, Pennsylvania

JOHN F. PERKINS Westinghouse Electric Corporation, East Pittsburgh, Pennsylvania

J. FRANKLIN ROACH Westinghouse Research and Development Center, Pittsburgh, Pennsylvania

PAUL G. SLADE Westinghouse Research and Development Center, Pittsburgh, Pennsylvania

DAVID T. TUMA† Carnegie-Mellon University, Pittsburgh, Pennsylvania

DAVID VALLO Westinghouse Electric Corporation, East Pittsburgh, Pennsylvania

ROSWELL C. VAN SICKLE* Westinghouse Electric Corporation, Trafford, Pennsylvania

ROY E. VOSHALL Westinghouse Research and Development Center, Pittsburgh, Pennsylvania

CHARLES L. WAGNER Westinghouse Electric Corporation, Pittsburgh, Pennsylvania

ROY E. WOOTTON Westinghouse Research and Development Center, Pittsburgh, Pennsylvania

---

*Retired
†Deceased

1
Introduction

THOMAS E. BROWNE, Jr.[*]/Consultant to Westinghouse Research and Development Center, Pittsburgh, Pennsylvania

1.1 History 1

1.2 Mechanics 5

1.3 Progress in Understanding 6

1.4 The Future 7

References 8

1.1 HISTORY

The interruption of electric power circuits has always been an essential function, especially in cases of overloads or short circuits when immediate interruption of the current flow becomes necessary as a protective measure. In earliest times, circuits could be broken only by separation of contacts in air followed by drawing the resulting electric arc out to such a length that it could no longer be maintained. As the voltage and current capacity of power systems grew, however this means of interruption soon became inadequate and special devices called *circuit breakers* had to be developed. Circuit breakers continually developed more interrupting ability as the systems expanded, but by the early 1920s the shortcomings of available circuit breakers had become a limiting factor in the further growth of electric power systems. This situation led to a marked worldwide increase in research and development effort on circuit-interrupting devices which has continued to the present. This research and development has led to improved understanding of the basic principles, which has enabled circuit breaker development to keep pace with the requirements of our still expanding power systems.

---
[*]Retired

The means for improving the interrupting ability of circuit breakers have been many and varied. The basic problem has been to control and quench, or extinguish, the high-power arc which necessarily occurs [1] at the separating contacts of a breaker when opening high-current circuits. The arc-quenching ambients have included air, compressed air, oil, strong magnetic fields, sulfur hexafluoride (SF_6), and a vacuum environment.

Compressed Air One of the earliest schemes [2] for "blowing out" the arc used air under pressure blown through the arc space between the opening contacts. Sometimes the air pressure is also used to open the contacts. For interrupting very high currents at moderate voltages cross-blast arrangements are used, forcibly lengthening the arc by blowing it against or around insulating splitters [3]. The more usual arrangement, suitable for higher voltages, confines the air blast by a nozzle, either conducting or insulating, so as to direct the flow axially through the arc space. Both single flow and double flow (blasting in opposite directions through two nozzles facing each other) are used, and multiple breaks are operated in series for the higher voltages. To increase the current-interrupting ability, in addition to larger nozzles, shunting resistors are used to reduce the amplitude and rate of rise of the circuit recovery voltage (see Chap. 2). In this case, auxiliary interrupters for the resistor current are required. Sometimes a succession of interrupters with increasing values of shunting resistance are employed to handle very large currents. Maximum ratings are also achieved by using very high air pressures, up to 50 or 80 atm, and opening of the contacts synchronously with the current has been tried [4]. This technology is now being superseded by that of SF_6 puffer breakers.

Oil Of many interrupting schemes tried in the period of active development before 1900, the use of mineral oil for arc quenching was included, but this medium did not come into general use until the early 1900s, when increasingly high transmission voltages demanded it. In the United States, Kelman [5] in California appears to have built the first commercial oil circuit breaker for high-voltage lines by simply immersing a switch, or a series of switches, in a steel tank filled with oil. These tank-type or *bulk oil* breakers have been used extensively in the United States for many years. Since the late 1920s these have been greatly improved by providing arc and pressure control structures, sometimes called *explosion pots* or *deion grids*, at the contact breaks [6]. These structures, of various forms, serve to limit the growth of the gas bubble surrounding the arc and to utilize the pressure generated by the gas formation to force a flow of oil and gas through the arc space.

Tank-type oil breakers were aided by improvement in the bushings needed to bring the current leads into the grounded metal tanks

Introduction

(see Chap. 8), increasing the ease with which relatively low-cost ring-type current transformers around the bushings could be provided as integral parts of the breakers.

In Europe, the trend after World War II has been toward "minimum oil" construction in which the arc control structures are surrounded by closely fitting oil-filled casings of insulating material. These *live-tank* breakers require no bushings but in use must be provided with separately insulated and so relatively expensive current transformers.

In the United States, oil circuit breakers have been used up to voltages of only 345 kV. At higher voltages compressed air has been used, but SF_6 gas is now becoming the medium of choice.

Water It should be mentioned that pure water, usually with a non-ionizing antifreeze material added, has had some use in Europe [7,8] as a nonflammable substitute for oil. The arc-produced gases, steam and hydrogen, are as effective as the vapor and hydrogen from oil in quenching the arc, but insulation problems have limited the use of this medium and at present no breakers are being built that use this technique.

Solid Materials Gas-blast generation by ablation of an arc-enclosing solid material such as hard fiber has been used extensively in load-break switches and power fuses. Boric acid is especially effective as the gas source in "expulsion" fuses [9,10], but this medium has not found use in power circuit breakers.

Magnetic Fields Another arc control means, usually limited to relatively low-voltage service, has been the use of *magnetic blowout* structures. Either by arrangement of leads or by series-connected coils, magnetic fields transverse to the arc serve to lengthen the arc and usually to force it into a quenching chamber. This scheme has been employed under oil but generally is a feature of air-break devices, from low-voltage contactors to breakers for circuit voltages up to 15 kV (see Chap. 11).

An early Westinghouse development was the *air deion* breaker, which utilized in the quenching chamber the minimum reignition voltage of multiple series-connected short arcs magnetically rotated at very high speed between copper plates. It was found, however, that air-break breakers for voltages up to 15 kV or so could be built more cheaply by magnetically driving the arc into labyrinths of insulating plates. These plates, or sometimes fins, are of ceramic or other heat-resisting materials, and various arrangements are used to lengthen the arc and force it into close contact with the plate surfaces. Arrangements of insulated metal plates without the arc-rotating feature are used in some low-voltage breakers (Chap. 14).

Gas Blast (SF_6) Early in 1950 it was demonstrated (see Chap. 9) that the electronegative gas sulfur hexafluoride, previously known for its superior dielectric properties [11], had extraordinary ability to interrupt ac arcs. In its first application to a power circuit breaker, the required gas flow for arc quenching was produced by thermal expansion of arc-heated gas. The "puffer" principle, piston-driven flow energized by the opening mechanism, was also used in early load-break switches. However, for the highest breaker interrupting ratings, early practices followed that of compressed-air designs in which separate compressors and storage tanks were provided to furnish the gas blast. Since an elevated gas pressure of a few atmospheres on the discharge side of the interrupting nozzles was also needed to maintain dielectric strength comparable to that of oil, these were called *two-pressure* breakers. Both dead-tank and live-tank breakers of this type were built and are in use. In recent years further development of puffer or *single-pressure* designs (Chap. 10) has achieved the highest needed ratings and these devices are now becoming the standard for the industry.

Vacuum (See Chaps. 12 and 13). The conceptual advantages of switching in a vacuum environment, the absence of a medium to break down dielectrically or to sustain arcing, and other practical advantages have long attracted experimenters. Some partial success in power circuit interruption in vacuum was achieved as early as 1926 [12], but performance was not consistent. By 1962 vacuum and metal purification technologies had advanced to the point that reliable interruption of high currents in vacuum could be demonstrated [13]. This required extensive development, particularly of contact materials and forms. Vacuum interrupters are now used successfully at medium voltages in breakers, reclosers, and enclosed switchgear, and also in lower-voltage contactor applications. Recent research and development [14,15] shows promise for ultrahigh current interruption by these breakers and also for possible use at transmission voltages.

Direct-Current Breakers The interruption of direct or continuous currents presents a special problem in that the current must be forced to zero and the magnetically stored energy must be dissipated. At voltages up to a few kilovolts air-break breakers with strong series magnetic fields and insulated metal or nonmetallic arc "chutes" are employed to lengthen and cool the arc, thus raising its sustaining voltage above that of the circuit. For switching the high-voltage dc transmission systems now coming into increasing use, such methods are inadequate, although high-pressure transverse oil blasts have been tried to achieve very high arc voltages. Two methods for producing current zeros are used: (1) discharging a precharged capacitor through the breaker arc in a direction opposite to that of the circuit current; and (2) providing a large enough capacitor shunt to cause the breaker arc to become unstable or to cause its current to oscillate with enough amplitude to reach zero. Method 1 has been employed with force-flow oil [16] or vacuum switches [17] and method 2 with compressed air [18], vacuum

Introduction

[19], or SF_6 breakers [20]. With all high-voltage dc interrupters, absorbing the magnetically stored energy in the circuit requires an additional element, which is preferably a nonlinear resistor of the type used as lightning surge arresters [21].

Synchronous Operation In alternating-current interruption it is apparent that switching very close to the moment when the current is passing through zero should be helpful. However, the very short time of 1 or 2 ms available for this makes for very difficult mechanical problems, so that there are only a few practical examples of this method. A few breakers using high-pressure compressed air [4,22] or SF_6 [23] have been developed. Special difficulties are presented by the asymmetry of short-circuit currents and by the need for immediate reclosure in case of a failure to interrupt if the quenching means are sized for low currents only. In some cases, synchronous operation is not required for interruption but serves only to reduce wear on the contacts by arcing. Synchronous *closing* of high-voltage breakers has been employed to minimize switching transient overvoltages on transmission lines [24].

1.2 MECHANICS (CHAPTER 16)

Circuit interrupters are primarily mechanical devices for closing and opening electrical circuits. Because breakers are depended on to protect power systems, high reliability is an essential requirement. Breakers must remain closed until called on to open, even when short-circuit currents exert strong magnetic opening forces. Upon receiving an opening signal, usually from system protective relays, they must open and finally interrupt their connected circuits in a very short time, 33 ms for 60 Hz or 40 ms for 50 Hz, often in only two cycles.

These requirements present many special difficulties. A common problem is that breakers may remain closed for long periods without operation or maintenance and still must be ready at all times for reliable high-speed response. During their closed periods they must continue to carry normal load currents without excessive power loss or heating. They must be able to close against very high momentary currents and remain closed, or latched. Large closing forces may be required for this and also to maintain low contact resistance (see Chap. 15). High opening speeds call for very high accelerations of necessarily somewhat massive contacts and of often fairly complex and long linkages, especially for high-voltage breakers. In the case of puffer breakers, a high rate of gas compression represents a very substantial increase in duty on the mechanism. With vacuum breakers, much smaller strokes are needed, but the contacts must still be held closed against the strong magnetic "popping" tendency of high momentary currents, and on opening, significant force is needed to break the contact welds that tend to form under vacuum conditions.

All of these requirements and difficulties, which are highly challenging to circuit breaker designers, have led to the use of various types of operating means. These have included springs and latches, magnetic solenoids, compressed air or gas with valves, and hydraulic mechanisms. Electromagnetic repulsion with energy stored in capacitors has found use for very high accelerations [22]. Maintenance of circuit breaker mechanisms is an essential duty of the users of breakers.

1.3 PROGRESS IN UNDERSTANDING

Dielectric Recovery It was pointed out by Slepian [25] that circuit interruption by an arc requires a favorable outcome of a kind of "race" between the reappearance of circuit voltage at switch contacts and the recovery of dielectric strength by the arc space between the contacts. In dc interruption this can occur only after the current has been forced to zero by the switch. In ac interruption the circuit itself provides the current-zero moments at which interruption may result from a successful race.

Circuit Recovery Favoring interruption is the unavoidable delay in circuit recovery, or the finite rate of rise of the recovery voltage (RRRV), caused by combinations of inductive, resistive, and capacitive effects and finite wave velocities in the power supply system (see Chaps. 2 and 3).

Arc Gap Recovery Unlike circuit recovery, which can be clearly understood and calculated with some precision by classical linear methods, arc gap recovery depends on properties of the relatively complex arc and so is fundamentally more difficult to deal with theoretically. In its broad aspects we can say that recovery requires loss of conductance of the previously ionized arc space, or deionization, followed by cooling of the previously very hot (many thousands of degrees) arc path gas so that its density may approach that in its normally insulating state. Even in the case of oil breakers, initial dielectric recovery must occur in the gas bubble surrounding the arc space. The limiting phenomena involved in arc gap recovery in practical circuit breakers are dealt with in detail in many of the following chapters.

Depending on circumstances, limitations in rate of recovery may occur either at contact, or electrode, surfaces or in the arc column region remote from these surfaces. In the case of high-voltage breakers, the column region is primary because only it can be made long enough to withstand high voltage during and after recovery. Since most modern high-voltage breakers employ forced flow of high-pressure gas to "blow out" the arc, the dielectric recovery of gas-blasted arc spaces, discussed in Chap. 7, is of basic importance. In interrupting the highest currents in the presence of very high circuit recovery rates, interaction near current zero of the still conducting

Introduction

arc with the circuit, discussed in Chap. 6, becomes limiting. Here we must consider an *energy balance* race, the loss of which may prevent the loss of arc conductance, a precondition for dielectric recovery. The physical aspects of the arc in a gas blast are discussed in Chap. 5.

In magnetic air breakers, the subject of Chaps. 11 and 14, the gas blast is magnetically induced and dielectric recovery of solid surfaces adjacent to the arc may be limiting.

Of growing importance are vacuum circuit breakers, in which the physical aspects of arcing and recovery are very different from those in gas-blast breakers. Here, recovery of dielectric strength depends on *reduction* of gas (vapor) density between the contacts.

The ability of circuit breakers to withstand open circuit and test voltages in the absence of arcing is dealt with in Chap. 8.

1.4 THE FUTURE

Based on the history and present state of development of circuit-interrupting devices, we can expect to see further development, mostly refinement and optimization, of single-pressure SF_6 breakers, possibly making more effective use of arc-pressure-induced gas flow. This effect is similar to that employed in the oil breakers, which are still in extensive use but which are being superseded. Similarly, further refinement of vacuum breakers may be expected together with their increased application, especially in the medium-voltage range.

There is a demand for current-limiting breakers in the high-voltage field, but their necessary cost may be expected to limit such breakers to special-purpose applications such as joining sections of power networks. Similarly, synchronous switching will be limited to special uses, such as the make switching of high-voltage transmission lines and the reduction of contact wear of frequently operated high-current breakers.

The need for switching in high-voltage dc transmission systems should encourage further development of special circuit-interrupting arrangements for this purpose. These will almost certainly employ presently available ac breakers with some modification of the arc-interrupting parts.

Another possibility is the further development of solid-state devices to replace the arc we now depend on for synchronization of interrupting effort with current zero and rapid dielectric recovery. Again, however, for some time to come cost and the high continuous power loss in these devices should continue to favor arc-employing switches for power interruption.

Based on my observations of circuit breaker developments over the last half century, I expect that continuing research, especially on properties of materials and the use of modern instrumentation and computer-aided theory, will eventually bring about radically new developments in circuit interruption practice not now foreseen.

REFERENCES

1. J. Slepian, The electric arc in circuit interrupters, J. Franklin Inst. *214*:413-442, October 1932.
2. R. H. Read, Extinguishing electric arcs, U.S. Pat. 716,848, December 23, 1902.
3. L. R. Ludwig, H. S. Rawlins, and B. P. Baker, New pneumatic circuit interrupter, Electr. Eng. *59*:528-533, 1940.
4. K. Kriechbaum, A half cycle air blast generator breaker for high power testing fields, IEEE Trans. Power Appar. Syst. *PAS-91*(3):747-753, 1972.
5. J. N. Kelman, Switch for high-potential circuits, U.S. Pat. 874,601, December 24, 1907.
6. B. P. Baker and H. M. Wilcox, The use of oil in arc rupture, Trans. AIEE *49*:297-301, 1930.
7. F. Kesselring, Expansion breaker, Elektrotech. Z. *51*:499-508, 1930.
8. W. M. Leeds, A high power oiless circuit interrupter using water, Electr. Eng. *60*:85-88, 1941.
9. J. Slepian and C. L. Denault, The expulsion fuse, Trans. AIEE *51*:157-165, 1932.
10. A. P. Strom and H. L. Rawlins, The boric acid fuse, Trans. AIEE *52*:1020-1026, 1932.
11. F. S. Cooper, Gas dielectric media, U.S. Pat. 2,221,671, November 12, 1940.
12. R. W. Sorensen and H. E. Mendenhall, Vacuum switching experiments at California Institute of Technology, AIEE Trans. *45*:1102-1105, 1926.
13. T. H. Lee, A. N. Greenwood, D. W. Crouch, and C. H. Titus, Development of power vacuum interrupters, IEEE Trans. Power Appar. Syst. *PAS-81*:629, 1963.
14. S. Yanabu, E. Kaneko, H. Okumura, and F. Aiyoshi, Novel electrode structure of vacuum interrupter and its practical application, IEEE Trans Power Appar. Syst. *PAS-100*(4):1966-1973, 1981.
15. J. G. Gorman, C. W. Kimblin, R. E. Voshall, R. E. Wien, and P. G. Slade, The interaction of vacuum arcs with magnetic fields and applications, IEEE Trans. Power Appar. Syst. *PAS-102*(2):257-266, 1983.
16. Å. Ekström, H. Härtel, H. P. Lips, W. Schultz, P. Joss, H. Holfeld, and Q. Kind, Design and testing of an HVDC circuit-breaker, CIGRE, Rep. 13-06, Paris, 1976.
17. A. N. Greenwood, P. Barkan, and W. C. Kracht, HVDC vacuum circuit breaker, IEEE Trans. Power Appar. Syst. *PAS-91*:1575-1588, 1972.
18. M. Sakai, Y. Kato, S. Tokuyama, H. Sugawara, and K. Arimatsu, Development and field application of metallic return protecting

breaker for HVDC transmission, IEEE Trans. Power Appar. Syst. *PAS-100*: 4860-4868, December 1981.
19. A. L. Courts, J. J. Vithayathil, N. G. Hingorani, J. W. Porter, J. G. Gorman, and C. W. Kimblin, A new dc breaker used as metallic return transfer breaker, IEEE Trans. Power Appar. Syst. *PAS-101*:4112-4121, October 1982.
20. A. Lee, J. E. Heidrich, and L. S. Frost, A HVDC breaker concept based on SF_6 puffer technology, 7th Int. IEE Conf. Gas Discharges Appl., London, 1982, p. 116.
21. G. D. Breur, L. Csuros, R. W. Flugum, J. Kauferle, D. Povh, and A. Schei, HVDC surge diverters and their application for overvoltage protection of HVDC schemes, CIGRE, Rep. 33-14, Paris, 1972.
22. Y. Nitta and N. Hiyokuni, Synchronous air blast circuit breaker for one cycle interruption, Fuji Electr. Rev. (Japan) *11*(4):95-104, 1965.
23. L. D. McConnell and R. D. Garzon, Development of a new synchronous circuit breaker, IEEE Trans. Power Appar. Syst. *PAS-92*:673-681, 1973.
24. W. M. Leeds, R. E. Friedrich, C. L. Wagner, and T. E. Browne, Jr., Application of switching surge, arc and gas flow studies to the design of SF_6 breakers, CIGRE, Rep. 13-11, Paris, 1970.
25. J. Slepian, Extinction of an a.c. arc, Trans. AIEE *27*:1398-1407, 1928.

General References

Flurscheim, C. H., ed., *Power Circuit Breaker Theory and Design*, rev. ed., Peter Peregrinus, London, 1982.
Garrard, C. J. O., High-voltage switchgear, Proc. IEE (Lond.) *123* (10R):1053-1080, 1976.
Lafferty, J. M., ed., *Vacuum Arcs: Theory and Application*, Wiley-Interscience, New York, 1979.
Lee, T. H., *Physics and Engineering of High Power Switching Devices*, MIT Press, Cambridge, Mass., 1975.
Maller, V. N., and M. S. Naidu, *Advances in High Voltage Insulation and Arc Interruption in SF_6 and Vacuum*, Pergamon Press, Elmsford, N.Y., 1981.
Ragaller, K., ed., *Current Interruption in High-Voltage Networks*, Plenum Press, New York, 1978.
Schalter-Lichtbögen (switch arcs), Elektrotech. Z. Ausg. A *93*, July 1972.
Special Issue on Arc Plasmas, IEEE Trans. Plasma Sci. *PS-8*, December 1980.
Special Issue on Atomic and Molecular Plasmas, Proc. IEEE, *59*, April 1971.

2
Electrical and System Aspects

R. GERALD COLCLASER, Jr./University of Pittsburgh, Pittsburgh, Pennsylvania

2.1 Electrical Environment   11
2.2 Transient Switching Laws   12
2.3 Arc Characteristics   16
    2.3.1 Static characteristic   17
    2.3.2 Dynamic characteristic   18
2.4 Direct-Current Interruption   21
    2.4.1 Arc instability   21
    2.4.2 Current injection   25
2.5 Alternating-Current Interruption   25

References   37

2.1 ELECTRICAL ENVIRONMENT

A circuit-interrupting device (breaker) operates in an electrical environment which imposes a unique set of criteria on the device. There are three major operating conditions: closed, open, and the transition from closed to open. In the closed position the device must conduct the continuous rated current without exceeding standardized total temperatures for the conductor and insulation structures. While closed, the complete insulation system to ground is continuously stressed by the sinusoidal power frequency voltage and also by transient overvoltages caused by lightning, switching, and system changes. If the device is in the open position, insulation across the open contacts is stressed in addition to the insulation to ground. Any transient voltage surges will produce increased voltages on the stressed side of the device.

If a fault occurs, the closed device is expected to react to a trip signal and to interrupt the fault current within the rated interrupting time, minimizing any disturbances to the system. At some point during the opening operation the current is interrupted, resulting in an electrical separation of the system at the breaker location. Immediately following current zero, the contacts are stressed by transient voltages produced by the system as it reacts to its new operating state.

In addition to the rated fault current, a high-voltage power circuit breaker is required to interrupt inductive currents from the maximum fault value down to the exciting current of unloaded transformers. Currents limited by capacitance elements must also be interrupted. Overvoltages must be limited or controlled to predetermined values during these various types of switching operations. This wide variety of operating conditions imposes often-conflicting constraints on the breaker design.

2.2 TRANSIENT SWITCHING LAWS

The connected network includes components that can be approximated by both lumped and distributed elements. The time frame of interest dictates the network simulation, possibly requiring circuit components usually considered lumped to be modeled using distributed elements, or vice versa.

Digital computer simulations can be used to calculate voltage and current transients in the connected network. Simplifications can often be used to obtain first-order approximations of critical wave shapes. If the network can be satisfactorily simulated with linear lumped and distributed parameters, Laplace transform techniques [1] can be used to provide solutions to the describing integrodifferential equations.

When constructing the set of transform equations that describe a particular circuit, initial conditions related to current and voltage are required. Equivalent to the law of conservation of energy, flux linkages in an inductive circuit prior to switching must be equal to the flux linkages after switching [2]. This is formally written as

$$L^- i^- = L^+ i^+ \qquad (2.1)$$

where L^- and i^- refer to the inductance and current prior to switching, and the plus signs denote current and inductance just after switching has occurred but before any time has elapsed. The corresponding conditions for the capacitive circuit require the conservation of charge, or

$$C^- v^- = C^+ v^+ \qquad (2.2)$$

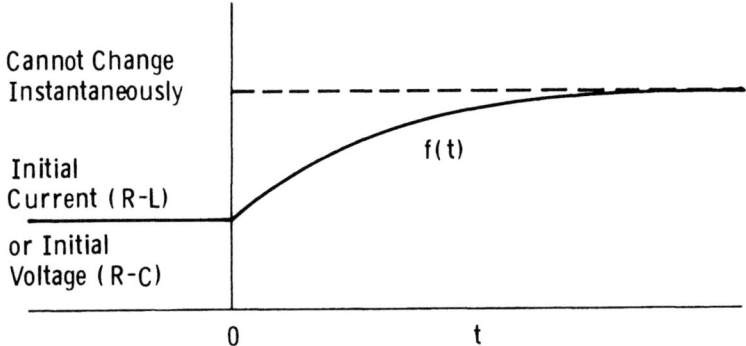

Figure 2.1 Initial conditions require continuity of charge or flux linkage at t = 0.

Here C and v are the capacitance and voltage, respectively. Although there may be situations where the inductance or capacitance changes during the switching operation, most applications will be based on the assumption that L^- is identical to L^+ and C^- is identical to C^+. The net result is that in almost all cases, the current through an inductance and the voltage across a capacitor cannot instantaneously change at the instant of switching. This is shown in Fig. 2.1, where, at t = 0, the current or voltage must be continuous.

If we consider the case of capacitor energization, an equivalent circuit is shown in Fig. 2.2. At t = 0, the switch closes and the defining equation is

$$E = Ri(t) + \frac{1}{C} \int i \, dt \qquad (2.3)$$

The corresponding Laplace transform equation is then

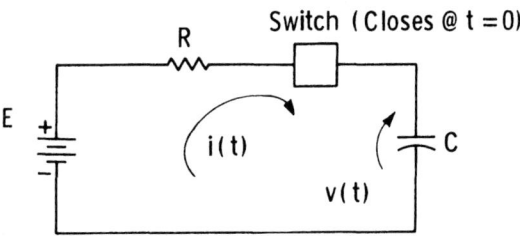

Figure 2.2 RC circuit (dc source).

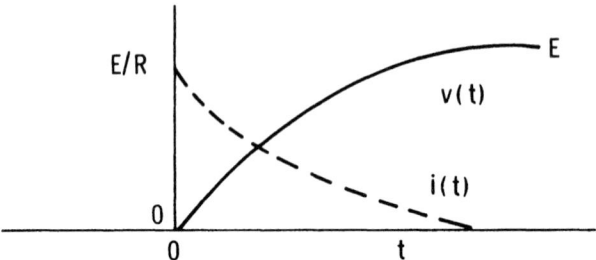

Figure 2.3 Current and voltage relationships in the RC circuit.

$$\frac{E}{s} = RI(s) + \frac{1}{Cs} I(s) + \frac{V_0}{s} \tag{2.4}$$

where V_0 is the initial value of capacitor voltage at the instant the switch is closed. Assuming that V_0 is zero, the current as a function of time is obtained by rearranging equation (2.4), and the inverse is

$$i(t) = \frac{E}{R} \varepsilon^{-t/RC} u(t) \tag{2.5}$$

where u(t) is the standard "unit step function." The transform of the voltage across the capacitor is

$$V_c(s) = \frac{1}{Cs} I(s) \tag{2.6}$$

or
$$v(t) = E[1 - \varepsilon^{-t/RC}]u(t) \tag{2.7}$$

Both current and voltage relations are shown in Fig. 2.3. Note that since the voltage on the capacitor prior to switching was zero, the voltage function begins at zero and builds up to E. On the other hand, the current through a capacitor can change instantaneously, and

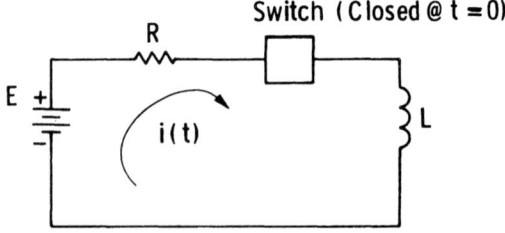

Figure 2.4 RL circuit (dc source).

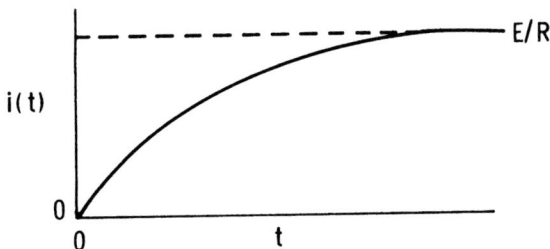

Figure 2.5 Current transient in the RL circuit.

in this case, assumes an initial value for the driving voltage E divided by the circuit resistance R. The inductive circuit, shown in Fig. 2.4, can easily be shown to have a current

$$i(t) = \frac{E}{R} [1 - \varepsilon^{-(R/L)t}] u(t) \qquad (2.8)$$

which is shown as a function of time in Fig. 2.5. Initially, since there was no current flowing in the inductance prior to switching, the current must remain zero at the instant of switch closure.

Although these concepts appear simple on the surface, they have far-reaching consequences when switching operations are attempted in the power system. If we consider an RL circuit with an alternating voltage source, as shown in Fig. 2.6, closure of the switch simulates the initiation of a fault on the network. The initial current through the inductance prior to switching is assumed to be zero. The defining differential equation is

$$E_m \sin(\omega t + \phi) = R i(t) + L \frac{di(t)}{dt} \qquad (2.9)$$

where ϕ denotes the point on the voltage wave when the switch is closed (t = 0). The current wave that satisfies this equation can be determined using Laplace transforms, and is written as

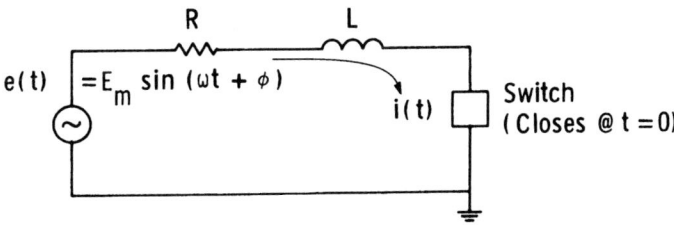

Figure 2.6 RL circuit (ac source).

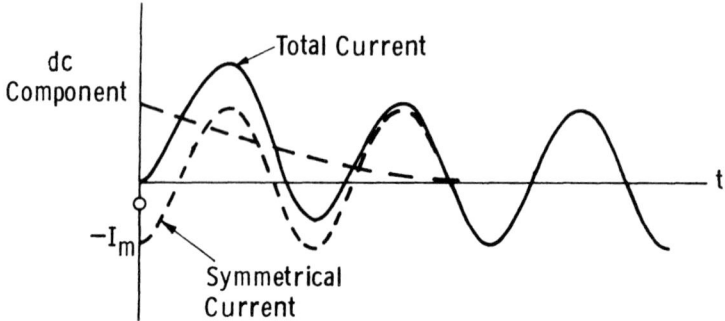

Figure 2.7 Asymmetrical current.

$$i(t) = \frac{E_m}{X} \frac{1}{\sqrt{(R/X)^2 + 1}} \left\{ \sin(\omega t + \phi - \theta) \right.$$

$$\left. - [\sin(\phi - \theta)] \varepsilon^{-(R/X)\omega t} \right\} u(t) \quad (2.10)$$

where $X = \omega L$ and $\theta = \tan^{-1}(X/R)$.

The solution for current includes a steady-state component and a transient component which is required to satisfy the conservation of flux linkages and results, as shown in Fig. 2.7, in an offset (dc) term which initially displaces the sinusoidal current. Since this type of transient occurs during fault initiation, breaker designs must account for the additional mechanical stresses imposed by the peak of the offset current, and the interrupting function must consider the requirements imposed by the unequal spacing of current zeros and the changes in current slope. The significance of the X/R ratio to breaker application can be seen in equation (2.10).

2.3 ARC CHARACTERISTICS

During an opening operation, the net result of the conservation of flux linkages is that the current which is flowing in the system inductances prior to switching continues to flow following the parting of the contacts. In the simplified RL circuit shown in Fig. 2.8, the switch opens at t = 0. The final contact point in the contacts has an extremely high current density, and the contact surface literally "explodes," setting the stage for a gaseous conduction arc.

As a circuit element, the arc behaves like a nonlinear resistor with the arc voltage in phase with the arc current. The arc voltage can be considered a function of arc current, but its exact shape and value

Electrical and System Aspects

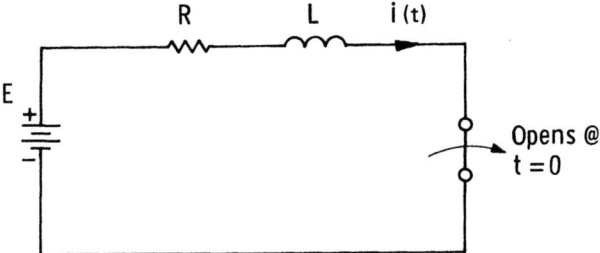

Figure 2.8 RL opening circuit (dc source).

also depend on the cooling and deionizing effects of the arc-extinguishing methods being used. Other factors include the length of the arc and the properties of the surrounding medium, especially its heat transfer characteristics.

2.3.1 Static Characteristic

A static, fixed-length dc arc has an arc voltage vs. current characteristic similar to that shown in Fig. 2.9. At low currents, of the order of 10 A, the arc has a falling volt-ampere characteristic, tending toward a constant power requirement. At higher currents the arc voltage or voltage gradient levels out, becoming practically independent of current but with a slight tendency to rise at very high currents. The important concept to note is that the arc voltage is not a function

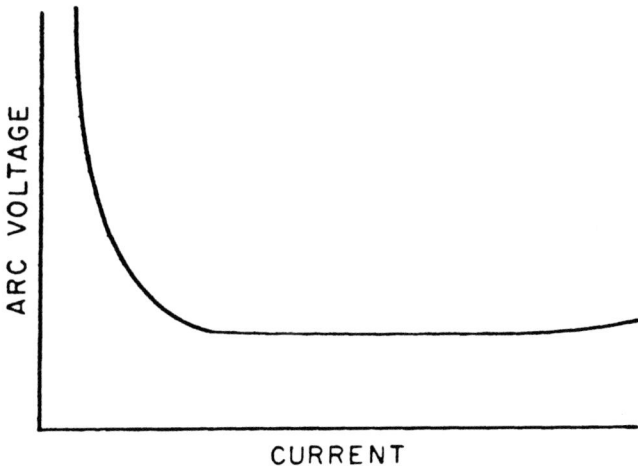

Figure 2.9 Static volt-ampere characteristic of the arc column.

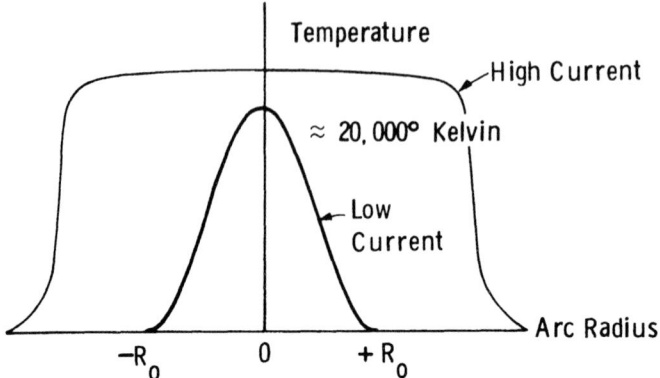

Figure 2.10 Arc temperature vs. radius profile.

of the system voltage, but instead is determined by the power input required to sustain the arc.

Although arc physics will be discussed in more detail in Chap. 4, it may be pointed out here that, together with the different volt-ampere behavior of low- and of high-current arcs illustrated in Fig. 2.9, there is distinct difference in the physical states of the arc column. Figure 2.10 shows roughly the different types of radial temperature distributions between low- and high-current arcs. Very approximately, the low-current arc, with its approach to constant power loss, changes its temperature and conductivity primarily in response to current changes. At high currents, on the other hand, the maximum arc temperature, on the order of 20,000 K, tends to be limited by radiant heat transfer so that it and the electrical conductivity have nearly constant values within the column. This column, then, responds to current changes by changes only in cross-sectional area, thus accounting for the near independence of current shown by the arc voltage at high current.

2.3.2 Dynamic Characteristic

Because of energy storage in the arc associated with its conductance and finite rates of energy flow, the arc conductance cannot respond to current changes instantaneously but together with the arc voltage, lags behind the current change. This is illustrated by the volt-ampere plot of Fig. 2.11 for the case of a low-current dc arc with alternating currents of limited amplitude superimposed. The degree of lag depends on both the ac frequency and the electrical inertia of the arc, characterized by its "time constant." In the figure, successively higher ac

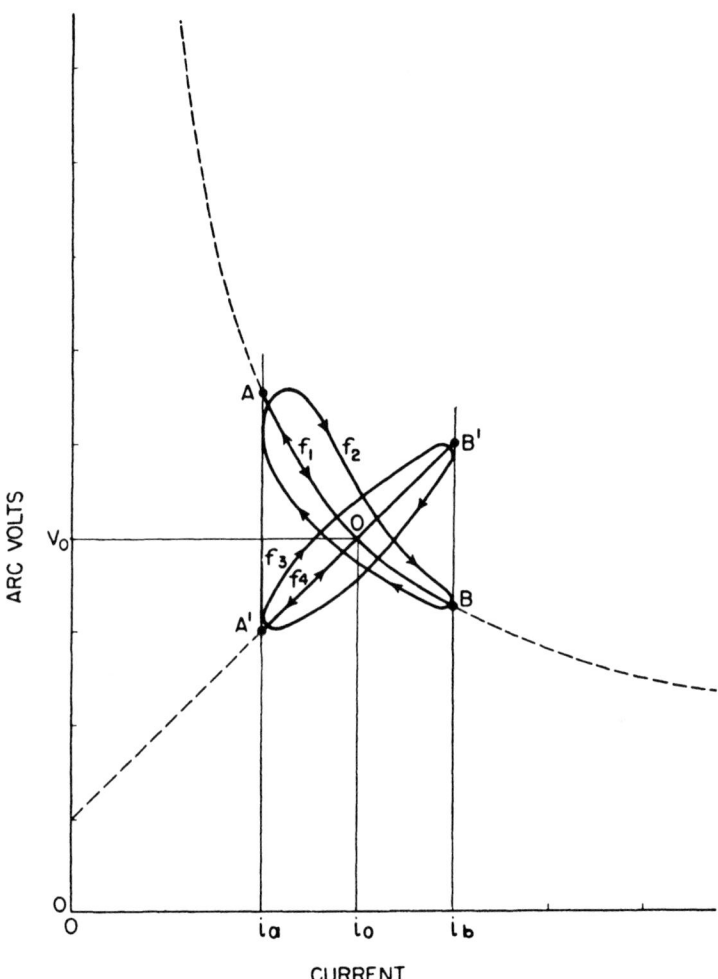

Figure 2.11 Dynamic characteristics of an arc with current pulsating at successively higher frequencies, $f_1 < f_2 < f_3 < f_4$. (From Ref. 3.)

frequencies are illustrated. Frequency f_1 is so low that the static characteristic is followed, while f_4 is so high that the arc conductance cannot follow at all but remains constant at the midvalue, i_0/V_0. Frequencies f_2 and f_3 have intermediate values between these two extremes.

In the case of large alternating-current power frequency arcs, the arc usually has a time constant much smaller than the half-cycle time, so the arc voltage tends to be nearly constant during each current half cycle, with only a slight hysteresis near each current-zero moment.

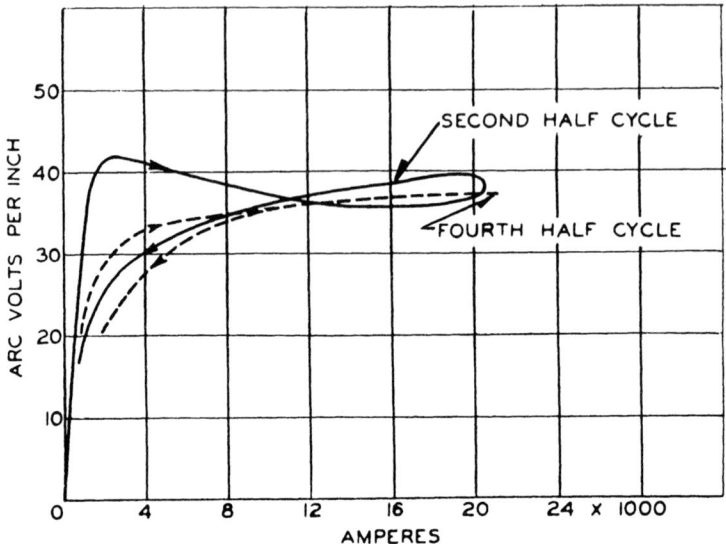

Figure 2.12 Volt-ampere cyclograms of a vertical 60-Hz arc 12 in. long in air, 15,000 A rms. (From Ref. 3.)

This is illustrated by the cyclogram, Fig. 2.12, showing observed behavior of a 15-kA 60-Hz vertical 12-in. arc in the open air. Figure 2.13 is a time plot for a high-current power frequency switch arc between gradually separating contacts. Arc properties are discussed in greater detail in Chaps. 4 and 5.

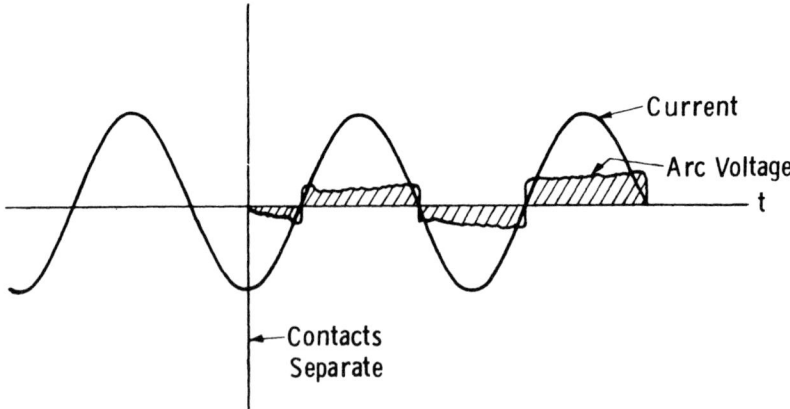

Figure 2.13 Arc voltage characteristics.

Electrical and System Aspects

2.4 DIRECT-CURRENT INTERRUPTION

Accepting the fact that the arc exists, what are the criteria for interruption in a dc circuit as shown schematically in Fig. 2.14? It is possible for the conservation of flux linkages to be satisfied and the current through the breaker to be interrupted instantaneously if the capacitance in parallel with the switch, shown dashed, is relatively large. The conservation of charge requires that the current be diverted into the capacitance initially, but normally the rapid rise in voltage would cause a reignition across the initially small switch gap even if the switch device path were noninductive. Neglecting the capacitance, the basic differential equation that describes the current relationship is

$$E = L\frac{di}{dt} + Ri + e_a \qquad (2.11)$$

Rearranging, the expression for di/dt, or rate of change of current with time, is

$$\frac{di}{dt} = \frac{1}{L}[(E - iR) - e_a] \qquad (2.12)$$

2.4.1 Arc Instability

Since initially $E - iR$ must be identically equal to zero, the di/dt at $t = 0$ must have a negative value equal to the arc voltage divided by the inductance L, indicating that the current initially decreases. If, however, the $E - iR$ term at some point is equal to the arc voltage, the di/dt is equal to zero and the current ceases to decrease. This is a stable point and in the circuit of Fig. 2.14, the arc will exist in a stable configuration. The basic criterion for arc interruption, therefore, is that the arc voltage term must continue to exceed the $E - iR$ term so that the current continues to decrease. This requires that at some time, the arc voltage exceed the system driving voltage. This can occur because the arc voltage is not a function of system voltage,

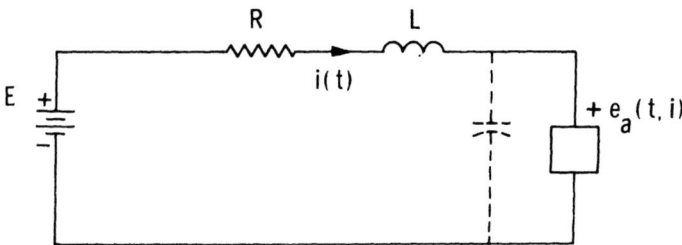

Figure 2.14 Dc switching circuit.

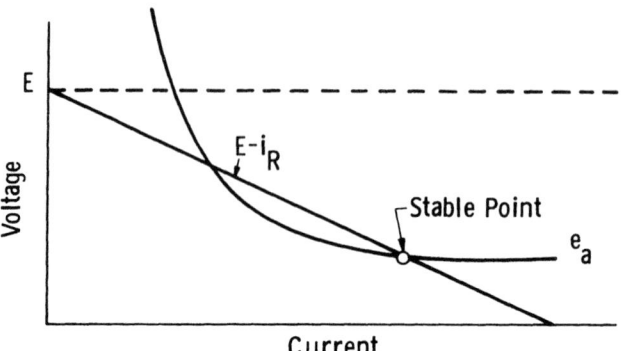

Figure 2.15 Current-voltage relations in the circuit of Fig. 2.14.

but is instead determined by the arc characteristics. As an example, Fig. 2.15 shows an arc voltage characteristic superimposed on the E − iR line for Fig. 2.14. Notice that, according to equation (2.12), a stable point occurs where the arc voltage and E − iR lines intersect at the higher current value circled. Figure 2.16, on the other hand, shows a second arc voltage characteristic, e_{a1}, which ensures interruption. As before, the arc voltage e_{a2} would result in a stable arc.

Since the arc is subject to wide variations of length and voltage gradient, insight can be gained by considering the arc voltage characteristic as a function of time, ignoring temporarily the current dependency. Figure 2.17 shows two possible idealized variations, a constant arc voltage and a linearly increasing arc voltage. These characteristics can be used to calculate the current characteristics for the circuit of Fig. 2.14. If we assume that the system has reached

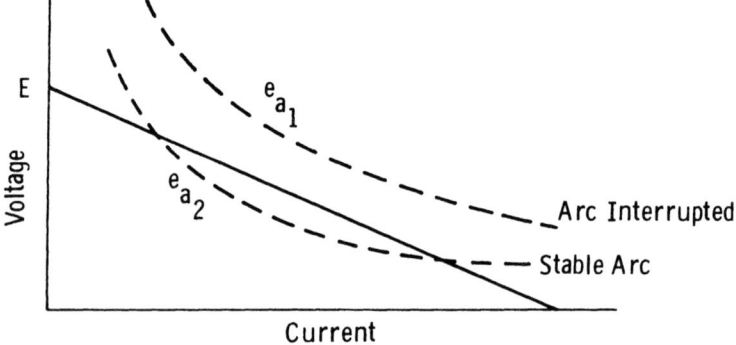

Figure 2.16 Criterion for interruption.

Electrical and System Aspects

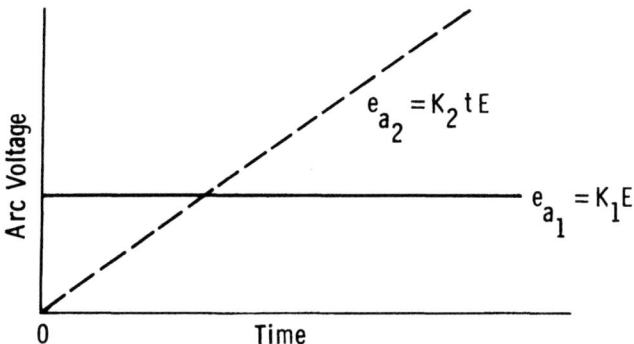

Figure 2.17 Arc voltage variations with time.

steady state prior to switching, the describing differential equation for the constant-arc-voltage case is

$$E = L \frac{di}{dt} + Ri + K_1 E \qquad (2.13)$$

where K_1 is a constant. The transformed equation is

$$\frac{E}{s} = LsI - L\zeta + RI + \frac{K_1 E}{s} \qquad (2.14)$$

where ζ is the initial value of the current, E/R. The resulting expression for current as a function of time is

$$i(t) = \frac{E}{R} [(1 - K_1) + K_1 \varepsilon^{-(R/L)t}] u(t) \qquad (2.15)$$

By setting this equation equal to zero, an estimate of the time required to interrupt the current as a function of K_1 can be obtained. The higher K_1 is, the shorter will be the interrupting time, keeping in mind that K_1 should be greater than 1.0 to ensure interruption.

It is also possible to consider a current-limiting function for the arc voltage by having the contacts part before the current has reached a steady state. This is shown schematically in Fig. 2.18, and the expression for the current following the contact part is obtained by substituting the value of the current at the contact part for the initial condition. In this example, the maximum value of fault current is determined by the slope KE of the arc voltage.

The arc voltage requirements discussed in the preceding section hold regardless of the level of system voltage. When high-voltage systems are considered, dc breaker design becomes difficult. Since it is impractical to force a current zero using feasible arc voltages in a

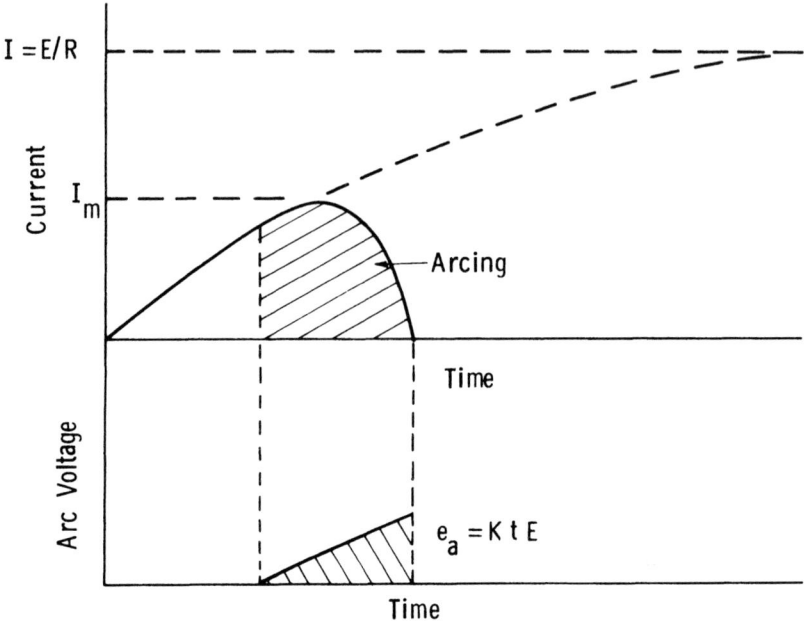

Figure 2.18 Current limited by arc voltage.

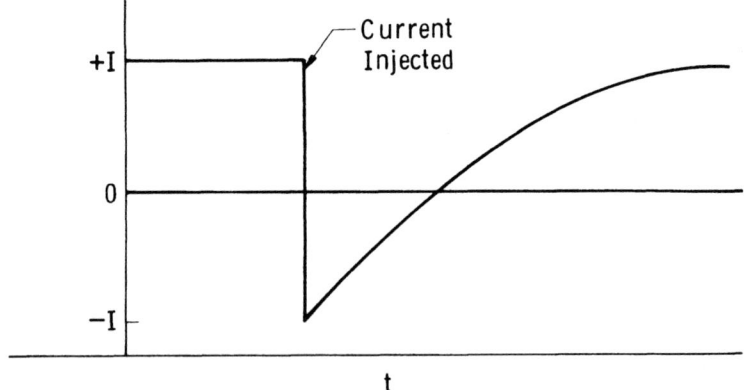

Figure 2.19 Forcing current zero with injected current.

Electrical and System Aspects

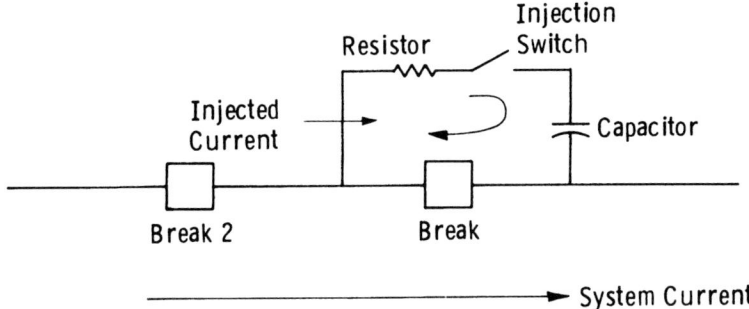

Figure 2.20 Simplified current injection circuit.

500-kV dc circuit, alternative methods have been explored. A brief discussion of the current injection concept follows.

2.4.2 Current Injection

If a current zero can be artificially produced by injecting a current of opposite polarity, a new basis for interruption can be developed. This is shown in Fig. 2.19. Passage through the initial current zero is too rapid for interruption, but the second current zero, as the current returns more slowly to the steady-state value, shows promise. A simplified current injection circuit is shown in Fig. 2.20. The capacitor is charged with a polarity such that the injected current is opposite in direction from the system current through the breaker. The resistor is used to limit the value of injected current to somewhat more than the system value. Such a scheme can produce an artificial zero, although the actual circuit design is considerably more involved than indicated. As might be expected, the circuit branch inductances and the arc dynamic characteristic complicate the switching problem considerably since it is now similar to that of the ac case. Another means for bringing the current to zero is the connection of a parallel capacitor, shown dashed in Fig. 2.14, of a large enough value that a growing oscillation of the arc and capacitor current is produced [3].

2.5 ALTERNATING-CURRENT INTERRUPTION

The alternating-current interruption process does not require the forcing of a current zero since two zeros naturally occur each cycle. When a current zero occurs, the magnetically stored energy in the circuit is zero and the arc may be prevented from reigniting. If arc reignition

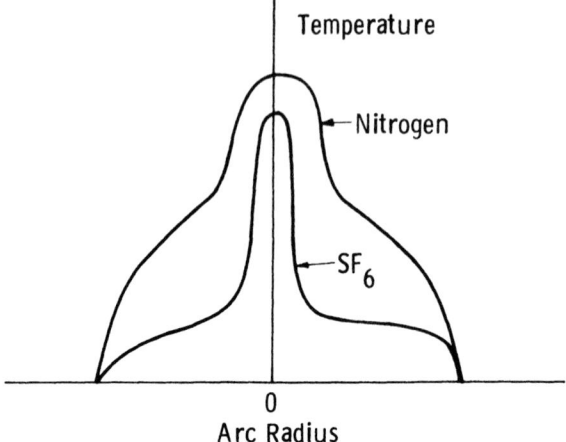

Figure 2.21 Arc temperature as a function of radius.

is not to occur, the recently conducting plasma must be cooled below the thermally conductive temperature.

Figure 2.21 shows a pictorial representation of arc temperature as a function of radius for near-zero-current arcs drawn in the molecular gases nitrogen and sulfur hexafluoride (SF_6). If the hot central core is removed from both the SF_6 and nitrogen arcs by cooling at current zero, a broader-based temperature distribution remains, as shown in Fig. 2.22. In this case, with the arbitrary minimum conducting temperature level shown, the arc in nitrogen would continue to conduct while the cooler arc in SF_6 would be extinguished. The rate of cool-

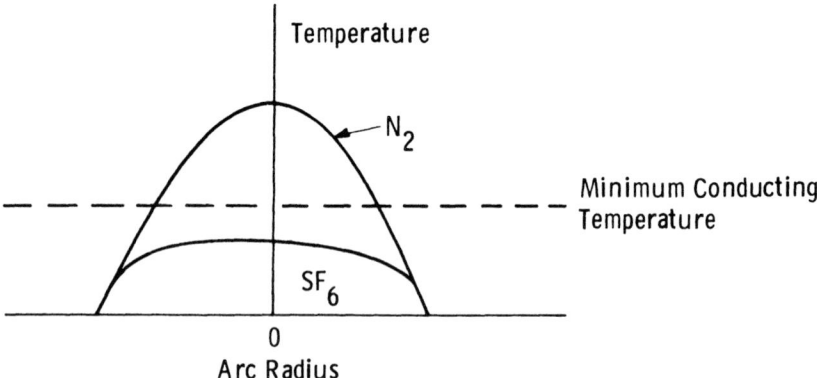

Figure 2.22 Temperature distribution after current zero.

Electrical and System Aspects

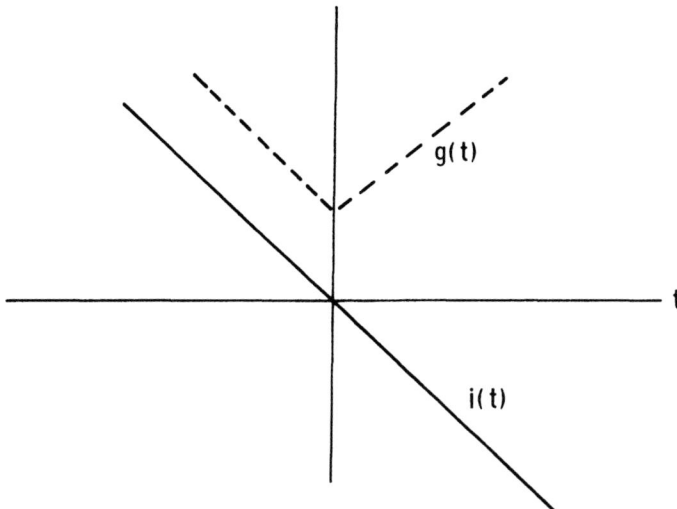

Figure 2.23 Ideal reignition.

ing required to obtain essentially the nonconducting state is on the order of 1000 K per microsecond. Although the arc current has reached a zero value, the arc column conductivity still has a finite value and so some "post arc" current can continue to flow in the opposite direction with a corresponding tendency for the arc conductance $g(t)$ to increase again, as shown in Fig. 2.23, causing a failure to interrupt the circuit. In a successful interruption, insufficient power input to the arc takes place and the conductance can be assumed to decay as shown in Fig. 2.24.

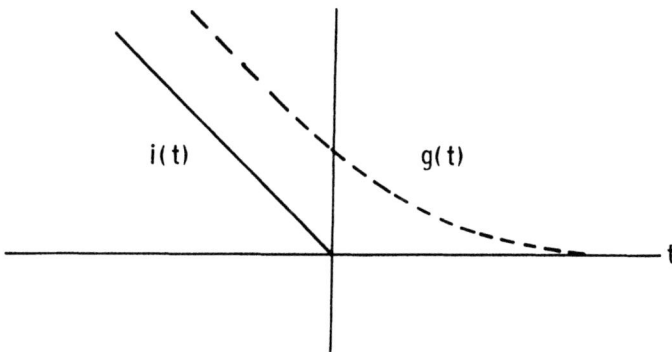

Figure 2.24 Ideal interruption, conductivity delay.

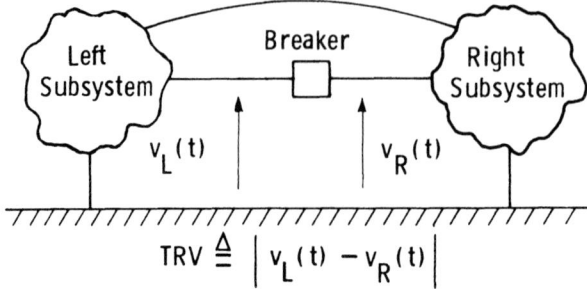

Figure 2.25 Definition of transient recovery voltage.

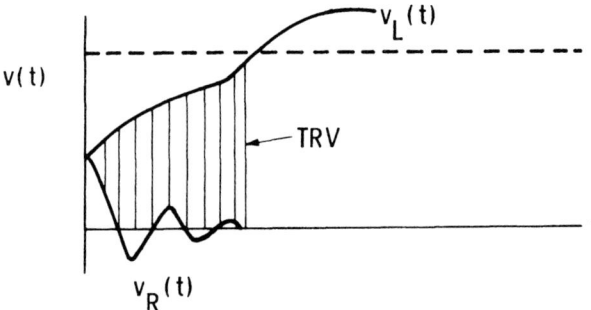

Figure 2.26 TRV components.

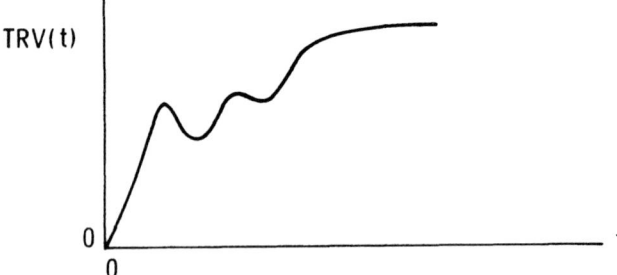

Figure 2.27 TRV across a breaker.

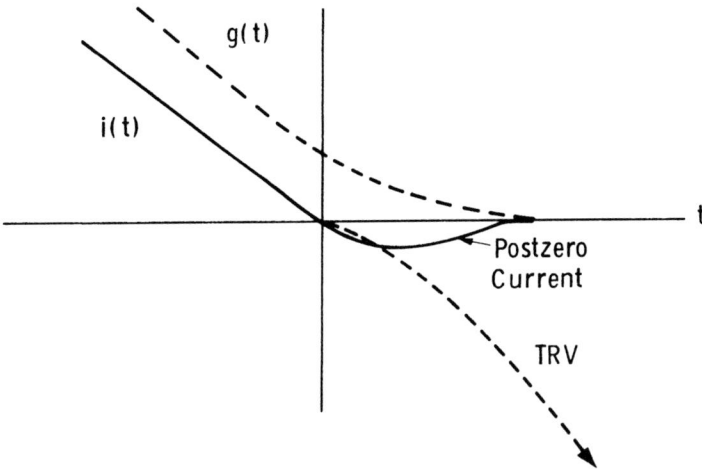

Figure 2.28 Effect of TRV on current following current zero.

The breaker, as mentioned previously, is embedded in the system as shown in Fig. 2.25. When the current through the breaker is interrupted, the entire system adjusts to its new operating state. The voltage transients produced separately by the left subsystem, $V_L(t)$, and the right subsystem, $V_R(t)$, stress the recently conducting arc plasma. The transient recovery voltage (TRV) is defined as the difference between the voltage on the left and right sides of the breaker. This is shown in Fig. 2.26, where the TRV is the difference between V_L and V_R. The difference voltage is shown in Fig. 2.27. The transient recovery voltage contributes, for arc space residual conductance $g(t)$, a power input i^2/g. Integrated over time, this is an energy input that may raise the temperature and the corresponding conductivity while the arc cooling effects are attempting to reduce the conductivity by removing energy. This is the "thermal region," in which interruption depends on the energy balance between the arc and the connected circuit, as discussed quantitatively in Chap. 6.

The effect of the transient recovery voltage is shown in Fig. 2.28, where postzero current exists during the residual conductance period as a result of the transient recovery voltage stress. It is possible for the TRV to overstress the conductance, resulting in an increase instead of the exponentially decaying decrease described previously. In such a case, shown in Fig. 2.29, the postzero current may extinguish or may, if the energy input exceeds the rate of energy loss, reignite the arc for another half cycle. This energy balance region is thus critical in the decision to attempt interruption. If the energy balance

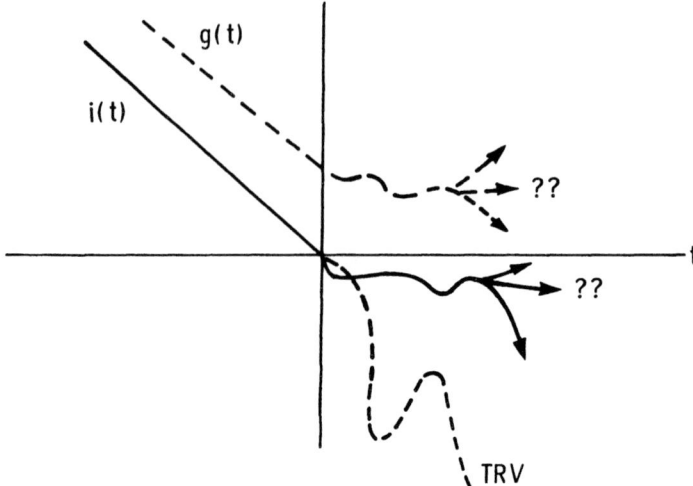

Figure 2.29 Conductivity variations.

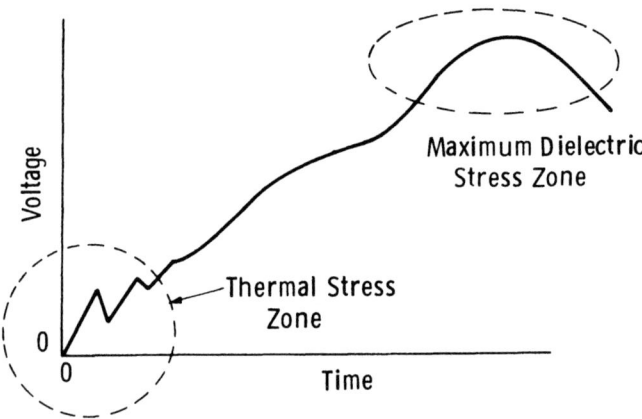

Figure 2.30 Thermal and dielectric breakdown regions.

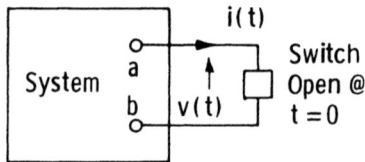

Figure 2.31 System representation.

Electrical and System Aspects 31

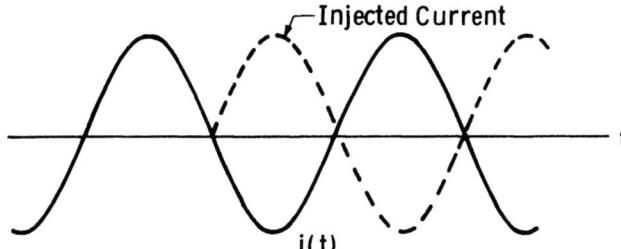

Figure 2.32 Fault current and injected current.

decision is favorable, the recently conducting arc space must still withstand the crest values of transient recovery voltage, which produce a dielectric stress tens or hundreds of microseconds later. The two regions are shown in Fig. 2.30, recognizing that this is oversimplified in that transition periods also exist.

For complete current interruption the energy balance decision must be favorable and the gap must not fail dielectrically. The distinguishing characteristic between an energy balance reignition and a dielectric failure is a discontinuity in the current flow which occurs prior to a dielectric failure. The transient recovery voltage in a dielectric situation is controlled by the circuit, whereas in an energy balance contest the TRV may be jointly controlled by both the circuit and the conductance of the arc.

Because of the extreme importance of the transient recovery voltage in the arc interruption process, system TRVs must be considered as a part of the design and application process. If the system is represented as a box connected to the breaker terminals at a and b, as shown in Fig. 2.31, the current i(t) that must be interrupted flows between terminals a and b. If a current is injected at the switch equal and opposite to i(t), the resulting current will be zero after the time of injection [4]. This is shown in Fig. 2.32 and the net result in Fig. 2.33. Note that the injected current must exactly duplicate the fault

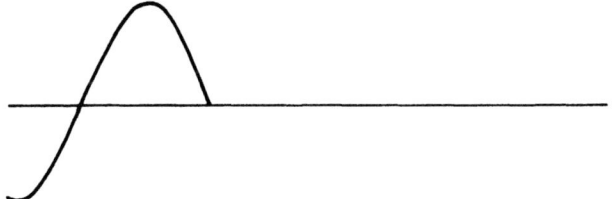

Figure 2.33 Resultant current: sum of i(t) and $i_{injected}$.

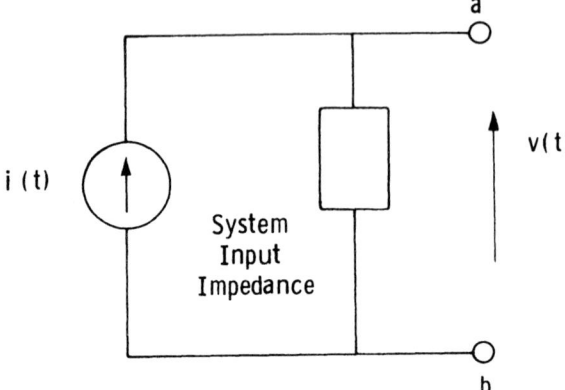

Figure 2.34 Norton equivalent circuit.

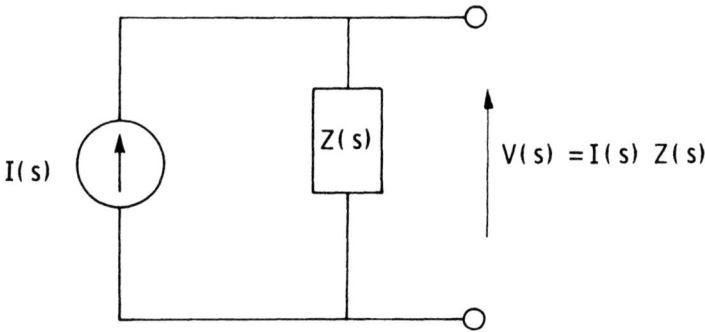

Figure 2.35 Equivalent Laplace transform representation.

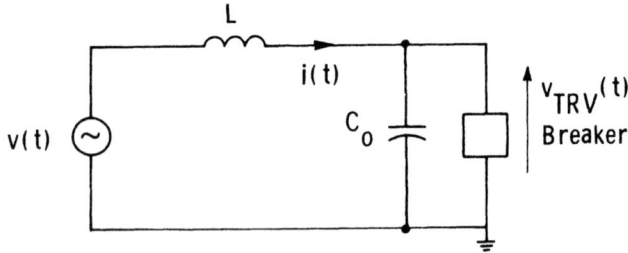

Figure 2.36 Equivalent inductance-dominated circuit.

Electrical and System Aspects

current over time. If the variable of interest, in this case the recovery voltage, is the transient that takes place because of the injection of current, a Norton equivalent circuit can be constructed as shown in Fig. 2.34. The current source is equivalent to the current being injected, and the system input impedance is the impedance of the system seen from terminals a-b with voltage sources shorted and current sources opened in the network. Using operational notation, an equivalent Laplace transform representation for the Norton equivalent circuit is shown in Fig. 2.35, where I(s) is the transform of the injected current, Z(s) is the operational impedance of the system viewed from terminals a-b, and the voltage V(s) is the product of I(s) and Z(s). The circuit elements used in the operational impedance form are R for resistance, LS for inductance, and 1/CS for capacitance.

As an example, a simple LC circuit can be used as shown in Fig. 2.36. The inductance determines the fault current prior to interruption, and at current zero the distributed capacitance of both the breaker and the inductance, shown lumped together as C_0, completes the circuit. Because the TRV frequency is high relative to 50 or 60 Hz, a linear ramp of current representing the current in the neighborhood of zero can be injected with a slope equal to that of the sinusoidal current at t = 0. In such a case the equation of i(t) is

$$i(t) = I_m \omega t \tag{2.16}$$

and the transform is

$$I(s) = \frac{I_m \omega}{s^2} \tag{2.17}$$

In the circuit shown in Fig. 2.36, the impedance looking into the terminals of the breaker is

$$Z(s) = \frac{1}{C} \frac{s}{s^2 + 1/LC} \tag{2.18}$$

The transform for the transient recovery voltage is

$$V(s) = I(s)Z(s) = \frac{I_m \omega}{C} \frac{1}{s(s^2 + 1/LC)} \tag{2.19}$$

and the corresponding time expression is

$$v(t) = I_m \omega L \left(1 - \cos \frac{1}{\sqrt{LC}} t\right) u(t) \tag{2.20}$$

or

$$v(t) = E_m \left(1 - \cos \frac{1}{\sqrt{LC}} t\right) u(t) \tag{2.21}$$

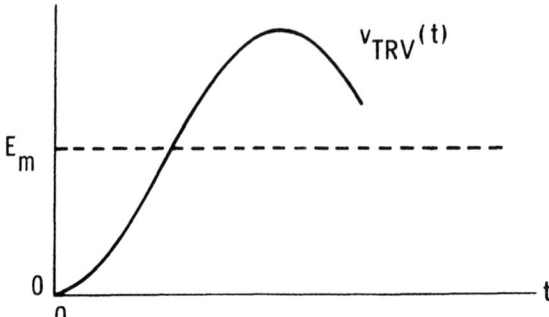

Figure 2.37 Estimated TRV (no damping).

Equation (2.21) is plotted in Fig. 2.37. Note that in this circuit no resistance has been included. The resistance inherent in the circuit, increased by skin effect because of the relatively high frequencies involved, will damp the overshoot to some extent, resulting in a lower peak voltage, as shown in Fig. 2.38. The key point is that the simplified method of analysis has resulted in an estimated transient recovery voltage of calculated frequency and shape. An actual power system is much more complicated than that shown in Fig. 2.36, as multiple sources and distributed parameters must be included.

To illustrate the effect of circuit parameters on transient recovery voltage, resistive, inductive, and capacitive circuits will be described separately. In the resistive circuit, the current is in phase with the system voltage and the transient recovery voltage is simply the sinusoidal power frequency voltage, as shown in Fig. 2.39. In the induc-

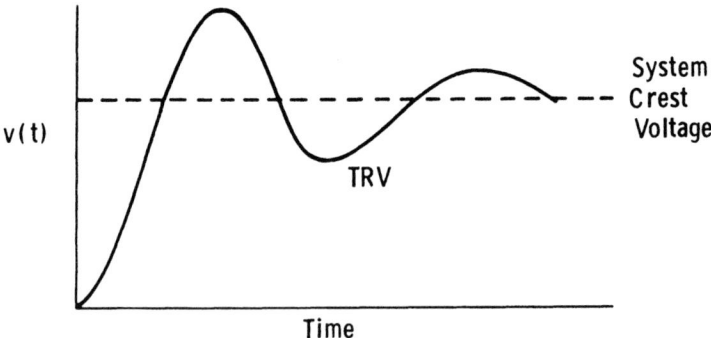

Figure 2.38 Damping of TRV.

Electrical and System Aspects

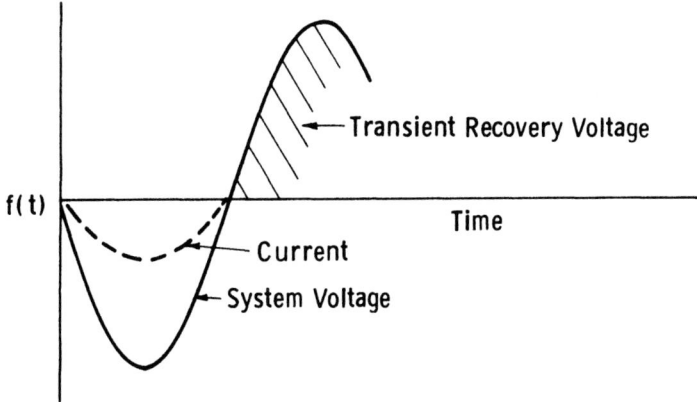

Figure 2.39 Resistive circuit.

tive case, which has previously been calculated, the current lags the system voltage by approximately 90° and the transient recovery voltage has a high-frequency component and an overshoot, as shown in Fig. 2.40. This is significantly more difficult to interrupt than the resistive circuit, with currents ranging from amperes to tens of kiloamperes. The capacitive circuit, shown in Fig. 2.41, has the current through the breaker limited by a capacitive element C. In this circuit

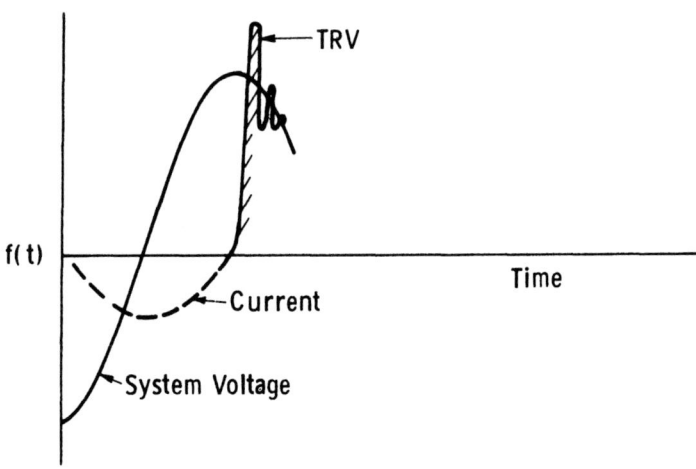

Figure 2.40 Inductive circuit.

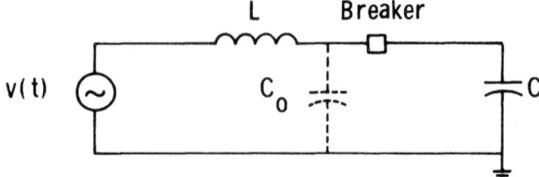

Figure 2.41 Capacitive circuit.

the current leads the system voltage by approximately 90°, and when the current is interrupted, the voltage on the capacitor bank is trapped. This is shown in Fig. 2.42.

One of the problems that arises with capacitor switching is the two-per-unit voltage stress that occurs across the contacts one-half cycle after interruption. Because the voltage across the contacts at current zero is very small and changes only at power system frequency, it is possible to interrupt the circuit if the current zero occurs approximately at the contact part. The later two-per-unit voltage then stresses a small gap distance and if a current reignition should occur, as shown in Fig. 2.42, theoretical overvoltages can reach three per unit, leading

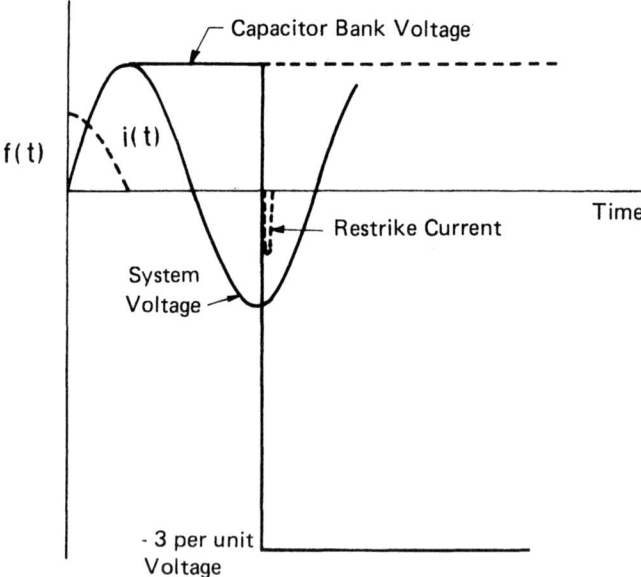

Figure 2.42 Capacitive circuit restrike.

Electrical and System Aspects

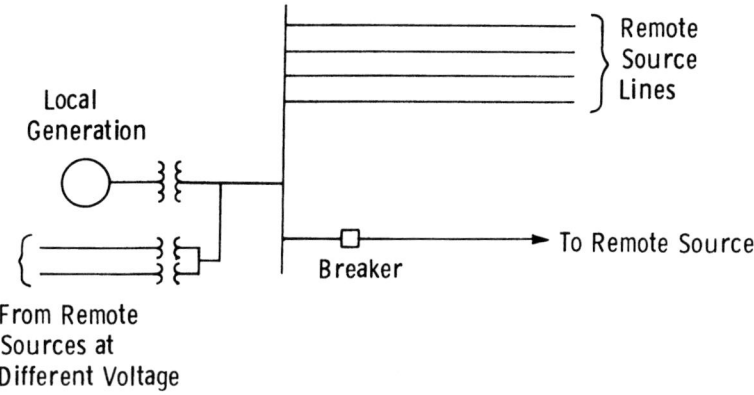

Figure 2.43 System representation.

to a four-per-unit voltage difference one-half cycle later. Multiple restrikes with reignitions occurring at later times can increase the stresses even further. This is discussed in detail in Chap. 3.

As is evident from this brief introduction, the process of interrupting an alternating current is complex and requires a knowledge of both arc physics and the stresses imposed by the system. A possible system configuration is shown in Fig. 2.43. Contributions to the fault current are supplied through local transformers and from remote sources over distributed lines. This combination of lumped and distributed parameters determines the shape of the transient recovery voltage and forms the basis for the ANSI standard rating and application requirements discussed in Chap. 3. Other more detailed electrical aspects of circuit interruption are dealt with in other chapters.

REFERENCES

1. F. E. Nixon, *Handbook of Laplace Transformation*, Prentice-Hall, Englewood Cliffs, N.J., 1960.
2. D. K. Cheng, *Analysis of Linear Systems*, Addison-Wesley, Reading, Mass., 1959.
3. T. E. Browne, Jr., The electric arc as a circuit element, J. Electrochem. Soc. *102*(1):27-37, 1955.
4. M. F. Gardner and J. L. Barnes, *Transients in Linear Systems*, Wiley, New York, 1956.

3
Circuit Breaker Application

CHARLES L. WAGNER /Westinghouse Electric Corporation, Pittsburgh, Pennsylvania

3.1 Circuit Breaker Standards   40
3.2 Standard Rating Tables   42
3.3 Rated Voltage   46
3.4 Insulation Levels   47
    3.4.1 Power frequency overvoltages   47
    3.4.2 Lightning surge voltages   48
    3.4.3 Switching surge overvoltages   55
    3.4.4 Special considerations   62
3.5 Continuous Current Rating   62
3.6 Interrupting Current Ratings   65
    3.6.1 Classical method   66
    3.6.2 Total current method   67
    3.6.3 Symmetrical current method   74
3.7 Momentary or Closing and Latching Rating   82
3.8 Reclosing Duty Factors   83
3.9 Rated Interrupting Time   84
3.10 Transient Recovery Voltage   85
    3.10.1 Standard TRV envelopes   87
    3.10.2 Calculation method   90
3.11 Capacitance Switching   105
    3.11.1 Interruption currents   107
    3.11.2 Transient recovery voltage   109
    3.11.3 Transient overvoltages   113
    3.11.4 Inrush currents   116
    3.11.5 Effect on other equipment   122

3.12 Other Application Considerations   122
    3.12.1 Live tank vs. dead tank   123
    3.12.2 Ganged vs. IPO breakers   125
    3.12.3 Open-breaker capacitance resonance   129

3.13 Summary   131

References   132

To apply a circuit breaker, like any other piece of electrical equipment, one must determine the operating requirements imposed by the system and select a device whose capabilities will meet or exceed these requirements. Circuit breaker capabilities are expressed in terms of rating values specified by industry standards. The application process therefore consists of system calculations to ensure that the various rated values will not be exceeded.

3.1 CIRCUIT BREAKER STANDARDS

Standards are a set of minimum requirements that all pieces of equipment such as circuit breakers must meet to perform their desired functions. Standards cover equipment design requirements to which the manufacturers must conform. They also cover the application requirements, which dictate the rules that the user must follow to ensure that the equipment will operate satisfactorily on the system.

    The first electrical standards were written by the AIEE (American Institute of Electrical Engineers), which is now the IEEE (Institute of Electrical and Electronic Engineers), in 1899 and were originally published in one volume covering all electrical equipment. In the early 1920s, separate standards on the various individual equipments were published. AIEE Standard No. 19 was the first standard pertaining specifically to circuit breakers. Today there are over a dozen separate standards on high-voltage circuit breakers alone. Reference 1 is an interesting treatise on the history of circuit breaker standards from 1899 to the present.

    It is important to note that practically each standard change or revision was precipated by a new system application problem or requirement. For example, the first circuit breaker standard defined the interrupting capability rating as "a given rms current at normal voltage to be interrupted two times in a 2 minute interval." It neglected to mention that there should be no oil or flame discharge or that interruption should occur in a specific time. This was, of course, changed in a later revision.

    More recently, when switching surges were found to be a limitation on extra-high-voltage (EHV) line insulation, methods of controlling

Circuit Breaker Application

Table 3.1 Present C37 High-Voltage Circuit Breaker Standards

Subject	Total current standard	Symmetrical current standard
Basis of Rating	C37.4	C37.04
Preferred Ratings	C37.6	C37.06
Reclosing Factors	C37.7	(Included in C37.06)
Test Code	C37.9	C37.09
Application Guide		
General	C37.5	C37.010
Transient Recovery Voltage		C37.011
Capacitance Current Switching		C37.012
Electrical Control Requirements	C37.11	C37.11
Guide Specifications	C37.12	C37.12
Synthetic Testing		C37.081
Sound-Level Measurement		C37.082
Definitions	C37.100	C37.100

these surges by circuit breakers were required, so the standards were revised to specify these control measures. In all these cases, the users and manufacturers got together to define the problem and the solution and document both in the appropriate standard.

Table 3.1 shows a listing of all the present high-voltage circuit breaker standards by subject and ANSI C37 Power Switchgear Committee number. Note that for most of the subject areas there are two documents. The first column (C37.4, C37.6, etc.) lists the documents for circuit breakers rated on a Total Current Basis. As will be described later, this was the original method of rating circuit breakers. In 1964, a new "Symmetrical Current" rating system was established and a new series of standards documents (C37.04, C37.06, etc.), shown in the second column, was written. The differences between the two rating systems are described in a later section. Most circuit breakers are manufactured today under the new Symmetrical Current Basis of rating, but there are still some of the old type being manufactured, so both sets of documents are still necessary.

Table 3.2 IEC High-Voltage Circuit Breaker Standards

Subject	Standard number
Direction of Motion of Operating Devices and Indicating Lamps	54
Part 1: General and Definitions	56-1
Part 2: Rating	56-2
Part 3: Design and Construction	56-3
Part 4: Type Tests and Routine Tests	56-4
Part 5: Rules for Selection of Circuit Breakers for Service	56-5
Part 6: Rules for Transport, Erection, and Maintenance	56-6
Guide to Testing with Respect to Switching of Shunt Capacitor Banks	56-7
HV Test Techniques and Measuring Devices	60
Out-of-Phase Switching	267
Synthetic Testing	427

In addition to the ANSI (American National Standards Institute) documents listed in Table 3.1, there are a series of International Electrotechnical Commission (IEC) documents on circuit breakers. These documents are listed in Table 3.2. Most of the requirements of the ANSI and IEC documents are the same, but there are a few differences. These differences will be discussed. In general, however, the application problems are similar, so the ANSI standards will be used here as the reference documents. For applications involving circuit breakers designed to IEC specifications, the respective IEC documents should be consulted.

3.2 STANDARD RATING TABLES

Tables 3.3 through 3.5 show typical rating tables taken from ANSI standards. These particular sample tables refer to outdoor circuit breakers rated 121 kV and above are taken from C37.06-1979 (the reference tables are shown in the table headings). There are similar tables in C37.06-1979 for indoor and outdoor circuit breakers rated 72.5 kV and below and in C37.6-1971 for the total current-rated circuit

Circuit Breaker Application

Line No.	Voltage		Insulation Level		Rated Values			Transient Recovery Voltage (16)				Rated Interrupting Time (7) Cycles	Rated Permissible Tripping Delay Y Seconds	Rated Max Voltage Divided by K kV, rms	Related Required Capabilities		
			Rated Withstand Test Voltage		Current										Current Values		
	Rated Max Voltage (1)a kV, rms	Rated Voltage Range Factor, K (2)	Low Frequency kV, rms	Impulse (3) kV, Crest	Rated Continuous Current at 60 Hz (4) Amperes, rms	Rated Short-Circuit Current at Rated Max kV) (5)(6) kA, rms		Rated Time to Point P T_2 μs	Rated Rate R kV/μs	Rated Delay Time T_1 μs					Max Symmetrical Interrupting Capability (8)	3-Second Short-Time Current Carrying Capability (9)	Closing and Latching Capability 1.6 K Times Rated Short-Circuit Current (9)(10) kA, rms
															K Times Rated Short-Circuit Current		
	Col. 1	Col. 2	Col. 3	Col. 4	Col. 5	Col. 6		Col. 7	Col. 8	Col. 9	Col. 10	Col. 11	Col. 12	Col. 13	Col. 14	Col. 15
	kV, rms					kA, rms		μs	kV/μs	μs	Cycles	Seconds	kV, rms	kA, rms	kA, rms	kA, rms
1	121	1.0			1200	20		276	1.7	2.9	3	—	121	20	20	32
2	121	1.0			1600	40		258	1.8	2.9	3	—	121	40	40	64
3	121	1.0			2000	40		258	1.8	2.9	3	—	121	40	40	64
4	121	1.0			2000	63		252	1.8	2.9	3	—	121	63	63	101
5	121	1.0			3000	40		252	1.8	2.9	3	—	121	40	40	64
6	121	1.0			3000	63		252	1.8	2.9	3	—	121	63	63	101
7	145	1.0			1200	20		331	1.7	3.2	3	—	145	20	20	32
8	145	1.0			1600	40		310	1.8	3.2	3	—	145	40	40	64
9	145	1.0			2000	40		310	1.8	3.2	3	—	145	40	40	64
10	145	1.0			2000	63		302	1.8	3.2	3	—	145	63	63	101
11	145	1.0			2000	80		298	1.8	3.2	3	—	145	80	80	128
12	145	1.0			3000	40		310	1.8	3.2	3	—	145	40	40	64
13	145	1.0		See Table 3.4	See Table 3.4	3000	63		302	1.8	3.2	3	—	145	63	63	101
14	145	1.0			3000	80		298	1.8	3.2	3	—	145	80	80	128
15	169	1.0			1200	16		396	1.7	3.4	3	—	169	16	16	26
16	169	1.0			1600	31.5		369	1.8	3.4	3	—	169	31.5	31.5	50
17	169	1.0			2000	40		361	1.8	3.4	3	—	169	40	40	64
18	169	1.0			2000	50		357	1.8	3.4	3	—	169	50	50	80
19	242(15)	1.0			1600	31.5		529	1.8	4.1	3	—	242	31.5	31.5	50
20	242(15)	1.0			2000	31.5		529	1.8	4.1	3	—	242	31.5	31.5	50
21	242(15)	1.0			3000	31.5		529	1.8	4.1	3	—	242	31.5	31.5	50
22	242(15)	1.0			2000	40		517	1.8	4.1	3	—	242	40	40	64
23	242(15)	1.0			3000	40		517	1.8	4.1	3	—	242	40	40	64
24	242(15)	1.0			3000	63		503	1.8	4.1	3	—	242	63	63	101
25	362(15)	1.0			2000	40		773	1.8	4.9	3	—	362	40	40	64
26	362(15)	1.0			3000	40		773	1.8	4.9	3	—	362	40	40	64
27	550(15)	1.0			2000	40		1325	1.6	5.4	2	—	550	40	40	64
28	550(15)	1.0			3000	40		1325	1.6	5.4	2	—	550	40	40	64
29	800(15)	1.0			2000	40		1531	1.9	7.9	2	—	800	40	40	64
30	800(15)	1.0			3000	40		1531	1.9	7.9	2	—	800	40	40	64

aSee notes, page 17 of Ref. 2. *Source:* Ref. 2.

Table 3.4 Schedule of Dielectric Test Values for Outdoor Circuit Breakers and External Insulation[a]

			Insulation Withstand Test Voltages							
	Low-Frequency		Impulse Test 1.2 × 50 Microsecond Wave					Switching Impulse		Minimum Creepage Distance of External Insulation to Ground, Inches
Rated Max Voltage kV, rms	1 Minute Dry rms kV	10 Second Wet rms kV	Full Wave Withstand kV Crest	Interrupter Full Wave kV, Crest	Chopped Wave,[b] kV Crest Minimum Time to Sparkover			Withstand Voltage Terminal to Ground with Circuit Breaker Closed kV, Crest	Withstand Voltage Terminal to Terminal on One Phase with Circuit Breaker Open kV, Crest	
					2 Microseconds Withstand	3 Microseconds Withstand				
Col. 1	Col. 2	Col. 3	Col. 4	Col. 5	Col. 6	Col. 7		Col. 8	Col. 9	Col. 10
15.5	50	45	110	—	142	126				9 (0.229 m)
25.8	60	50	150	—	194	172				15 (0.381 m)
25.8[c]	60	50	125	—	—	—				15 (0.381 m)
38.0	80	75	200	—	258	230		Not Required		22 (0.559 m)
38.0[c]	80	75	150	—	—	—				22 (0.559 m)
48.3	105	95	250	—	322	288				28 (0.711 m)
72.5	160	140	350	—	452	402				42 (1.067 m)
121	260	230	550	412	710	632				70 (1.778 m)
145	310	275	650	488	838	748				84 (2.134 m)
169	365	315	750	552	968	862				93 (2.489 m)
242[d]	425	350	900	675	1160	1040				140 (3.556 m)
362[d]	555	Not Required	1300	975	1680	1500		825	900	209 (5.309 m)
550[d]	860		1800	1350	2320	2070		1175	1300	318 (8.077 m)
800	960		2050	1540	2640	2360		1425	1550	442 (11.227 m)

[a]Refer to Tables 3, 4, and 5 of Ref. 2.
[b]1.2 × 50 microsecond positive and negative wave. All impulse values are phase-to-phase and phase-to-ground and across the open contacts.
[c]These circuit breakers are intended for application on grounded wye distribution circuits equipped with surge arresters.
[d]These circuit breakers are intended for application only on systems effectively grounded, as defined in *Neutral Grounding Devices*, IEEE Publication No. 32, 1972.
NOTE: For dielectric test voltage values for indoor ac high-voltage circuit breakers, see Tables 1 and 2 of Ref. 2.
Source: Ref. 2.

Circuit Breaker Application

Table 3.5 Preferred Capacitance Current Switching Ratings for Outdoor Circuit Breakers

			General-Purpose Circuit Breakers					Definite-Purpose Circuit Breakers				
			Rated Capacitance Switching Current[a]					Rated Capacitance Switching Current[a]				
					Shunt Capacitor Bank or Cable					Shunt Capacitor Bank or Cable		
						Back-to-Back					Back-to-Back	
							Inrush Current[a]					Inrush Current[a]
Line No.	Rated Max Voltage kV, rms	Rated Short-Circuit Current kA, rms	Rated Continuous Current[a] Amperes, rms	Overhead Line Current Amperes, rms	Isolated Current[a] Amperes, rms	Current[a] Amperes, rms	Peak Current kA	Frequency Hz	Overhead Line Current Amperes, rms	Isolated Current[a] Amperes, rms	Current[a] Amperes, rms	Peak Current kA	Frequency Hz
1	121	20	1200	50	50				160	315	315	16	4250
2	121	40	1600	50	50				160	315	315	16	4250
3	121	40	2000	50	50				160	315	315	16	4250
4	121	63	2000	50	50				160	315	315	16	4250
5	121	40	3000	50	50				160	315	315	16	4250
6	121	63	3000	50	50				160	315	315	16	4250
7	145	20	1200	63	63				160	315	315	16	4250
8	145	40	1600	80	80				160	315	315	16	4250
9	145	40	2000	80	80				160	315	315	16	4250
10	145	63	2000	80	80				160	315	315	16	4250
11	145	80	2000	80	80				160	315	315	16	4250
12	145	40	3000	80	80				160	315	315	16	4250
13	145	63	3000	80	80				160	315	315	16	4250
14	145	80	3000	80	80	No ratings are established since these breakers should not be applied for back-to-back switching			160	315	315	16	4250
15	169	16	1200	100	100				160	400	400	20	4250
16	169	31.5	1600	100	100				160	400	400	20	4250
17	169	40	2000	100	100				160	400	400	20	4250
18	169	50	2000	100	100				160	400	400	20	4250
19	242	31.5	1600	160	160				200	400	400	20	4250
20	242	31.5	2000	160	160				200	400	400	20	4250
21	242	31.5	3000	160	160				200	400	400	20	4250
22	242	40	2000	160	160				200	400	400	20	4250
23	242	40	3000	160	160				200	400	400	20	4250
24	242	63	3000	160	160				200	400	400	20	4250
25	362	40	2000	250	250				315	500	500	25	4250
26	362	40	3000	250	250				315	500	500	25	4250
27	550	40	2000	400	400				500	500	—	—	—
28	550	40	3000	400	400				500	500	—	—	—
29	800	40	2000	500	500				500	500	—	—	—
30	800	40	3000	500	500				500	500	—	—	—

[a]See note and footnotes to Table 1A of Ref. 2.
Source: Ref. 2.

breakers. Although the numbers may be different, the format for all the tables is similar, so Tables 3.3 through 3.5 will be used to demonstrate the various rating quantities involved, their meaning, and the system calculations required in the application to ensure that these rated values are not exceeded.

3.3 RATED VOLTAGE

Circuit breakers have a rated maximum voltage, shown in column 1 of Table 3.3, which is the highest voltage at which the breaker is designed to operate satisfactorily and perform at its rated capability. Note that this is a *maximum* voltage as distinguished from a normal or *nominal* system voltage. Power systems may be called 110- or 115-kV systems and normally operate at approximately this voltage. In applying circuit breakers, however, one must ensure that the system voltage does not exceed the rated maximum voltage, which in this case is 121 kV, for any normal or abnormal condition under which the circuit breaker will be called upon to operate.

For the circuit breakers of Table 3.3, the maximum short-circuit currents that can be interrupted are independent of operating voltage. For some ratings, however, the fault current capability is inversely proportional to operating voltage up to a certain limit, which is called the Rated Voltage Range Factor K, shown as column 2 of Table 3.3. While Table 3.3 shows a K of 1.0, some circuit breakers rated 72.5 kV and below have K factors as high as 3.75. For these ratings, the inverse relationships hold up to a limit of K times the rated short-circuit current.

For example, assume a circuit breaker having a Rated Maximum Voltage of 15 kV, a K factor of 3.75, and a Rated Short-Circuit Current of 9.8 kA. If this circuit breaker were applied on a system that had a maximum operating voltage of 12 kV, its capability would be

$$\text{Interrupting capability} = \text{Rated Short-Circuit Current} \times \left(\frac{\text{Rated Maximum Voltage}}{\text{Operating Voltage}} \right)$$

$$= 9.8 \times \frac{15}{12} = 12.25 \text{ kA}$$

Since this is less than K × Rated Short-Circuit Current (= 9.8 × 3.75 = 36.75 kA), the circuit breaker could be applied on this system provided that the fault current at this point did not exceed 12.25 kA. If this unit had a K factor of 1.1 rather than 3.75, however, its limit would be 1.1 × 9.8 = 10.78 kA rather than 12.25 kA. This factor is discussed further in a later section.

Circuit Breaker Application

3.4 INSULATION LEVELS [2a]

Like other system and equipment insulation, circuit breaker insulation is exposed to three types of system overvoltages: power frequency (50 or 60 Hz) overvoltages and lightning and switching surges. Table 3.4 defines the required insulation strength of circuit breakers for these three types of overvoltages. In the following sections the origin of these overvoltages is discussed, together with the methods commonly used to limit their magnitudes and any special application problems encountered in meeting the required circuit breaker insulation ratings.

3.4.1 Power Frequency Overvoltages

As discussed above, the rated maximum voltage is the upper limit for normal operation of the breaker. During abnormal conditions the breaker insulation systems may be subjected to higher line-to-ground and open-gap voltages. Under line-dropping conditions, the Ferranti effect of the line and overspeed conditions can cause as much as a 20 to 30% rise at the far end of the open line. A line-to-ground fault on the system can cause the line-to-ground voltage on the unfaulted phase to exceed normal by as much as 73%, depending on the system grounding. The voltage across the open breaker can be as high as 2.2 times normal if the two systems connected to the open breaker are operating out of synchronism.

Column 2 of Table 3.4 shows the low-frequency test voltage in root-mean-square (rms) kilovolts. Note that in all cases these values are higher by a large margin than the power frequency overvoltages that can occur on utility systems. The reason for this apparently excessive margin goes back to the history of insulation testing. Before much was known about lightning and switching surges, the only test given circuit breakers as well as other electrical apparatus was the low-frequency 1-min 50- or 60-Hz test. The test values were selected based on experience and were shown to be sufficient proof that the breaker would adequately withstand all the overvoltages, which we now know as lightning and switching surges, that occur on operating systems.

A question often raised is whether the low-frequency test could not now be eliminated since the power frequency overvoltages are so much lower than the test values, and since lightning and switching surge performance is now verified by direct testing. The main reason it has not been eliminated is that the low-frequency test is the only production test performed on breakers or breaker components, and is thus a quality control measure for the insulation systems.

Column 3 of Table 3.4 shows that for breakers rated 242 kV and below, a low-frequency test voltage somewhat lower than the dry test value should be applied to the breaker when it is exposed to simulated rainfall. Again, these test values are higher than the power frequency overvoltages encountered on systems, but were established to demonstrate the switching impulse withstand of the breakers. At the extra high voltages, 362 kV and above, separate wet and dry switching impulse tests have been established, so low-frequency wet tests are no longer required at these ratings.

In recent years contamination testing has become of increasing interest, but to date no uniformly acceptable test procedure has been established. Circuit breaker standards recognize this requirement by specifying a minimum creepage distance required for all external insulation paths to ground. These minimum distances are shown in column 10 of Table 3.4. Examination of the values of column 10 will show that each listed distance in inches divided by the maximum rated voltage, expressed as an rms line-to-ground value, equals 1.0. This 1.0-in./kV value has been shown by experience to be adequate for light to moderate contamination. Some transmission lines and stations have operated satisfactorily with values as low as 0.68 and 0.83 in./kV. For heavily contaminated areas such as those exposed to salt, fog, or chemical fumes, longer creepage distances are required. Transmission lines having a 2.0-in./kV criterion are in service under these conditions. Special breaker bushings may be required in these areas.

From a power frequency standpoint, therefore, the only application problem to consider is contamination. If heavy contamination is possible, special high-creepage bushings or support columns may be required. In most cases, however, the power frequency insulation levels are considerably higher than the overvoltage encountered on utility systems.

3.4.2 Lightning Surge Voltages

The ability of a circuit breaker to withstand lightning impulse voltages is defined by ANSI by the three test values shown in Table 3.4. Column 4 shows the Full-Wave Withstand level, often called the Basic Impulse Level (BIL). This is based on a 1.2 × 50 impulse wave, which is a wave whose front time is 1.2 μs and decreases to half value in 50 μs. As will be discussed subsequently, this is not fully descriptive of all surges, so columns 6 and 7 specify the 2- and 3-μs Chopped Wave Withstand levels. These tests use the same 1.2 × 50 test wave except that the tail of the wave decreases to zero, or is "chopped" at times of 2 and 3 μs, respectively. Note that the values of columns 6 and 7 are 1.29 and 1.15 times the BIL of the circuit breaker shown in column 4. This relationship reflects the inherent characteristics of external line and equipment insulation and of the rod gaps that are sometimes used as protection. The volt-time curve of sulfur hexafluoride (SF_6) equipment is relatively flat, however, and for this equipment it is often necessary to raise the internal BIL to meet the chopped wave levels.

Circuit Breaker Application

IEC lightning impulse standards are somewhat different from the ANSI standards. At 242 kV and below, the Full-Wave Withstand levels (BILs) are essentially the same as ANSI, but IEC does not specify any Chopped Wave Withstand levels, so for these ratings the ANSI requirements are more severe. At 300 kV and above, the IEC BILs are lower. For example, the BIL for 550 kV breakers are 1550 kV as opposed to the ANSI level of 1800 kV. Again, no chopped wave levels are specified. IEC, however, specifies a "bias test" for the open breaker. In this test, a voltage equal to 70% of the normal line-to-neutral rated voltage is applied to one terminal of the open breaker at the same time that the full-wave impulse test level is applied to the other terminal. For the 550-kV breaker this bias voltage is 300 kV, making the total applied voltage across the open breaker equal to 1550 kV plus 300 kV, or 1850 kV. This is still less than the 2320-kV chopped wave level of ANSI, however, so the ANSI test is more severe.

Whereas in the preceding section it was shown that there were very few application problems from a low-frequency insulation standpoint, since the power frequency insulation strength was so much greater than the maximum system power frequency overvoltages that could occur, quite the opposite is true for lightning impulse voltages. The potential overvoltages generated by lightning far exceed any practical insulation level that could be designed into the circuit breaker. The application problem, therefore, is to select the proper protective devices, such as surge arresters, to limit overvoltages to the proper levels. This requires knowledge of the surge origin and the protective levels that can be obtained.

Lightning surges can originate in a station by either a direct stroke in the station or by a stroke to a transmission line feeding the station [3]. In a well-designed station the former is extremely rare, so only line strokes are usually considered in lightning insulation studies.

Surges originating on the line can be caused by direct stroke to the conductor (shielding failure) or by a stroke to the ground wire or tower which then flashes over to the conductor (back flashover). Again, with a properly shielded line, the latter strokes are of primary concern.

The magnitude and wave shape of the lightning surge entering a station depend on the insulation level of the line and the distance between the point of stroke origin and the station. Figure 3.1 shows a typical surge voltage entering a 550-kV station as a function of stroke terminating distance. For a stroke just outside the station (0.0 mi) the wave has a high short voltage peak and then drops off at a relatively slower rate. For strokes originating farther from the station, this peak is attenuated and distorted by line losses and corona. The 0.5-, 1.0-, and 2.0-mi curves of Fig. 3.1 show this attentuation and front distortion.

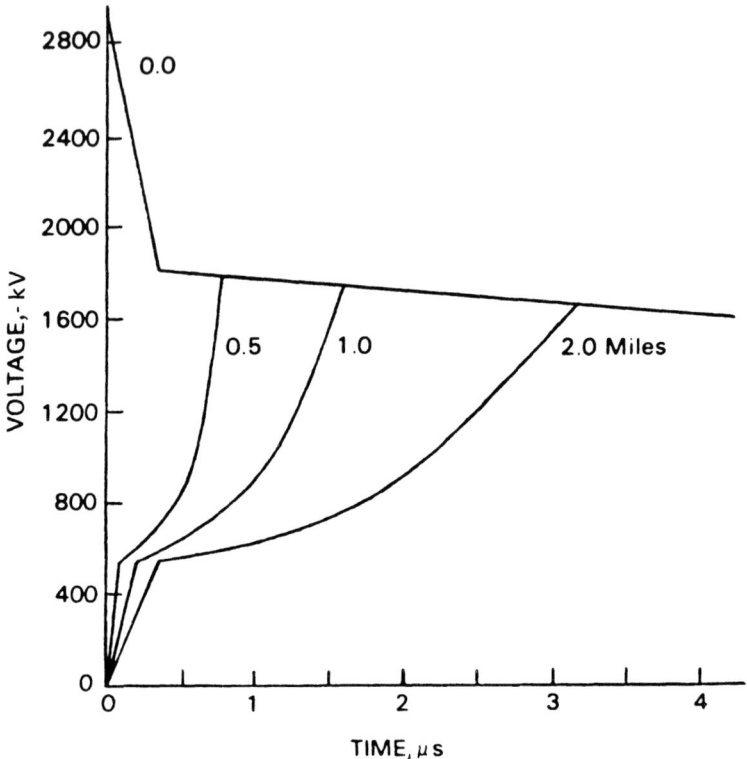

Figure 3.1 Typical incoming surge voltages for strokes terminating on a 550-kV line 0.0, 0.5, 1.0, and 2.0 mi from the station. (From C. L. Wagner, Insulation considerations for ac high-voltage circuit breaker, IEEE Tutorial Course on Application of Power Circuit Breakers, IEEE Spec. Publ. 75CH0975-3-PWR, pp. 52-61, © 1975, IEEE.)

When the surge arrives it is modified by the terminating impedance of the station. Taking the extreme condition of the stroke terminating at an open breaker, the surge would double in amplitude. Thus even for the 2.0-mi stroke of Fig. 3.1, the surge seen by the breaker would exceed 3000 kV. Obviously, 550-kV equipment insulation levels of this magnitude are impractical, so some means of protection is required.

The required impulse insulation levels for substation equipment depend on the protective characteristics of the voltage-limiting device and the number and location of the devices that are applied in the substation [4-7]. In the station itself surge arresters are the normal protective device used, while arresters or rod gaps are used on the line entrance terminals.

Circuit Breaker Application

Closed-Breaker Protection

Although a complete dissertation on arrester application is beyond the scope of this chapter, a brief summary of the process will be given. References 4-7 can be consulted for a more complete description.

The first step in the process is to select an arrester rating. This requires a study of the maximum power frequency system voltages that can appear across the arrester terminals during faults or other system abnormalities. This factor is a function of the system grounding and depends of the X_0/X_1 and R_0/X_1 ratios at the arrester location during various system operating conditions [3,5].

Once the arrester rating is selected and its protective characteristics specified, the number and location of the arresters must be determined. Surge arresters are normally applied directly on the terminals of the power transformers, so transformer insulation is determined by adding a suitable margin (usually 15%) to the arrester protective characteristics and using this for the transformer BIL.

Usually, however, circuit breakers must rely on surge arresters located remotely from their terminals for protection. Traveling waves are produced which develop higher voltages at these remote locations. Reference 4 shows a means for determining these voltages and computer programs are available for even the most complex stations. Figure 3.2 shows a typical output of such a program. The solid curve is the voltage at the arrester, with the dashed curve the voltage at the termination point, which in this case is a transformer rather than a breaker. Note that whereas the arrester limits its voltage to less than 500 kV, the voltage 200 ft away at the transformer is over 600 kV.

The wave shape of Fig. 3.2 is typical of these types of surges: fronts of 1 to 2 μs and sharply falling tails. Since this shape does not agree with the standard 1.2 × 50 test wave shape used to determine the BIL in column 4 of Table 3.4, two methods have been used to analyze the severity of the overvoltage. The first method, *time-point comparison*, compares the peak of the overvoltage with the magnitude of the standard surge taken off the 1.2 × 50 time-lag curve (obtained by plotting the BIL and chopped wave levels from Table 3.4) at a time at which the odd-shaped surge is effectively chopped to zero. For example, the wave of Fig. 3.2 would be compared to a standard 1.2 × 50 wave chopped at approximately 1.5 μs. Since some judgment is required to select this time value, a margin of at least 10% is used in the evaluation.

The second method of analysis is the *severity index method*, described in Ref. 8. This method equates the Destructive Energy (DE) of the nonstandard wave with that of the standard 1.2 × 50 wave to make the comparison. The DE is the time integral of the voltage wave raised to the K power, where K depends on the shape of the time-lag curve for the insulation being stressed. Again, a margin of at least 10% should be used in the evaluation.

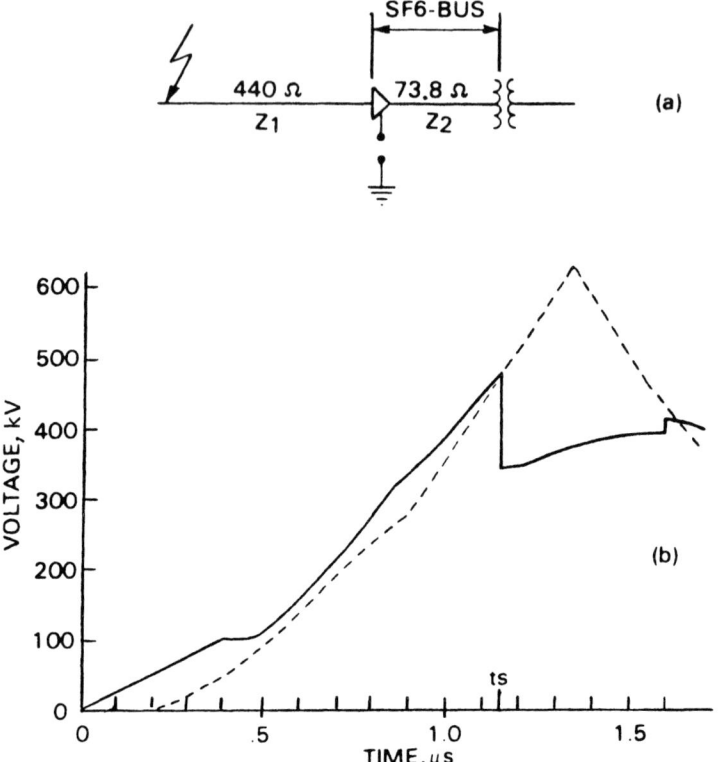

Figure 3.2 Voltage buildup in a 242-kV gas insulated system with a 192-kV arrester. Solid curve, voltage at the arrester; dashed-curve, voltage at the transformer. (From C. L. Wagner, Insulation considerations for ac high-voltage circuit breaker, IEEE Tutorial Course on Application of Power Circuit Breakers, IEEE Spec. Publ. 75CH0975-3-PWR, pp. 52-61, © 1975, IEEE.)

The next step in the investigation depends on the purpose of the particular study. References 5 and 6 describe studies of new equipment where the purpose of the study was to determine what BIL the new breaker should have. Here the problem is to balance the cost of additional surge arresters against the savings in breaker cost due to the lower BIL. Figure 3.3 shows what is involved in such a determination. The figure shows a typical 550-kV substation layout with two transformers and nine breakers. The breaker positions and bus lengths in feet are shown on the diagram. Surge arresters would definitely be located on the transformers and it remained to be determined where and how many additional arresters would be needed to

Circuit Breaker Application

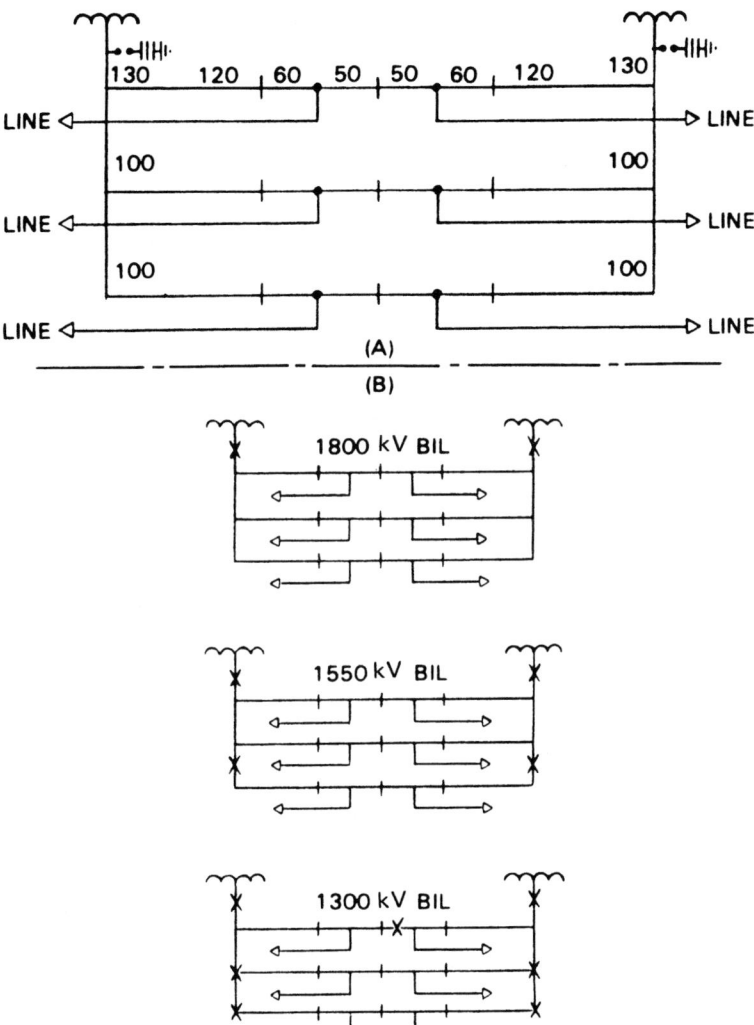

Figure 3.3 Typical station layout. Arrester locations indicated by X. Short cross lines indicate breaker locations, with all distances in feet. (From C. L. Wagner, Insulation considerations for ac high-voltage circuit breaker, IEEE Tutorial Course on Application of Power Circuit Breakers, IEEE Spec. Publ. 75CH0975-3-PWR, pp. 52-61, © 1975, IEEE.)

protect the breakers. Using the methods described above, the voltages throughout the substation were determined for the worst breaker-open surge origination combination involved. Then the arrester locations to protect the various breaker BILs considered were determined. In this case, use of two transformer arresters was all that was needed to protect 1800-kV BIL breakers. If the breaker BIL was reduced to 1550 kV, two additional arresters were needed. For 1300-kV BILs a total of five additional arresters were required. Balancing the cost of the arresters against the savings in breaker insulation led to the selection of 1550 kV in this particular case [5].

The more usual application of this method is to start with an existing breaker insulation level and simply determine the number and location of arresters to protect the breakers. Using Fig. 3.3 again as the example, since the standard 550-kV breaker has an 1800-kV BIL, the application study would show that the two transformer surge arresters provide adequate protection.

Open-Breaker Protection [7]

The preceding discussion showed that the station arresters provide protection for all closed breakers on the bus and even the bus side of open breakers. However, if a breaker is open, these arresters cannot provide protection to the line side of the breaker.

Three methods of solution to this problem are: (1) apply arresters on the line side of all breakers; (2) apply rod gaps on the line side of all breakers; or (3) do nothing. The selection of the proper method requires an evaluation of the economics and the risks in any given application.

Some of the factors to be considered are listed below.

1. *Surge arresters*: Surge arresters applied to the line side of every circuit breaker give absolute protection for the circuit breaker against all lightning and switching surges. The obvious disadvantage to this scheme is its high cost. The low probability of breaker flashover, discussed below, often negates justification of this large expenditure.

2. *Rod gaps*: Rod gaps can be applied on the line side of the breaker at about a 30:1 decrease in cost over the use of surge arresters. The quality of protection is somewhat poorer since a gap spacing must be selected that will protect the open breaker but still not flash over on lightning and switching surges when the breaker is closed. Reference 7 describes in detail the problems and compromises required to provide this protection. The gap spacing must be large enough so that the bus arresters will prevent flashover of the gap due to lightning or switching surges when the breaker is closed. The gap spacing must be small enough to provide protection for the breaker in the open position. Meeting these two requirements is often a delicate proposition.

Circuit Breaker Application

3. *No protective device*: The third approach is to do nothing on the line terminals and to rely on the low probability that surges of high enough magnitude to cause damage will impinge on the open breaker. This low probability can be demonstrated by examining the situations that can occur to cause this condition.

The first situation occurs when the breaker is standing open on the line. Normal operating practice, however, dictates that when a breaker is to be open for a prolonged period of time, the breaker disconnecting switch is also opened, so that no surge would reach the breaker. Departures from this practice are extremely rare, but in those few cases where it is normal operating practice, some form of protection should be employed.

The second situation, which is of primary concern, is caused by the second component of a lightning stroke. Consider a lightning stroke hitting a transmission line and causing a flashover to the phase conductor. The resulting surge voltage is transmitted to the station, and since the breaker has not yet opened, the breaker is protected by the arrester within the station. Should the lightning stroke exhibit a second component, a second surge will be transmitted toward the station. If the breaker has now opened or is in the process of opening, the bus-side arrester will not now provide protection to the line side of the breaker and flashover could occur. Reference 7 discusses the probabilities of this occurrence. Based on an 1800-kV BIL 550-kV breaker, it is estimated that the probability of this second component causing a flashover is once in 40 to 800 years. For a single breaker, this appears to be a low probability. The counterargument is that on a utility system there are many such breakers, which increases the probability that a flashover will occur.

3.4.3 Switching Surge Overvoltages

The third type of overvoltage to which circuit breakers are subjected is ironically usually caused by the circuit breaker itself, as evidenced by the name "switching surges." There are many types of switching surges that occur on a system. There are surges caused by faults occurring on the system and surges caused by circuit breakers clearing the fault. There are surges caused by line dropping and load rejection due to opening load-carrying lines. Out-of-step switching causes system surges. All of these surges cause problems to lines and certain electrical apparatus, but the major source of switching surges that cause the most concern to circuit breaker insulation and application are surges caused by closing and reclosing transmission lines [9-11].

Figure 3.4 shows the simplified theory of how switching surges are generated. Since the maximum switching surge usually occurs when reclosing into an uncompensated transmission line, Fig. 3.4(a) shows a

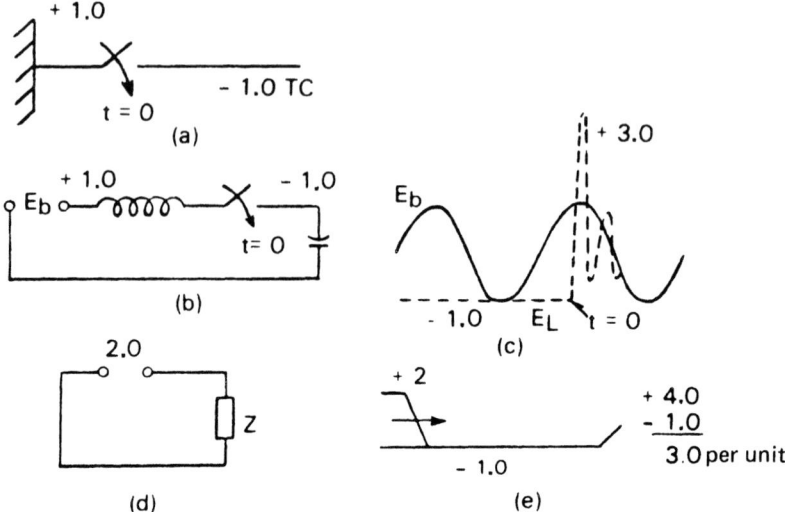

Figure 3.4 Simplified illustration of the switching surge generated by the reclosing of the circuit breaker: (a) system condition represented; (b) and (c) 60-Hz solution; (d) and (e) traveling-wave solution. (From C. L. Wagner, Insulation considerations for AC high-voltage circuit breaker, IEEE Tutorial Course on Application of Power Circuit Breakers, IEEE Spec. Publ. 75CH0975-3-PWR, pp. 52-61, © 1975, IEEE.)

breaker closing into a line having a −1.0 per unit trapped charge at the instant that the bus voltage is +1.0 per unit.

Figure 3.4(b) and (c) show how this condition can be analyzed from a power frequency standpoint. In this case the line is represented by a capacitor and the system is represented by an inductance. Note that with no damping, the line voltage starts from a −1 per unit, approaches the bus voltage and overshoots to a maximum of +3 per unit, and then oscillates around the bus voltage. Figure 3.4(d) and (e) show how this condition can be analyzed from a traveling-wave standpoint. In this case the line is represented by a surge impedance and the voltage traveling down the line is equal to the voltage across the breaker just prior to closing (equal to 2 per unit). When the wave hits the end of the line it doubles to a value of 4 per unit, but since it started at −1, the net value is 3 per unit. Thus the two methods of analysis give the same value, 3 per unit.

Figure 3.4 is an extremely simplified description of the phenomenon and neglects such factors as source impedance, other lines, coupling, and so on. Many computer studies have been run examining the effects

Circuit Breaker Application

of these factors, and many papers are available on the subject [10,11]. It does illustrate the cause of the problem and can be used to describe the methods of protection and control.

Protection

Surge arresters again act as the ultimate protection of circuit breakers and other apparatus, so the comments given previously for lightning impulses also apply for switching surges, with one exception. The exception is that the wavefronts of switching surges are usually much longer than lightning surges. For example, while the standard lightning impulse test uses a 1.2 × 50 wave, the standard switching impulse test wave is 250 × 2500. This slower wavefront means that the traveling-wave effect of separation distance no longer applies, and a lightning arrester located anywhere in a situation holds the switching surge level to essentially a constant value [5]. When surge arresters are applied for breaker protection, therefore, there is no problem with switching surges.

For transmission lines and the line terminals of open breakers, there may be a problem. At 242 kV and below, transmission lines and circuit breakers designed to give satisfactory outage performance from a lightning standpoint will not experience flashovers due to switching surges since the switching impulse insulation strength is inherently well above the maximum surges that can occur on the system. For EHV systems, 362 kV and above, this is not the case. The surges that can occur with conventional breakers can exceed the switching impulse critical flashover level of the line. Some means of control in the circuit breaker must therefore be utilized.

Control

The most common means of control used today consists of one or more steps of resistance inserted in the contact circuit of the circuit breaker prior to closing of the main contacts [10,11]. Figure 3.5(a) shows schematically how this is accomplished. During a closing operation (at t = 0) an auxiliary contact of the breaker closes and inserts a resistor R in series with the bus and line. After a short time delay (t = t_1) the main contacts of the breaker close, shorting out the resistor, and the breaker is therefore in its normal closed current-carrying condition. Figures 3.5 and 3.6 show in a manner similar to that of Fig. 3.4 how this resistor reduces the magnitude of the switching surges.

Figure 3.5 shows the single resistor breaker using the traveling-wave analysis. Here we have two transients, the first [represented by Fig. 3.5(b) and (c)] caused by inserting the resistor in the circuit, and the second [shown in 3.5(d) and (e)] caused by shorting out the resistor. In Fig. 3.5(b), the circuit is represented by the resistor in the breaker being in series with the surge impedance of the line. The

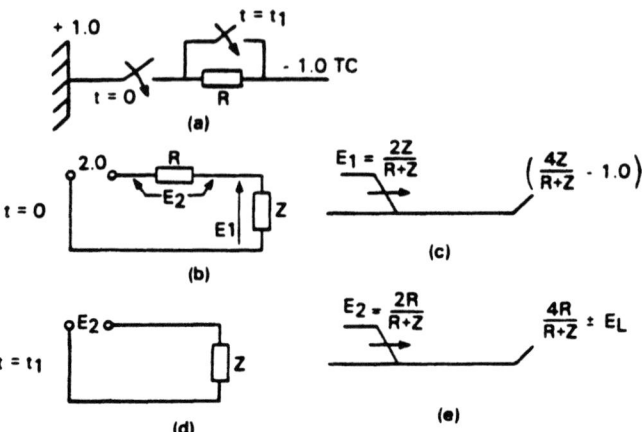

Figure 3.5 Traveling-wave analysis for a single-resistor circuit breaker. (From C. L. Wagner, Insulation considerations for AC high-voltage circuit breaker, IEEE Tutorial Course on Application of Power Circuit Breakers, IEEE Spec. Publ. 75CH0975-3-PWR, pp. 52-61, © 1975, IEEE.)

voltage surge traveling down the line, E_1, is equal to $2Z/(R + Z)$, which is the 2 per unit voltage across the breaker impressed across the potential divider circuit of R and Z in series. Note in Fig. 3.5(c) this E_1 surge doubles at the end of the line and the resultant voltage is this value minus the 1 per unit trapped charge voltage. From this circuit it can be seen that the higher the value of R, the lower the value of the insertion transient.

Figure 3.5(d) and (e), although not strictly correct, show in general terms that the higher the resistor, the higher the transient caused when the resistor is shorted. E_2 in Fig. 3.5(d) and (e) is equal to the voltage E_2 in Fig. 3.5(b), which of course increases as the resistor R increases. Figure 3.5(e) shows that this value of E_2, equal to $2R/(R + Z)$, doubles at the end of the line, $4R/(R + Z)$, and is superimposed on the power frequency voltage E_L at the particular instant of closure.

Figure 3.6 is a better way of showing the latter effect. It shows the power frequency equivalent of the circuit after the resistor has been inserted and all the transients due to inserting the resistor have decayed. Note that the voltage E_R is now the driving voltage when the resistor is shorted out at time t_1. If the main contact should close at the time the resistor voltage E_R is equal to 0 (shown at t_{1a}), there

Circuit Breaker Application

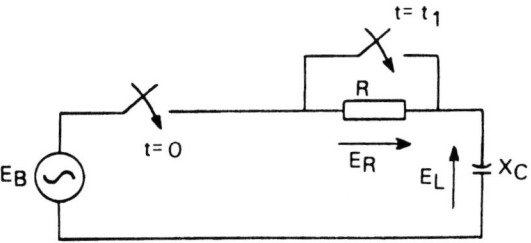

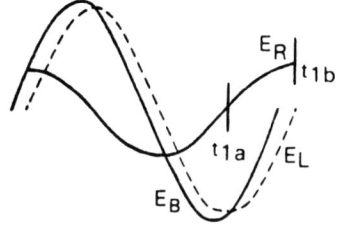

Figure 3.6 60-Hz analysis for a single-resistor circuit breaker. (From C. L. Wagner, Insulation considerations for AC high-voltage circuit breaker, IEEE Tutorial Course on Application of Power Circuit Breakers, IEEE Spec. Publ. 75CH0975-3-PWR, pp. 52-61, © 1975, IEEE.)

would be no shorting out transient in the circuit. The maximum surge occurs, however, when this E_R is a maximum (shown at t_{1b}). Note here that the larger the value of R, the higher the value of E_R and therefore the higher the maximum shorting-out transient.

While Figs. 3.5 and 3.6 describe how a single-step resistor breaker functions, other forms of control are possible and are used [11]. Two or three steps of resistance can further reduce the insertion and shorting-out transients. Controlled closing of the contacts has also been used. For example, if the breaker of Fig. 3.6 were controlled always to close the main contacts at t_{1a}, the shorting-out transient would be reduced or eliminated, allowing a higher value of R and thus a lower insertion transient. These and other schemes have been used to some extent, but use of the single inserted resistor remains the primary method of control.

Many computer studies have been run to determine the maximum switching surges that can occur on systems. The results of one study [11] show the maximum switching surges with and without control in the breaker as follows:

Control means	Maximum switching surge voltage (per unit)
No control	3.9
One resistor	2.05
Two resistors	1.7
Three resistors	1.5
Two resistors with controlled closing	1.5

In addition, many field tests have been run to measure surges that actaully occur on power systems. Figure 3.7 shows the results of many switching operations on a 362-kV system. Comparing Fig. 3.7 with the table above shows that, in general, computer studies are conserva-

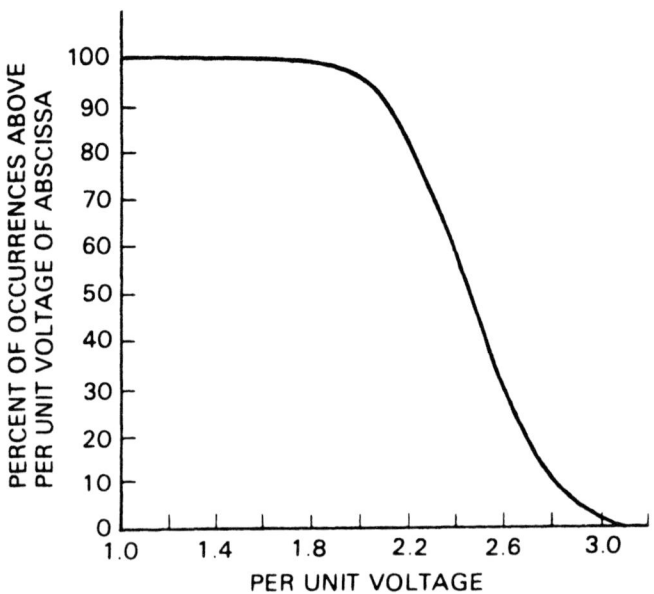

Figure 3.7 Cumulative frequency distribution curve for field test measurements of switching surge magnitudes due to reclosing on unterminated transmission lines. (From C. L. Wagner, Insulation considerations for AC high-voltage circuit breaker, IEEE Tutorial Course on Application of Power Circuit Breakers, IEEE Spec. Publ. 75CH0975-3-PWR, pp. 52-61, © 1975, IEEE.)

Circuit Breaker Application

tive. For purposes of insulation coordination studies, therefore, a value of 3.00 per unit is usually used as the maximum uncontrolled switching surge that will be encountered on a system.

Switching Impulse Insulation Levels

Columns 8 and 9 of Table 3.4 give the ANSI switching impulse withstand levels for line-to-ground and terminal-to-terminal or open gap insulation in circuit breakers. Note that no values are given for units rated at 242 kV and below. The reason for this is that the inherent switching impulse strength of this insulation is greater than 3.0 per unit, so tests at these ratings are not necessary.

For circuit breakers rated 362 kV and above, Table 3.4 shows that two switching impulse test voltages are specified. The first is the terminal-to-ground withstand level measured with the breaker in the closed position. All these values are less than the maximum uncontrolled surge voltage of 3.0 per unit that can occur on power systems. This is not a problem, however, for with the breaker closed, the breakers are protected by the bus-side arresters, whose switching surge protective levels are well below the values given in the tables.

With the circuit breaker open, however, the bus surge arresters are of no assistance and if line entrance arresters are not used, the open-gap insulation must exceed the maximum surge that can occur. Table 3.4 shows that the 362-kV rating has a 3.05 per unit withstand, which is greater than the maximum uncontrolled surge. This level was established, since some 362-kV breakers use switching surge control measures and some do not.

At 550 kV all breakers use some method of control. Table 3.4 shows a withstand level of 2.9 per unit, which is still considerably above the 2.05 per unit level given above for a "one-resistor" breaker and the 2.2 per unit switching surge factor specified in C37.06 for 550-kV circuit breakers. The withstand level for 800-kV units is 2.48 per unit from Table 3.4, which again is above the 1.7 and 2.05 per unit given in the table above and the 2.0 per unit specified in C37.06.

The IEC line-to-ground switching impulse withstand levels are essentially the same as ANSI's (i.e., 1175 kV for a 550-kV breaker). For the open-gap insulation, however, they again specify a bias test. In this case a reduced voltage (900 kV for the 550-kV breaker) is applied to one terminal of the open breaker while full line-to-neutral rated crest voltage (430 kV) is applied to the other terminal. This gives 1330 kV total voltage across the breaker instead of the 1300 kV specified by ANSI.

From an application standpoint, however, there is no problem in protecting circuit breakers against switching surges. The only consideration for switching surges is to specify the proper control measures for 362 kV and higher ratings, although even here, except in rare cases, the standard breakers are satisfactory for all systems.

3.4.4 Special Considerations

The preceding sections have discussed the usual insulation factors that must be considered in the application of circuit breakers. There are other factors that must be considered in special applications. For example, standard breakers are designed to operate at altitudes of up to 3300 ft (1000 m). For altitudes greater than this, a derating factor on the rated maximum voltage and the basic impulse insulation level must be applied [12]. These factors are 1.0 for up to 3300 ft (1000 m), 0.95 for 5000 ft (1500 m), and 0.80 for 10,000 ft (3000 m) with interpolation of these factors for intermediate altitudes.

For breakers that use high-pressure air or gas as the insulating medium, consideration must be given to the possible loss of insulation strength should this pressure be reduced. In some breakers, a stuck blast valve will cause the high-pressure gas or air to be exhausted to atmosphere, with the result that little or no insulation strength remains. With some other breakers, this particular malfunction would not be a problem, since the high-pressure gas flows into the low-pressure system, causing an increase in insulation strength. However, should a rupture in the low-pressure system also occur, the gas would escape, again with a reduction in insulation strength.

Present-day insulation coordination studies for equipment and most transmission lines have used a nonprobabilistic approach. The basic concept, as illustrated in the preceding sections, is to select an insulation level that is greater by a suitable margin than the maximum overvoltage that can impinge on it. This concept is, of course, sound for nonself-restoring insulation, that is, insulation that once it is punctured will not heal itself, and must therefore be taken out of service for repair.

Self-restoring insulation, such as transmission lines and external insulation of circuit breakers and other equipment, need not be repaired after a flashover. The resulting system fault is cleared by the circuit breaker, and following a suitable arc deionization time, the insulation strength of the flashover path is restored. Since the consequence of such a failure of self-restoring insulation is so much reduced, there has been considerable activity in the industry toward the use of probabilistic techniques in insulation coordination studies, especially in the studies of ultrahigh-voltage (UHV) systems rated 1000 kV and above.

3.5 CONTINUOUS CURRENT RATING

The total temperature of the various internal and external parts of circuit breakers, like other electrical equipment, must be limited to certain values to prevent degradation of their mechanical and insulating properties. Since current flowing in the circuit breaker represents

Circuit Breaker Application

a heat source, limits on the magnitude of this current must be established. One such limit is the rated continuous current shown in column 5 of Table 3.3. This is the maximum value of current that the circuit breaker can carry for sustained periods at an altitude above sea level of 3300 ft (1000 m) or less and when the ambient temperature does not exceed 40°C.

Normally, the application problem is simply to determine the maximum load current that will flow through the circuit breaker and verify that it is less than the value given in column 5 of table 3.3. Occasionally, there are special applications that require additional consideration. Some of these are:

1. *Shunt capacitor banks*: The rated current of capacitor banks should be multiplied by 1.25 for ungrounded and 1.35 for grounded neutral banks. This is to account for operation of the capacitors at up to 10% above rated voltage, tolerances in the capacitance of the banks, and the additional heating caused by harmonic currents flowing in the banks.

2. *High-altitude applications*: For applications at altitudes above 3300 ft (1000 m), a derating factor must be applied to the continuous current ratings [12]. These factors are 1.0 for up to 3300 ft (1000 m), 0.99 for 5000 ft (1500 m), and 0.96 for 10,000 ft (3000 m), with interpolation of these factors for intermediate altitudes.

3. *Ambients above or below 40°C*: Since *total temperature* is the critical limit for the materials in electrical apparatus, circuit breakers have been designed so that the *temperature rise* caused by current flow in the breaker does not exceed a value equal to this total temperature minus the maximum standard ambient temperature of 40°C. If the actual ambient temperature is above or below 40°C, the temperature rise due to the current flow in the circuit breaker could be more or less than that caused by the rated continuous current as given in the standards. This allowable current rating is given by the formula [13]

$$I_a = I_r \left(\frac{\theta_{max} - \theta_a}{\theta_r} \right)^{1/1.8}$$

where

I_a = allowable continuous current, at the actual ambient temperature (not to exceed two times I_r), amperes

I_r = rated continuous current, amperes

θ_{max} = allowable hottest-spot total temperature ($\theta_{max} = \theta_r + 40°C$), degrees Celsius

θ_a = actual expected ambient temperature, degrees Celsius

θ_r = allowable hottest-spot temperature rise at rated current, degrees Celsius

The values for θ_r and θ_{max} for the various components of circuit breakers are given in Refs. 12 and 13.

4. *Short-time load current capability*: If the circuit breaker has been carrying a current less than its rated continuous value for a long time (greater than 4 h), its actual hottest-spot total temperature will be less than θ_{max}. It is, therefore, permissible to operate at a current above rated for a short time and still not exceed θ_{max}, and thus still not adversely affect the components and materials in the circuit breaker. The allowable time for this increased current is given by the formula [13]

$$t_s = \tau \left(-\ln \left\{ 1 - \frac{\theta_{max} - Y - \theta_a}{Y[(I_s/I_i)^{1.8} - 1]} \right\} \right)$$

where

$$Y = (\theta_{max} - 40°C)(I_i/I_r)^{1.8}$$

τ = thermal time constant of the circuit breaker, hours (presently 0.5 h)

I_s = short-time load current, amperes

I_i = initial current, prior to I_s, amperes

$\theta_{max}, \theta_a, I_r$ = same as in preceding section

A simplified graphical solution of this equation with examples is given in Ref. 13.

5. *Emergency overload conditions*: The conditions above assume that θ_{max} is not exceeded, and therefore the useful life of the circuit breaker is not affected. There may be certain emergencies where to maintain system security overloads above these limits must be accepted. Reference 13 specifies factors to be applied to the rated continuous current that will allow operation at 15°C above θ_{max} for an emergency period of 4 h or 10°C above θ_{max} for an emergency period of 8 h. Rules similar to those above for conditions of ambient temperatures above and below 40°C and short-time loadability are also given. Other limitations, such as number of occurrences before maintenance, are also specified. Note that operation at these higher temperatures may cause a reduction in the operating life of the circuit breaker.

Circuit Breaker Application

3.6 INTERRUPTING CURRENT RATINGS

One of the primary purposes of circuit breakers is to protect the system and equipment on the system from short circuits that occur on the system. Thus one of the primary problems in the application of circuit breakers is the calculation of the maximum short-circuit currents that the circuit breaker must carry and interrupt.

Figure 3.8 shows the first few cycles of a typical short-circuit current that could be seen by a circuit breaker. Note that it is made up of a decaying sinusoidal ac component superimposed on a decaying unidirectional or dc component. The circuit breaker must withstand mechanically the forces caused by the first peak of this current (point 1 on the figure) and must interrupt the energy in the arc caused by the current at the time when the circuit breaker contacts part (that could be point 2 on the figure). The application problem is, therefore, to calculate the first peak current and the total current at the time of contact part for the worst short-circuit conditions to which the circuit breaker would be subjected and select a circuit breaker rating that would withstand this duty.

The most accurate method of calculating these currents is to solve the classical machine short-circuit equations [14]. This will be called the *classical method* in the following sections. This method can be quite laborious, so several simplified methods have been developed: the total current method of Ref. 15 for Total Current Rated circuit breakers and the symmetrical current method of Ref. 13 for Symmetrical Current Rated circuit breakers. These three methods are discussed in the following sections.

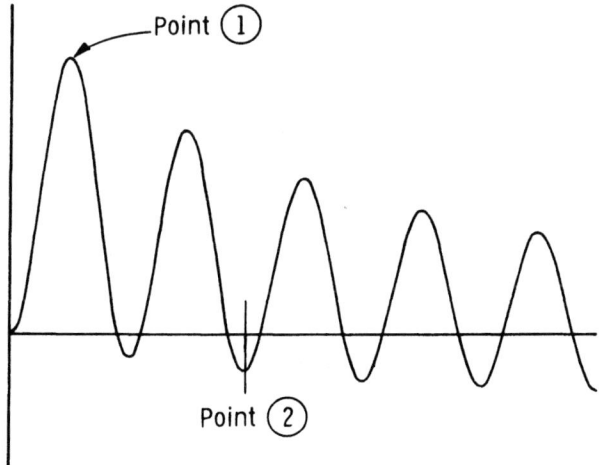

Figure 3.8 Typical short-circuit current.

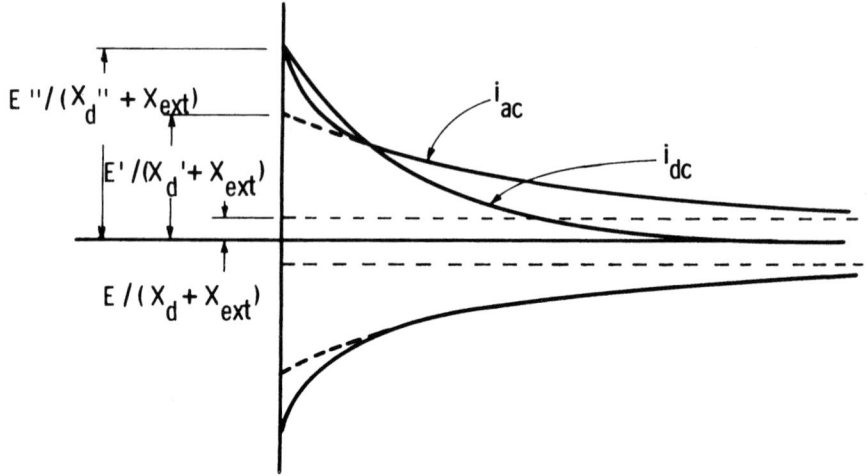

Figure 3.9 Envelope of ac and dc components of a three-phase fault on a generator through an external impedance.

3.6.1 Classical Method

Figure 3.9 shows the dc component and the envelope of the symmetrical ac component for a three-phase fault on a generator having a simple external impedance. Expressed in equation form, the ac component is

$$i_{ac} = \left(\frac{E''}{X_d'' + X_{ext}} - \frac{E'}{X_d' + X_{ext}}\right)\varepsilon^{-t/T''}$$
$$+ \left(\frac{E'}{X_d' + X_{ext}} - \frac{E}{X_d + X_{ext}}\right)\varepsilon^{-t/T'} + \frac{E}{X_d + X_{ext}}$$

where X_d'', X_d', and X_d are, respectively, the subtransient, transient, and synchronous reactances of the generator; X_{ext} is the external reactance; and E'', E', and E are, respectively, the voltages behind subtransient and transient reactance and the air-gap voltage for the loading condition of the generator prior to the short circuit.

$$T'' = \frac{X_d'' + X_{ext}}{X_d' + X_{ext}} T_{do}''$$

$$T' = \frac{X_d' + X_{ext}}{X_d + X_{ext}} T_{do}'$$

Circuit Breaker Application

where T_{do}'' and T_{do}' are the open-circuit subtransient and transient time constants, respectively, of the generator.

The magnitude of the dc component depends on the point on the wave at which the fault is initiated, but the maximum value is

$$i_{dc} = \sqrt{2}\left(\frac{E''}{X_d'' + X_{ext}} - I_L\right)\varepsilon^{-t/T_a}$$

where I_L is the load current prior to the fault, and both I_L and $E''/(X_d'' + X_{ext})$ are vector quantities.

$$T_a = \frac{X_2 + X_{ext}}{2\pi f(r_a + r_{ext})}$$

where X_2 and r_a are the negative sequence reactance and the dc armature resistance of the generator and r_{ext} is the external resistance.

For line-to-ground faults, the equations above can be used with r_{ext} and X_{ext} being equal to the sum of the negative- and zero-sequence impedances of the generator and external impedance plus the positive-sequence external impedance. The faulted-phase current is, of course, equal to three times the resulting currents calculated above.

The i_{ac} calculated above is a rms quantity with i_{dc}, of course, an instantaneous quantity. To convert these values into a total rms value in which the breaker duties are expressed, the following expression should be used:

$$i_{rms} = \sqrt{i_{ac}^2 + i_{dc}^2}$$

where i_{ac} and i_{dc} are the values calculated above at the time t which is under consideration. If t is 1/2 cycle, the i_{rms} calculated would be the value that would correspond to the Momentary or Close and Latch rating of the breaker. If t were the contact parting time of the breaker, the i_{rms} calculated would be compared to the interrupting rating of the circuit breaker.

3.6.2 Total Current Method

The classical method can be quite laborious, especially when there are a number of generating sources and meshed networks, so the industry has continually examined new methods to simplify the calculations. In the early 1950s, the total current method of rating and applying circuit breakers was developed. Circuit breakers were given an MVA rating that was equivalent to the total rms current (i_{rms} above) that it could interrupt multiplied by rated voltage. At other operating voltages, the interrupting capability was

Table 3.6 Reactance Quantities and Multiplying Factors for Application of Circuit Breakers

		Reactance quantity for use in X_1		
	Multi-plying factor	Synchronous generators and condensers	Synchronous motors	Induction machines

A. Circuit breaker interrupting duty

1. General case

8-cycle or slower circuit breakers[a]	1.0	Subtransient[b]	Transient	Neglect
5-cycle circuit breaker	1.1			
3-cycle circuit breaker	1.2			
2-cycle circuit breaker	1.4			

2. Special case for circuit breakers at generator voltage only. For short-circuit calculations of more than 500,000 kVA (before the application of any multiplying factor) fed pre-dominantly direct from generators, or through current-limiting reactors only

8-cycle or slower circuit breakers[a]	1.1	Subtransient[b]	Transient	Neglect
5-cycle circuit breakers	1.2			
3-cycle circuit breakers	1.3			
2-cycle circuit breakers	1.5			

Circuit Breaker Application

3. Air circuit breakers rated 600 V and less	1.25	Subtransient	Subtransient	Subtransient
B. Mechanical stresses and momentary duty of circuit breakers		
1. General case	1.6	Subtransient	Subtransient	Subtransient
2. At 5000 V and below, unless current is fed predominantly by directly connected synchronous machines or through reactors	1.5	Subtransient	Subtransient	Subtransient

[a] As old circuit breakers are slower than modern ones, it might be expected a low multiplier could be used with old circuit breakers. However, modern circuit breakers are likely to be more effective than their slower predecessors, and therefore the application procedure with the older circuit breakers should be more conservative than with modern circuit breakers. Also, there is no assurance that a short circuit will not change its character and initiate a higher current flow through a circuit breaker while it is opening. Consequently, the factors to be used with older and slower circuit breakers well may be the same as for modern 8-cycle circuit breakers.

[b] This is based on the condition that any hydroelectric generators involved have amortisseur windings. For hydroelectric generators without amortisseur windings, a value of 75% of the transient reactance should be used for this calculation rather than the subtransient value.

Source: These application factors derived from ANSI C37.5-1953. Table from *Electrical Transmission and Distribution Reference Book,* © 1950, Westinghouse Electric Corporation, East Pittsburgh, Pa.

$$I = I_{rated}\left(\frac{\text{Rated Voltage}}{\text{Operating Voltage}}\right)$$

but not greater than a stated maximum value. Thus the MVA was constant over a range of values. The mechanical limit or the maximum peak value of current at 1/2 cycle was called the Momentary Rating and was approximately 1.6 times the rated interrupting current expressed in rms amperes.

For example, a 5000-MVA 69-kV oil breaker has a rated voltage interrupting capability of 42 kA and a maximum interrupting capability at 44 kA. If it were used on a system with a maximum operating voltage of 68 kV, it could interrupt 42 × 69/68 = 42.62 kA. Its momentary rating would be 70 kA, which is approximately 1.6 × 44 kA.

The initial version of C37.5, the Application Guide for High-Voltage Circuit Breakers, issued in 1953, contained a table similar to Table 3.6. To apply a circuit breaker, one had only to divide the system voltage by the net system impedance (using the appropriate sequence impedances and connections for unbalanced faults) and multiply the results by the factors in Table 3.6 to obtain the total rms current at the time of contact part and the 1/2-cycle peak rms duty for the circuit breaker.

For example, assume that the 5000-MVA 69-kV breaker just discussed was a 5-cycle breaker applied out on the system, away from local generation, when the system impedance was 3% based on 100 MVA and 68 kV. The E/X value at this point is

$$\frac{100}{\sqrt{3} \times 68 \times 0.03} = 28.3 \text{ kA}$$

Multiplying this by the 1.1 and 1.6 factors from Table 3.6 gives an interrupting current of 31.13 kA and a momentary current of 45.28 kA, which are both less than the 42.62- and 70-kA capability values of the circuit breaker.

The "general case" factors of 1.0 through 1.4 of Table 3.6 recognized the decrement of the dc component of the fault current based on the average system X/R values existing in the early 1950s (approximately X/R = 15). The slower the breaker, the more the dc had decayed, and, in fact, for the 8-cycle breaker having a 4-cycle contact parting time, the dc was assumed to be zero, and thus the multiplying factor was 1.0. For this general case, it was assumed that the impedance external to the generator was so high that there was no ac decrement. For the "special case" of Table 3.6, for circuit breakers connected directly to generator buses, the X/R ratios were greater than 15, so the multiplying factors were raised; but also faults on these buses would have some ac decrement, so the factors were modified accordingly.

This was a relatively simple calculation and application method, but as systems grew, the X/R of systems increased. As a result, the cur-

Circuit Breaker Application

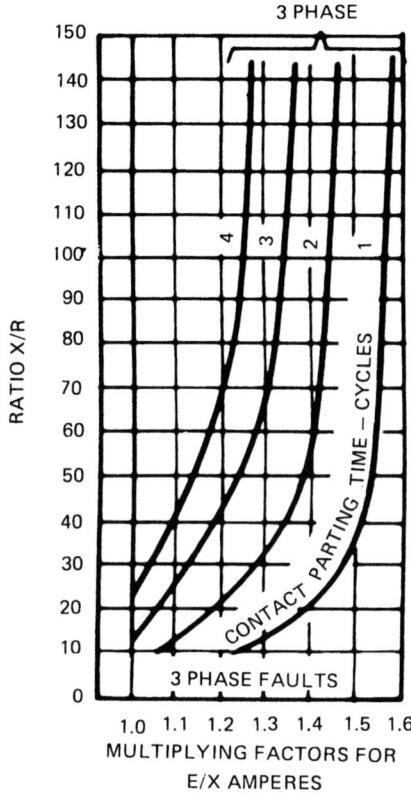

Figure 3.10 Three-phase-fault multiplying factors which include effects of ac and dc decrement. (From ANSI/IEEE C37.5-1979, Guide for calculation of fault currents for application of ac high-voltage circuit breakers rated on a total current basis, © 1979, IEEE.)

rents calculated by this method were lower than the actual current. In 1969, a new method was developed and published in C37.5-1969. This method replaced the factors of Table 3.6 with the three curves of Figs. 3.10, 3.11, and 3.12. Figures 3.10 and 3.11 are used for three-phase and line-to-ground faults, respectively, for fault locations where both ac and dc decrement must be considered. This is where there is not more than one transformer between the fault and the generators, or if the external impedance between the fault and the generator is less than 1.5 times the subtransient reactance (X_d'') of the generator. For all other three-phase and line-to-ground faults, only the dc decrement must be considered, and Fig. 3.12 should be used.

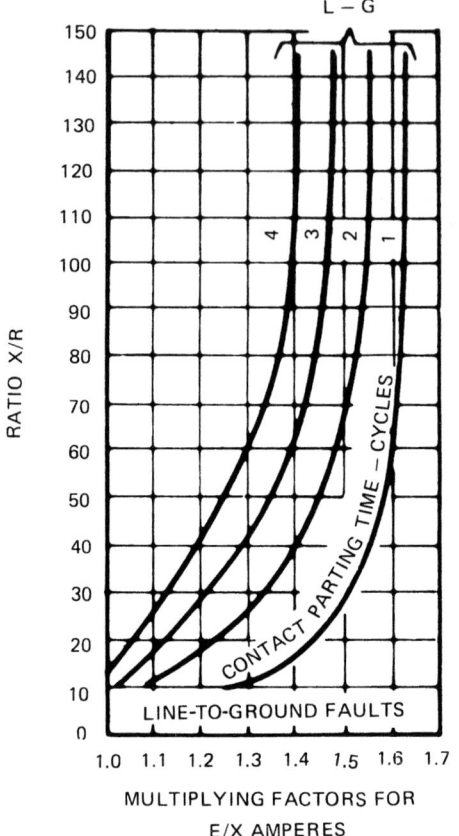

Figure 3.11 Line-to-ground-fault multiplying factors which include effects of ac and dc decrement. (From ANSI/IEEE C37.5-1979, Guide for calculation of fault currents for application of ac high-voltage circuit breakers rated on a total current basis, © 1979, IEEE.)

To use this method, you must now calculate the system X/R at the fault point. In most digital short-circuit studies, an impedance (R + JX) network reduction is made of all the series-parallel connections in the network. This resulting impedance can be used to obtain the fault current magnitude, but should *not* be used for the equivalent X/R ratio, as it will give a value lower than the actual value. C37.5, as well as C37.010 (as will be discussed later), states that the X/R ratio should be calculated by making a reduction of the network using only reactances in the circuits. This will give the equivalent X. A similar

Circuit Breaker Application

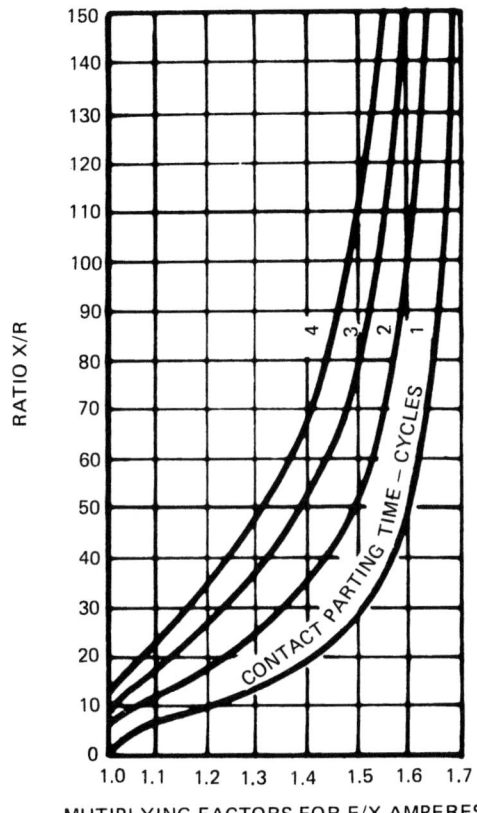

Figure 3.12 Three-phase and line-to-ground-fault multiplying factors which include effects of dc decrement only. (From ANSI/IEEE C37.5-1979, Guide for calculation of fault currents for application of ac high-voltage circuit breakers rated on a total current basis, © 1979, IEEE.)

reduction of the network using only resistances is then made to obtain an equivalent R. These equivalent X's and R's are then used for the X/R ratio to enter Figs. 3.10 through 3.12. This will give a fault-current value higher than the actual, but for circuit breaker applications, it is preferred to be conservative, considering the risks of breaker failure.

Returning to our 69-kV 5000-MVA breaker application, assume that the 3% system impedance has an X/R ratio of 20. Since the breaker is applied away from a generation source, Fig. 3.12 should be used for the multiplying factor. For and X/R of 20, the factor is 1.13 for a 5-cycle breaker. Note that this is greater than the 1.1 factor from Table

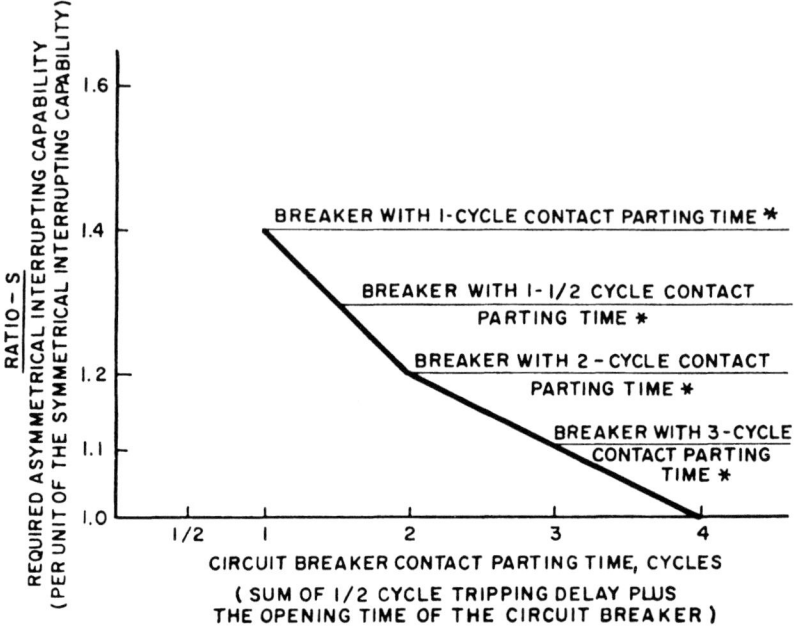

Figure 3.13 Power circuit breaker design requirements. (From ANSI/IEEE C37.010-1979, Application guide for ac high-voltage circuit breakers rated on a symmetrical current basis, © 1979, IEEE.)

3.6. The breaker duty is now 31.98 kA, which, however, is still less than the 42.62-kA capability of the circuit breaker.

3.6.3 Symmetrical Current Method

No sooner had the old total current method of rating and applying circuit breakers been established (1952) than the standards committees began working on "simplifying" them. Since most of the circuit breakers fell into the general case category of Table 3.6, why not build the multiplying factors of 1.0 through 1.4 into the breakers, and then give the breakers only a symmetrical rating? In applying the breakers, then, only a simple E/X calculation would have to be made.

This suggestion was accepted by the manufacturers, and the so-called *S factors* were built into the breakers. Figure 3.13 shows these factors, which are the ratios of the asymmetrical current (total rms current) to the symmetrical current at breaker contact part. Note that this is the same as Table 3.6, except that the 2-cycle breaker (which has a contact parting time of 1.5 cycles) now has a 1.3 factor instead of 1.4.

Circuit Breaker Application

Although this was a simplification, certain problems soon became apparent. As mentioned previously, these factors were based on a fixed X/R ratio (approximately 15). Curves of additional factors to account for different X/R values had to be developed. The S factors built into the circuit breaker were based on no ac decrement in the fault, so two sets of X/R curves had to be developed to cover with and without ac decrement. The old method was based on a constant MVA rating, but the new method was based on a constant "current."

As it turned out, this was not a "simple" change in rating, and it took 12 years to accomplish. In 1964, however, the new Symmetrical Current Basis of Rating, Standards C37.04 through C37.010, were established. Although not simpler to apply, they were more accurate, and even the old standard C37.5 was changed, as discussed previously, to increase its accuracy.

The new breaker standards (C37.06) now have three quantities that define their interrupting capability: Rated Maximum Voltage, Rated Voltage Range Factor K, and Rated Short-Circuit Current. These are given in columns 1, 2, and 6 of Table 3.3. The first two factors and how they influence the third were discussed in Sec. 3.3 and will not be repeated here. How they define the interrupting capability of the breaker so far as what fault currents the circuit breaker must interrupt will be discussed.

C37.04-1979 [12] states that the circuit breaker must interrupt three-phase and phase-to-phase faults up to its Symmetrical Current Capability, which is equal to

$$\text{Rated Short-Circuit Current} \times \frac{\text{Rated Maximum Voltage}}{\text{Maximum Operating Voltage}}$$

at contact parting time (which is equal to 1/2 cycle relay time plus breaker minimum opening time), but this current is not to exceed K times Rated Short-Circuit Current. For line-to-ground faults the Symmetrical Current Capability is 1.15 times the three-phase fault value, but not more than K times Rated Short-Circuit Current. As discussed in Sec. 3.3, K in Table 3.3 is equal to 1.0, so this increased capability does not exist except for lower-voltage breakers which have K factors greater than 1.0. This can be advantageous.

The three-phase and phase-to-phase fault and the line-to-ground fault asymmetrical (or total current) fault capability is S times the symmetrical values, where the S factors are shown in Fig. 3.13 for the various contact parting times. (Note: 2-, 3-, 5-, and 8-cycle breakers have contact parting times of 1.5, 2, 3, and 4 cycles, respectively.)

The Momentary Rating of the Total Current breakers has been replaced by the Close and Latch Current Rating shown in column 15 of Table 3.3. The rms value of this rating is equal to 1.6 times K times Rated Short-Circuit Current, and the peak (or crest) value is equal to 2.7 times K times Rated Short-Circuit Current.

As an example, a 46-kV 5-cycle breaker has a Rated Maximum Voltage of 48.3 kV, a Rated Short-Circuit Current of 17 kA, and a K factor of 1.21. KI is equal to 21 kA. If operated at 43 kV, its three-phase Symmetrical Current Capability is 17(48.3/43) = 19.1 kA. Its Line-to-Ground Symmetrical Current Capability is 21 kA, since 1.15 × 19.1 = 22 kA, which is greater than KI. Its three-phase Asymmetrical Capability is 1.1 × 19.1 = 21 kA (since the S factor for a 5-cycle breaker is 1.1), and its Line-to-Ground Asymmetrical Capability is 1.1 × 21 = 23.1 kA.

Table 3.7

Type of rotating machine	Positive sequence reactances for calculating:	
	Interrupting duty per unit	Closing and latching duty per unit
All turbo-generators, all hydro-generators with amortisseur windings, and all condensers (see note 2)	$1.0X_d''$	$1.0X_d''$
Hydro-generators without amortisseur windings (see note 2)	$0.75X_d'$	$0.75X_d'$
All synchronous motors (see notes 1, 4, and 5)	$1.5X_d''$	$1.0X_d''$
Induction motors (see notes 3, 4, and 5)		
Above 1000 hp at 1800 r/min or less	$1.5X_d''$	$1.0X_d''$
Above 250 hp at 360 r/min		
From 50 to 1000 hp at 1800 r/min or less	$3.0X_d''$	$1.2X_d''$
From 50 to 250 hp at 3600 r/min		
Neglect all three-phase induction motors below 50 hp and all single-phase motors		

Circuit Breaker Application

Table 3.7 (Continued)

Notes:
(1) X_d'' of synchronous rotating machines is the rated-voltage (saturated) direct-axis subtransient reactance.
(2) X_d' of synchronous rotating machines is the rated-voltage (saturated) direct-axis transient reactance.
(3) X_d'' of induction motors equals 1.00 divided by per unit locked-rotor current at rated voltage.
(4) The current contributed to a short circuit by induction motors and small synchronous motors may usually be ignored on utility systems except station service supply systems and at substations supplying large industrial loads. At these locations, as well as in industrial distribution systems or locations close to large motors, or both, the current at 1/2 cycle will be increased by the motor contribution to a greater degree, proportionately, than the symmetrical current will be increased at minimum contact parting time. In these cases, an additional calculation of 1/2 cycle current should be made using the methods of Sec. 5.3.1 or 5.3.2 of Ref. 13 and the appropriate reactance values given above under the heading "closing and latching duty." A 1.6 multiplying factor should be used for asymmetry and this result must not exceed the closing and latching capability of the circuit breaker being used.
(5) These rotating machine reactance multipliers and the E/X ampere multipliers of Figs. 8 and 9 of Ref. 13 include the effects of ac decay. However, the methods for calculation of system short-circuit current described in Secs. 5.3.1 and 5.3.2 of Ref. 13 incorporate sufficient conservatism to permit the simultaneous use of a rotating machine reactance and an E/X ampere multiplier from Fig. 8 or 9 of Ref. 13.
(6) When the contribution of large individual induction motors is an appreciable portion of the short-circuit current, substitution for the tabulated multiplying factors of more accurate multipliers based on manufacturer's time constant data is appropriate. Using $I = (E/X_d'') \times \varepsilon^{-t/T''}$, as the expression for the exponential decay of induction motor symmetrical current to a terminal short circuit, the reactance multiplying factor is $\varepsilon^{+t/T''}$, where t is the proper time after initiation of the short circuit and T'' is the motor short-circuit time constant. (Both should be in the same time units.) For example, using manufacturers' motor data for T'', the reactance multiplying factor for determining the interrupting duty may be found using t equal to the circuit breaker minimum contact parting time. For a circuit breaker with a 5-cycle rated interrupting time, t = 3 cycles (0.05 s). For determining the closing and latching duty, t = 0.5 cycle (0.00833 s) in the reactance multiplying factor calculations.
Source: Ref. 13.

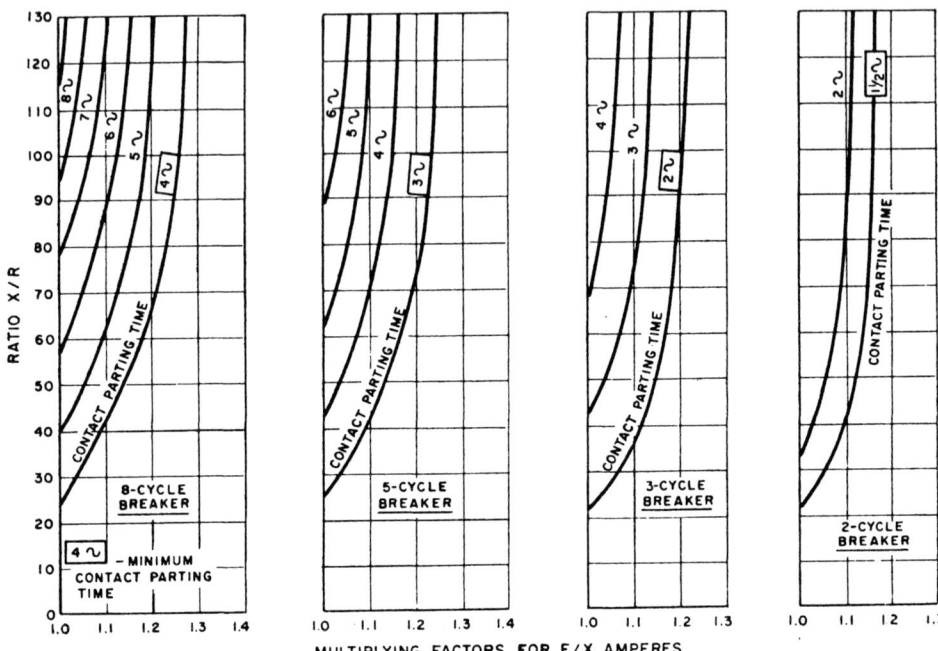

Figure 3.14 Three-phase-fault multiplying factors which include effects of ac and dc decrement. (From ANSI/IEEE C37.010-1979, Application guide for ac high-voltage circuit breakers rated on a symmetrical current basis, © 1979, IEEE.)

To apply this breaker, C37.010 decribes two methods: the E/X Simplified Method and the E/X Method with Adjustment for AC and DC Decrement. In the Simplified Method, the actual voltage at the fault point (E) is divided by the reactance (X) of the system obtained by a network reduction using only reactances in the sequence networks. The reactances to use for the system elements are given in Table 3.7. Note that for motors, different reactances are used when calculating the Breaker Interrupting duty and the Closing and Latching duty.

For the three-phase fault, the fault current is E/X. If this value is equal to or less than 80% of the breaker three-phase fault capability, the breaker application is satisfactory.

For the line-to-ground fault, the fault current is $3E/(2X_1 + X_0)$. If this value is equal to or less than 70% of the breaker line-to-ground fault capability, the breaker application is satisfactory. The reasoning behind these 80 to 70% values will be discussed later.

Circuit Breaker Application

If the three-phase and line-to-ground fault currents are not less than 80 to 70%, respectively, the more refined E/X Method with Adjustment for AC and DC Decrement must be used. This method is very similar to the method already discussed for the C37.5-1979 Total Current Rated Circuit Breakers. In fact, the C37.5-1969 method was developed after this new C37.010 method. In this method, after calculating the E/X vaules, one must determine an equivalent R for the circuit. Again, this is done by performing a network reduction from the fault point using only the resistances of the circuit elements in the network to be reduced. (Note that the same multipliers used for the reactances in Table 3.7 should be used for the resistances.) Dividing the equivalent X by the equivalent R gives the X/R of the system at the fault point.

Since the S factors of Fig. 3.13 were based on a system X/R of approximately 15, if the X/R of the system under consideration is equal to or less than 15, the E/X and $3E/(2X_1 + X_0)$ values calculated previously can be assumed to be the correct values for the circuit breaker duty. If they are less than the circuit breaker capabilities, the application is satisfactory.

For X/R ratios greater than 15, the correction factor curves of Figs. 3.14 through 3.16 must be used. Figures 3.14 and 3.15 are to be used for three-phase and line-to-ground faults, respectively, where both ac and dc decrement must be considered. This occurs when there is not more than one transformation between the fault and any generators, or where the external impedance between the generator and fault is equal to or less than 1.5 times the subtransient reactance of the generator. If neither of these conditions are met, ac decrement can be neglected, and Fig. 3.16 is used for both three-phase and line-to-ground faults.

Note that these curves are similar to those of Figs. 3.10 through 3.12 for Total Current Rated Circuit Breakers. The values of Figs. 3.14 through 3.16 are less, since they include the S factors of Fig. 3.13 that have been built into the breakers. The other difference is that they include curves for times other than the normal contact parting time of each breaker. This is done to allow the use of time delays in the relay system to reduce the interrupting duty on the circuit breaker.

As with the C37.5-1969 method for Total Current Rated Breakers, the multiplying factors of Figs. 3.14 through 3.16 are used with the E/X and $3E/(2X_1 + X_0)$ values to determine the interrupting duty on the breakers. If these duties are less than the capability of the breaker, the application is satisfactory.

To demonstrate the use of this method, assume that the 46-kV 5-cycle breaker discussed previously is applied on a 43-kV system whose three-phase fault E/X value is 16.5 kA and line-to-ground fault $3E/(2X_1 + X_0)$ value is 17 kA. Since both of these values are greater than

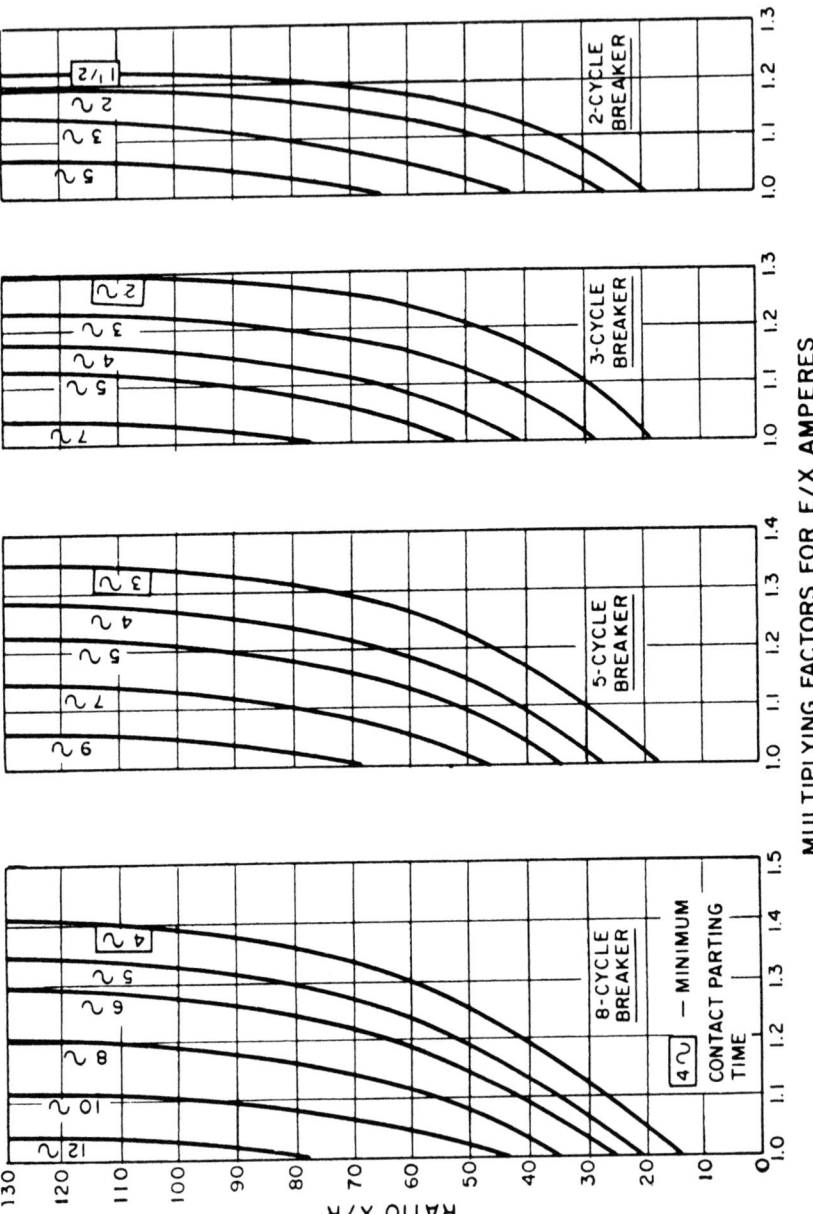

Figure 3.15 Line-to-ground-fault multiplying factors which include effects of ac and dc decrements. (From ANSI/IEEE C37.010-1979, Application guide for ac high-voltage circuit breakers rated on a symmetrical current basis, © 1979, IEEE.)

Circuit Breaker Application

If the three-phase and line-to-ground fault currents are not less than 80 to 70%, respectively, the more refined E/X Method with Adjustment for AC and DC Decrement must be used. This method is very similar to the method already discussed for the C37.5-1979 Total Current Rated Circuit Breakers. In fact, the C37.5-1969 method was developed after this new C37.010 method. In this method, after calculating the E/X vaules, one must determine an equivalent R for the circuit. Again, this is done by performing a network reduction from the fault point using only the resistances of the circuit elements in the network to be reduced. (Note that the same multipliers used for the reactances in Table 3.7 should be used for the resistances.) Dividing the equivalent X by the equivalent R gives the X/R of the system at the fault point.

Since the S factors of Fig. 3.13 were based on a system X/R of approximately 15, if the X/R of the system under consideration is equal to or less than 15, the E/X and $3E/(2X_1 + X_0)$ values calculated previously can be assumed to be the correct values for the circuit breaker duty. If they are less than the circuit breaker capabilities, the application is satisfactory.

For X/R ratios greater than 15, the correction factor curves of Figs. 3.14 through 3.16 must be used. Figures 3.14 and 3.15 are to be used for three-phase and line-to-ground faults, respectively, where both ac and dc decrement must be considered. This occurs when there is not more than one transformation between the fault and any generators, or where the external impedance between the generator and fault is equal to or less than 1.5 times the subtransient reactance of the generator. If neither of these conditions are met, ac decrement can be neglected, and Fig. 3.16 is used for both three-phase and line-to-ground faults.

Note that these curves are similar to those of Figs. 3.10 through 3.12 for Total Current Rated Circuit Breakers. The values of Figs. 3.14 through 3.16 are less, since they include the S factors of Fig. 3.13 that have been built into the breakers. The other difference is that they include curves for times other than the normal contact parting time of each breaker. This is done to allow the use of time delays in the relay system to reduce the interrupting duty on the circuit breaker.

As with the C37.5-1969 method for Total Current Rated Breakers, the multiplying factors of Figs. 3.14 through 3.16 are used with the E/X and $3E/(2X_1 + X_0)$ values to determine the interrupting duty on the breakers. If these duties are less than the capability of the breaker, the application is satisfactory.

To demonstrate the use of this method, assume that the 46-kV 5-cycle breaker discussed previously is applied on a 43-kV system whose three-phase fault E/X value is 16.5 kA and line-to-ground fault $3E/(2X_1 + X_0)$ value is 17 kA. Since both of these values are greater than

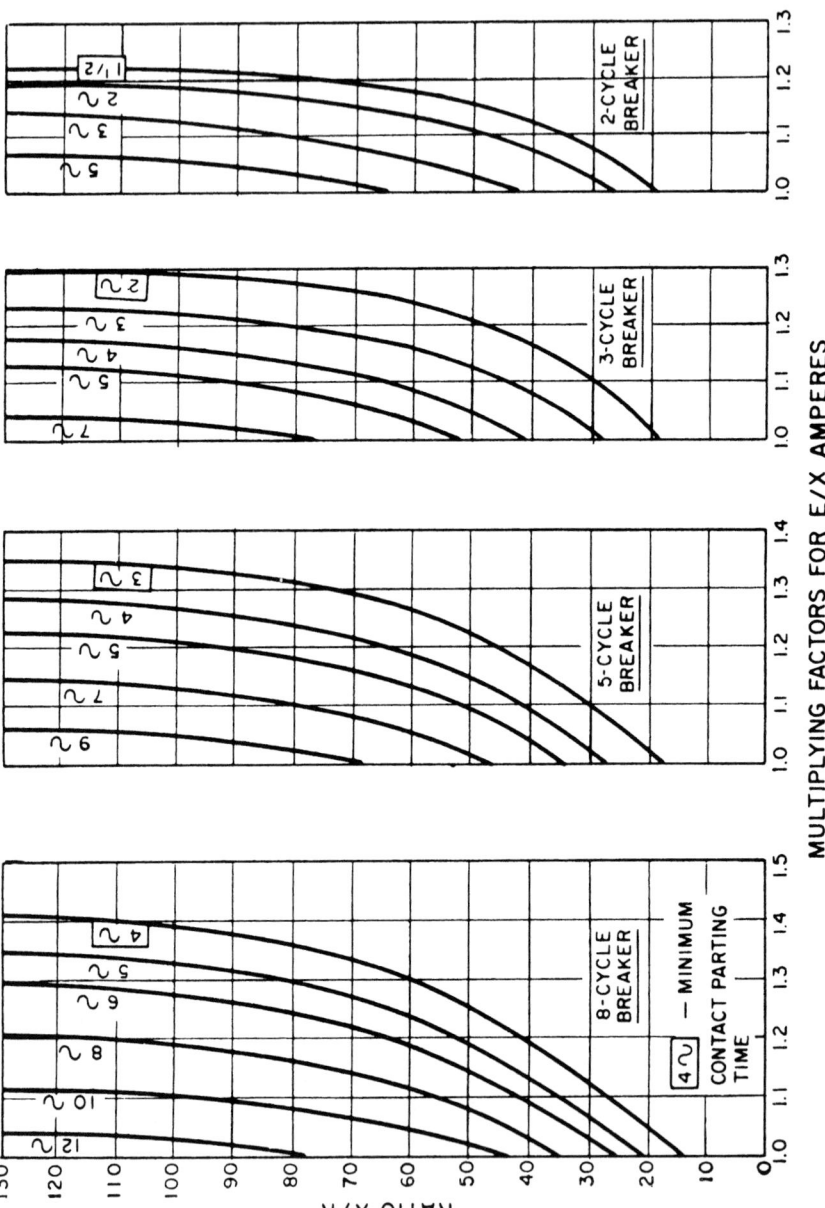

Figure 3.15 Line-to-ground-fault multiplying factors which include effects of ac and dc decrements. (From ANSI/IEEE C37.010-1979, Application guide for ac high-voltage circuit breakers rated on a symmetrical current basis, © 1979, IEEE.)

Circuit Breaker Application

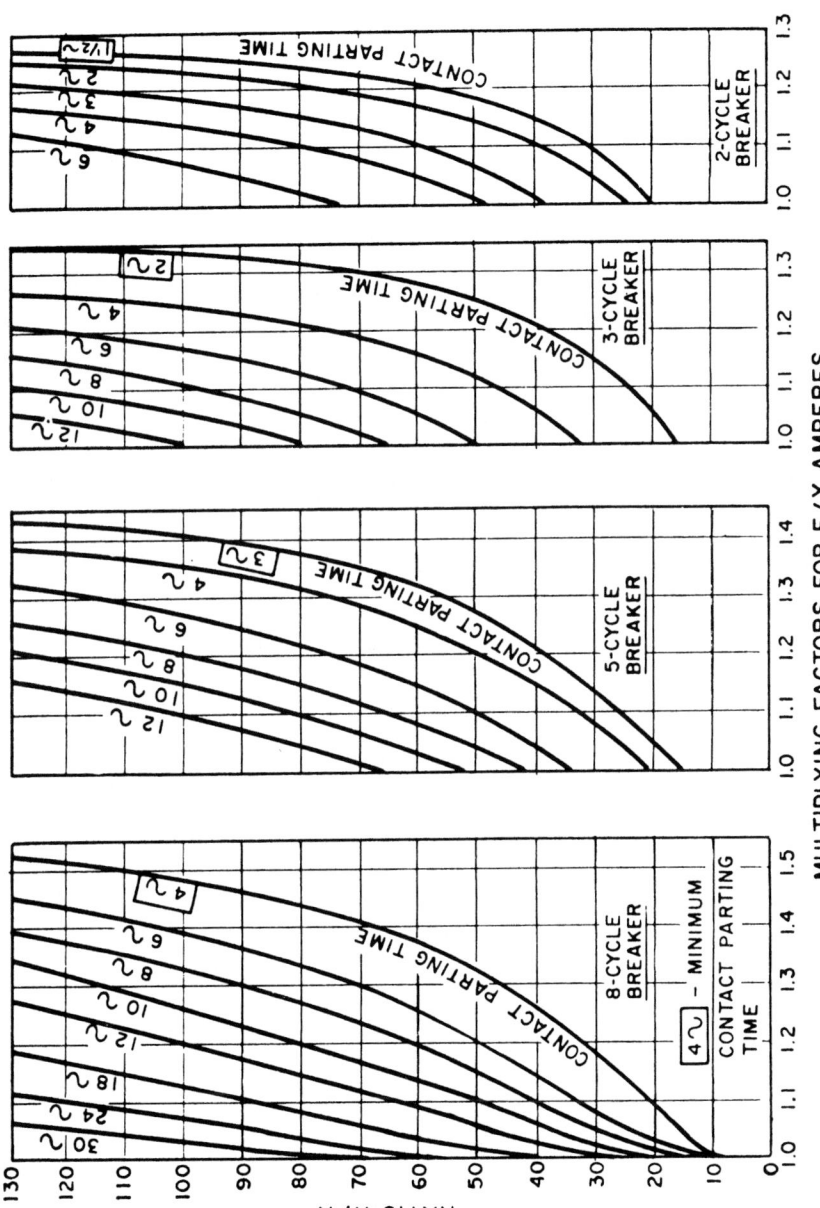

Figure 3.16 Three-phase- and line-to-ground-fault multiplying factors which include effects of dc decrement only. (From ANSI/IEEE C37.010-1979, Application guide for ac high-voltage circuit breakers rated on a symmetrical current basis, © 1979, IEEE.)

80 and 70% of the 19.1- and 21-kA capabilities of the breaker that was previously determined, Figs. 3.14 through 3.16 must be used in the calculations. Assume that the three-phase fault X/R is 33 and the line-to-ground fault X/R is 35. Also, assume that there is no more than one transformation between the fault and the local generators. From Fig. 3.14, the multiplying factor for a 5-cycle breakers and an X/R ratio of 33 is 1.05. 1.05 times the 16.5-kA E/X value gives 17.325 kA, which is less than the 19.1-kA capability of the breaker. For an X/R ratio of 35, Fig. 3.15 gives a factor of 1.13 for the 5-cycle breaker. This value times the 17-kA $3E/(2X_1 + X_0)$ value gives 19.21 kA, which is less than the 21-kA capability of the breaker. Thus the application is satisfactory.

An examination of the curves of Figs. 3.14 through 3.16 shows where the "magic" 80% to 70% values given above developed. In Fig. 3.14, the maximum factor for any breaker or X/R ratio is approximately 1.25. The reciprocal of 1.25 is 0.8. Similarly, for Fig. 3.15, the maximum factor is 1.42, whose reciprocal is 0.7. In Fig. 3.16, the factors are greater than these limits for the higher X/R values, but when the breaker is located so far out on the system that ac decrement can be neglected, the X/R values are also lower, so the practical limits for 80 to 70% also apply here.

It should be noted that most circuit breakers designed and built since 1964 are defined and should be applied using the symmetrical current method of calculation described here and in C37.010-1979. There are, however, still a number of circuit breakers in service and being built under the old standards, and these should be applied under the total current method of calculation described in Sec. 3.6.2 and given in C37.5-1969. Of course, the classical method can be used for both types.

3.7 MOMENTARY OR CLOSING AND LATCHING RATING

As mentioned in the preceding section, circuit breakers must be able to close, latch, and withstand the mechanical forces of the first half-cycle peak current of any short circuit up to its rating. In Total Current rated circuit breakers, this current is called its Momentary Rating. In Symmetrical Current rated circuit breakers, it is called the Closing and Latching rating. The rms value of this rating is equal to 1.6 times its interrupting rating. The peak (crest) value is equal to 2.7 times its rms interrupting rating.

In most transmission systems, the fault current at 1/2 cycle will be less than the Momentary or Closing and Latching Current rating if the Interrupting rating is satisfactory. On systems with a fairly large motor contribution, however, this may not be so. In this case, a new X network reduction must be made using the reactances listed in Table 3.7 for Closing and Latching duty. The resulting E/X value

Circuit Breaker Application

after this reduction should be multiplied by 1.6 and then compared with the Momentary or Closing and Latching capability of the circuit breaker being considered.

3.8 RECLOSING DUTY FACTORS

The standard-duty cycle for a circuit breaker is CO + 15 s + CO, which means that it can close into a fault and immediately open, clearing the fault. After waiting 15 s, it can repeat this sequence. It can perform this duty cycle up to its rated interrupting current capability. Deviations from this standard-duty cycle may require derating of some circuit breakers. For example, with oil breakers, if reclosing should occur in less than 15 s, and/or if there are more than two operations in the duty cycle, the interrupting medium (in this case, oil) will not have recovered to its fullest extent to allow a full-rated, short-circuit interruption.

Circuit breaker standards [2,12,13] recognize this limitation by specifying reduction factors for each operation over two and each reclosing time interval less than 15 s. The equations for determining this derating factor are

$$R = 100 - D\ (\%)$$

and

$$D = d_1(n - 2) + d_1 \frac{15 - t_1}{15} + d_1 \frac{15 - t_2}{15} + \cdots$$

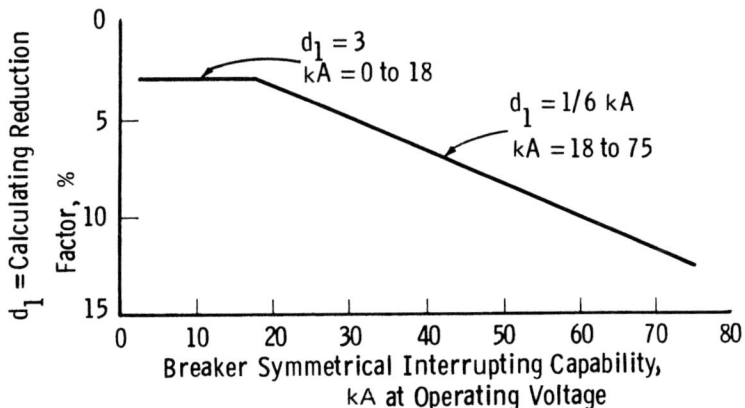

Figure 3.17 Interrupting capability factors for reclosing service. (From Ref. 2, © 1979, American National Standard Institute, Inc.)

where

    R = reclosing capability factor

    D = total reduction factor, %

    d_1 = calculating reduction factor of symmetrical interrupting capability (see Fig. 3.17), %

    n = total number of openings

    t_1 = first time interval less than 15 s

    t_2 = second time interval less than 15 s

    t_3 = ...

To illustrate this procedure, assume that the 46-kV 5-cycle breaker used in the example in Sec. 3.6.3 again is to be used on a 43-kV system, but with a duty cycle of 0 + 0 s + CO + 15 s + CO + 60 s + CO. As calculated previously, its Symmetrical Current Capability at 43 kV is 19.1 kA. From Fig. 3.17, d_1 is equal to 3.2%. From the equations above,

$$D = 3.2(4 - 2) + 3.2\left(\frac{15 - 0}{15}\right) + 3.2\left(\frac{15 - 15}{15}\right) + 0 = 9.6$$

and

    R = 100 − 9.6 = 90.4%

The symmetrical interrupting capability of this circuit breaker for this duty cycle at 43 kV is therefore 19.1 × 0.904 = 17.27 kA.

Note that not all breakers need to be derated for reclosing duties. For example, SF_6 gas breakers are not derated. Each manufacturer should be consulted for the applicability of derating for its particular designs.

3.9 RATED INTERRUPTING TIME

Rated interrupting time of a circuit breaker is defined as the maximum interval between energization of the trip coil and interruption of the current in all poles of the breaker. Standard ratings are 2, 3, 5, and 8 cycles. This rating applies to an initial opening of currents equal to 25% or more of rated short-circuit current. For a close-open operation the opening times may exceed rated by up to 1 cycle for 5- and 8-cycle breakers and by up to 1/2 cycle for breakers rated 3 cycles or less. For currents less than 25% of rated, these times can also be exceeded by as much as 50% for 5 and 8 cycles, and as much as 1 cycle for breakers rated 3 cycles or less. For those breakers using

Circuit Breaker Application

opening resistors, the interrupting time of the resistor current may be longer.

Column 10 of Table 3.3 shows that all outdoor circuit breakers rated 121 kV through 362 kV are rated 3 cycles, while all 550- and 800-kV circuit breakers are rated 2 cycles. Actually, designs of all circuit breakers from 121 kV through 800 kV are available with 2-cycle ratings. In general, the 2-cycle ratings are more expensive than the 3-cycle designs, so the application engineer must examine the system to see if the 2-cycle clearing time is required.

Reducing the fault-clearing time from 3 to 2 cycles will, of course, reduce the damage at the point of fault, but this effect is minor. For applications where system stability is critical, this reduction of fault-clearing time by one cycle may be important, but of greater importance is the reduction of 2 cycles (one cycle in the primary breaker and one cycle in the backup breaker) in the breaker failure fault-clearing time should the primary fault breaker fail to operate.

3.10 TRANSIENT RECOVERY VOLTAGE

When a circuit breaker attempts to interrupt a current, whether it be load current or fault current, a voltage is generated across the circuit breaker contacts to oppose this change in current. This voltage is called the *transient recovery voltage* (TRV) and is equal to the difference in the voltages on the load side (V_y) and source side (V_x) of the circuit breaker after the breaker contacts have parted. The wave shape and magnitudes of V_x and V_y depend on the system configuration and constants both before and after the contacts open. Figure 3.18 shows the TRV for several simple single-phase circuits.

Figure 3.18(a) shows the interruption of a simple resistive circuit. Prior to opening, V_x and V_y are, of course, equal, and the current, i, is in phase with the voltage. At current zero, the circuit is opened, and i and V_y remain equal to zero. The bus voltage V_x continues as a sinusoid, and the TRV ($V_x - V_y$) is simply a 1/2 loop of the original sinusoidal system voltage with a maximum value of 1.0 per unit occurring 1/4 cycle after circuit opening.

Figure 3.18(b) shows a similar condition for a capacitive circuit. In this case, when the current passes through zero, the voltage V_y is trapped on the capacitor at its maximum negative value. As the bus voltage V_x continues along its normal sinusoid, the TRV ($V_x - V_y$) will appear as a (1 − cos) wave with a small value immediately following current zero, but reaching a value of 2.0 per unit 1/2 cycle after circuit opening. This is discussed further in Sec. 3.11.

For both of the cases above, the rate of rise of TRV is rather low immediately after current zero. Such is not the case when switching inductive circuits, as shown in Fig. 3.18(c). Here at current zero, V_x continues along its sinusoidal path, but V_y immediately heads for

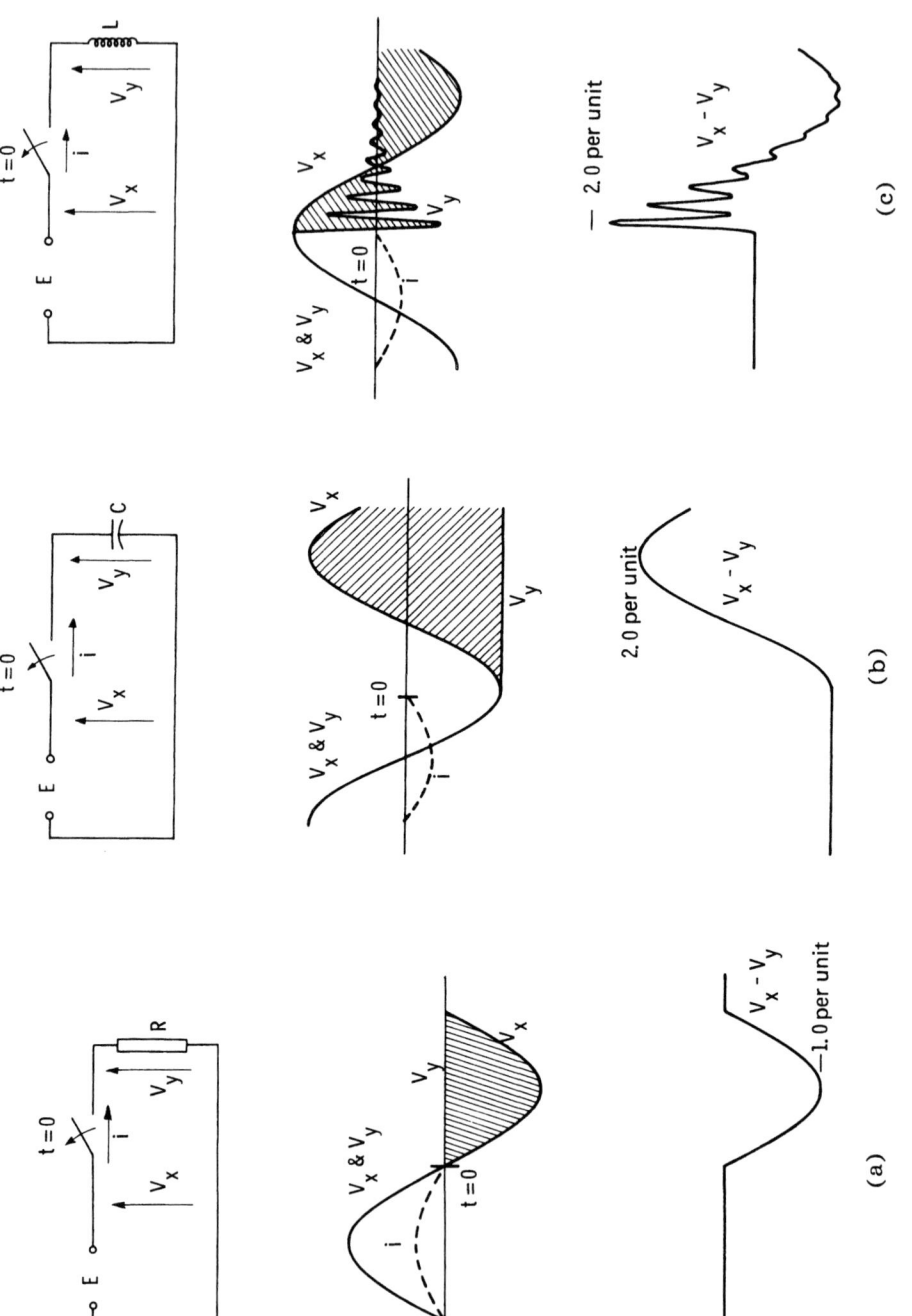

Figure 3.18 Transient recovery voltages of simple (a) resistive, (b) capacitive, and (c) inductive circuits.

Circuit Breaker Application

zero, and with the inherent capacitance, overshoots to -1.0 per unit, or some value less than this depending on the damping in the circuit. With no damping, the TRV ($V_x - V_y$) would be an oscillatory wave ($1 - \cos$) having a peak amplitude of 2.0. The frequency, and thus the time to crest of the wave peak, depend on the L and C of the circuit, but it is apparent that the rate of rise of TRV immediately after current zero is significantly higher than in the other two cases.

The shape of the TRV wave affects the circuit interruption process in two important areas. In the short-time (10 to 20 µs) area, called the *energy balance region*, failure to interrupt can be caused by thermal considerations. When the contacts of the circuit breaker part, a plasma arc is developed. The current times the arc voltage (= power input) causes temperatures in this arc to approach 50,000 K. To extinguish the arc, it must be cooled to such a level that the space between the contacts can act as an insulator. In a gas-blast circuit breaker, high-pressure gas is forced through the arc to cool it. In this regard, we have two opposing forces. In the vicinity of current zero, the gas is attempting to cool the arc, while the arc voltage, or TRV, is generating heat in the arc space attempting to reestablish the arc. Thus there is an energy balance race between the gas cooling and the TRV heating. If the gas wins, we have a successful interruption, at least in the short-time region.

Assuming that the gas won the energy balance race, there is now a second race—a dielectric race. Is the contact separation sufficient to cause the dielectric strength of the gap to exceed the voltage applied across the gap? This is the dielectric region of the TRV and usually occurs hundreds of microseconds after current zero. If the breaker contacts also win this race, we have a truly successful interruption.

With this perhaps oversimplified description of TRV and its effect on the circuit interruption process, the application process will now be discussed.

3.10.1 Standard TRV Envelopes

ANSI C37.06-1979 [2] defines two TRV envelopes. For circuit breakers rated 72.5 kV and below, the envelope is defined as a ($1 - \cos$) wave with a crest voltage (E_2) equal to 1.88 times the rated maximum voltage of the circuit breakers. The time to crest (T_2) of this wave varies as a function of the voltage. These T_2 values are given in Tables 3 and 4 of C37.06-1979 [2].

For circuit breakers rated 121 kV and above, the envelope is defined in terms of an exponential-cosine (ex-cos) wave. This is shown in Fig. 3.19. The exponential is defined by an initial rate of rise R shown in column 8 of Table 3.3 and a final crest value of 1.5 $\sqrt{2}/\sqrt{3} = 1.225$ times rated maximum voltage. The ($1 - \cos$) wave is

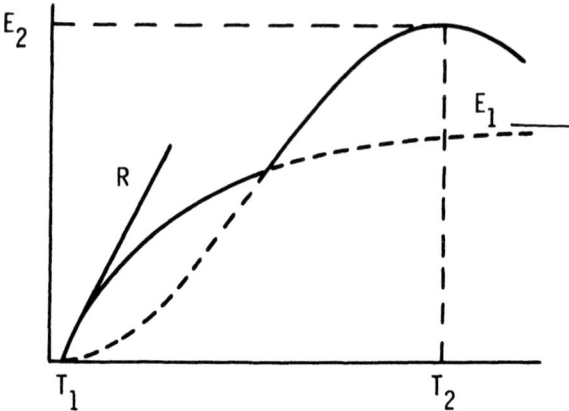

Figure 3.19 ANSI standard TRV test wave.

defined by a crest value (E_2) of 1.76 times rated maximum voltage and a time to crest (T_2), which varies as a function of voltage (see column 7 of Table 3.3). There is also a delay of T_1 (column 9) in the initial portion of the exponential to account for the effect of bus and breaker capacitance.

The values given above are for interruption at rated short-circuit current. For values less than rated, the multipliers of Fig. 3.20 are used.

IEC standards also define two TRV envelopes. For circuit breakers rated 72.5 kV and below, the envelope is defined by the *two-parameter method* shown in Fig. 3.21. u_c is equal to 1.715 times the rated maximum voltage of the circuit breaker and t_3 varies as a function of the voltage, similar to the T_2 ANSI value. The t_d, u', t' characteristic shown is to allow for the bus and circuit breaker capacitance effect similar to the T_1 delay in ANSI.

For circuit breakers rated 100 kV and above, the TRV envelope is defined by the *four-parameter method* shown in Fig. 3.22. For breakers rated 100 through 170 kV applied on ungrounded systems, u_c is equal to 1.715 times rated maximum voltage. For 100- to 170-kV

Figure 3.20 TRV rate (K_r) and voltage (K_t) multipliers for fractions of rated current. K_t (note 1) is for 121 kV and above K_t (note 2) is for 72.5 kV and below. (From Ref. 2, © 1979, American National Standards Institute, Inc.)

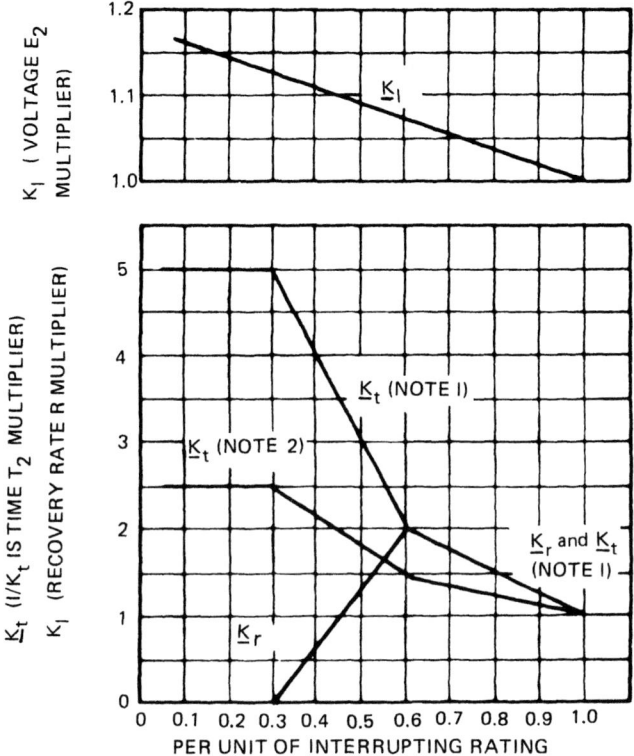

Related Required Transient Recovery Voltage Capabilities of Circuit Breakers at Various Interrupting Levels for Terminal Faults

Percent of Interrupting Rating*	Multipliers for Rated Parameters				
	72.5 kV and Below		121 kV and Above		
	E_2	T_2	R†	E_2	T_2
100	1.00	1	1	1.00	1
60	1.07	0.67	2	1.07	0.5
30	1.13	0.4	0	1.13	0.2
7	1.17	0.4	0	1.17	0.2

NOTE: Interpolation between the above given points is linear, as shown in the above curves.

*Ratio of the symmetrical current component of the current being considered to the related required symmetrical interrupting capability (defined in 5.10.2.1 of American National Standard Rating Structure for AC High-Voltage Circuit Breakers, ANSI/IEEE C37.04-1979), is stated in percent.

†Applies only to circuit breakers rated 121 kV and above since the rated transient recovery voltage is defined as the envelope of a one minus cosine curve which has R = 0.

Figure 3.20

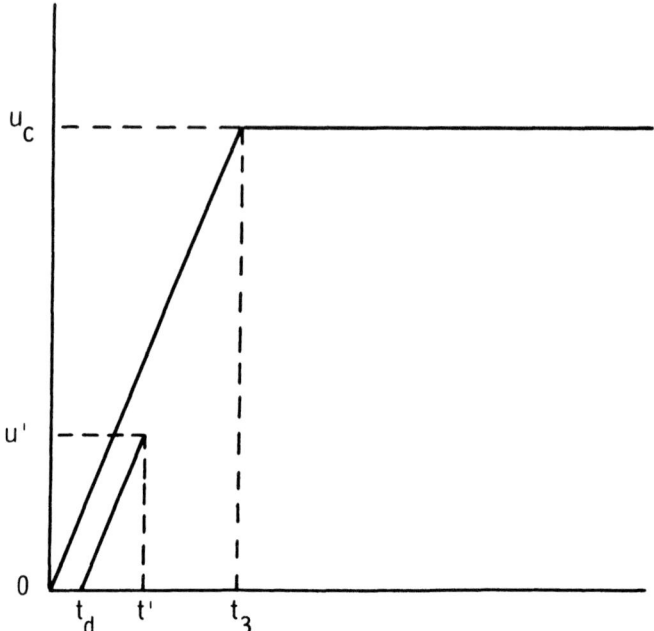

Figure 3.21 IEC two-parameter TRV test characteristic.

circuit breakers on grounded systems and all circuit breakers rated 245 kV and above, u_c is equal to 1.486 times rated maximum voltage. The reason for these different multipliers will be discussed subsequently. In all cases, u_1 is equal to 0.7 times u_c. t_1 varies with the rated voltage of the circuit breaker, similar to the T_2 ANSI value. t_2 is always three times the t_1 value. Again the t_d, u', t' characteristic allows for the effect of bus and circuit breaker capacitance. As with ANSI, higher TRV values are specified for interruption at currents less than rated short-circuit current. Both ANSI and IEC standards also specify that all breakers must withstand the sawtooth wave shape of short-line faults. This will be discussed later in this section.

3.10.2 Calculation Method

The TRV to which a circuit breaker is subjected depends on the type of fault, the location of the fault, and the type of circuit switched. By means of the following examples, the effect of these parameters on the TRV will be shown, as will the method of TRV calculation and their correlation with the standard TRV test waves.

Circuit Breaker Application

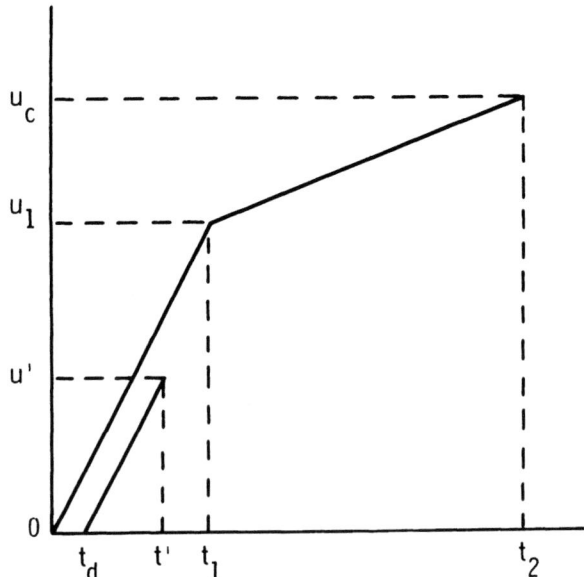

Figure 3.22 IEC four-parameter TRV test characteristic.

Three-Phase Grounded Transformer Terminal Fault

Figure 3.23 shows the simplest type of fault: a three-phase-to-ground fault on the low-side terminals of a delta-wye transformer. As shown in the symmetrical component solution of Fig. 3.23(b), the steady-state voltage on the fault side of the first pole of the circuit breaker to open is zero, while on the transformer side it is 1.0 per unit. Figure 3.23(c) and (d) show the transient involved when this first pole is opened. Note that the transformer is not a pure inductance; it has capacitances to ground and between and across the windings. As shown in Refs. 16 and 17 this distributed capacitance can be represented as a lumped capacitance shown as C_t in Fig. 3.23(c). Prior to the breaker pole opening, the voltages on both sides of the breaker, E_x and E_y, are equal to zero. Immediately after the pole opens, E_y remains equal to zero, but E_x increases to E (which is 1.0 per unit), overshoots, and oscillates around E. With no damping in the circuit, E_x would reach a peak of 2.0 per unit, but with damping it would be somewhat less than 2.0. Since the TRV is E_x minus E_y, it would also be an oscillatory wave with a peak value of something less than 2.0 per unit. The frequency of the oscillation is a function of the inductance and the lumped capacitance of the transformer.

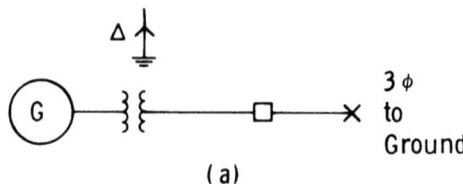

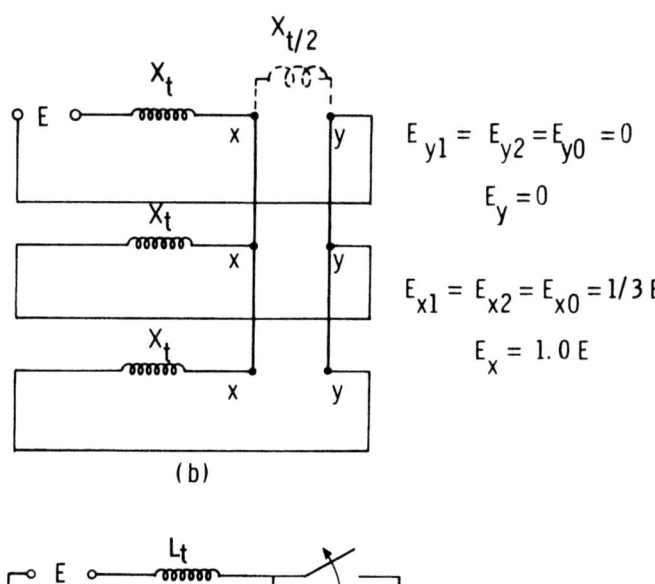

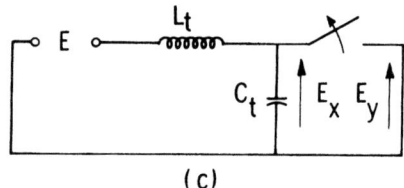

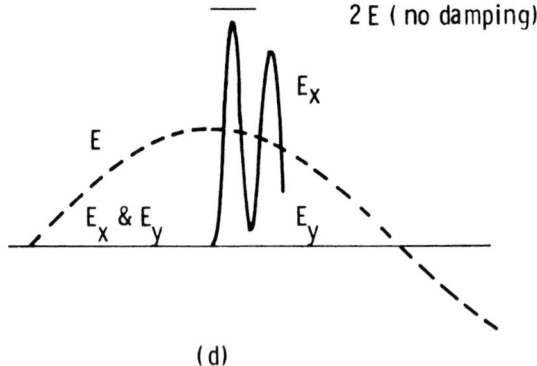

Figure 3.23 TRV for a three-phase grounded fault: (a) system represented; (b) sequence network connection; (c) transient network; (d) system voltages.

Circuit Breaker Application

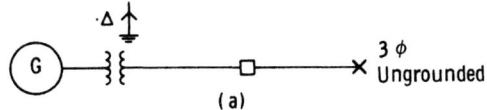

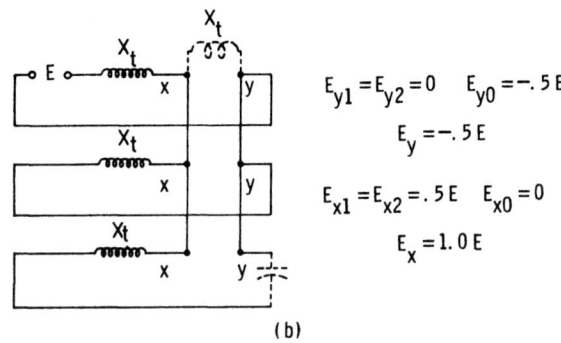

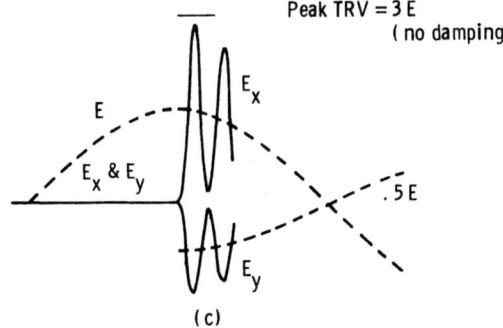

Figure 3.24 TRV for a three-phase ungrounded fault: (a) system represented; (b) sequence network connection; (c) system voltages.

Three-Phase Ungrounded Transformer Terminal Fault

Figure 3.24 shows a similar calculation for an ungrounded three-phase fault. Note that again the steady-state voltage on the transformer side of the first pole of the circuit breaker to open is 1.0 per unit, but now the fault-side voltage is minus 0.5 per unit. Figure 3.25 shows this phenomenon in vector form. The triangle a, b', c' represents the unfaulted three-phase voltages. During the fault, this triangle collapses to zero. With one pole (phase a) of the breaker open, E_a rises

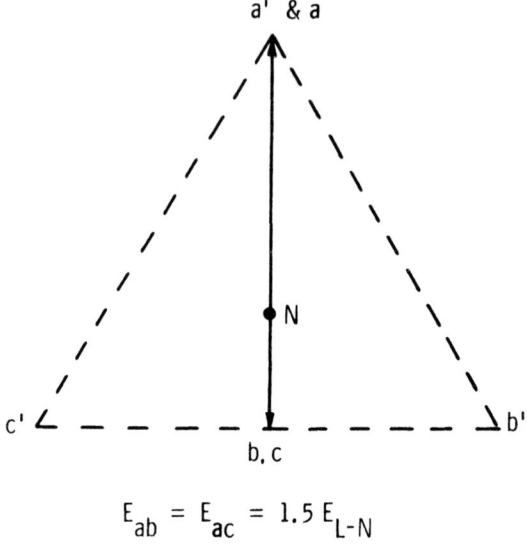

$E_{ab} = E_{ac} = 1.5 E_{L-N}$

Figure 3.25 Vector representation of voltage across the first pole to open a three-phase ungrounded fault.

to point a. E_{bc} is zero, so points b and c lie at the midpoint of vector $E_{b'c'}$. Since all three phases are still tied together at the fault point, the voltage on the fault side of the opened phase a is $E_b = E_c$ or minus 0.5 per unit, which agrees with Fig. 3.24(b). The voltage on the transformer side of the opened phase is E_a or plus 1.0 per unit, which again agrees with part (b). The net steady-state voltage across the breaker pole is 1.5 per unit. Figure 3.24(c) shows the transient involved in this transition. Again, E_x starts at zero and oscillates around E at the frequency of the transformer, which is a function of L_t and C_t. Without damping, E_x would reach a peak of 2.0 per unit. E_y, on the other hand, starts at zero and oscillates around minus 0.5E at the same transformer natural frequency. The TRV across the breaker is again E_x minus E_y and is an oscillatory wave with a peak value of 3.0 per unit with no damping or something less than 3.0 peak unit depending on the amount of damping present.

Three-Phase-to-Ground Fault on an Ungrounded System

If the transformer of Fig. 3.23(a) was ungrounded, the zero-sequence source circuit would be open, so that E_{X1}, E_{X2}, and E_{X0} in Fig. 3.23(b) would now all equal $(1/2)E$. The steady-state voltage on the source side of the circuit breaker and thus across the circuit breaker would again be 1.5 per unit. The TRV would again be an oscillatory

Circuit Breaker Application 95

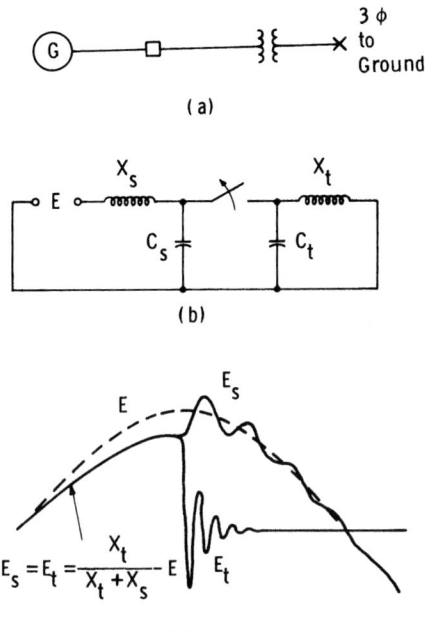

Figure 3.26 TRV for a high-side circuit breaker for a low-side three-phase grounded fault: (a) system represented; (b) transient network; (c) system voltages.

wave with a peak value of 3.0 per unit assuming no damping. Thus a three-phase grounded fault on an ungrounded system produces the same TRV as a three-phase ungrounded fault on a grounded system.

High-Side Breaker with a Three-Phase Fault on the Low Side of the Transformer

Figure 3.26 shows a case similar to Fig. 3.23, except now the circuit breaker is on the high side of the transformer. In this case, we must know the equivalent system reactance X_s and capacitance C_s, as well as the transformer reactance X_t and capacitance C_t.

Prior to the breaker opening, the voltage on both sides of the breaker is equal to $X_t/(X_t + X_s)$ times E. After the first breaker pole opens, the voltage on the system side of the breaker oscillates around E at a frequency based on the inductance and capacitance of the system. The voltage on the transformer oscillates around zero at a frequency based on the inductance and capacitance of the transformer.

(Note that the transformer frequency is usually higher than the system oscillation frequency.) The breaker TRV is again equal to E_s minus E_t, which in this case is the difference between two oscillatory waves.

Several interesting conclusions can be drawn from the examples above. First, in all cases, the TRV for the first pole to open is a (1 − cos) wave. This is true for all applications where the fault-current source is primarily through transformers. Since breakers rated 72.5 kV and below are generally applied in this manner, ANSI standards have specified the (1 − cos) wave as the standard test wave for this class of equipment. Since the natural frequency of transformers generally decrease as the voltage and MVA ratings increase, C37.06-1979 recognizes this by increasing T_2 (time to peak of the oscillation) as the voltage rating increases, and decreasing T_2 (with the T_2 multiplier of Fig. 3.20) as the percent of interrupting rating decreases.

The second point of interest is that the three-phase ungrounded fault produces the highest TRV. (The power frequency driving voltage is 1.5 per unit compared to 1.0 per unit for the three-phase grounded fault.) Although this is an extremely rare condition, it can occur, so ANSI Standards have specified this fault to define the required TRV capability of the breaker. For this condition, Fig. 3.24 shows that the peak voltage of the oscillation is 3.0 per unit, assuming no damping. This is equivalent to $3 \times \sqrt{2}/\sqrt{3} = 2.45$ times rated maximum voltage. E_2 (see Fig. 3.19) in C37.06-1979 is given as 1.88 per unit, which allows for approximately 25% decrement of the oscillatory wave.

IEC considers that a three-phase ungrounded fault is too rare a condition on which to base their standards. However, they do consider that a number of systems rated 170 kV and below can be operated effectively ungrounded. Since for this case the power frequency driving voltage is again 1.5 per unit, the undamped peak of the oscillatory voltage is 3.0 per unit or 2.45 times rated maximum voltage. u_c in Figs. 3.21 and 3.22 is equal to 1.715 for these ratings, which allows for approximately 30% damping for the oscillatory wave.

IEC also assumes that all systems of 245 kV and above will be effectively if not solidly grounded. A three-phase-to-ground fault on these systems will produce a peak of the oscillatory voltage of less than 3.0 per unit or 2.45 times rated voltage. The 1.486 value for u_c instead of 1.715 reflects this condition. Also, as stated earlier, IEC specifies two values for circuit breakers rated 100 through 170 kV: one for u_c equal to 1.715 times rated voltage and one for u_c equal to 1.486 times rated voltage. This reflects the fact that some of these systems will be ungrounded and therefore use circuit breakers with the 1.715 factor, whereas some will be grounded and use breakers with the 1.486 factor.

Circuit Breaker Application

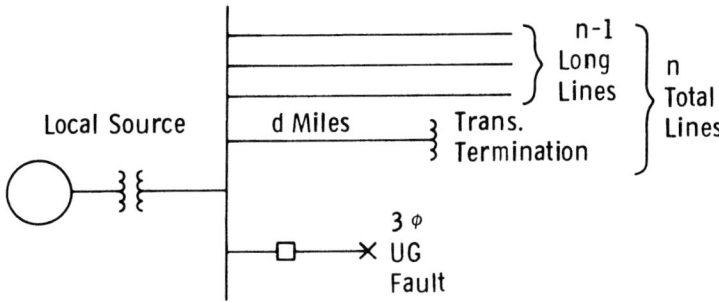

Figure 3.27 Typical high-voltage switching station.

The application problem for these circuit breakers is, therefore, to determine the L and C of the transformer (using the methods of Refs. 16 and 17) and calculate the natural frequency (f_n) of the circuit. The maximum TRV will be a (1 − cos) wave having a peak of 2.45 times rated maximum voltage and a time to crest of ($1/2f_n$) seconds. The time-to-crest value should be compared to the T_2 specified in C37.06-1979 times the multiplier from Fig. 3.20 for the particular fault-current magnitude (in percent of rated interrupting current) that exists at the station. The maximum crest value of 2.45 per unit is greater than the 1.88 E_2 value in the standards, so a computer study using the

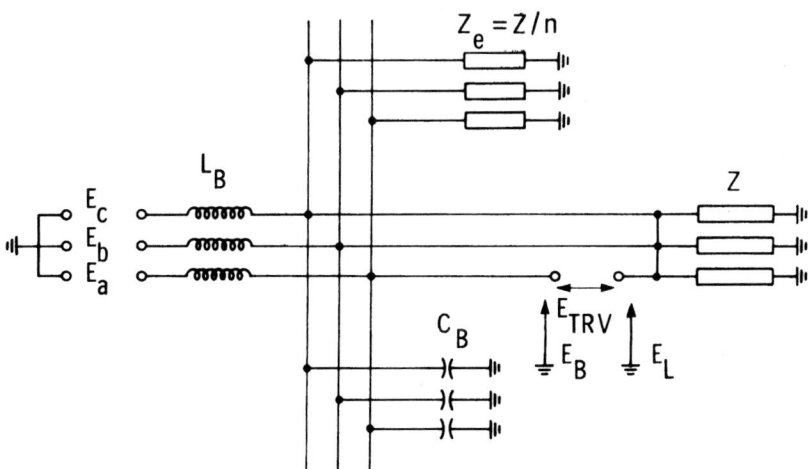

Figure 3.28 Three-phase transient network equivalent of Fig1 3.27. E_a = 1.0 per unit and E_b = E_c = −0.5 per unit when phase a breaker pole clears.

actual damping constants should be made to determine the actual crest value. If both values are less than rated, the application is satisfactory.

Composite Circuits

Circuit breakers rated 121 kV and above are usually applied in composite circuits where the sources of fault current are from transmission lines, as well as transformers. This complicates the calculation process. Figure 3.27 shows such a circuit for a typical high-voltage station. Part of the fault current I_F is supplied from the local source, I_L, through a transformer, and the remaining part, I_R, is supplied over (n) transmission lines also connected to the faulted bus.

Figure 3.28 shows the three-phase transient network for this system at the time that the first breaker pole interrupts the three-phase ungrounded fault current. Note that E_a at this instant is equal to 1.0 per unit, with E_b and E_c both equal to -0.5 per unit. The transformer is represented by its inductance L_B. The transformer and bus capacitance is lumped into one C_B. The transmission lines are represented by their surge impedances Z_e, which is equal to the individual surge impedance Z divided by the number of lines (n). The faulted line also has its surge impedance Z connected to the fault point, since this is an ungrounded fault. The transient recovery voltage E_{TRV} for this circuit is the difference between the bus-side transient E_B and the line-side transient E_L.

The simplest method of solving this circuit is by the *current injection method*. This method consists of setting all the voltage sources equal to zero and inserting or injecting a current pulse into the open-pole path that is equal and opposite to the current that would have flowed if the breaker pole had remained closed. Thus a half cycle of the three-phase fault current is injected into the opened phase of Fig. 3.28. The resulting transient voltages E_B and E_L can then be determined.

To ease the calculation, the circuit of Fig. 3.28 can be reduced to the simpler equivalent circuit of Fig. 3.29. On the bus side, the voltage E_B is determined by the voltage drop across the transformer inductance L_B, the bus capacitance C_B, and the equivalent surge impedance Z_e of the (n) transmission lines all in parallel. This is the top half of the circuit of Fig. 3.29(a). On the fault side of the breaker pole, the injected current path is through the phase b and phase c circuit elements in parallel. Thus the bottom half of Fig. 3.29(a) shows $L_B/2$, $2C_B$, and $Z_e/2$ as the paralleled impedances. The Z/3 element represents the three surge impedances of the faulted line in parallel. The total recovery voltage E_{TRV} is the total drop across this series-parallel circuit.

In the majority of applications, the circuit of Fig. 3.29(a) is overdamped by the parallel resistance of the line-surge impedances, so the

Circuit Breaker Application

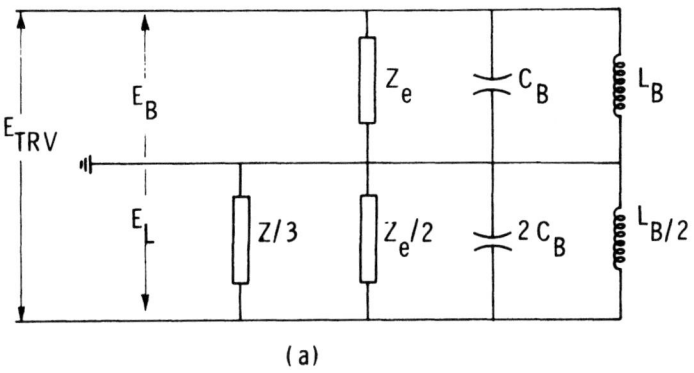

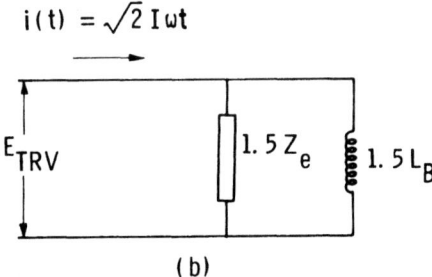

Figure 3.29 Equivalent transient network of Fig. 3.28: (a) complete equivalent; (b) neglecting faulted line surge impedance and bus capacitance.

capacitance C_B can be neglected. Also, $Z/3$ is usually so much larger than $Z_e/2$ that it, too, can be neglected. With these two simplifications, the circuit of Fig. 3.29(a) reduces to the simple LR parallel circuit of Fig. 3.29(b). Also, since the time of interest of the TRV wave is short compared to a half cycle of a power frequency wave, a ramp current of $\sqrt{2}\,I\omega t$ can be used to represent the injected fault current at current zero. This is also shown on Fig. 3.29(b).

The solution of Fig. 3.29(b) is

$$E_{TRV} = 1.5\sqrt{2}\,I\omega L_B \left[1 - \exp\left(-\frac{Z_e}{L_B}\right)\right]$$

where I is the rms symmetrical value of the total fault current I_F in Fig. 3.27. As shown in Fig. 3.30, this is an exponential wave having a crest value equal to $(1.5\sqrt{2}\,I\omega L_B)$ and an initial rate of rise equal to $1.5\sqrt{2}\,I\omega Z_e$. This is the initial wave that appears across the breaker

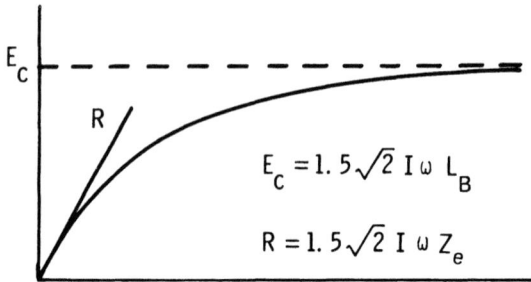

Figure 3.30 Initial transient component.

pole. It also appears as traveling waves on each of the transmission lines. When one of these waves reaches a discontinuity on the line such as another bus or a transformer termination, a reflected wave is produced, which travels back toward the faulted bus. The value of this reflected wave is K_r times the initial wave, where [18]

$$K_r = \frac{Z_t - Z}{Z_t + Z}$$

Z_t is the effective impedance of the termination, and Z is the surge impedance of the line. A transformer termination appears as an open circuit, so $Z_t = \infty$ and $K_r = 1.0$. Thus the reflected wave is approximately equal to the initial wave. If the termination is another bus with a large number of additional lines, Z_t would approach zero and K_r would approach -1.0. The reflected wave in this case would be negative and with a magnitude equal to the initial wave.

When the reflected wave reaches the faulted bus, it again sees a terminating impedance of the $1.5L_B$ and $1.5Z_e$ of Fig. 3.29. The effective wave that appears across the breaker pole is K_t times the reflected wave, where [18]

$$K_t = \frac{2Z_t}{Z_t + Z}$$

where Z_t in this case is the $1.5L_B$ and $1.5Z_e$ in parallel, and Z is again the surge impedance of the line.

The total effective reflected wave that appears at the breaker is $K_r K_t$ times the initial wave. This wave adds to the initial wave as shown in Fig. 3.31. Note that it adds at a time equal to 10.7 × d microseconds, where d is the distance to the termination in miles, as shown in Fig. 3.27. This is the travel time for the wave to travel to the termination and back.

Circuit Breaker Application

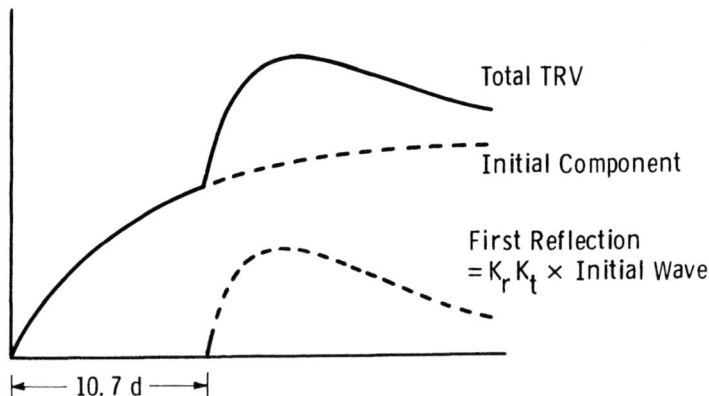

Figure 3.31 Total TRV, including first reflection. d, distance in miles to first line discontinuity. Time is in microseconds.

While Fig. 3.31 shows only the effect of one reflection, additional reflections do occur. These can be calculated as above if the application warrants this additional complexity.

The discussion above gives the general method of calculating the TRV for these composite circuits. Further details on this method are given in Refs. 16 and 19. Note that the wave of Fig. 3.31 is very similar to that of the standard (ex-cos) wave of Fig. 3.19. The application problem for breakers on these types of circuits is, therefore, to calculate the R of Fig. 3.30 and the wave of Fig. 3.31 and compare them with the R and T_2 specified in C37.06-1979 (see Table 3.3) times the multipliers from Fig. 3.20 for the particular fault-current magnitude (in percent of rated interrupting current) that exists at the station. If the calculated value is less than rated, the application is satisfactory.

Short-Line Faults

The preceding examples have assumed that the faults have occurred very near the breaker terminals. This, of course, gives the highest fault current that the breaker must interrupt, but it does not necessarily give the most difficult TRV for the breaker. Figure 3.32 shows a condition where with the fault a short distance out on a transmission line, the TRV produced, while relatively small in magnitude, can have an extremely high rate of rise. This is called the short-line-fault TRV.

Figure 3.32(a) shows the fault location ℓ miles from the breaker. Figure 3.32(b) shows the steady-state condition. Note that the fault current has been reduced from E/X_s to $E/(X_s + X_\ell)$. The voltage at the circuit breaker is

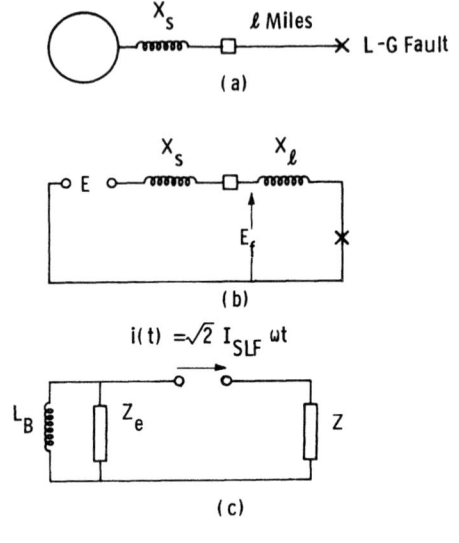

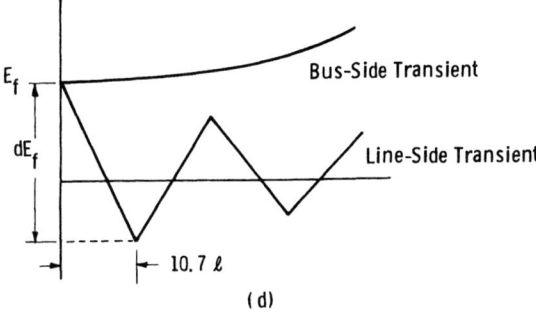

Figure 3.32 Short-line-fault TRV: (a) system represented; (b) 60-Hz equivalent; (c) transient equivalent; (d) transient voltages.

$$\frac{X_\ell}{X_s + X_\ell} E$$

Figure 3.32(c) shows the equivalent transient network. On the bus side of the breaker is the LR circuit—similar to Fig. 3.29, but the 1.5 factor is missing, since the fault being considered is a line-to-ground fault, not a three-phase ungrounded fault. On the line side of the breaker, the section of transmission line to the fault point is represented by the surge impedance of the line. For single-conductor lines, C37.04-1979 specifies an ohmic value of Z of 450 Ω. For bundled-con-

Circuit Breaker Application

ductor lines, Z is specified to equal 360 Ω. Again, the injected current is the ramp function $\sqrt{2}I\omega t$, where I now is the value of the fault current reduced from the bus fault current by the additional line impedance to the fault, X_ℓ.

The injected ramp current times Z gives the value of the traveling-wave voltage that is impressed on the line. This, too, is the ramp function equal to $\sqrt{2}\,I_{SLF}Z\omega t$. When this traveling wave arrives at the fault point, it sees a short circuit or a terminating impedance of zero. K_r is thus -1.0, so a negative wave equal in magnitude and wave shape is produced and travels back toward the circuit breaker. When it arrives back at the circuit breaker, it sees an open circuit, so it causes a reversal in the overall voltage; thus the sawtooth wave shape shown in Fig. 3.32(d). The time of the reversal is $10.7 \times \ell$ microseconds, as shown. This is the time it takes the original wave to travel the ℓ miles to the fault and then back to the circuit breaker.

Note the effect of the distance ℓ. As ℓ increases, X_ℓ increases, and thus the fault-current magnitude I_{SLF} and the rate of rise decrease. The turnaround time (10.7ℓ), however, increases, which increases the magnitude of the sawtooth. As ℓ decreases, the opposite occurs; the rate of rise increases, but the magnitude decreases. It has been found in practice that the worst combination occurs at lengths near 1/2 mi or 1 km, so this condition was originally called the kilometric fault.

On a completely lossless sytem, the sawtooth wave would have a starting point of plus E_f as shown in Fig. 3.32(d) and reach a negative peak of minus E_f and continue the sawtooth with these same peak values. Since there are losses, ANSI standards specify a damping factor d, which for single-conductor lines is 1.8 and for bundled conductors is 1.6. Thus the amplitude of the first peak is only dE_f rather than $2E_f$. With small amplitudes, a further reduction is caused by the prezero arc voltage.

Whereas Fig. 3.32(d) shows the initial sawtooth ramp function starting at the time of current extinction, there is actually a short time delay of 0.2 and 0.5 µs caused by the capacitance of apparatus on the line side of the circuit breaker. This has no effect on the magnitude of the recovery voltage but does affect the initial rate of rise of the transient.

As shown in Fig. 3.32(d), the sawtooth wave occurs on the line side of the circuit breaker. There is still the bus-side TRV which must be added to the sawtooth for the overall circuit breaker TRV. The normal bus-side TRV discussed previously is usually so slow compared to the line-side TRV, however, that it can usually be ignored. For example, with ℓ equal to 1/2 mi, the peak of the sawtooth occurs in approximately 5 µs. Table 3.3 shows T_1 values of 2.9 to 7.9 µs and T_2 values of 276 to 1531 µs; so it is obvious that this component of the *bus-side* recovery voltage can be ignored for worst-case line lengths.

There is an additional component of the bus-side transient for the higher current ratings, however, called the *initial transient recovery voltage* (ITRV), which appears in the very short time period immediately after current interruption. This component is a sawtooth wave similar to the line-side wave caused by the traveling wave transmitted from the circuit breaker back toward the substation bus. Since the distance from the breaker to the first discontinuity of the bus is usually short, the peak magnitude of this wave is much smaller than the magnitude of the line-side wave; however, it cannot be ignored.

Although not included in the present version of ANSI standards, proposed revisions specify two alternative test methods for short-line faults. The first methods consists of modifications to the test circuit to produce the ITRV on the source side of the test breaker and the delayed sawtooth wave on the line side of the breaker. Since this results in a complex test setup, an alternative test method is to apply only the unmodified line-side sawtooth wave across the breaker. The

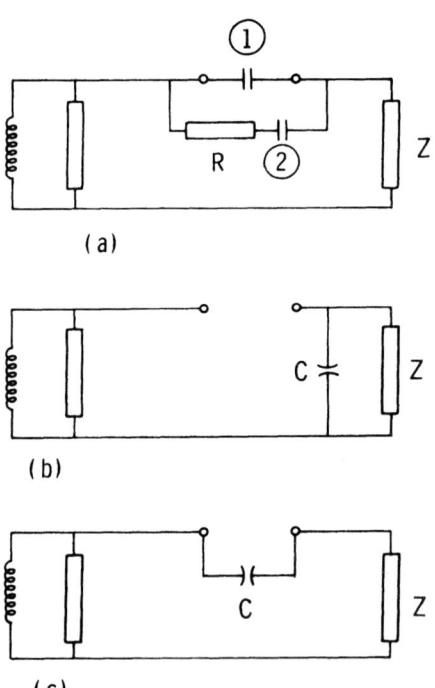

Figure 3.33 Circuit breaker modifications for short-line-fault TRV control: (a) opening resistor; (b) shunt capacitor; (c) open-gap grading capacitor.

Circuit Breaker Application

latter method recognizes that removing the delay in the line-side component compensates for the ITRV bus-side component.

Application engineers need not be concerned with calculating the short-line-fault condition, since the standards state that breaker manufacturers must design their equipment to withstand all short-line-fault conditions. Engineers should be aware of the phenomenon, however, to understand the methods used by the designers to meet these requirements.

In most cases, the basic interrupter of the circuit breaker is adequate to meet the short-line-fault duty. In some cases, however, auxiliary devices are used to modify the TRV, so that the interrupter can perform satisfactorily. Some of these means are shown in Fig. 3.33.

Some circuit breakers use a two-stage opening sequence for TRV control. In the first stage, the main contacts, shown as 1 in Fig. 3.33(a), open, which leaves a resistor in the main fault circuit. Approximately one cycle later, contact 2 opens, completely isolating the circuit. Comparing this operating condition with that of Fig. 3.32(c). it can be seen that the injected current ($\sqrt{2}\, I_{SLF}\omega t$) now has a parallel path through the resistance R, which reduces the rate of rise of the sawtooth voltage wave. If R = Z, the rate of rise is reduced to approximately one-half. With R in the circuit, the fault current has been drastically reduced, so there is no problem when contact 2 opens.

Another means of reducing the short-line-fault TRV is by means of a shunt capacitor connected to the line side of the circuit breaker. This is shown in Fig. 3.33(b). Note that this causes a delay in the initial buildup of the sawtooth wave due to the ZC time constant of the combination. Reference 20 gives more details on this application. It is also treated in Chap. 6.

A third method, shown in Fig. 3.33(c), is to apply a capacitor across the open contacts of the breaker. Most multibreak circuit breakers already use grading capacitors across the contacts to obtain the proper voltage division for interruption and surge voltage control. By increasing the value of the capacitance, the short-line-fault TRV performance can also be improved. As can be seen in Fig. 3.33(c), this capacitance acts the same as the shunt capacitor in that the sawtooth voltage wave is delayed by the ZC time constant.

3.11 CAPACITANCE SWITCHING

In addition to switching fault and load current, circuit breakers must be capable of switching capacitance currents. This may take the form of switching shunt capacitor banks, or unloaded cables or overhead transmission lines. Capacitance switching presents different problems than that of fault switching. Some of these are:

1. *Interruption currents*: Capacitance currents are smaller than fault currents. They are thus easier to interrupt, and there is less arc energy.

2. *Transient recovery voltage*: As was shown in Fig. 3.18(b), the TRV is a (1 − cos) wave. The initial TRV is thus much less than for fault currents. This, coupled with the low current magnitude, leads to an easy interruption and, in fact, causes early arc extinction at small contact separation distances. One-half cycle later, however, there is 2.0 per unit or higher voltage across this smaller contact spacing, which has the potential for possible restrikes of the arc.

3. *Transient overvoltages*: If restrikes occur, they can cause high line-to-ground overvoltages in the LC circuit of the capacitance and the system.

4. *Inrush currents*: Switching of the LC circuit can produce high-magnitude and high-frequency inrush currents that can damage the circuit breaker or other equipment on the system. It can also produce high transient voltages in the substation control wiring.

These problems are recognized by the standards by defining two classifications of circuit breakers: General-Purpose and Definite-Purpose Circuit Breakers. The former are to be used in general applications and for switching only isolated capacitor banks. The Definite-Purpose Circuit Breaker can switch capacitor banks subject to certain limitations on inrush current and frequency. Table 3.5 shows the rating values for these two breaker types.

Table 3.5 shows that each type of breaker is given a rated overhead line and isolated capacitor bank current rating. The values are the same for the General-Purpose breaker. For the Definite-Purpose breaker, the isolated bank current rating is approximately twice that of the overhead line rating, and both of these are higher than those for the General-Purpose breaker.

The table also shows what was previously mentioned: that General-Purpose breakers should not be used for back-to-back switching of capacitor banks, and no rating values are given. For Definite-Purpose breakers, a continuous current rating is given (which is equal to the isolated bank rating), but there are also limits specified for the inrush current magnitudes and frequencies.

The significance of these rating values will be discussed in the following application sections, but it should be mentioned here that the values in Table 3.5 are minimum required levels. Certain breaker capabilities exceed these values. For example, SF_6 breakers, either of the two-pressure or puffer design, and certain vacuum breakers have rated capacitor switching currents equal to their rated continuous current. Also, their inrush current capabilities are equal to their peak closing and latching current ratings with no limit on frequency. Thus the manufacturer should be consulted if the values calculated for the application in question exceed the values of Table 3.5.

Circuit Breaker Application

In the following sections, the four problem areas mentioned above are discussed as they pertain to the application of circuit breakers for switching of capacitance circuits.

3.11.1 Interruption Currents

The first step in the application process is to determine the magnitude of the continuous capacitive current that must be switched or interrupted. For overhead transmission lines, this means that we must divide the maximum system line-to-ground voltage E by the capacitive reactance X_c of the transmission line. Where the exact parameters of the line are not available, it is sufficiently accurate to use

$$X_c = \frac{0.18}{\ell} \times 10^6 \ \Omega \quad \text{for single-conductor lines}$$

or

$$X_c = \frac{0.14}{\ell} \times 10^6 \ \Omega \quad \text{for bundled-conductor lines}$$

where ℓ is the line length in miles.

In determining the E value, the voltage rise on the open-ended line should be considered. Using the ABCD constants of the line and system is the most exact method for this determination. A simpler method is to calculate the bus voltage rise, which is equal to the line capacitance current times the source impedance, and then add the average voltage rise along the line, which is equal to 1/4 times the line capacitive current times the line inductive reactance (X_L). X_L for single-conductor lines is approximately equal to 0.8 Ω/mi, and for bundled-conductor lines approximately 0.6 Ω/mi.

For cable circuits, the same type of calculation is made using the cable capacitive and inductive reactances. For both overhead line and cable applications, if there are shunt reactors permanently connected to the circuit switched, their currents should be subtracted from the capacitive current.

When lines or cables are dropped or deenergized from the low side of transformers connected to the line, the breaker current is equal to the line capacitive current times the turns ratio of the transformer. This can result in higher currents, but as discussed in the next section, the TRV is considerably lower, so the overall interruption process is less severe.

For capacitor banks, the breaker capacitance current is determined by multiplying the rated bank current by 1.25 for ungrounded banks or 1.35 for grounded neutral banks. Rated bank current is equal to the rated kVA of the bank divided by $\sqrt{3}$ times the rated (or nominal) voltage of the bank. The 1.25 and 1.35 factors come from the fact that capacitors can be operated up to 10% above their rated voltage, so the actual system voltage could be up to 1.1 times the rated

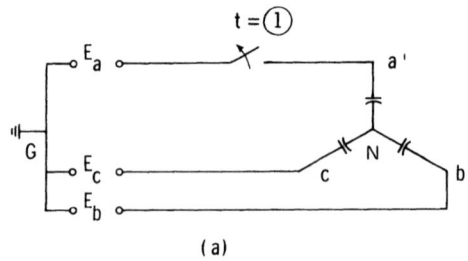

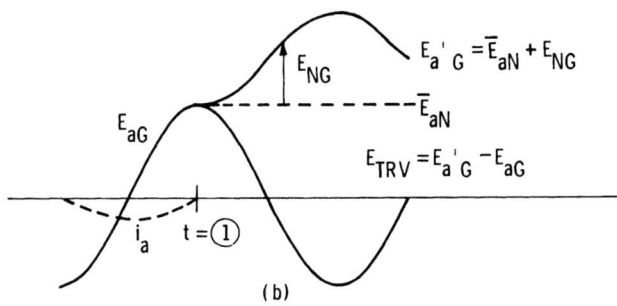

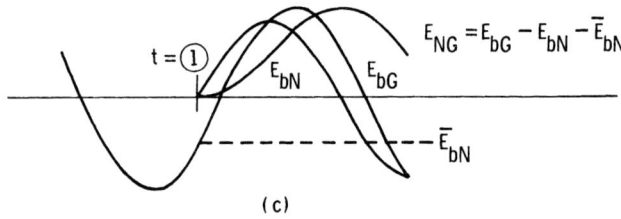

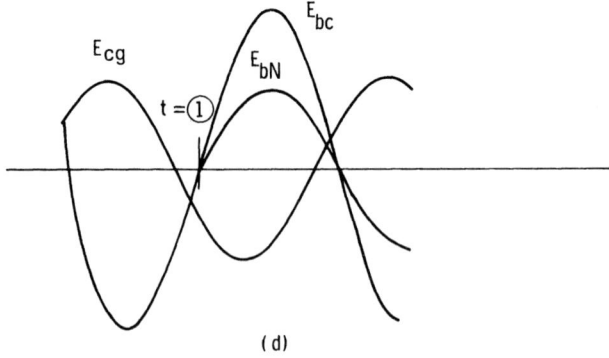

Figure 3.34 TRV for ungrounded capacitor bank: (a) equivalent circuit; (b), (c), and (d) voltage wave shapes.

Circuit Breaker Application

value. Also, the capacitor manufacturing tolerance is between -0 and $+15\%$ with an average of 5%. This adds another multiplier of 1.05 to 1.15. The final consideration is that capacitors on a system provide a low impedance for system harmonics. Grounded capacitor banks provide a path for all harmonics, so a multiplier of 1.1 is used for this contingency. Ungrounded banks block the flow of triple harmonics (third, sixth, ninth, etc.), so a factor of only 1.05 is used for these banks. Combining all these factors gives the 1.25 and 1.35 overall factors specified above.

3.11.2 Transient Recovery Voltage

The transient recovery voltage (TRV) across the circuit breaker when switching a capacitive circuit depends on the ratio of the positive-sequence capacitance (C_1) to the zero-sequence capacitance (C_0) of the circuit. For grounded capacitor banks and cable circuits, $C_1/C_0 = 1.0$, and the circuit can be considered a single-phase capacitance switching circuit. Figure 3.18(b) shows that the TRV for this circuit is a $(1 - \cos)$ wave with a peak value of 2.0 per unit.

For an ungrounded capacitor bank $C_1/C_0 = \infty$. This circuit is slightly more complicated to analyze, but will be described using Fig. 3.34. Figure 3.34(a) is the circuit being analyzed and defines the voltages used in the remaining part of the figure. Figure 3.34(b) through (d) show the voltages appearing in phases a, b, and c, respectively, of the circuit. E_{aG}, E_{bG}, and E_{cG} are the three system voltages to true ground. These voltages are equal and displaced 120° from each other.

Figure 3.34(b) shows that at $t = 1$, the current in phase a (i_a) passes through zero, and the circuit is opened. This traps a voltage $\overline{E}_{aN}$ across the capacitance in phase a. Note that this is the voltage to the neutral (N) of the bank and not to true ground (G).

The instantaneous voltage across phase b and c capacitances at $t = 1$ is -0.5 per unit. Thus the trapped voltages on these two capacitors are -0.5 per unit. This is shown as $\overline{E}_{bN}$ in Fig. 3.34(c), again to the neutral, N, not ground, G. $\overline{E}_{cN}$ is not shown in Fig. 3.34(d), since it is not needed in this explanation.

After $t = 1$, the power frequency voltage across the phase b and c capacitors is the line-to-line voltage E_{bc} shown on Fig. 3.34(d). This voltage is divided equally across the two capacitors, so the voltage E_{bN} across the phase b capacitor is shown as one-half E_{bc} on Fig. 3.34(d).

The voltage of the neutral of the bank (E_{NG}) is equal to the line-to-ground voltage of the system (E_{bG}) minus the power frequency voltage across the phase b capacitor (E_{bN}) minus the trapped charge on the capacitor ($\overline{E}_{bN}$). This is shown in Fig. 3.34(c). E_{NG} is a $(1 - \cos)$ wave starting at zero at $t = 1$ and reaching a crest of 1.0 per unit one-half cycle later.

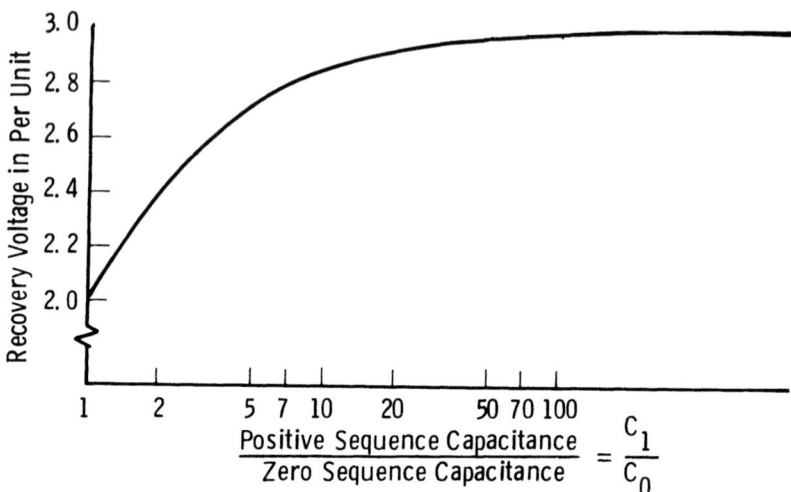

Figure 3.35 TRV for the first pole to open a capacitive circuit. [From I. B. Johnson, A. J. Shultz, N. R. Schultz, and R. B. Shores, Some fundamentals on capacitance switching, *AEEE Trans.* 74(Pt. III): 727-736, © 1953, AIEE—now IEEE.]

Returning to Fig. 3.34(b), the voltage on the capacitor side of the breaker to true ground (E_{aG}) is equal to the trapped voltage $\overline{E}_{aN}$ plus the voltage of the neutral to ground E_{NG}. Since the voltage on the source side of the breaker is E_{aG}, the TRV of the breaker is equal to $E'_{aG} - E_{aG}$. Note that this again is a (1 − cos) wave reaching a peak of 3.0 per unit one-half cycle after current extinction.

For grounded neutral capacitors and cables ($C_1/C_0 = 1$), therefore, the peak TRV is 2.0 per unit, and for ungrounded capacitors ($C_1/C_0 = \infty$), the peak TRV is 3.0 per unit. For overhead transmission lines C_1/C_0 is equal to 1.5 to 2.0, therefore, the TRV is between 2.0 and 3.0 per unit. Figure 3.35 shows how this TRV varies as a function of C_1/C_0. For C_1/C_0 between 1.5 and 2.0, this figure shows a TRV of between 2.2 and 2.4. Standards, therefore, specify a TRV of 2.4 per unit for overhead line switching.

As shown in Fig. 3.36, shunt reactors reduce the severity of the TRV on line switching. The reactors drain the trapped charge from the line capacitance and cause an oscillating voltage on the line. If the shunt reactor MVA exactly equaled the charging MVA of the line, the frequency of the oscillation would equal power frequency and the TRV would be essentially zero. The example of Fig. 3.36 is for 25% compensation ($X_L = 4X_C$). Ths causes a line oscillation of 30 Hz as shown. Note here that the peak voltage is still 2.0 per unit, but as

Circuit Breaker Application

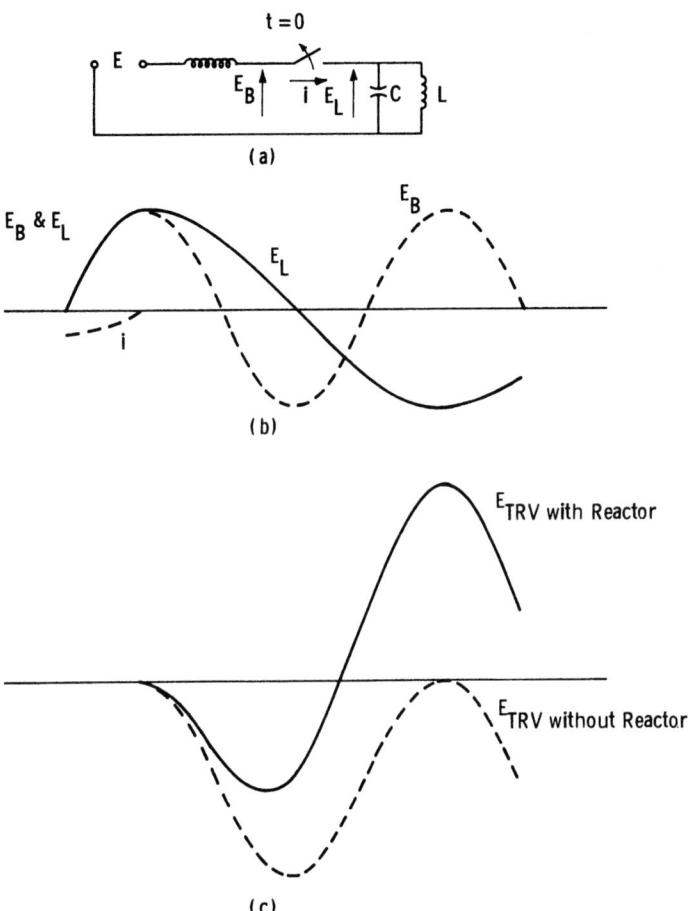

Figure 3.36 Effect of a shunt reactor on TRV: (a) equivalent circuit; (b) circuit voltages; (c) TRV with and without shunt reactor.

shown in Fig. 3.36(c), the peak occurs one cycle after current interruption, instead of one-half cycle in the uncompensated case.

As mentioned previously, switching a transmission line through a transformer increases the capacitive current, but decreases the severity of the TRV. This is demonstrated in Fig 3.37. Figure 3.37(a) is the same as Fig. 3.36, except that the inductance (L_t) is now nonlinear. In the unsaturated region, it is a very high value. When it saturates, it becomes a very low value. Figure 3.37(b) shows that when the current is interrupted at a current zero, a voltage is trapped on the LC circuit. Assuming that L_t is infinite at this moment (unsaturated), a dc voltage

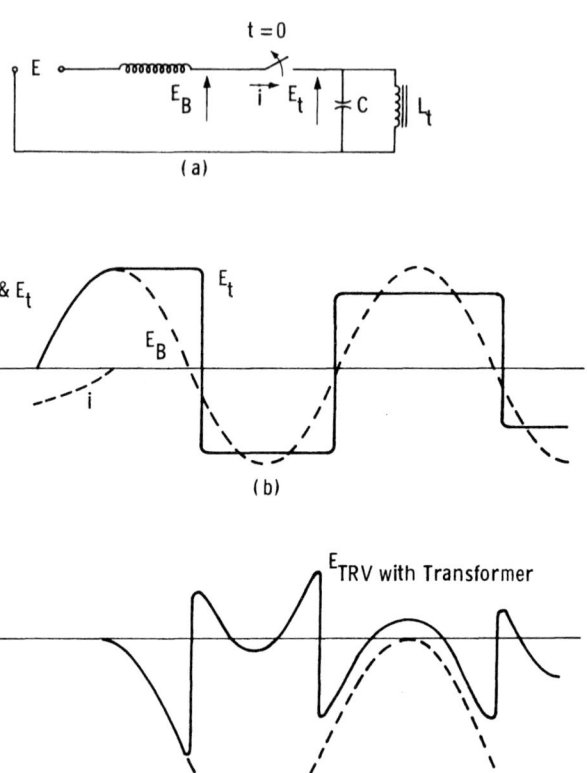

Figure 3.37 Effect of a transformer on TRV: (a) equivalent circuit; (b) circuit voltages; (c) TRV with and without transformer.

exists across L_t. This dc voltage ultimately causes the transformer to saturate, and L_t becomes a low value. The low L_t discharges C at a high frequency and causes the E_t to reverse itself. When it reaches a negative peak, the transformer comes out of saturation, and L_t again becomes infinite. The voltage E_t becomes a negative dc value and remains as such until the transformer saturates again. This causes L_t to become a low value, and again the voltage oscillates, this time to a positive value. This process continues until the losses in the circuit completely discharge the energy from the capacitance. Note that due to the losses, the peaks of the square waves continually decrease.

Circuit Breaker Application

Since the flux in the transformer is proportional to the area under the voltage wave, decreasing the magnitudes of the square waves means that the duration of each wave must increase to make the total area under the square wave equivalent to the saturation flux density. Thus E_t is a series of decreasing-magnitude, increasing-duration square waves. The resulting TRV is shown in Fig. 3.37(c). Note that it is less severe than that when a transformer is not involved.

3.11.3 Transient Overvoltages

C37.04-1979 [12] specifies that the voltage produced across the capacitor (with respect to the capacitor neutral, not to ground) during an opening operation should not exceed 3.0 per unit for a General-Purpose breaker, 2.5 per unit for a Definite-Purpose breaker rated 72.5 kV and below, and 2.0 per unit for a Definite-Purpose breaker rated 121 kV and above. For the 121 kV and above applications, it is assumed that the system and capacitor banks are grounded. These voltage limits cannot be exceeded more than once in 50 switching operations.

Figure 3.18(b) showed that on a normal opening operation the voltage on the capacitor did not exceed 1.0 per unit, so this meets the standard limits. But, as mentioned above, the interruption is so easy that current is interrupted at small contact separation distances. One-

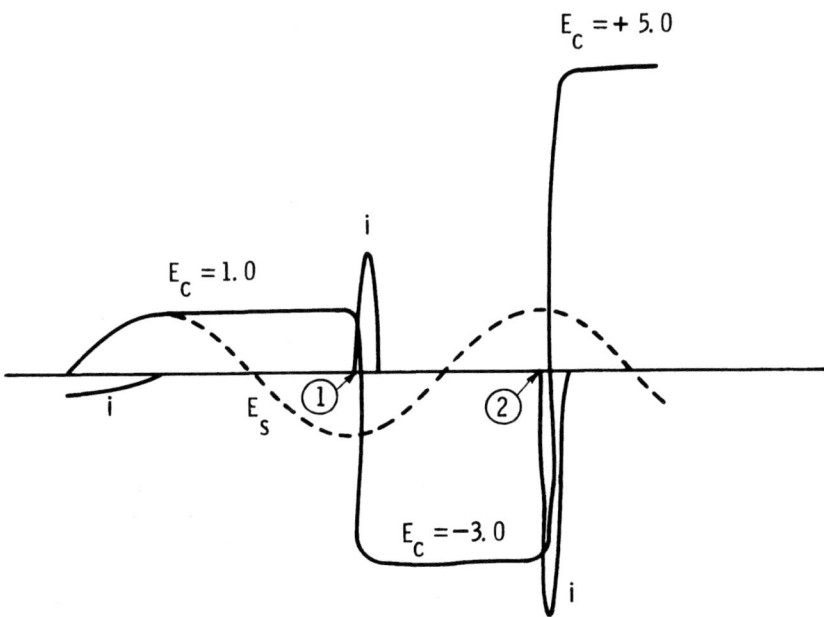

Figure 3.38 Effect of circuit breaker restrikes.

half cycle later, when the recovery voltage is 2.0 per unit, the contact gap may not be great enough to withstand this voltage, and a restrike may occur. Figure 3.38 shows what occurs if this should happen.

Assume that the restrike occurs at the peak of the recovery voltage (point 1). At this time the capacitor voltage (E_c) is equal to +1.0 per unit, and the system voltage (E_s) is equal to -1.0 per unit. When the circuit is reestablished, the capacitor voltage approaches E_s and overshoots. With no damping, it reaches a peak of minus 3.0 per unit. The inrush current at this time is also shown (on an expanded scale). Note that it is also an oscillatory wave passing through zero when the capacitor voltage reaches its negative peak. If the breaker extinguishes the arc at this current zero, a -3.0 per unit voltage will remain trapped on the capacitor.

Although not shown, the system voltage also has an oscillation during this period, but it will shortly return to its steady-state value. The recovery voltage will, therefore, continue its (1 − cos) waveform until one-half cycle later (point 2) it will reach 4.0 per unit. Suppose that the breaker restruck at that time. Now the capacitor voltage again approaches E_s and overshoots reaching a peak of +5.0 per unit. This would certainly cause equipment damage.

The discussion above shows that since a General-Purpose breaker is permitted to produce a 3.0 per unit overvoltage, it is allowed to restrike once on capacitor switching, but not more than once. Definite-Purpose breakers are only allowed 2.0 or 2.5 per unit overvoltages, so they must either not restrike or have some means of controlling the voltages if a restrike should occur.

SF_6 breakers meet this requirement, since they do not restrike, at least not more than once in 50 switching operations. Other breaker types, such as oil breakers, generally cannot meet this requirement, so for those breakers given a Definite-Purpose rating, voltage control methods are used. In this case, opening resistors are used. Figure 3.39 shows the operation of such a breaker using opening resistors.

Figure 3.39(a) shows that the breaker has two sets of contacts. The first set, when they open at t = 1, inserts a resistor R in series with the capacitance circuit. The second set, when they open at t = 2, then opens the RC circuit, completely clearing the circuit. Figure 3.39(b) through (d) show how this sequence controls the transient overvoltages.

Figure 3.39(b) shows the system voltage and the current involved. Prior to t = 1, the current is equal to the normal capacitor current. At t = 1, the resistor is inserted. The resistor ohms is normally selected to equal $0.4X_c$, so that when the resistor is inserted, there is very little change in current magnitude or phase angle. At t = 2, the second set of contacts open, and the current goes to zero.

Figure 3.39(c) shows the capacitor voltage. Again at t = 1, there is very little change in capacitor voltage due to the small ohmic value of R. At t = 2, the circuit is opened, and a voltage equal to -1.0 per unit is trapped on the capacitor.

Circuit Breaker Application

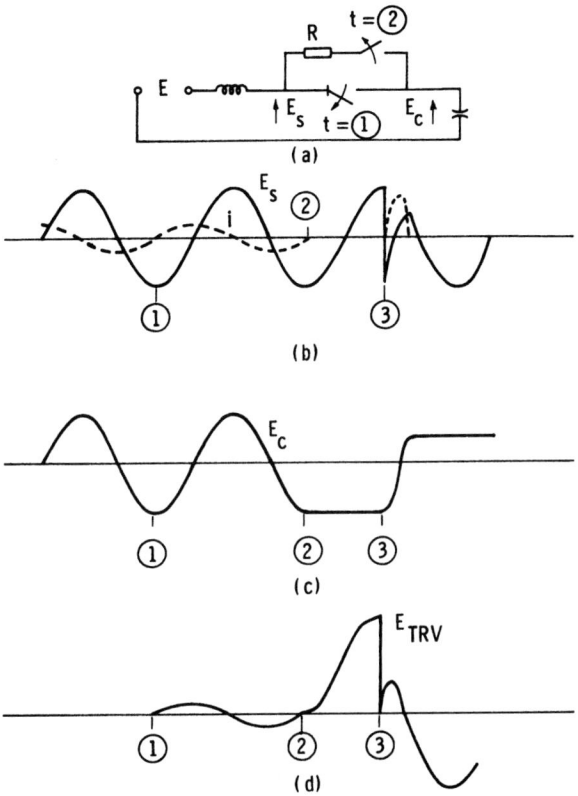

Figure 3.39 Capacitor switching with opening resistor: (a) equivalent circuit; (b) system voltage; (c) capacitor voltage; (d) TRV.

In the meantime, the recovery voltage, shown in Fig. 3.39(c), is very low after $t = 1$, since it is only the voltage drop across the resistor. Even though extremely rare, should a restrike occur during this period, the driving voltage (the voltage across the resistor) is so low that only a very small system disturbance would result. At $t = 2$, the normal $(1 - \cos)$ recovery voltage would occur.

Should a restrike occur, say at point 3, the circuit would be overcritically damped through the resistor R. Figures 3.39(b) through (d) show that the resulting overvoltages would be less than 2.0 per unit, and thus the requirements for a Definite-Purpose Circuit Breaker would be satisfied.

3.11.4 Inrush Currents

Switching of capacitance circuits causes not only transient voltages, but also transient currents called *inrush currents*. The magnitude and frequency of the inrush current is a function of the system voltage at the point on the wave at breaker contact closing, the inductance and capacitance of the circuit, the charge on the capacitor at time of closing, and any control impedances, such as closing resistors, that are used in the circuit breaker. Switching of single-capacitor banks usually does not cause problems, since the inrush currents are less than the normal short-circuit current at that location. When two or more banks are switched back to back, the high magnitude and high frequency of the inrush currents can cause problems in the circuit breaker or the associated system, or both.

In the circuit breaker it is the rate of change of current (di/dt) that produces the problem. The high rate of energy released in the arc, especially at the small contact separation distances involved, generates explosive forces in the interrupting chamber that can cause physical damage to the interrupter. The oil breaker is especially vulnerable to this condition. The standards [2] (see Table 3.5) therefore provide upper limits of inrush current magnitude and frequency that define the di/dt capability of these breakers to withstand this switching duty.

Note that SF_6 breakers are relatively unaffected by di/dt, so the limits expressed in Table 3.5 do not apply to these breakers. The only restriction on inrush currents for SF_6 breakers is that the peak of the inrush current does not exceed the peak of the rated Closing and Latching Current.

Table 3.5 shows that both General-Purpose and Definite-Purpose Circuit Breakers have Isolated Current ratings, but only Definite-Purpose Circuit Breakers have Back-to-Back Current ratings. The latter is further limited by a current magnitude and frequency rating. An Isolated Bank condition is said to exist when the di/dt of the inrush current is equal to or less than the di/dt of the normal rated short-circuit current of $\sqrt{2}\, 2\pi f_s I_{sc} = 533 I_{sc}$ for $f_s = 60$ Hz. Thus, even if there are two capacitor banks in a station, if the inductance between the banks is large enough to limit the di/dt to this value, it could be considered an Isolated Bank condition, and a General-Purpose Circuit Breaker could be applied.

Figures 3.40 and 3.41 show the methods of calculating the inrush currents for capacitor bank switching. On the left side of the figures, the circuit equations are given, and on the right side are simplified expressions that were developed in Ref. 21 and given in Table 1 of that reference.

For single-bank switching, Fig. 3.40 shows that inrush current peak is equal to the crest value of the line-to-ground voltage times the quantity $\sqrt{C/L}$. The frequency is $1/2\pi\sqrt{LC}$, and the maximum di/dt

Circuit Breaker Application

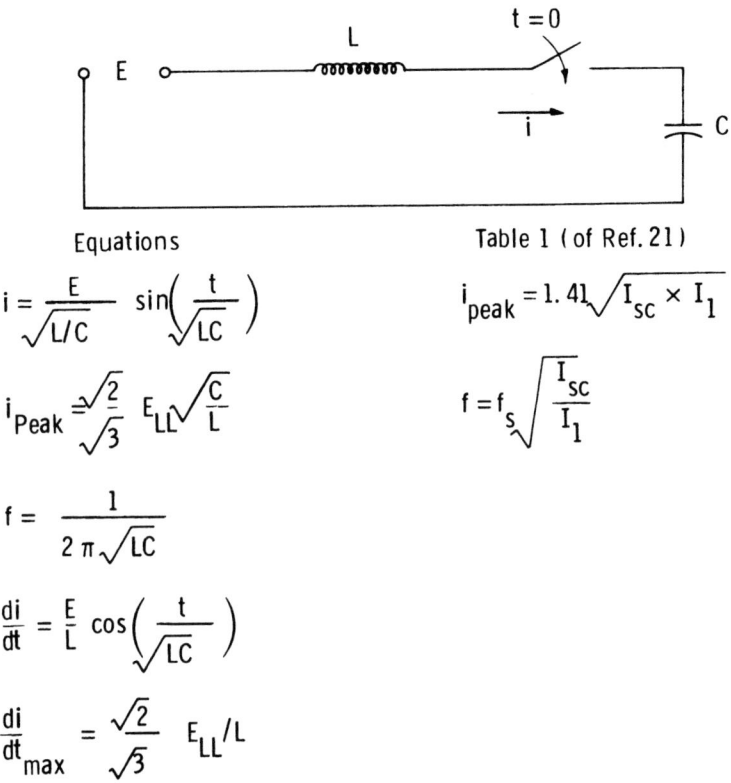

Figure 3.40 Inrush current for single-capacitor-bank switching.

is the crest line-to-ground voltage divided by the inductance (E_{LG}/L). Note that the di/dt for a short circuit on the substation bus is E_{LG}/L_S. Since L in Fig. 3.40 must always be greater than L_S, this confirms that if the switching device can interrupt the system short circuits, it will also be satisfactory for switching an isolated capacitor bank.

The quantity I_{sc} on the right-hand side of Fig. 3.40 is defined as the symmetrical short-circuit current available at the breaker location and is equal to $E_{LL}/(\sqrt{3}\,\omega L)$. I_1 is the current rating of the capacitor bank being switched and is equal to $E_{LL}/(\sqrt{3}\,X_c) = E_{LL}\omega C/\sqrt{3}$. Substituting these quantities in the expression for i_{peak} gives

$$i_{peak} = 1.41\sqrt{I_{sc} I_1} = 1.41\sqrt{\frac{E_{LL} E_{LL} \omega C}{\sqrt{3}\,\omega L \sqrt{3}}} = \frac{\sqrt{2}}{\sqrt{3}} E_{LL}\sqrt{\frac{C}{L}}$$

which is the same as shown on the left side of Fig. 3.40. A similar substituion in the expression for f (where f_s is the system frequency or $\omega/2\pi$) gives

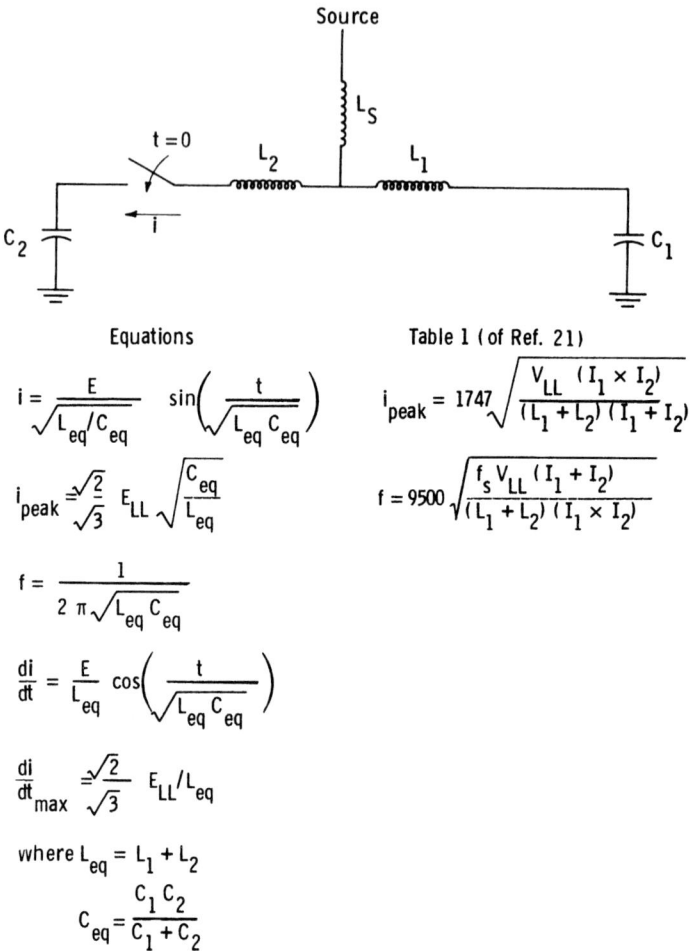

Figure 3.41 Inrush current for back-to-back capacitor-bank switching.

$$f = f_s \sqrt{\frac{I_{sc}}{I_1}} = \frac{\omega}{2\pi} \sqrt{\frac{E_{LL}\sqrt{3}}{\sqrt{3}\,\omega L E_{LL}\,\omega C}} = \frac{1}{2\pi\sqrt{LC}}$$

which again is the same as the equation on the left.

Figure 3.41 shows the condition where a capacitor bank C_2 is being energized from a bus that has another energized bank or banks of capacitors already connected to it. As shown, there are three inductances involved in this circuit. L_s is the inductance of the source,

Circuit Breaker Application

which is usually much larger than L_1 and L_2 and so is usually ignored. L_1 and L_2 are the inductances of the capacitors and bus work connected to the two capacitor banks C_1 and C_2, respectively. L_2 includes the inductance of the bus, which is approximately 0.25 µH/ft, plus the inductance of the capacitor bank itself. For banks rated 46 kV and below, this bank inductance is approximately 5 µH. For banks above 46 kV, this value is approximately 10 µH. If C_1 is a single bank, L_1 is determined the same as L_2. If it is more than one bank, an equivalent L must be determined based on the series-parallel combination of the actual inductances. Note that the equations on the left side of Fig. 3.41 are the same as those on Fig. 3.40 if L_1 and L_2 are combined into L_{eq} and C_1 and C_2 into C_{eq}. Also, the simplified expressions on the right-hand side now use the continuous current ratings of both banks, I_1 and I_2, in their formula.

The expressions in Figs. 3.40 and 3.41 assume that the capacitors being switched have no voltage or charge existing on them at the time of energization. For normal switching, this is a valid assumption, since all capacitors have built-in discharge resistors which drain their charge between switching operations. Should a restrike occur (as in Fig. 3.38) during an opening operation, however, this will not be true. There could be a negative voltage of 1.0 per unit trapped on the capacitor from the first opening, and if the restrike occurred when the system voltage was +1.0 per unit, the net driving voltage in Figs.

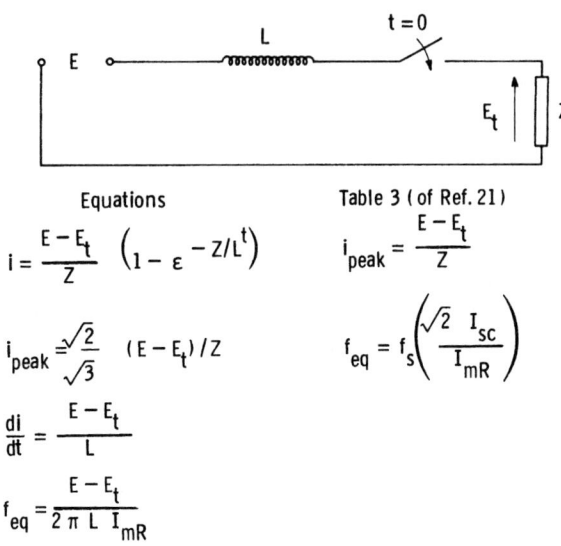

Figure 3.42 Inrush current for single-cable switching.

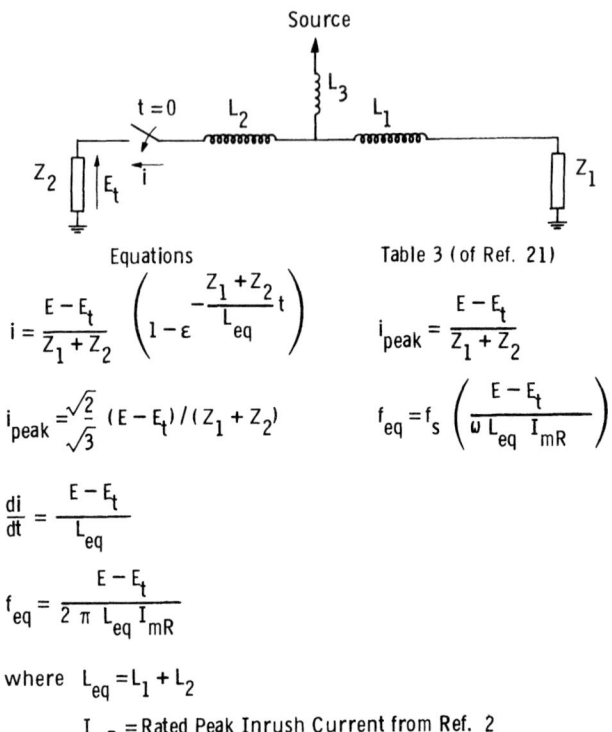

Figure 3.43 Inrush current for back-to-back cable switching.

3.40 and 3.41 would be 2.0 per unit. The i_{peak} and di/dt values from the equations of these figures should, therefore, be doubled for the restrike condition.

Switching of cable circuits is very similar to capacitor switching, as they too produce high rates of rise of current. Figures 3.42 and 3.43 show the circuits and equations for these conditions. The circuit of Fig. 3.42 for switching a single cable is the same as Fig. 3.40 except that now C is replaced by the cable surge impedance Z. The equation for the current is now an exponential having a magnitude of $(E - E_t)/Z$ and a di/dt of $(E - E_t)/L$. Note that now we have to include the voltage E_t as the trapped voltage on the cable, since there are no built-in discharge resistors in cable circuits.

The breaker capability in Table 3.5 is expressed as an inrush current peak I_{MR} and frequency f_R. Since the current wave for cable switching is an exponential, for the application the exponential wave must be converted to an equivalent frequency. Since di/dt is the real limit on the breaker duty, this conversion is accomplished by converting

Circuit Breaker Application

the rated inrush current I_{MR} and rated frequency f_R into a rated di/dt. Thus

$$\text{Rated } i = I_{MR} \sin 2\pi f_R t$$

$$\text{Rated } \frac{di}{dt} = 2\pi f_R I_{MR}$$

If the actual di/dt $\leq$ rated di/dt, the application is satisfactory. Thus

$$\text{Actual } \frac{di}{dt} = \frac{E - E_t}{L}$$

$$\text{Rated } \frac{di}{dt} = 2\pi f_R I_{MR}$$

or

$$\frac{E - E_t}{L} = 2\pi f_R I_{MR}$$

$$feq = \frac{E - E_t}{2\pi L I_{MR}}$$

If feq $\leq f_R$, the application is satisfactory.

Figure 3.43 shows the back-to-back cable switching circuit, and is similar to the back-to-back capacitor switching circuit of Fig. 3.41. Again, the capacitances C_1 and C_2 are replaced by the surge impedances Z_1 and Z_2, and the trapped charge voltage E_t cannot necessarily be neglected.

The breaker application problem for capacitor and cable switching is, therefore, to calculate the inrush current magnitude and frequency using the methods of Figs. 3.40 through 3.43. These values are compared to the rated values of Table 3.5. If the calculated values are less than the rating, the application is satisfactory. If they are above the Table 3.5 values, the manufacturer should be consulted to see if the capability of the breaker under consideration exceeds the Table 3.5 values. For example, as mentioned previously, SF_6 breakers are limited only by the Closing and Latching Current rating of the breaker.

If the capability of the circuit breaker is exceeded, means of controlling the inrush currents are required. One method is to use resistors in the closing and opening sequences of the breaker. This was discussed briefly in conjunction with Fig. 3.39. A more common method, however, is to install current-limiting reactors in the capacitor circuits. In effect, these reactors increase the inductances L_1 and L_2 in Figs. 3.40 through 3.43 to decrease the magnitude and frequency of the inrush currents.

3.11.5 Effect on Other Equipment

The preceding discussion has been concerned primarily with the effect of capacitance switching on the circuit breaker performing the switching. Other equipment in the station could also be affected, however. For example, other breakers connected to the bus will be subjected to the discharge currents of the banks should a fault occur on the load side of these breakers. If the breaker is closed at the time of the fault, this should not be a problem, provided that the discharge current is less than the Closing and Latching Current rating of the breaker, since the discharge will have attenuated before contact part of the breaker. Should the breaker in question close into the fault, however, the discharge current should be checked against its capacitive switching capability.

The high magnitude and frequency of the inrush currents can induce voltages in the control wiring and can cause failures in the relaying and control equipment. Reference 22 provides recommendations for grounding of capacitor banks and layout and shielding of control circuits to minimize these problems. In addition, many users install current-limiting reactors to reduce the inrush currents for this reason, even though they are not needed for the circuit breakers.

The high-magnitude high-frequency inrush currents can cause problems in the instrument transformer circuits. The burdens in current transformer secondaries are made up of reactance as well as resistance. At power frequency the voltage drop across this reactance is usually very low. In the kilohertz range of inrush currents, these voltages can be appreciable, and voltage limiters are usually required.

Linear couplers are especially susceptible to these problems, since their secondary voltage is directly proportional to the product of the magnitude and frequency of the input current. For example, the output of a linear coupler is 5 V for each 1000 A of 60-Hz input current. If the inrush current is 12,000 A at a frequency of 4250 Hz, the secondary voltage is

$$E_s = 5 \times \frac{12,000}{1000} \times \frac{4250}{60} = 4250 \text{ V}$$

which is above the insulation level of the secondary circuits. Voltage limiters are certainly required in this application.

3.12 OTHER APPLICATION CONSIDERATIONS

Although the preceding sections have discussed the major application problems as they apply to the standard rating factors pertaining to circuit breakers, there are other choices to consider when selecting a certain breaker for a given application. Some of these choices are discussed in the following sections.

Circuit Breaker Application

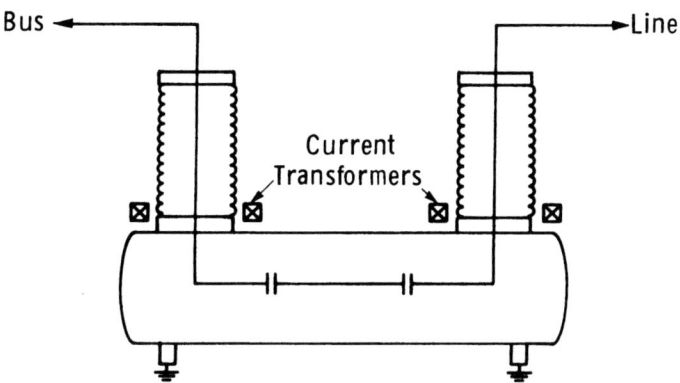

Figure 3.44 Schematic of a dead-tank circuit breaker.

3.12.1 Live Tank vs. Dead Tank

For many years in the United States, all high-voltage circuit breakers were of the *dead-tank* design, where the interrupting elements were enclosed in a grounded metal tank, and the line and bus conductors entered the interrupting chamber through entrance bushings. A schematic diagram of such a breaker design is shown in Fig. 3.44.

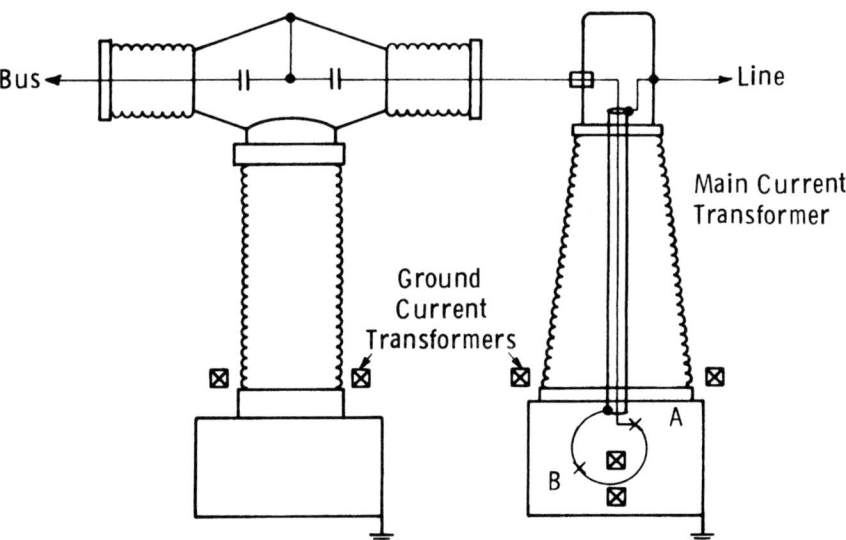

Figure 3.45 Schematic of a live-tank circuit breaker.

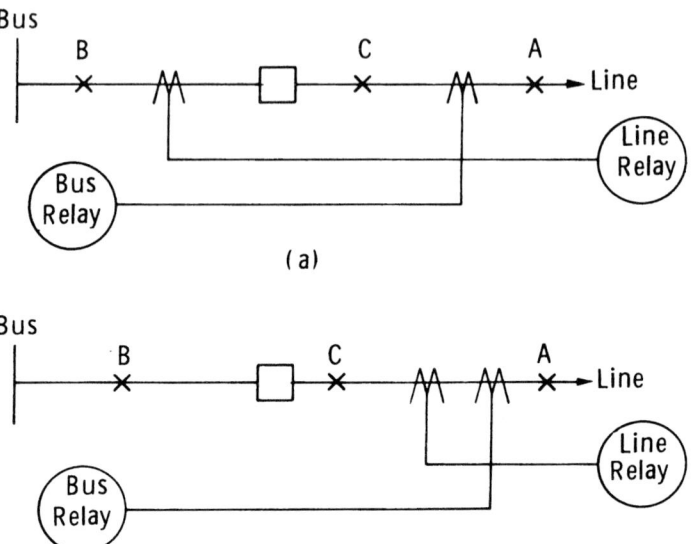

Figure 3.46 Single current transformer: (a) per circuit breaker; (b) versus two current transformers.

With the advent of EHV circuit breakers, the concept of the *live-tank* circuit breaker, shown schematically in Fig. 3.45, has evolved. In this design, the breaker interrupters are mounted in a container at line potential and insulated from ground potential by porcelain insulating columns. The major advantage of the live tank is its lower cost, especially at the higher voltage ratings. The major disadvantage is that it requires externally mounted current transformers, which are more expensive and require more substation space than the bushing current transformers, shown in Fig. 3.44, on dead-tank designs.

Due to their low cost, it is usual practice on dead-tank breakers to mount current transformers on both the line-side and bus-side bushings of the breaker. As shown in Fig. 3.46(a), the current transformers on the bus-side bushings are connected to the line relays; thus line faults, such as A, trip the breaker and the far end termination to clear the fault. The current transformers on the line-side bushings are connected to the bus relays, so that bus faults such as B trip the breaker and the other bus breakers to clear the fault. Faults inside the breaker, such as C are seen by both the line and bus relays, so both the line and bus breakers are tripped immediately to clear the fault. This type current transformer arrangement is called *overlapping protection*.

Circuit Breaker Application

To obtain pure overlapping protection for a live-tank breaker, two externally mounted current transformers would be required, one on each side of the breaker. Due to the high cost and large space requirements, this is almost never done. Instead, the usual practice is to install multisecondary current transformers only on one side (usually the line side) of the breaker. Figure 3.46(b) shows that for faults at A and B, the tripping sequence is the same. Only for an internal breaker fault (fault C) is there a difference. In this case, the fault appears as a bus fault, and all the bus breakers trip, *but* the fault is not cleared until the breakers at the far end of the line are tripped by their backup relays.

Some users accept this limitation, since the probability of such a fault is extremely rare. For those who want complete protection, however, ground current transformers, shown in Fig. 3.45, are installed around each breaker column. These current transformers are arranged so that any internal breaker fault will result in current flow inside the column current transformer. Auxiliary relays connected to the secondary of these current transformers will instantaneously trip all bus breakers and the far-end line breakers, and therefore accomplish the same function of overlapping protection of the dead-tank breakers.

Although Fig. 3.45 shows a ground current transformer around the main current transformer, this is felt not to be justified, due to the small probability of faults not being detected by the normal relays. For example, since the metal head of the main current transformer is solidly connected to the line terminal and is insulated from the breaker terminal, any external flashover will appear as a line fault and be cleared by normal relaying. Also, since the line lead of the bushing completely encloses the return lead, all internal bushing faults must again appear as a line fault. In the base of the current transformer, any faults in the current transformer "eye" from the line lead to point B on the figure will also appear as a line fault. The only section not covered, therefore, by the line relays is the section of the eye from point B to point A. The probability of faults in this section is so remote that the cost and complexity of the bushing ground current transformer on the main current transformer column is felt not to be justified.

3.12.2 Ganged vs. IPO Breakers

In Sec. 3.9 it was mentioned that reducing the breaker interrupting time from 3 to 2 cycles could, in certain instances, be important to system stability. Similarly, reducing the clearing time to 1 cycle would be a further aid to stability. There is another breaker concept, however, that provides greater gains in system stability, and that is *independent pole operation* (IPO) circuit breakers.

As mentioned previously, it is not the 1 or more cycle saving in primary clearing time of the breaker that affects system stability; rather, it is the 2 or more cycles in backup clearing time that is important. Reference 23 shows that the critical fault clearing time for systems using modern turbine generators is 6 to 8 cycles for a three-phase fault condition. If all relays and circuit breakers operate correctly, there is no problem with either a 2- or 3-cycle breaker in meeting this criterion. If for some reason the breaker should fail to operate, the breaker failure relays and backup breakers must operate to clear the fault. For this case, even with 1- or 2-cycle breakers, the 6- to 8-cycle total clearing time would be difficult to obtain.

Reference 23 also showed that if the three-phase fault were downgraded into a line-to-ground fault after 3 cycles, the total critical fault clearing time would be raised to 12 to 14 cycles. Any breaker, even a 3-cycle breaker, would allow a backup clearing time of this value. The problem, therefore, is how to accomplish this.

The first criterion is met by designing the protection system so that no piece of electrical or mechanical equipment will cause all poles of a circuit breaker to fail to open on a three-phase fault. This requires duplicate relays, instrument transformers, battery circuits, and trip coils in the circuit breakers. It also requires separate mechanisms for each phase or pole of the circuit breaker, with no common elements in the gas or air systems that could cause misoperation of all three mechanisms.

This is the concept of IPO or independent pole operation. Rather than having one mechanism that operates all three poles together (which is called gang operation), an IPO breaker has three separate mechanisms, with associated elements, which ensures that each phase pole of the circuit operates independently. It is virtually impossible, therefore, for all breaker poles to fail to operate on a three-phase fault. With this concept, even a 3-cycle breaker will be satisfactory for most system stability considerations.

All present 500- and 800-kV breakers automatically are IPO breakers, since for mechanical reasons they require three operating mechanisms. For breakers rated 362 kV and below, the standard breaker is ganged, but IPO can be provided at added cost.

In addition to cost, the IPO breaker causes an additional problem to the application engineer. Since the three breaker poles are mechanically independent, they can misoperate independently. For example, during a closing operation, two of the phases can close and latch correctly, but the third may fail to latch and drift open. This causes a system unbalance and could cause system or circuit breaker damage. There are two means to detect and correct this situation: a circuit breaker control circuit scheme and a new protective relay. A functional schematic of the breaker control circuit scheme is shown in Fig. 3.47. The scheme consists of a series-parallel connection of the breaker a and b switches of the three breaker poles in series with the three breaker trip coils. The a contacts are closed when the breaker is closed, and the b contacts are closed when the breaker is open. Thus, when all three breaker poles are closed, all three a contacts are

Circuit Breaker Application

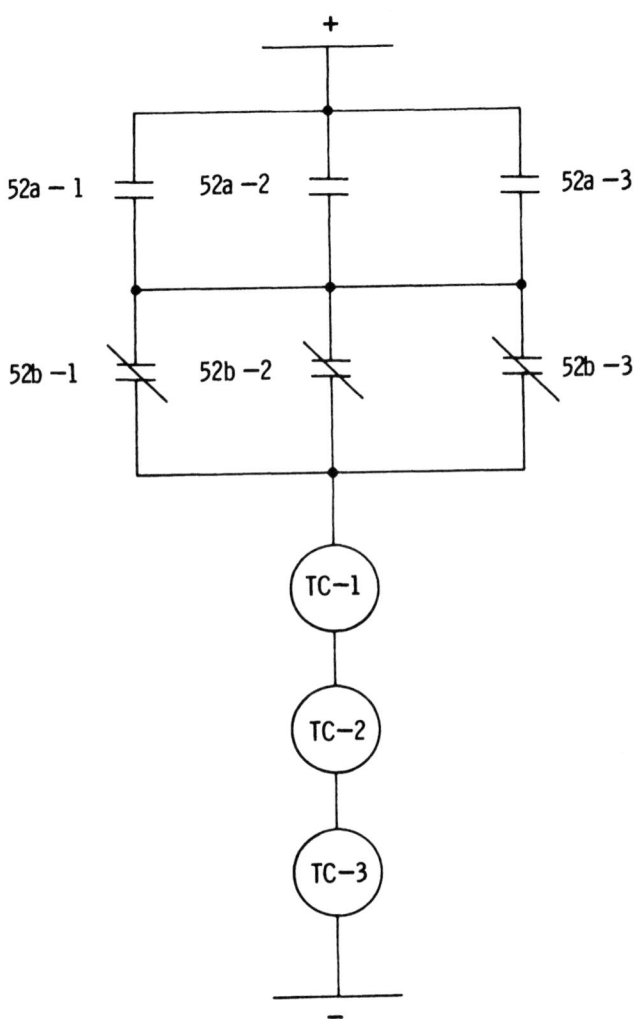

Figure 3.47 Functional schematic of a circuit breaker pole disagreement circuit.

closed, but all b contacts are open and tripping does not occur. If the three breaker poles are open, all three b contacts are closed, but all a contacts are open, and tripping does not occur. If, however, poles 2 and 3 are closed and pole 1 drifts open, 52a-2 and 52a-3 are closed, and when 52b-1 closes, the trip coils are energized. This will trip poles 2 and 3 and prevent operation with the open pole.

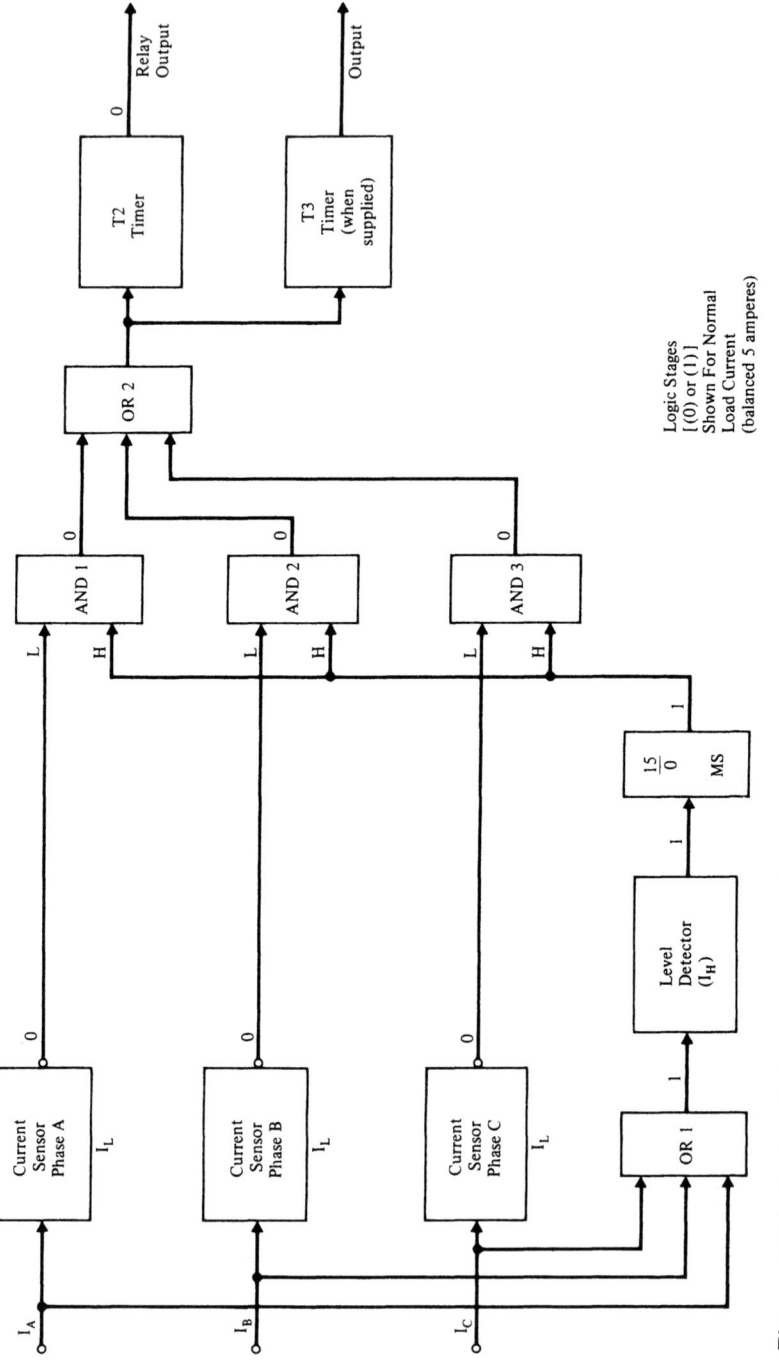

Figure 3.48 SLB pole disagreement relay logic diagram.

Circuit Breaker Application

This breaker disagreement circuit protects against mechanical misoperations of the circuit breaker. There are certain system electrical conditions that can cause phase disagreements. For example, an electrical flashover can occur through one phase of a mechanically open breaker. Also, one of the trip coils may be inoperative, which would prevent all three breaker poles from clearing. To provide for these contingencies, a new relay, the SLB pole disagreement relay, was developed. A functional schematic of this relay is shown in Fig. 3.48.

The basic principle of this relay is measurement of the magnitude of the currents in all three phases of the circuit breaker. If one or two of the currents are below a certain level I_L while the other one or two currents are above a certain level I_H, a phase disagreement is indicated, and the breaker involved plus its backup breakers are tripped. Normally, I_L is set for 20 mA secondary current and I_H for 60 mA. These levels are used to detect charging current on open-ended lines. In certain cases, under light load conditions, these settings can cause false tripping. For these applications, an SLB-1 relay is available which uses zero-sequence voltage to supervise the I_H setting. This modification is not shown in Fig. 3.48.

Reference 24 contains more details on this relay and its application. It should be noted here, however, that this relay, although developed primarily for use with IPO breakers, is also advisable for use with ganged breakers to protect the breaker and system against electrical flashover of the contacts of open breakers.

3.12.3 Open Breaker Capacitance Resonance

Multibreak circuit breakers use capacitors across their individual contacts to maintain the proper voltage gradients during open and opening conditions. As mentioned in Sec. 3.10, these capacitors can also be used to decrease the high rate of rise of recovery voltage under short-line-fault conditions. In both cases, the larger the capacitance value, the greater the improvement in circuit breaker operation.

Unfortunately, this capacitance can cause ferroresonance problems with potential transformers connected to otherwise deenergized buses. If one or more open breakers remain connected to the bus with their disconnect switches closed, the open breaker capacitance and the potential transformer magnetizing impedance can constitute a resonance circuit. This phenomenon is illustrated in Fig. 3.49.

Figure 3.49(a) shows the circuit breaker with its grading capacitors. C_B represents the bus capacitance. Also shown is a potential transformer connected to the bus. Neglecting the transformer for the moment, Fig. 3.49(b) shows the equivalent circuit with only the bus capacitance. The grading capacitors and the bus capacitance act as a voltage divider when the breaker is open, and the voltage on the bus is $C_0/(C_0 + C_B)$. Thus the "dead" bus is not at zero potential, but is also never greater than 1.0 per unit.

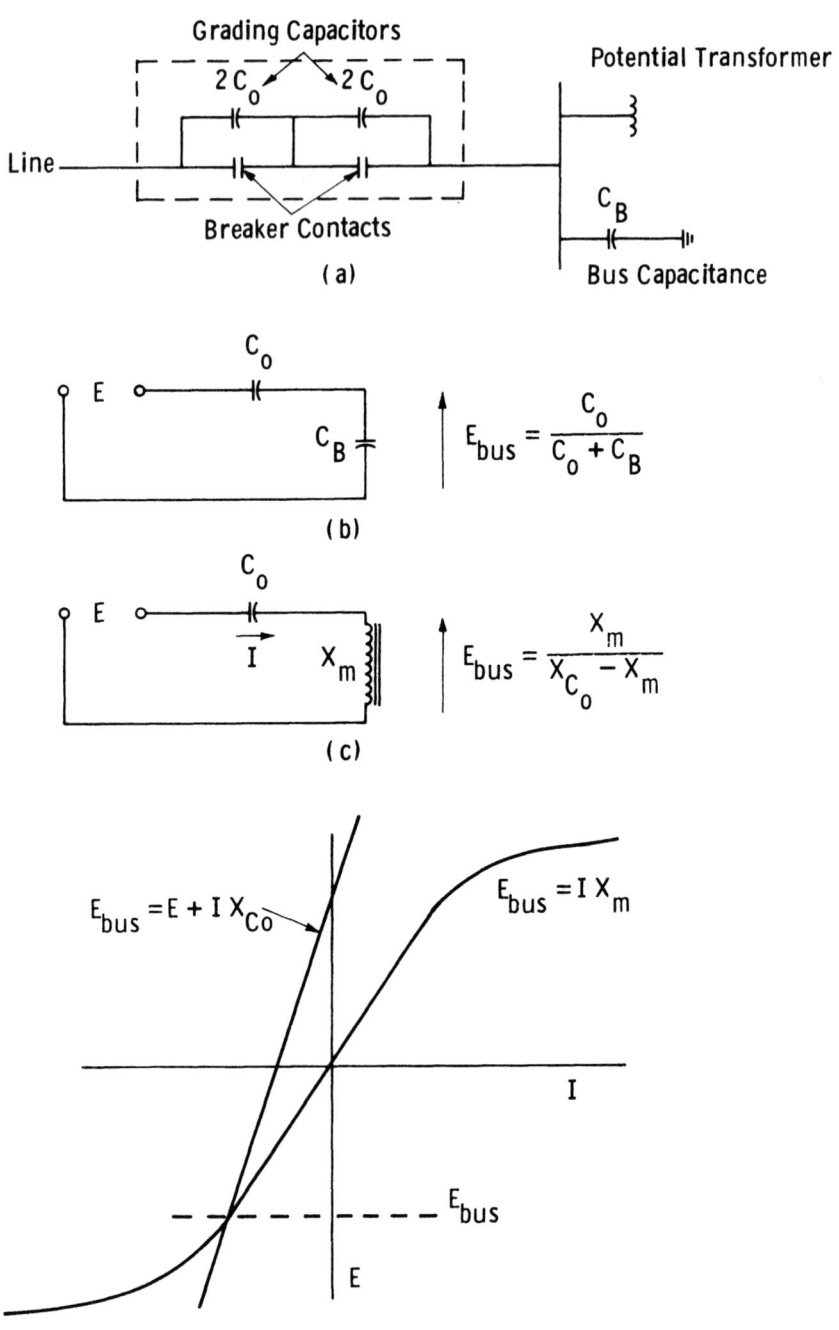

Figure 3.49 Breaker capacitance-potential transformer resonance phenomena: (a) system condition; (b) equivalent circuit without transformer; (c) equivalent circuit with transformer; (d) saturation solution.

Circuit Breaker Application

Figure 3.49(c) shows, however, that with a potential transformer connected to the bus, the bus voltage is $X_m/(X_{c0} - X_m)$, where X_m is the magnetizing impedance of the potential transformer. If X_m is equal to X_{c0}, it would appear that the bus voltage would become infinite. This, of course, cannot occur because the transformer is a nonlinear impedance. Figure 3.49(d) shows that the bus voltage is determined by the intersection of the transformer saturation curve and the capacitance line. On the figure, if the capacitance in the breaker were increased, the slope of this line would decrease, and the voltage on the bus would increase. If the intersection point occurs above the knee of the saturation curve, the potential transformer can be damaged.

The cure for this resonance condition is to insert a low-ohmic-value resistance in the secondary circuit of the potential transformers. From Ref. 25, the value of this resistance should be equal to or less than

$$R_c = \frac{X_c X_s}{X_c - X_s} \sqrt{\left(\frac{X_c E_q}{X_s E_q}\right)^2 - 1}$$

where

X_c = Thévenin's equivalent capacitance of the breaker and bus, secondary ohms

X_s = air-core reactance of transformer, secondary ohms

E_s = equivalent saturation voltage of transformer, secondary volts

E_q = Thévenin's equivalent voltage, secondary volts

This resistor should be inserted any time that the bus is completely deenergized by opening all the circuit breakers feeding the bus. This can be accomplished by the b contacts of the circuit breakers or by an overvoltage relay in the secondary circuit of the potential transformer. If the value of the resistance is such that the loading on the potential transformer exceeds its thermal rating, a time-delay relay is also required to cut the resistor out of the circuit before its short-time rating is exceeded.

3.13 SUMMARY

An attempt has been made to define and explain the major problem areas that must be considered when applying circuit breakers. In most, if not all cases, references have been cited in which the reader can obtain more detailed information on the pertinent subject area.

REFERENCES

1. G. N. Lester, History of circuit breaker standards, IEEE Tutorial Course on Application of Power Circuit Breakers, IEEE Spec. Publ. 75CH0975-3-PWR, 1975, pp. 8-17.
2. ANSI C37.06-1979, Preferred ratings and related required capabilities for ac high-voltage circuit breakers rated on a symmetrical current basis.
2a. C. L. Wagner, Insulation considerations for ac high voltage circuit breakers, *IEEE Tutorial Course on Application of Power Circuit Breakers*, IEEE Spec. Publ. 75 CH 0975-3-PWR, 1975, pp. 52-61.
3. A. A. Johnson, Insulation coordination, in *Electrical Transmission and Distribution Reference Book*, Westinghouse Electric Corporation, Pittsburgh, Pa., 1950, Chap. 18.
4. J. M. Clayton and R. W. Powell, Application of arresters for complete lightning protection of substations, *AIEE Trans.* 77 (Pt. III):1608-1614, February 1959.
5. C. L. Wagner, J. M. Clayton, F. S. Young, and C. L. Rudasill, Insulation levels for VEPCO 500-kV substation equipment, *IEEE Trans. Power Appar. Syst. PAS-83*:236-241, March 1964.
6. H. W. Anderl, C. L. Wagner, and T. H. Dodds, Insulation coordination for gas insulated substations, *IEEE Trans. Power Appar. Syst. PAS-92*:1622-1630, September/October 1973.
7. A. R. Hileman, C. L. Wagner, and R. B. Kisner, Jr., Open breaker protection of EHV systems, *IEEE Trans. Power Appar. Syst. PAS-88*:1005-1014, July 1969.
8. A. R. Jones, Evaluation of the integration method for analysis of non-standard surge voltages, *AIEE Trans.* 73(Pt. III-B):984-990, August 1954.
9. A. J. McElroy, H. M. Smith, W. S. Price, and D. F. Shankle, Field measurement of switching surges on unterminated 345 kV transmission lines, *IEEE Trans. Power Appar. Syst. PAS-82*: 465-487, August 1963.
10. C. L. Wagner and J. W. Bankoske, Evaluation of surge suppression resistors in high-voltage circuit breakers, *IEEE Trans. Power Appar. Syst. PAS-86*:698-707, June 1967.
11. R. G. Colclaser, C. L. Wagner, and E. P. Donohue, Multi-step resistor control of switching surges, *IEEE Trans. Power Appar. Syst. PAS-88*:1022-1028, July 1969.
12. ANSI/IEEE C37.04-1979, Rating structure for ac high-voltage circuit breakers rated on a symmetrical current basis.
13. ANSI/IEEE C37.010-1979, Application guide for ac high-voltage circuit breakers rated on a symmetrical current basis.
14. C. F. Wagner, Machine characteristics, in *Electrical Transmission and Distribution Reference Book*, Westinghouse Electric Corporation, Pittsburgh, Pa., 1950, Chap. 6.

15. ANSI/IEEE C37.5-1979, Guide for calculations of fault currents for application of ac high-voltage circuit breakers rated on a total current basis.
16. ANSI/IEEE C37.011-1979, Application guide for transient recovery voltage for ac high-voltage circuit breakers rated on a symmetrical current basis.
17. W. Wanger and J. K. Brown, Calculation of the oscillations of the recovery voltage after the rupture of short circuits, Brown Boveri Rev., November 1937, pp. 284-302.
18. C. F. Wagner, Wave propagation on transmission lines, in *Electrical Transmission and Distribution Reference Book*, Westinghouse Electric Corporation, Pittsburgh, Pa., 1950, Chap. 15.
19. R. G. Colclaser, Jr., and D. E. Buettner, The traveling wave approach to transient recovery voltage, *IEEE Trans. Power Appar. Syst.* PAS-88:1028-1035, June 1969.
20. R. G. Colclaser, Jr., L. E. Berkebile, and D. E. Buettner, The effect of capacitors on the short-line fault component of transient recovery voltage, *IEEE Trans. Power Appar. Syst.* PAS-90:660-669, March/April 1971.
21. ANSI/IEEE C37.012-1979, Application guide for capacitance current switching of ac high-voltage circuit breakers rated on a symmetrical current basis.
22. ANSI/IEEE C37.99-1980, Guide for protection of shunt capacitor banks.
23. C. L. Wagner and H. E. Lokay, Independent-pole circuit breakers improve system stability performance, Westinghouse Eng., September 1973, pp. 130-137.
24. C. L. Wagner and E. A. Udren, Pole disagreement relaying for independent-pole circuit breakers, Westinghouse Eng., November 1973, pp. 168-173.
25. E. D Price, Voltage transformer ferroresonance in transmission substations, Texas A&M Conf. Protective Relay Eng., College Station, Tex., April 1977.

Additional ANSI and IEEE Standards of Interest

ANSI/IEEE C37.09-1979, Test procedure for ac high-voltage circuit breakers rated on a symmetrical current basis.
ANSI/IEEE C37.081-1981, Guide for synthetic fault testing of ac high-voltage circuit breakers rated on a symmetrical current basis.
ANSI/IEEE C37.082-1982, Sound level measurements for ac high-voltage circuit breakers rated on a symmetrical current basis.
ANSI/IEEE C37.4-1953, (R 1981) Definitions and rating structure for ac high-voltage circuit breakers rated on a total current basis.
ANSI C37.6-1971, (R 1976) Schedules of preferred ratings for ac high-voltage circuit breakers rated on a total current basis.

ANSI C37.7-1960, (R 1976) Interrupting rating factors for reclosing service for ac high-voltage circuit breakers rated on a total current basis.

ANSI/IEEE C37.9-1953, (R 1981) Test code for ac high-voltage circuit breakers rated on a total current basis.

ANSI C37.11-1979, Requirements for electrical control for ac high-voltage circuit breakers rated on a symmetrical current basis and a total current basis.

ANSI C37.12-1981, Guide specification for ac high-voltage circuit breakers rated on a symmetrical current basis and a total current basis.

ANSI/IEEE C37.100-1981, Definitions for power switchgear.

4
Nature of the Electric Arc

JOACHIM V. R. HEBERLEIN, CLIVE W. KIMBLIN, and ANTHONY LEE / Westinghouse Research and Development Center, Pittsburgh, Pennsylvania

4.1 Introduction   135
4.2 Physical Description   136
    4.2.1 Overview   136
    4.2.2 The cathode   140
    4.2.3 Anode phenomena   142
    4.2.4 Arc column   143
4.3 Description of the Recovery Phenomena   152
4.4 Summary   154
References   154

4.1 INTRODUCTION

The arc has always existed in transient form in nature as the lightning stroke, and as such it has always been feared because of its destructive and incendiary power. It was not until soon after sources of continuous electric current became available that it was discovered in 1808 as a manageable laboratory curiosity [1]. After this it found, and continues to find, considerable use for illumination and as a high-temperature source for welding and some chemical processes [2]. The electrical characteristics of arcs in the near-10-A range, as used in lighting, have been studied extensively since the early part of this century [3,4], and indeed continue to receive investigation [5,6].

In this chapter we summarize the important features of electric arcs in circuit interruption devices and provide an overview of the literature and recent work in this area. For readers who wish to go further into this subject, other chapters of this book as well as textbooks on this subject [7-12] and specific references given in this chapter will form a good basis. The types of discharge we are primarily concerned with form when two current-carrying contacts are physically moved apart in an insulating medium. The current levels at the instant of contact part typically exceed the ampere level, and we are therefore concerned with understanding arcing phenomena, in particular the change from a conducting to an insulating medium at current zero. Other types of discharges are certainly of importance, including recent applications of non-selfsustained discharges to pulse power switching [13,14], but the arc remains of paramount importance to circuit breaker applications.

The modeling and design of arcing and interruption devices are complicated by the wide variety of breaker applications. Electrical power control is required for a variety of ac circuits with voltages from several tens of volts, such as aircraft distribution systems, to several hundreds of kilovolts, such as major transmission lines. Furthermore, the current to be controlled can vary from amperes to hundreds of kiloamperes. Finally, electrical power control is also required in both low-voltage and high-voltage dc circuits. The reader can gain an impression of the broad scope of electrical power control devices involving arcs from Tables 4.1 and 4.2. These tables are not intended to cover every circuit breaker rating and type, but are intended to indicate the diversity of arcing media and arcing current requirements for the broad categories of utility power control, and for industrial plus commercial power control. Circuit breaker arcing phenomena have been studied for all of these media, and the results of some of these studies are detailed in Chaps. 5, 6, 7, 9, 10, 11, 12, and 14. In general, the modeling and theoretical understanding of high-pressure gas-blast arcs are the most complete. For these arcs, the conditions of local thermodynamic equilibrium (LTE) are closely approximated, and approaches used in modeling the LTE arc column are described in this chapter.

4.2 PHYSICAL DESCRIPTION

4.2.1 Overview

The arc can be generally divided into three regions: the column, the cathode region, and the anode region (see Fig. 4.1). The measurable variables with which each region of the arc can be characterized are electric field and temperature distribution. Figure 4.2 shows a typical distribution of the potential along the axis of an electric arc. The potential gradient in the column is dependent on the arc current and the energy exchange with the surroundings, including the gas type and pressure, flow velocity, and solid boundaries. The column cross section tends to adjust itself automatically so that the potential gradient assumes the lowest possible value [15]. This potential gradient can vary by more than two orders of magnitude depending on the

Table 4.1 Overview of Switching Applications

	Applications	Media	Pressure	Fault I	System V	Continuous I
Utility Power Control	• Distribution Apparatus	Oil	>1 atm	12/25kA	5/38kV	0.6/1.2kA
		Vacuum	10^{-6} Torr			
	• Switchgear	Oil / SF_6 / Air	>1 atm	20/48kA	5/15kV	0.6/3kA
		Vacuum	10^{-6} torr			
	• Power Circuit Breaker	SF_6 / Air	>5 atm	5/90kA	24/800kV	1.2/4kA
		Oil	>1 atm			
	• Fuses	Silver Wire/Sand	>1 atm	0.01/50kA	2.7/38kV	0.01/3kA
	• Current Limiter (Under Development)	SF_6 / Air / Vacuum		15/100kA	15/145kV	3kA
	• Generator Breaker	SF_6 / Air	~15 atm	300kA	38kV	25kA
	• dc Switching — Metallic Return Transfer Breaker	SF_6 / Air / Oil	>5 atm / >1 atm	3kA	80kV	3kA
	• Line Breaker	Vacuum	10^{-6} torr	3/10kA	600kV	3kA

Table 4.2 Overview of Switching Applications

	Applications		Media	Pressure	Fault I	System V	Continuous I
Industrial and Commercial Power Control	ac Switching	• Circuit Breakers • Switches • Current Limiters	Air	atm	10/200kA	0.1/0.6kV	0.02/4kA
		• Fuses	Silver Wire/Sand Sodium	1 atm	0.3/3kA	0.1/7.5kV	0.01/1kA
		• Controllers	Air	atm	1/18kA	2.2/7.2kV	0.2/0.8kA
			Vacuum	10^{-6} torr			
	dc Switching	• Contactors	Air	atm	10/100kA	2/8kV	0.2/6kA
		• Low Voltage Switches	Air	atm			
			Vacuum Mercury	$\sim 10^{-2}$ torr	6/100kA	6/100V	5/100kA
Examples of New Switching Systems (Under Development)	• Tokamak Ohmic Heating Interrupter		Vacuum	10^{-6} torr	d.c	30/60 kV	20/100kA
	• Pulse Power		Variety of Media for • Erosion Switches (Vacuum) • E Beam Switches (Gas Mixtures) • Launcher Switches (Air) • Triggered Gaps (Vacuum, Gas)		Switch Currents Vary from kA to MA with Pulse Durations of ns to ms and Load Voltages Vary from kV to MV		

Nature of the Electric Arc

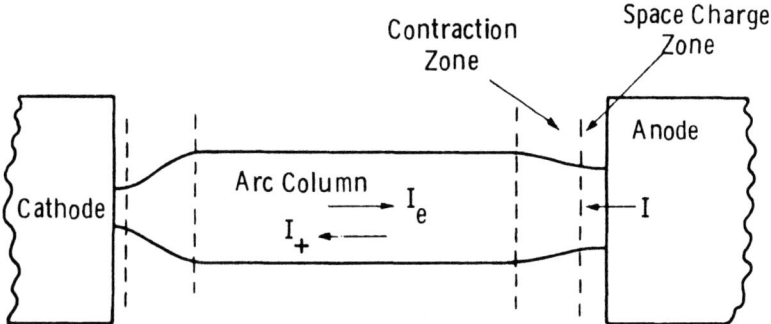

Figure 4.1 Schematic of an arc with uniform column constricted at the electrodes (thickness of sheath regions exaggerated).

design of the arcing device. There are essentially no space charges associated with the arc column (positive column) region, because the electron charges are balanced by an equal number of ionic charges, even though the current flow is essentially determined by the electron mobility. The mechanism which provides the required charge carrier densities depends strongly on the pressure in the discharge, and ranges from ejection of electrons and ions from the cathode in a vacuum arc (see Chap. 12) to "thermal ionization" in high-pressure arcs. Here, the peak temperature in the arc column typically ranges from 7000 to 25,000 K, depending on arcing medium and configuration.

The electrode regions adjacent to the arcing contacts perform two functions for the arc. They serve as the transition from a gaseous conductor with variable conductivity, the arc column, to a solid con-

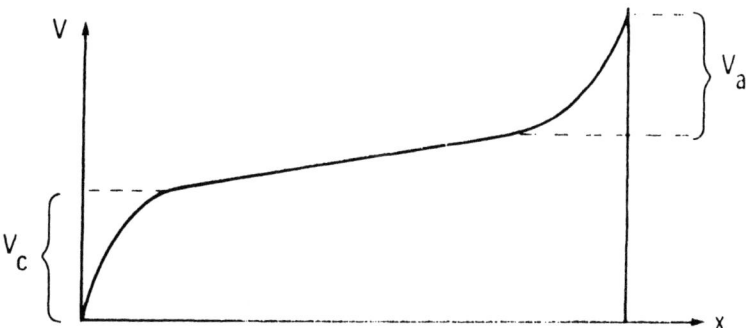

Figure 4.2 Schematic of potential distribution in the arc (length of sheath regions exaggerated).

ductor with essentially constant conductivity. They also deliver charge carriers to the column. Whereas the first role is passive, the second role can be considered active, and is therefore strongly material dependent. The cathode is always active, providing electrons from the material which are accelerated across a high-field region, the electrode sheath, until they have enough energy to ionize neutral particles.

The potential drop in the cathode region is primarily a function of the cathode material and can be divided [16,17] into the cathode fall and the drop in the cathode boundary layer. The cathode fall is typically 10 to 20 V and extends over a region of 10^{-3} to 10^{-2} mm, whereas the cathode boundary layer can be a few millimeters thick.

The potential drop in the anode region is again a function of geometry and can vary widely. Again, it is divided into an actual anode fall of 0 to 5 V, extending over a region of ~10^{-3} mm, and an anode boundary layer [16,18-21]. This boundary layer forms because of the axial energy loss from the arc column to the cold anode, and it is again typically a few millimeters thick.

4.2.2 The Cathode

The role of the cathode in an electric arc is to deliver the current-carrying electrons to the arc column. Cathodes are generally characterized according to the mechanism by which they release the electrons. Refractory materials with high boiling points, such as tungsten, carbon, molybdenum, or zirconium, emit electrons when they are heated to a temperature less than the evaporation temperature (thermionic emission). There is typically one stationary hot spot with a temperature in excess of 3500 K, which is heated predominantly through bombardment of ions accelerated in the cathode sheath. The current densities j, which can be obtained from thermionic cathodes, are given by the Richardson-Dushman equation:

$$j = AT^2 \exp\left(\frac{-e\phi}{kT}\right) \tag{4.1}$$

where

 T = cathode temperature

 ϕ = work function of the cathode material (i.e., the energy required to release one electron from a metal surface, typically 4 to 5 V for electronegative metals and 1.5 to 4 V for electropositive metals such as thorium)

 e = electronic charge

Nature of the Electric Arc

k = Boltzmann constant

A = a materials constant with a value ~60 A/cm^2 K^2 for most metals

Typical current densities are 10^4 A/cm^2. From the standpoint of ac interruption, the cooling of the heated spot is slow compared to the time scale associated with the rate of rise of recovery voltage. On the other hand, thermionic materials mixed with a good conductor such as copper or silver (Chap. 15) display a desirable low erosion characteristic under high-current arcing conditions [22] and can therefore be used in breaker applications where the arc column provides the interruption function.

For cold cathode arcs, where the cathode material experiences significant evaporation at temperatures too low for significant thermionic emission, the electron emission is more complex. The general energy balance for the surface spot [16,17,23] involves energy supply terms (ion bombardment plus joule heating) and energy loss terms (neutral particle loss, electron cooling, and heat conduction). Quantitative values for the individual terms in this energy balance equation, together with overall modeling of the cathode spot region, have received a great deal of attention but remain the subject of dispute. The reader is directed elsewhere [23,24] for detailed analyses of nonrefractory cathode spot emission. From a qualitative standpoint, we can state that field emission contributes to the nonrefractory cathode spot operation. Here the electrical field, probably enhanced by local surface imperfections, becomes sufficiently high to draw electrons from the metal. Typical current densities are 10^6 to 10^8 A/cm^2. Many investigators consider that cathode spot electron emission involves a combination of thermionic and field emission, called TF emission.

The emission process has been studied particularly extensively for vacuum arcs, where transient explosive phenomena may also be involved [25-28]. For vacuum arcs (see Chap. 12) the cathode not only provides electrons, but also metal ions which contribute to the quasineutral interelectrode plasma. Electrons and ions emanate [29,30] from individual small spots on the cathode surface, each carrying a current between ~15 and ~150 A, depending on material. Because of the extremely small size of these spots, cooling is almost instantaneous when the alternating current decreases to zero.

The current density at the cathode is usually at least two orders of magnitude higher than the current density in the arc column. The consequence is a magnetic pressure gradient due to the change of the self-magnetic field which accelerates the gas flow away from the cathode (Maecker effect [31]). The resulting entrainment of cold gas from the surrounding areas leads to the formation of a thermal boundary layer in front of the cathode in which the cold gas is heated to the plasma column temperatures. A simplified quantitative description of this effect is given in Sec. 4.2.4.

4.2.3 Anode Phenomena

The anode can be either active or passive. In its passive mode, it serves only to collect electrons carrying current from the arc column to the surface. A space-charge region in front of the anode accelerates the electrons from the column to permit sufficient ionization in this region, where the temperature is too low to assure thermal ionization (anode sheath region). If, however, the thermal boundary layer between arc column and anode surface is small enough, and the electron density gradients are high enough so that a substantial electron diffusion flux exists, this space-charge region is unnecessary [21]. The interaction between the arc column and a passive anode has been studied extensively, and the different physical mechanisms of energy transfer have been quantitatively described [18-21,32,33].

In its active mode, the anode evaporates and this evaporated material can be ionized in the anode region to supply ions to the plasma. A typical example of an active anode is found in vacuum arcs, where, for high anode current densities, anode spots can form and contribute to the supply of plasma needed to conduct current between the contacts [29,34,35]. The operating temperature of the anode spot is determined by the power balance at the surface, where the power input q_{an} is given by the heat flux density from the plasma to the anode surface:

$$q_{an} = q_{cg} + q_{Rg} + j\left(\frac{5}{2}\frac{kT_e}{e} + V_a + \phi_a\right) \quad (4.2)$$

where

q_{cg} = heat flux by conduction from the plasma to the anode

q_{Rg} = heat flux by radiation from the plasma to the anode

$j\left(\frac{5}{2}\frac{kT_e}{e} + V_a + \phi_a\right)$ = heat flux associated with the electron current from the plasma into the anode: (1) electron enthalpy, (2) directed electron energy gained in the anode fall region, (3) chemical energy of the electrons as expressed by the anode work function

Nature of the Electric Arc

The anode surface loses heat q_{loss} according to the relationship

$$q_{loss} = q_{ev} + q_{cs} + q_{Rs} \tag{4.3}$$

where

q_{ev} = heat loss by material evaporation

q_{cs} = heat conducted from the surface to the bulk material

q_{Rs} = heat radiated by the heated surface

Which term is dominant depends strongly on the specific configuration. For example, for vacuum arcs the most important terms in the energy balance are heating by the current flow and heat loss by anode material evaporation [36,37]. For a high-pressure arc with a refractory anode, the heat flux by conduction to the anode can be comparable to the heat flux associated with the current, and these power inputs are largely balanced by radiation from the anode surface. It should be noted that in the sheath region as well as in the thermal boundary layer, nonequilibrium prevails [38], making theoretical treatment difficult. Analytical descriptions have been derived assuming a two-temperature fluid model for the boundary layer: a set of conservation equations for the electrons, assuming a maxwellian energy distribution according to a temperature T_e and a second set of equations for the heavy particles (ions and atoms) with a maxwellian distribution according to a temperature T_h [21,32,33] (see the Appendix for the description of the maxwellian energy distribution).

In general, the anode spot current density is significantly lower than the current density at the cathode spot. However, magnetic effects cannot be neglected because of a reinforcing effect. A small flow away from the anode due to a magnetic pressure gradient results in the suction of cold gas into the anode region, which cools the arc and leads to a stronger constriction. This results in an increase in jet action until the increased heat losses in the constriction region stabilize the arc [39].

4.2.4 Arc Column

All circuit breaker arcs which are initiated by contact separation will pass through an initial phase where electrode vapor is the dominant source of charged particles [40,41]. With increasing contact separation, the influence of electrode vapor remains dominant for vacuum arcs (see Chap. 12), and this vapor can also influence the properties of the intercontact arc column in molded-case breakers (see Chap. 14). For long arcs in high-pressure gas-blast circuit breakers, however, the effects of contact material are minimal, and the intercontact arc column is dependent primarily on the gaseous ambient. The properties of such an arc column are the subject of this section.

Over the last 50 years, many arc column models have been developed, and recently, powerful computers have allowed considerable sophistication of these models. Although our understanding of the physical processes in the arc and their relative importance has been considerably enhanced by these models, it must be said that a quantitative model that would directly aid the engineer in all aspects of the design of a high-pressure circuit breaker has not yet been formulated.

The plasma in the arc column can be described by the equations of fluid dynamics and the laws of thermodynamics. However, the special properties of the plasma make it necessary for several terms in the equations to be included because the usual approximations (e.g. constant properties) cannot be made. Also, the following processes have to be considered:

1. The plasma is chemically reacting; in addition to the conservation of mass equation one needs the rate equations for the different reactions. For local thermodynamic equilibrium, the rate equations become the equilibrium mass action laws. In the simplest case when the reaction is the ionization of a monatomic gas, this mass action law becomes the Saha equation, which gives the degree of ionization in the gas (see the Appendix).

2. Since the plasma is electrically conducting, one has to consider in the momentum equation the terms describing the interaction with magnetic fields. These magnetic fields are either externally applied or are generated by the current flowing in the arc (self-magnetic fields). As an example of the influence of the self-magnetic fields, we will consider the axial acceleration of the plasma due to a radial constriction as can occur, for example, in front of the electrode or in a nozzle (Maecker effect [31]). The change in current density between the contricted and nonconstricted arc results in different magnetic pressures p at different axial locations z. This results in an axial pressure gradient at the arc center:

$$p(z) = \int_0^{R_a} j(r,z) B_\phi(r) \, dr$$

$$B_\phi(r) = \frac{\mu_0}{r} \int_0^r jr \, dr$$

Nature of the Electric Arc

Assuming a uniform current density over the arc cross section, we obtain

$$\frac{\partial p}{\partial z} = \frac{\mu_0 I}{4\pi} \frac{\partial j}{\partial z} \qquad (4.4)$$

where

B_ϕ = circumferential magnetic field component

μ_0 = magnetic permeability

I = total arc current

R_a = arc radius

Resultant velocities of more than 100 m/s have been observed.
3. There is volumetric heat generation in the plasma which is described by a corresponding term in the energy equation. This term is obtained from Ohm's law for resistive heat dissipation of electric energy.
4. The plasma is strongly radiating, and the radiation transport term in the energy equation becomes very important. In many cases, a considerable portion of this radiative energy is being reabsorbed in the plasma, and one has to add to the energy equation the radiation transfer equation [42,43], or a substitute, such as a tabulated value for the net emission coefficient (see Chap. 5) [44].
5. There is one thermodynamic equation of state for every component present in the plasma (e.g., electrons, ions, atoms, and molecular species).

Taking all these considerations into account, we arrive at the very comprehensive set of equations given below. It should be noted that solution of this set of general equations is a formidable task, but for all practical purposes considerable simplifications can be made (see, e.g., Chap. 5). For readers who have limited interest in the detailed equations, we demonstrate at the end of this section that some qualitative predictions of the arc behavior can be made by using the very simplest form of the energy conservation equation [see equation (4.15)].

Conservation Equations in Vector Notation

[See Eqs. (4.5), (4.6), and (4.7) on pages 146 and 147.]

Conservation Equations in Tensor Notation

[See Eqs. (4.5a), (4.6a), and (4.7a) on page 148.]

Conservation Equations in Vector Notation

Conservation of Mass (Continuity Equation):

$$\frac{\partial \rho}{\partial t} + \vec{\nabla} \cdot (\rho \vec{v}) = 0 \qquad (5)$$

- $\frac{\partial \rho}{\partial t}$: Density Change in Control Volume
- $\vec{\nabla} \cdot (\rho \vec{v})$: Difference Between Mass Flows Entering and Leaving the Control Volume. Note that the Density is in General a Function of Location

Conservation of momentum (Navier-Stokes Equation):

$$\rho \frac{\partial \vec{v}}{\partial t} + \rho (\vec{v} \cdot \vec{\nabla}) \vec{v} = -\vec{\nabla} p + \frac{4}{3}\vec{\nabla}\mu(\vec{\nabla}\cdot\vec{v}) - \vec{\nabla}\times\mu(\vec{\nabla}\times\vec{v}) + \vec{j}\times\vec{B} + \rho\vec{g} \qquad (6)$$

- $\rho \frac{\partial \vec{v}}{\partial t}$: Time Rate of Change of Momentum in Control Volume
- $\rho(\vec{v}\cdot\vec{\nabla})\vec{v}$: Difference of Momentum of Flows Entering and Leaving the Control Volume
- $-\vec{\nabla}p$: Accelerating Pressure Gradient. This Term is Important in Arcs with Superimposed Flow
- $\frac{4}{3}\vec{\nabla}\mu(\vec{\nabla}\cdot\vec{v}) - \vec{\nabla}\times\mu(\vec{\nabla}\times\vec{v})$: Surface Forces (Stresses) Acting on Control Volume. The First Term is Important for Compressible Fluid Flow, i.e. when the Flow Velocities Approach or Surpass the Sound Velocity. The Second Term is Important when Strong Velocity Gradients are Present, Which is Usually the Case in Arcs with Superimposed Flow.
- $\vec{j}\times\vec{B}$: Accelerating Magnetic Force. This Term is Important Only when a) External B-fields are Applied, or b) When a Strong Change in Arc Channel Diameter Results in Strong Interaction with the Self-Magnetic Field
- $\rho\vec{g}$: Accelerating Gravity Force (Buoyancy). This Term is Usually Only Important when there are no or Small Superimposed Flows (Free-Burning Arcs).

Nature of the Electric Arc

Conservation of Energy:

$$\rho \frac{\partial H}{\partial t} + \rho (\vec{v} \cdot \vec{\nabla})H - \frac{\partial p}{\partial t} - (\vec{v} \cdot \vec{\nabla}p) = \vec{\nabla} \cdot \frac{\lambda}{c_p} \vec{\nabla}H - \vec{\nabla} \cdot \vec{q}_R + \Phi + \vec{j} \cdot \vec{E}$$

- $\rho \frac{\partial H}{\partial t}$: Time Rate of Enthalpy Change in Control Volume
- $\rho (\vec{v} \cdot \vec{\nabla})H$: Difference of Enthalpy of Fluids Entering and Leaving the Control Volume (Gas Heating)
- $\frac{\partial p}{\partial t}$: Work Needed for (Produced by) Pressure Change. This Term is Usually Small Compared to the Enthalpy Change
- $(\vec{v} \cdot \vec{\nabla}p)$: Work Performed by the Pressure Gradient. This Term has to be Considered in Nozzle Flow Situations with Strong Pressure Gradients
- $\vec{\nabla} \cdot \frac{\lambda}{c_p} \vec{\nabla}H$: Heat Loss by Conduction. This is Usually a Dominant Loss Term
- $\vec{\nabla} \cdot \vec{q}_R$: Heat Loss by Radiation (Change of Radiation Flux). This Term is Important at High Currents
- Φ: Viscous Dissipation Function (See Below)
- $\vec{j} \cdot \vec{E}$: Electric Energy Dissipation in Control Volume. This is the Most Important Heat Addition Term

With the Viscous Dissipation Function:

$$\Phi = \sum_{i,j,k=1}^{3} \left[\mu \left(\frac{\partial v_i}{\partial x_j} + \frac{\partial v_j}{\partial x_i} \right) - \frac{2}{3} \mu \frac{\partial v_k}{\partial x_k} \delta_{ij} \right] \frac{\partial v_i}{\partial x_j}$$

This Heating Term is Usually Neglected

Conservation Equations in Tensor Notation

Continuity Equation: $\dfrac{\partial \rho}{\partial t} + \dfrac{\partial \rho v_i}{\partial x_i} = 0$ (4.5a)

Momentum Equation in the Direction of i-Coordinate:

$$\rho \frac{\partial v_i}{\partial t} + \rho v_j \frac{\partial v_i}{\partial x_j} = -\frac{\partial p}{\partial x_i} + \frac{\partial}{\partial x_j}\mu\left(\frac{\partial v_i}{\partial x_j} + \frac{\partial v_j}{\partial x_i}\right) - \frac{2}{3}\frac{\partial}{\partial x_i}\left(\mu \frac{\partial v_j}{\partial x_j}\right) + j_j B_k \delta_{ijk} + \rho g_i$$

(4.6a)

Energy Equation

$$\rho \frac{\partial H}{\partial t} + \rho v_i \frac{\partial H}{\partial x_i} - \frac{\partial p}{\partial t} - v_i \frac{\partial p}{\partial x_i} = \frac{\partial}{\partial x_i}\left(\frac{\lambda}{c_p}\frac{\partial H}{\partial x_i}\right) - \frac{\partial q_{Ri}}{\partial x_i} + \Phi + j_i E_i$$

(4.7a)

With $\Phi = \mu\left(\dfrac{\partial v_i}{\partial x_j} + \dfrac{\partial v_j}{\partial x_i}\right)\dfrac{\partial v_i}{\partial x_j} - \dfrac{2}{3}\mu \dfrac{\partial v_k}{\partial x_k}\delta_{ij}\dfrac{\partial v_i}{\partial x_j}$

Note: In Tensor Notation Every Term in which a Subscript Appears Twice has to be Summed Over All Three Coordinates, e.g.:

$$\frac{\partial \rho v_i}{\partial x_i} \equiv \frac{\partial \rho v_1}{\partial x_1} + \frac{\partial \rho v_2}{\partial x_2} + \frac{\partial \rho v_3}{\partial x_3}$$

and $v_j \dfrac{\partial v_i}{\partial x_j} \equiv v_1 \dfrac{\partial v_1}{\partial x_1} + v_2 \dfrac{\partial v_1}{\partial x_2} + v_3 \dfrac{\partial v_1}{\partial x_3}$

The following additional equations are needed for the solution of the conservation equations.

The radiation flux density across the surface having the surface normal vector $\hat{n}$ is given by

$$q_R = \int_\nu \int_\omega i_\nu \hat{n} \cdot d\underline{\omega}\, d\nu$$

(4.8)

where one integral extends over the entire solid angle ω and the other over the entire frequency range in which there is significant radiation. The spectral radiation intensity is given by the radiation transport equation:

$$\frac{di_\nu}{ds} = \epsilon_\nu - \kappa_\nu i_\nu$$

(4.9)

ds being an arbitrary path differential, and ϵ_ν and κ_ν the emission

Nature of the Electric Arc

and absorption coefficients, respectively (i.e., local properties of the plasma).

The enthalpy H is defined as

$$H = H_0 + \int_{T_0}^{T} c_p \, dT$$

Ohm's law can be expressed

$$I = \int_A \underline{j} \cdot d\underline{A} = E \cdot \int_A \sigma \, d\underline{A} \tag{4.10}$$

where $\underline{j}$ is the current density integrated over the arc cross-sectional area.

The charge neutrality

$$en_e = \sum_{i=1}^{r+1} q_i n_i \tag{4.11}$$

where we equate the total electron charge to the total ion charge of all i stages of ionization from the singly ionized particle (i = 1) to the highest ionized particle (i = r + 1).

The Saha equation (r mass action laws, one for each level of ionization) is

$$\frac{n_{r+1} n_e}{n_r} = \frac{2Z_{r+1}}{Z_r} \frac{(2\pi m_e kT)^{3/2}}{h^3} \exp\left(\frac{-\chi_{r+1}}{kT}\right) \tag{4.12}$$

The equation of state for a perfect gas is

$$p = \sum_{i=0}^{r+1} (n_e + n_i) kT \tag{4.13}$$

where i = 0 designates the nonionized atom. Finally, we need the relation

$$\rho = \frac{kML}{R} \left(n_e + \sum_{i=0}^{r+1} n_i \right) \tag{4.14}$$

List of Symbols

- A area
- B magnetic field strength
- c_p specific heat at constant pressure
- E electric field strength
- e electronic charge
- g gravitational acceleration

H	gas enthalpy	V	voltage
H_0	enthalpy at a reference temperature T_0	V_a, V_c	anode and cathode fall voltages, respectively
h	Planck's constant	v	velocity
I	current	x_i, x_j, x_k	set of orthogonal coordinates
i_ν	spectral radiation intensity	Z_r, Z_{r+1}	partition function of the rth ion and the (r + 1)th ion, respectively
j	current density		
k	Boltzmann constant		
L	Loschmidt's number		
M	molar mass of gas		
m_e	electron mass	z	axial coordinate
n_e	electron number density	ε_ν	spectral emission coefficient
n_r, n_{r+1}	density of the ions in the rth and (r + 1)th ionized state, respectively	κ_ν	spectral absorption coefficient
		λ	thermal conductivity
		μ	viscosity
P_R	radiative power density for optically thin radiation	μ_0	permeability of free space
		ν	frequency of radiation
p	pressure		
q	heat flux density	ρ	gas density
q_i	charge of i times ionized ion	σ	electrical conductivity
		ϕ	viscous dissipation function
R	universal gas constant		
r	radial coordinate	ϕ_a	anode work function
s	arbitrary path in radiating volume	χ_r	ionization potential of the rth ion
T	temperature	ω	solid angle
t	time		

Subscripts

e	electron	r	designating state of ionization
i, j, k	relating to the direction of the coordinates		
		R	radiation

If we consider for the moment only single ionization (r = 0) we have 10 equations for 12 parameters (p, v, H, ρ, j, E, I, q_R, i_ν, n_e, n_1, n_0). Consequently, measurement of two of these parameters allows us to solve this coupled set of differential equations. The most frequently measured parameters are arc current I and pressure p(z) or field strength E(z).

As shown in Chap. 5, the gas properties μ, λ, σ, and c_p are strongly and nonlinearly varying functions of temperature, making these equations nonlinear and requiring a numerical solution procedure. Chapters 5 and 7 will describe solution procedures.

Nature of the Electric Arc

For one special case, the conservation equations become decoupled: for the case of a thermally and hydrodynamically fully developed, wall-constricted arc with negligible axial pressure drop. For this case $\partial v_z/\partial z = 0$; $v_r = 0$; $\partial H/\partial z = 0$, and we arrive at an energy equation which is similar in form to the equation derived in 1935 by Elenbaas [45] and Heller [46], frequently called the Elenbaas-Heller equation:

$$-\frac{1}{r}\frac{\partial}{\partial r}\left(\lambda r \frac{\partial T}{\partial r}\right) + P_R = \sigma E^2 \qquad (4.15)$$

In many instances, this equation will permit a quick and qualitative evaluation of effects of increased current or reduced channel diameter on a wall-constricted arc.

As an example, let us assume that the cooling rate at the arc boundary is increased while the current is kept constant. The arc will respond by reducing its diameter, that is, the geometric location where the temperature has a value associated with an appreciable electrical conductivity occurs at a smaller radius. From the Elenbaas-Heller equation we can see that if the temperature gradient at the arc boundary had the same value at the reduced diameter as it formerly had for the larger diameter, the peak temperature would have to be lower. On the other hand, the arc current remains constant, so the current density j (the current divided by the cross-sectional area) must increase. According to Ohm's law $j = \sigma E$, the volumetric power dissipation σE^2 in the arc column must increase as well. In the center of the arc column, this increased power dissipation will increase the arc temperature. Consequently, in order to satisfy the steady-state energy balance (Elenbaas-Heller equation), the temperature gradient at the arc boundary must increase. The increased conductive and radiative heat losses will counteract the center temperature rise until a new equilibrium condition is established. Thus one has the paradoxical effect that increased cooling of the arc column can lead to higher peak temperatures.

An increase in arc current will in general lead to an increase in arc diameter. The current density will increase with arc current only when further increases in arc diameter increase the heat loss at the arc boundary. Such circumstances occur when the arc boundary approaches a cold wall, or when convection or radiation losses become dominant. Then increased current density will lead to higher peak temperatures.

Shown in Chap. 5 are the strong differences of some of the properties between different gases. In general, the high-temperature electrical conductivity is similar for all gases. The thermal conductivity of different gases, however, can vary by orders of magnitude (e.g., those of hydrogen and argon). The reason for this is that (1) gases consisting of lighter molecules have in general a higher thermal conductivity, and (2) that molecular gases have very high thermal conductivities in the temperature region where they dissociate. Similar strong differences can be found in the radiation properties of different gases.

From the Elenbaas-Heller equation, we can again see how the arc will respond to a different gas. A higher thermal conductivity will increase the heat loss from the arc, and the arc responds with a smaller diameter and higher peak temperatures. Consequently, an arc in hydrogen will typically have a smaller diameter and higher peak temperatures than an arc in nitrogen. A somewhat different effect results from an increase in radiative power density. As long as the radiation is not reabsorbed in the arc (optically thin radiation), it acts to reduce the temperature of the radiating volume element. In general, higher radiation losses tend to limit peak temperatures.

4.3 DESCRIPTION OF THE RECOVERY PHENOMENA

The preceding section has provided a brief description of arcing characteristics without considering the effect of time-dependent current changes. The behavior of arc discharges under ac circuit conditions is, however, of prime importance for current interruption and control. In particular, the focus for many chapters in this book is the critical period around current zero when the discharge medium changes from a conductor to an insulator.

At current zero, the column temperature rapidly reduces due to continued power loss via conduction, convection, and radiation. Physical insight into the initial arc column recovery process (electrical conductance decay) can be obtained by considering an idealized cylindrically symmetric arc discharge where the losses are solely by radial thermal conduction. We will make the further simplification that the current is initially steady state, and examine the consequence of suddenly removing the intercontact voltage. At this point the input power vanishes. The initial condition is described by the Elenbaas-Heller equation [equation (4.15) with $P_R = 0$], and the subsequent free recovery condition for a volume element of the arc is given by

$$\frac{1}{r}\frac{\partial}{\partial r}\left(r\lambda \frac{\partial T}{\partial r}\right) = \rho c_p \frac{\partial T}{\partial t} \qquad (4.16)$$

where the left-hand side is the heat loss due to radial thermal conduction, and the right-hand side of the equation refers to the change in energy content of the volume element. Here ρ is the gas density, T the gas temperature, c_p the specific heat at constant pressure, and λ the thermal conductivity.

Expressed in terms of heat flux potential $S = \int_0^T \lambda \, dT$, equation (4.16) reduces to the form

Nature of the Electric Arc

$$\frac{1}{r}\frac{\partial}{\partial r}\left(r\frac{\partial S}{\partial r}\right) = \frac{1}{K}\frac{\partial S}{\partial t} \tag{4.17}$$

where K, the thermal diffusivity, = $\lambda/\rho c_p$.

Frind [47] considered λ constant and solved equation (4.17), whence

$$S(r,t) = S(r,0)e^{-t/\theta} \tag{4.18}$$

where θ, the thermal time constant, has the value

$$\theta = \frac{R_a^2}{(2.4)^2 K} \tag{4.19}$$

This solution is valid if the conduction radius R_a remains stationary during the free recovery. Assuming a linear relationship between the electrical conductivity σ and the heat flux potential S, we obtain

$$\sigma(r,t) = \sigma(r,0)e^{-t/\theta} \tag{4.20}$$

so the plasma has an exponential decay of conductance [48-50] with a thermal time constant θ.

For many arcs this thermal time constant, θ, has values ranging from submicroseconds to microseconds, implying an extremely rapid change of gap conductance following removal of the intercontact voltage.

The description of recovery phenomena above gives useful insight into the dependence of θ on arc radius and thermal diffusivity, but is grossly simplified for application to circuit breaker arcs. The complications are listed below, with appropriate references to chapters providing more accurate descriptions:

1. In ac circuits the current essentially approaches current zero linearly.
2. A rapidly rising recovery voltage, of opposite polarity to the arc voltage, appears across the recovering arc column within microseconds of current zero. This circuit reapplied voltage can cause post-arc currents to flow through the decaying arc column with consequent power input to offset the power loss (see Chaps. 5 and 6).
3. Arc gap recovery involves dielectric recovery in addition to reduction of arc gap conductance (see Chap. 7).
4. In practice, convection, conduction, and radiation all contribute to the energy balance around current zero (see Chaps. 5, 6, and 7).
5. Electrode region recovery also plays a major role in air breakers, where the arc interacts with deion arc chutes (see Chaps. 11 and 14), and in vacuum circuit breakers, where electrode effects are dominant (Chap. 12).

4.4 SUMMARY

In this century there have been major advances in the physical understanding of arcing and interruption phenomena. It is now possible to provide broad descriptions of both steady-state and transient arc column and electrode phenomena. However, detailed descriptions of certain aspects of arcing phenomena remain the subject of much discussion. Examples include details of cathode and anode spot operation, the role of turbulence in interruption, the factors that affect arc stability, and many others.

At the present time it is impractical to provide quantitative descriptions of arcing and interruption phenomena which encompass the broad range of currents, voltages, and media encountered in today's switching devices. However, the improved understanding of these phenomena certainly contributes to the continued evolution of such devices.

REFERENCES

1. H. Davy, *Elements of Chemical Philosophy*, Vol. 1, Smith and Elder, London, 1812, p. 152.
2. W. S. Weedon, The electric arc, *Trans. Am. Electrochem. Soc.* 5:171, 1904.
3. H. Ayrton, The electric arc, *The Electrician,* London, 1902.
4. W. B. Nottingham, Normal arc characteristic curves: dependence on absolute temperature of anode, *Phys. Rev. 28*:764, 1926.
5. J. F. Waymouth, *Electric Discharge Lamps,* MIT Press, Cambridge, Mass., 1971.
6. R. J. Zollweg, The modeling of modern high pressure arc lamps, *NBS Spec. Publ. 561, Proc. 10th Mater. Res. Symp.,* September 1978, Gaithersburg, Md., 1979.
7. T. H. Lee, *Physics and Engineering of High Power Switching Devices,* MIT Press, Cambridge, Mass., 1975.
8. J. D. Cobine, *Gaseous Conductors,* Dover, New York, 1958.
9. E. Pfender, Electric arcs and arc gas heaters, in *Gaseous Electronics* (M. N. Hirsh and H. J. Oskam, eds.), Vol. I, Academic Press, New York, 1978.
10. G. R. Jones and M. T. C. Fang, The physics of high-power arcs, *Rep. Prog. Phys. 43*:1415-1465, 1980.
11. J. M. Somerville, *The Electric Arc,* Methuen, London, 1959.
12. K. Ragaller, ed., *Current Interruption in High-Voltage Networks,* Plenum Press, New York, 1978.
13. R. O. Hunter, Electron beam switching, *Proc. First Int. Pulsed Power Conf.,* Lubbock, Tex., November 1976, Paper IC8.
14. L. E. Kline, Performance predictions for electron beam con-

trolled on/off switches, *IEEE Trans. Plasma Sci.* PS-10:224, 1982.
15. M. Steenbeck, Untersuchungen am Luftlichtbogen im schwerefreien Raum, *Z. Tech. Phys.* 33:593, 1937.
16. G. Ecker, Electrode components of the arc discharge, *Ergeb. Exakt. Naturwiss.* 33:1-104, 1961.
17. A. E. Guile, Arc-electrode phenomena, *Proc. IEE (Lond.) IEE Rev.* 118(9R):1131-1154, 1971.
18. G. Busz-Peuckert and W. Finkelnburg, Die Abhängigheit des Anodenfalles von Stromstärke und Bogenlänge bei Hochtemperaturbögen, *Z. Phys.* 140:540-546, 1955.
19. P. Schoeck and F. Maisenhälder, Zur Druckabhängigkeit der Anodenfallspannung von Hochdrucklichtbögen, *Beitr. Plasmaphys.* 5:345, 1966.
20. T. K. Bose and E. Pfender, Direct and indirect measurements of the anode fall in a coaxial arc configuration, *AIAA J.* 71:1643, 1969.
21. H. A. Dinulescu and E. Pfender, Analysis of the anode boundary layer of high intensity arcs, *J. Appl. Phys.* 51:3149, 1980.
22. W. R. Wilson, High-current arc erosion of electric contact materials, *AIEE Trans.* 74(Part III):657-664, August 1955.
23. J. M. Lafferty, ed., *Vacuum Arcs, Theory and Applications*, Wiley-Interscience, New York, 1980. See Chap. 7 by G. Ecker, "Theoretical aspects of the vacuum arc."
24. T. H. Lee et al.: Voltage distribution, ionization and energy balance in the cathode region of an arc, *Aeronaut. Res. Lab. Rep. 64-152*, Wright-Patterson Air Force Base, Ohio, 1964.
25. G. A. Mesyats, Electron explosive emission and electrical discharge in vacuum, *Proc. 6th Int. Symp. Discharge Electr. Insul. Vacuum*, Swansea, Wales, 1974, pp. 21-47.
26. J. E. Daalder, A cathode spot model and its energy balance for metal vapor arcs, *J. Phys. D.: Appl. Phys.* 11:1667-1682, 1978.
27. B. Jüttner, Formation time and heating mechanism of arc cathode craters in vacuum, *J. Phys. D.: Appl. Phys.* 14:1265-1275, 1981.
28. E. Hantzsche, Theory of cathode spot phenomena, *Physica*, 104C:3-16, 1981.
29. M. P. Reece, The vacuum switch—Part I: Properties of the vacuum arc, *Proc. IEE (Lond.)* 110:793-802, April 1963.
30. C. W. Kimblin, Erosion and ionization in the cathode spot regions of vacuum arcs, *J. Appl. Phys.* 44:3074-3081, July 1973.
31. H. Maecker, Plasmaströmungen infolge eigenmagnetischer Kompression, *Z. Phys.* 141:198, 1955.
32. T. K. Bose, Anode heat transfer for a flowing argon plasma at elevated electron temperature, *Int. J. Heat Mass Transfer* 15:1745-1763, 1972.

33. N. Sanders et al., Studies of the anode region of a high-intensity argon arc, *J. Appl. Phys.* 53(6):4236, 1982.
34. C. W. Kimblin, Anode voltage drop and anode spot formation in dc vacuum arcs, *J. Appl. Phys.* 40:1744, 1969.
35. G. R. Mitchell, High current vacuum arcs, *Proc. IEE (Lond.)* 117:2315, 1970.
36. J. D. Cobine and E. E. Burger, Analysis of electrode phenomena in the high-current arc, *J. Appl. Phys.* 26:895, 1955.
37. C. W. Kimblin, Anode phenomena in vacuum and atmospheric pressure arcs, *IEEE Trans. Plasma Sci.* PS-2:310-319, December 1974.
38. J. V. R. Heberlein and E. Pfender, Investigation of the anode boundary layer of an atmospheric pressure argon arc, *IEEE Trans. Plasma Sci.* PS-5:171, 1977.
39. D. M. Chen and E. Pfender, Modeling of the anode contraction region of high intensity arcs, *IEEE Trans. Plasma Sci.* PS-8:252, 1980.
40. R. Holm and E. Holm, *Electric Contacts: Theory and Application*, Springer-Verlag, New York, 1967.
41. P. G. Slade and M. F. Hoyaux, The effect of electrode material on the initial expansion of an arc in vacuum, *IEEE Trans. Parts Hybrids Packag.* PHP-8:35, 1972.
42. C. H. Church et al., Studies of highly radiative plasmas using the wall-stabilized pulsed arc discharge, *AIAA J.* 4:1947, 1966.
43. J. J. Lowke and E. R. Capriotti, Calculation of temperature profiles of high pressure electric arcs using the diffusion approximation for radiation transfer, *J. Quant. Spectrosc. Radiat. Transfer* 9:207, 1969.
44. J. J. Lowke, R. J. Zollweg, and R. W. Liebermann, Theoretical description of ac arcs in mercury and argon, *J. Appl. Phys.* 46:650, 1975.
45. W. Elenbaas, Ähnlichkeitsgesetze der Hochdruckentladung, *Physica* 2:169, 1935.
46. G. Heller, Dynamical similarity laws of the mercury high pressure discharge, *Physics* 6:389, 1935.
47. G. Frind, On the decay of arcs, Part I: Theoretical considerations; Part II: Experimental tests of the theory, *Z. Angew. Phys.* 12:231-237, 515-521, 1960.
48. C. W. Kimblin and H. Edels, Electrical conductance decay of interrupted arc columns, *Br. J. Appl. Phys.* 17:1607-1619, 1966.
49. H. Christmann, W. Frie, and W. Hertz, Experimentelle und theoretische Untersuchungen des Leitwertabklingens in Stickstoffkaskadenbogen, *Z. Phys.* 203:372-388, 1967.
50. G. R. Jones and H. Edels, Electrical conductance decay of arc columns in some common gases, *Z. Phys.* 229:14-32, 1969.

5
Physical Theory of the Arc in a Gas Blast

JOHN J. LOWKE / Commonwealth Scientific and Industrial Research Organization, Sydney, New South Wales, Australia

5.1 Introduction   157

5.2 Arcs at High Current   159
    5.2.1 Two-dimensional calculations   159
    5.2.2 Channel model   162

5.3 Transient Arcs at Low Current   169
    5.3.1 Slepian's race theory and electrode sheath effects   169
    5.3.2 Prince's wedge model   173
    5.3.3 Cassie model: Convective effects   173
    5.3.4 Mayr model: Conduction effects   174
    5.3.5 Detailed energy balance calculations   175
    5.3.6 Nonequilibrium models   180

5.4 Summary   183

References   183

5.1 INTRODUCTION

It is self-evident that whether a circuit breaker succeeds or fails to interrupt a given current in a circuit is a result of many complex phenomena. Of necessity the engineer has to attempt to have some conception of the dominating physical processes to have some guidance in circuit breaker design. The earliest circuit breaker designers viewed the arc as similar to a flame that needed to be extinguished by immersion in oil or water [1]. This chapter is concerned with later designs where the arc is extinguished primarily by convective axial flow, as is the case with modern oil, air, sulfur hexafluoride (SF_6), or puffer breakers.

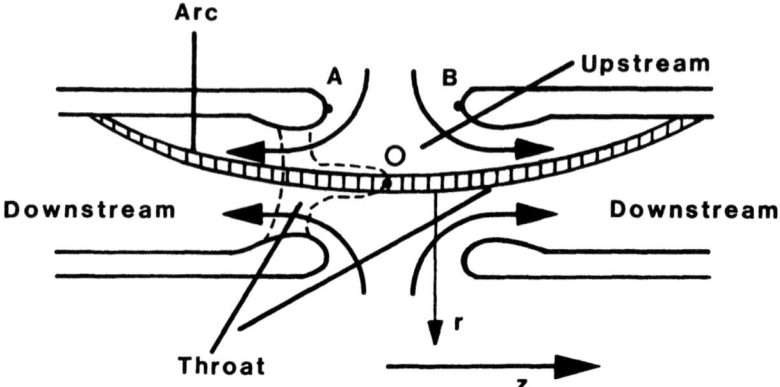

Figure 5.1 Representation of the electrode configuration of a circuit breaker. (From Elektrotech. Z. Archiv. 2:267-272, 1980.)

In Fig. 5.1 a representation is shown of the basic electrode arrangement of such circuit breakers with axial flow. In a double-flow breaker the arc is initiated at two points of contact (e.g., A and B) when the two electrodes separate. The arc is then swept by an imposed gas flow to the position shown. In a single-flow breaker, the flow is in only one direction, with, for example, the left-hand electrode solid, as indicated by the dashed lines of Fig. 5.1.

There are two fundamental physical models that are used to describe what happens whenever an electrical discharge flows in a gas. The first is the arc model, where the arc plasma is treated as a simple hot fluid having an electrical conductivity which depends uniquely on temperature. The local temperature depends on heat transfer effects—a balance between ohmic electrical heating and cooling by conduction, convection, and radiation.

The second gas discharge model is the glow model, which applies when the electrons are much hotter or more energetic than the heavy particles. The heavy particles can even be at room temperature, as in a fluorescent lamp. In the glow model, detailed equations are solved for electron properties such as energy, recombination, ionization, or attachment.

Our discussion of arcs in gas flow can be conveniently divided into two sections. Arcs at high current are discussed in Sec. 5.2 using the arc or fluid model with the arc plasma assumed to be in local thermodynamic equilibrium. The properties of the Cassie model follow from this arc theory: arc diameters can be predicted for any given nozzle shape, and phenomena of clogging and even reverse gas flow can be predicted, at least in principle. These arcs can generally be regarded as being in the steady state, as even for time-varying arcs the percentage change in current is small in the plasma transit time in the nozzle of ~20 μs.

Time-varying arcs of low current are discussed in Sec. 5.3, and both arc and glow models need to be used. Generally, arc physics is used to explain thermal breakdown, and glow physics is used for dielectric breakdown. In glow physics, properties of the electrons are evaluated as distinct from properties of the background heavy particles.

The number of completely different physical models that have been proposed to account for circuit interruption is remarkable. Historically, the focus of attention has moved downstream in the circuit breaker. In the early 1930s there was much discussion about the upstream region of the circuit breaker. Slepian's race theory [2] originally involved sheath effects near the cathode. Prince's wedge theory [3] postulated that convective effects introduced a wedge of liquid or cold gas at the upstream stagnation point and that this growing wedge caused circuit interruption. From the late 1930s thermal conduction effects, presumably strongest where the arc is narrowest at the nozzle throat, were analyzed usually using mathematical formulations initiated by Cassie [4] and Mayr [5]. Now, in the 1980s, as suggested earlier by Slepian [6.7], turbulence in the nozzle region downstream of the throat has been proposed as being crucial to interruption [8,9].

The advent of computers has greatly increased our predictive capabilities. We now have detailed predictions of arc temperature, velocity, pressure, and arc radius as a function of both time and axial position. Account is taken of the material functions of the gas, and also details of the nozzle shape. Despite this progress, there is still a lack of agreement among workers as to such basic questions as the importance of turbulence—and indeed the relative sensitivity of the upstream, throat, and downstream regions of the breaker.

5.2 ARCS AT HIGH CURRENT

One of the simplest physical representations of an arc is to consider the arc as a channel, isothermal with radius, which can be represented mathematically in one dimension, as is shown in Sec. 5.2.2. First, however, we consider results of two-dimensional calculations.

5.2.1 Two-Dimensional Calculations

It is assumed that the arc can be treated as a hot plasma in local thermodynamic equilibrium. The electron density and arc conductance at all positions are determined from the equilibrium fraction of ionization corresponding to the local plasma temperature. The local plasma temperature is determined by the local energy balance equation, which is

$$\sigma E^2 = \nabla(-k\nabla T) + \rho C_p \bar{v} \cdot \nabla T + U \qquad (5.1)$$

The electrical energy input due to the electric field E is equated with losses due to conduction, convection, and radiation, respectively. In equation (5.1), $\bar{v}$ is the plasma velocity, σ the electrical conductivity, k the thermal conductivity, ρ the density, C_p the specific heat, and U the radiation emission coefficient. The material functions k, ρ, C_p, and U are functions of temperature. For example, thermal and electrical conductivity as functions of temperature for common gases are shown in Figs. 5.2 and 5.3.

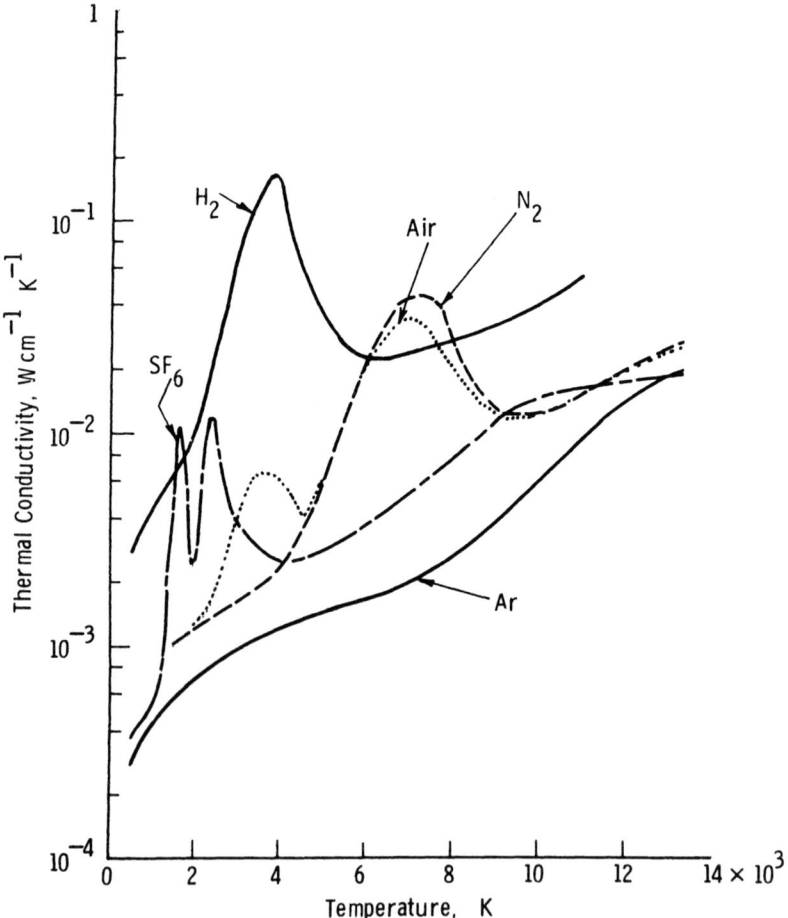

Figure 5.2 Thermal conductivity as a function of temperature. (From Ref. 10.)

Physical Theory of the Arc in a Gas Blast

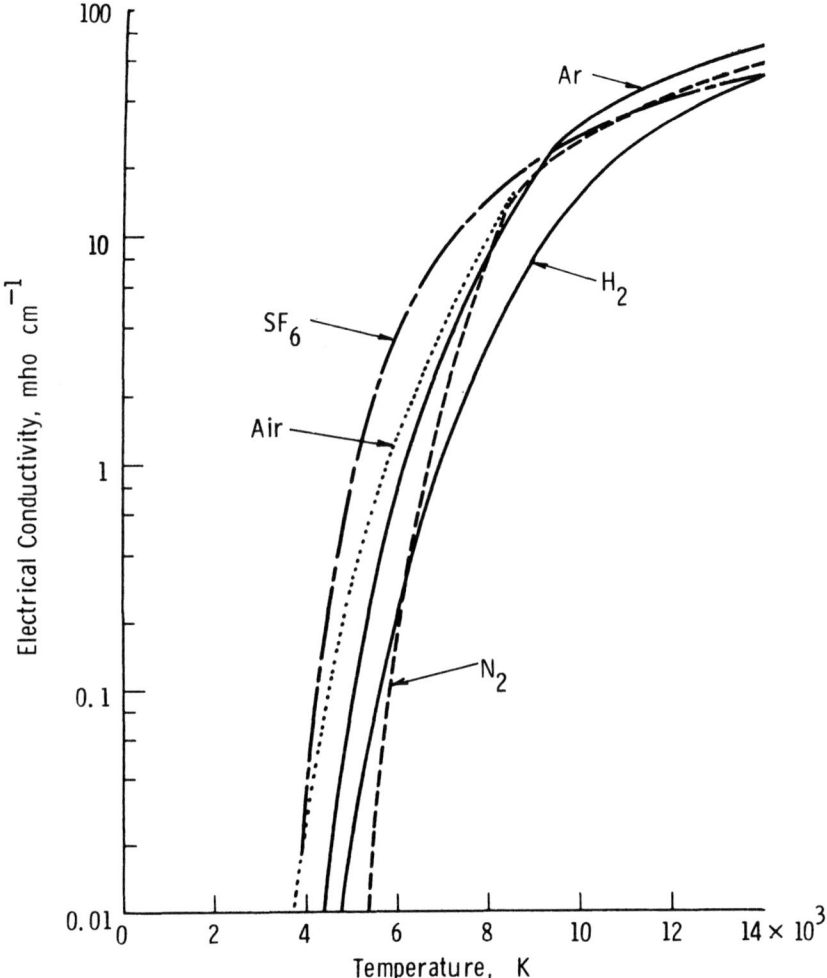

Figure 5.3 Electrical conductivity as a function of temperature. (From Ref. 10.)

Equation (5.1) can be solved to give temperature as a function of r and z after eliminating E and $\bar{v}$. Electric field E can be eliminated using Ohm's law for any given current I. Thus E is given by

$$E = \frac{I}{\int 2\pi r \sigma \, dr}$$

Plasma velocity $\bar{v}$ can be eliminated using gas dynamic relations which define Mach number as a function of nozzle shape [11].

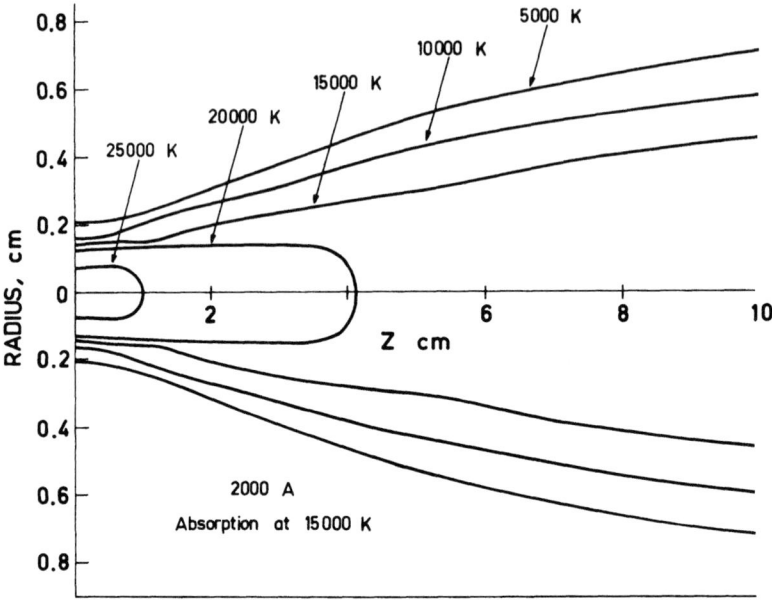

Figure 5.4 Derived contours of temperature for a 2000-A arc in nitrogen. (From Ref. 12.)

A feature of recent research has been the elucidation of radiation as an energy dissipation mechanism. Central arc temperatures at high currents are greater than 2000 K so that a major fraction of the radiation is in the ultraviolet portion of the spectrum. Photons of this ultraviolet radiation are of sufficient energy that they photoionize the colder particles at the edge of the arc. For high-current arcs it is a good first approximation to ssume that all radiation is absorbed in the cooler outer regions of the arc.

In Fig. 5.4 are shown derived contours of temperature from solutions of equation (5.1) for a 2000-A arc in nitrogen. The central radiation was assumed to be absorbed in a short distance near the 15,000-K contour. The effect of this self-absorption is to produce a shelf in the temperature profile as a function of radius. Such shelves are shown in Fig. 5.5, where calculations are compared for the three cases of absorption at 15,000 K, 10,000 K, and no absorption at all.

5.2.2 Channel Model

Although the two-dimensional solutions of Figs. 5.4 and 5.5 indicate that arcs are far from isothermal, analysis that considers the arc as a simple channel leads to significant insights. We view the arc as a

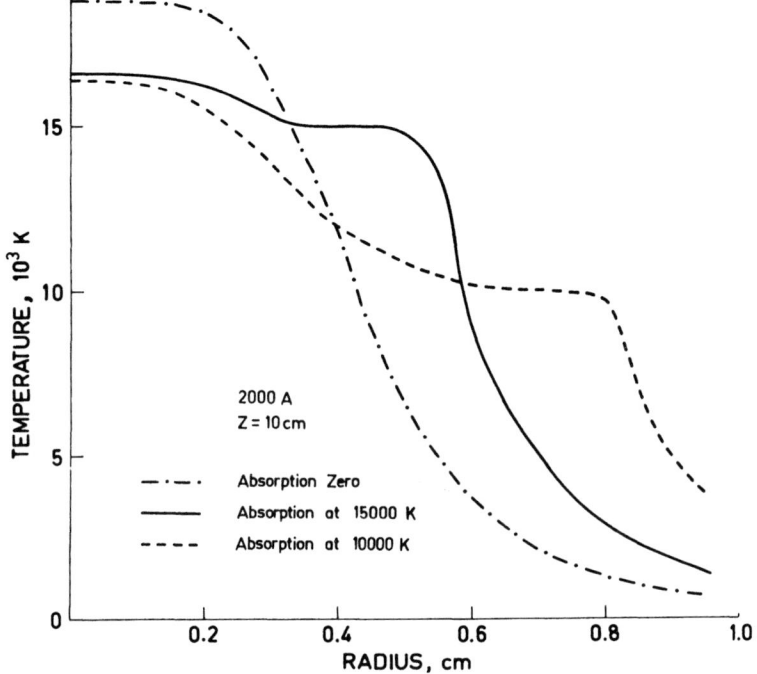

Figure 5.5 Effect of self-absorption of radiation on the radial profiles of temperature. (From Ref. 12.)

channel of area A and temperature T which is isothermal with respect to radius. Equations then give the properties of the Cassie arc model analytically and also an expression for arc diameter.

Cassie Model

Conservation equations for the channel model [13] are functions only of z. From equation (5.1), at the arc center, temperature will be defined from

$$\sigma E^2 = -\rho C_p v_z \frac{\partial T}{\partial z} + U \qquad (5.2)$$

Ohm's law gives

$$I = \sigma E A \qquad (5.3)$$

If we assume that all radiation is reabsorbed at the edge of the arc, all electrical energy produces plasma, and we can use another energy equation,

$$\sigma E^2 A = \frac{\partial}{\partial z} \rho h v_z A \tag{5.4}$$

where h is the enthalpy of the arc plasma.

Using equation (5.2) to eliminate E and equation (5.3) to eliminate A, we obtain from equation (5.4) an equation in temperature in which I cancels [13]. Thus temperature T(z) is independent of current, provided that the axial pressure and Mach number distributions are independent of current, which is true if the nozzle is not clogged. Furthermore, from equation (5.2), if T(z) is independent of current, E or arc voltage is independent of current. Finally, from equation (5.3), as E and T are independent of I, arc area will vary proportionally with I. These properties are those of the Cassie model.

The coupled equations, (5.2) through (5.4), can be solved to give numerical solutions of the three variables T, A, and E as functions of z. The theoretical curves for a 2000-A arc in nitrogen compare very well with the experimental points of Hermann et al. [14], as shown in Figs. 5.6 and 5.7.

A further consequence of equations (5.3) and (5.4) is an approximate expression for arc diameter in the Cassie regime. On eliminating E and integrating with respect to z, we obtain [13]

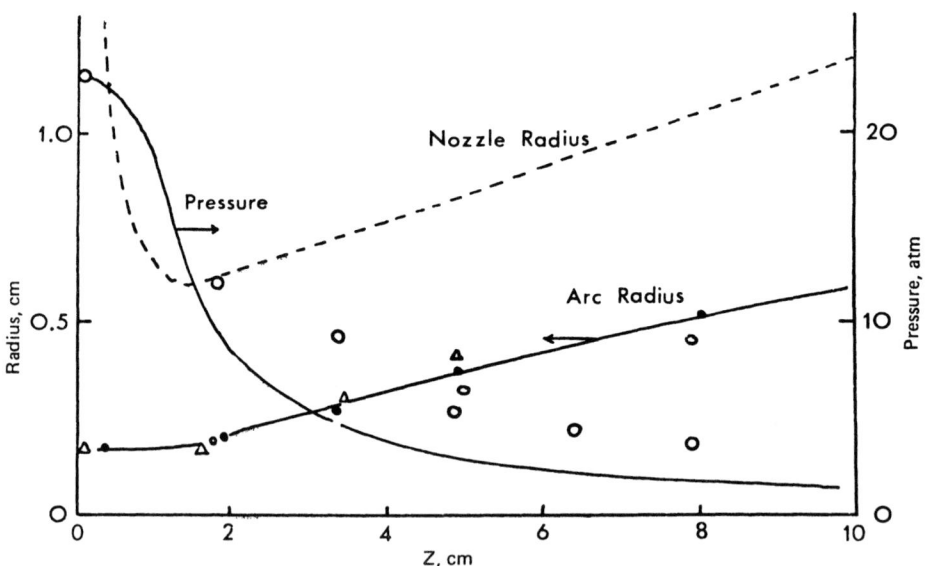

Figure 5.6 Theoretical curves of radius and pressure from the channel model compared with the experiment for a 2000-A arc in nitrogen. (From Ref. 15.)

Physical Theory of the Arc in a Gas Blast

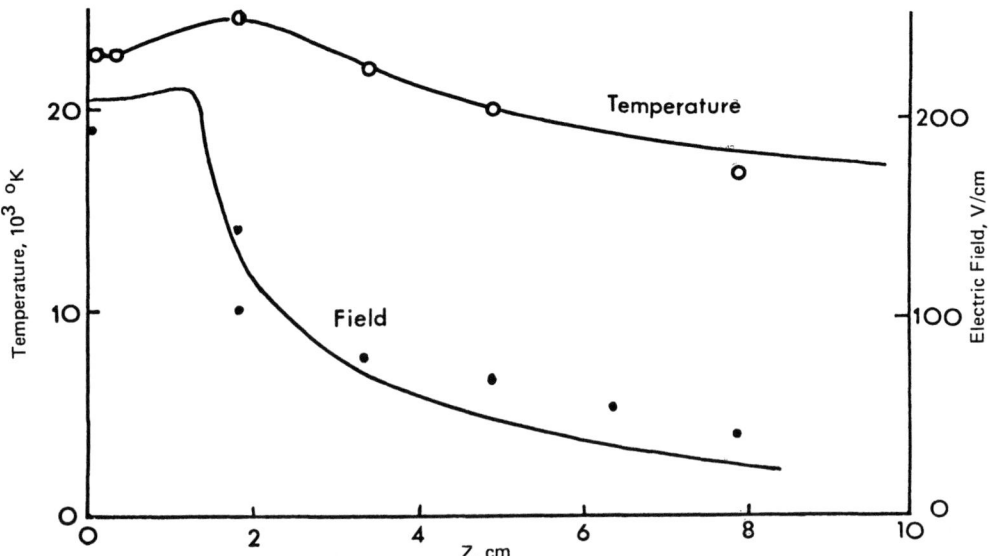

Figure 5.7 Theoretical curves of temperature and electric field from the channel model compared with the experiment for a 2000-A arc. (From Ref. 15.)

$$A = \sqrt{\frac{2z}{\rho h v_z \sigma}}\, I$$

if it be assumed that $\rho h v_z$ is independent of z. In Fig. 5.8 are shown predictions of arc diameter at the nozzle throat for both air and SF_6 obtained from the equation above, compared with experimental points. It is assumed that $\sigma \sim 100$ S/cm, that v_z, the sonic velocity, is 5×10^5 cm/s, and ρh at 100 kPa (1 atm) is ~ 0.8 J/cm^3. Predicted diameters D are insensitive to the approximations because D depends on only the one-fourth power of $\rho h v_z \sigma$.

Nozzle Clogging

At high currents, the area of cross section of the arc becomes significant compared with the nozzle cross section and the gas flow is impeded. The axial distributions of pressure and Mach number then differ from the isentropic gas relations and we say that the nozzle is partially clogged.

The channel model can be modified to include these effects by including the equations of conservation of mass and momentum, which have to be solved together with equations (5.2) to (5.4). The mass conservation equation is

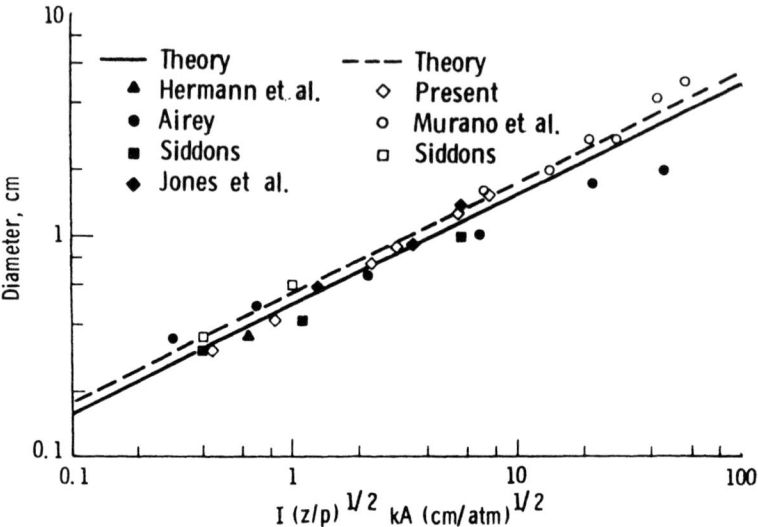

Figure 5.8 Comparison of theoretical and experimental arc diameters at the nozzle throat. (From Ref. 13.)

$$\frac{\partial(\rho v_z A)}{\partial z} + \frac{\partial[\rho_c v_{zc}(Q - A)]}{\partial z} = 0 \qquad (5.5)$$

where $Q(z)$ is the nozzle area and the subscript c refers to the cold gas surrounding the arc. The momentum conservation equation is

$$\rho v_z \frac{\partial v_z}{\partial z} = -\frac{\partial p}{\partial z} \qquad (5.6)$$

where p is the gas pressure. These additional equations define $p(z)$ and $v_z(z)$, provided we assume that the Mach number of the plasma equals the Mach number of the cold gas (i.e., $v_z/a = v_{zc}/a_c$, where a is sonic velocity).

Equations (5.2) through (5.6) define the five variables $T(z)$, $A(z)$, $E(z)$, and $v_z(z)$ from the Cassie regime up to very high currents when the arc completely fills the nozzle and $A(z) = Q(z)$. In the latter case, with an insulating nozzle the input energy is dissipated largely by ablation of the nozzle wall. This ablation increases the pressure inside the nozzle.

Solutions of equations (5.2) through (5.6) have been investigated numerically [16] for various currents in the nozzle of Ref. 14. In the solutions we assume that U and ρ are proportional to p, and that k, σ, C_p, and h are independent of p. In Fig. 5.9 we show calculated arc radii as functions of current. It is seen that the arc first touches the

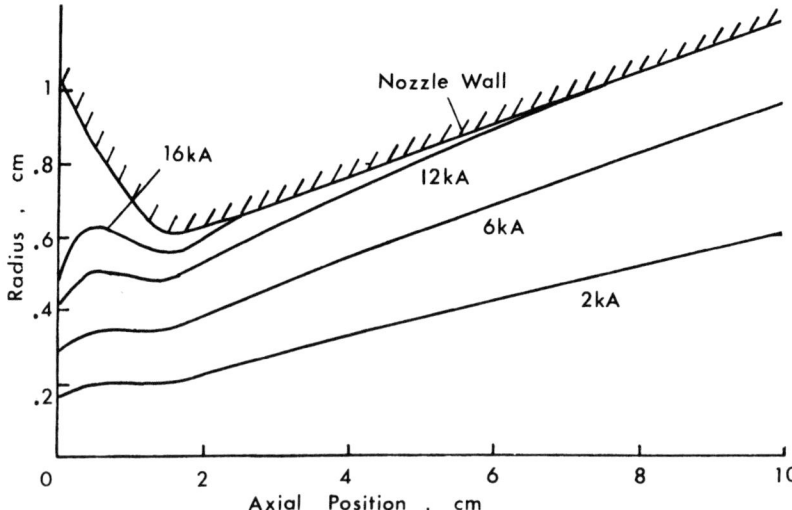

Figure 5.9 Calculated arc radii as functions of current extending to the clogging regime. (From Ref. 16.)

nozzle wall downstream of the throat. In Fig. 5.10 are shown the calculated Mach number distributions $M(z)$. At low currents $M(z)$ is approximately the isentropic distribution for the given nozzle shape with Mach 1 at the throat. At very high currents flow is dominated by the very high pressures inside the nozzle produced by ablation and $M(z)$ tends to a different limit, with Mach 1 at both ends of the nozzle.

The pressure distribution inside the insulating nozzle as a function of current is shown in Fig. 5.11. Again at low currents there is the conventional isentropic distribution corresponding to the nozzle shape $Q(z)$. At very high currents the equations predict that the distribution of pressure $p(z)$ is proportional to the square of the current.

The derived volt-ampere characteristic and temperature at the nozzle throat as functions of current are shown in Fig. 5.12. The voltage changes from being constant with current at low current, corresponding to the Cassie model, to being proportional to the current at high current, corresponding to the plasma acting like a resistance. The axial temperature distribution changes from its distribution in the Cassie regime to a different limit in the fully clogged regime.

Figures 5.8 through 5.12 are instructive in giving a qualitative picture of the transition from the Cassie regime to a clogged and

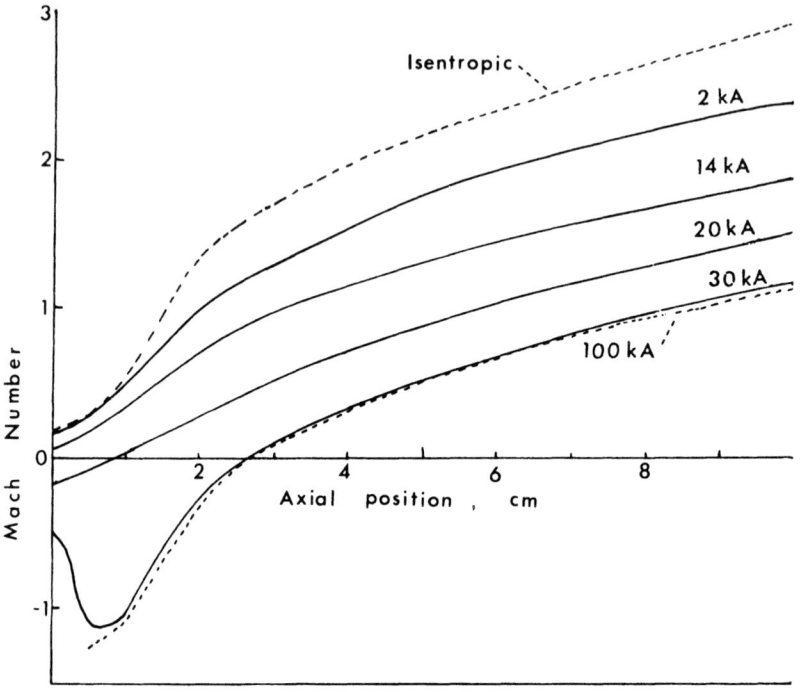

Figure 5.10 Mach number distributions as functions of current. (From Ref. 16.)

finally ablation stabilized arc in a nozzle. At high currents, however, two assumptions of equations (5.2) through (5.6) are invalid. First, not all of the radiation from the center of the arc is of wavelength short enough to ionize the ablated material from the wall. As a consequence, a zone of vapor usually exists between the nozzle wall and the arc plasma. This vapor is transparent to some of the radiation. The mass flow in this vapor can be significant compared with the plasma mass flow, and as a consequence the calculated pressure and Mach number distributions are varied from those of Figs. 5.10 and 5.11. Second, as the pressure inside the nozzle becomes large, the radiation emission coefficient U is no longer proportional to pressure because the radiation intensity attains the blackbody limit. To be within this limit it is necessary that DU be less than bT^4, where b is the constant for blackbody radiation. These effects are considered for cylindrical nozzles in Ref. 17.

Physical Theory of the Arc in a Gas Blast

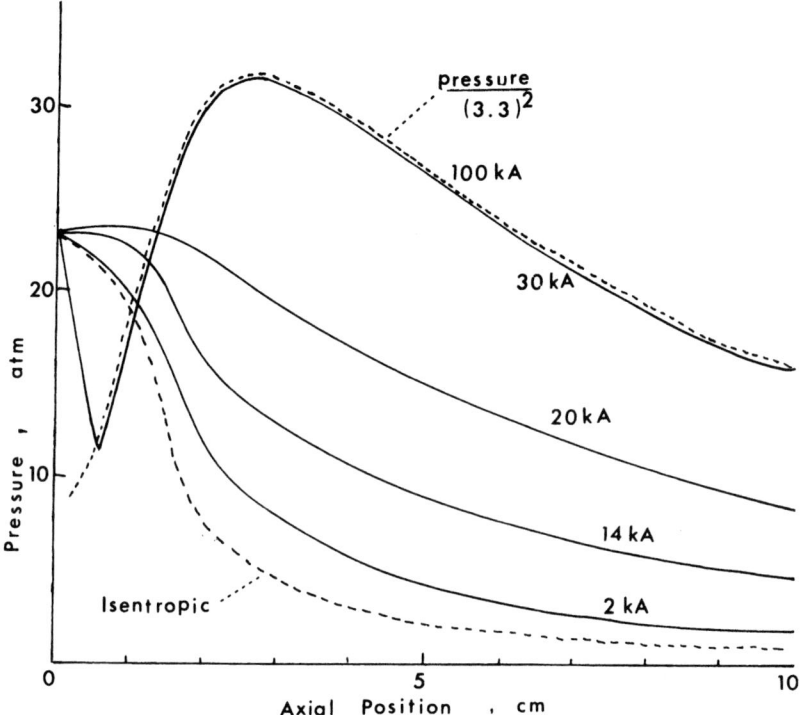

Figure 5.11 Pressure distributions as functions of current. (From Ref. 7.)

5.3 TRANSIENT ARCS AT LOW CURRENT

Although high-current arcs are of importance to the circuit breaker engineer to assess the high-current limit of the breaker and the onset of clogging, the prime consideration for any breaker is the mechanism of circuit interruption. Explanations of circuit interruption have used both arc physics and glow physics.

5.3.1 Slepian's Race Theory and Electrode Sheath Effects

Historically, the first theory of circuit interruption was given by Slepian in 1928 and has become known as the *race theory* [2]. Immediately after current zero, electrons are drawn away from the cathode by the reapplied field, leaving a sheath of positive ions as shown in Fig. 5.13. Slepian postulated that the field produced in this sheath needed to be always less than the critical breakdown field for air, which is ~30,000 V/cm. Thus success or failure depended on a race

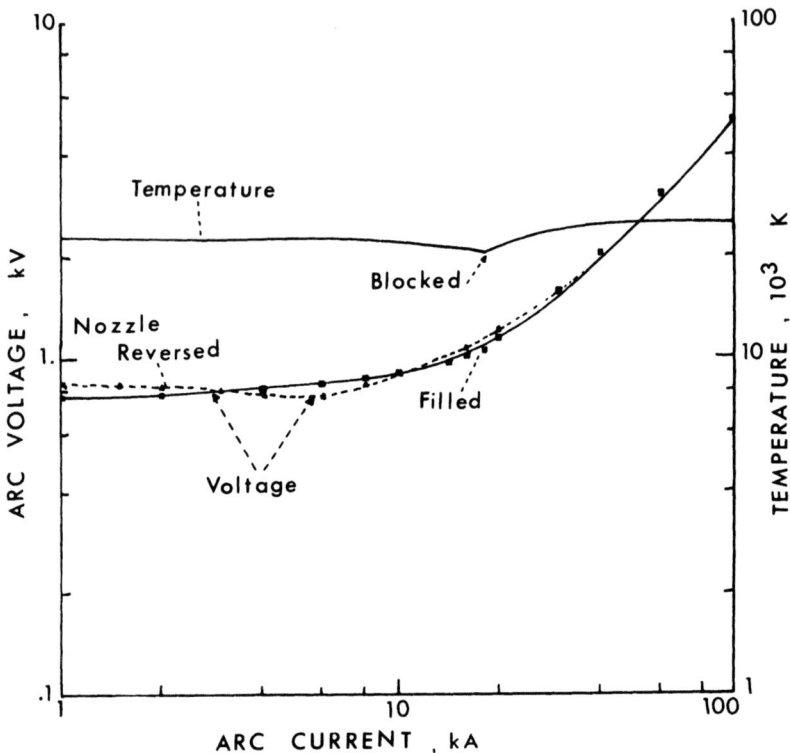

Figure 5.12 Calculated V-I characteristics and plasma temperature at the nozzle throat. (From Ref. 16.)

between the rate of recombination of the ions, which increases the sheath thickness for a given voltage and so reduces the sheath field, and the rate of rise of recovery voltage, which increases both the thickness of the space-charge sheath and the electric field in the sheath.

The recombination rate of the ion density n_+ is normally given by $dn_+/dt = -\gamma n_+^2$, where γ is the recombination coefficeint and t the time. On integration, $n_+ \sim 1/\gamma t$. The electric field E in the ion sheath of thickness d can be found on integration of Poisson's equation, which gives $E = en_+(z - d)/\varepsilon$, where e is electronic charge, ε the permittivity constant, and z the distance from the electrode. Further integration of E gives an expression for the voltage in terms of d. The maximum field in the sheath, E_m, is en_+d/ε and can be used to eliminate d in the expression for V to give $V = \varepsilon E_m^2/2en_+ = \varepsilon\gamma E_m^2 t/2e$. The rate of rise of recovery voltage thus needs to be less than $\varepsilon\gamma E_m^2/2e$.

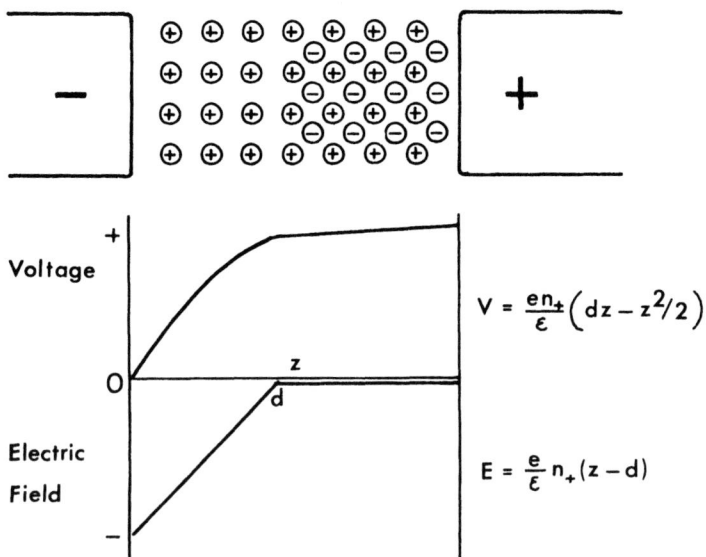

Figure 5.13 Electrode sheath effects. (From Elektrotech. Z. Archiv. 2:267-272, 1980.)

Slepian obtained fair agreement with experiment using this expression with $E_m = 30,000(273/T)$ V/cm, where T is the absolute temperature.

Although the mainstream of thinking has moved away from a consideration of sheath effects to energy balance analyses, there has been a recent revival of interest in space-charge effects. Hoyaux [18] believed that after current zero the reapplied electric field could not enter the arc because the arc is a plasma. Rather, the reapplied field falls entirely over the sheath regions near the electrodes, where the plasma is cooled by the electrodes. Arc reignition is determined by whether this cold sheath is broken down by the reapplied field. Dzierzbicki and Tarocinski [19] have a series of papers analyzing sheath effects in circuit interruption.

A recent sophisticated model of the sheath is that of Eliasson and Schade [20] applied to dielectric breakdown. Space-charge effects of positive and negative ions are analyzed near circuit breaker electrodes in SF_6 to make predictions of dielectric breakdown. It is generally accepted that sheath effects are important for predictions of dielectric breakdown, which may occur several hundreds of microseconds after current zero when ion densities are low. However, for thermal breakdown when electron and ion densities are much higher, the space-charge and thermal sheath regions are so thin that their resistances can usually be neglected compared with the column resistance.

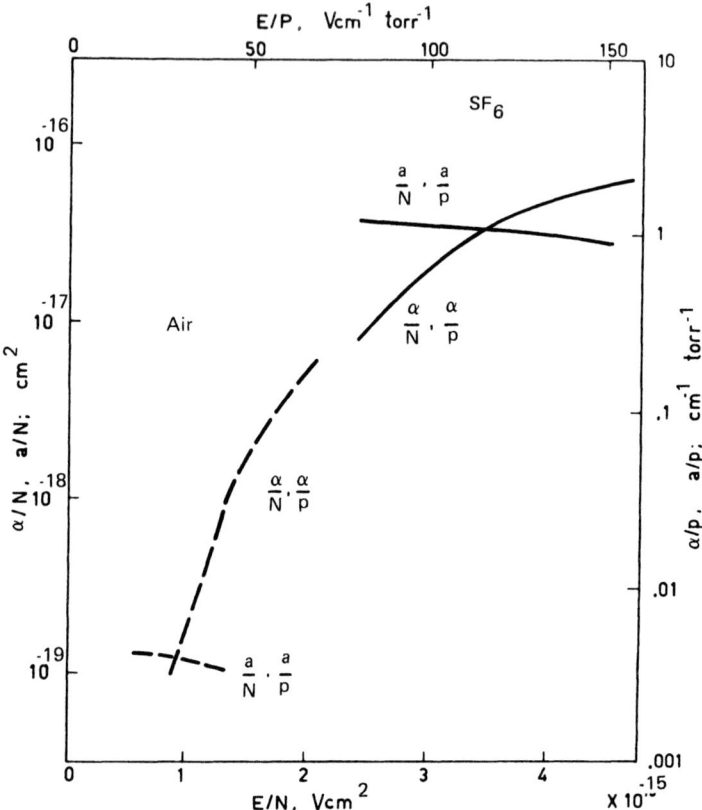

Figure 5.14 Ionization (α) and attachment (a) coefficients in air (broken curves) and SF_6 (unbroken curves).

An overriding consideration in any sheath analysis which uses glow discharge physics rather than arc physics is the relative importance of electron ionization and attachment. For both air and SF_6, electron attachment is dominant at low electric fields, whereas ionization is dominant at high fields. In Fig. 5.14 we show ionization and attachment coefficients, α and a, for both air and SF_6 as functions of E/N, where N is the gas number density [21,22]. The value of E/N for which ionization equals attachment is ~10^{-15} V cm^2 for air and 4×10^{-15} V cm^2 for SF_6, indicating that SF_6 is greatly superior to air in withstanding dielectric breakdown.

5.3.2 Prince's Wedge Model

In contrast to Slepian's race theory, Prince and co-workers proposed a simple *wedge model*. It was proposed that at current zero when there is no electrical power input, the arc is cut by a wedge of cold gas entering the upstream region between the electrodes. For single-flow breakers the wedge was believed to be at the upstream electrode, whereas for double-flow breakers the wedge would be at the stagnation point, marked 0 in Fig. 5.1. Whether an electric current continued after current zero was thought to depend on whether electrical breakdown occurred through this wedge of cold gas when the circuit voltage is reapplied after current zero. The current interruption capability on a circuit breaker is obviously aided by a large thickness of the wedge that would be produced by high convective flow velocities, and by a low rate of rise of recovery voltage. Initially, this wedge theory was proposed for oil breakers by Prince and Skeats [3], where the wedge was thought to consist of liquid oil Prince obtained experimental correlations between the estimated rate of rise of dielectric strength from the derived velocity of the oil in the nozzle and the measured rate of rise of recovery voltage. Later the wedge theory was applied to gas-blast breakers [23].

Slepian [6] argued convincingly against the wedge theory on the basis that immediately when a region of the interelectrode plasma starts to cool to form a wedge, the resistance R of this region of the arc increases. However, as the current I is the same for all axial positions, most of the input power I^2R is deposited in the cold region of high resistance, preventing the formation of any wedge. He also pointed out that fluid continuity in a region of accelerating flow results in stretching of an arc column, rather than cutting.

Despite the decline of the wedge theory, there is no doubt that the limiting rate of rise of recovery voltage for arc breakdown is much more sensitive to the position of the upstream electrode than for that of the downstream electrode [24]. Any model of arc interruption must explain these experimental results. Effects of the stagnation point, the electrode wake, and electrode vapor are still at the forefront of discussion.

5.3.3 Cassie Model: Convective Effects

As an alternative to the cold wedge theory, Cassie postulated that the arc area was continuously reduced by the convective flow. Cassie [4] said: "The deformation of the column due to the gas flow is something like that of a piece of elastic of circular cross section fixed at one end to the upstream contact while the other end is made to move at a speed equal to that of sound in the gas at arc temperature."

The variation of such a plasma column of area of cross section A is given by the time-dependent form of equation (5.4):

$$\frac{\partial(\rho hA)}{\partial t} = \sigma E^2 A - \frac{\partial}{\partial z}\rho h v_z A \tag{5.7}$$

where t is the time. To a first approximation $\rho \propto 1/T$ and $h \propto T$ and ρh is constant. Thus, if A and $\rho h v_z$ are approximately independent of z, equation (5.7) becomes

$$\frac{1}{A}\frac{\partial A}{\partial t} = \frac{\sigma E^2}{\rho h} - \frac{\partial v_z}{\partial z} \tag{5.8}$$

Arc conductance G is approximately proportional to A if arc temperature T is constant, so that

$$\frac{1}{G}\frac{\partial G}{\partial t} = \frac{\sigma E^2}{\rho h} - \frac{\partial v_z}{\partial z} \tag{5.9}$$

which is the Cassie equation, except that $\sigma/\rho h$ and $\partial v_z/\partial z$ are undetermined constants in the Cassie equation.

Experimental results are in very good agreement with the Cassie model at high currents and even down to zero current for high rates of current fall. However, from equation (5.8), for zero electric field, arc area is predicted to decay exponentially with characteristic time constant. Thus even for zero reapplied field, the arc area never decays to zero to produce an arc interruption. The filament of high-temperature plasma that remains will always be restored to an arc by reapplication of a sufficiently high voltage after current zero. For a complete model of arc interruption we need to consider effects such as thermal conduction to reduce the arc temperature at low currents.

5.3.4 Mayr Model: Conduction Effects

Mayr [5] considered the decay in arc temperature by thermal conduction and the temperature dependence of electrical conductivity. The equation for G derived by Mayr is

$$\frac{1}{G}\frac{dG}{dt} = \alpha EI - \beta$$

where β is a constant giving decay of G by thermal conduction and α is a further empirical constant which from equation (5.7) is $\sim 1/\rho hA$.

With this interpretation, arc interruption is determined principally by the thermal and electrical conductivity, suggesting that we attempt to correlate these material functions with the interruption ability of various gases. The very large peak in the thermal conductivity of hydrogen (see Fig. 5.2) suggests that this gas would be a good interrupting medium, as is borne out by the success of oil circuit breakers, where the dissociated arc products of oil are largely hydrogen. However, from Fig. 5.2, the thermal conductivity peak for SF_6

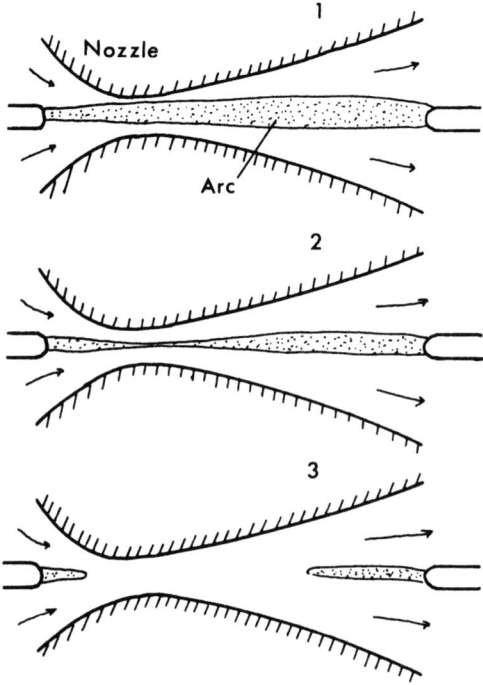

Figure 5.15 Arc decay dominated by conduction at the throat. (From Elektrotech. Z. Ausg. A 2:267-272, 1980.)

is much smaller than for nitrogen on air, yet SF_6 is acknowledged as a better circuit breaker medium. As the action of thermal conduction will be greatest where the radius is smallest (i.e., at the throat), the progressive decay of the arc is pictured as in Fig. 5.15.

5.3.5 Detailed Energy Balance Calculations

A recent significant development has been computer predictions of success or failure of arc interruption using various forms of the time-dependent energy balance equation. Because of the significant self-absorption of radiation energy, the input energy of the arc at high currents is largely dissipated by the axial convective flow. At high currents, arcs are approximately isothermal radially and conduction has a relatively small effect. It is only at very small currents, when the arc radius is small, that conduction is significant in reducing the central temperature and thus increasing arc resistance. Conservation equations of energy, momentum, and mass in principle define temperature, velocity, and pressure as a function of position and time. However, there are important differences among different workers as to how to represent the energy transfer process mathematically. There have been three groups of papers solving conservation euqations for arc properties.

A first group of papers uses the simple channel model of Sec. 5.2.2 to represent the arc. The time-dependent forms of equations (5.2) through (5.4) are solved to obtain the axial distribution of arc area and temperature as a function of time, for various external circuits and values of dI/dt before current zero [25,26].

However, calculations of current interruption are dependent on the boundary conditions chosen at the upstream electrode. Also, it seems unlikely that an isothermal model will be adequate at low currents when even a slightly nonisothermal arc will have a significant effect on the derived electric field for a given current.

The second group of papers, by Fang and co-workers [27,28], uses the integral method of Cowley [29]. This approach takes a more accurate account of radial variations at low current than the channel model by defining effective areas for thermal density, enthalpy flux, and electrical conduction, together with two shape factors, which are ratios of the latter two areas to the thermal arc area. The shape factors are represented as functions of local dynamic power loss. These functions are determined empirically, it being claimed that they are reasonably independent of the type of arc being studied. The integral methods have been particularly successful in predicting properties of an efflux arc near current zero. What is of special interest is that agreement is obtained between theory and experiment "without resorting to turbulence and without adjustment of parameters" [28]. However, significant differences exist between theoretical predictions and the experimental results of Frind et al. [24], which may be due to the lack of any account of thermal conduction losses in the theory. Furthermore, the pressure dependence of arc-interrupting properties is significantly underestimated by the model.

Of considerable impact to the subject has been the third group of papers, the experimental and theoretical papers of Ragaller and co-workers [7,8,14]. In the paper of Hermann and Ragaller [8], a theoretical method is outlined giving predictions for both air and SF_6 in reasonable agreement with experimentally found arc interruption properties as functions of both maximum current and tank pressure of the circuit breaker. The principal physical process of circuit interruption is claimed to be turbulence operating just downstream of the nozzle throat, where it dominates over thermal conduction. In Fig. 5.16 is shown their picture of the arc near current zero, with the tenuous turbulent region downstream of the throat. It even shows some breakup of the arc column into parallel filaments as first suggested by Slepian [30].

Physical Theory of the Arc in a Gas Blast

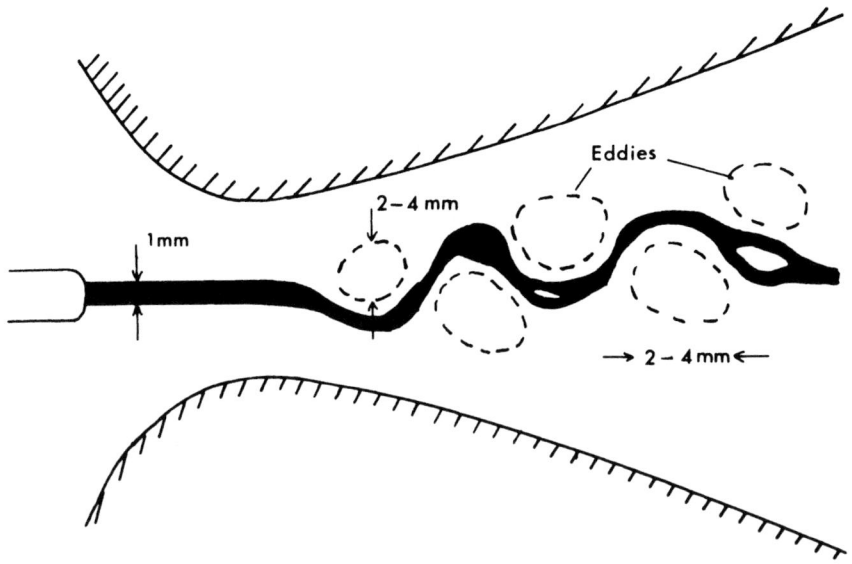

Figure 5.16 Turbulent picture of arcs from Ragaller et al. (From Elektrotech. Z. Ausg. A 2:267-272, 1980.)

In the theoretical model, arc temperature is considered to be independent of axial position. Three regions are considered for the temperature profile; (1) the arc core, (2) the surrounding thermal layer, and (3) the external cold gas, as shown in Fig. 5.17. Equations are solved for the mass, momentum, and energy balance between these regions. Material functions appropriate to the circuit breaker medium are required for the computer program, together with other parameters, such as nozzle dimensions.

The paper of Hermann and Ragaller [8] contains most impressive agreement between theoretical and experimental results for both air and SF_6. Nevertheless, no detail of the axial variation of arc properties is retained in the theory, and it is assumed that the axial pressure gradient is proportional to position. Also, there are some parameters representing turbulence, radiation absorption, and the shape of the temperature profiles which are evaluated from experimental results. Frind et al. [31] are skeptical that the influence of turbulence is dominant at current zero for their own gas-blast arcs and claim that their measured values of maximum current after current zero for circuit interruption are significantly less than predicted by theory when account is taken of turbulence.

Special mention must be made of the papers of Swanson [9]. An extensive series of papers originating with the two-dimensional treat-

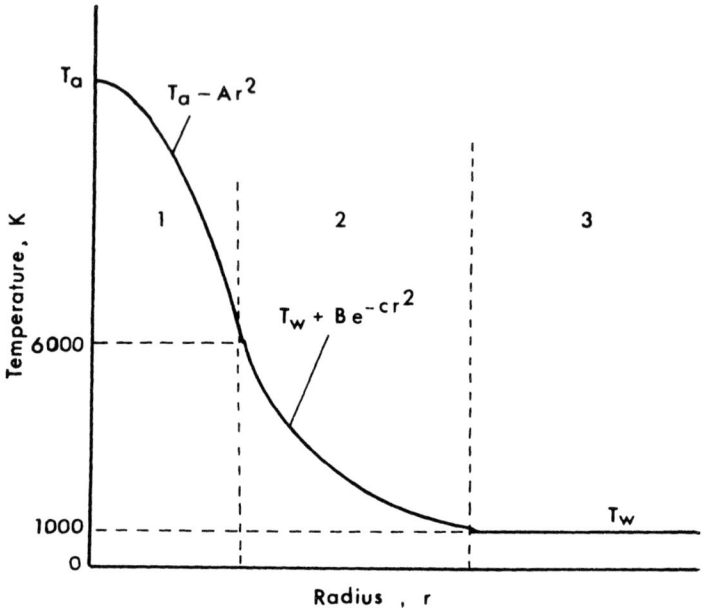

Figure 5.17 Radial zones of the theoretical model of Hermann and Ragaller. (From Elektrotech. Z. Archiv. 2:267-272, 1980.)

ment of Swanson and Roidt was the first theoretical treatment to include turbulence and amounts to a first-order integral method [32].

The Swanson an Roidt model, similar to those of Lowke [12], Lowke and Ludwig [13], Tuma and Lowke [15], and El-Akkari and Tuma [26], was derived solely from the integrated energy equation and is based on the assumption that two limiting forms of this equation, for arc area and temperature, can be solved simultaneously to define a reasonable arc model around current zero. It uses the adiabatic approximations for nozzle pressure and Mach number and assumes that the radial distribution of the heat flux potential is a Bessel function.

Figure 5.18 shows calculated temperature distributions along the previously shown nozzle at current zero marked $t = 0$, and at three postzero times as the arc cools with the application of a near-critical rate of rise of recovery voltage. Swanson has shown that in the temperature range above about 8000 K at which the electrical conductivity of the gas increases nearly linearly with temperature, the dynamic behavior of the arc approaches that of the Cassie equation. Conversely, at temperatures below 6000 K the arc behavior approaches that of the Mayr equation. As the arc cools it passes successively from the near-Cassie regime through a 2000-K range of intermediate

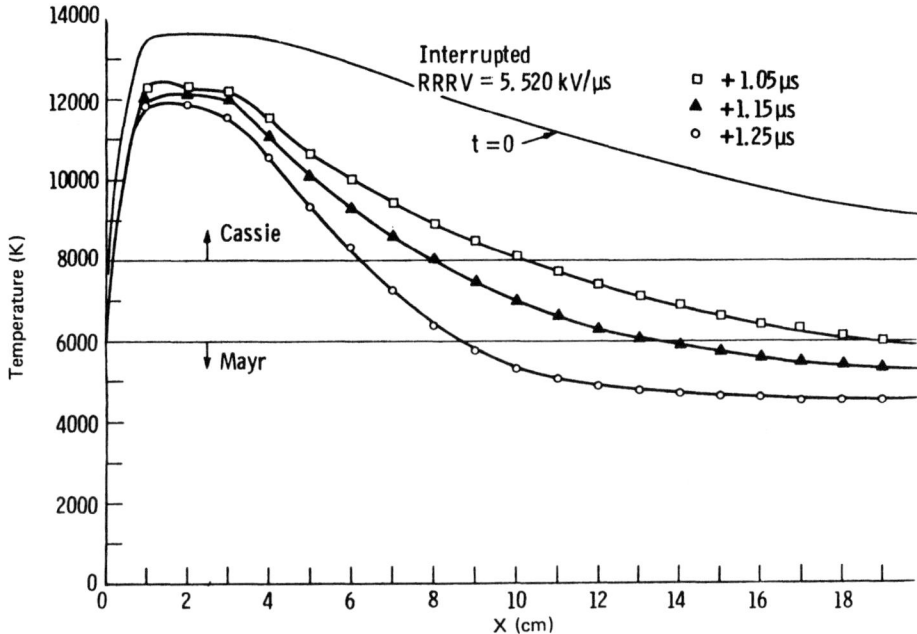

Figure 5.18 Postzero arc temperature as a function of axial distance calculated by Swanson. (From Ref. 9.)

behavior to the near-Mayr regime. The calculations indicate that this transition occurs first at the downstream end of the nozzle and proceeds upstream toward the nozzle constriction. A later model by Tuma [33] shows a somewhat similar behavior with nitrogen gas in this nozzle but with SF_6 shows cooling into the Mayr regime, starting at the constriction.

The usefulness of Swanson's arc model as a circuit element is illustrated by Figs. 5.19 and 5.20 showing arc model voltages and currents around current zero for two prezero rates of current change (the smaller near critical) with arc-circuit interaction included in the calculation. The resemblance to observed behavior of reigniting SF_6-blasted arcs in striking (see Fig. 14 of Ref. 9).

It is evident that careful work is now necessary to investigate the sensitivity of results to the various terms and parameters in these mathematical models. It will be only after different research groups test each others' models that definitive conclusions can be made on such questions as the importance of turbulence or the dominant physical processes in arc interruption.

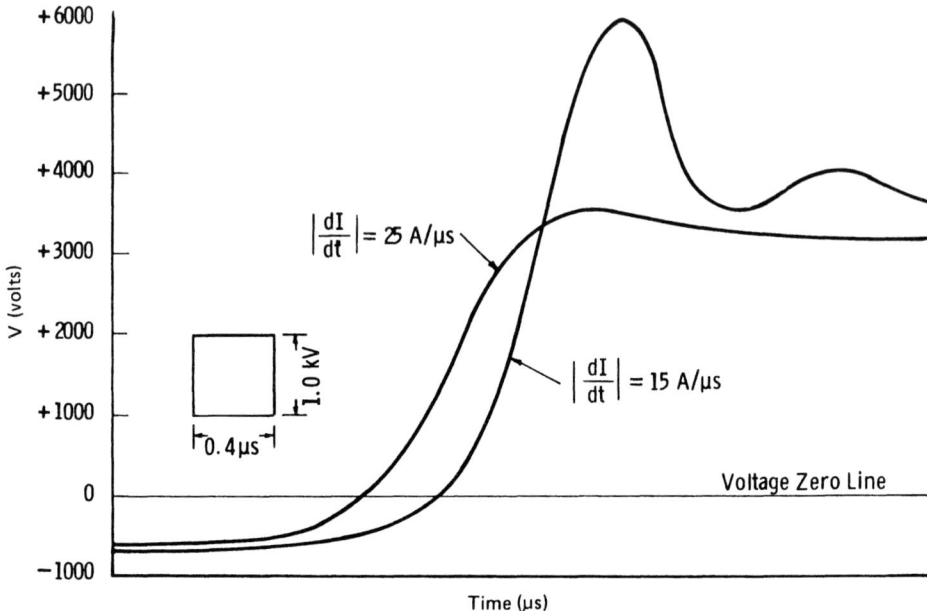

Figure 5.19 Calculated arc voltage with circuit interaction from Swanson. (From Ref. 9.)

5.3.6 Nonequilibrium Models

There is a final possibility that nonequilibrium effects are dominant at current zero. There are two major possibilities of nonequilibrium.

Electron Nonequilibrium In the arc models discussed above, the plasma is treated as a fluid and it is assumed that the electron density is appropriate to the plasma temperature. However, Rieder and Urbanek [34] have pointed out that immediately after current zero, when the electron density is low, the high reapplied electric field is likely to accelerate electrons to high energies sufficient to cause enhanced ionization. The instantaneous electron density will then be much higher than the equilibrium value appropriate to the gas temperature. No attempts have been made to account for this ionization in a rigorous way.

Some experimental evidence of such nonequilibrium is shown in Fig. 5.21 for a wall-stabilized ac arc in argon. At the current of 20 A, radiation losses should be negligible, and as there is no imposed flow, there should be no turbulence losses. Argon material functions are thought to be known accurately so that good agreement is expected between theory and experiment. Yet in the region of current zero, the low experimental arc voltages indicate an abundance of electrons

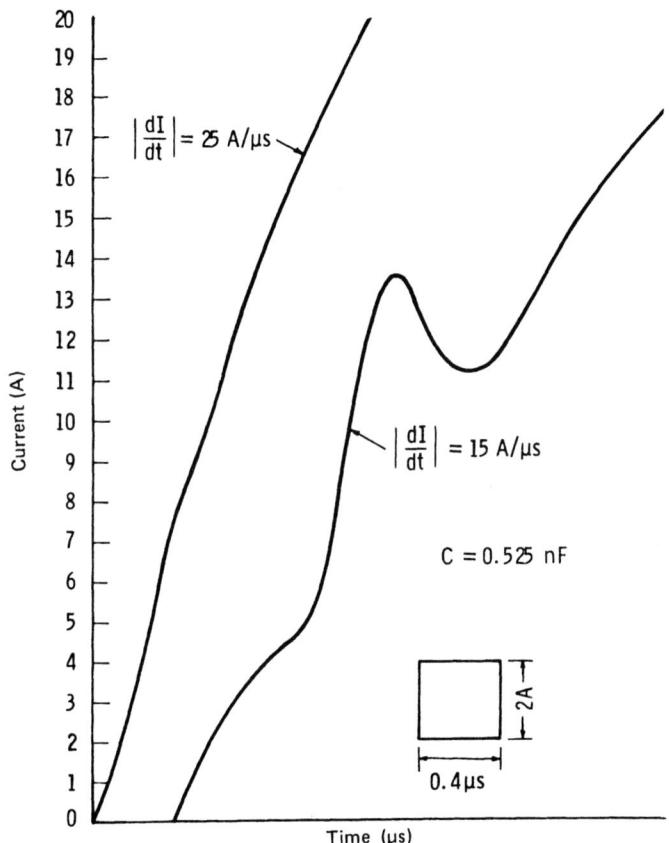

Figure 5.20 Calculated postarc current from Swanson. (From Ref. 9.)

and conductance above that appropriate to the equilibrium electrical conductivity [36]. It is possible that such effects also occur for arcs in circuit breakers.

Molecular Nonequilibrium A further nonequilibrium process results from the slow equilibration times of atoms to reform molecules when the dissociated atoms cool at current zero [37]. Thus time constants for $N + N \rightarrow N_2$ and $S + 6F \rightarrow SF_6$ can be many milliseconds. This effect was first cited by Hertz [38] to explain the strong decay of electrical conductance of wall-stabilized arcs in SF_6 on removing the dc electric field.

A significant cooling process in SF_6 at current zero is the blowing into the arc of the cold SF_6 molecules, which on dissociation absorb heat. This heat of dissociation is not reliberated because recombination

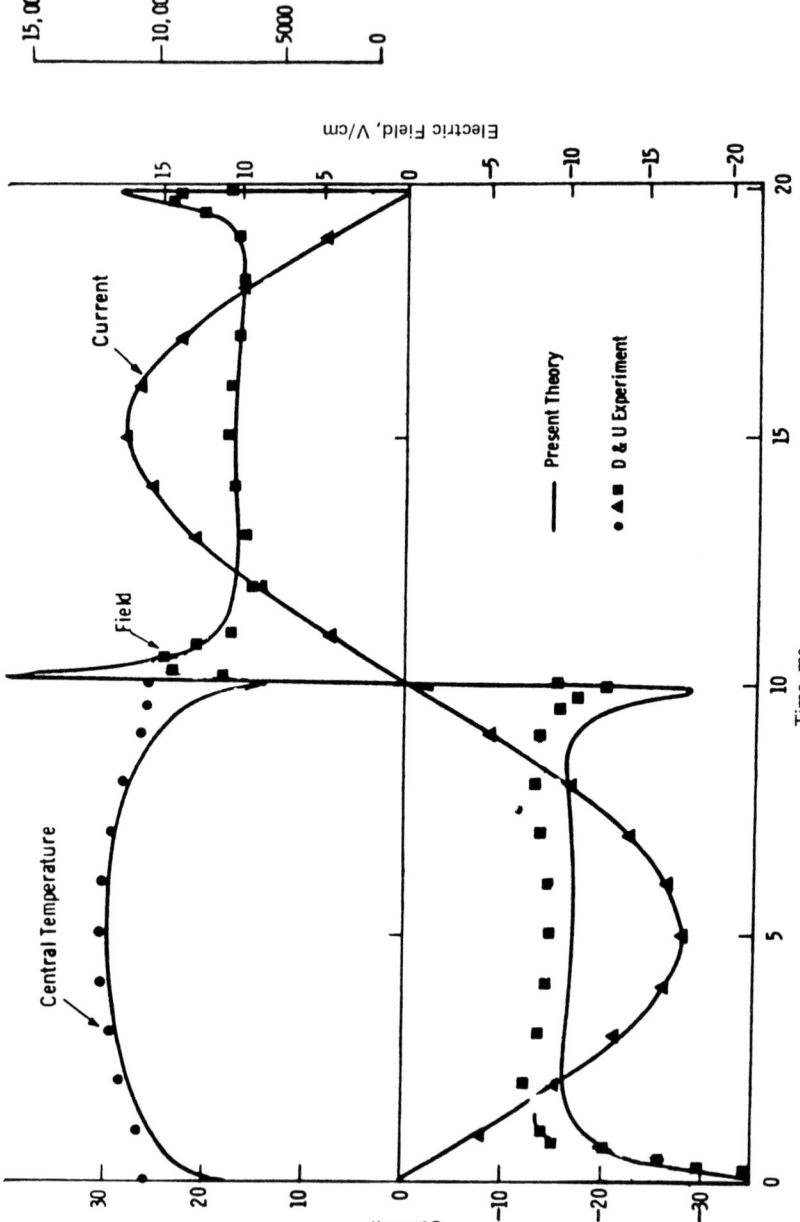

Figure 5.21 Experimental and theoretical V-I oscillograms for a wall-stabilized ac arc of 30 A in argon. (From Ref. 36.)

times of the molecules are large compared with the microsecond time scale in which arc interruption is determined.

This effect operates to cool a plasma to the temperature at which molecules dissociate, which roughly corresponds to the temperature at which there are peaks in the thermal conductivity. From Fig. 5.2, this temperature is ~2000 K for SF_6 and 7000 K for air. This effect causes large values of ρh, which at 2000 K are much larger in SF_6 than for air and so may give SF_6 superior interrupting ability. Thus the effect causes SF_6 plasma to be cooled to 2000 K, at which temperature the electrical conductivity σ is negligible (see Fig. 5.3). For air, however, the value of σ at 7000 K is still significant.

5.4 SUMMARY

It is generally agreed that properties of arcs at high current can be represented by the energy balance equation, which assumes that the arc plasma is in local thermodynamic equilibrium. However, for low currents completely different physical models of arc interruption have been proposed by different authors. In particular, the role and representation of turbulence effects is a matter of keen discussion. It is necessary for detailed tests to be made of the sensitivity of models to various parameters before any unanimity can be achieved on the relative importance of the many possible physical processes. Nonequilibrium effects may also be important.

REFERENCES

1. V. Zajic, *High Voltage Circuit Breakers*, Artia. Prague, 1957.
2. J. Slepian, Extinction of an a.c. arc, Trans. AIEE *47*:1398, 1928.
3. D. C. Prince and W. F. Skeats, The oil blast circuit breaker, Trans. AIEE *50*:506-512, 1931.
4. A. M. Cassie, Introduction to the theory of circuit interruption, in *Circuit Breaking* (H. Trencham, ed.), Butterworth, London, 1953, p. 46; see also CIGRE 10, Rep. 102, Paris, 1939.
5. O. Mayr, Contributions to the theory of the static and the dynamic arc, Arch. Elektrotech. 37:588, 1943.
6. J. Slepian, Displacement and diffusion in fluid flow arc extinction, Trans. AIEE *60*:162-167, 1941.
7. K. Ragaller, Physics of arcs in circuit breakers, Proc. 13th Int. Conf. Phenom. Ionized Gases, Berlin, 1977, Invited Papers.
8. W. Hermann and K. Ragaller, Theoretical description of the current interruption in gas blast breakers, Trans. IEEE Power Appar. Syst. *PAS-96*:1546-1555, 1977.

9. B. W. Swanson, Theoretical models for the arc in the current zero regime, in *Current Interruption in High Voltage Networks* (K. Ragaller, ed.), Plenum Press, New York, 1978, p. 137.
10. J. J. Lowke, R. E. Voshall, and H. C. Ludwig, Decay of electrical conductance and temperature of arc plasmas, J. Appl. Phys. 44:3513-3523, 1973.
11. H. W. Liepmann and A. Roshko, *Elements of Gas Dynamics*, Wiley, New York, 1957, p. 126.
12. J. J. Lowke, Two-dimensional properties of arcs in high speed flow, Proc. 7th Int. Conf. Gas Discharges Appl., London, 1982, pp. 20-23.
13. J. J. Lowke and H. C. Ludwig, A simple model for high current arcs stabilized by forced convection, J. Appl. Phys. 46: 3352-3360, 1975.
14. W. Hermann, U. Kogelschatz, L. Niemyer, K. Ragaller, and E. Schade, Study of a high current arc in a supersonic nozzle flow, J. Phys. D.: Appl. Phys. 7:1703-1722, 1974.
15. D. T. Tuma and J. J. Lowke, Prediction of properties of arcs stabilized by forced convection, J. Appl. Phys. 46:3361-3367, 1975.
16. P. Kovitya, J. J. Lowke, and A. D. Stokes, Theory of arc clogging in nozzles, Inst. Eng. (Aust.), Elect. Eng. Trans. EE16:172-175, 1980.
17. P. Kovitya and J. J. Lowke, Theoretical predictions of ablation stabilized arcs confined in cylindrical tubes, J. Phys. D: Appl. Phys. (in press) 1984.
18. M. F. Hoyaux, *Arc Physics*, Springer-Verlag, New York, 1968, p. 24.
19. S. Dzierzbicki and Z. Tarocinski, Influence on cathode phenomena of the strength of the cathode sheath, Int. Conf. Phenom. Ionized Gases, Bucharest, 1969, p. 297.
20. B. Eliasson and E. Schade, Electrical breakdown in SF_6 at high temperatures, Int. Conf. Phenom. Ionized Gases, Berlin, 1977, p. 409.
21. J. Dutton, F. M. Harris, and F. Llewellyn Jones, The determination of attachment and ionization coefficients in air, Proc. Phys. Soc. 81:52-64, 1963.
22. M. S. Bhalla and J. D. Craggs, Measurement of ionization and attachment coefficients in sulphur hexafluoride in uniform fields, Proc. Phys. Soc. 80:151-160, 1962.
23. D. C. Prince, J. A. Henley, and W. K. Rankin, The cross-airblast circuit breaker, Trans. AIEE 59:510-517, 1940.
24. G. Frind, R. E. Kinsinger, R. D. Miller, H. T. Nagamatsu, and H. O. Noeske, Fundamental investigation of arc interruption in gas flows, Final EPRI Rep. EL-284, Electric Power Research Institute, Palo Alto, Calif., 1977.

25. D. T. Tuma and E. Fong, Current zero deformation by interaction of the air blast arc with the test circuit, IEEE Trans. Power Appar. Syst. *PAS-99*:976-981, 1980.
26. F. R. El-Akkari and D. T. Tuma, Simulations of transient and zero current behavior of arcs stabilized by forced convection, IEEE Trans. Power Appar. Syst. *PAS-96*:1784-1788 (1977).
27. S. K. Chan, M. T. C. Fang, and M. D. Cowley, The dc arc in a supersonic nozzle flow, IEEE Trans. Plasma Sci. *PS-6*:394-405, 1978.
28. M. T. C. Fang and D. Brannen, Performance prediction of gas blast circuit breakers, Proc. IEEE Conf. Gas Discharges, Liverpool, England, September 1978, p. 9.
29. M. D. Cowley, Integral method of arc analysis, J. Phys. D: Appl. Phys. *7*:2218-2245, 1974.
30. J. Slepian, The electric arc in circuit interrupters, J. Franklin Inst. *214*:413-442, 1932.
31. G. Frind, L. E. Prescott, and J. H. Van Noy, Measurements of postzero current and of power loss in air blast arcs, J. Phys. D: Appl. Phys. *12*:133-137, 1979.
32. B. W. Swanson and R. M. Roidt, Numerical solutions for an SF_6 arc, Proc. IEEE *59*:493-501, 1971.
33. D. T. Tuma, A comparison of the behavior of SF_6 and N_2 blast arcs around current zero, Trans. IEEE Power Appar. Syst. *PAS-99*:2129-2137, 1980.
34. W. Rieder and J. Urbanek, A theory of thermal non-equilibrium arc conditions, CIGRE, Rep. 107, Paris, 1966.
35. L. Detloff and J. Uhlenbusch, Measurement and calculation of temperature profiles and characteristics of cascade arcs, Z. Angew. Phys. *28*:205-210, 1970.
36. J. J. Lowke, R. J. Zollweg, and R. W. Liebermann, Theoretical description of ac arcs in mercury and argon, J. Appl. Phys. *46*:650-660, 1975.
37. K. P. Brand and J. Kopainsky, Particle densities in a decaying SF_6 plasma, Appl. Phys. *16*:425-432, 1978.
38. W. Hertz, Measurement and explanation of conductivity decay of cylindrical arcs, Z. Phys. *245*:105-125, 1971.

6
Calculation of Arc-Circuit Interaction

LESLIE S. FROST*/Westinghouse Research and Development Center, Pittsburgh, Pennsylvania

THOMAS E. BROWNE, Jr.†/Consultant to Westinghouse Research and Development Center, Pittsburgh, Pennsylvania

Part 1. COMPUTER SOLUTION OF ARC MODELS WITH CONNECTED CIRCUIT

6.1 Mathematical Models   188
    6.1.1 Mayr equation   188
    6.1.2 Cassie equation   189
    6.1.3 Hochrainer model   190
    6.1.4 Other models   191
    6.1.5 Cassie-Mayr-Cassie model   191

6.2 Combined Model Solution   192
    6.2.1 Resistance shunted arc   192
    6.2.2 Capacitance shunting   199
    6.2.3 Combined resistance and capacitance shunting   206

Part 2. APPROXIMATE ANALYTICAL MODEL SOLUTIONS

6.3 Impressed Current and Voltage   211

6.4 Interaction of the Arc with the Circuit   218

6.5 Relation of Model Parameters to Current Rate of Change   223
    6.5.1 Application to short-line fault   224

6.6 Use of the Cassie Criterion   227
    6.6.1 Uses of the interrupting limit equation   229
    6.6.2 Determination of arc model parameters   233

6.7 Modeling of the Puffer Breaker   235

6.8 Nomenclature   236

References   239

---

*Currently Consulting Engineer, Pittsburgh, Pennsylvania
†Retired

Part 1 Computer Solution of Arc Models with Connected Circuit

6.1 MATHEMATICAL MODELS

Since the electric arc does not offer a constant resistance to a changing current, to enable us to calculate the transient arc current and voltage relations near current zero we must employ mathematical models of the arc in the form of differential equations. As discussed in Chap. 5, much progress has been made in recent years in physical analysis of the circuit breaker arc, resulting in rather elaborate sets of partial differential equations which can be dealt with only by modern high-speed computers. However, such equations still retain necessary approximations and in use must start with some experimental data from actual breaker tests. Of more practical use for engineering calculations which treat the arc as a purely electrical circuit element are simpler models devised from greatly simplified physical considerations or as logical oversimplifications of the more elaborate physical models.

A basic concept of the simplified dynamic arc models is that of a stored energy Q in the arc associated with its electrical conductance G; that is

$$G = F(Q) = F\left[\int (W - N) \, dt\right] \tag{6.1}$$

where W is the power input and N is the power loss. This stored energy and its necessarily finite rate of change accounts for the thermal inertia of the arc—that which limits the rate of change of G. In differential form equation (6.1) becomes

$$\frac{dG}{dt} = (W - N) \frac{dF(Q)}{dQ} \tag{6.2}$$

6.1.1 Mayr Equation

An early and much used model by O. Mayr [1] is based on simplified physics of an arc column of fixed cross-sectional area losing energy by radial thermal conduction only and having its temperature in the range for which conductivity of the gas varies nearly exponentially with temperature as given by the Saha relation for thermal ionization.*
This analysis led Mayr to the approximate relations

$$F(Q) = K \exp\left(\frac{Q}{Q_0}\right) \tag{6.3}$$

and

$$N = N_0 \tag{6.4}$$

---

*See Appendix.

Calculation of Arc-Circuit Interaction

where N_0 is a constant power loss and Q_0 is a constant quantity of stored energy. Also defining a characteristic time quantity,

$$\theta = \frac{Q_0}{N_0} \tag{6.5}$$

leads to Mayr's differential equation,

$$\frac{dG}{dt} = \frac{G}{\theta}\left(\frac{W}{N_0} - 1\right) \tag{6.6}$$

in which both θ and N_0 are constant parameters. The meaning of θ is clarified by integration of equation (6.6) with $W = 0$, leading to

$$G = G_0 \exp\left(\frac{-t}{\theta}\right) \tag{6.7}$$

showing that without power input the arc conductance decays exponentially from its initial value G_0 with the time constant θ.

6.1.2 Cassie Equation

A. M. Cassie [2] based his slightly earlier model on physics of an arc column in a gas blast. His idealized arc had a fixed temperature and was cooled by forced convection. In this case, for relatively large currents, conductivity, stored energy density, and power loss per unit volume may all be taken as constant to a first approximation so that G, Q, and N in equation (6.1) or (6.2) are all proportional to a cross-sectional area A and so to each other. In the steady state, with $N = W = Ge^2$, we have the relation for the arc voltage:

$$e = \sqrt{\frac{N}{G}} = E_0 \tag{6.8}$$

a constant value independent of the cross-sectional area and hence of the current. With nonsteady conditions, $N \neq Ge^2$, G varies with the changing A and the arc voltage e departs from E_0. Again, defining a time constant as

$$\theta = \frac{Q}{N} \tag{6.9}$$

independent of A, leads to the equation (6.2) relations

$$N = \frac{Q}{\theta}$$

$$F(Q) = \frac{Q}{E_0^2 \theta} \tag{6.10}$$

$$\frac{dF(Q)}{dQ} = \frac{1}{E_0^2 \theta}$$

Thus equation (6.2) becomes

$$\frac{dG}{dt} = \frac{G}{\theta}\left(\frac{e^2}{E_0^2} - 1\right) \tag{6.11}$$

In terms of the arc current $i = Ge$, this equation can be conveniently written [3,4]

$$\frac{d}{dt}(G^2) + \frac{2G^2}{\theta} = \frac{2}{\theta}\left(\frac{i}{E_0}\right)^2 \tag{6.12}$$

With applied voltage or current equal to zero, these equations also integrate to equation (6.7), showing that θ has the same meaning in both models. More recently it has been shown [5] that the physical considerations underlying the Mayr model lead also to the Cassie equation if arc temperatures and corresponding levels of ionization are so high that the gas conductivity increases nearly linearly with temperature rather than exponentially. It is known that this high-temperature range occurs in high-current arc columns subjected to intense gas blasts and may exist also in such arcs near current zero.

6.1.3 Hochrainer Model

A somewhat different and purely empirical approach has been taken by A. Hochrainer and his students [6]. They start with the "generalized" equation

$$\frac{dG}{dt} = \frac{G^* - G}{\theta} \tag{6.13}$$

where G^* is a steady-state value of arc conductance equal to i/e for the same current on an assumed static volt-ampere characteristic of the arc. This equation leads to the Mayr equation if θ and the arc power loss N are both constant, and with other assumptions also to the Cassie equation. However, in the use of this approach it has been usual practice to employ the Mayr equation, or some modification of it, for the whole arcing period but to take N and θ not as constants but as empirical functions of either time or the instantaneous arc conductance G. The latter course has had reasonable success in matching near-zero arc current and voltage behavior to CRO observations but it suffers from two problems:

1. Determination of model parameters for a given breaker arc depends on experimentally difficult measurements, especially near current zero,

Calculation of Arc-Circuit Interaction

and also on the assumption that the arc actually follows the simplified differential equation in detail.

2. The nonlinear equation does not lend itself to even approximate analytical solutions but requires step-by-step computer solutions, using the empirical relations for $N(G)$ and $\theta(G)$, which, as noted in problem 1, are difficult to determine with precision near current zero.

6.1.4 Other Models

Other variations in approximate arc modeling have included simultaneous combinations of Cassie-like and Mayr-like terms. Examples are that by Cassie and Mason [7] and a somewhat similar but more complex model by Urbanek [8] which includes the dielectric breakdown regime. Still other models, mentioned in Chap. 5 and Ref. 9, have been based more closely on the physics of the arc in a gas blast but require the extensive use of modern high-speed computers. It is our opinion that at present these models, which are necessarily still approximate, are rather difficult for general engineering use.

6.1.5 Cassie-Mayr-Cassie Model

Based on early observations of gas-blast breaker arc characteristics [10,11] and more recent physical analyses [5], a successively combined model has been used [12]. In this the Cassie equation for the high-current arc is assumed to apply prior to current zero and also following a "thermal" reignition shortly after a current zero. The Mayr equation is employed in the intermediate time region between the two Cassie regimes, or simply following the Cassie regime for a successful interruption (neglecting the possible case of later dielectric breakdown). In this combination, the constant power loss N_0 in the Mayr equation is determined by a solution of the preceding Cassie equation. Similarly, a subsequent Cassie equation has its new steady-state arc voltage E_0' determined by the Mayr equation solution. A common time-constant value θ is assumed throughout the calculation.

Justification

Although physical analyses and careful measurements both indicate that the parameters E_0, N_0, and θ in the model equations are never strictly constant for an actual arc, these analyses and observations [11,13] do indicate that during the brief critical time around current zero these parameters vary so much more slowly than the arc current or voltage that we are justified in assuming them to be momentarily constant. Since in actual use they are evaluated by comparison with the behavior of only the *integrated* forms of the model equations, they are actually effective *average* values during the critical time of interaction.

6.2 COMBINED MODEL SOLUTION

Because the Cassie and Mayr differential equations are nonlinear, step-by-step computer solutions are required unless the equations are linearized (see Sec. 6.3).

6.2.1 Resistance Shunted Arc

The most important application of these calculations is to the case of the short-line fault discussed in Chap. 3. An equivalent circuit diagram for such a fault from a single line to ground is shown in Fig. 6.1. When dealing with the initial rapid rise of the postzero arc voltage $e(t)$, the bus-side voltage V_0 may be assumed to be constant, the C_{bus} value may be assumed infinite (negligible transient impedance to ground), and the line section may be replaced by the purely resistive surge impedance Z_0 to ground. V_g is the rms single-line-to-ground bus voltage and L_G is the effective system inductance on the bus side. It can be shown [12] that the pre-current-zero momentarily constant bus-side voltage is

$$V_0 = \sqrt{2}\, V_g(1 - f) - f|E_0| \tag{6.14}$$

where $f = I/I_0$ is the fault fraction ($f < 1$ because of the current limiting effect of the line) and E_0 is the prezero nearly constant arc voltage. Thus the simplified circuit for this calculation becomes that of Fig. 6.2, where L is now the inductance ℓL_1 of the line section. For lumped constant test circuits, V_0 becomes simply $\sqrt{2}\, V_g$ and the shunting element is R_s, representing Z_0.

For convenience, the arc conductance G is normalized in terms of a current- and time-zero quantity G_0 resulting from solution of the Cassie equation (6.12) for an impressed current ramp as described in Sec. 6.3. By equation (6.56) of Part 2, this is

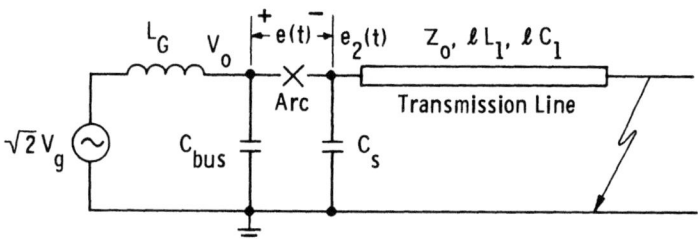

Figure 6.1 Schematic circuit of a capacitance shunted circuit breaker subjected to short-line fault. (From L. S. Frost, Dynamic arc analysis of short-line fault tests for accurate circuit breaker performance specification, IEEE Trans. Power Appar. Syst. PAS-97:478-484, ©1978, IEEE.)

Calculation of Arc-Circuit Interaction

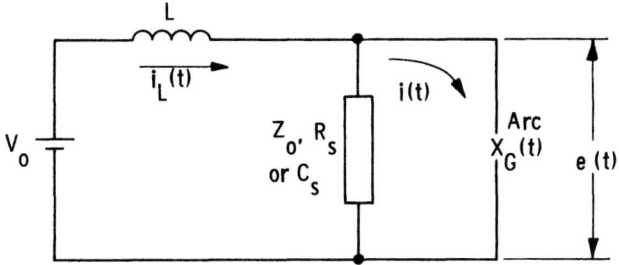

Figure 6.2 Equivalent diagram of a shunted arc, short-line-fault circuit. (From L. S. Frost, Dynamic arc analysis of short-line fault tests for accurate circuit breaker performance specification, IEEE Trans. Power Appar. Syst. *PAS-97*:478-484, © 1978, IEEE.)

$$G_0 = \frac{\omega \theta I}{E_0} \tag{6.15}$$

and correspondingly, the current-zero ramp-produced arc power loss is

$$N_0 = \omega \theta I E_0 \tag{6.16}$$

both expressed in terms of the steady-state Cassie model arc voltage E_0 (readily estimated from test oscillograms), the rate of decrease of the short-circuit current $\dot{I} = \sqrt{2}\omega I$ and the arc time constant θ. With shunting, $\dot{I}$ is not constant, causing the actual current-zero conductance $G(0)$ to be somewhat less than G_0. Here I is the rms value of the circuit fault current of angular frequency $\omega = 2\pi f$ radians per second (f is the power circuit frequency). During the Mayr equation period just after $t = 0$, the arc power loss is assumed to have the constant value N_0. In the arc reignition case, return to the Cassie equation occurs if and when $dG/dt = 0$. At this time, G has fallen to the value G_{min}, less than $G(0)$. Note that combining equations (6.15) and (6.16) shows that $E_0 = \sqrt{N_0/G_0}$. Similarly, with constant N_0 and the reduced conductance value G_{min}, we have a new higher value of steady-state Cassie arc voltage:

$$E'_0 = \sqrt{\frac{N_0}{G_{min}}} \tag{6.17}$$

Using equation (6.16), this becomes

$$E'_0 = \sqrt{\frac{\omega \theta I E_0}{G_{min}}} \tag{6.18}$$

Thus the three successive differential equations for the reigniting arc case may be written

$$\frac{dG}{dt} = \frac{G}{\theta} \begin{cases} \left(\dfrac{e^2}{E_0^2} - 1\right) & \text{(Cassie until } t = 0\text{)} \\ \left(\dfrac{ei}{\omega \theta I E_0} - 1\right) & \text{(Mayr until } \dfrac{dG}{dt} = 0 \text{ at } t_{min}\text{)} \\ \left(\dfrac{e^2 G_{min}}{\omega \theta I E_0} - 1\right) & \text{(Modified Cassie after } t_{min}\text{)} \end{cases}$$

(6.19a)

(6.19b)

(6.19c)

The associated circuit equations are

$$i_L(t) = \dot{I}t \tag{6.20}$$

$$e(t) = \frac{i_L(t)}{G(t) + 1/R_s} \tag{6.21}$$

$$i(t) = G(t)e(t) \tag{6.22}$$

In short-line-fault cases in which E_0 is an appreciable fraction of $\sqrt{2}V_g$, the prezero $\dot{I}$ in (6.20) becomes

$$\left|\frac{di}{dt}\right|_L = \frac{\sqrt{2}V_g + |E_0|}{L_G + L_\ell} \tag{6.23}$$

where $L\ell$ is the line section inductance in series with L_G.

Figure 6.3 shows the postzero arc conductance behavior resulting from computer calculations, including interaction of the arc model with a series of values of the shunting conductance, $G_s = 1/R_s$. Here the quantities are shown normalized according to the relations

$$\tau = \frac{t}{\theta}$$

$$g(\tau) = \frac{G}{G_0} \tag{6.24}$$

$$\beta = \frac{G_s}{G_0} = \frac{R_0}{R_s}$$

By trial a critical value β_c of the shunting ratio β was found to be 1.427, or close to $\sqrt{2}$. The corresponding calculated arc currents and voltages are shown in Fig. 6.4 in which the normalizing relations

Calculation of Arc-Circuit Interaction

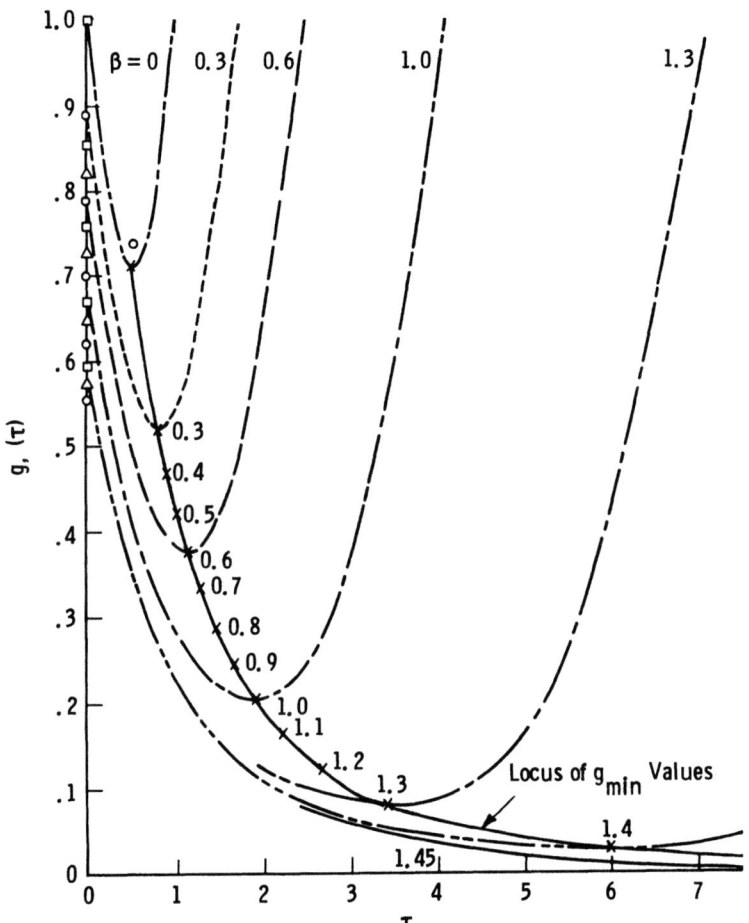

Figure 6.3 Postzero arc conductance vs. time, normalized. (From L. S. Frost, Dynamic arc analysis of short-line fault tests for accurate circuit breaker performance specification, IEEE Trans. Power Appar. Syst. *PAS-97*: 478-484, © 1978, IEEE.)

$$j_a(\tau) = \frac{i}{\omega \theta I} \tag{6.25a}$$

$$u(\tau) = \frac{e}{E_o} \tag{6.25b}$$

are used.

Figure 6.5 shows the good fit obtainable between the model-calculated arc current and voltage behavior near zero and CRO observa-

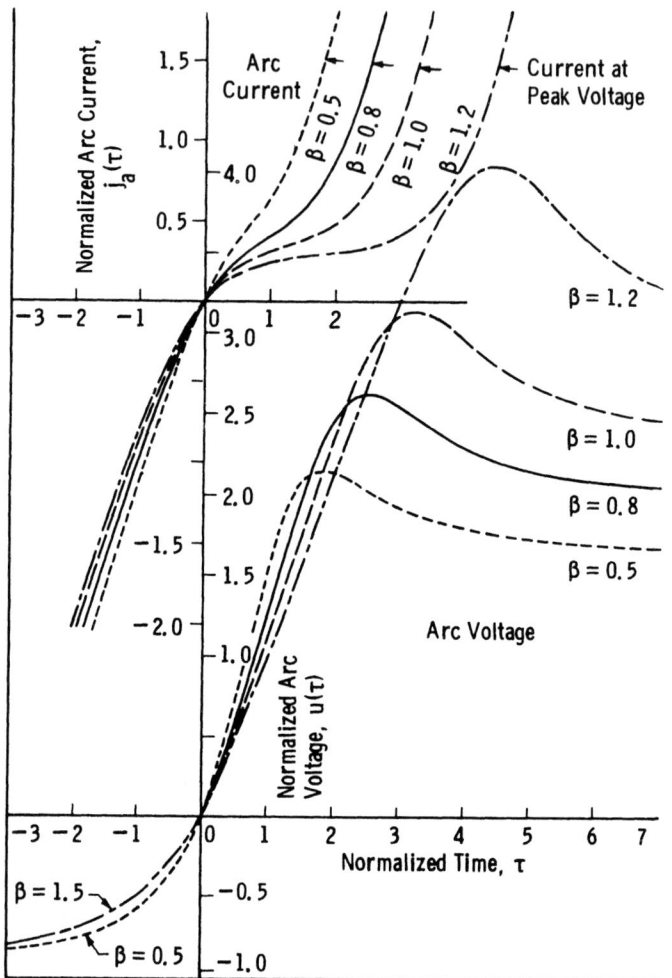

Figure 6.4 Normalized arc voltage and current vs. time for various values of $\beta = G_s/G_0$, model solution. (From L. S. Frost, Dynamic arc analysis of short-line fault tests for accurate circuit breaker performance specification, IEEE Trans. Power Appar. Syst. *PAS-97*:478-484, © 1978 IEEE.)

tions of an actual SF_6 interrupter arc shunted by a conductance slightly lower than the critical value $\beta_c G_0$. The model parameters E_0 and θ were adjusted to give the best fit between the calculated solid curves and the circled points from the test oscillogram.

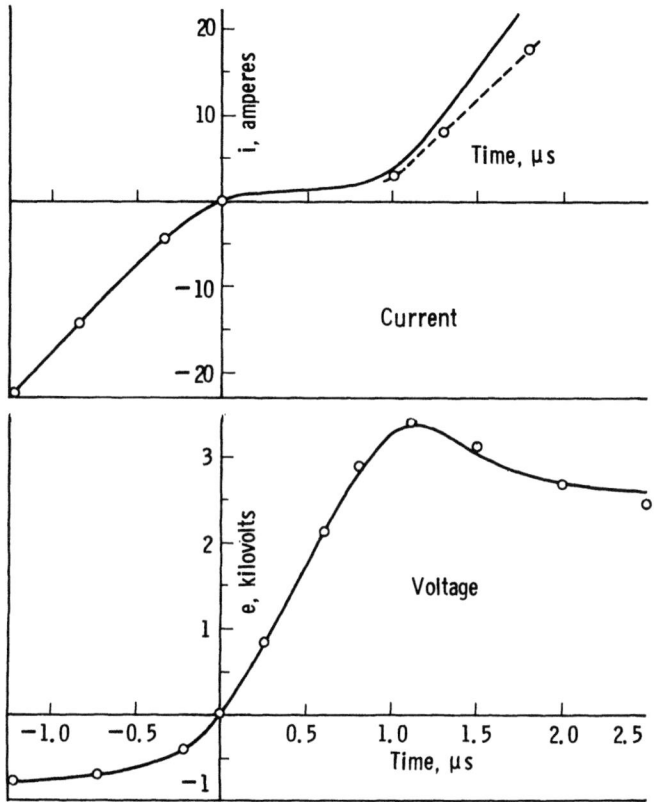

Figure 6.5 Arc voltage and current for model SF$_6$ breaker test. Solid curves: theory for $E_0 = 0.91$ kV, $\theta = 0.282$ μs, $R_S = 0.193$ kΩ, $\dot{I} = 20.9$ A/μs, $\beta = 1.15$. Circles and dashed curve: observed data as read from test oscillogram. (From L. S. Frost, Dynamic arc analysis of short-line fault tests for accurate circuit breaker performance specifications, IEEE Trans. Power. Appar. Syst. PAS-97:478-484, © 1978, IEEE.)

Critical Rate of Voltage Rise

With purely resistive shunting the nearly linear rate of rise of recovery voltage $\dot{V}_R$ after the postarc current has become negligible is to a good approximation simply the inductance-limited rate of change of current $\dot{i}_L$ times R_s, or

$$\dot{V}_R = R_s \dot{i}_L = \sqrt{2}\omega R_s I \tag{6.26}$$

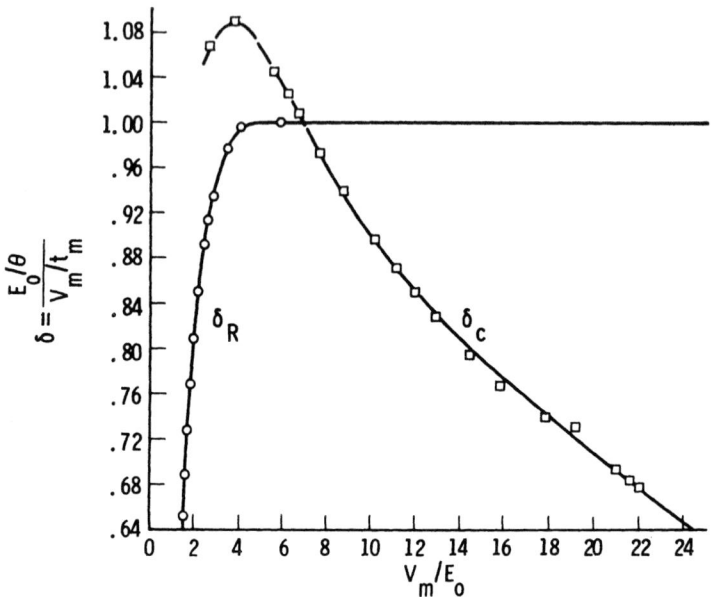

Figure 6.6 Performance multiplier for resistance and capacitance shunts. (From L. S. Frost, Dynamic arc analysis of short-line fault tests for accurate circuit breaker performance specification, IEEE Trans. Power Appar. Syst. PAS-97:478-484, © 1978 IEEE.)

From equation (6.24) this can be written

$$\dot{V}_R = \frac{\sqrt{2}\,\omega R_0 I}{\beta} \qquad (6.27)$$

The *critical* value of the resistance-limited rate of voltage rise may be obtained by substituting for R_0 the value from equation (6.15) and taking the critical value of $\beta = \beta_c$ as exactly $\sqrt{2}$ to give the surprisingly simple result [12]

$$\dot{V}_{Rc} = \frac{E_0}{\theta} \qquad (6.28)$$

This will be used later in Part 2.

The approximations leading to equation (6.28) break down for peak recovery voltages V_m not large compared with E_0. The curve marked δ_R in Fig. 6.6 shows the departure from equation (6.28) at smaller values of V_m/E_0. V_m/t_m is the average slope of the computed recovery voltage. The departure is less than 5% for $V_m/E_0 > 3$.

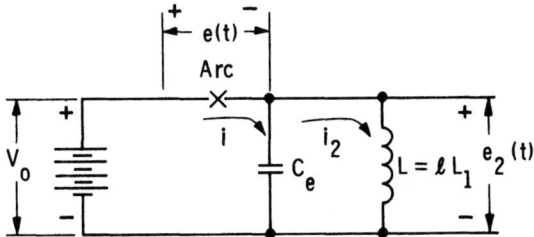

Figure 6.7 Stationary bus equivalent circuit for C_s-shunted short-line fault. (From L. S. Frost, Dynamic arc analysis of short-line fault tests for accurate circuit breaker performance specifications, IEEE Trans. Power Appar. Syst. PAS-97:478-484, © 1978, IEEE.)

6.2.2 Capacitance Shunting

Although it can be shown that Fig. 6.2 with C_S as the shunting element can represent this case, Fig. 6.7 for the short-line-fault case will be referred to here since modification of line-fault transients is the principal use of capacitance shunting. Analysis [12] of the effect of the shunting capacitance C_S at the line-side breaker terminal shows that when C_S is half or more of the distributed line capacitance ℓC_1, the recovery voltage across the breaker is very nearly of $(1 - \cos)$ form with an angular frequency

$$\upsilon = \frac{1}{\sqrt{\ell L_1 C_e}} \qquad (6.29)$$

where

$$C_e = C_s + k\ell C_1 \qquad (6.30)$$

The line section of length ℓ has uniformly distributed inductance values of L_1 and capacitance values of C_1 per unit length, yielding a surge impedance of $Z_0 = \sqrt{L_1/C_1}$. Computer solutions of recovery voltages for an idealized line shunted by capacitance show that for best fits k varies between 0.3 and 0.4 over a reasonable range of $C_s/\ell C_1$. Thus k can be taken as roughtly 1/3.

The circuit equations to be solved simultaneously with those of the arc model, equation (6.19) are

$$e(t) = V_0 - e_2(t) \qquad (6.31)$$

$$i(t) = G(t)e(t) \qquad (6.32)$$

$$\frac{de_2}{dt} = \frac{i - i_2}{C_e} \qquad (6.33)$$

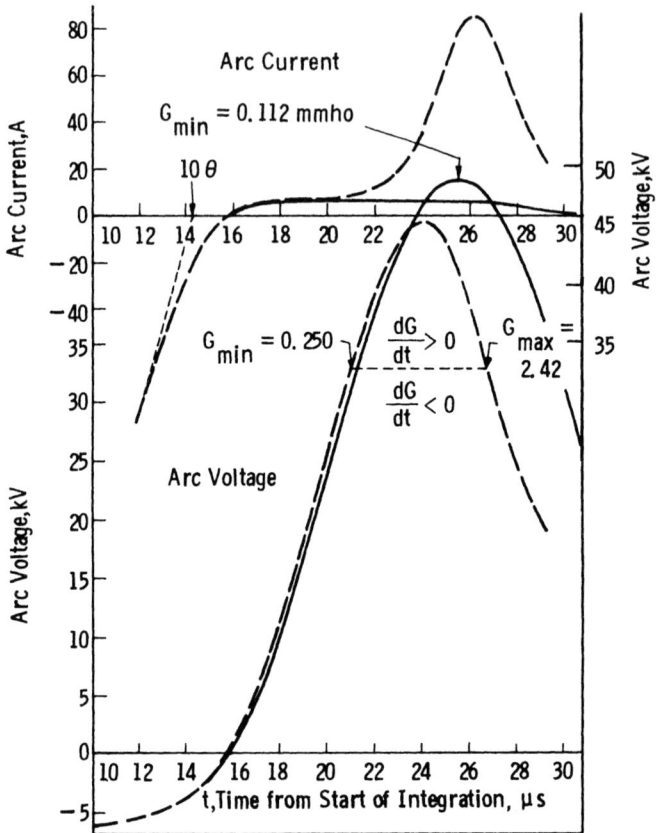

Figure 6.8 Calculated arc voltage and current vs. time for $f = 0.85$, $I_0 = 80$ kA, $\theta = 1.4305$ µs, $E_0 = 7.319$ kV, $V_0 = 23.48$ kV. Dashed curves: $C_S = 12$ nF, $(\theta/G)(dG/dt)|_{max} = 0.913$; solid curves: $C_S = 13.443$ nF, $(\theta/G_0)(dG/dt)|_{max} = -0.0017$. (Courtesy of Westinghouse Electric Corporation, Pittsburgh, Pa.)

$$\frac{di_2}{dt} = \frac{e_2}{L} \tag{6.34}$$

Currents and voltages around zero with near-critical values of C_S are shown in Fig. 6.8 for a particular example. Here, the bus-side voltage $V_0 = 23.48$ kV, the 60-Hz short-circuit current $I_L = 0.85 \times 80 = 68$ kA, the arc model parameters were $E_0 = 7.32$ kV (for two series breaks) and $\theta = 1.43$ µs. The line-surge impedance Z_0 was 0.4 kΩ. The dashed curves are for $C_S = 12$ nF and the solid curves for $C_S = 13.44$ nF. Points at which G_{min} occurred are shown on the voltage

Calculation of Arc-Circuit Interaction

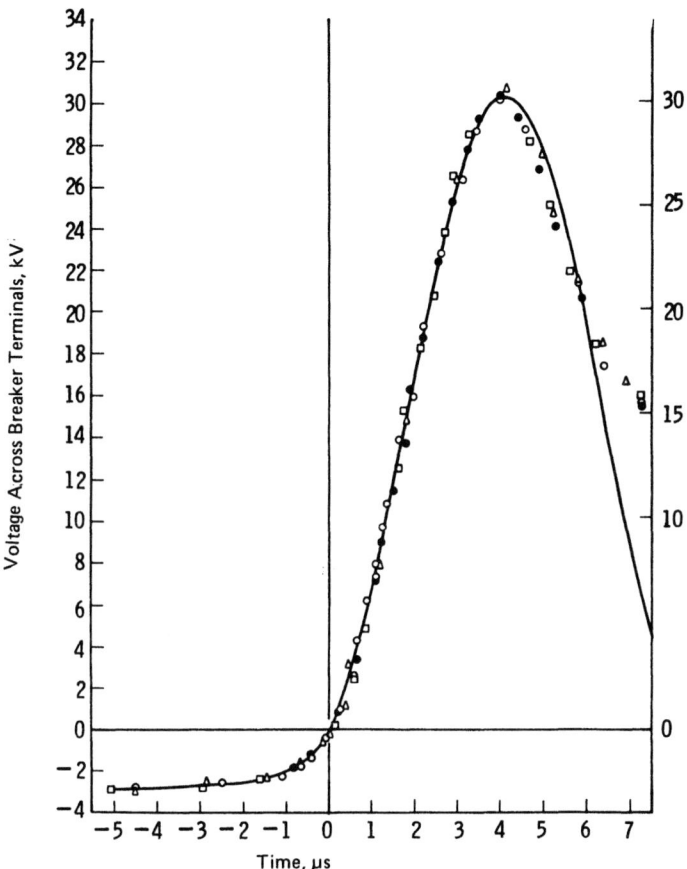

Figure 6.9 Comparison of successful tests with model solution. (Data from Mitsubishi Electric Corporation courtesy of Westinghouse Electric Corporation, Pittsburgh, Pa.)

plots, together with G_{max} for the lower C_S value. In this calculation, computation was started at $t = -10\theta$ with respect to the projected time of current zero (from the unmodified current slope). Figure 6.9 illustrates the excellent fit between such a model-calculated restored voltage curve for four successful tests of a single-pressure sulfur hexafluoride (SF_6) breaker under nominally identical conditions and the arc voltage points from test cathode-ray oscillograms [14]. Incidentally, this figure also illustrates the consistent behavior of the type of breaker dealt with in Chap. 10.

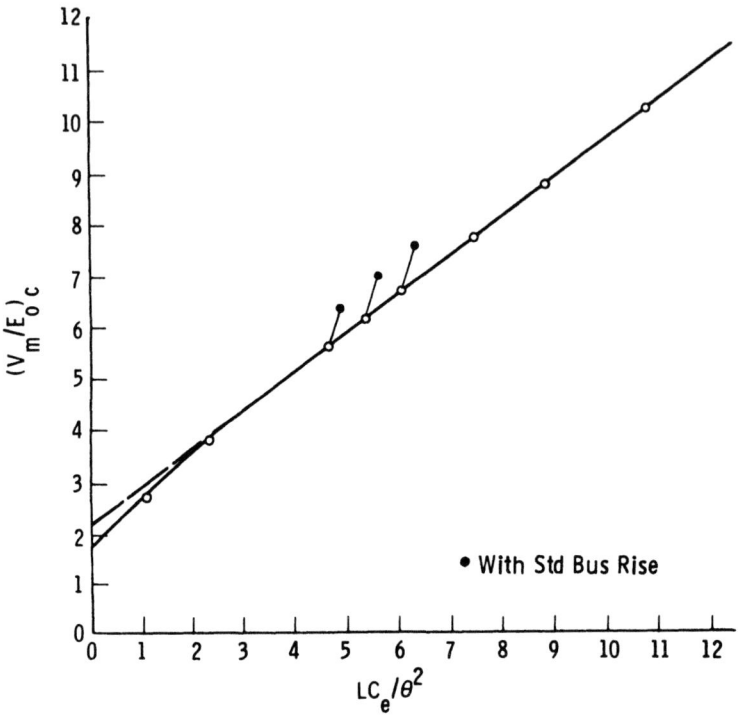

Figure 6.10 Normalized critical peak arc recovery voltage vs. interaction parameter LC_e/θ^2. $I_0 = 80$ kA, $V_g = 140$ kV. (Courtesy of Westinghouse Electric Corporation, Pittsburgh, Pa.)

Critical Capacitance Shunting

In Fig. 6.8 for $C_s = 12$ nF the arc conductance G decays after its maximum at 26.8 μs, pointing to its eventual disappearance and so to a successful interruption in spite of considerable arc reheating. However, for simplicity, and allowing for some error in the modeling, the slightly higher C_s value of 13.44 nF was taken to be the fully adequate shunting value and the corresponding relation

$$\left.\frac{dG}{dt}\right|_{V_m} = 0 \qquad (6.35)$$

was adopted as the general criterion for adequacy, or for the "critical" case.

Following the example of Mayr [1], the results of these calculations can be plotted in dimensionless form as $(V_m/E_0)_c$ vs. $(v_r)^{-1}$, or vs.

Calculation of Arc-Circuit Interaction

$(\upsilon_r)^{-2} = LC_e/\theta^2$ in Fig. 6.10. The equation for the linear part of the Fig. 6.10 plot is very nearly

$$\left(\frac{V_m}{E_0}\right)_c = 2.17 + \frac{0.75 LC_e}{\theta^2} \tag{6.36}$$

or alternatively,

$$\frac{LC_e}{\theta^2} = \frac{4}{3}\left(\frac{V_m}{E_0}\right)_c - 2.9 \tag{6.37}$$

These linear equations are also useful for finding the indicated value of θ from critical test data when solved for θ to give

$$\theta = \sqrt{\frac{LC_e}{(4/3)(V_m/E_0)_c - 2.9}} \tag{6.38}$$

In the case of a critical line-fault test, equation (6.30) may be used with equation (6.38) to obtain approximately

$$\theta = \sqrt{\frac{L(C_s + L/3Z_0^2)}{4V_m/3E_0 - 2.9}} \tag{6.39}$$

in terms of CRO measured voltages and the line inductance and surge impedance and the shunting capacitance.

Another useful result from the computer calculations, convenient for determining the required "effort" E_0/θ of the breaker from the test data, is the curve marked δ_c in Fig. 6.6, using the average rate of rise of the recovery voltage. From this curve, the arc time constant θ may be obtained in terms of E_0. In another approximate form of the same data, Fig. 6.11 shows a useful characteristic of the normalized critical-case recovery voltage curves. Here the parameter is $t_m/\theta = \pi/\upsilon\theta$, the ratio of time to peak (approximated by the half-cycle duration) to the arc time constant. In the range of t_m/θ values from 3 to 8, the peak voltages may be seen to fall very close to the straight line for the simple relation

$$\frac{V_m}{E_0} = \frac{t_m}{\theta} \tag{6.40}$$

which is equivalent to the approximation $\delta_c = 1$ in Fig. 6.6. Thus a rough estimate of the effort ratio E_0/θ becomes simply the observed ratio V_m/t_m as long as t_m/θ is in the range 3 to 8. From this ratio the estimated arc time constant is

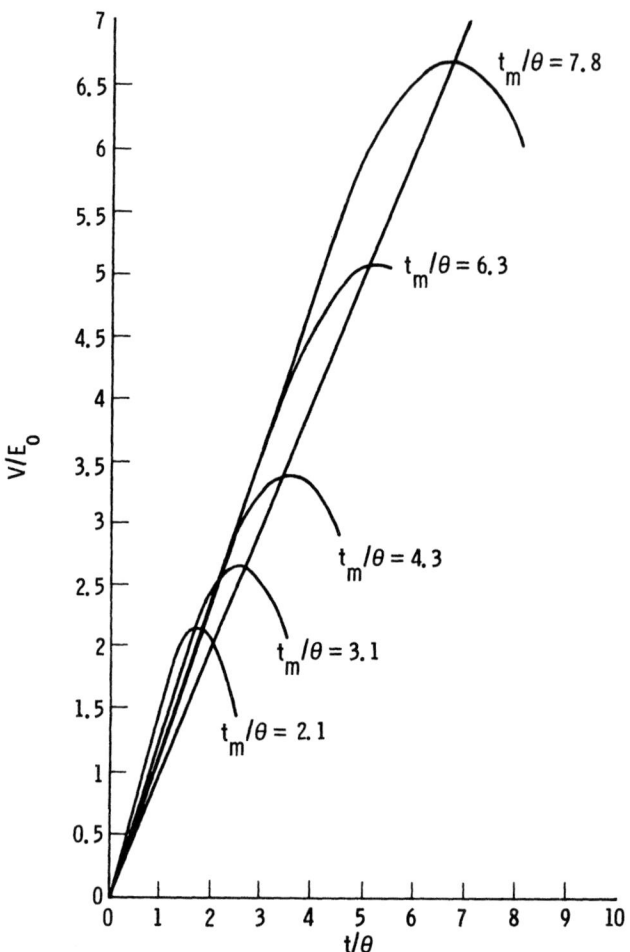

Figure 6.11 Normalized critical voltage rise curves with t_m/θ as the parameter. Points where $V_m/t_m = E_0/\theta$ lie on straight line. (From T. E. Browne, Jr., Practical modeling of the circuit breaker arc as a short line fault interrupter, IEEE Trans. Power Appar. Syst. *Pas-97*: 838-847, © 1978, IEEE.)

$$\theta \cong \frac{t_m}{V_m/E_0} \qquad (6.41)$$

Estimating the Cassie Model Arc Voltage

When relatively slow-speed oscillograms are available, E_0 may be taken as the peak arc voltage occurring just before current zero. If only

Calculation of Arc-Circuit Interaction

high-speed CRO's suitable for measuring V_m/t_m are available, a computer-obtained arc voltage equation can be expressed in terms of the earliest available arc voltage V_e and the corresponding time t_e to give

$$E_0 = V_e \sqrt{1 - \frac{1.028}{t_e} + \frac{1}{t_e^2(0.534/\theta + 0.263\upsilon)^2}} \quad (6.42)$$

based on a first estimate of θ. Since θ estimates depend on E_0, final values of E_0 and θ require iteration, but convergence is usually rapid. Note that in this use, t_e values are negative. For somewhat rougher estimation, a simpler expression, applying strictly to critical resistance shunting, is

$$E_0 = V_e \sqrt{1 - 0.76\,\frac{\theta}{t_e} + 2.24\left(\frac{\theta}{t_e}\right)^2} \quad (6.43)$$

Agreement between these equations is reasonable for t_m/θ values between 1.5 and 8, and improves for larger values of $-t_e/\theta$. If t_e/θ is near -10, both equations show that V_e/E_0 is near 0.95.

Estimation of Required Capacitance Shunting

Equation (6.37) together with (6.30) can be used to determine critical C_s values where the other terms in equation (6.37) are known. In the short-line-fault case, V_m and L can be related to line length and current and circuit voltage, and empirical relations between E_0 and θ and the current can be used. Results of such calculations are shown in Fig. 3 of Ref. 15. This subject is covered more completely in the alternative approach of Sec. 6.6.1 of Part 2.

Effect of Bus Recovery Voltage

In the calculations described so far, the rate of voltage rise on the bus or system side of the breaker has been neglected since in larger systems it is generally small compared with the short-line component of rate of rise. The three solid circles in Fig. 6.10 show calculated values of the normalized added voltage from the bus according to the American National Standards* for this example. An increase in LC_e/θ^2, or of C_e at the same L/θ^2, to compensate for this additional rise is indicated as the horizontal displacement of the points from the critical case line in the figure. A required increase in C_s of about 7% is illustrated in the example shown in Fig. 3 of Ref. 15.

---

*"American National Standard Requirements for Transient Recovery Voltage for Ac High-Voltage Circuit Breakers Rated on a Symmetrical Current Basis" (IEEE Std. 327-1971 and ANSI C37.072-1971).

6.2.3 Combined Resistance and Capacitance Shunting

In the short-line-fault case with finite lumped capacitance C_S at the breaker line terminal, the distributed L and C values of the line affect the initial rate of voltage rise "seen" by the breaker in the same manner as a purely resistive shunt with the value Z_0 (see Chap. 2). When $C_S/\ell C_1$ is relatively small, the restored voltage wave shape is between the triangular wave for pure resistance shunting and the nearly $(1 - \cos)$ wave for large C_S values. A method has been developed [12] for combining the results described above of the Cassie-Mayr-Cassie computer solutions for resistance shunts, 6.2.1, and capacitance shunts, Sec. 6.2.2. We give a brief description of the method here, and illustrate its usefulness in accurate characterization of circuit breaker performance by means of an example.* The equivalent circuit is shown in Fig. 6.12.

Consider a series of SLF tests carried out in a high-power laboratory on an SF_6 gas-blast circuit breaker. From each cathode-ray oscillogram, represented schematically in Fig. 6.13, the indicated parameters are evaluated. In general, the slope to peak voltage, V_m/t_m, will be increasingly lowered compared to the steepest slope, r_v, as shunt capacitance is increased. The relative values of these two slopes is thus a measure of the relative shunting effect of the line impedance Z_0, acting as a resistance shunt, and the lumped capacitance C_S.

The results of the C-M-C computer solutions for resistance and capacitance shunting described above can be succinctly represented in

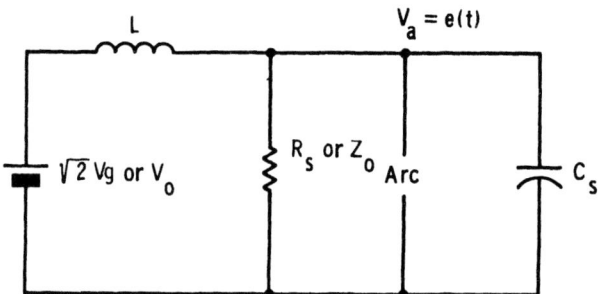

Figure 6.12 Equivalent near-current-zero circuit for arc-circuit interaction with parallel resistance and capacitance shunting. (From T. E. Browne, Jr., Practical modeling of the circuit breaker arc as a short line fault interrupter, IEEE Trans. Power Appar. Syst. *PAS-97*: 838-847, © 1978, IEEE.)

---

*This work was done at the same time as that reported in Ref. 12 but was not published.

Calculation of Arc-Circuit Interaction

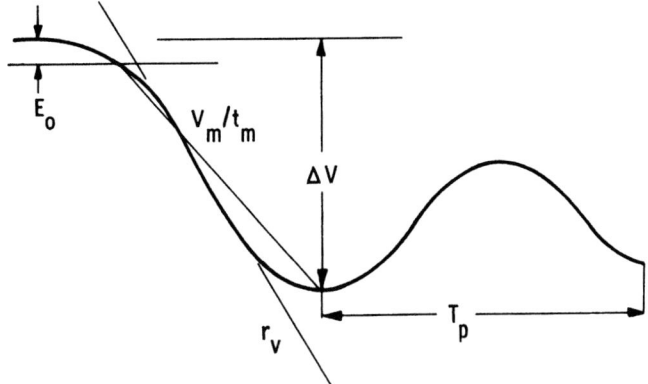

Figure 6.13 Schematic of a fast voltage oscillogram showing the four measurements needed to solve the arc-circuit interaction for each test. (From L. S. Frost, Dynamic arc analysis of short-line fault tests for accurate circuit breaker performance specification, IEEE Trans. Power Appar. Syst. *PAS-97*: 478-484, © 1978, IEEE.)

two plots, each with one curve for each type of shunt. The steepest slope of the recovery voltage rise can be expressed in the resistance shunt case as

$$r_v = \left.\frac{de(t)}{dt}\right|_{max} = \alpha_R(U) \dot{\imath} Z_0 \tag{6.44}$$

where $\alpha_R(U)$ is shown in Fig. 6.14 plotted against $U = V_m/E_0$; it is unity except for low values of U. Defining the effective impedance with shunt capacitance C_s,

$$Z' = \frac{2}{\pi}\sqrt{\frac{L}{C_e}} \tag{6.45}$$

where C_e is the equivalent lumped capacitance at the breaker [see equation (6.30)]. With this definition, an equation similar to (6.44) describes the voltage slope to peak in the capacitance shunt case,

$$\frac{V_m}{t_m} = \alpha_c(U) \dot{\imath} Z' \tag{6.46}$$

$\alpha_c(U)$ is also plotted in Fig. 6.14.

The other master parameters from the C-M-C computer results describe the ratio of the arc characteristic E_0/θ to the appropriate rate of rise of voltage:

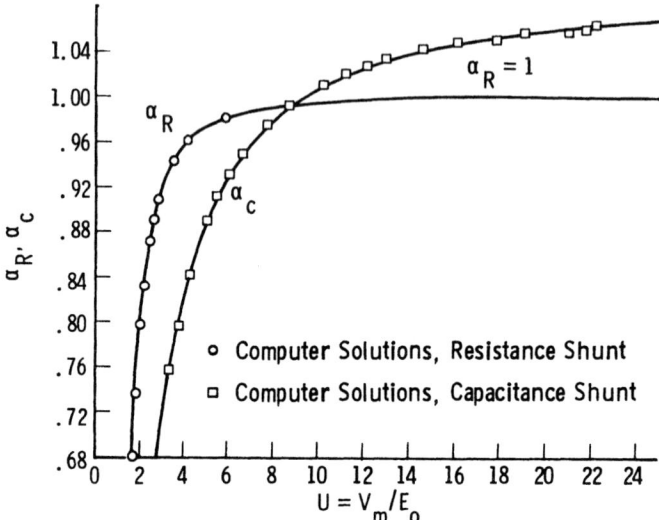

Figure 6.14 Voltage slope coefficients for resistance and capacitance shunts. (From L. S. Frost, Dynamic arc analysis of short-line fault tests for accurate circuit breaker performance specification, IEEE Trans. Power Appar. Syst. *PAS-97*:478-484, © 1978, IEEE.)

$$\delta_R(U) = \frac{E_0/\theta}{r_V} \qquad (6.47)$$

$$\delta_C(U) = \frac{E_0/\theta}{V_m/t_m} \qquad (6.48)$$

These are shown in Fig. 6.6. δ_R is unity except at very low U values, and δ_C shows more effective protection due to capacitance shunting as the voltage ratio increases.

Over the whole range it is found that the following inequality holds:

$$1 \leq \frac{r_V}{V_m/t_m} \leq 1.45 \qquad (6.49)$$

with the lower limit representing zero capacitance and the upper limit representing very large capacitance. We define the mix factor

$$f_d = \frac{[r_V/(V_m/t_m)] - 1}{0.45} \qquad (6.50)$$

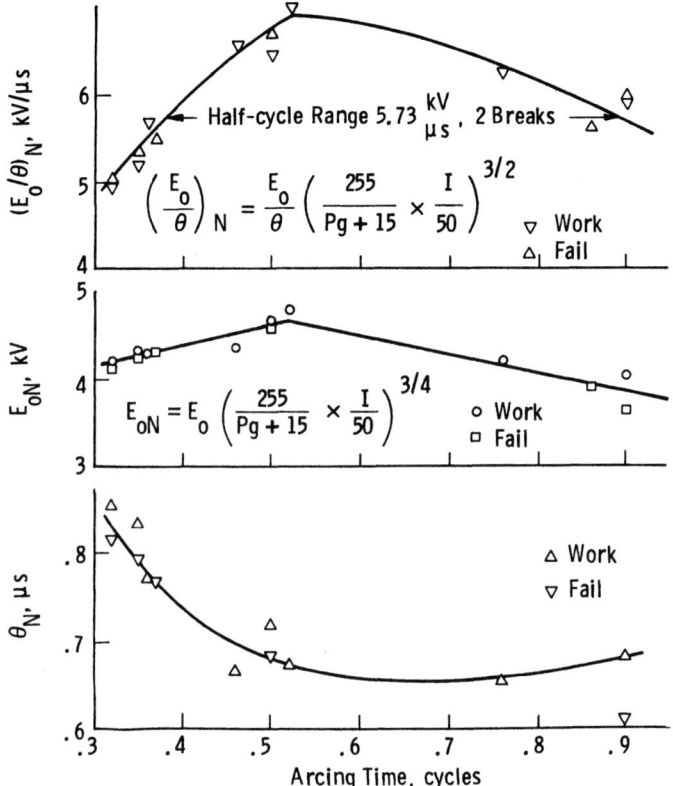

Figure 6.15 Standard curves for an SF_6 breaker of normalized E_0/θ, E_0, and θ vs. arcing time t_a. (Courtesy of Westinghouse Electric Corporation, Pittsburgh, Pa.)

which can be experimentally determined for each SLF test by measurements indicated in Fig. 6.13. It is used for interpolating between the R_S and C_S limits for α and δ.

$$\alpha(U) = 1 - f_d + f_d \alpha_c(U) \quad U \geq 8 \quad (6.51)$$

$$\delta(U) = 1 - f_d + f_d \delta_c(U) \quad U \geq 4 \quad (6.52)$$

These coefficient values used in equations (6.46) and (6.48) enable the evaluation of Z' and E_0/θ for each SLF circuit breaker test.

We illustrate the usefulness of this method in unfolding the effect of differing C_S and Z_0 values in order to obtain clear values of the inherent rate of rise of recovery voltage (RRRV) for each test, by

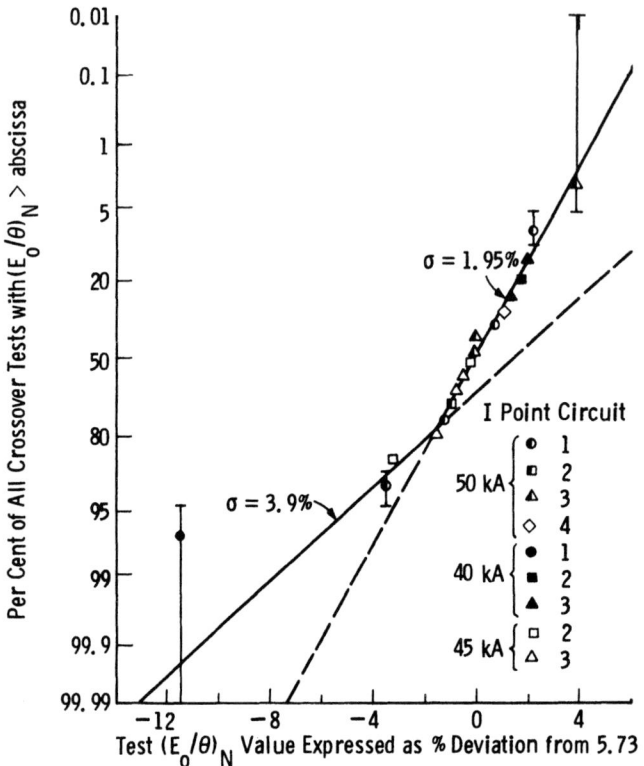

Figure 6.16 Probability plot of the 18 tests which worked with stress above E_{0s}/θ_s of Fig. 6.15, or which failed with stress below E_{0s}/θ_s. (Courtesy of Westinghouse Electric Corporation, Pittsburgh, Pa.)

applying it to a group of 80 SLF tests on a Westinghouse developmental two-pressure circuit breaker. These tests were at 40, 45, and 50 kA, with a range of C_S values.

The best information on breaker capability is of course provided by those tests closest to the interrupting limit. Figure 6.15 shows the final results, based primarily on 18 "crossover" tests, ones that cleared stresses comewhat higher than critical, or failed on stresses below critical. Note the small scatter and the expanded scales. The top curve shows inherent RRRV capability E_0/θ vs. arcing time. It increases at first as electrode gap and pressure difference increase, reaches a peak and falls off at long arcing times due to the onset of clogging effects. The extinction voltage E_0 and time constant θ, the components of this performance ratio E_0/θ, show similar trends.

Calculation of Arc-Circuit Interaction

The 18 crossover tests provide the data for a useful probability plot, shown in Fig. 6.16. The probability of working at stresses above the critical value fall off with a standard deviation σ of about 2%, a remarkable indication that circuit breakers are quite deterministic, when the relative stress in different test circuits is taken into account. The few data below a deviation (on the abscissa) of about −2% show about twice the standard deviation of the main body of the data.

The examples in Part 1 show by comparison with experiment the practical utility of calculations employing a combination of the relatively simple Cassie and Mayr arc model equations.

Part 2 Approximate Analytical Model Solutions

6.3 IMPRESSED CURRENT AND VOLTAGE

To make analytical solutions of the Cassie or Mayr arc model equations possible, following Mayr [1,3,10] we linearize them by assuming either an applied current or an applied voltage (i.e., functions of time only). Thus they become more easily solved linear first-order differential equations. As shown in equation (6.12), before current zero we use Cassie's equation for the arc conductance $1/R$ in the convenient form

$$\frac{d}{dt}\frac{1}{R^2} + \frac{2}{\theta}\frac{1}{R^2} = \frac{2}{\theta}\frac{i^2(t)}{E_0^2} \qquad (6.53)$$

The simplest useful case, approximating a sinusoidal current near zero, is the applied current ramp

$$i(t) = \dot{I}t = \sqrt{2}\omega I t \qquad (6.54)$$

where $\dot{I} = di/dt$ (assumed constant; I = rms current) and $\omega = 2\pi f$ is the angular frequency of the corresponding power system, approximately 377 rad/s for a 60-Hz system and 314 rad/s for a 50-Hz system.

Solution [10] of equation (6.53) for the current of equation (6.54) yields the relation

$$\frac{1}{R} = \frac{1}{R_0}\sqrt{1 - 2\frac{t}{\theta} + 2\left(\frac{t}{\theta}\right)^2} \qquad (6.55)$$

where

$$R_0 = \frac{E_0}{\dot{I}\omega\theta} \qquad (6.56)$$

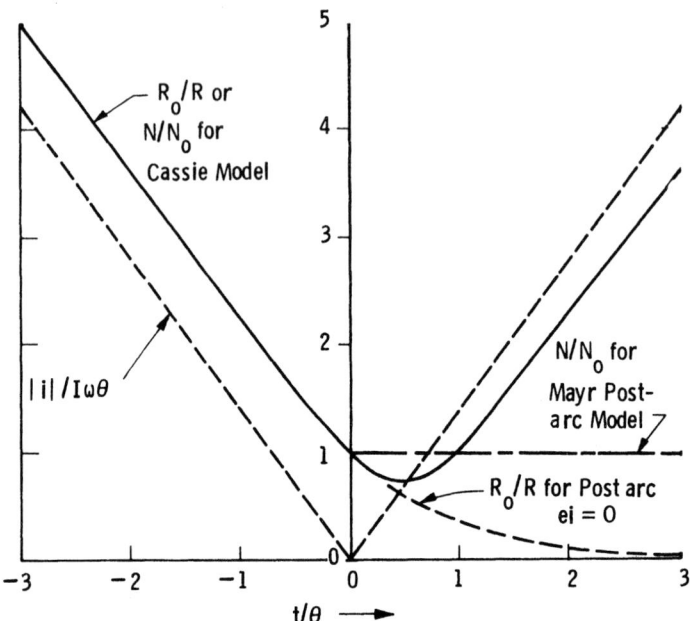

Figure 6.17 Lag of arc conductance and power loss behind current magnitude, quantities normalized. (From T. E. Browne, Jr., An approach to mathematical analysis of a-c arc extinction in circuit breakers, AIEE Trans. (Pt. III): 1508-1517, © 1959, IEEE.)

is the arc resistance at current zero. Since the simplest physical model fitting the Cassie equation has both conductance and power loss simply proportional to the arc column area, we deduce that the equation for power loss N is of the same form as equation (6.55) and find that the current-zero value is

$$N_0 = E_0 I \omega \theta \qquad (6.57)$$

Correspondingly, the near-current-zero steady-state arc voltage is related to R_0 and N_0 by

$$E_0 = \sqrt{R_0 N_0} \qquad (6.16)$$

Figure 6.17 shows in dimensionless form the assumed current ramp magnitude (dashed lines) as the current passes through zero and the calculated response of both the arc conductance $1/R$ and the arc power loss N according to the Cassie model. The postzero heavy dashed extension shows the exponential further decay of arc conduc-

Calculation of Arc-Circuit Interaction

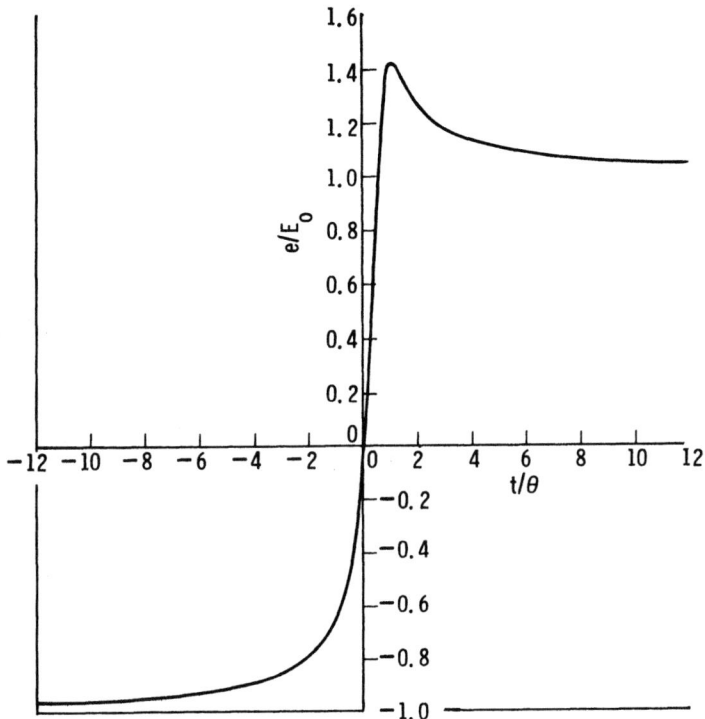

Figure 6.18 Arc voltage vs. time by the Cassie equation for linear passage of current through zero. (From T. E. Browne, Jr., A study of a-c arc behavior near current zero by means of mathematical models, AIEE Trans. 67:141-153, © 1948, IEEE.)

tance or power loss for the case of *no* postzero current or power input. In Fig. 6.18 is plotted the normalized arc voltage

$$e = Ri(t) \tag{6.58}$$

according to the Cassie equation solution for the current ramp extending through zero.

In the combined Cassie-Mayr model used for near-critical conditions, we assume a transition to the Mayr model at current zero, using the postzero form of the equation with the restored voltage $V_a(t)$ applied,

$$\frac{dR}{dt} - \frac{R}{\theta} = -\frac{V_a^2(t)}{\theta N_0} \tag{6.59}$$

Because the Cassie and Mayr equations are identical for zero power input, the time of transition is not critical so long as it occurs while $V_{ai} \ll N_0$. Again, the simplest useful form of the restored voltage, approximating that for resistance shunting of the arc in an inductive circuit (initial part), is a ramp of voltage

$$v(t) = \dot{V} t \qquad (6.60)$$

with the constant rate of rise $\dot{V}$. The solution of (6.59) with (6.60) applied may be written

$$\frac{R}{R_0} = \left(\frac{\dot{V}}{\dot{V}_c}\right)^2 \left[1 + \frac{t}{\theta} + \frac{1}{2}\left(\frac{t}{\theta}\right)^2\right] + \left(1 - \frac{\dot{V}^2}{\dot{V}_c^2}\right) \exp\left(\frac{t}{\theta}\right) \qquad (6.61)$$

where R_0 is the Cassie current ramp solution value, equation (6.56), taken as an initial condition for solution of equation (6.59), together with N_0 (constant in the Mayr equation) and with the same arc time constant θ. The quantity

$$\dot{V}_c = \frac{E_0}{\sqrt{2\theta}} \qquad (6.62)$$

In equation (6.61) it may be seen that R approaches infinity (the arc is extinguished) only if

$$\dot{V} \leqslant \dot{V}_c \qquad (6.63)$$

Therefore, $\dot{V}_c$ is the *critical* rate of rise of recovery voltage. This solution with its result, equation (6.62), is useful where the arc extinction or reignition is effectively decided during the initial ramplike rise of the restored voltage, which is actually finite. Figure 6.19 shows calculated "postarc" currents for such a case with several values of the ratio

$$\alpha = \frac{\dot{V}}{\dot{V}_c} \qquad (6.64)$$

In this plot $I_1 = \sqrt{2} I$ is the peak value of an equivalent prezero sinusoidal current with angular frequency ω. Except for the exactly critical case ($\alpha = 1$), the work-fail decision generally occurs within the first five or six arc time constants.

Exceptions to the infinite ramp solutions occur for recovery voltages which depart from linearity or reach peak values within the first few multiples of θ. One case of principal importance, approximating the voltage recovery in the case of a short-line fault, is that of a symmetrical triangular wave which can be represented by the relations

Calculation of Arc-Circuit Interaction

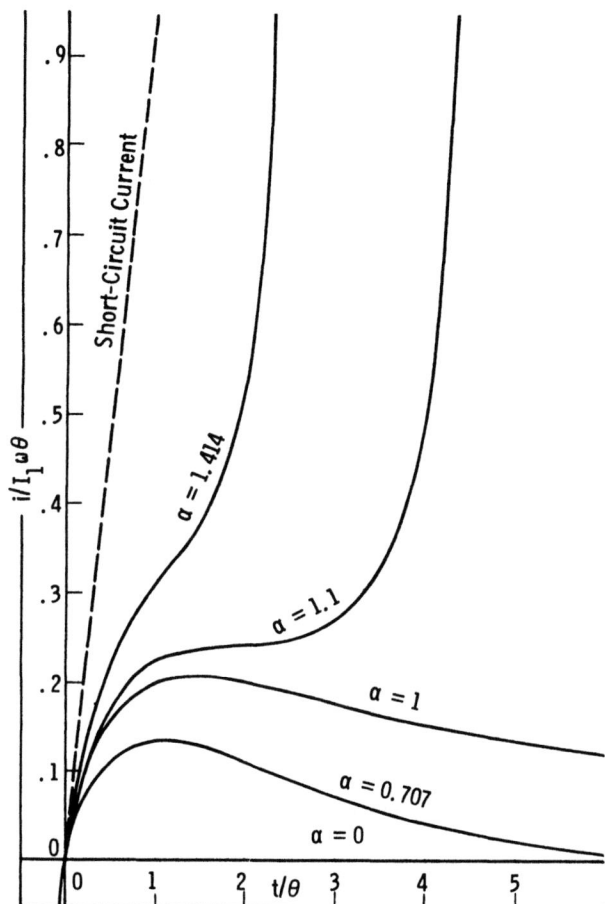

Figure 6.19 Arc current vs. time by the Mayr equation with linear impressed voltage rise ($\alpha = \dot{V}/\dot{V}_c$). (From T. E. Browne, Jr., A study of a-c arc behavior near current zero by means of mathematical models, AIEE Trans. 67:141-153, © 1948, IEEE.)

$$v_1 = \frac{\hat{V}}{\tau} t \qquad \text{for } 0 < t < \tau$$

and (6.65)

$$v_2 = \frac{\hat{V}}{\tau}(2\tau - t) \qquad \text{for } \tau < t < 2\tau$$

Here $\hat{V}$ is the peak value and τ the time to peak of the wave, replacing V_m and t_m used in Part 1. By integration in two steps of equation

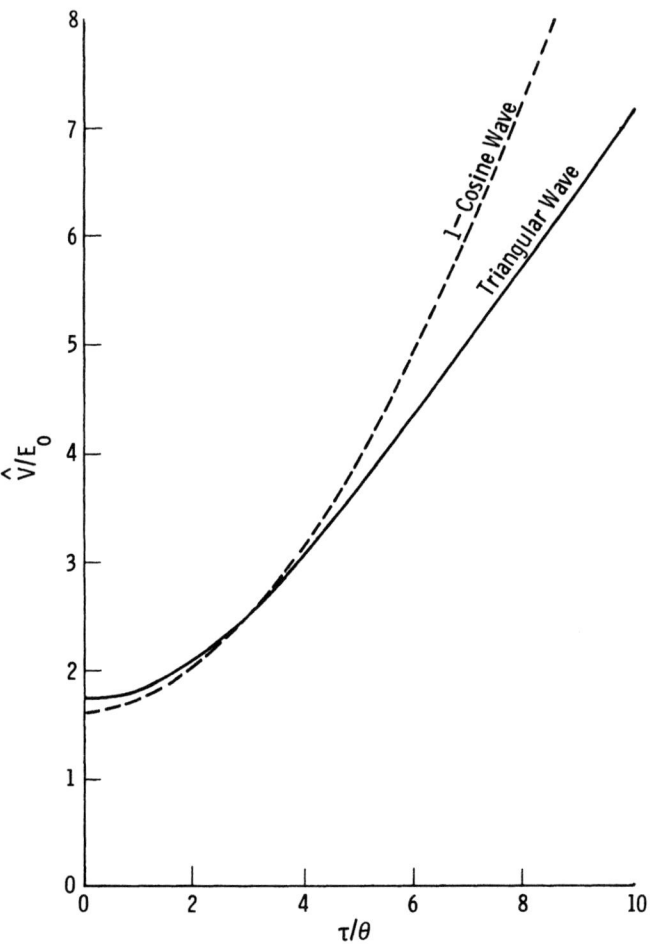

Figure 6.20 Critical peak recovery voltages by the Mayr equation for impressed triangular and $(1 - \cos)$ waveforms vs. time to peak τ. (From T. E. Browne, Jr., Practical modeling of the circuit breaker arc as a short line fault interrupter, IEEE Trans. Power Appar. Syst. PAS-97:838-847, © 1978, IEEE.)

(6.59) with these two voltage ramps successively applied and equating the final arc resistance at $t = 2\tau$ to R_0 for the critical case, the *relation for criticality* is found to be

$$\frac{\hat{V}}{E_0} = \frac{\tau}{\sqrt{2\theta}} \left[1 - \frac{\tau/\theta}{\sinh(\tau/\theta)} \right]^{-1/2} \qquad (6.66)$$

Calculation of Arc-Circuit Interaction

This dimensionless equation is plotted as the solid curve in Figure 6.20. Beyond the time of peak $\tau = 6\theta$, the straight line shows that, as with the infinite ramp, only the slope $\hat{V}/\tau$ of the wave is critical. The curvature at shorter times show that here the limited amplitude becomes important and near zero the curve becomes flat, showing that *only* the amplitude $\hat{V}$ is critical at this limit. Here the root-mean-square value of the voltage is simply equal to E_0.

Another important recovery voltage waveform, applying to the capacitance shunted arc in an inductive circuit, is the oscillatory voltage represented by the equation

$$v = \frac{\hat{V}}{2}(1 - \cos \upsilon t) \tag{6.67}$$

where

$$\upsilon = \frac{1}{\sqrt{LC_s}} \tag{6.68}$$

is the angular frequency of the oscillation between the current-limiting inductance L and the shunting capacitance C_S. The *critical relation* for this applied waveform is found to be

$$\frac{\hat{V}}{E_0} = \sqrt{\frac{8}{3}\left[1 + \frac{5}{4\pi^2}\left(\frac{\tau}{\theta}\right)^2 + \frac{1}{4\pi^4}\left(\frac{\tau}{\theta}\right)^4\right]} \tag{6.69}$$

where now

$$\tau = \frac{1}{2f_0} = \frac{\pi}{\upsilon} \tag{6.70}$$

is the time to peak of this wave. The curve for this equation is shown dashed in Fig. 6.20. For longer times to peak it approaches a square-law curve rather than a straight line, but for times to peak less than 5θ it is very close to the critical curve for the peak of the triangular wave. Again, at times to peak approaching zero the rms value of this somewhat flatter voltage wave is simply equal to E_0. For both wave shapes, it may be seen that as τ approaches zero the critical rate of rise approaches infinity (i.e., the arc is not then sensitive to rate of rise).

Using relations (6.70) and (6.68), (6.69) can be rearranged to give the *critical value* of shunting capacitance,

$$C_s = \frac{5\theta^2}{2L}\left[\frac{3}{5}\sqrt{1 + \frac{2}{3}\left(\frac{\hat{V}}{E_0}\right)^2} - 1\right] \tag{6.71}$$

required to prevent arc reignition.

6.4 INTERACTION OF THE ARC WITH THE CIRCUIT

Although the solutions above are useful as first approximations, if we are to include all of the pertinent quantities in our calculations we must deal with the complete arc-circuit interaction. Rather than by computer solution with the nonlinear equations as in Part 1, in this approach we still use the analytic solutions of the linearized equations but combine them with additional approximate relations to account for the interaction. In the circuit assumed for this analysis, shown in Fig. 6.12, the arc is shunted by resistance R_s and capacitance C_s in parallel. This corresponds to many laboratory test circuits and to the near-current-zero situation for short-line-fault interruption. The circuit voltage is assumed constant at its peak value for the short period considered in this inductive circuit. The capacitance may include that added to reduce the rate of voltage rise but is always present at least as the residual capacitance of the interrupter and other circuit elements as "seen" by the arc. It will be shown later that the residual or stray capacitance, although small, has an appreciable effect on arcs which have very small time-constant values near current zero. Common examples of such arcs are those in SF_6-blast circuit breakers. The assumed relations for determining arc-circuit interaction are illustrated in Fig. 6.21. A first simplifying assumption is that the rate of current change

$$\dot{i}_L = \frac{\sqrt{2}V_g - V_a}{L} \cong \frac{\sqrt{2}V_g}{L} \tag{6.72}$$

through the inductance is driven by the peak circuit voltage $\hat{V} = \sqrt{2}\,V_g$ alone, neglecting the arc voltage drop V_a, and so remains constant. It is also assumed that the postzero arc current and its rate of change are so small that we may neglect their effects on the restored voltage, a fair approximation for most circuit breakers near critical conditions, especially those employing SF_6. The current component through the resistance shunt,

$$i_R = \frac{V_a}{R_s} \cong \frac{E_0}{R_s} \tag{6.73}$$

far enough ahead of current zero that V_a is nearly equal to E_0, causes the arc current i_a to lead i_L by the time increment Δt_R, as shown by the dashed extension of the early i_a to zero. This lead-time component is

$$\Delta t_R = \frac{i_R}{\dot{i}_L} \tag{6.74}$$

Calculation of Arc-Circuit Interaction

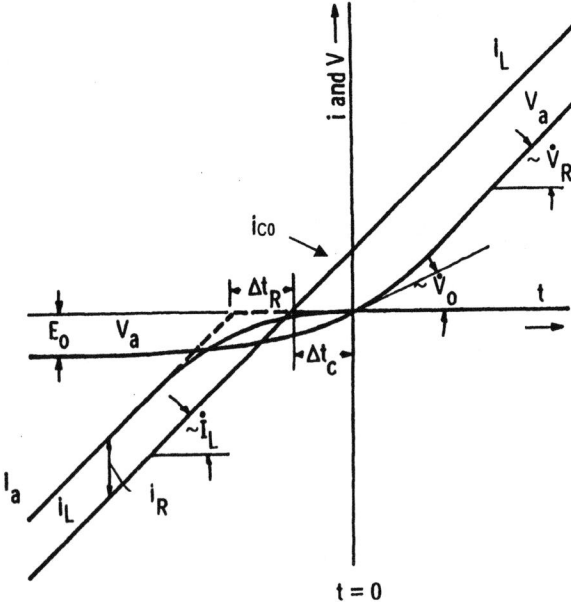

Figure 6.21 Time variation of model currents and voltage near zero. (From T. E. Browne, Jr., Practical modeling of the circuit breaker arc as a short line fault interrupter, IEEE Trans. Power Appar. Syst. PAS-97:838-847, © 1978, IEEE.)

There is an additional time component delaying current zero which is caused by the capacitance current. At arc current zero, $i_L = i_{C0}$ alone and

$$i_{C0} = C_s \dot{V}_0 \qquad (6.75)$$

The corresponding additional time delay is

$$\Delta t_C = \frac{i_{C0}}{\dot{i}_L} = \frac{C_s \dot{V}_0}{\dot{i}_L} \qquad (6.76)$$

Thus the arc shunting circuit elements cause a total time lag between the time of the linear extension of the arc current to zero and the instant of actual current zero,

$$\Delta t = \Delta t_R + \Delta t_c = \frac{E_0/R_s + C_s \dot{V}_0}{\dot{i}_L} \qquad (6.77)$$

Equation (6.56) gives us the value R_0 of the arc resistance R at $t = -\Delta t$ if the linear extrapolation of i_a to zero is assumed. If we also assume, as shown by the dashed line, that i_a remains at zero from $-\Delta t$ onward, solutions of either Cassie's or Mayr's equation yield the value of arc resistance at the actual current and voltage zero as

$$R'_0 = R_0 \exp\left(\frac{\Delta t}{\theta}\right) \qquad (6.78)$$

Making the further assumption, corresponding to the Mayr equation, that the arc power loss remains at the value N_0 after $-\Delta t$, the power loss relation $N = V_a^2/R$ shows us that the Cassie arc model must have a new steady-state arc voltage at the actual current zero given by

$$E'_0 = \sqrt{R'_0 N_0} \qquad (6.79)$$

which by (6.78) and (6.16) is

$$E'_0 = E_0 \exp\left(\frac{\Delta t}{2\theta}\right) \qquad (6.80)$$

By analysis of the circuit, neglecting postarc current, it can be shown that the equation of the postzero modified voltage "ramp" in Fig. 6.21 is

$$V_a = \dot{V}_R \left\{ t - R_s C_s (1 - a) \left[1 - \exp\left(\frac{-t}{R_s C_s}\right) \right] \right\} \qquad (6.81)$$

where

$$\dot{V}_R = R_s \dot{i}_L \qquad (6.82)$$

is the rate of rise of the final straight-line part of the recovery voltage and

$$a = \frac{\dot{V}_0}{\dot{V}_R} \qquad (6.83)$$

is the ratio of the initial to the final slope of the volt-time curve. The reduction in initial slope is determined by the arc-circuit interaction before current zero. The solution of the Mayr equation (6.59) with equation (6.81) voltage impressed yields [15] the *critical* resistance-limited rate of voltage rise

$$\dot{V}_R = \frac{E'_0}{\sqrt{2\theta} S(a,x)} \qquad (6.84)$$

where

Calculation of Arc-Circuit Interaction

$$S(a,x) = \sqrt{1 - \frac{(1-a)(1+2x)x}{(1+x)^2} + \frac{(1-a)^2 x^2}{(1+x)(2+x)}} \quad (6.85)$$

and

$$x = \frac{R_s C_s}{\theta} \quad (6.86)$$

To correct for our neglect of the actual power input during Δt in deriving equation (6.80) for E'_0, we use the result of the nonlinear equation solution from Part 1, showing that with interaction and resistance shunting alone, the critical rate of voltage rise is very nearly

$$\dot{V}_R = \frac{E_0}{\theta} \quad (6.87)$$

This is similar to equation (6.62) but without the factor $1/\sqrt{2}$. It can be shown [15] that this correction is equivalent to shortening the resistance-shunt-produced time delay of current zero by the factor ln 2. Thus the modified form of equation (6.80) is

$$E'_0 = E_0 \exp\left[\frac{(\ln \sqrt{2})\Delta t}{\theta}\right] \quad (6.88)$$

Substituted in equation (6.84), this gives the criticality relation

$$\dot{V}_R = \frac{E_0}{\sqrt{2}\,\theta S(a,x)} \exp\frac{(\ln \sqrt{2})E_0/R_0 + C_s \dot{V}_0}{i_L \theta} \quad (6.89)$$

For further simplification we use the dimensionless ratio

$$y = \frac{\dot{V}_R \theta}{E_0} \quad (6.90)$$

and also using equations (6.86), (6.82), and (6.83), (6.89) becomes

$$y = \frac{\exp[(\ln\sqrt{2})(1/y + ax)]}{\sqrt{2}S(a,x)} \quad (6.91)$$

As shown in Ref. 16,* this relation can be plotted as y vs. a with x as a parameter. The resulting curves have relatively flat minima at values of $a = a_m$ which are related to the ratio x as shown in Fig. 6.22. Comparison with many cathode-ray oscillograms of near-critical tests of a model SF_6-blast interrupter, using various R_S and C_S values,

---

*In equation (11) of Ref. 16 the coefficient of the last term should have been written 1/ln 2 instead of 1/ln $\sqrt{2}$.

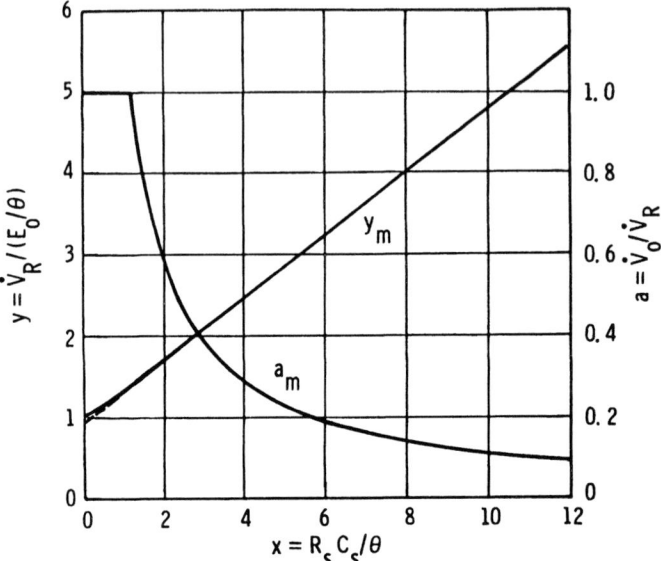

Figure 6.22 Plots vs. x of y_m and a_m. (From Fig. 3 of T. E. Browne, Jr., Simplified estimation of critical quantities for short-line fault interruption, IEEE Trans. Plasma Sci. *PS-8*:400-405, © 1980, IEEE.)

shows that actual observed a values are close enough to a_m that we are justified in using $a = a_m$ as a basis for estimation. When this is done, the minimum y values, y_m, plot vs. x as almost exactly the straight line,

$$y_m = 0.927 + 0.380x \tag{6.92}$$

shown in Fig. 6.22. Very close to this is the slightly simplified line with the equation

$$y_m = 1 + 0.4x \tag{6.93}$$

Thus we are justified in using this unexpectedly simple relation for estimation where the restored voltage amplitude and duration are great enough to justify the assumption of an infinite voltage ramp with its initial slope reduced by the presence of the arc shunting circuit elements.

Returning to arc and circuit quantities, the equation (6.93) relation for criticality can be expressed in various useful ways as

$$\dot{V}_R = \frac{E_0}{\theta}\left(1 + \frac{0.4 R_s C_s}{\theta}\right) \tag{6.94}$$

Calculation of Arc-Circuit Interaction

$$\dot{I} = \sqrt{2}\omega I = \frac{E_0}{\theta}\left(\frac{1}{R_s} + \frac{0.4C_s}{\theta}\right) \tag{6.95}$$

$$C_s = 2.5\theta\left(\frac{\dot{I}\theta}{E_0} - \frac{1}{R_s}\right) \tag{6.96}$$

and

$$\theta = \frac{E_0}{2\dot{V}_R}\left(1 + \sqrt{1 + \frac{1.6\dot{V}_R R_s C_s}{E_0}}\right) \tag{6.97}$$

In these expressions, $\dot{V}_R$, $\dot{I}$, I, C_s, R_s and θ are critical values for energy-balance-limited interruption with an "infinitely" extending recovery voltage ramp having the form of equation (6.81).

6.5 RELATION OF MODEL PARAMETERS TO CURRENT RATE OF CHANGE

For practical use of the equations above, we need to relate the model parameters E_0 and θ to the interrupted current I, or its maximum rate of change $\dot{I}$. From an extensive study of unclogged stored pressure SF_6-blast interrupters near their interrupting limits, relatively simple power-law relations have been found to be good approximations. These are

$$E_0 = nE_1^*\dot{I}^{-1/2} = nE_1^*(\sqrt{2}\omega I)^{-1/2} \tag{6.98}$$

and

$$\theta = \theta_1^*\dot{I}^{3/2} = \theta_1^*(\sqrt{2}\omega I)^{3/2} \tag{6.99}$$

For convenience, at a given power system frequency or ω, we prefer relations to the rms current values rather than to their rates of change. These are

$$E_0 = nE_1 I^{-1/2} \tag{6.100}$$

and

$$\theta = \theta_1 I^{3/2} \tag{6.101}$$

in which the normalized quantities E_1 and θ_1, characterizing one break of a given interrupter under prescribed conditions of gas medium and pressure, are also functions of ω. The factor n is the number of identical series breaks in each phase of the complete breaker. *Thus the*

Cassie-Mayr arc model for a given circuit breaker is completely defined by n, E_1, θ_1, *and* ω, *together with equations (6.100) and (6.101).*

It should be pointed out here that other current dependencies are found for other breaker types, particularly for the single-pressure or puffer type of breaker, where clogging of the nozzle is generally present and the gas pressure applied to the arc interrupting nozzle is therefore a function of the arc current. For these, the arc model can still be used in limited ranges with empirically determined exponents differing from those in equations (6.100) and (6.101).

As illustrative examples here, we shall use the stored pressure breaker relations above, with which equation (6.94) becomes

$$\dot{V}_R = \frac{nE_1}{\theta_1 I^2}\left(1 + \frac{0.4 R_s C_s}{\theta_1 I^{3/2}}\right) \qquad (6.102)$$

and, combining values of I, equation (6.95) thus becomes

$$I = \sqrt[3]{\frac{nE_1}{\sqrt{2}\omega R_s \theta_1}\left(1 + \frac{0.4 R_s C_s}{\theta_1 I^{3/2}}\right)} \qquad (6.103)$$

We cannot directly solve for the current I, but the equation correctly shows the observed fact that with a resistance shunted stored pressure breaker the current-interrupting limit varies as the cube root of the shunting conductance $1/R_s$. With iteration, equation (6.103) can also be used to calculate the critical current where shunting capacitance is present. For the capacitance required for interruption of a given current, equation (6.95) can be combined with (6.100) and (6.101) to give

$$C_s = 2.5\theta_1\left(\frac{\sqrt{2}\omega\theta_1 I^{9/2}}{nE_1} - \frac{I^{3/2}}{R_s}\right) \qquad (6.104)$$

Note that this equation applies strictly only to the case where rate of rise control is primarily by resistance shunting, but with capacitance as a small additional aid, and also where the peak restored voltage $\hat{V}$ is much greater than E_0.

6.5.1 Application to Short-Line Fault

For relatively long line faults the equations derived above can be used by simply substituting the line surge-impedance,

$$Z_0 = \sqrt{\frac{L_1}{C_1}} \qquad (6.105)$$

Calculation of Arc-Circuit Interaction

for R_s. Here, as before, L_1 and C_1 are the per-unit-length values of line inductance and capacitance. For the line (assumed lossless) we also have the relations

$$L = L_1 \ell = \frac{Z_0 \ell}{c_0} \tag{6.106}$$

and

$$C_L = C_1 \ell = \frac{\ell}{Z_0 c_0} \tag{6.107}$$

where ℓ is line length and c_0 is the wave velocity, close to the velocity of light, 0.186 mi/μs (0.299 km/μs).

For *short*-line faults, voltage transients of limited amplitude must be dealt with in the important cases. Therefore, the triangular wave solution result, equation (6.66), must be employed but modified now to account for interaction. This modification, to make equation (6.66) agree with (6.87) as τ/θ approaches infinity, simply eliminates the $1/\sqrt{2}$ factor in equation (6.66). Rearranged, the critical rate of rise for the *short* line without capacitance shunting but with arc-circuit interaction becomes

$$\dot{V}_Z = \frac{\hat{V}}{\tau} = \frac{E_0/\theta}{\sqrt{1 - [(\tau/\theta)/\sinh(\tau/\theta)]}} \tag{6.108}$$

Comparison with equation (6.94) for the capacitance modified *infinite* voltage ramp shows that to account for the limited amplitude with the short line we should simply replace the 1 within the parentheses by

$$\left[1 - \frac{\tau/\theta}{\sinh(\tau/\theta)}\right]^{-1/2}$$

a quantity greater than 1 for τ/θ finite. Since, as shown in Part 1, the transient voltage amplitude is determined by both the line length and the prezero arc voltage E_0, as shown in Ref. 16, an additional term θ/τ must also be added. As limited by the surge impedance, the critical rate of rise then becomes

$$\dot{V}_Z = \frac{E_0}{\theta}\left\{\frac{\theta}{\tau} + \left[1 - \frac{\tau/\theta}{\sinh(\tau/\theta)}\right]^{-1/2} + \frac{0.4 Z_0 C_s}{\theta}\right\} \tag{6.109}$$

Making use of equation (6.95) similarly modified yields for the critical current

$$I_c = \frac{E_0}{\sqrt{2}\,\omega\theta Z_0}\left\{\frac{\theta}{\tau} + \left[1 - \frac{\tau/\theta}{\sinh(\tau/\theta)}\right]^{-1/2} + \frac{0.4 Z_0 C_s}{\theta}\right\} \tag{6.110}$$

Upon putting in the stored pressure interrupter relations (6.100) and (6.101), we find that for the short-line fault, equation (6.103) has been modified to

$$\frac{I_c}{I_\infty} = \sqrt[3]{\frac{\theta}{\tau} + \left[1 - \frac{\tau/\theta}{\sinh(\tau/\theta)}\right]^{-1/2} + I_c^{3/2}} \qquad (6.111)$$

where

$$I_\infty = \sqrt[3]{\frac{nE_1}{\sqrt{2}\,\omega Z_0 \theta_1}} \qquad (6.112)$$

is the critical current for an unshunted line ($C_s = 0$) approached when the length, and therefore τ, becomes large enough that the combined functions of τ/θ in equation (6.111) approach unity. In equation (6.111), θ rather than $\theta_1 I^{3/2}$ has been retained in the τ/θ ratio because in numerical treatment it is most convenient to evaluate the equation for a series of values of τ/θ and then, using the line-length relation

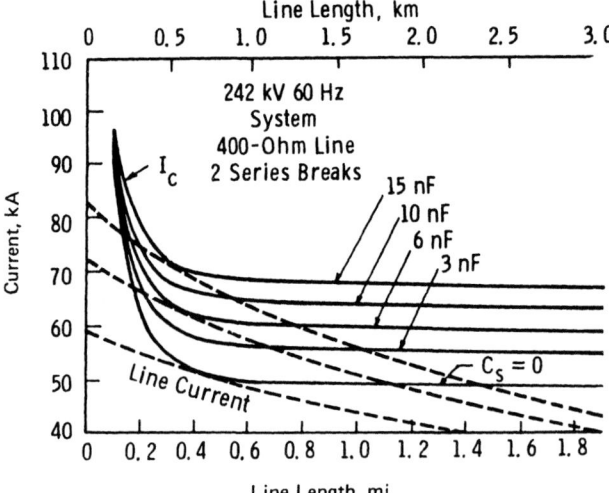

Figure 6.23 Example of critical line-fault currents for interruption and permissible line currents with capacitance shunting, plotted vs. line length. (From T. E. Browne, Jr., Simpified estimation of critical quantities for short-line fault interruption, IEEE Trans. Plasma Sci. PS-8: 400-405, © 1980, IEEE.)

Calculation of Arc-Circuit Interaction

$$\ell = \frac{\tau c_0}{2} \equiv \frac{1}{2} \theta_1 I^{3/2} \frac{\tau}{\theta} c_0 \tag{6.113}$$

to plot points on an I vs. ℓ curve for each τ/θ value. Figure 6.23 shows results of such calculations for the stated system example, using E_1 and θ_1 values determined for a particular SF_6 circuit breaker. These values in convenient units are $E_1 = 30.2$ kV(kA)$^{1/2}$ and $\theta_1 = 0.00255$ μs/(kA)$^{3/2}$. In the figure are drawn three curves of short-circuit line current I_L as functions of line length. These are adjusted to be tangent to the particular critical current curves for shunting capacitances of 0, 6, and 15 nF. The line currents are calculated by the equation

$$\frac{1}{I_L} = \frac{1}{I_0} + \frac{\omega Z_0 \ell}{c_0 V_g} \tag{6.114}$$

in which I_0 is the "bus fault" current (for zero line length) and V_g is the system phase voltage, $1/\sqrt{3}$ times the line-to-line voltage, which is 242 kV in this example.

In agreement with experiment (see, e.g., Fig. 17 of Ref. 17), the critical current curves of Fig. 6.23 are asymptotic to constant values at greater line lengths, or line inductances L in the reference, but turn up sharply for small line lengths. Calculated improvements by capacitance shunting are illustrated in Fig. 6.23. Also illustrated is an outstanding characteristic of short-line faults, the existence of a "worst" line length, that for which the line current and the critical current curves just touch each other in the figure. The calculation, using our derived critical current equation, also shows that this critical line length is a function of the arc characteristics of a particular breaker as well as the properties of the power system and of the line.

6.6 USE OF THE CASSIE CRITERION

In his 1939 paper [2], A. M. Cassie derived an effective impedance Z(k) of the circuit connected to an arc where k, a rate constant for the arc, was found [10] to be equivalent to $1/(2\theta)$ in the Mayr equation. During the near-current-zero period, if the arc resistance exceeded Z(k), decreasing power input would cause the resistance to increase without limit, so the arc would be extinguished. For the parallel shunting circuit of Fig. 6.12, Cassie's impedance (or admittance as expressed) is given by

$$\frac{1}{Z(k)} = \frac{1}{kL} + \frac{1}{R_s} + kC_s = \frac{2\theta}{L} + \frac{1}{R_s} + \frac{C_s}{2\theta} \tag{6.115}$$

which is treated as an effective pure shunting conductance with the two terms, $2\theta/L$ and $C_S/2\theta$, added to $1/R_S$. If, as was done in Ref. 3, we replace R_S by $Z(k)$ in the voltage ramp solution of Mayr's equation, (6.62), and use (6.58), we get an equation for critical current for interruption,

$$I = \frac{E_0}{2\omega\theta}\left(\frac{1}{R_S} + \frac{C_S}{2\theta} + \frac{2\theta}{L}\right) \qquad (6.116)$$

Except for the last term within the parentheses, this is rather similar to equation (6.95) derived previously for the capacitance modified voltage ramp. For the short line with characteristic impedance Z_0, capacitance shunting, and use of the dimensionless ratios, we find that

$$y = \frac{1}{\sqrt{2}} + \frac{x}{2\sqrt{2}} + \frac{\sqrt{2}\,\theta Z_0}{L} \qquad (6.117)$$

where L is the inductance of the line section. Numerically, the equation is

$$y = 0.707 + 0.354x + \frac{1.414\theta Z_0}{L} \qquad (6.118)$$

For a line section long enough to produce effectively an infinite voltage ramp, the last term can be neglected and the equation may be seen to be that of a line close to and nearly parallel with the y_m vs. x line in Fig. 6.22, which was previously derived for the capacitance modified infinite voltage ramp. For a finite line length, the last term in equation (6.118) simply raises the y vs. x line without changing its slope, just as the function of τ/θ does in our previously derived equation (6.109). By trial, it is found that the relatively simple third term in equation (6.118) has so nearly the same effect as the more complex function of τ/θ in the previously derived equations that, as a practical matter, *we are justified in using the Cassie-criterion-derived equations instead of the previous more elaborate ones for short-line-fault calculations.* This is a suprising conclusion because Cassie's theory was derived for lumped-constant circuits only and in our manipulations we have treated the line as having both an inductance L and a surge impedance $Z_0 = \sqrt{L/C_L}$. However, since both the Cassie and Mayr equations and many of our previous simplifying assumptions are only approximate, we may assume that either of these approaches to estimating equations is equally valid so long as the calculated results are nearly identical, and they have been found to be so for all the conditions dealt with so far.

Using the unclogged stored pressure arc model relations (6.100)

Calculation of Arc-Circuit Interaction

and (6.101), the critical short-line-fault current relation, using the Cassie criterion, becomes

$$\frac{I}{I_{\infty c}} = \sqrt[3]{1 + \frac{Z_0 C_s}{2\theta_1 I^{3/2}} + \frac{2Z_0 \theta_1 I^{3/2}}{L}} \qquad (6.119)$$

where now

$$I_{\infty c} = \sqrt[3]{\frac{nE_1}{2\omega Z_0 \theta_1}} \qquad (6.120)$$

is only slightly different (by 12%) from equation (6.112) for I_∞. Equation (6.119) can be numerically solved for the critical current I by iteration, but it is more convenient to solve it for L and use equation (6.106) to get an equation for critical line length as a function of current,

$$\ell = \frac{2c_0 \theta_1 I^{3/2}}{2\omega Z_0 \theta_1 I^3 / nE_1 - Z_0 C_s / 2\theta_1 I^{3/2} - 1} \qquad (6.121)$$

Curves of I_c vs. line length by this equation, computed for the example of Fig. 6.23, are found to be in such good agreement with those of Fig. 6.23 that equations (6.119) and (6.120) or (6.121) may be used with confidence as replacements for the more awkward previously derived equations (6.111) and (6.112), also requiring the use of equation (6.113).

6.6.1 Uses of the Interrupting Limit Equation

Since short-line-fault interruption is frequently the most difficult for a high- or medium-voltage circuit breaker, equation (6.121) with (6.114) is very useful to the designer as a tool for extrapolation from limited test data. It also permits quantitative evaluation of proposed design changes since the formula takes account of all the principal interactions of the pertinent variables. Especially useful is the relatively simple algebraic form of equation (6.121), permitting quick and easy calculation with the aid of only a pocket calculator.

Effect of Gas Pressure

Using the approximate relations, based on observations of SF_6-blast

interrupters,

$$E_0 \sim \left(\frac{P}{P_1}\right)^{1/2} \quad \text{and} \quad \theta \sim \frac{P_1}{P} \tag{6.122}$$

where P is the absolute upstream stored gas pressure and P_1 is the reference pressure for E_1 and θ_1, equation (6.121) becomes

$$\ell = \frac{2c_0 \theta_1 I^{3/2}}{2\omega Z_0 \theta_1 I^3 / nE_1 \sqrt{P/P_1} - Z_0 C_s (P/P_1)^2 / 2\theta_1 I^{3/2} - P/P_1} \tag{6.123}$$

With this equation a series of critical I_c vs. fault fraction or line length curves may be plotted, similar to those of Fig. 6.23 but with the pressure ratio P/P_1 as the parameter.

Number of Series Breaks

A set of curves plotted against line length and with the number of breaks n as parameter is shown for the chosen example in Fig. 6.24. These have been drawn with the reasonable assumption that 1 nF of stray shunting capacitance is present. The curve for two series breaks is in agreement with the rating of this particular breaker, 63 kA bus-fault short-circuit current.

It would appear that four breaks should be able to handle a 90-kA bus fault, but this presumes that the E_1 and θ_1 values would still

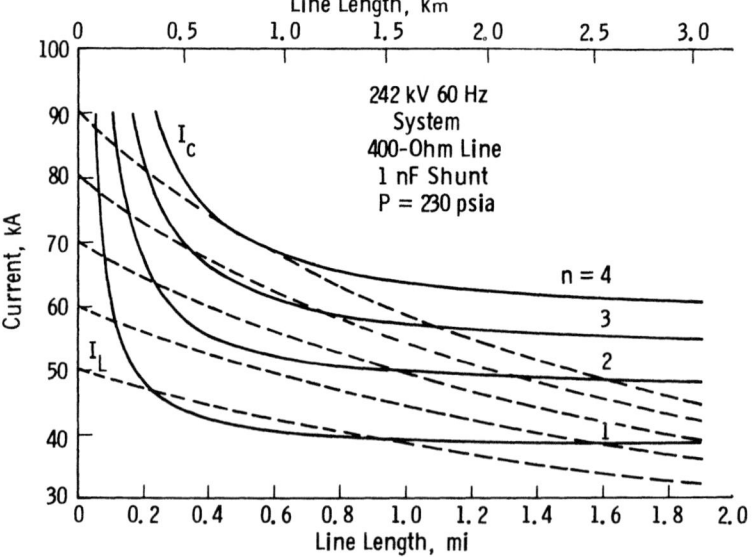

Figure 6.24 Critical line-fault currents I_c and line currents I_L for various bus faults, plotted vs. line length. Numbers n of series breaks as the parameter.

Calculation of Arc-Circuit Interaction

hold good at the higher critical current level of 70 kA. This presumption would require proof since nozzle clogging might be present at this current.

Effect of Scale

Although the size of a gas-blast interrupter nozzle is usually related primarily to clogging limits (see Chap. 5), it may be of interest to designers to determine the effect of changes in size of similar nozzles (diameter and all other associated dimensions changed in the same ratio) upon interrupted current limits according to these energy balance calculations. For this purpose we make a plausible assumption that arc voltage varies as the 0.6 power of the nozzle diameter D and that the arc time constant increases directly with D. (Note that for the same pressure, the pressure *gradient* varies inversely as D.) With these assumptions, based at least partly on observations, and also presuming use of 6 nF of shunting capacitance, the plots of Fig. 6.25 show a considerable advantage obtainable by *decreasing* rather than increasing the nozzle size. Of course, practical limitations, including

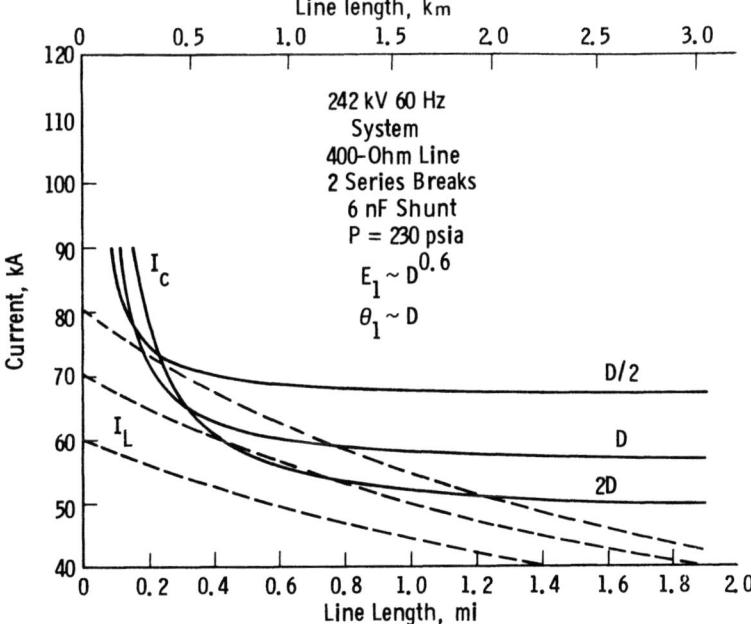

Figure 6.25 Critical line currents I_c and line currents I_L for various bus faults, plotted vs. line length. Diameter D of similar nozzles as the parameter.

clogging limits, would undoubtedly restrict permissible changes to much less than the 2:1 ratios illustrated here. It is interesting to observe that the most difficult line lengths are also considerably affected by these nozzle-size changes.

Required Capacitance Shunting

One of the most useful possible calculations is of the amount of shunting capacitance that may be required to meet a given rating. Thus, the capacitance required to lift the critical I vs. ℓ curve to or above the line current curve for a given I_0, can be estimated from a plot like Fig. 6.23, but it can also be expressed as a formula. Equation (6.121) can be solved for C_s and the current expressed in terms of the fault fraction f. Expressing ℓ also in terms of f and I_0 by means of equation (6.114) with $I_L = fI_0$, we obtain

$$C_s = 4\omega \theta_1^2 (fI_0)^4 \left[\frac{\sqrt{fI_0}}{nE_1} - \frac{1}{V_g(1-f)} \right] - \frac{2\theta_1 (fI_0)^{3/2}}{Z_0} \quad (6.124)$$

Figure 6.26 is a plot of C_s vs. f for our example with the bus-fault current I_0 equal to 63, 70, and 80 kA. The critical C_s can also be plotted against line length by use of the relation

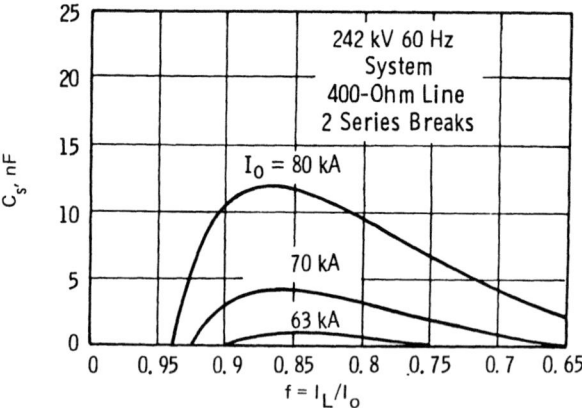

Figure 6.26 Critical capacitance shunts vs. fault fraction for some bus-fault currents I_0. (From T. E. Browne, Jr., Simplified estimation of critical quantities for short-line fault interruption, IEEE Trans. Plasma Sci. PS-8:400-405, © 1980, IEEE.)

Calculation of Arc-Circuit Interaction

$$\ell = \frac{c_0 V_g (1 - f)}{\omega Z_0 I_0 f} \tag{6.125}$$

which is equation (42) of Ref. 15.

Since calculation [12] shows that with C_s values 0.5 or more times the line capacitance ℓC_1, the restored voltage transient approaches very closely the $(1 - \cos)$ shape, we can also use equation (6.71) as a basis for estimating the critical C_s in such cases. The oscillatory voltage is closely matched by an equivalent oscillation between the line inductance ℓL_1 and a lumped capacitance $C_e = C_s + 0.346 \ell C_1$, or approximately by assuming that C_s is paralleled by one-third of the line capacitance. On this basis an alternative formula for critical shunting capacitance has been derived [15] and adjusted so as to agree exactly with computer calculations reported in Part I. This formula may be written

$$C_s = 2.75 \omega \theta_1^2 (fI_0)^4 \left[\frac{\sqrt{2fI_0}}{nE_1} - \frac{1+f}{V_g(1-f)} \right] - \frac{V_g(1-f)}{3Z_0^2 \omega fI_0} \tag{6.126}$$

Being fairly similar to equation (6.124), it gives nearly the same results for f values which are near the most difficult.

Calculations of critical I vs. f or ℓ curves by equation (6.126) also agree well with those shown here in the region of significance. For greater line lengths or smaller f values, however, these curves do not flatten out but rise with increasing line lengths [5]. This false rise occurs because the calculation assumes lumped circuit values and so does not recognize the return of the voltage transient to nearly triangular shape as the ratio $C_s/\ell C_1$ becomes small with increasing ℓ values.

6.6.2 Determination of Arc Model Parameters

Since our approximate arc modeling is quantitatively usable only for extrapolation from test results, evaluation of model parameters from results of near-critical tests is essential.

Cassie Model Arc Voltage

As discussed in Sec. 6.2.1, E_0 is readily obtainable from test oscillograms. Log-log plots of E_0 vs. current, like Fig. 9 of Ref. 15, can be used to obtain the E_1 value in equation (6.100) for a given breaker if a straight line can be fitted to the data.

Arc Time Constant

For tests with essentially linear restored voltage having amplitudes large compared with E_0, a first estimate of θ may be obtained using the computer-determined relation from Part 1, equation (6.87), in the form $\theta = E_0/\dot{V}_R$, where $\dot{V}_R$ is estimated to be the *critical* linear rate of voltage rise. Since even with pure resistance shunting some stray parallel capacitance is always present, a more accurate value is obtainable from equation (6.97) with C_s measured or estimated. C_s values are seldom much less than 1 nF. Possibly more accurate values of θ may be obtained from the much more elaborate equation (6.89) together with (6.85) and (6.86) where both $\dot{V}_R$ and the initial rate of rise $\dot{V}_0$ are known, yielding the ratio a.

Equations (6.100) and (6.101) give the ratio

$$\frac{E_0}{\theta} = \frac{nE_1}{\theta_1 I^2} \qquad (6.127)$$

In cases of interrupters for which these relations apply, it is convenient to plot on log-log coordinates, as in Fig. 10 of Ref. 15, the E_0/θ ratio vs. the current I, and then using the critical case equations to calculate θ. Actual tests have a "work" or "fail" result, so the calculated E_0/θ values must be either above or below the true critical values. A straight line drawn with a slope of -2 serves to separate almost completely the "work" from the "fail" points. The position of this fitted line determines E_1/θ_1 in equation (6.127).

Use of Short-Line-Fault Test Data

For a first and somewhat rough estimate, if test data without shunting capacitance and for sufficiently long lines are available so that the critical I_∞ can be estimated, and relations of the form of equations (6.100) and (6.101) can be assumed, equation (6.112) or (6.120) can be used to get a first estimate of E_1/θ_1. Then, with E_1 estimated from test oscillograms, we also have an estimated value of θ_1 for the test breaker. More accurate estimates, taking account of shunting capacitance and known line length, may be obtained from equation (6.121). For example, if as before E_1 can be estimated from oscillograms, rearrangement of equation (6.121) and solution as a quadratic in θ_1 gives us the equation

$$\theta_1 = \frac{Z_0 C_s}{I_c^{3/2}[\sqrt{1 + 4Z_0 C_s (\omega Z_0 I_c^{3/2}/nE_1 - c_0/\ell)} - 1]} \qquad (6.128)$$

where I_c is the critical current for a line of length ℓ shunted by the capacitance C_s.

When at least two points on a curve of limiting current vs. line length are available, as in Fig. 17 of Ref. 17 for example, both E_1 and θ_1 can be obtained by solving simultaneously either equation (6.121) or

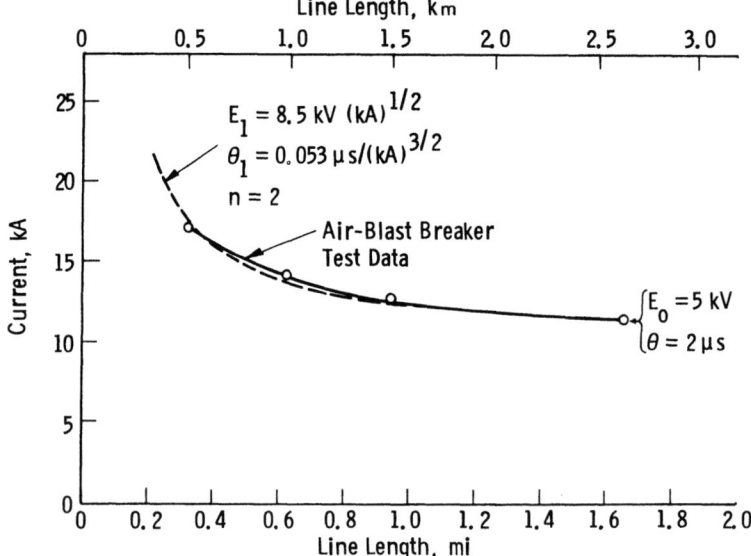

Figure 6.27 Critical currents for interruption by an air-blast breaker vs. line length from the short-line-fault test data in Ref. 17. Dashed curve by equation (6.121).

(6.128) for the two points. Figure 6.27 shows the same data plotted (solid curve) vs. line length instead of line inductance and the closely approximating dashed curve plotted by equation (6.121) with the E_1 and θ_1 indicated values obtained in this way. Since the effective shunting capacitance was not stated, a value of 1 nF was assumed in the calculation, which was relatively insensitive to C_S in this case. At the maximum tested line length of 1.65 mi (2.65 km) for which the critical current was 11.3 kA, the indicated model pre-zero arc voltage was found to be $E_0 = 2 \times 8.5/\sqrt{11.4} = 5.0$ kV and the arc time constant was $\theta = 0.053(11.4)^{3/2} = 2.0$ μs.

It should be noted that these values were obtained for an air-blast breaker. The good fit of the calculated to the experimental curve strongly suggests that the relations of arc voltage and time constant to current magnitude first observed for an SF_6-blast interrupter are of similar form for this stored pressure air-blast breaker.

6.7 MODELING OF THE PUFFER BREAKER

As mentioned in Sec. 6.5 (see also Chap. 10) modeling of the single-pressure or puffer-type circuit breaker arc is made more difficult by the presence of nozzle clogging and of pressure variation as a function of arc current. This pressure variation, pressure rising with current magnitude, is determined by the dynamic characteristics of the piston and contact driving mechanism as well as by nozzle flow and ablation relations. Thus complete modeling requires a rather complete computer program.

However, if the gas pressure at a given current zero moment is known, an equation of the form of equation (6.123) should apply for short-line-fault calculations here also. A possible approach is to account for the effect of current magnitude on the pressure and inflowing gas temperature by assuming relations like equations (6.100) and (6.101) for the arc model parameters but with other current exponents. Studies indicate that relations of this simplicity do not apply over a wide current range, but in limited ranges near the current-interrupting limits, such modeling may still be useful for puffer breakers.

6.8 NOMENCLATURE

Symbol	Meaning	Convenient units
a	$\dot{V}_0/\dot{V}_R$	
a_m	value of a for minimum y	
A	arc cross-sectional area	
c_0	electromagnetic wave velocity	mi/μs or km/μs
C_L	line-to-ground capacitance of line section	nF
C_1	line capacitance per unit length	nF/mi or nF/km
C_e	effective shunt capacitance including the line	nF
C_s	shunting capacitance	nF
e	instantaneous arc voltage	kV
E_0	Cassie model near-current-zero steady-state arc voltage	kV
E_0'	Cassie arc voltage after reignition	kV
E_1	per break Cassie arc voltage for unit arc current at given frequency	kV(kA)$^{1/2}$
E_1^*	per break Cassie arc voltage for unit rate of change of circuit current	kV(A/μs)$^{1/2}$

Calculation of Arc-Circuit Interaction

Symbol	Meaning	Convenient units
f	system frequency or fault fraction, I_L/I_0	Hz
f_0	frequency of restored voltage oscillation	MHz
g	normalized arc conductance, G/G_0	
G	arc conductance, i/e	$(k\Omega)^{-1}$
G*	steady-state arc conductance	$(k\Omega)^{-1}$
G_0	arc conductance at current zero by Cassie equation with impressed current ramp	$(k\Omega)^{-1}$
i	instantaneous arc current	A
i_L	instantaneous inductance limited current	A
i_C	instantaneous capacitance current	A
i_{C_0}	capacitance current at current zero	A
I	rms circuit current	kA
I_L	rms line-fault current	kA
I_1	peak value of circuit current	kA
I_0	bus-fault current (for $\ell = 0$)	kA
$\dot{I}$	rate of change of I, dI/dt	A/µs
j_a	normalized arc current, $i/\omega\theta I$	
k	ratio of effective to actual line capacitance in Part 1	
k	rate constant for the arc conductance in Cassie model theory [i/(2θ)] in Part 2	$(\mu s)^{-1}$
ℓ	line length	mi or km
L	current-limiting inductance or line inductance to fault	mH
L_1	line inductance per unit length	mH/mi or mH/km
L_G	inductance of connected power system	mH
n	number of series breaks per phase of breaker	

Symbol	Meaning	Convenient units
N	arc power loss	kW
N_0	constant power loss of Mayr arc model	kW
P	absolute upstream gas pressure	atm
P_1	reference absolute gas pressure corresponding to E_1 and θ_1	atm
Q	stored arc energy associated with arc conductance	kWs
Q_0	stored energy term in Mayr equation	kWs
R	arc resistance, $1/G$	kΩ
R_0	arc resistance at current zero for impressed current ramp	kΩ
R'_0	arc resistance at current zero with current distortion	kΩ
R_s	arc shunting resistance	kΩ
S	function of a and x by equation (2.34)	
t	time after current zero	μs
t_a	arcing time at 60 Hz	cycles
t_m	time to peak recovery voltage in Part 1	μs
Δt	time delay of actual current zero behind projected current zero	μs
u	normalized arc voltage, e/E_0	
V_a	arc voltage (same as e)	kV
V_g	rms power circuit phase voltage	kV
V_0	bus-side system voltage for short-line fault	kV
$\dot{V}_0$	initial rate of rise of restored voltage	
$\dot{V}_R$	resistance limited RRRV = $R_s \dot{i}_L$	kV/μs
$\dot{V}_c$	critical value of $\dot{V}_R$ or $\dot{V}_z$	kV/μs
V_m	peak restored voltage in Part 1	kV
$\hat{V}$	peak restored voltage in Part 2	kV
W	arc power imput = $V_a i$	kW

Symbol	Meaning	Convenient units
x	normalized time delay, $R_s C_s/\theta$	
y	normalized rate of voltage rise, $\dot{V}_R \theta/E_0$	
y_m	minimum value of y with respect to a	
Z_0	line surge impedance	$k\Omega$
α	ratios of actual to critical RRRV	
β	ratio of actual shunting conductance G_s to G_0	
δ	ratio of E_0/θ to average rate of rise to peak voltage	
θ	arc time constant	μs
θ_1	value of θ for unit I in Part 2.	$\mu s/(kA)^{3/2}$
θ_1^*	arc time constant for unit rate of change of circuit current	$\mu s/(A/\mu s)^{3/2}$
υ	angular frequency of oscillatory restored voltage	$(\mu s)^{-1}$
υ_p	normalized angular frequency	$\upsilon\theta$
τ	normalized time, t/θ, in Part 1	
τ	time to peak recovery voltage in Part 2	μs
ω	angular frequency of power circuit = $2\pi f$	$(ms)^{-1}$

REFERENCES

1. O. Mayr, Beiträge zur Theorie des statischen und des dynamischen Lichtbogens, Arch. Elektrotech. 37:588-608, 1943.
2. A. M. Cassie, Arc rupture and circuit severity: a new theory, CIGRE, Rep. 102, Paris, 1939.
3. T. E. Browne, Jr., An approach to mathematical analysis of a-c arc extinction in circuit breakers, AIEE Trans. 78(Part III): 1508-1517, 1959.
4. T. E. Browne, Jr., The electric arc as a circuit element, J. Electrochem. Soc. 102(1):27-37, 1955.
5. B. W. Swanson, R. M. Roidt, and T. E. Browne, Jr., A thermal model for short-line fault interruption, Elektrotech. Z. Archiv. 93:375-380, July 1972.

6. A. Grütz and A. Hochrainer, Rechnerische Untersuchung von Leistungsschaltern mit Hilfe einer verallgemeinerten Lichtbogentheorie, Elektrotech. Z. Archiv. 92:185-191, 1971.
7. A. M. Cassie and F. O. Mason, Post-arc conductivity in gas-blast circuit-breakers, CIGRE, Rep. 103, Paris, 1956.
8. J. Urbanek, Zur Berechnung des Schaltverhaltens von Leistungsschaltern, eine erweiterte Mayr-Gleichung, Elektrotech. Z. Archiv. 93:381-385, 1972.
9. Y. Nakamichi and D. T. Tuma, Analysis of electric arcs in ac circuits, IEEE Trans. Power Appar. Syst. PAS-102:586-595, March 1983
10. T. E. Browne, Jr., A study of a-c arc behavior near current zero by means of mathematical models, AIEE Trans. 67:141-153, 1948.
11. J. Urbanek, The time constant of high voltage circuit breaker arcs before current zero, Proc. IEEE 59:502-508, April 1971.
12. L. S. Frost, Dynamic arc analysis of short-line fault tests for accurate circuit breaker performance specification, IEEE Trans. Power Appar. Syst. PAS-97:478-484, March/April 1978.
13. E. Schwalb, U. Habedank, H. H. Schramm, E. Slamecka, and K. Zückler, Investigation of unit testing and full-pole testing of single pressure type double nozzle SF_6 circuit-breaker under short-line fault conditions, CIGRE, Rep. 13-03, Paris, 1978.
14. M. Murano, H. Nishikawa, A. Kobayashi, T. Okazaki, and S. Yamashita, Current zero measurement for circuit breaker phenomena, IEEE Trans. Power Appar. Syst. PAS-94:1890-1900, 1975.
15. T. E. Browne, Jr., Practical modeling of the circuit breaker arc as a short line fault interrupter, IEEE Trans. Power Appar. Syst. PAS-97:838-847, May/June 1978.
16. T. E. Browne, Jr., Simplified estimation of critical quantities for short-line fault interruption, IEEE Trans. Plasma Sci. PS-8:400-405, December 1980.
17. S. Yamazaki, M. Hosokawa, T. Goto, N. Nakanishi, and J. Tomiyama, Synthetic testing and interrupting phenomena under short line fault conditions, IEEE Trans. Power Appar. Syst. PAS-91:773-782, May/June 1972.

7
Postarc Dielectric Recovery in a Blast Arc

DAVID T. TUMA†/Carnegie-Mellon University, Pittsburgh, Pennsylvania

7.1 Introduction   241
7.2 Experimental Investigations   243
    7.2.1 General characteristics of the recovery   244
    7.2.2 Effect of pressure   246
    7.2.3 Effect of the nozzle geometry   247
    7.2.4 Effect of arc current   249
    7.2.5 Effect of the gas   250
    7.2.6 Effect of turbulence   252
    7.2.7 Effect of space charge   252
    7.2.8 Location of the high-dielectric-strength region   257
7.3 Theoretical Investigations   257
    7.3.1 Physical processes   257
    7.3.2 Early models of arc channel dielectric recovery   262
    7.3.3 Models for the gas-blast arc channel in free recovery   263
7.4 Summary   271

References   271

7.1 INTRODUCTION

After the gas-blast arc current reaches the zero value, the arc channel is subjected to a rate of rise of recovery voltage (RRRV) induced by the rest of the electric network to which the arc gap is connected. A postarc current of the order of a few amperes may flow in response to the applied reverse voltage. The resulting electric power input into

†Deceased

the arc channel counteracts the drive by the energy-loss mechanisms to reduce the arc temperature—and therefore electrical conductivity—leads to a race [1] that may conclude in either reignition or interruption of the arc channel. The arc channel may survive reignition during the next several microseconds (thermal reignition), in which case it continues to experience an ever-increasing transient recovery voltage (TRV) across it while its resistance increases dramatically. The arc channel current—if any—now becomes so small that the electric energy input into the channel can be ignored. Recombination among the electrons and the heavy-particle species in the arc channel reduces the number of free electrons fairly rapidly, but negative and positive ions remain in the channel for several hundred microseconds [2]. The channel may eventually break down several tens of microseconds to several milliseconds after current zero when the TRV has reached very high values. This breakdown is termed a dielectric breakdown, owing to the paucity of mobile charge carriers (i.e., free electrons) in the extinguished arc channel prior to breakdown and to the similarity of the suddenness of this breakdown to that observed in cold gases. However, there are differences between the physical mechanisms of this dielectric breakdown and that occurring in cold gases, owing mainly to the presence of high and nonuniform gas temperatures, residual ionic charges, and various gas subspecies in the extinguished arc channel. The channel susceptibility to dielectric breakdown, then, depends on its spatial temperature distribution, the concentrations of the various gas species (both charged and uncharged), and the distribution of the electric field along the channel as determined by the nozzle geometry, the TRV, and the spatial distribution of the remaining charged particles in the channel. Both the channel temperature and its particle concentrations depend on the channel properties at current zero, the energy loss mechanisms cooling the channel, the cold gas pressure distribution, and the ongoing diffusion processes of and chemical reactions among the various species of the channel.

A clear distinction between thermal and dielectric breakdowns in the extinguished arc channel in the regime where one ends and the other begins is not universally agreed upon; but it is accepted that thermal breakdown is characterized by a slow collapse of the channel voltage and a high power input into the channel, while dielectric breakdown is precipitated by a sudden and power efficient process [3]. It is suspected that the few electrons present in the channel undergoing dielectric breakdown do not possess a maxwellian energy distribution [4], which may account for the high efficiency of the dielectric breakdown process.

Most studies of dielectric breakdown have been experimental in nature, motivated, most probably, by the need for industry to develop a better interrupting device. Although some of the work has been performed on actual circuit breakers, much of it has been done on model circuit breakers that allow measurement of relevant parameters under controlled conditions. Scaling model results to actual circuit breakers requires, however, proper theoretical understanding of the underlying physical processes. Recently, some rigorous theoretical treatments of the free dielectric recovery process in the gas-blast interrupter have been reported [5-7], with quite encouraging results. Further theoretical work is still necessary for a better understanding of many aspects of the dielectric recovery.

7.2 EXPERIMENTAL INVESTIGATIONS

Two experimental methods for investigating dielectric recovery in the extinguished arc channel are reported in the literature [8]. The first—the ac method—involves the use of a synthetic test circuit that impresses a TRV across the channel gap directly after current zero [9-11]. The second—the free recovery method—employs a variable-length dc arc which is usually, but not necessarily, terminated suddenly (in <1 µs) followed by the application of a fast-rising voltage step at a predetermined time after current zero to test for breakdown [7,8,12,13] (Fig. 7.1). The ac method for testing for dielectric recovery suffers from integrating the effects of the current decay to zero, the influence of residual currents after current zero, a fixed polarity of the TRV, and current chopping prior to current zero [10]. While these effects are usually present in the actual operation of a circuit breaker, their simultaneous presence does not allow the investigation of the contribution of each separate mechanism

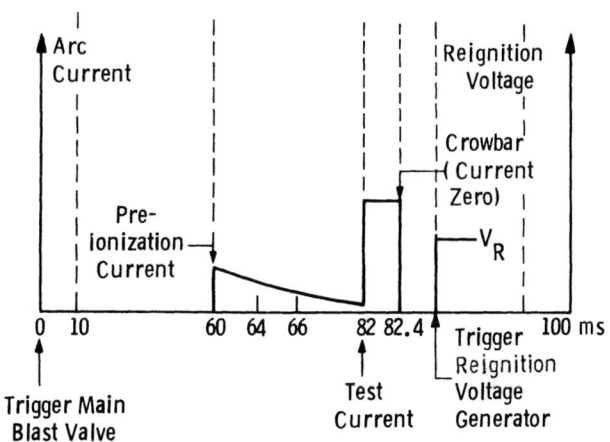

Figure 7.1 Sequence of operations for conducting a free dielectric recovery test on a gas-blast interrupter. (From J. F. Perkins and L. S. Frost, Dielectric recovery and predicted ac performance of blown SF_6 arcs, IEEE Trans. Power Appar. Syst. *PAS-91*:368, © 1972, IEEE.)

to the dielectric recovery of the breaker. The free recovery method, however, allows the investigation of the effect of each mechanism to the recovery process under controllable conditions. For example, the sustenance of the discharge by a known dc current for a determined amount of time followed by its abrupt interruption provides initial conditions at current zero that retain many of the arc plasma properties during the current phase. An attempt to relate dielectric recovery data obtained under free recovery conditions to the recovery in an ac circuit interrupter has been made; however, the suggested approach has met with some criticism [13].

Usually, measurements are made of the recovery of the channel gas temperature and its electrical conductance in addition to its dielectric strength. Differential interferometry and classical Schlieren techniques are employed to determine variation in the gas index of refraction [14]. The use of a pulsed argon laser could freeze motion for periods of 70 ns. The index of refraction measurement allows a determination of the radial density profile for temperatures below the gas dissociation temperature, from which a radial temperature profile may be calculated. Unfortunately, this technique is usually valid for temperatures below 2000 K for nitrogen and 1500 K for sulfur hexafluoride (SF_6) [14]. Electrical conductance is measured through the application of a voltage across the gap—below the breakdown value—and determining the resulting current. This measurement is made difficult by the need to detect very small values of current [14,15]. However, values of channel resistance and capacitance of the order of 10^{10} Ω and 0.1 pF, respectively, can be measured [14,15].

7.2.1 General Characteristics of the Recovery

Initial studies of the dielectric recovery of the arc channel after current zero—in both free-burning and blast arcs—confirmed a general behavior that can be grossly described as a fast linear increase of recovery voltage with delay time of the application of the recovery voltage followed by a slow phase [8,9,10,13,16] (Fig. 7.2). However, closer analysis of the data shows [8,10,12,14] what may be better described as a fast-slow-fast-slow rate of recovery (Fig. 7.3). An initial explanation of the middle pause in the rate of recovery rested on the observation that it occurs at applied recovery voltages which are gas specific. Thus it was suggested [8] that the pause, in the case of a free-burning arc, is associated with the breakdown of a neutral cathode sheath at the minimum spark breakdown voltage of the working gas. Recently, however, it has been suggested that the pause associated with the recovery of the channel of a blast arc is caused by the delay of temperature decay of the channel owing to the release of the heat of dissociation of the parent gas [11].

A consistent result obtained in dielectric recovery tests of all arcs, particularly blast arcs, is the wide scatter in the recovery volt-

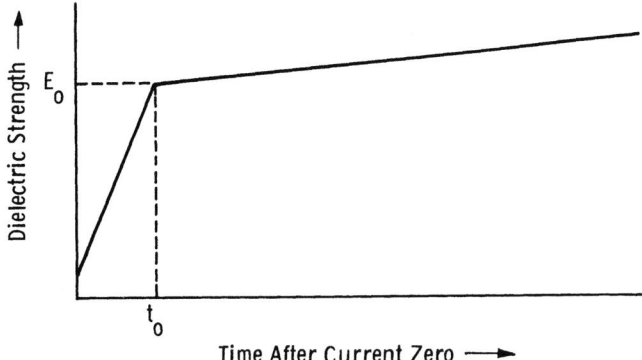

Figure 7.2 General characteristic of the dielectric recovery of an arc. (From T. E. Browne, Jr., Extinction of a-c arcs in turbulent gases, Trans. AIEE 51:185, © 1932, IEEE.)

ages (Figs. 7.3 and 7.4), even though experimental conditions and precurrent zero arc properties are reproduced quite well [9,14,15,17]. Furthermore, the amount of scatter is observed to be especially pronounced in the slow recovery period. This result emphasizes the significance to the recovery process of a physical mechanism which has a wide statistical variation. It has been pointed out [11] that the statis-

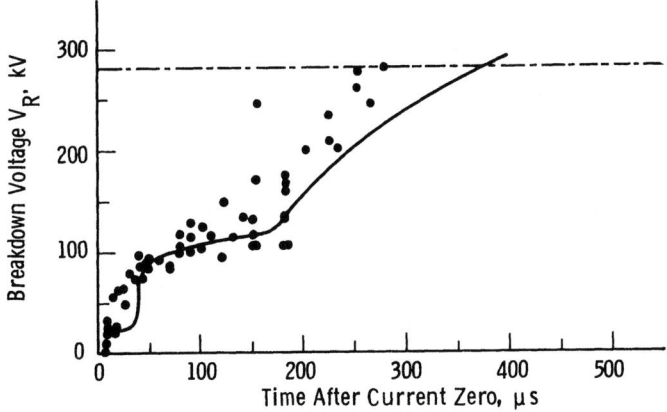

Figure 7.3 Calculated (solid curve) and measured (dots) dielectric recovery characteristics for an SF_6-gas-blast model interrupter. (From K. P. Brand, W. Egli, L. Niemeyer, K. Ragaller, and E. Schade, Dielectric recovery of an axially blown SF_6-arc after current zero, Part III: Comparison of experiment and theory, IEEE Trans. Plasma Sci. PS-10:162, © 1982, IEEE.)

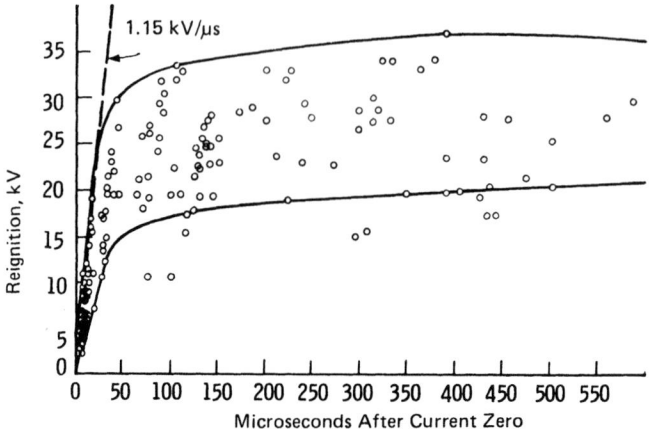

Figure 7.4 Dielectric recovery characteristics for a 60-Hz 1200-A rms air-blast arc in a metal orifice with 20 psig upstream pressure. Open circles represent measured reignition voltages. The two curves represent envelopes enclosing the estimated regions of uncertainty within which most of the breakdowns occurred. (From T. E. Browne, Jr., Dielectric recovery by an a-c arc in an air blast, Trans. AIEE 65:169, © 1946, IEEE.)

tics of the progression of a streamer toward full dielectric breakdown in the arc channel may be the cause for the scatter in the results. It is also possible that the statistics associated with turbulent eddy generation and destruction may also contribute, since turbulent mixing and energy transport seem to be so important in the recovery process.

7.2.2 Effect of Pressure

Dielectric recovery rate is found to increase with gas pressure [9,10, 15]. In particular, the early fast linear recovery with delay time [15] in a SF_6-blast model interrupter depends on the ratio of upstream to downstream pressure, being 11, 3.4, 2.1, and 1.8 kV/μs for pressure ratios of 6.8/1, 3.4/1, 1.8/1, and 1.3/1, respectively. However, the plateau in the dielectric recovery of an air-blast interrupter reached at about 300 μs after current zero (Fig. 7.5) is rather insensitive to the tank pressure used [9], but the upstream tank pressure needed to ensure the dielectric strength of the arc channel [9] increases with the interrupted rms ac current (Fig. 7.6). These results seem to indicate that not only the pressure but also its axial gradient strongly influence the recovery process.

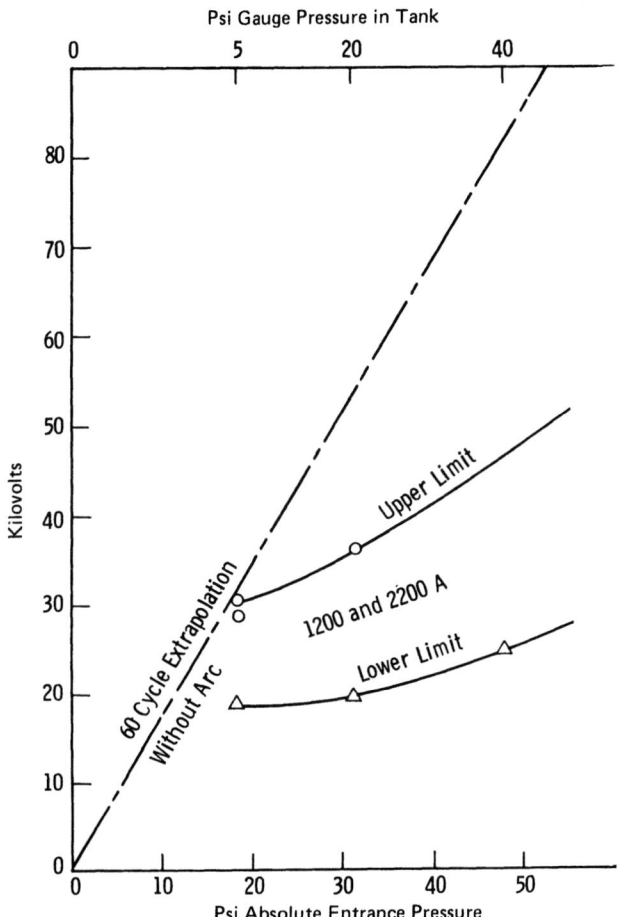

Figure 7.5 Effect of the upstream pressure on the dielectric strength, 300 μs after current zero, of 1200-A and 2200-A rms air-blast arcs. (From T. E. Browne, Jr., Dielectric recovery by an a-c arc in an air blast, Trans. AIEE 65:189, © 1946, IEEE.)

7.2.3 Effect of the Nozzle Geometry

The geometrical features of the discharge chamber that are expected to influence arc channel recovery are the nozzle convergence and divergence angles, nozzle throat dimensions, nozzle length, and position of the electrodes. These parameters determine axial pressure gradients, cold gas mass flow rates, nozzle clogging possibilities during the peak current phase, and total discharge channel length. An extensive

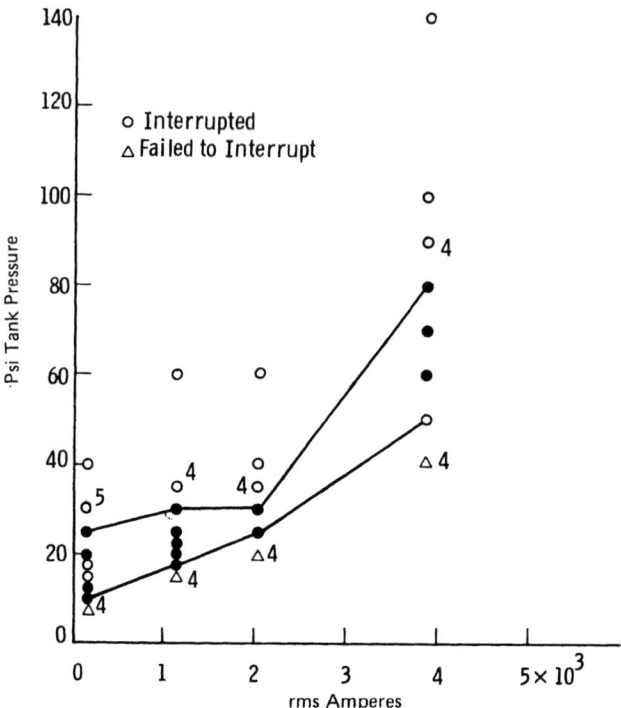

Figure 7.6 Interrupting pressure range as a function of arc rms current in an air blast. The upper line is connected through points representing tests at the maximum pressure at which a failure to interrupt occurred, and the lower line through points of minimum pressure for a successful interruption. (From T. E. Browne, Jr., Dielectric recovery by an a-c arc in an air blast, Trans. AIEE 65:169, © 1946, IEEE.)

parametric study of these features for an SF_6-blast interrupter [18] resulted in specific recommendations for optimum operation. For example, it is found that optimum upstream electrode distance from the nozzle throat is about one-half the throat diameter, and that the dielectric recovery is insensitive to nozzle divergence angle, although a cylindrical nozzle dramatically diminishes the recovery capability of the interrupter. A more recent study [14] finds that the geometry affects the dielectric recovery mainly through its influence on the axial profile of both the pressure and the net electric field.

7.2.4 Effect of Arc Current

The magnitude of the interrupted arc current is expected to influence the dielectric strength of the extinguished arc channel through three mechanisms. First, radiation emission from the arc during the high-current regime creates a hot nonconducting gas mantle surrounding the core [19], which is usually slow to cool down, persists long into the postarc period, and acts as a thermal barrier to radial energy transport from the arc channel. The radial extent of this mantle is expected to increase with the magnitude of the arc current [19,20]. Second, the initial conditions of the extinguished arc channel at current zero depend on the magnitude of the interrupted ac arc current. For example, both the arc channel radius and temperature at current zero increase with the arc current magnitude [21,22]. Third, in the event that the arc clogs the nozzle throat near its peak value, ablation products from the nozzle surface would contaminate the arc channel plasma and, therefore, alter its electrical conductivity and chemical composition for a period corresponding to the transit time of the gas in the channel.

It was observed early that the recovery rate of free-burning [10] and blast ac arcs [9] decreases with increasing interrupted current. To retain a constant value of dielectric strength for the arc channel as the ac current is increased, it is found [9] that the upstream tank pressure needs to be raised (Fig. 7.6). When the free recovery method is used to study the dependence of dielectric recovery of a SF_6-blast model interrupter on the arc current [13], it is observed that the time delay needed to provide a specific dielectric strength for the channel increases with arc current (Fig. 7.7). This trend is seen, however, to be violated for the higher currents where the required delay is higher at a lower current (350 A compared to 1000 A). A possible explanation for this reversal [13] is the larger contribution of self-magnetic-field pumping and upstream gas heating at the higher current toward the forced axial flow, which, effectively, translates to the imposition of a higher upstream pressure. A repetition of the experiment at a much higher upstream pressure rendered the relative contribution of this effect to the axial flow nonsignificant, and the delay needed to attain the required dielectric strength again increased with the arc current [13].

An investigation of the influence of the arc current on the dielectric recovery of the interrupter over a wide range of current values shows a negligible effect at rather low values, but a significant effect when the current exceeds some critical value [23] (Fig. 7.8). A phenomenological analysis [20] of the dependence of the interruption limiting curve of an SF_6 circuit breaker on ac current has delineated the regimes of failure of the breaker under both dielectric and thermal modes.

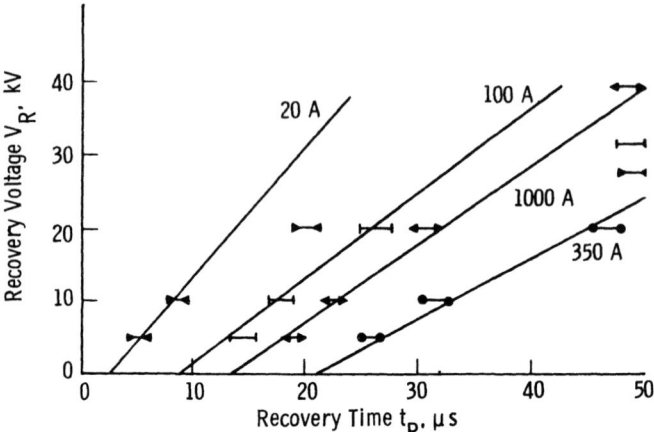

Figure 7.7 Dielectric recovery of an SF_6 gas-blast arc in a free recovery test with arc current as a parameter. Upstream pressure is 35 psia. Straight lines are empirical fits to the measured data. (From J. F. Perkins and L. S. Frost, Dielectric recovery and predicted ac performance of blown SF_6 arcs, IEEE Trans. Power Appar. Syst. PAS-91: 368, © 1972, IEEE.)

7.2.5 Effect of the Gas

High-voltage circuit breakers ususally employ air, SF_6, or oil as an interrupting medium. The latter contributes hydrocarbon gases and hydrogen for the interruption process when the arc dissociates the oil molecules. Early tests [10,24] show a dramatic dependence of the dielectric recovery on the filling gases with hydrogen being the most

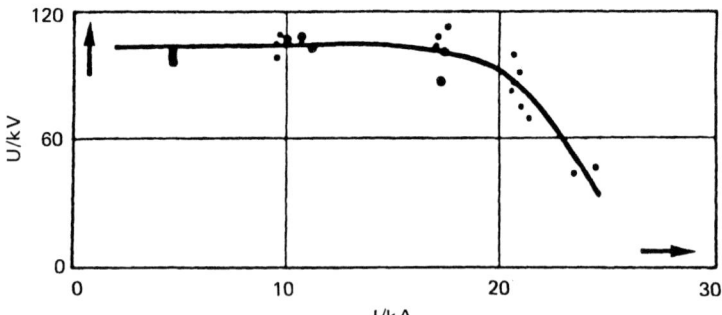

Figure 7.8 Dielectric breakdown voltage as a function of arc current in a double-nozzle SF_6 circuit breaker. (From Ref. 23.)

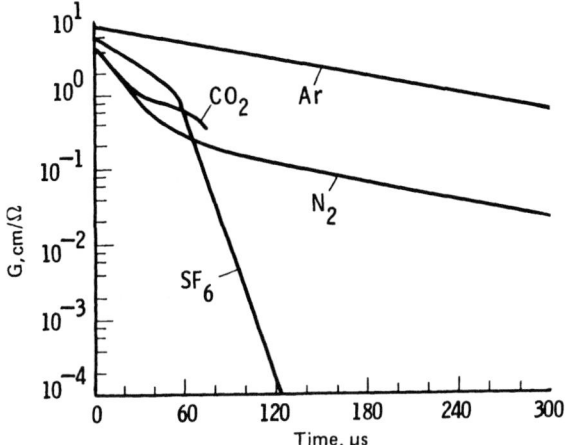

Figure 7.9 Electrical conductance decay in interrupted cascade arcs. (From W. Hertz, H. Motschmann, and H. Wittel, Investigations of the properties of SF_6 as an arc quenching medium, Proc. IEEE 59:485, © 1971, IEEE.)

effective followed by carbon dioxide, oxygen, helium air, nitrogen, and finally argon, in that order. Tests on the dielectric recovery in SF_6, oil, and N_2 interrupters show a wide variation among them, with SF_6 being superior at very high recovery voltages (see Fig. 9.2). Investigation of the free recovery of cascade arcs in SF_6 and N_2 indicate [25] (Fig. 7.9) that nitrogen has a slightly faster electrical conductivity decay rate while the arc channel temperature is above 10,000 K, but at lower temperatures the SF_6 arc channel electrical conductivity drops precipitously below that of the N_2 arc channel. The investigation [25] also confirms that efficient energy-loss mechanisms are responsible for the faster electrical conductivity decay rate rather than electron attachment by atoms and molecules of the SF_6 gas.

Interferograms of the arc channel prior to current zero and during the dielectric gas recovery period show a distinct difference in the extent of the hot gas mantle surrounding the channel core between air- and SF_6-blast arcs. In a 2000-A air-blast arc [26] a well-defined mantle extending several millimeters is clearly evident, while in a 1200-A SF_6-blast arc [14], no mantle can be distinguished within the resolution of the measuring system. Since the hot gas mantle acts as a bottleneck to radial energy transport from the channel core to the surrounding cold gas during the dielectric periods, the absence of such a mantle in the SF_6 arc channel clearly provides a faster cooling mechanism than in a corresponding air arc channel.

7.2.6 Effect of Turbulence

Turbulent flow in the gas blast arc is observed in all phases of the arc discharge [27]. In the high-current regime, turbulent flow is evident in the downstream region but is absent from the upstream part [22], a situation that persists as the arc current is reduced to the zero value. About 60 µs after current zero in an SF_6-gas-blast model interrupter, turbulent eddies are observed to form at the nozzle entrance and to progress upstream until, some 10 µs later, the whole channel is engulfed in it [14], as can be seen in the series of interferograms of Fig. 7.10. Why turbulence starts in the upstream region at such a delayed time is not understood now, but it can be speculated that propitious conditions for the development of an instability leading to turbulent flow in that region are met at that time.

A correlation study between the incipience of turbulent flow and the dynamics of the channel diameter in the upstream region shows [14] that the channel diameter increases when turbulent flow is established. This is consistent with the hypothesis that turbulent eddies are very effective in reducing the core temperature while broadening its radius through their efficient radial energy transport.

Because turbulent flow is conducive to accelerating the cooling rate of the channel core, the employment of an upstream grid to enhance turbulence intensity in the upstream region has been investigated [14]. Results show that the grid improves the dielectric recovery rate during the first phase of the recovery, but degrades it during the later phase. It is not clear what aspect of the induced turbulent flow causes the modified behavior. But such results point out the complexity of the physical processes associated with turbulent flow.

7.2.7 Effect of Space Charge

Because it distorts the externally applied electric field, space charge in a hot gas is found to modify the breakdown behavior of the gas, causing it to depart from Paschen curve predictions [3,28]. Investigations into the influence of space charge in a gas on its breakdown behavior have been performed in shock tubes [3] and in the wake of a cross-flow arc [28]. Breakdown in a hot nitrogen gas produced in a shock tube has been studied as a function of its electron density [3]. The two regions of breakdown—the thermal and the dielectric—are defined on the basis of the values of the electron densities present in the gas (Fig. 7.11). A later investigation [4] modifies this classification to take account of the actual electron energy distribution, which is suspected of having at least as important a role as the value of the electron density on the breakdown voltage.

Figure 7.10 Interferograms in the upstream region of a double-nozzle SF_6-gas-blast arc ramped from a 1200-A dc value to zero at the rate of 27 A/µs for the following times after current zero: (a) −40 µs; (b) 4.8 µs; (c) 23 µs; (d) 34 µs; (e) 54 µs; (f) 85 µs; (g) 112 µs; (h) 320 µs; (i) 512 µs. (From K. Ragaller and E. Schade, Dielectric recovery of an axially blown SF_6-arc after current zero, Part 1: Experimental investigation, IEEE Trans. Plasma Sci. *PS-10*:141, © 1982, IEEE.)

Figure 7.10 (Continued)

Postarc Dielectric Recovery in a Blast Arc 255

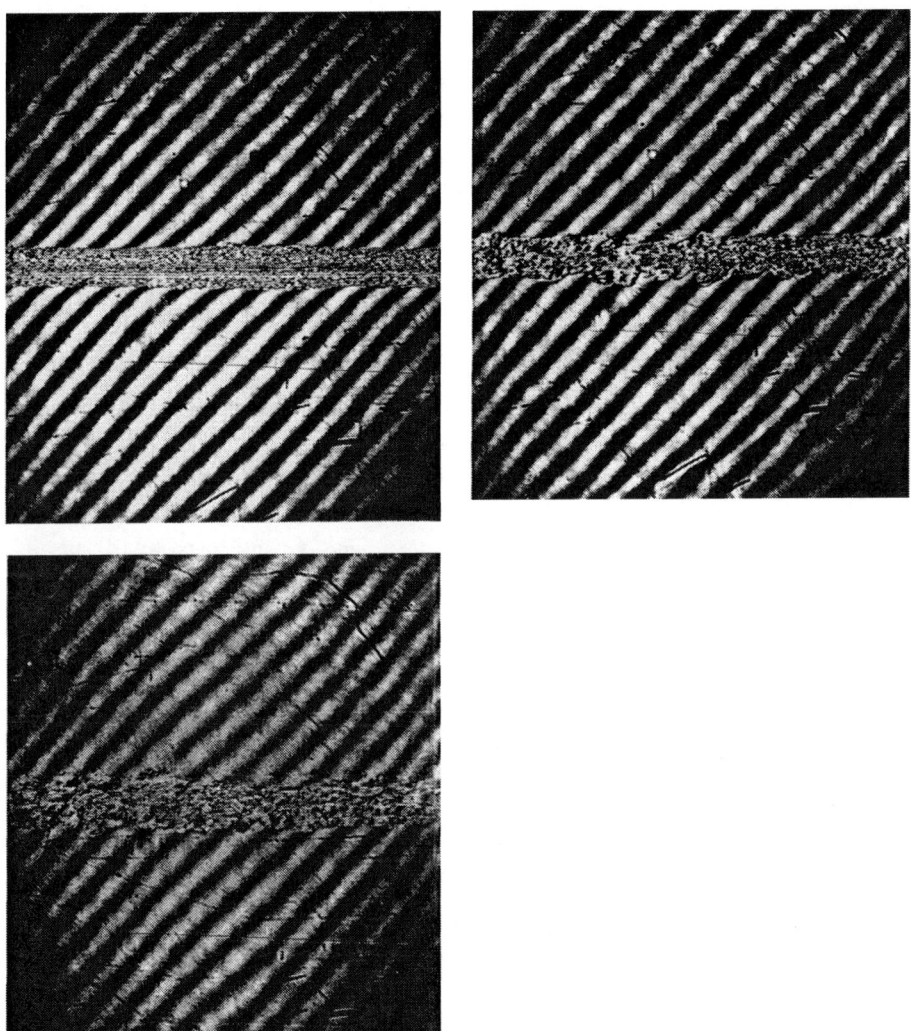

Figure 7.10 (Continued)

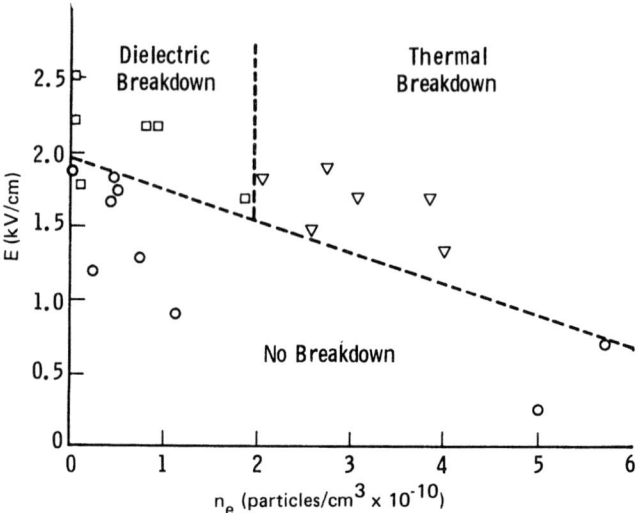

Figure 7.11 Regimes for dielectric and thermal breakdown in nitrogen gas as a function of the electron density in the gas. (From T. H. Lee, A. N. Greenwood, and D. R. White, Electrical breakdown of high-temperature gases and its implications in post-arc phenomena in circuit breakers, IEEE Trans. Power Appar. Syst. *PAS-84*:1116, © 1965, IEEE.)

In a gas-blast interrupter the space charge distributes itself in a characteristic time along the channel so as to even out the peaks in the externally applied field [23]. In an SF_6-blast arc channel the charges, 10 μs after current zero, consist mainly of F^- and S_2^+ ions of density $\simeq 10^{14}$ cm^{-3} at the stagnation point [11]. The space-charge concentration decays thereafter by three body recombination. When the space-charge concentration falls below 10^{10} cm^{-3}, around 100 μs after current zero, its influence on the channel dielectric properties can be ignored [11].

The dielectric strength of the arc channel at times larger than 100 μs is found to depend on whether a recovery voltage is applied right after current zero or not. This is caused by the axial redistribution of the space charge in response to the applied voltage and therefore the modification of the axial profile of the electric field [11]. Thus instead of the upstream region withstanding the total applied voltage, the redistributed space charge allows the rest of the channel to share in the voltage drop. As a consequence, the free recovery method is expected to yield a lower value for the dielectric strength of the channel than would be obtained by the ac method [11].

7.2.8 Location of the High-Dielectric-Strength Region

In arcs not subject to forced convection, probe measurements indicate that dielectric strength of the extinguished arc channel resides mainly in the space next to the electrodes [24]. It is suggested that the TRV appears initially across sheaths at the electrodes, whose thickness increases with the applied voltage. When the applied voltage exceeds the sheath breakdown voltage, breakdown of the whole gap occurs. However, when turbulent flow is introduced to the arc channel, by rotating the arc in a radial magnetic field, dielectric strength increases substantially and becomes positively correlated with the overall arc length and the intensity of the generated turbulent flow [24].

Experiments in an air-blast model interrupter [9] suggest that the region which supports the dielectric recovery starts initially near the nozzle entrance, then moves downstream to the nozzle throat. More exhaustive investigations [2,11,15] on an SF_6-blast model interrupter indicate that dielectric recovery commences in the turbulent downstream region during the first 100 μs after current zero, while the upstream region remains in a relatively conducting state. Later, when turbulent flow engulfs the upstream region and the influence of the space charge on the field distribution vanishes, the upstream region, particularly the stagnation point area, assumes most of the voltage drop.

7.3 THEORETICAL INVESTIGATIONS

7.3.1 Physical Processes

After current zero is reached, the extinguished arc channel experiences a process of temperature decay, chemical reactions, and partial diffusion whose ultimate result is the return of the gas to its initial cold, totally recombined state. This process is expected to occur in a state that is characterized by some deviation from local thermodynamic equilibrium. This deviation may take the form of one or more of the following: nonequality of the electron and heavy particle temperatures, nonmaxwellian electron energy distribution, chemical nonequilibrium, and demixing of the various particle species.

In the dielectric breakdown regime the electric energy input from the externally applied electric field is negligible compared to the energy loss from the arc channel. The major energy-loss mechanisms are radial and axial convection, expansion cooling, and radial conduction. The latter consists of classical conduction, including chemical energy diffusion, and the more important radial energy exchange induced by free shear turbulence. Radiative emission loss is considered to be much smaller than these other mechanisms, owing to the usually low channel temperature (<10,000 K).

Nonequal Electron and Heavy-Particle Temperatures

The externally applied electric field gives energy mainly to the free electrons in the channel, owing to their high mobility. In turn, the electrons transfer part of this energy to the heavy particles through elastic and inelastic collisions at a rate related to the difference between the temperatures of the electrons and the heavy particles. For this temperature difference to stay small compared to the electron or heavy-particle temperature, the Maecker criterion [29,30] should hold:

$$\frac{T_e - T_g}{T_e} = \frac{M}{m_e} \left(\frac{e\lambda E}{4kT_e}\right)^2 \ll 1 \qquad (7.1)$$

where T_e, m_e, and T_g, M are the temperature and mass of the electrons and the heavy particles, respectively; e the electronic charge; λ the electron mean free path; E the electric field; and k the Boltzmann constant. Thus equation (7.1) suggests that small mean free paths (e.g., high pressures), small electric fields, and high temperatures are conducive to minimizing temperature differences between electrons and heavy particles. When the temperature difference is small, it is feasible to obtain approximate values for the plasma material and transport properties from those applicable to the LTE plasma—a procedure that is rather uncomplicated to implement, but could also lead to serious errors if not properly utilized.

It has been suggested [31] that the difference between the electron and heavy-particle temperatures could explain the inability of Mayr's theory to explain experimental thermal breakdown results. A modified Mayr-type equation has thus been derived, which accounts for a higher electron density and temperature, which in turn predicts results in better agreement with experiment. However, in this derivation a multitemperature Saha equation [32,33] is utilized. This form of the Saha equation has been shown [34] to yield erroneous results; therefore, the conclusions drawn from the modified Mayr equation [31] should be interpreted carefully. When the electron and heavy-particle temperatures differ substantially, it is appropriate [34] that the kinetic equations for each of the channel plasma species be solved together with a proper set of momentum and energy conservation equations to determine the channel plasma properties.

Nonmaxwellian Electron Energy Distribution

When the externally applied electric field exceeds some critical value for a given pressure and electron density, the electron energy distribution starts to depart from a maxwellian, and the concept of an electron temperature becomes invalid. The high electric field causes some electrons to gain excessive energies, making them very efficient ionizing agents. This process may explain the sudden and highly efficient

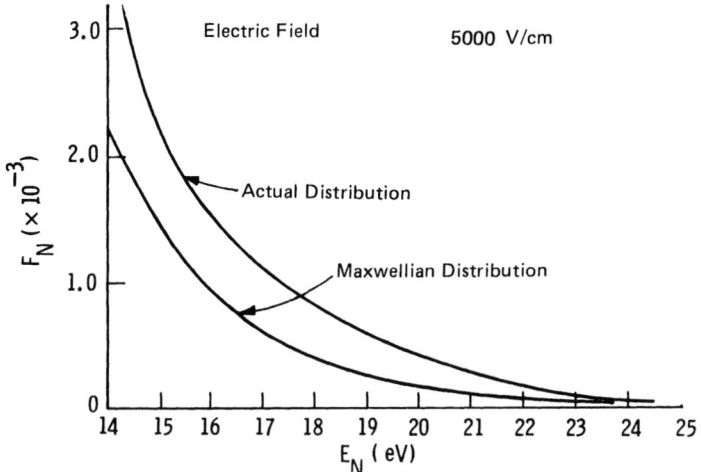

Figure 7.12 High-energy electron population in a maxwellian energy distribution and in an actual distribution calculated from the Boltzmann equation in the presence of a 5000-V/cm electric field in a hot (3000 K) nitrogen gas of density 3.35×10^{18} particles per cubic centimeter and electron density of 10^{11} electrons per cubic centimenter. (From Ref. 4.)

nature of dielectric breakdown [4]. In this situation the arc channel plasma would be far removed from an equilibrium state, and the calculation of the material and transport properties of the plasma cannot follow the equilibrium-based approach. A solution of the Boltzmann equation to obtain basic plasma properties becomes necessary in this case [4]; and the results show that when a rather moderate electric field of 5000 V/m is present in the channel, the population of high energy ($\geqslant 14$ eV) electrons substantially exceeds that calculated from an equivalent maxwellian energy distribution (Fig. 7.12). The enhanced high-energy electron population would greatly increase the rate of ionization and in turn boost both the channel electrical conductivity and its space-charge concentration.

Chemical Nonequilibrium

During the recovery period, the various arc constituents undergo a series of chemical reactions whose ultimate result is the reconstitution of the original gas state in existence prior to the initiation of the arc. The rates of progress of the different reactions depend in a complex way on the concentrations and energies of the various species at any position and time during the recovery period, and the rate of temperature decay of the arc channel. When the temperature decay

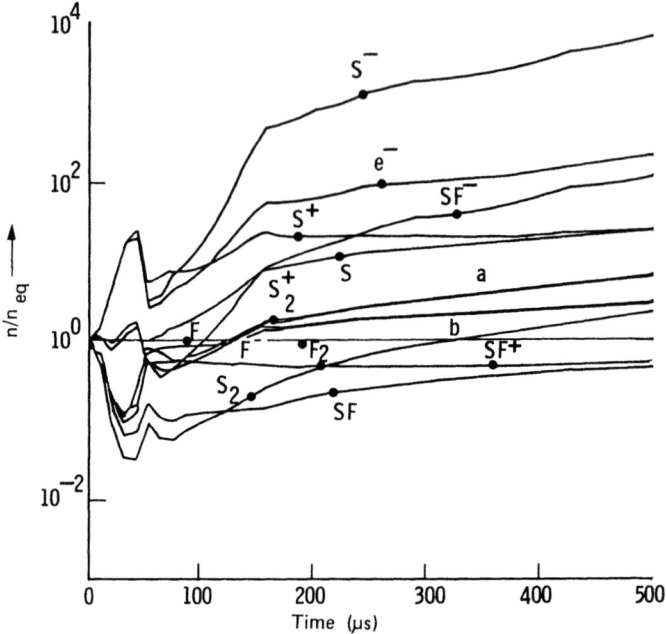

Figure 7.13 Decay of relative particle concentrations after current zero in the channel of an SF_6-blast arc. n_{eq} is the concentration obtained under partial diatomic equilibrium conditions. (From Ref. 35.)

rate of the channel is small compared to the ongoing reaction rates, the concept of local thermodynamic equilibrium may be invoked and the concentrations of the various species, as well as the plasma material and transport properties, can easily be calculated from statistical mechanics considerations. However, when the channel temperature decay rate becomes comparable to or faster than some reaction rates, a state of chemical nonequilibrium develops within the channel plasma that requires special treatment to obtain valid plasma properties. This may involve a perturbation approach or possibily a solution of the kinetic equations for the plasma particle concentrations, in conjunction with momentum and energy conservation equations. It can be done, for example, under the assumption of the existence of partial equilibrium in the channel [23,35] during its free recovery. Partial equilibrium is described as an equilibrium state in which only a subset of the possible particle species of the plasma are considered to exist. For example, in a decaying SF_6 plasma, a partial diatomic equilibrium state has been considered [23,35] in which electrons, ions, atoms, diatomic molecules, and diatomic molecular ions exist in equilibrium, while molecules and ions with more than two atoms are neglected. When the plasma temperature is changing rapidly, reaction rates among the various species may not be fast enough to preserve the partial equilibrium state. Results of calculations [35] of particle concentrations from detailed rate equations in a decaying SF_6 plasma whose temperature drops from 10,000 K to 3000 K in 150 µs show substantial deviation, in some instances reaching two orders of magnitude, from values obtained under partial diatomic equilibrium conditions (Fig. 7.13). It is apparent, then, that a state of chemical nonequilibrium is not uncommon whenever the arc plasma temperature is changing rapidly.

Demixing of Species

In the presence of steep temperature gradients such as those found in the decaying arc channel, the radial diffusion rates of the various channel species are not the same, leading to a deviation of the particle concentrations from those dictated by the stoichiometric ratios of the parent gas [36]. Moreover, this demixing effect causes the concentration ratios of the various species to vary in space. When this effect becomes significant, the determination of particle concentrations cannot proceed according to the assumption that they depend only on the local values of temperature and pressure. It now becomes necessary to solve the continuity equation for each species with account taken of its own diffusion and, perhaps, convective velocities.

Turbulence

The fluid-flow properties in the nozzle of the gas-blast arc are paramount in shaping arc dielectric recovery. The role of turbulence in enhancing energy and particle radial transport was noted [1,16,24] during the early development of the gas-blast breaker. Recent work [14] has added significant experimental evidence of the presence of large and small turbulent eddies in the arc channel throughout the dielectric recovery period. Unfortunately, a rigorous mathematical description of turbulent transport is not available at present to relate experimental observations unequivocally to the effect of turbulence. However, phenomenological models of turbulent transport based on the Prandtl mixing length [37] have been able to predict results consistent with experiment [11,38,39]. It is also reasonable [7] to apply a sensitivity analysis to assess the range of possible results obtainable from a set of input parameters.

According to the Prandtl mixing-length hypothesis [37], the effect of free shear turbulence on the radial transport of momentum and energy can be accounted for by modifying the plasma viscosity η and thermal conductivity κ to include laminar and turbulent contributions:

$$\eta = \eta_\ell + 0.5\varepsilon_t \tag{7.2}$$

$$\kappa = \kappa_\ell + C_p \varepsilon_t \qquad (7.3)$$

where η_ℓ and κ_ℓ are the laminar parts and ε_t represents the turbulent eddy diffusivity of momentum, which is given by

$$\varepsilon_t = \lambda \rho \delta \, \Delta V \qquad (7.4)$$

with λ a constant determined from experiment and ΔV an appropriate difference in axial velocity between the hot and cold gases. δ is an effective radial dimension of the arc channel, usually taken as the channel radius; ρ is the mass density; and C_p is the specific heat at constant pressure.

Recent investigations into the nature of turbulent shear flows [40] have revealed the existence of both large-scale and small-scale eddies. The former are coherent, exist for significant lifetimes, and have important bearings on the mixing of the two regions. This character can be clearly distinguished from the photographs of Fig. 7.14. The large-scale properties of the three flows are very much the same even though the Reynolds numbers span a large range extending from the laminar to the highly turbulent. The large, coherent eddies tend to entrain large amounts of one of the fluids, which is then acted upon by turbulent small eddies, causing complete mixing of the fluids of the two regions and, in effect, resulting in efficient energy and momentum transport [40].

7.3.2 Early Models of Arc Channel Dielectric Recovery

During the early development of gas-blast interrupters, two theories for dielectric recovery were proposed: the displacement theory [41, 42] and the diffusion theory [1]. In the displacement theory it is hypothesized that a cool nonionized section of gas replaces a part of the extinguished arc channel and progressively displaces more and more of the arc channel until the whole channel is eventually replaced with nonionized gas. It is further hypothesized that the applied voltage across the channel stresses mainly this nonionized section of the channel and therefore the dielectric strength of the channel becomes dependent on the length of the nonionized section, which increases with time proportionately to the axial velocity of the nonionized gas. The diffusion theory proposes an enhanced radial diffusion of both ions and energy because of the free shear turbulence generated by the axial flow of the ionized and nonionized gases. The high degree of turbulence is presumed to enhance the rate of deionization and cooling of the arc channel by two orders of magnitude over classical processes. Such an enhanced rate is found to be consistent with observation. The arguments put forth to support either theory are rather qualitative in nature and any calculations presented to support either model are not rigorous. While some fundamentally sound arguments are presented to dispute the contentions of the displacement theory [1], there do not exist today definitive experiments that completely and totally support either of the two models. This stems, in part, from our lack of understanding of many aspects of turbulent flow. It may be possible, for example, that large eddies coexisting with small eddies (Fig. 7.14), may give some support to the displacement model, in that they may lead, when the channel radius is very small, to the entrainment of large amounts of cold gas into the channel, effectively severing it.

7.3.3 Models for the Gas-Blast Arc Channel in Free Recovery

In the absence of a TRV, it can be safely assumed that all channel species, including free electrons, have a maxwellian energy distribution. The decaying arc channel temperature can, then, be calculated under the assumption of various degrees of deviation from the equilibrium state. As a first approximation, the conservation equations for mass, momentum, and energy in differential form [5,7] may be taken to describe the channel dynamic behavior:

$$\frac{\partial \rho}{\partial t} + \frac{\partial}{\partial z}(\rho V_z) + \frac{1}{r}\frac{\partial}{\partial r}(r\rho V_r) = 0 \tag{7.5}$$

$$\rho \frac{\partial V_z}{\partial t} + \rho V_z \frac{\partial V_z}{\partial z} + \rho V_r \frac{\partial V_z}{\partial r} = -\frac{dp}{dz} + \frac{1}{r}\frac{\partial}{\partial r}\left(r\eta \frac{\partial V_z}{\partial r}\right) \tag{7.6}$$

$$\rho \frac{\partial h}{\partial t} + \rho V_z \frac{\partial h}{\partial z} + \rho V_r \frac{\partial h}{\partial r} = V_z \frac{dp}{dz} + \frac{1}{r}\frac{\partial}{\partial r}\left(\frac{r\kappa}{C_p}\frac{\partial h}{\partial r}\right) + \eta \left(\frac{\partial V_z}{\partial r}\right)^2 \tag{7.7}$$

where ρ is the mass density, V_r and V_z the radial and axial velocities, respectively; p the pressure imposed by the cold gas flow, usually determined from either the isentropic or isothermal approximations [43] for gas flow in a laval nozzle; h the enthalpy, related to temperature T by $dh = C_p dT$; η and κ the viscosity and thermal conductivity, respectively, and include both laminar and turbulent flow contributions.

Equation (7.6) represents momentum conservation for the axial directions; radial momentum conservation is not of interest since the radial pressure gradient is assumed to be negligible. The externally applied axial pressure gradient is balanced by viscous, inertial, and axial momentum transport forces and establishes the axial velocity profile in the channel. Equation (7.7) describes the energy balance for the arc channel. Energy-loss mechanisms consist of axial and radial convection, mechanical expansion cooling, radial conduction, and viscous dissipation [6]. Other mechanisms, such as axial conduction and thermoelectric effects, are found to be orders of magnitude smaller in the channel. Energy loss by radiation can also be neglected

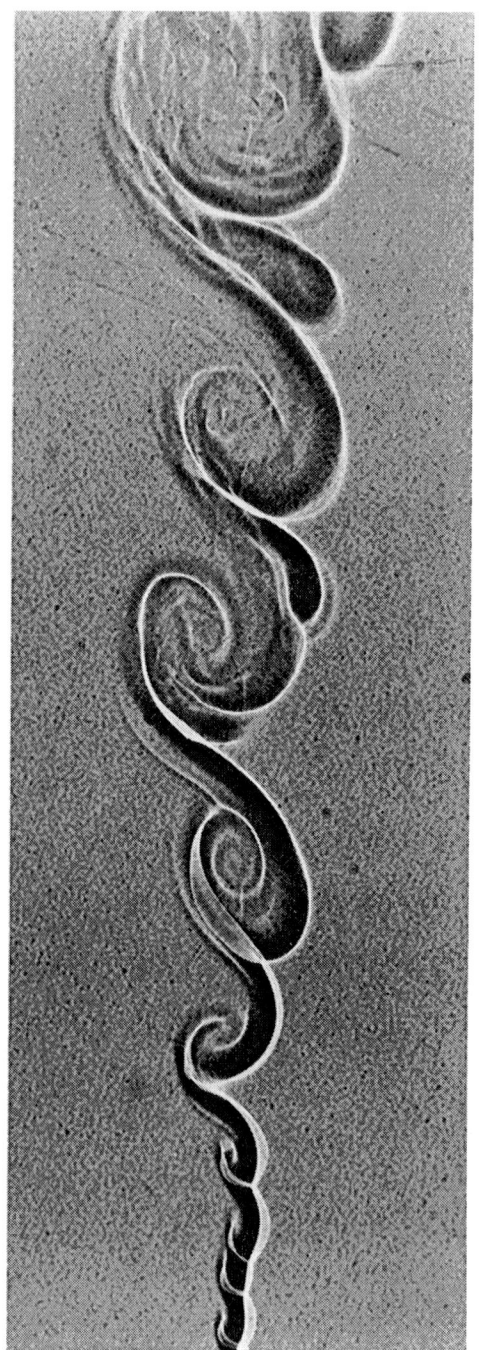

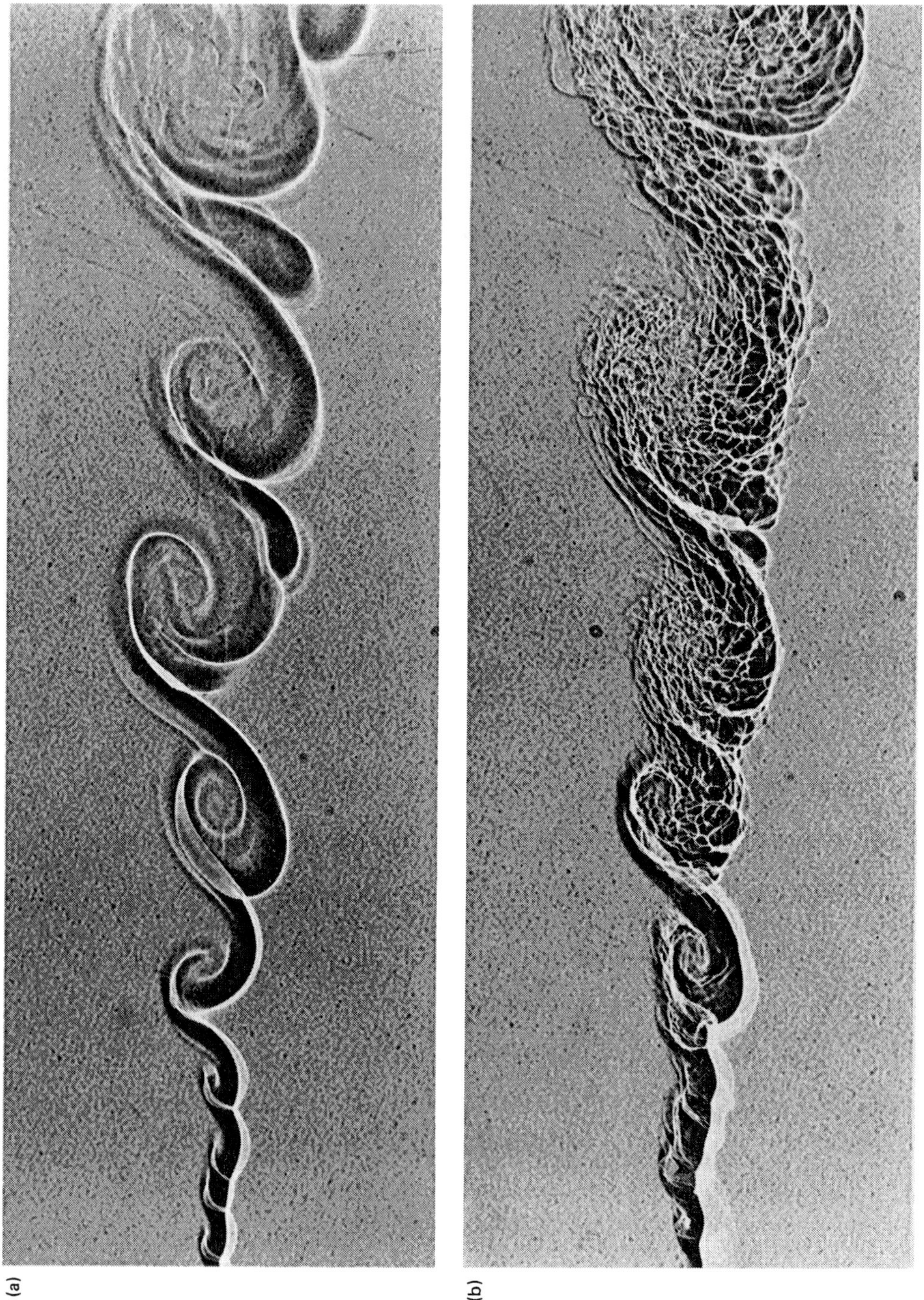

Postarc Dielectric Recovery in a Blast Arc

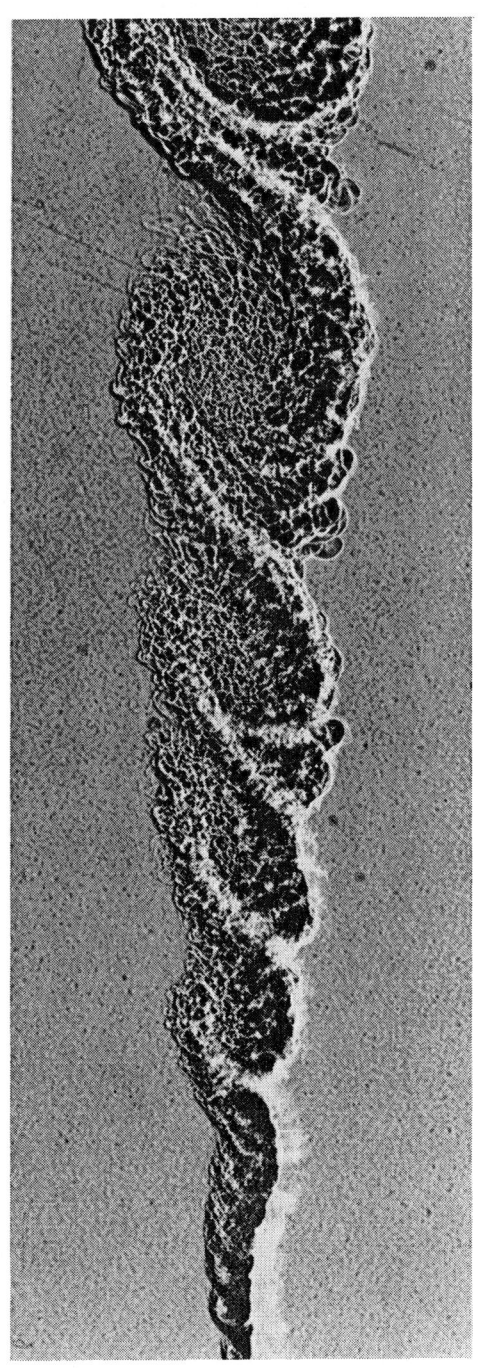

Figure 7.14 Mixing layers between two streams of helium and nitrogen gases with velocity ratio of 0.38 and density ratio of 7 for three Reynolds numbers in the helium stream: (a) 3×10^4; (b) 6×10^4; (c) 1.2×10^5. (From Ref. 40.)

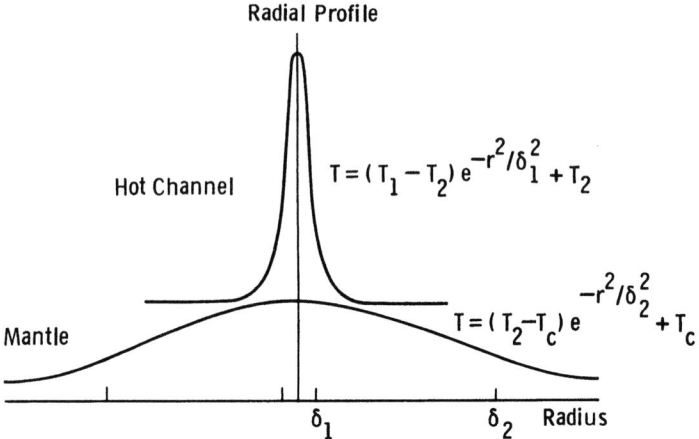

Figure 7.15 Double gaussian radial temperature profile for the arc channel and hot gas mantle of the two-zone model. (From E. Richley and D. T. Tuma, Mechanisms for temperature decay in the freely recovering gas blast arc, IEEE Trans. Plasma Sci. *PS-10*:2, © 1982, IEEE.)

for channel temperatures below 10,000 K. In free recovery an external electric field is not present, and therefore an electric power input term is not included in the energy balance equation.

One-Dimensional Model

The decay of the temperature, velocity, and radius of the extinguished gas-blast arc channel in SF_6 and N_2, in free recovery, has been theoretically investigated [5,6] by numerically solving a one-dimensional model. The reduction of the problem to a one-dimensional situation markedly reduces the computational time needed to obtain a simulation. The axial and temporal decay of the arc channel parameters are simulated by solving equations (7.5) through (7.7) in z and t, while the functional radial dependencies of the channel parameters are taken to be known. The radial temperature and velocity profiles are taken to have a double gaussian dependence: one for the channel core, the other for the hot gas mantle (Fig. 7.15). Thus for the channel core (region I)

$$T_I(r) = (T_1 - T_2) \exp\left(\frac{-r^2}{\delta_1^2}\right) + T_2 \qquad (7.8)$$

$$V_{zI}(r) = (V_1 - V_2) \exp\left(\frac{-r^2}{\beta_1^2}\right) + V_2 \qquad (7.9)$$

and for the hot gas mantle (region II)

$$T_{II}(r) = (T_2 - T_c) \exp\left(\frac{-r^2}{\delta_2^2}\right) + T_c \qquad (7.10)$$

$$V_{zII}(r) = (V_2 - V_c) \exp\left(\frac{-r^2}{\beta_2^2}\right) + V_c \qquad (7.11)$$

Initial conditions on the arc channel and the hot gas mantle are obtained either from previous calculations of the thermal period or from experiment.

The conservation equations (7.5) through (7.7) are then solved in the channel core and the hot gas mantle for the variables T_1, T_2, V_1, V_2, δ_1, δ_2, β_1, and β_2 as functions of t and z for a given set of initial conditions at current zero, subject to the boundary conditions at the upstream stagnation point and at the interface between the channel core and the hot gas mantle. The six equations needed to determine the parameters above are obtained from the conservation equations evaluated at the channel axis (i.e., at r = 0) and their integral form across the radial extent of the channel and the hot gas mantle.

With the assumption of existence of LTE, the material and transport properties of the channel plasma and mantle are uniquely determined once the pressure and temperature are specified. These have been calculated by Frost and Liebermann [44] for SF_6 and Yos [45] for N_2.

The contribution of turbulent radial energy transport and viscosity and radial gas flow cooling to the dielectric recovery of the extinguished arc channel are found to be most significant. More important, it is determined that the size of the hot gas mantle is paramount in controlling the channel cooling rate once the central temperature declines below the 5000 K value (Fig. 7.16). This is so because the slowly cooling hot gas mantle limits the radial temperature gradient that controls radial heat flow from the channel. Only in the event that the hot gas mantle becomes very thin, thus losing its large heat capacity and therefore allowing its temperature to decline rapidly in response to energy loss, does it permit the channel to efficiently loose energy by radial heat flow. Initial conditions in the channel core at current zero, however, are found to have negligible influence on the cooling rate of the channel during the dielectric period [5,6].

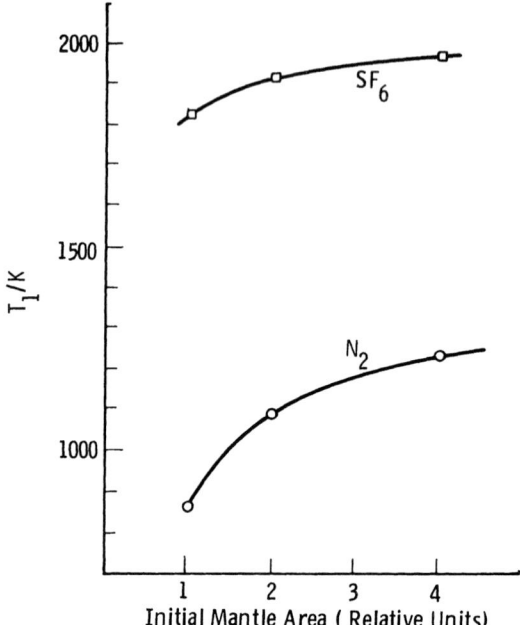

Figure 7.16 Dependence of the channel temperature T_1 at the nozzle exit 100 μs after current zero on the initial hot gas mantle area. (From E. Richley and D. T. Tuma, Mechanisms for temperature decay in the freely recovering gas blast arc, IEEE Trans. Plasma Sci. PS-10:2, © 1982, IEEE.)

The influence of scaling up the nozzle pressure on the temperature decay of the channel is also found to be negligible (Fig. 7.17). This result could be explained by noting that both the heat content of the channel as well as the cooling mechanisms scale up in the same fashion with pressure, thus nullifying each other's effect.

The one-diemensional model suffers from having to specify a priori the radial functional forms of the channel and mantle temperature and velocity profiles. In addition, the assumption of the existence of LTE in the channel, although not necessitated by the use of a one-dimensional model, limits the range of validity of the results. Nonetheless, the predictions of this model agree with several experimental observations; and many of the disagreements may be mainly a result of applying inappropriate initial conditions such as to the radius of the hot gas mantle at current zero.

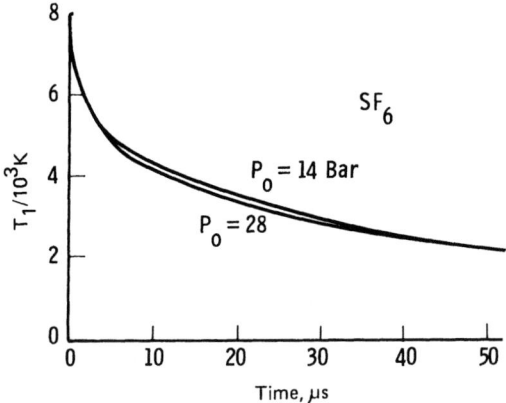

Figure 7.17 Time evolution of the channel temperature T_1 at the nozzle exit for two upstream pressures. (From E. Richley and D. T. Tuma, Mechanisms for temperature decay in the freely recovering gas blast arc, IEEE Trans. Plasma Sci. *PS-10*:2, © 1982, IEEE.)

Two-Dimensional Model

Recently, a two-dimensional model for the free recovery of a gas blast arc channel based on the conservation equations (7.5) through (7.7) has been numerically simulated [7] and its predictions evaluated in the light of experimental data [11]. Model predictions are found to be in disagreement with experiment unless turbulent energy and momentum transport are included in the model. Turbulent shear flow contributions to radial energy and momentum transport are introduced using the Prandtl mixing-length hypothesis [37] with the usual provisions for adjusting the constant of equations (7.2) through (7.4). Moreover, turbulent eddies are considered to exist at and near the upstream stagnation point, despite the lack of significant flow velocities in the area. The justification for this action is the experimental observation of their presence there (Fig. 7.10). The intensity of turbulent eddies at the stagnation point is taken to be equal to that calculated one characteristic eddy length (1 to 4 mm) downstream of it. While channel velocity is quite low in the upstream region, turbulent diffusivity is found to be nearly as significant as in the downstream region, owing to its dependence on the high upstream pressure [equation (7.4)].

Results of the simulation of the free recovery of an SF_6-gas-blast arc channel show that when turbulent shear flow effects are introduced in the upstream region, the stagnation-point temperature declines dramatically in a very short period of time. Under such conditions, the formation of molecules larger than diatomic is retarded and the channel plasma departs from a state of chemical equilibrium. Consequently, the transport and material properties cannot all be taken as pertaining to an LTE plasma. As an approximation, only the density and the specific heat are considered to require non-LTE values. An argument, based on the kinetics of the most important chemical reactions in the channel, is made to derive modified values for these two properties from LTE values. In essence, the modified values amount to a shift of the LTE values to lower temperatures in addition to some distortion in the shape of their functional dependence on temperature. The shift in the values of density and specific heat to lower temperatures is found to have a strong effect on the channel recovery behavior; in particular, they allow the prediction of the fast-slow-fast recovery characteristic of the channel (Fig. 7.3).

The two-dimensional model predicts the channel temperature recovery rate to be insensitive to the value of the upstream tank pressure, a finding that is also predicted by the one-dimensional model. However, because the upstream pressure affects the temperature radial profile of the channel core at current zero—higher pressure leads to a narrower temperature profile—the temperature decay rate is found to increase with pressure. Similarly, the rms arc current affects the recovery rate through its influence on the channel initial conditions at current zero—the larger the rms current, the wider the channel core. Interestingly, a sensitivity analysis for the model shows the temperature recovery rate to be rather insensitive to the value of the peak core temperature at current zero.

The nozzle geometry affects the ultimate dielectric strength of the channel through its determination of the axial pressure gradient and electric field in the upstream region, a prediction consistent with experimental observation [7,18]. The model further predicts that the channel temperature decay rate increases with the square of the upstream pressure gradient.

The two-dimensional model convincingly demonstrates the ability of a rigorous theoretical treatment of the dielectric recovery of a gas blast to elucidate the physical mechanisms determining its behavior, it correctly predicts many aspects of its characteristics, provides the means by which test results on a model interrupter may be scaled to a working prototype, and yields guiding principles for further tests that may achieve valuable results.

The present model, however, has limitations imposed by some of the assumptions made in its derivation. A more reliable methodology than the ad hoc approach presently employed is needed for determining critical material and transport properties in a channel not in a state of chemical equilibrium. Provisions are needed for analyzing cases in which the electron temperature deviates appreciably from that of the heavy particles in the channel core. Modifications to the model should be added to account properly for the effects of applying a TRV to the channel at current zero. Criteria for determining the onset of turbulent flow in the upstream region from basic principles need to be established.

These and other considerations should certainly be attempted in future theoretical investigations.

7.4 SUMMARY

Experimental and theoretical investigations into the post-arc recovery of gas blast interrupters have contributed substantially to our understanding of the basic mechanisms controlling the recovery process and to our ability to predict and design more desirable interrupter performance. There are still many areas that require further investigation and which are now amenable to experimental and computational tools and techniques. Paramount attention in any future investigation should be concerned with exploring the conditions that precipitate turbulent flow in the upstream part of the arc channel. A more detailed study of turbulent energy and momentum transport in the channel core is also needed. The various modes of deviation from equilibrium in the channel should be more rigorously treated and their influence on the recovery process assessed. A well-defined theoretical criterion needs to be formulated for the onset of dielectric breakdown in the channel of the extinguished arc.

ACKNOWLEDGMENTS

The author wishes to acknowledge valuable discussions with the Editor. This work was supported by NSF grant CPE-8111625.

REFERENCES

1. J. Slepian, Displacement and diffusion in fluid-flow arc extinction, Trans. AIEE *60*:162, 1941.
2. R. Moll and E. Schade, Dielectric recovery of axially blown SF_6 arcs, Proc. 6th Int. Conf. Gas Discharges Appl.(Edinburgh), IEE Conf. Publ. 189, Vol. 1, 1980, p. 13.
3. T. H. Lee, A. N. Greenwood, and D. R. White, Electrical breakdown of high-temperature gases and its implications in post-arc phenomena in circuit breakers, IEEE Trans. Power Appar. Syst. *PAS-84*(12):1116, 1965.
4. T. H. Lee and A. N. Greenwood, The mechanism of thermal breakdown of a high temperature gas and its implication to circuit breakers, CIGRE, Rep. 13.06, Paris, 1968.
5. E. Richley and D. T. Tuma, Free recovery of the gas blast arc column, IEEE Trans. Plasma Sci. *PS-8*(4):405, 1980.
6. E. Richley and D. T. Tuma, Mechanisms for temperature decay in the freely recovering gas blast arc, IEEE Trans, Plasma Sci. *PS-10*:2, 1982.

7. K. Ragaller, W. Egli, and K. P. Brand, Dielectric recovery of an axially blown SF_6-arc after current zero, Part II: Theoretical investigation, IEEE Trans. Plasma Sci. *PS-10*:154, 1982.
8. F. W. Crawford and H. Edels, The reignition voltage characteristics of freely recovering arcs, Proc. IEE (Lond.) *107*:202 1960.
9. T. E. Browne, Jr., Dielectric recovery by an a-c arc in an air blast, Trans. AIEE *65*:169, 1946.
10. W. O. Kelham, The recovery of electric strength of an arc discharge column following rapid interruption of the current, Proc. IEE (Lond.) *101*:321, 1954.
11. K. P. Brand, W. Egli, L. Niemeyer, K. Ragaller, and E. Schade, Dielectric recovery of an axially blown SF_6-arc after current zero, Part III: Comparison of experiment and theory, IEEE Trans. Plasma Sci. *PS-10*:162, 1982.
12. W. M. Bauer and J. D. Cobine, Gap recovery strength of ac arcs at high pressures, Gen. Electr. Rev. *44*:315, 1941.
13. J. F. Perkins and L. S. Frost, Dielectric recovery and predicted ac performance of blown SF_6 arcs, IEEE Trans. Power Appar. Syst. *PAS-91*:368, 1972.
14. K. Ragaller and E. Schade, Dielectric recovery of an axially blown SF_6-arc after current zero, Part I: Experimental investigation, IEEE Trans. Plasma Sci. *PS-10*:141, 1982.
15. R. Moll and E. Schade, Investigation of the dielectric recovery of SF_6 blown high-voltage switchgear arcs, Proc. 3rd Int. Symp. High Voltage Eng., Milan, 1979, Tech. Rep. 32.07.
16. T. E. Browne, Jr., Extinction of a-c arcs in turbulent gases, Trans. AIEE *51*:185, 1932.
17. S. S. Attwood, W. C. Dow, and W. Krausnick, Reignition of metallic a-c arcs in air, Trans. AIEE *50*:854, 1931.
18. J. F. Perkins and L. S. Frost, Effect of nozzle parameters on SF_6 arc interruption, IEEE Trans. Power Appar. Syst. *PAS-92*: 961, 1973.
19. J. J. Lowke, Radiative energy transfer in circuit breaker arcs, in *Current Interruption in High-Voltage Networks* (K. Ragaller, ed.), Plenum Press, New York, 1978, pp. 299-328.
20. K. Ragaller and K. Reichert, Introduction and survey: physical and network phenomena, in *Current Interruption in High-Voltage Networks* (K. Ragaller, ed.), Plenum Press, New York, 1978, pp. 1-28.
21. F. R. El-Akkari and D. T. Tuma, Simulation of transient and zero current behavior of arcs stabilized by forced convection, IEEE Trans. Power Appar. Syst. *PAS-96*:1784, 1977.
22. W. Hermann, U. Kogelschatz, L. Niemeyer, K. Ragaller, and E. Schade, Investigation on the physical phenomena around current zero in HV gas blast breakers, IEEE Trans. Power Appar. Syst. *PAS-95*:1165, 1976.

23. J. Kopainsky, Influence of the arc on breakdown phenomena in circuit breakers, in *Current Interruption in High-Voltage Networks* (K. Ragaller, ed.), Plenum Press, New York, 1978, pp. 329-354.
24. T. E. Browne, Jr., Dielectric recovery of a-c arcs in turbulent gases, Physics 5:103, 1934.
25. W. Hertz, H. Motschmann, and H. Wittel, Investigations of the properties of SF_6 as an arc quenching medium, Proc. IEEE 59:485, 1971.
26. W. Hermann, U. Kogelschatz, K. Ragaller, and E. Schade, Investigation of a cylindrical, axially blown, high-pressure arc, J. Phys. D: Appl. Phys. 7:607, 1974.
27. G. R. Jones, The influence of turbulence on current interruption, in *Current Interruption in High-Voltage Networks* (K. Ragaller, ed.), Plenum Press, New York, 1978, p. 95.
28. B. Eliasson and E. Schade, Electrical breakdown of SF_6 at high temperatures, Proc. 13th Int. Conf. Phenom. Ionized Gases, Berlin, Vol. 1, 1977, p. 409.
29. W. Finkelburg and H. Maecker, Elektrische Bogen und thermische Plasmen, in *Handbuch der Physik* (S. Fluegge, ed.), Vol. 22, Springer-Verlag, Berlin, 1956, p. 254.
30. H. Maecker, Theory of thermal plasma and application to observed phenomena, in *Introduction to Discharge and Plasma Physics* (S. C. Haydon, ed.), University of New England, Armidale, Australia, 1964, p. 245.
31. W. Rieder and J. Urbanek, New aspects of current-zero research on circuit-breaker reignition. A theory of thermal nonequilibrium arc conditions, CIGRE, Rep. 107, Paris, 1966.
32. I. Prigogine, Extension de l'equation de Saha au plasma nonisotherme, Bull. Cl. Sci. Acad. R. Belg. 26(5):53, 1940.
33. A. V. Potapov, Chemical equilibrium of multitemperature systems, High Temp. 4(1):48, 1966.
34. E. Richley and D. T. Tuma, On the determination of particle concentrations in multitemperature plasmas, J. Appl. Phys. 53:8537, 1982.
35. K. P. Brand and J. Kopainsky, Particle densities in a decaying SF_6 plasma, Appl. Phys. 16:425, 1978.
36. W. Frie, Berechnung der Gaszusammensetzung und der Materialfunktionen von SF_6, Z. Phys. 201:269, 1967.
37. H. Schlichting, *Boundary Layer Theory*, 6th ed., McGraw-Hill, New York, 1968.
38. B. W. Swanson, Theoretical models for the arc in the current zero regime, in *Current Interruption in High-Voltage Networks* (K. Ragaller, ed.), Plenum Press, New York, 1978, pp. 137-184.
39. D. T. Tuma, A comparison of the behavior of SF_6 and N_2 blast arcs around current zero, IEEE Trans. Power Appar. Syst. PAS-99:2129, 1980.

40. A. Roshko, Structure of turbulent shear flows: a new look, AIAA J. *14*:1349, 1976.
41. D. C. Prince and W. F. Skeats, The oil-blast circuit breaker, Trans. AIEE *50*:506, 1931.
42. F. Kesselring and F. Koppelmann, Das Schaltproblem der Hochspannungstechnik, Arch. Elektrotech. *30*:71, 1936.
43. D. T. Tuma and J. J. Lowke, Prediction of properties of arcs stabilized by forced convection, J. Appl. Phys. *46*:3361, 1975.
44. L. S. Frost and R. W. Liebermann, Composition and properties of SF_6 and their use in a simplified enthalpy flow arc model, Proc. IEEE *59*:474, 1971.
45. J. Yos, Revised transport properties for high temperature air and its components, AVCO Space Systems Division, 1967, Tech. Rep. Z 220.

8
Dielectric Properties of Circuit Breakers

ALAN H. COOKSON, LYON MANDELCORN, ROY E. WOOTTON, and J. FRANKLIN ROACH/Westinghouse Research and Development Center, Pittsburgh, Pennsylvania

8.1 Introduction   275
8.2 Field Analysis   276
    8.2.1 Field plotting techniques   276
    8.2.2 Field utilization factors   277
8.3 Insulation Properties of Materials   278
    8.3.1 Compressed gas breakdown   278
    8.3.2 Insulating liquids   297
    8.3.3 Insulating solids   310
    8.3.4 Vacuum   325
8.4 Bushings   337
    8.4.1 Bushing external insulation strength   338
    8.4.2 Bushings designs   341
References   348

8.1 INTRODUCTION

The insulation characteristics of circuit breakers play a critical part in the insulation coordination required for breakers in transmission and distribution systems. The internal insulation (e.g., at the contacts) and external insulation (e.g., in air at the bushing) needs to be coordinated so that in the event of an excessive overvoltage, flashover will occur externally where it will generally be nondestructive and the breaker will not be damaged.

This chapter reviews the characteristics of the insulation media used in the breakers: compressed gases [primarily sulfur hexafluoride (SF_6), as air-blast breakers are now being rapidly replaced by SF_6

systems for new installations], insulating liquids (primarily oil), and vacuum. These characteristics are for the dielectric media insulation without arcing; the arc recovery characteristics across the contacts are discussed in the appropriate arc interruption chapters. Solid insulation is also reviewed, as of course dielectric supports are always required in breakers. The various types of bushings and characteristics of the external air insulation are also discussed.

Techniques for calculating the field distribution are described, together with the field analysis for practical geometries, so that with the appropriate section on the dielectric characteristics, the breakdown voltage can be calculated for typical breaker configurations.

8.2 FIELD ANALYSIS

8.2.1 Field Plotting Techniques

Accurate field analysis is required of the complex structures in circuit breakers in order to use the dielectric materials in an efficient and reliable manner. For most of these designs it is essential to be able to determine accurately the maximum field, in addition to knowing the field distribution.

There are several techniques available [1]. Field sketching is a technique for calculating the field by successive approximations. It requires a skilled, experienced person, who can get results rapidly, perhaps with an accuracy of 10%. It is difficult for problems with complex, asymmetrical geometries and mixed dielectrics, and is now usually only done for initial analyses.

Electrolytic tanks use current between electrodes as an analog for the study of the field distribution [2]. This technique permits rapid evaluation of changes in simple models. For complex models involving different media, wedge-shaped models need to be constructed. Typical accuracy of 3% is claimed, but this may be difficult in cases where the curvature of electrodes is sharp or rapidly changing. The tank technique is simple and quick for many applications.

Similarly, the conducting paper analog [3] is rapid and easy to use for simple geometries, but requires skill and time for multidielectric configurations, in which case the accuracy is limited.

Considerable advances in the application of the computer for field analysis for complex multidielectric configurations have been made in the past 10 years, using digital techniques for the solution of Laplace's and Poisson's equations with the boundary conditions satisfied. Some techniques have solved Laplace's equation by finite difference techniques [4]. Other approaches have used integrals of Laplace's or Poisson's equations either by using discrete charges or by dividing the electrode surface into subsections of charges [5]. Typically, the results are given either as the equipotential distribution (Fig. 8.1) or can be given as the numerical value of the field at discrete points. For

Dielectric Properties of Circuit Breakers

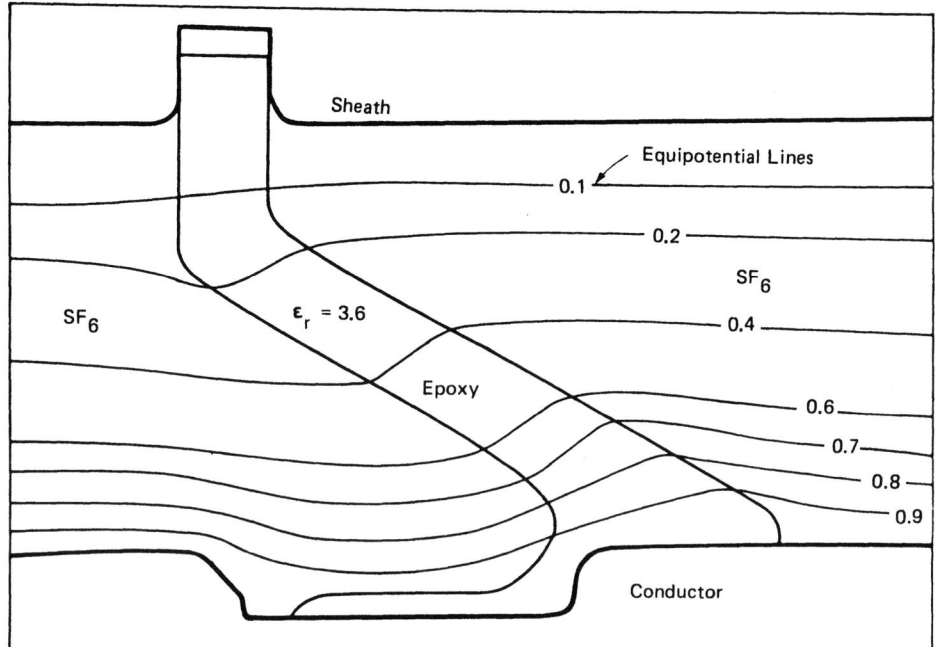

Figure 8.1 Equipotential distribution for a conical insulator in an SF_6 insulated bus structure.

many systems, such as SF_6 gaseous insulation where the breakdown is very sensitive to the maximum field, using the charge simulation technique enables the field to be accurately determined (e.g., ±2%) at the conductor and dielectric surfaces (Fig. 8.2). Techniques can be used to increase the sensitivity of the analyses at critical areas by decreasing the grid size or increasing the number of line charges.

8.2.2 Field Utilization Factors

For many applications, it is unnecessary to use field analysis, as the complex geometries can be approximated by a simple geometry such as a sphere-sphere, sphere-plane, internal cylinders, internal spheres, external cylinders, torus-plane, torus within cylinder, sphere within cylinder, rod-plane, edge-plane, and edge-edge.

Figures 8.3 through 8.10 give curves for the ratios of maximum to average fields (E_{max}/E_{av}) for the above geometries [6]. This is the reciprocal of the field utilization factor $\mu = E_{av}/E_{max}$. These can then be used to calculate the maximum field and the breakdown voltage for

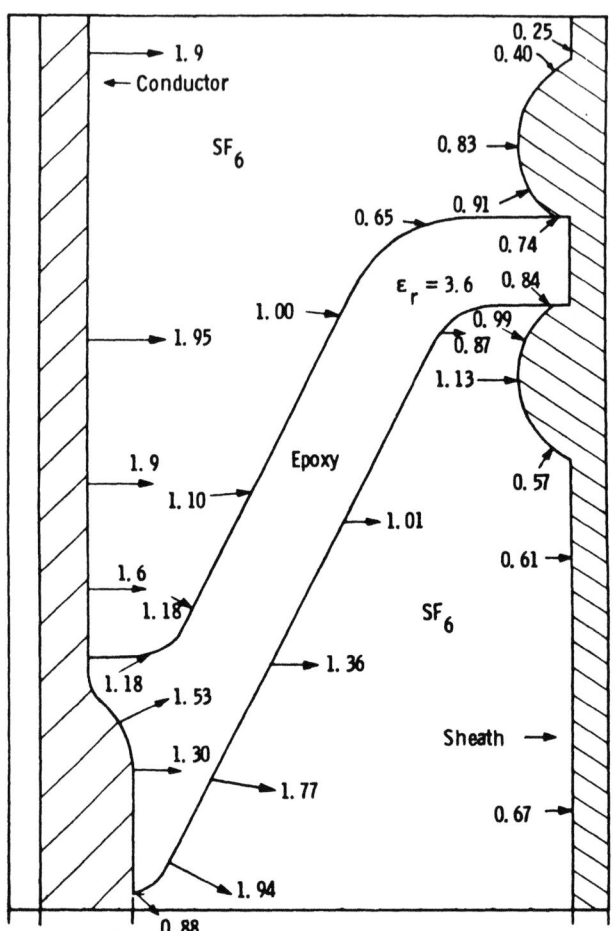

Figure 8.2 Field values on a conical insulator from line-charge calculation. Fields are normalized to E_{av}, set at unity by assuming 11.5 kV applied across an 11.5 cm gap in the system. Conductor field removed from insulator = 1.78 in these units.

the applied dielectric media of SF_6, oil, vacuum, or solid insulation, as described later.

8.3 INSULATION PROPERTIES OF MATERIALS

8.3.1 Compressed Gas Breakdown

Failure of Paschen's Law

If a voltage is applied across a uniform field gas gap of separation d at pressure p and gas density ρ, then for clean conditions the break-

Dielectric Properties of Circuit Breakers

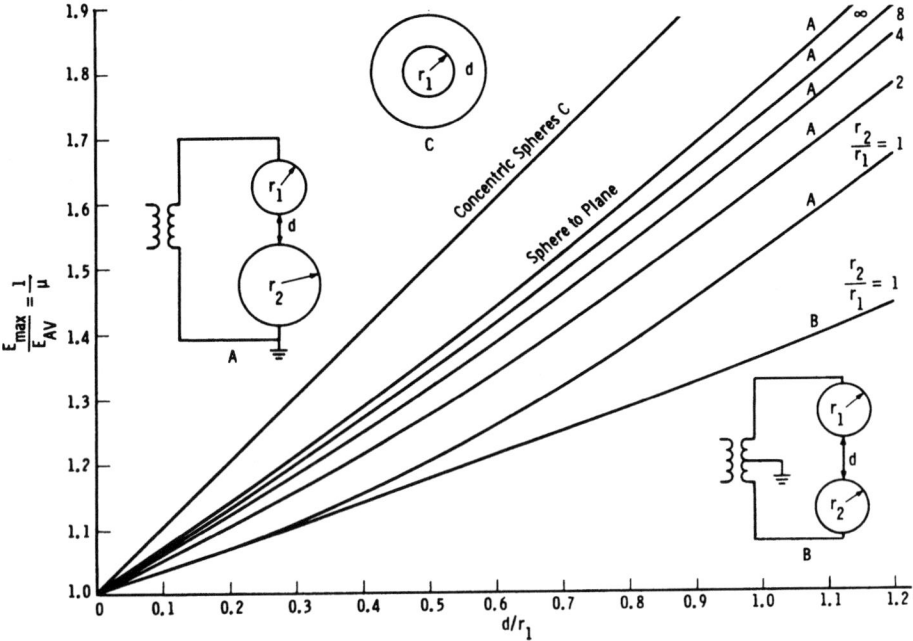

Figure 8.3 Stress concentration factor for spheres and sphere to plane.

down voltage V_s is a function of the product (ρd). This is known as Paschen's law [7]. Often the function pd is used instead of ρd, which is usually valid except at high pressures when p is no longer proportional to ρ.

In practice, deviations occur from Paschen's law as shown in Fig. 8.11 for SF_6 [8]. These deviations, which are such that V_s is lower than predicted, typically begin to occur at values of the field E of the order of 10 to 20 MV/m and at pressures of 2 or 3 atm [9]. The breakdown field decreases with increasing electrode surface roughness, contamination, electrode area, and electrode separation as described below. The data in Fig. 8.11 are for laboratory tests and should only be applied with care to practical systems as described later. The pressure unit kilopascals (kPa) is used: 1 atm = 1.013 bar, and 1 bar = 10^2 kPa.

The failure of Paschen's law is due to field nonuniformities at the electrodes, such as electrode microprojections or dust, which initiate enhanced ionization in the gas and cause a reduction of the breakdown voltage [9].

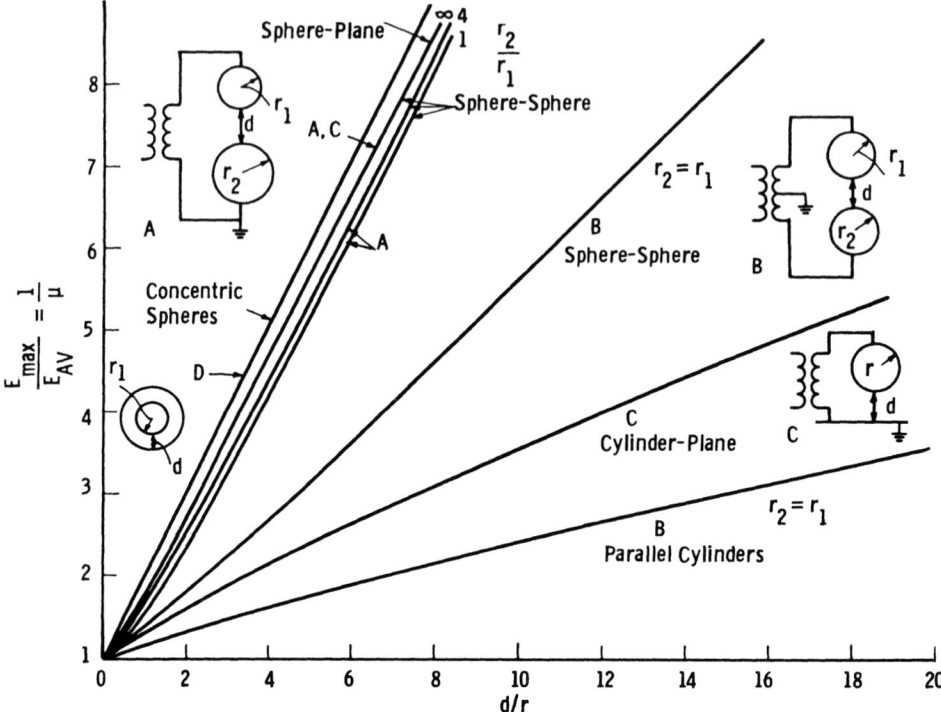

Figure 8.4 Stress concentration and utilization factor for spheres and cylinders.

Conditioning

At fields of the order 10 to 20 MV/m, it is observed that the breakdown voltage increases with repeated sparking of the electrodes (Fig. 8.12). This is attributed to progressive destruction of the high-field sites or microprojections on the electrodes referred to above. The spark conditioning effect becomes more significant with increasing field, pressure, and electrode area. At pressures of 1 or 2 atm, spark conditioning can result in the breakdown voltage increasing to the theoretical value. However, at higher pressures (e.g., SF_6 at 5 atm) even though the value of V_s may increase by 50% over the initial value, it is still below the theoretical breakdown voltage by Paschen's law.

Spark conditioning should be treated with some caution for practical systems. Some conditioning can be done in the testing laboratory and also in the field if the energy stored in the source is limited to prevent excessive damage to the conductor. However, practical systems are designed so that spark conditioning only rarely occurs. Also,

Dielectric Properties of Circuit Breakers

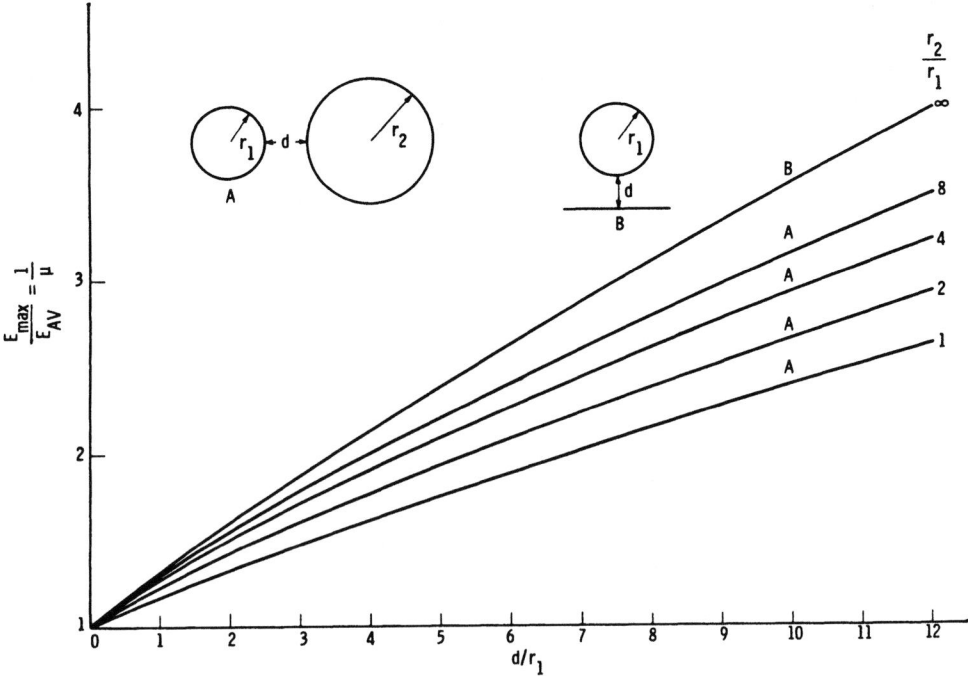

Figure 8.5 Stress concentration factor for external parallel cylinders.

it should be noted that most experimental data quoted in the literature are the conditioned values and not the first spark values.

Stress conditioning (without sparking) can also occur where the applied voltage is increased very slowly, resulting in an increase in the breakdown voltage [9]. This is attributed to the slow destruction of the high-field sites by ion bombardment, or to the movement of particles out of the high-field region. Controlled stress conditioning can be particularly important for practical systems where particles and contamination can be moved by the field into "particle traps," where they are deactivated [10].

Gap Dependence

The breakdown field E_s in compressed SF_6 (and other gases) increases with decreasing electrode separation, as typically shown in Fig. 8.13. This is because, for the smaller gaps and lower gas densities, higher fields are required to obtain the critical net ionization in the gap.

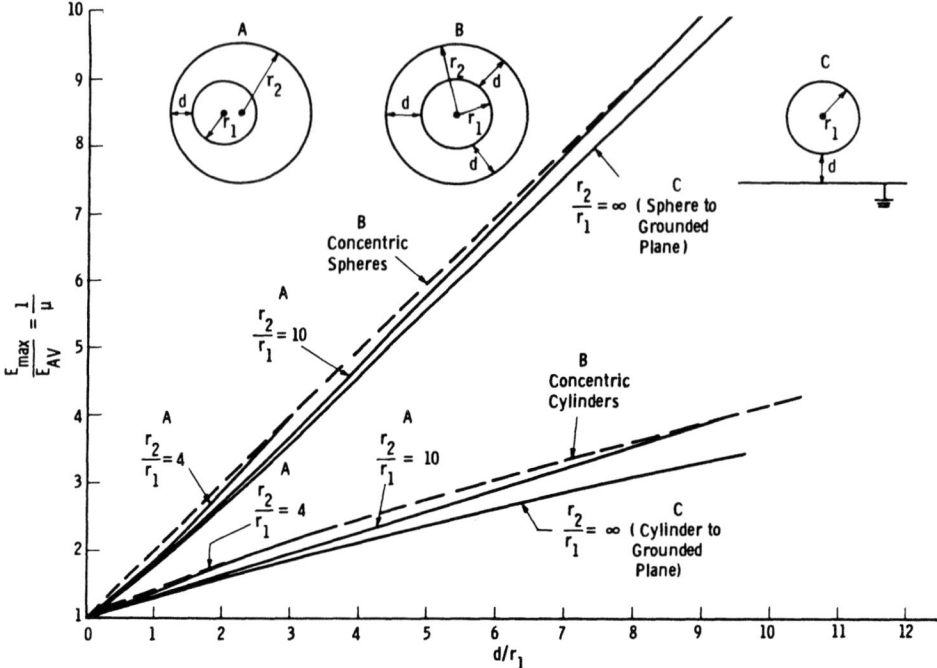

Figure 8.6 Stress concentration factor for internal spheres and cylinders.

Effect of Electrode Geometry

The experimental results described so far have been for uniform fields. Practical geometries of course are rarely uniform, but are usually coaxial cylinders or approximate to spherical electrodes. These geometries are designed so that the factor (E_{max}/E_{av}) is normally not more than 4 or 5.

For very nonuniform field geometries, unlike uniform field or quasi-uniform field systems, corona can be detected before breakdown; Fig. 8.14, for example, shows the corona onset and breakdown characteristics of a point-plane gap in SF_6. The corona onset voltage increases monotonically with pressure, but the breakdown voltage initially increases with pressure, reaches a maximum, and at a critical pressure then decreases with increasing pressure. This "corona stabilization" effect is due to a space-charge cloud produced in the gap, causing redistribution of the gap field. At pressures beyond the critical pressure, the space-charge cloud cannot effectively shield all the point, so that with increasing pressure this region is more constricted due to decreased range of photoionization or photodetachment

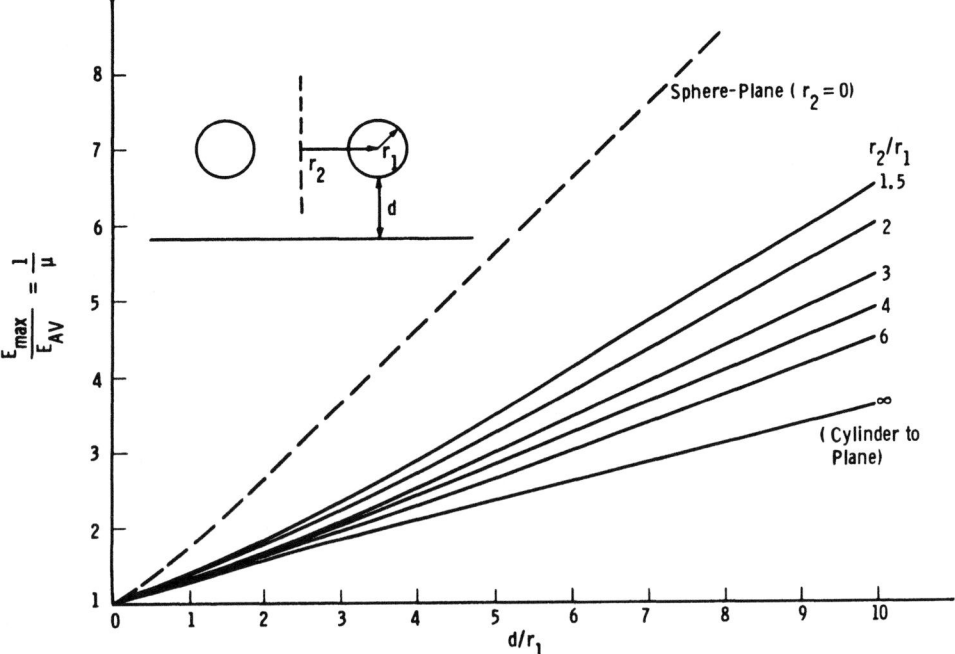

Figure 8.7 Stress concentration and utilization factor for a toroid parallel to a plane.

in the gas (which plays an important role in the production of the space charge cloud) and reduced electron diffusion. The breakdown sparks in this pressure range are no longer straight but are curved and occur at the sides of the point. Eventually, at a certain pressure there is no further corona stabilization and breakdown occurs almost immediately after corona. At higher pressures the breakdown voltage again increases with pressure. These are the general nonuniform field characteristics for ac, lightning, and switching impulse conditions.

Practical systems with SF_6 cannot utilize this corona stabilization effect because of degradation and decomposition of the SF_6. The practical designs using nonuniform field geometries therefore limit the field nonuniformity as much as possible; for these geometries, any ion cloud formation would lead to breakdown.

Typical SF_6 breakdown results for coaxial geometries are shown in Fig. 8.15, where the breakdown field (at the conductor) is given for ac, lightning, and switching impulse. At atmospheric and higher pressures and for the geometries where the field decreases rapidly

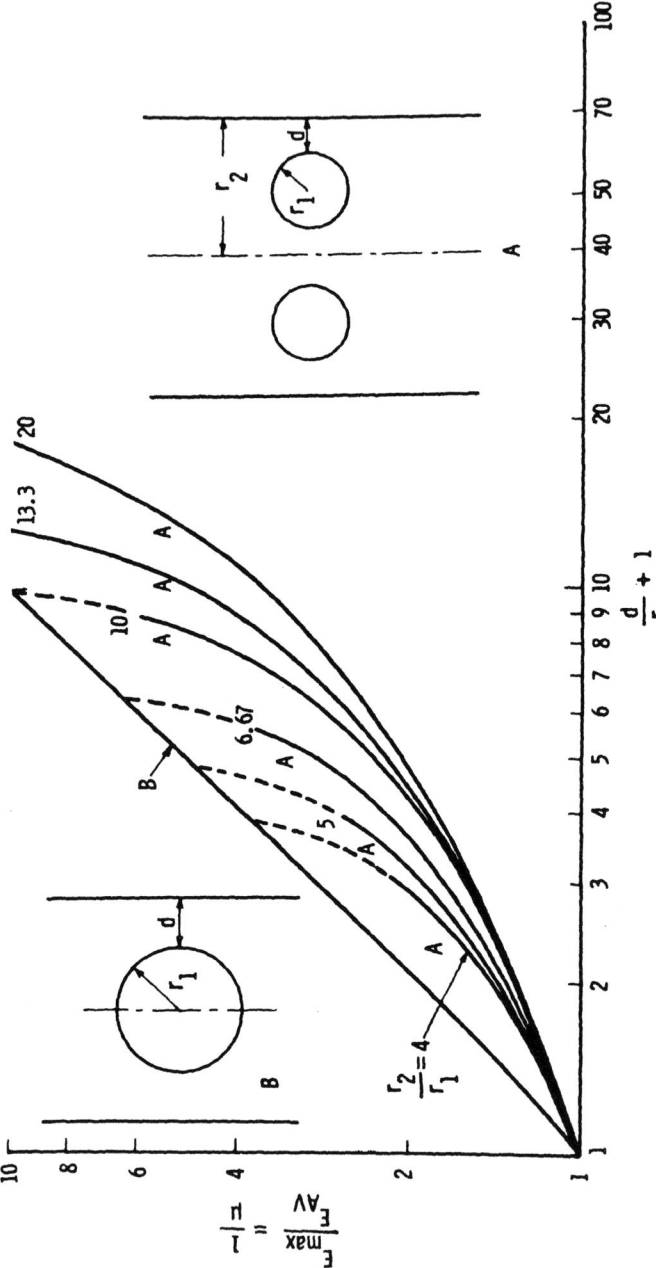

Figure 8.8 Stress concentration factor for a torus within a cylinder and a sphere within a cylinder.

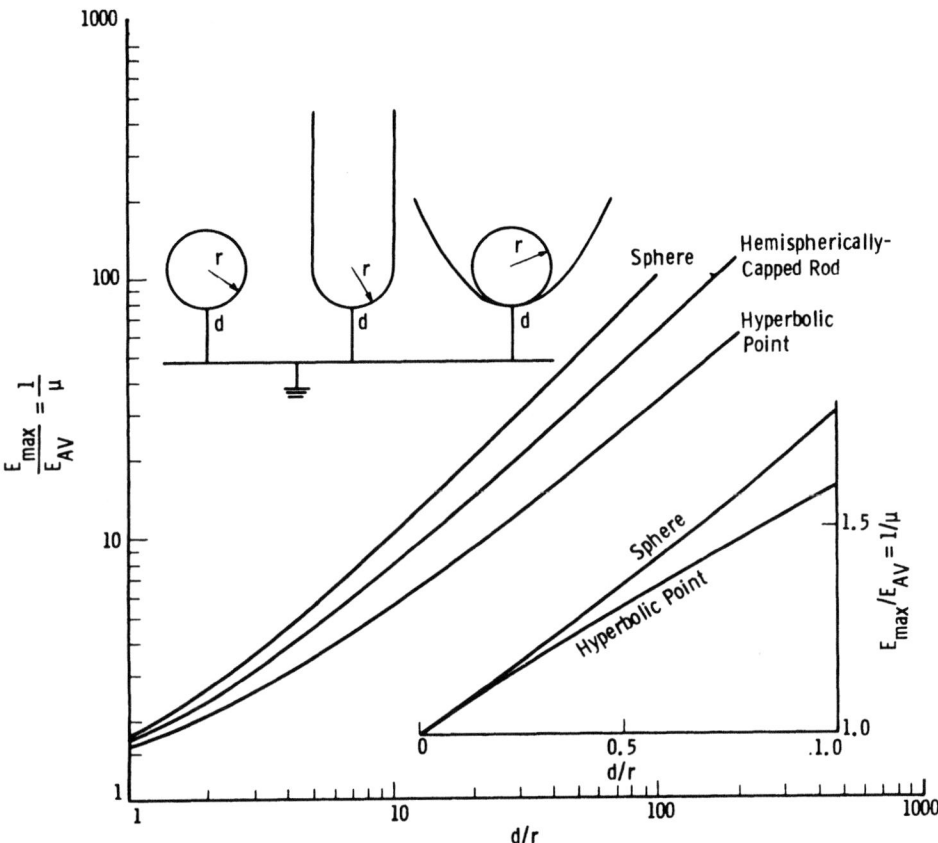

Figure 8.9 Stress concentration factor for hyperbolic point to plane, hemispherically capped rod to plane, and sphere to plane.

with distance from the high-field electrode, then, unlike the uniform field case, there is relatively little direct dependence of E_s on the electrode separation, although the separation of course enters into the calculation of the conductor field. In practice, this means that, for example, with a coaxial geometry in SF_6 with a particular conductor, the breakdown field at the conductor will be almost independent of the diameter of the outer electrode.

Area Effect

An electrode area effect has been observed in compressed SF_6 and other gases when the field exceeds values of the order 10 MV/m [12,13].

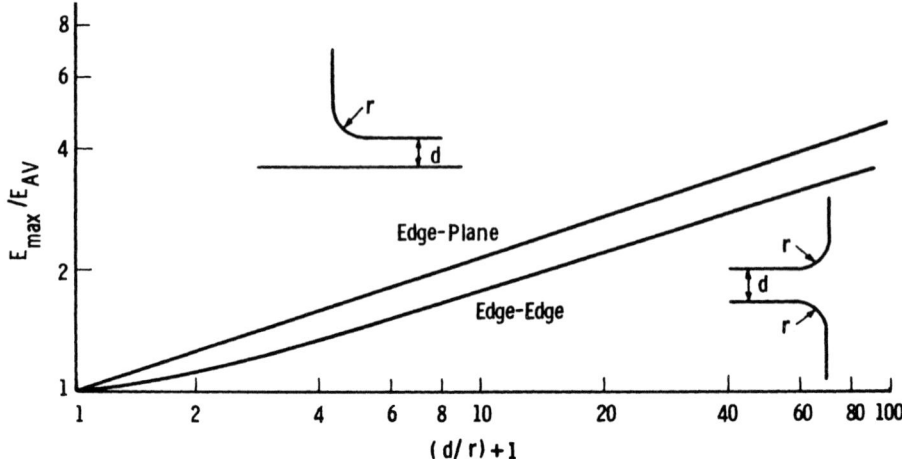

Figure 8.10 Edge-to-plane and edge-edge stress concentration factors.

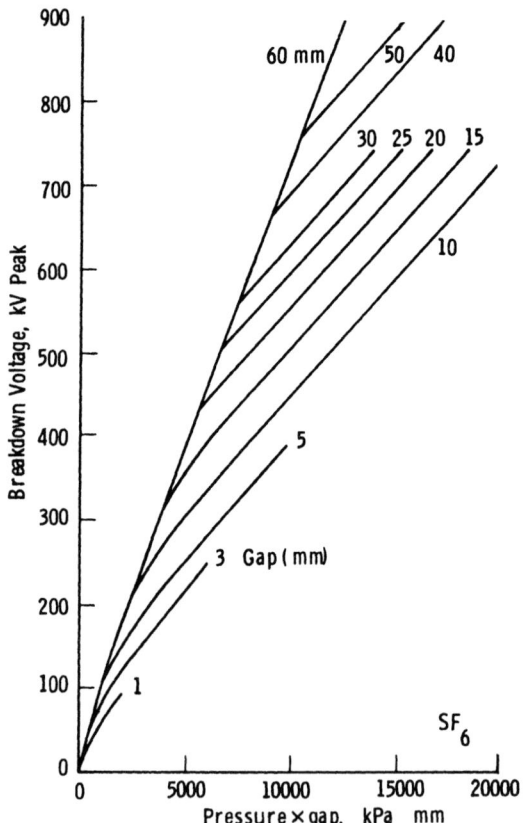

Figure 8.11 Paschen plot for ac uniform field breakdown in SF_6. (From T. W. Dakin et al., Electra. No. 32:61, and A. H. Cookson, Proc. IEE (Lond.) 128:303, © 1981, IEE.)

Dielectric Properties of Circuit Breakers

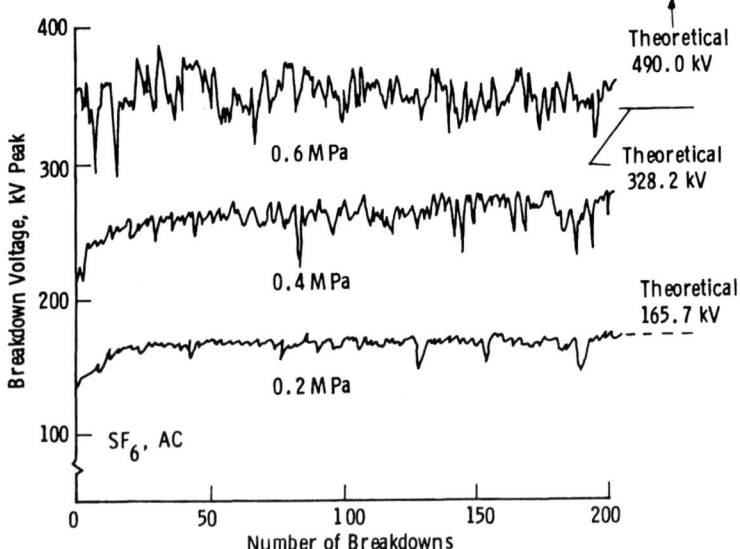

Figure 8.12 Spark conditioning in SF_6 for 132 mm/152 mm-diameter coaxial electrodes, 116 mm long, with ±0.5-μm finish. (From T. Nitta, N. Yamada, and Y. Fujiwara, IEEE Trans. Power Appar. Syst. *PAS-93*: 623, © 1974, IEEE.)

Here the field decreases with increasing electrode area. This effect, which becomes more pronounced with increasing pressure [13] (Fig. 8.16), can be calculated from extreme value statistics which relate to breakdown initiation at a weak point or flaw on an electrode surface. Some analyses indicate a lower or minimum breakdown field, whereas some others suggest a monotonical decrease with increasing area. The fact is that compressed-SF_6 systems with very large electrode area, as in long compressed-gas-insulated transmission lines, have been placed successfully in service.

Electrode Surface Finish

The finish of the cathode is critical for compressed-SF_6-insulated systems, as high-field sites can result in enhanced emission and ionization and can cause a significant reduction in the breakdown voltage. Figure 8.17 shows typical results of the effect of the surface finish in SF_6 compared with calculated values, assuming different values of the projection heights h and profile shape. Calculations indicate that in SF_6 for values of ph ≤ 4 kPa mm, no deviation should occur from the clean "intrinsic" value [14]. Practical systems usually have the

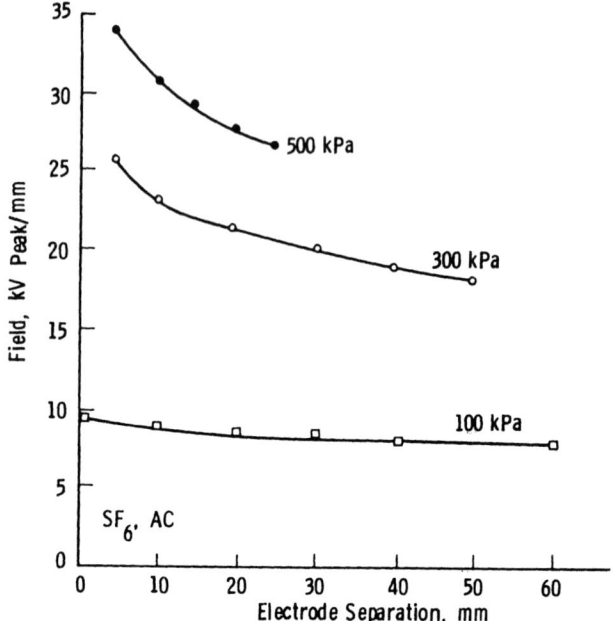

Figure 8.13 Breakdown field as a function of uniform field electrode separation. (From Ref. 8.)

surface finely sanded and then highly polished, resulting in an overall surface finish much better than 100 μm.

Insulating coatings on small electrode area systems can increase the SF_6 breakdown strength [9]. However, they have not been generally used in practical systems except at small selected critical areas. This is because for large, practical geometries, the improvements due to the coated surface may be masked by the effect of contamination.

Particle Effects

Particle contamination can be present, either remaining after assembly or produced by the operation of the equipment. Metallic particles are especially harmful, as they acquire a charge on the electrodes in the field and can then be elevated and move to the high-field conductor or onto an insulator. Here they can cause breakdown at a low value initiated from the enhanced fields at the particles. Figure 8.18 shows the effect of fine wire particles in reducing the ac breakdown voltage in SF_6. These effects, which become more pronounced with increasing pressure and increasing particle length, are especially severe under ac and dc conditions. Under impulse conditions the particle movement

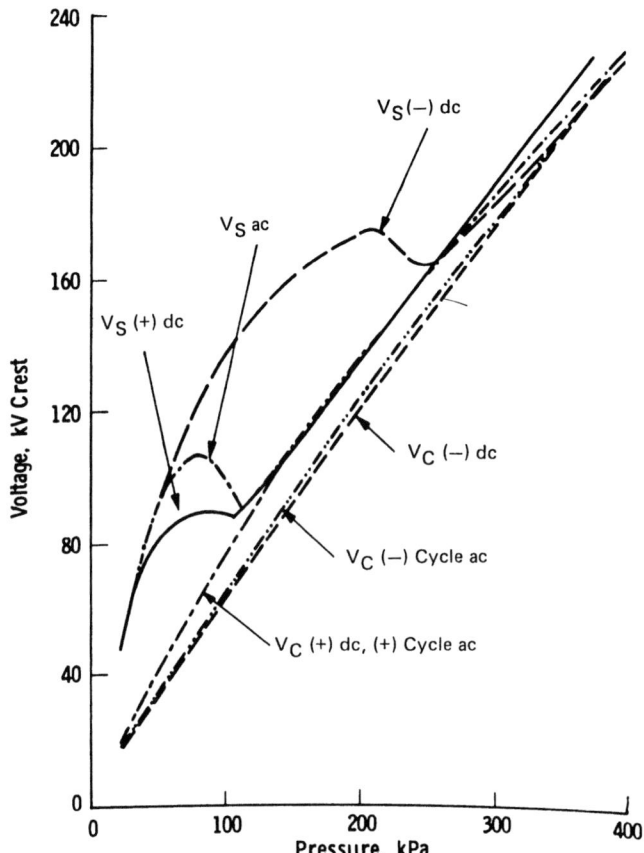

Figure 8.14 Corona onset V_C and breakdown voltage V_S for rod-plane gap in SF_6. Rod radius 5 mm, gap 40 mm. (From Ref. 11.)

is more restricted, but particles still result in lower breakdown voltages if they are already at critical high-field areas on the conductors or insulators.

To prevent this problem, it is necessary to ensure cleanliness in assembly and operation. For some systems it is advantageous to use particle traps, especially for SF_6 systems, where the stresses are very high [10]. Examples are bus and other components in SF_6-gas-insulated substations. The basic particle trap design consists of created regions of very low field at preferred areas, for example by using a slotted or perforated shield at the grounded enclosure (Fig. 8.19). Any metallic particles are elevated by electrostatic force in

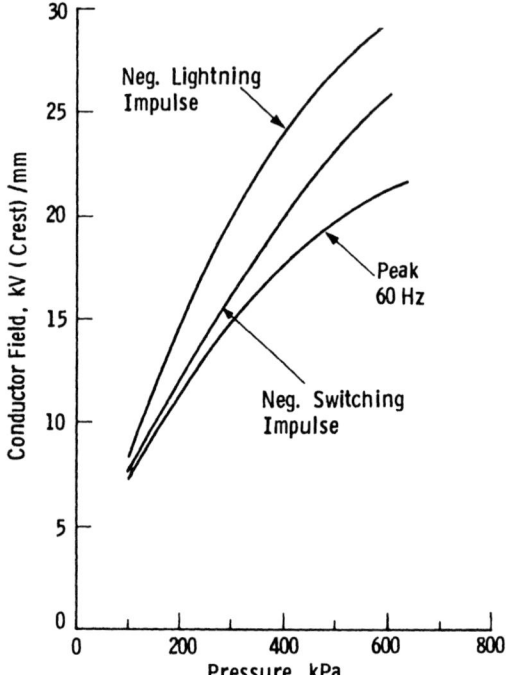

Figure 8.15 Breakdown fields in SF_6 from coaxial geometry 89/226-mm-diameter 0.9-m-long industrial CGIT finish. (From Y. Kawaguchi, K. Sakata, and S. Manju, IEEE Trans. Power Appar. Syst. *PAS-90*:1072, © 1971, IEEE.)

the field and move to the strategically placed traps, where they fall through the small openings into the zero-field region and so are electrically and mechanically trapped.

Voltage Waveform Effects

The effect of voltage waveform on breakdown in compressed SF_6 is shown typically in Fig. 8.15; the negative impulse breakdown value is on the order of 10 to 15% lower than the positive value, the negative lightning impulse E_s being typically 15% above the negative switching impulse value and about 20% above the peak ac value [15]. The dc breakdown field in SF_6 is equal to the peak ac value for very clean conditions. However, if contamination is present, the dc values can then be 20% lower than the ac values for the same conditions.

The voltage/time-to-breakdown characteristics are important for insulation coordination. These are determined by the statistical and

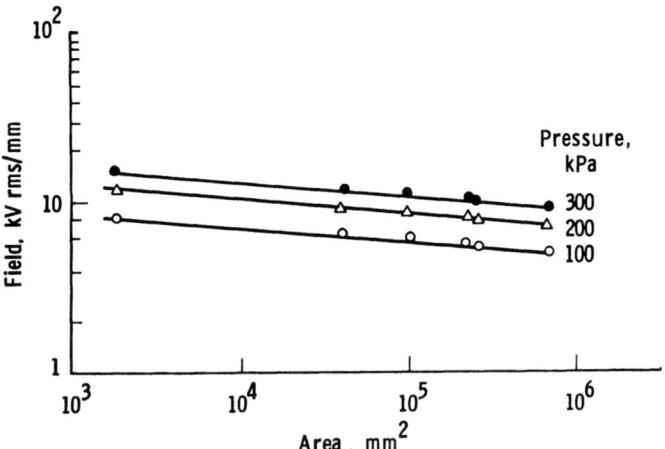

Figure 8.16 Breakdown field in SF_6 as a function of the equivalent electrode area. (From A. H. Cookson, Proc. IEE (Lond.) *128*:303, © 1981; and C. Masetti and B. Parmigiani, 3rd Int. Symp. High Voltage Engr., Paper 32.15, © 1979, IEEE.)

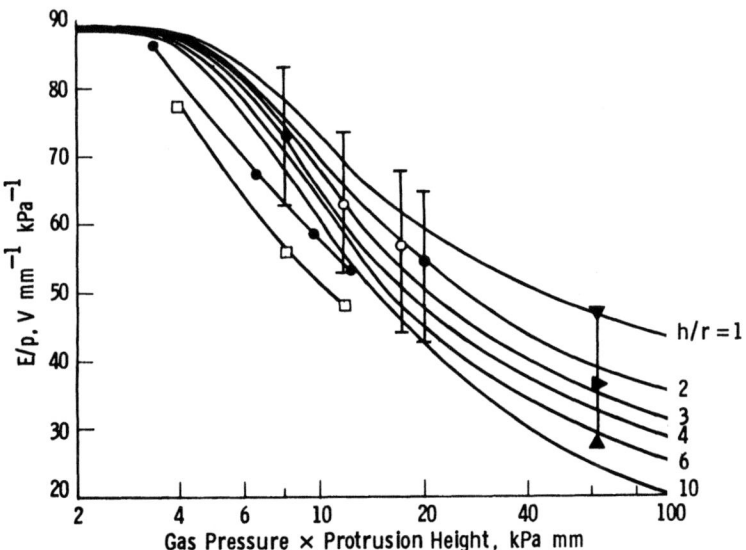

Figure 8.17 Comparison of theoretical and practical threshold values of the breakdown gradient in SF_6 assuming surface projection of height h and hemispherical radius r on the cathode. (From S. Berger, IEEE Trans. Power Appar. Syst. *PAS-95*:1073, © 1976, IEEE.)

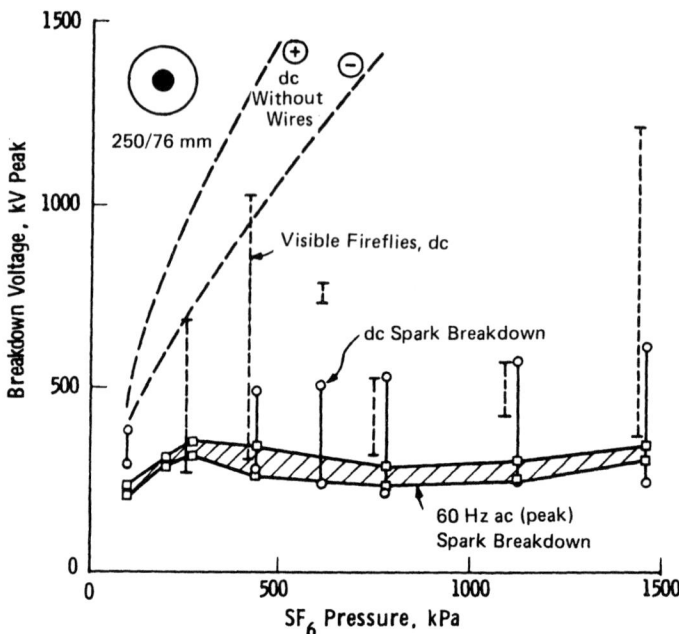

Figure 8.18 Breakdown in SF_6 with Al wires, 6.4 mm long, 0.45 mm diameter. (From C. M. Cooke, R. E. Wootton, and A. H. Cookson, IEEE Trans. Power Appar. Syst. *PAS-96*:768, © 1977, IEEE.)

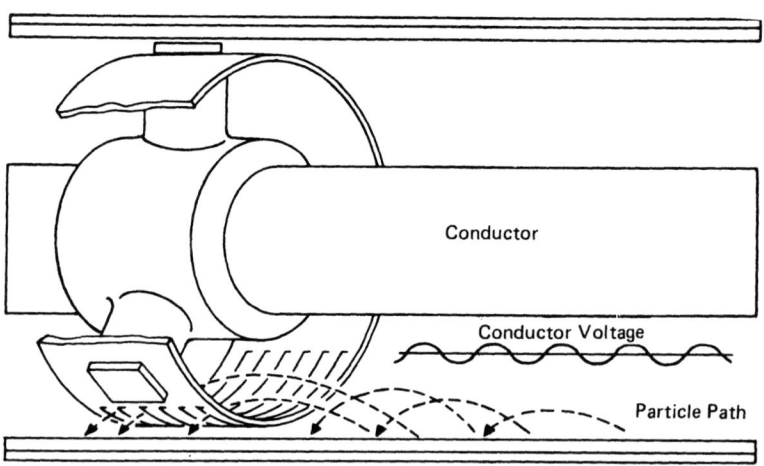

Figure 8.19 With Westinghouse tri-trap design, any particles "left over" from installation pass under the spacer and into the trap and are permanently immobilized in a low field region.

Dielectric Properties of Circuit Breakers

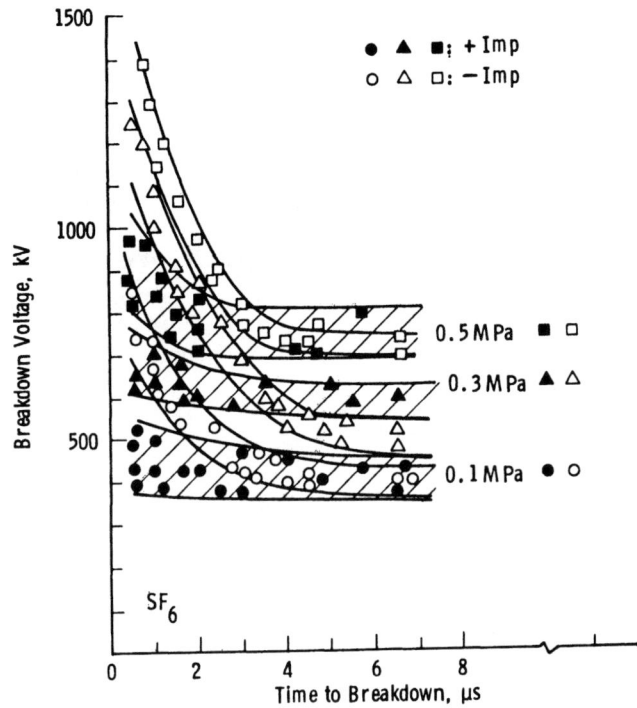

Figure 8.20 Volt-time characteristic for a 30-mm-diameter-hemisphere rod gap of 150 mm. (From Ref. 16.)

formative time lags. Typical characteristics are given in Fig. 8.20. The upcurving at the short times (<2 µs) is of particular concern for coordination purposes. Some results indicate that the positive and negative characteristics can cross over in this region; that is, the critical, lower value for short times is the positive impulse breakdown characteristic, and for long times it is the negative value.

Insulator Flashover

Insulators are critical in compressed-gas-insulated systems, since they support the conductors or other high-voltage components. The insulators used in gas circuit breakers and gas insulated substations are made of filled cast epoxy, usually of bisphenol A, cycloaliphatic, or hydantoin formulation and filled with quartz or hydrated alumina. They are cast under vacuum, either directly onto the conductor or onto metallic inserts. This is to avoid voids in the insulator or at the

Figure 8.21 Westinghouse Electric conical and tri-post cast epoxy insulators with particle trap used on Westinghouse compressed-gas-insulated transmission lines.

conductor interface, which could result in insulator puncture or surface flashover.

Several designs are available, depending on the application (Fig. 8.21). The conical and disk insulators, which can be support or gas-tight insulators, are captured between shielding rings or flanges at the grounded enclosure. The tri-post, which is a simple support insulator, has an encircling metallic ring and is fixed or moves inside the enclosure.

The shape of the insulators is critical. The profile is determined by field plotting, as shown earlier, so as to optimize the design; that is, minimize the resultant field inside the insulator, along the insulator surface and at the adjacent conductor and stress shields. Mechanical stress analysis and testing is also done to ensure adequate mechanical performance. With careful design the flashover voltage of the insulators can be made similar to the gas breakdown voltage of the practical geometries used in gas-insulated systems.

Insulating components are used in other parts in gas-insulated circuit breakers, such as operating rods. These can be cast epoxy, Micarta, or glass-fiber-epoxy-reinforced rods. As the space factor is not usually as critical in these applications as for the support insulator, simple shielding at the ends is adequate. When insulators are used in a circuit breaker or disconnect switch environment, where chemically active arced gases occur, the insulator material is critical to prevent excessive surface conduction leading to flashover. The steps taken to prevent this are to use efficient gas driers to keep the moisture level low (typically <100 ppm by volume in SF_6) and to use epoxy formulations or surface coatings especially formulated to stop the arc-product-induced surface conduction.

Calculation of Breakdown Voltage

Region Where Paschen's Law Is Satisfied The gas breakdown voltage in SF_6 for the region where Paschen's law is still satisfied may be calculated from the physics of gas breakdown. If α is the Townsend first ionization coefficient [the number of ionizing collisions per unit dimension (millimeters) in the field direction], and η is the attachment coefficient [the number of attachments per unit dimension (millimeters) in the field direction], ionization measurements have shown that for SF_6 [7,17] the following empirical equations are satisfied:

$$\frac{\alpha}{p} = 23\left(\frac{E}{p}\right) - 1.234 \text{ mm}^{-1}/\text{kPa}$$

$$\frac{\eta}{p} = -4\left(\frac{E}{p}\right) + 1.135 \text{ mm}^{-1}/\text{kPa}$$

and

$$\frac{\alpha - \eta}{p} = 27\left[\left(\frac{E}{p}\right) - \left(\frac{E}{p}\right)_c\right] \text{mm}^{-1}/\text{kPa}$$

where the field E is in kV/mm and the pressure in kPa. $(E/p)_c$ is the critical value of E/p of 87.8 V/mm kPa at which $\alpha = \eta$ (i.e., when the net ionization is zero) and is termed the limiting value of E/p.

The usual Townsend breakdown criterion [7] for a uniform field gap is

$$\gamma\{\exp[(\alpha - \eta)d] - 1\} = 1 - \frac{\eta}{\alpha} \qquad (8.1)$$

where γ is the second Townsend ionization coefficient, defined as the number of secondary electrons produced per electron in the gap. This secondary ionization could be by positive ions or photoemission at the cathode or by photoionization in the gas.

An alternative breakdown criterion has been developed [7] where an avalanche reaches a critical size to initiate a "streamer." The number of electrons and positive ions in the avalanche head is then sufficient to distort the field in the gap so that with photoionization in the gases, anode and cathode directed streamers are initiated to lead to breakdown. The streamer criterion in SF_6 is of the form [18]

$$\int_0^{Z_0} [\alpha(Z) - \eta(Z)]dZ = K \qquad (8.2)$$

where Z is the line element in the field direction and Z_0 is the critical avalanche length. There is some controversy over the value of K, the discharge constant. The value of 10.5 is most frequently used [18], but values up to 18 have also been used in the calculations. Other variations in this breakdown criterion have been proposed, which generally give similar calculated values to that determined from equation (8.2).

For the simple uniform field case in SF_6 where Paschen's law is still valid, the breakdown voltage may be calculated from the equation

$$V_s = \left(87.8 \frac{V}{mm\ kPa}\right) pd + 0.5\ kV \qquad (8.3)$$

where pd is in mm kPa. This is valid only for the region $100 \leqslant pd \leqslant 2000$ mm kPa.

Paschen's law is valid for uniform field conditions only. However, a similar mathematical analysis of breakdown can be made for nonuniform field breakdown (where there is no corona) to yield a "similarity law" [20,21]. For coaxial electrodes, this is of the form [21]

$$\frac{E_s}{p} = \left(\frac{E}{p}\right)_c \left(\frac{1 + 3.82}{\sqrt{pr}}\right) \qquad (8.4)$$

where p is in kPa, r is the radius in mm of the high-field conductor, and E_s is the breakdown field.

Region Where Paschen's Law Is Not Satisfied The failure of Paschen's law due to, for example, electrode surface effects and contamination, cannot yet be completely accommodated in a single theoretical model, although significant advances have been made [22].

The effect of surface roughness has been calculated using the streamer criterion, equation (8.2), for ionization occurring at a surface projection [14,23], as shown in Fig. 8.17. It should be noted that for SF_6, the coefficients α and η are very strongly dependent on the field,

Dielectric Properties of Circuit Breakers

and so are very sensitive to surface projections when the pressure is such that the projection height is similar in magnitude to the average distance between ionizing collisions, $1/\alpha$.

Similarly, there has been some success in using the streamer criterion in SF_6 for calculating the breakdown voltage with simple, free spherical particle contamination [24,25].

For the more complex, practical geometries with not completely controlled conditions of electrode finish and cleanliness, there has been some success in developing and applying empirical equations [22, 26]. A typical equation for the breakdown voltage V_s is of the form

$$V_s = B(\mu d)^C p^D \tag{8.5}$$

where μ is the stress concentration factor (E_{av}/E_{max}) as described in Sec. 8.2.2, d the electrode separation, B a constant, C a constant of the order 0.8, and D depends on the voltage waveform and averages 0.7.

An alternative empirical approach is to use the experimentally derived breakdown field (E_b) data for the different voltage waveforms for typical electrode and cleanliness conditions as shown in Fig. 8.15, and to then calculate V_s from

$$V_s = E_b \mu d F \tag{8.6}$$

Here F is a factor to take account, compared with the experimental reference data, of the electrode area effect (e.g., from Fig. 8.16), and different surface roughness conditions (e.g., from Fig. 8.17).

8.3.2 Insulating Liquids

Functions, Mineral Oil, and Requirements

Insulating oils are used in dead-tank and live-tank breakers, and perform two functions. One is to insulate the gap and the conductor. The other is to provide interruption of the arc generated on opening the breaker. Hydrogen is a principal product of the arc in the oil, and aids extinction of the arc. Because of the arcing and the arc products, which weaken the electric strength of the oil, the electrical stresses used in the oil in breakers are considerably lower than in cables and transformers.

Only mineral oils are used in oil breakers for the reason that they provide a large amount of hydrogen for arc extinction. Their property specifications are given in Table 8.1. In general, these are similar to mineral oil for transformers. A typical composition of such an oil is given in Table 8.2, the principal constituents being paraffinic, naphthenic, and aromatic hydrocarbons.

Table 8.1 Electrical, Physical, and Chemical Properties of Mineral Oil for Circuit Breakers and Transformers

Dielectric breakdown	
ASTM D877	30 kV
ASTM D1816	28 kV
Dissipation factor—ASTM D924	
At 25°C	0.05%
At 100°C	0.3%
Flash point—ASTM D92	145°C min.
Interfacial tension—ASTM D971	40 dyn/cm min.
Pour point, for circuit breakers	−50°F max.
Specific gravity at 60°F	0.91 g/cm^3
Viscosity	
At 0°C	75 cSt
At 40°C	12 cSt
At 100°C	3 cSt
Oxidation stability—ASTM D2440	
At 72 h	0.15% sludge
At 164 h	0.3% sludge
Corrosive sulfur—ASTM D1275	Noncorrosive
Water content	35 ppm max.
Neutralization number, Total acid number	
—ASTM D974	0.03 mg KOH/g

An essential requirement of circuit breaker insulating oil is that it should maintain its electrical insulating function throughout its operating life. Specifically, it must maintain the following:

1. Bulk long-term and impulse electric strength
2. Surface flashover electric strength

Table 8.2 Typical Composition of Mineral Oil Used in Circuit Breakers and Transformers

Aromatics	30% (may be less)
Naphthenics	5%
Paraffinics	65% (may be greater)

Table 8.3 Carbon and Gas Generation by Arcs in Mineral Oils

Mineral oil composition, % of various carbons[a]			Arc energy, kW-s	Carbon, mg	Gas, cm^3
Paraffinic	Aromatic	Naphthenic		(in 1200 cm^3 oil)	
73	12	15	10	114	560
63	7	30	11	114	572
—	—	—	11	124	587
—	—	—	10	125	583
64	11	25	11	116	586

[a]According to ASTM D2140.

Contaminants: Carbon

Contaminants have considerable effect on these properties, particularly the bulk electric strength. One of the principal contaminants is carbon, which is an arc product in oil circuit breakers. Table 8.3 shows that the production of carbon by arcs in a test oil breaker was not affected by the paraffinic/naphthenic/aromatic ratio of several transformer type oils that were tested [27].

Electric Strength of Technical Grade Oil: Statistical Representation

Electric strength data and relationships that pertain to mineral oil in circuit breakers, as well as other power equipment, are those obtained with technical-grade mineral oils. In particular, the power frequency electric strength of these oils is lower than that of highly purified oils, which are not economically feasible for practical power applications. Impurities or defects dominate the electric strength level of technical-grade oils even in their initial state before use. For a circuit breaker, the effects of generation of carbon and ingress of moisture to decrease the power frequency electric strength must be considered. Impulse strength, however, is not as much affected by impurities, as is power frequency strength. The effects of impurities in causing breakdown take longer than the time duration of impulse applications.

A characteristic of ac and impulse breakdown data for mineral oil under a given condition of test is great variability. Weber and Endicott [28,29] were among the first to report that minimum value statistics can be used to represent these breakdowns. It is implied that breakdown in oil is initiated at a randomly occurring defect or weak link. Breakdown levels and relationships given here were generally derived from statistical treatments of sufficient numbers of samples and tests to be meaningful for this analysis. Such relationships have also been used for the electrical breakdown of solid-liquid composites and solids.

The essential use of statistical analysis is to obtain a proper fit of the data and a measure of its expectancy. This does not give direct evidence for a given mechanism. Besides minimum-value statistics, Weibull and normal statistics have been used for electrical breakdown of various insulators and insulation systems, the choice of the statistic being predicated on the best straight line. Figure 8.22 is a minimum-value distribution for front-of-wave breakdown data for 60 Hz, and Fig. 8.23 is a normal distribution for 1.5/40 µs impulse data, which does not fit as well in Fig. 8.22.

AC, Impulse, and Front-of-Cycle Electric Strength

A study was made by Dakin et al. [30] on comparing the continuous 60-Hz, 60-Hz front-of-wave, and 1.5/40-µs electric strengths of mineral oil in the ASTM D1816 cell with VDE electrodes. Front-of-the-wave voltage is not much used, but it provided a long time impulse of 4×10^{-3} s, in the range of switching impulses. It was found (Table 8.4) that only the ac strength was affected beneficially by filtering, and adversely by a previous sequence of 60-Hz ac, 60-Hz front-of-wave, and impulse breakdowns. The impulse/60 Hz ratio changed with decreasing 60-Hz strength from about 3 to about 5. Impulse and front-of-wave breakdown levels were about the same, the maximum difference being 13%. Figure 8.24 shows the decrease of breakdown

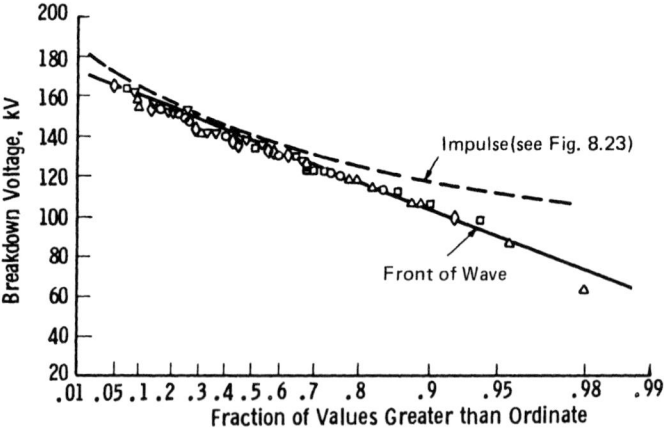

Figure 8.22 Distribution of the 60-Hz front-of-wave breakdown voltages for a 2-mm (0.08-in.) oil gap on extreme-minimum-value paper. ASTM D1816 cell, VDE electrodes. (From Ref. 30.)

Dielectric Properties of Circuit Breakers

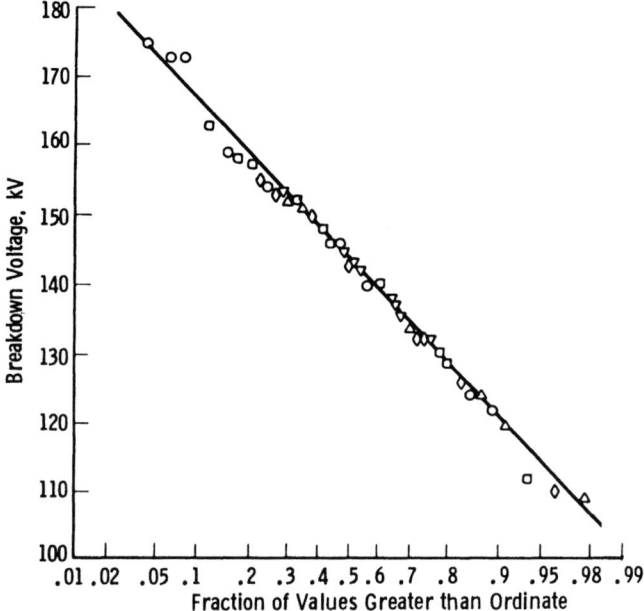

Figure 8.23 Distribution of the impulse breakdown voltage of a 2-mm (0.08-in.) oil gap on normal probability paper. ASTM D1816 cell, VDE electrodes. (From Ref. 30.)

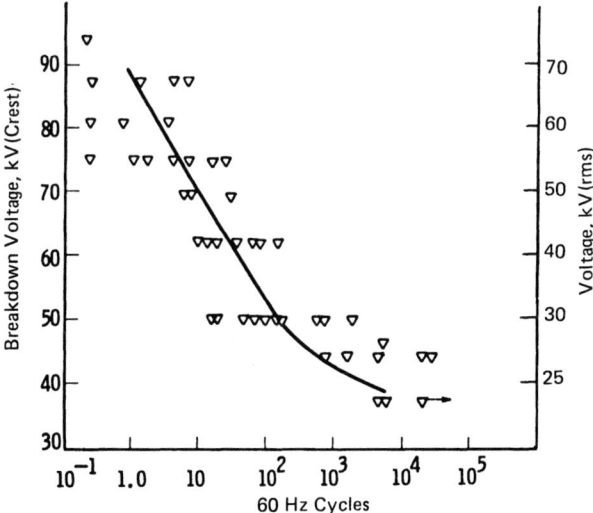

Figure 8.24 Time to breakdown of a 2.5-mm (0.1-in.) oil gap. ASTM D877, 2.5-cm (1-in.) square-edge electrodes. (From Ref. 30.)

Table 8.4 Breakdown Strengths of Mineral Oil with Different Applications of Voltage, in ASTM D1816 Cell, 2-mm Gap, VDE Electrodes

Oil treatment and water content, $\mu g/cm^3$	Applied voltage[a]			Ratio	
	60-Hz	Front of 60-Hz wave	1.5/40 μs impulse	F.W.[b]/ impulse	Impulse/ 60 Hz
		Breakdown voltage, kV			
	Crest				
—, 47	42 ⟶	119 ⟶	134	0.89	3.1
Filtered, 20	54 ⟶	127 ⟶	147	0.87	2.7
New, filtered, degassed, 18	52 ⟶	145	142	1.02	2.7
	34	139 ⟵			
	37 ⟶	⟶	⟶ 138	1.01	3.7
Above left overnight, 40	30 ⟶	⟶	⟶ 151		5.0
	29 ⟵	136 ⟵		0.90	

[a] Arrows indicate sequence of measurements.
[b] F.W., front of 60-Hz wave.

voltage with time of applied voltage from 60-Hz front-of-wave (i.e., 1/4 cycles) to more than 10^4 cycles (~4 ms to >3 min).

However, it was found in another study [31] that the breakdown voltage of mineral oil decreased with increasing time of voltage application from microseconds to milliseconds, giving a corresponding decrease in the impulse/power frequency ratio. A low impulse ratio of 1.3 was found for 500-μs impulses where uniform field electrodes were used [32].

Gap Length, Stressed Volume, and Area Effects

Relationships are needed to apply small-model oil breakdown data to full-size equipment. The simplest variable is the gap length, but the relationship of breakdown voltage to gap length depends on the geometry of the gap or the conductors and on the sample size. The quality of the mineral oil, notably the absence of impurities, is of course another variable. The stressed volume variable, specifically the volume in which 80 to 90% of the stress is contained, seems to be effective in representing breakdown voltage data for various electrode configurations and a large range of gap lengths, the breakdown voltage stress decreasing with increasing stressed volume. One difficulty

Dielectric Properties of Circuit Breakers

here is the calculation of the stressed volume in complicated gap configurations. However, it was found in various investigations [28,29, 33,34] that breakdown voltage stress of mineral oil essentially decreases with electrode area. A comparison of the use of the stressed volume and the electrode area variables with breakdown voltage stress data from nine studies indicated that, particularly for power frequency breakdown, the stressed volume variable represents the data better for technical-grade and contaminated oil and the area criterion is better for highly purified oil [34]. Since oil in a breaker is of the technical grade to begin with and becomes relatively contaminated with use, only the stressed volume criterion will be discussed here.

An extensive study was made by Kawaguchi et al. [35] on the effect of varying gap length and stressed oil volume over large ranges, and of different electrode configurations on the impulse and power frequency breakdown voltages of mineral oil. Calculations are given for determining the volume containing 90% of the stress for the various electrode configurations.

Relationships between power frequency breakdown voltage and gap length are given in Fig. 8.25 for gap lengths ranging from 2.5 to 250 mm. Each electrode configuration gave a different curve where the breakdown voltage increased less than linearly with gap length, and was generally higher with more uniform field geometries. These results are for ac voltage applied at 3 kV/s, and the breakdown levels were 10 to 20% higher than with voltages applied in a 1-min step sequence.

These data are plotted in Fig. 8.26 as ac breakdown voltage stress versus 90% stressed volume, one curve apparently being sufficient within the data variations. It can be seen that this curve has a significant discontinuity at a relatively high volume, where the breakdown continues to decrease but more slowly with increasing stressed volume. It is noteworthy that this curve is the same as one obtained with a uniform field distribution.

Figure 8.27 gives the impulse breakdown voltage stress as a function of stressed volume, and here also one curve represents the various electrode configurations used in this study. Although not stated, these breakdown data were probably obtained with $1/50$-μs impulse voltages as they agree well with $1/50$-μs data of another study [36]. Comparing Figs. 8.26 and 8.27, it can be seen that the impulse/power frequency breakdown ratio was 3.2 for the high and low stressed volumes, and 2.5 for an intermediate range of volumes.

Moisture and Carbon Contamination Effects

For circuit breaker design, it is necessary to use the electric strengths of mineral oil which is contaminated with moisture and arc products, principally carbon. These cause the electric strength to decrease significantly. Many breakers are not hermetically sealed, they are allowed

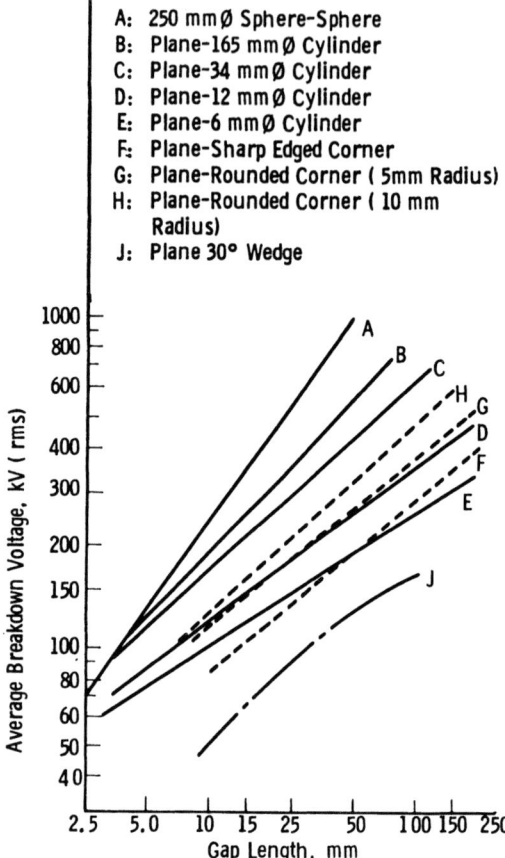

Figure 8.25 Ac breakdown voltage-gap length relations for various electrode configurations, at a 3-kV/s rate of voltage rise. (From Y. Kawaguchi, H. Murata, and M. Ikeda, IEEE Trans. Power Appar. Syst., *PAS-91*:9, © 1972, IEEE.)

to breath, and the oil will pick up moisture from the ambient. Arcing during opening of the contacts will generate carbon. Mineral oil with an ASTM D1816 or D877 electric strength which is reduced from ⩾35 kV to 20 kV by contamination is generally unacceptable for power equipment operation.

The electric strength decreases markedly with moisture content to the point of moisture saturation, beyond which the decrease is quite gradual. This is illustrated in Fig. 8.28. It is generally accepted that the effect of moisture in decreasing breakdown voltage in oil is associated with an interaction with particles. It was shown that a better

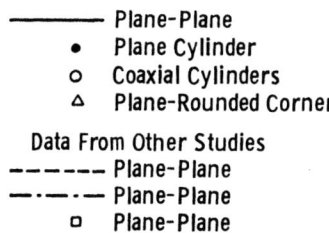

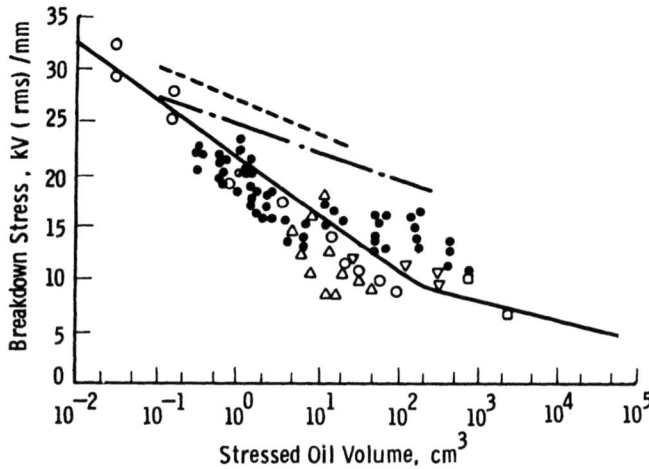

Figure 8.26 Ac breakdown voltage stress: 90% stressed volume relations for various electrode configurations, at a 3-kV/s rate of voltage rise. (From Y. Kawaguchi, H. Murata, and M. Ikeda, IEEE Trans. Power Appar. Syst. *PAS-91*:9, © 1972, IEEE.)

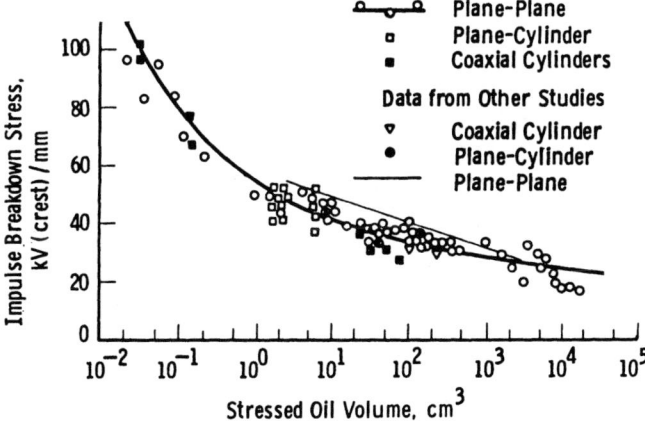

Figure 8.27 Impulse breakdown stress: 90% stressed oil volume relations for various electrode configurations. (From Ref. 26.)

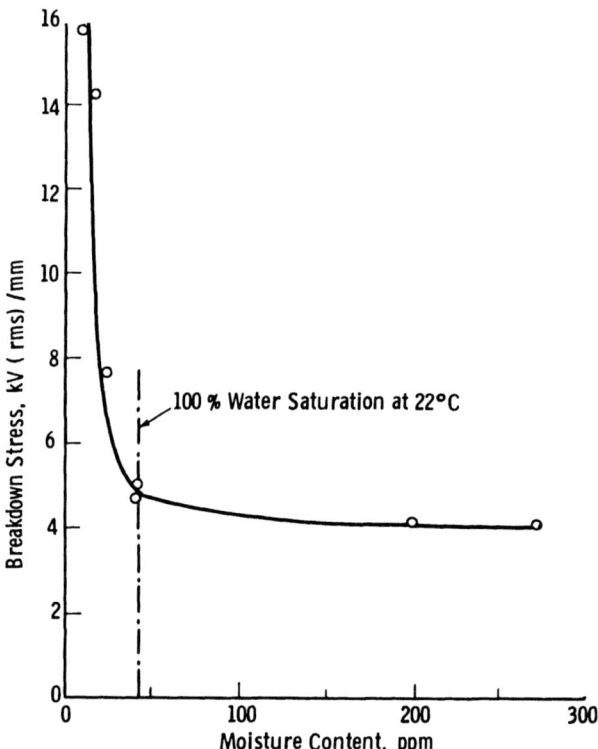

Figure 8.28 Effect of moisture on the breakdown strength of mineral oil, 5-mm-gap 127-mm-diameter aluminum flat-plate electrodes. (From Ref. 36.)

filtered oil has higher electric strength at equivalent moisture levels [37].

A number of studies have been made on the effect of particles on the breakdown of mineral oil, where particles were purposely introduced. The work of Darveniza [38] is particularly significant for oil circuit breakers because it deals with carbon (graphite) particles in mineral oil, as carbon is an unavoidable product of arcs which occur on opening these breakers. Of related interest is the work on the relationships that control the breakdown process initiated by a particle and the movement of a particle out of the field [39]. Other work that should be referenced here dealt with conducting and nonconducting particles, and how to characterize them [40].

Darveniza [38] showed that two modes of mineral oil breakdown occur when carbon particles are present, one at low voltage stresses and

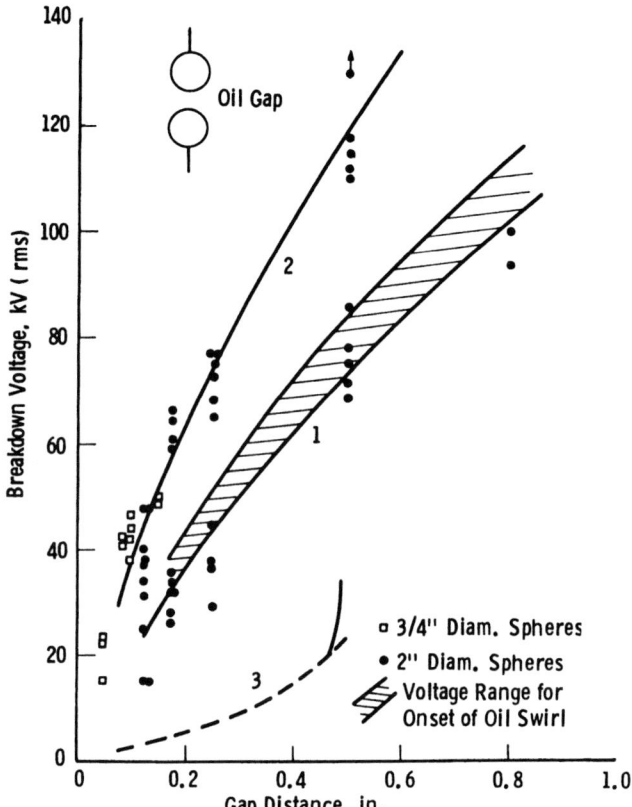

Figure 8.29 Individual breakdown voltages in 60-ppm graphite contaminated oil. Curves 1 and 2 at 2 kV/sec, curve 3 at 1 kV, 1 min steps. (From Ref. 38.)

one at high stresses. Figure 8.29 illustrates this where breakdown voltages are plotted against gap distance, for spherical gaps in oil contaminated with 60 ppm of carbon. At relatively low stresses the carbon particles lined up between the electrodes, forming a bridge across which breakdown could occur. If the voltage stress reached sufficiently high levels that caused fluid motion, or swirl, indicated by the hatched region in Fig. 8.29, the low-strength bridge was broken by the ejection of particles from the gap, and breakdown would occur at relatively high voltages. Thus two breakdown regions are delineated, one of low voltage which is below or at the region of oil swirl, and one of high voltage which is where there is strong fluid motion. A breakdown voltage histogram clearly shows these two regions. The effect of fluid motion is consistent with the observation of

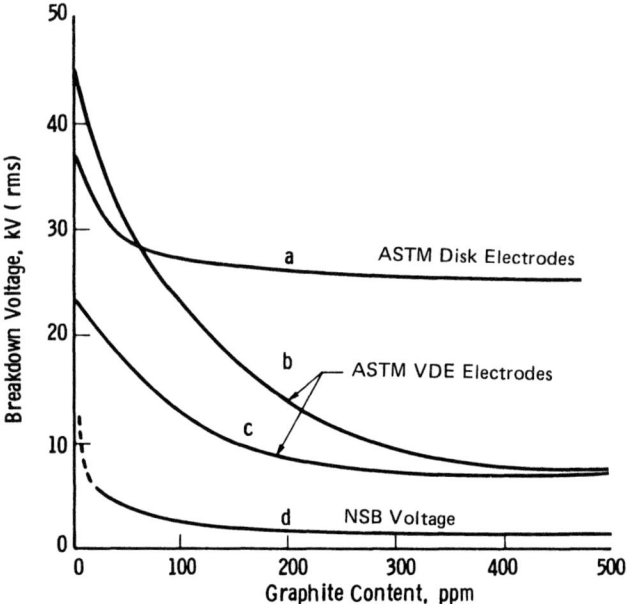

Figure 8.30 Electric strength of transformer oil containing carbon, graphite. (a) Determined according to ASTM D877-64 at 2 kV/sec; (b) determined according to ASTM D1816-60T at 1/2 kV/sec; (c) the lower limit of breakdown voltages recorded during tests with VDE electrodes; (d) average nonsustained breakdown voltage during 1 kV 1 min step tests with disc and VDE electrodes. (From Ref. 38.)

Nelson et al. [36] that the breakdown voltage of mineral oil was increased by causing it to circulate.

Figure 8.30 gives plots of breakdown voltage of mineral oil with various levels of carbon contamination, determined according to ASTM D877 and D1816 tests with disk and VDE electrodes. In these and other tests with carbon contaminations, it was noted that nonsustained breakdown (NSB) occurred, evidenced by audible clicks or crackles, below the usually observed sustained breakdown level [38]. These were taken to represent actual breakdown, which would be expected to occur with a high-power, low-impedance circuit. According to the NSB criterion, contaminated oil in both tests gave the same low breakdown values.

Carbon contamination, however, had much less effect on the breakdown voltages with needle-needle and needle-plane electrodes than with spherical and disk electrodes. A carbon bridge could not be established with a needle electrode because considerable liquid motion occurred in the vicinity of its point where the voltage stress became high at a relatively low voltage.

Flashover at Mineral Oil-Solid Interfaces

The flashover voltage stress at a mineral oil-solid interface can be considerably lower than the breakdown voltage of the bulk oil by a factor of about 5. Various investigators have found this to depend on numerous factors, such as the surface roughness and the dielectric constant of the solid, on the presence of gas and moisture, on particles, on capacitive current flow, on the normal voltage stress and on the tangential voltage stress distribution. In most investigations of flashover at a liquid-solid interface the liquid is mineral oil, but the solid is pressboard or paper, which do not appear together with mineral oil in oil breakers. However, a study by Weschler and Riccitiello [41] on flashover in mineral oil dealt with the effects of dielectric constant of the solid, which would include different solids. Flashover data obtained in this study are given in Fig. 8.31, which fall between a theoretically derived curve for a maximum effect of surface roughness, and one for completely smooth surfaces where there is no reduction in the breakdown voltage.

Mechanism of Breakdown

The foregoing observed breakdown voltage effects in mineral oil, particularly those associated with bulk breakdown, lead to the presently accepted concept that breakdown is initiated at a randomly occurring defect. This is derived from the statistical nature of the measured breakdown voltages. The fact that the breakdown strength of mineral oil is increased by filtration is indicative that particles may be a major

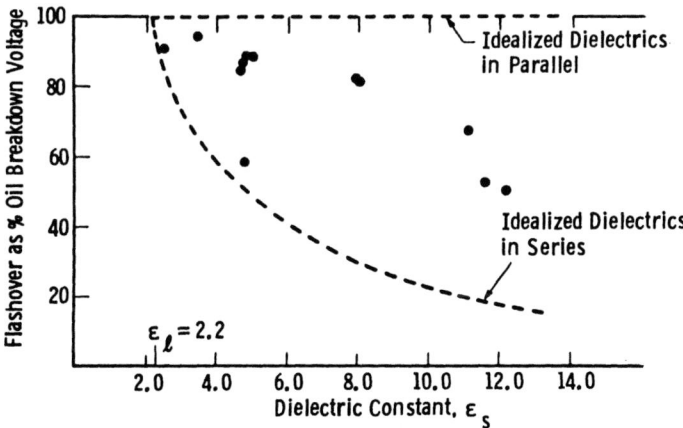

Figure 8.31 Flashover voltage in oil between cylinders as a function of the dielectric constant of the parallel solid. (From K. Weschler and M. Riccitiello, AIEE Trans., 80:365, © 1961, IEEE.)

source of defects. Particles are particularly associated with defects in technical-grade mineral oil, and may act as sites of high stress on an electrode. Krasucki [39] dealt with the relationships governing the motion of particles that lead to breakdown as well as their ejection from a high-stress region.

Very highly purified mineral oils show different characteristics of breakdown than technical-grade oils. Their breakdown voltages are higher, as shown earlier. Another significant difference found by Sletten and Dakin [42] was that technical-grade oil breaks down on the positive half cycle of applied 60 Hz and highly purified oil on the negative half cycle.

However, the effect of a defect in initiating a breakdown discharge, and the subsequent phenomena leading to propagation of discharges, apparently occur in times that are in the range of 60-Hz cycles, which are longer than with impulse voltage. This follows the observation that 60-Hz breakdown is much more sensitive to the quality of the oil than is impulse breakdown.

There are various suggestions regarding the actual way defects or particles initiate a breakdown streamer. According to theoretical and experimental models of Kok and Corbey [43], particles form a bridge between electrodes, across which thermal breakdown occurs. This is based on microscopic observation of particle behavior in a dielectric fluid, and that the presence of water was particularly effective in causing breakdown. Krazucki [39,44] suggests that vaporization occurs at the site of a defect, followed by a breakdown discharge in the bubble. The vaporization is said to occur when the pressure on the liquid due to the electric field at a particle exceeds the hydrostatic pressure and surface tension of the liquid on the surface of the particle. Dissolution and the formation of gas at a high stress point have also been considered as possible ways for bubbles to be generated, assuming that the initial breakdown occurs in bubbles.

8.3.3 Insulating Solids

Functions and Materials

Insulating solids provide several functions in circuit breakers, some concurrently with one component. The essential electrical isolations are provided by the bushing or bushings and by the contact supports. Insulating solids satisfy mechanical requirements of various designs, serving in part or completely as components for structure and pressure containment in some oil breakers, and for transfer of gas in some SF_6 breakers. Other functions of insulating solids in circuit breakers, which are auxiliary and generally do not entail high electrical or mechanical stresses, are to provide a low-friction surface for moving parts.

Dielectric Properties of Circuit Breakers

The materials used for the critical insulation are generally porcelain, or those based on thermosetting resins and on paper. These resins are usually epoxies and phenolics. Paper is used in composites with resin and with oil.

Various resin formulations with a powder and glass fiber filler are used to achieve required mechanical and electrical properties. Aluminum oxide trihydrate filler, and to some extent silica powder, improve resistance to tracking. Silica powder is added to epoxy resin to decrease the thermal expansion coefficient to match that of the conductor so that a good resin-to-metal bond is obtained, which is needed for both mechanical and electrical stability. Glass fiber is used to achieve structural strength for load bearing and for sustaining pressure.

Thermoplastics and elastomers are used for the low-stress auxiliary functions. Nylon (polyamide), polycarbonate, and polymethyl methacrylate are used for arc control. Teflon (polytetrafluoroethylene) is used as a liner in SF_6 breakers to resist arc products, and as a low-friction surface for moving parts.

Requirements

In developing criteria or requirements for long-term serviceability of the insulations, particularly those that are critical, it must be realized that their major deterioration effects can be due to the simultaneous application of electrical and mechanical stresses. These may synergistically promote deterioration. The main thrust of this presentation is on the electrical effects, and in considering them it will be noted where mechanical effects have a role.

Essentially, the effects of voltage are due to the continuously applied voltage stress, and to sporadically applied overvoltages due to switching or lightning impulses. The following effects have to be considered for reliable functioning under operational conditions:

1. Short-time and long-time electric strength, including treeing phenomena.
2. Discharges in voids that can cause progressive deterioration. Voids may be created by mechanical effects.
3. Thermal runaway.
4. Surface flashover and tracking, and also tracking within the insulation.

The effects listed above, except short-time breakdown and surface flashover, occur over a long period of time until failure ensues. All pertain to organic-based insulators, while the short-time effects pertain also to porcelain. The following discussion deals mainly with organic-based materials, generally on the basis of work done with epoxies. A good reference to this is a review by Dakin [45] on epoxies in applications related to electrical equipment.

Bulk Electric Stability

The bulk electric stability of a solid insulator can decrease with increasing time of voltage application; for design purposes it is necessary to know both the short-time electric strength that it can withstand under lightning and switching impulses, and the maximum voltage stress that would give long-term stability under rated voltage dictated by the usually required operating life to 20 to 40 years. The necessary short-time voltage withstand or breakdown data can be obtained easily and directly. Long-term bulk stability takes cognizance of deteriorating effects that are due to treeing, which may be associated with discharges, to discharges originating internally, in voids, or on the surface, and to heating of the insulation that could result in thermal runaway.

Short-Time and Impulse Electric Strengths

Electric strengths of filled and unfilled epoxy for short time, power frequency voltage application, and for lightning impulse (1/50 µs) conditions were obtained by using embedded sphere-to-sphere, and point-to-plane electrodes [46] (Figs. 8.32 and 8.33). The breakdown stress decreased with increasing electrode separation, from 0.5 to 5 mm, in these nonuniform fields, as the highest stresses behave in a less than linear way in these configurations. However, in uniform fields the breakdown voltage stress appeared to be constant with gap separations to 6 mm [45,47]. Also, the electric strength of unfilled epoxy is greater than that of filled epoxy [46,48,49] (Figs. 8.32 and 8.33). This is possibly due to the stress on the resin being higher in the filled material as the dielectric constant of the filler is higher, and also that the resin-filler interfaces might be a source of voids, which break down readily.

The impulse strengths are, as expected, greater than the power frequency strengths, where the voltage is ramped or applied in steps. However, Kind and Schiweck [50] found that on applying the voltage in single pulses or cycles, the breakdown strength in a 3-mm uniform field gap, and the partial discharge inception of a 20-mm nonuniform gap, are independent of the rate of voltage rise, from <1 kV/µs to about 10^3 kV/µs (Figs. 8.34 and 8.35). This was confirmed by observations that the single 60-Hz cycle and 1.5/40-µs breakdown voltages of a filled epoxy were similar [49].

Treeing

The long-time operating voltage strength of solid organic-based insulation is limited by the rather incompletely understood breakdown phenomenon known as treeing, which is not associated with thermal effects or with observable discharges. This has been studied considerably in polyethylene. Breakdown voltage-time data have been obtained by

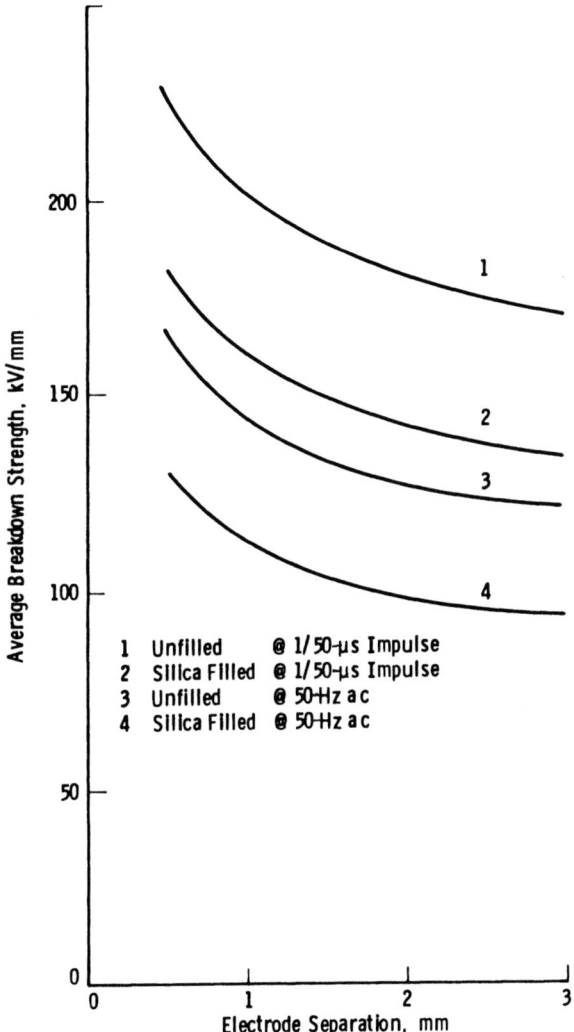

Figure 8.32 Breakdown strength of bisphenol epoxy resin system with embedded 20-mm-diameter spherical electrodes. (From Ref. 46.)

Studniarz and Dakin [49] for a quartz-filled epoxy in a uniform field gap of 2.5 and 3.2 mm. Figure 8.36 shows the voltage-time relation obtained where the specimen electrodes were silver paint, and Fig. 8.37 gives the data for specimens with cast-in electrodes. The shortest time to breakdown was $\sim 10^{-2}$ s, with one 60-Hz cycle, and the longest was the 60-Hz equivalent of about 10^5 h obtained by using 420 Hz and

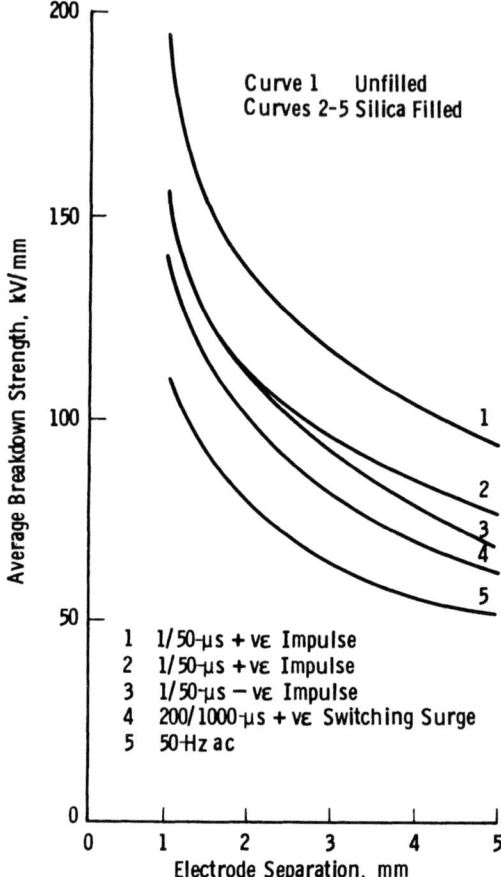

Figure 8.33 Breakdown strength of bisphenol epoxy resin system with embedded point-to-plane electrodes. (From Ref. 46.)

1440 Hz, which gave 7- and 24-fold accelerations. The breakdown times at equivalent voltages were shorter with the silver paint than with the cast-in electrodes. This may have been due to microbubbles between the paint and the resin, the bond being apparently better with the cast-in electrodes. It is difficult to extrapolate these curves to obtain a voltage at which the insulation would be stable for a required operating time of 20 to 40 years. A relationship proposed by these authors [51] for accelerated life data under discharge conditions may be applicable here.

$$E - E_s = -k \log [t_f(E - E_s)] + k_1 \tag{8.7}$$

Dielectric Properties of Circuit Breakers 315

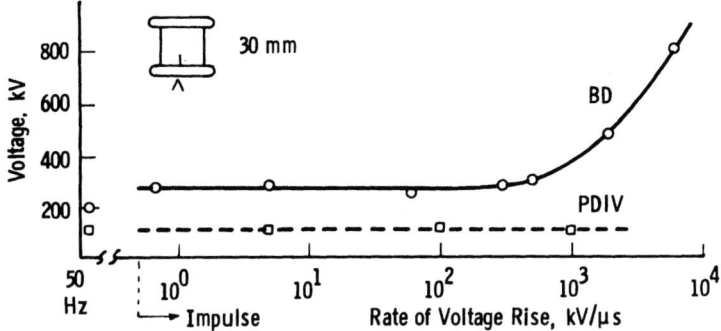

Figure 8.34 Partial discharge inception and breakdown voltage of epoxy resin at different rates of voltage rise, with point-to-plane electrodes. (From Ref. 50.)

Here E_s is the threshold voltage stress where breakdown would not occur, E the applied accelerating test voltage stress, t_f is the corresponding failure time, and k and k_1 are constants.

Effects of mechanical stress were checked on the 1-min breakdown voltage in this investigation [49]. In the absence of cracking, there was no reduction in breakdown voltage, but this was reduced by a factor of about 3 when cracking occurred.

Treeing effects are manifest by treelike formations which start at a sharp point on a conductor or even on a polished surface in a uniform field, possibly at an asperity, and grow quite sporadically through the insulator to the opposite conductor when failure occurs. This phenomenon in the absence of water in the insulation is specifically

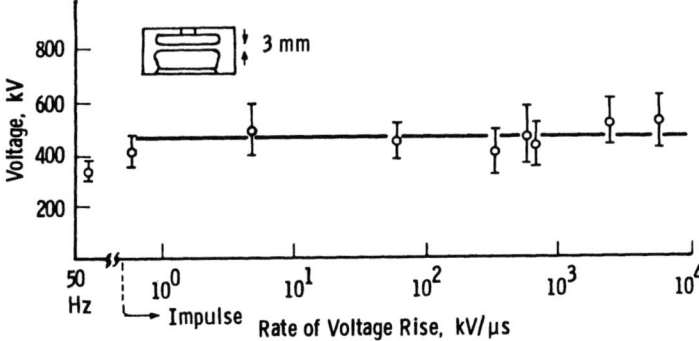

Figure 8.35 Breakdown voltage of epoxy resin at different rates of voltage rise with uniform-field electrodes. (From Ref. 50.)

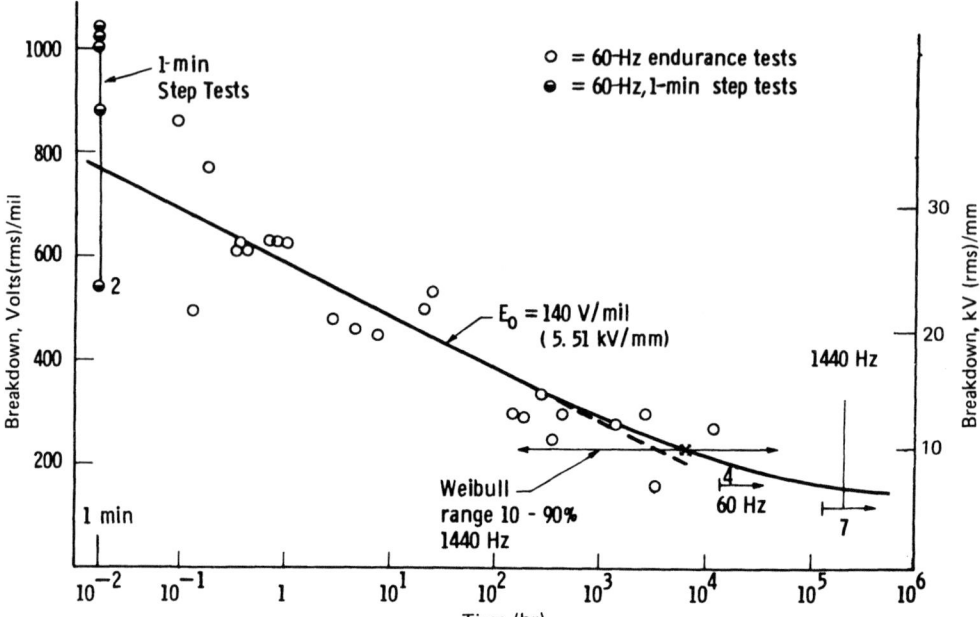

Figure 8.36 Voltage endurance of cast epoxy with silver paint electrodes at 25°C. (From Ref. 49.)

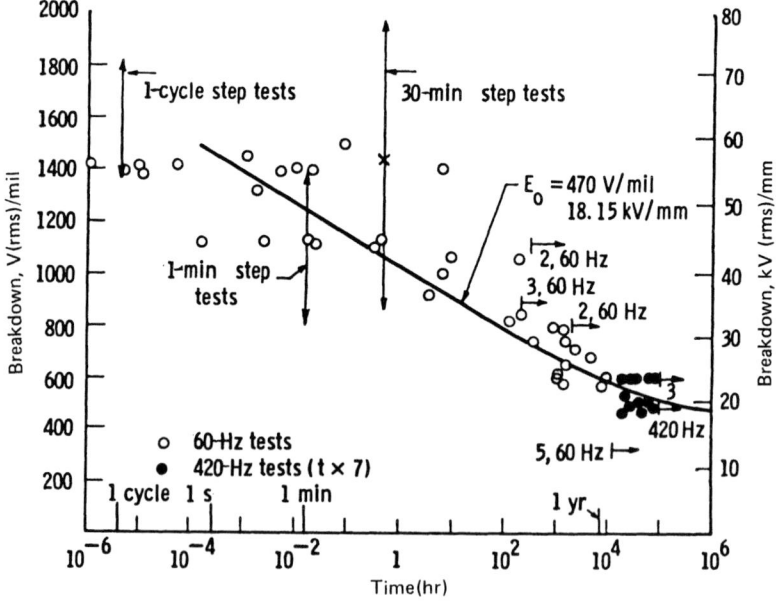

Figure 8.37 Voltage endurance of epoxy with cast-in electrodes at 25°C. (From Ref. 49.)

Dielectric Properties of Circuit Breakers

known as electrical treeing, and with water, which decreased the time to failure, it is known as electrochemical treeing. It is assumed to be initiated by the formation, or possible presence, of a microcavity.

Discharges in Voids, Discharges on the Surfaces, and Accelerated Testing

Discharges can occur on the surface of an insulator or within it and cause progressive erosion and sometimes conduction through its bulk, leading to failure. Surface discharges originate at an air (gas) insulator-conductor edge when the stress in the gas is high enough to cause a partial discharge. Such high stresses, which are localized, may be due to sharpness or roughness of the conductor. Discharges may occur in voids which can be present in the bulk of the insulator or at the conductor-insulator interface where the bond is poor or is broken.

Accelerated tests are available to determine the life expectancy of an insulator under discharge condition (e.g., ASTM D2275), and acceleration is achieved by increased voltage stress and frequency. Acceleration by applying higher frequencies than the power frequency is effective since it has been found that the equivalent power frequency life in cycles is independent of frequency to 1000 or 2000 Hz when the sample is not heated by the applied voltage [52]. However, the voltage stresses used should be as close as possible to the operating stress, as discontinuities have been observed in the volt-time curves, which may be overoptimistically extrapolated from higher voltages.

A voltage-time curve was developed for a silica-filled epoxy (Fig. 8.38) under conditions of surface discharges [49]. This curve is particularly amenable to analysis by equation (8.7). According to

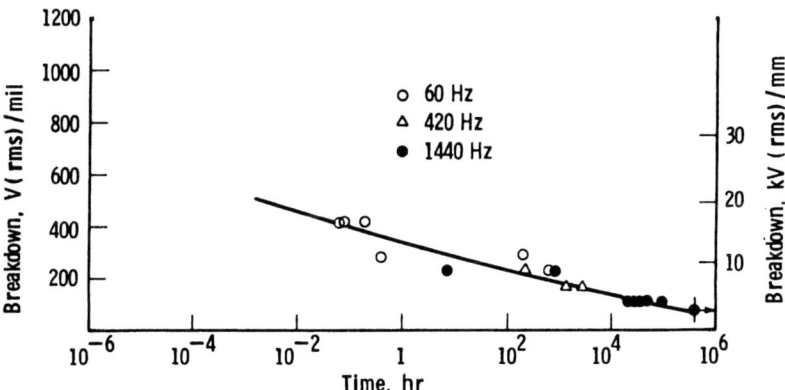

Figure 8.38 Voltage endurance of silica-filled cast epoxy with surface discharges. (From Ref. 49.)

this relation, E_S is obtained, which gives the stress below which the insulation is free of discharges.

One relation is used often in which the breakdown time t_f varies as $(E)^n$, where n is generally -7 to -5. Another relation is used where log t_f varies inversely with E.

A critical quantity needed for evaluating an insulation subject to discharges is E_S, which is just below the partial discharge inception voltage stress. This can be obtained directly by a partial discharge measurement, from extrapolation of the volt-breakdown-time curve, and from careful analysis of the voltage stresses at a conductor edge or in a void if the necessary geometrical parameters are known. For surface discharges, the partial discharge inception voltage stress depends on sharpness of the conductor edge, and was found to be proportional to $(1/d\varepsilon)^{\sim 1/2}$, where d is the thickness of the insulation and ε is its dielectric constant (Fig. 8.39). With voids, this depends on the geometry of the voids; the partial discharge inception voltage stress is εE for laminar voids and $[3\varepsilon/(1 + 2\varepsilon)]E$ for spherical voids, where the total void thickness is much smaller than the insulation thickness, and E is the stress in the solid insulation. Partial discharges occur in a void when the product of the pressure in the gas and the thickness exceeds the breakdown voltage according to the characteris-

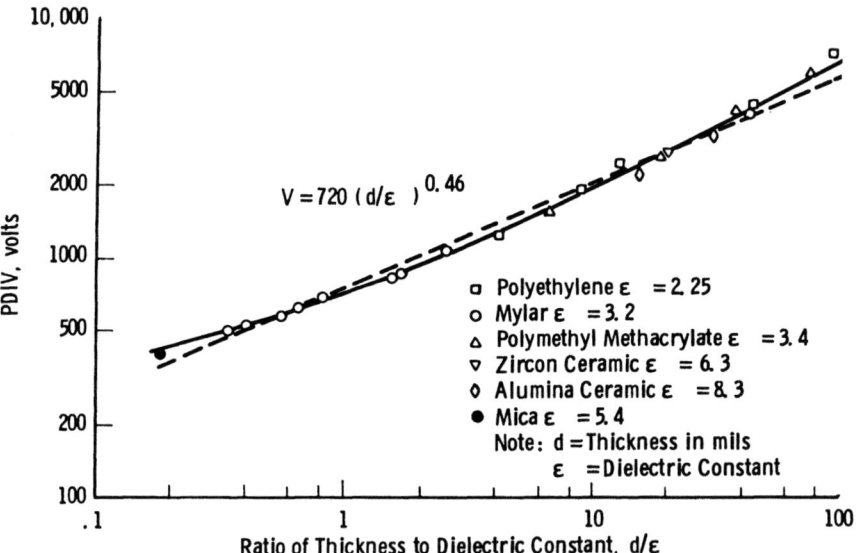

Figure 8.39 Partial discharge inception voltages on dielectric surfaces with 1/2-in.-diameter square-edge electrodes. (From Ref. 53.)

Dielectric Properties of Circuit Breakers

tic Paschen curve. There is a minimum breakdown voltage, which for air is 235 V rms at 0.007 bar-mm. Partial discharges would appear at 7.5 kV/mm in an epoxy, $\varepsilon = 4.5$, with 0.007-mm-thick laminar voids filled with air at 1 atm.

Surface discharge testing of an insulator may be sufficient to evaluate it for effects of discharges in voids, as the life with surface discharges has often been found to be similar to that with discharges in voids [51]. However, this may not always be true, as shorter lives than expected were obtained with certain voids with sharp edges, where there apparently was particularly high voltage stress concentration.

Thermal Runaway

Another form of bulk breakdown of solid insulation is thermal runaway, which occurs when the heat input due to the applied voltage exceeds the heat lost to the surroundings. This heat input is proportional to the capacitance C or the dielectric constant ε of the insulation under voltage stress and its dissipation factor, $\tan \delta$, and is equal to $2\pi f V^2 C \tan \delta$, or $2\pi f E^2 \varepsilon \tan \delta$ (per unit volume) where V and E are the applied voltage and voltage stress, and f is the frequency. Relationships for the heat loss are available, or may be calculated, and may be obtained for the heat input by determining the product of the capacitance and dissipation factor as a function of temperature. It is necessary to know if the dissipation factor increases during operational aging to the point of thermal runaway. This increase can be due to ambient and internal operating effects, such as moisture, thermal aging, discharges, and discharge products. These considerations apply to thermoset and paper-oil insulation systems, but hardly to porcelain, which is generally very stable and resistant to ambient and discharge effects and to high temperatures.

Pollution Effects

The effects that lead to failure over the surface are generally due to atmospheric pollution to which bushings are exposed. This occurs when dirt or salt settles on the surface, and their ions are combined with moisture from dew or rain to form a conducting film. The current through this film causes the water to evaporate, and as dry spots or bands are formed the current is interrupted locally, which causes arcing [54]. This effect leads to decreased flashover voltage and to tracking and erosion.

The electric strength of a surface, or flashover voltage, may also be decreased by conducting particles and by partial discharges on the surface. This is of concern for solid insulation under relatively high voltage stress, as in compressed-gas cables.

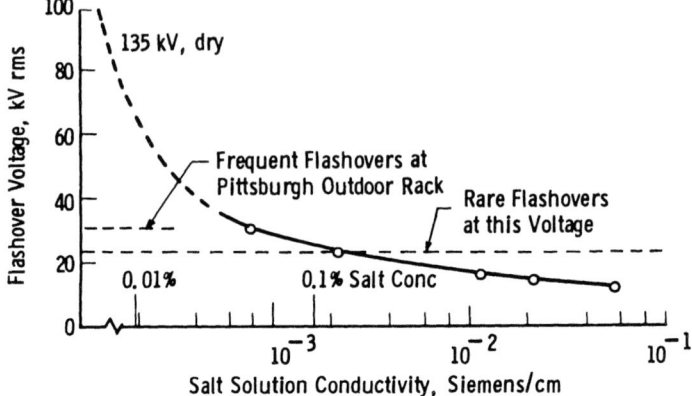

Figure 8.40 Flashover voltages of a 38-cm (15-in.) insulator strip in the salt fog chamber. (From T. W. Dakin and G. A. Mullen, IEEE Trans. Electr. Insul. EI-7:169, © 1972, IEEE.)

Flashover Voltage

The wet surface flashover voltage of outdoor insulators is a function of the conductivity of the film of water, which contains ions from the solid pollution. This is shown in Fig. 8.40 for porcelain insulation. The flashover voltage in an outdoor ambient with moderate pollution and rainfall is of the order of 25% of the flashover voltage obtained under dry conditions. A safe design voltage rating for outdoor bushings in most areas in the United States is about 0.03 kV/mm (0.75 kV/in.) of surface length, following the convolutions of the usual shedded structure.

Tests on filled epoxy insulators that had been aged under voltage in an outdoor test rack showed that this aging decreased the flashover voltage, measured with conducting films of water, by only 10% [55] (Table 8.5). The physical effect of the aging that could be associated with a decrease in flashover voltage was an increase in surface roughness.

Tracking

With organic-based insulation, the arcs that develop as a result of pollution and moisture can cause the underlying surface to track or to erode. Tracking is essentially carbonization of the surface that develops in a treelike formation which progresses from one conductor until it reaches the other, at which point failure occurs. Internal tracking can occur when the resin contains glass fiber. The ion containing moisture seeps into the resin-fiber interfaces, which become

Table 8.5 Flashover Voltage Stresses of Insulation Samples from the Westinghouse Research and Development Outdoor Test Rack

Specimen	Flashover voltage stress, kV/in. at salt concentration/conductivity[a]			Surface roughness, μ in. at 2 in. from ends	
	0.03% 600 μS/cm	0.03% 5000 μS/cm	3.0% 49,000 μS/cm	Top end	Bottom
2-in.-wide strip with a 15-in. gap					
New, unexposed sample (sandblasted)	1.8	1.07	0.9	150	100
Sample from rack, smoothest	1.9	1.1	0.9	38-42	34-40
Sample from rack	1.5	1.0	0.8	460	450
Sample from rack, roughest	1.5	1.0	0.87	250-280	230-250
New, unexposed 1-in.-diameter rod with a 6-in. gap (sandblasted)	2.7	1.8	1.3	100-120	105-120
Shed insulator, from rack, tested single shed					
Convex shed, 8.6 in. creepage	2.0	1.3	1.05	210-240	
Concave shed, 8.3 in. creepage	2.5	1.7	1.2	200-230	

[a]Flashover voltages were measured in a Salt Fog test chamber. The precipitation rate, into 7-cm-diameter beakers, was 5.8 cm^3/cm^2 h.
Source: Ref. 60.

conducting and the localized currents cause sufficient heating to carbonize the adjacent resin [54].

Phenolics and other aromatic materials track readily. Aluminum oxide trihydrate filler imparts track resistance, and is very effective when added to cycloaliphatic epoxy and some other epoxies. Only such materials that are very resistant to tracking should be considered for outdoor use. These can erode, however, and have to be evaluated for erosion resistance.

There are several tests for determining track and erosion resistance. The Westinghouse Differential Wet Tracking test (ASTM D2302) and the Inclined Plane test (ASTM D2303) are relatively fast screening tests, primarily for track resistance, but can also serve for erosion resistance. However, the test known as the Salt Fog test, which uses a controlled spray of salt and fog, gives realistic effects on insulator samples of sizes approaching those used in service [56]. This test was perfected at Westinghouse Research and Development Center with concurrent testing on outdoor racks which use similar specimens and voltages. These racks are located in different parts of the United States, subjecting the samples to coastal and industrial pollution effects. The Salt Fog test was found to accelerate the pollution effects by a factor of about 300 times those obtained from the racks (Fig. 8.41). This was based on the erosion of an aluminum oxide trihydrate-filled epoxy, which gave weight-loss rates of about 0.8% per day in the Salt Fog test for exposure times of 1 to 4 days.

In fact, the experience with the test racks has shown that with over 8 years of testing, various inorganic-filled epoxies showed good resistance to tracking under these conditions [57]. Even silica-filled bisphenol epoxy, which is an aromatic resin, showed only minor tracking effects which did not lead to failure; but with aluminum oxide trihydrate added to this composition, resistance to tracking was better. Cycloaliphatic epoxy with aluminum oxide trihydrate was, as expected, very resistant to tracking.

Insulation Composition and Designs for High-Voltage Operation

In the foregoing the major effects that limit insulation life were discussed, as was voltage stress for high-voltage operating conditions, which are required for high-power circuit breakers. Definite indications emerge from these regarding insulation formulations, processing, and designs, which have been and are being developed to optimize stability and performance.

For short-term high-voltage requirements, such as for lightning and switching impulses as well as power frequency electric strength, a design favoring uniform field distribution may be all that is needed. This, of course, would considerably aid long-term voltage stability.

Stability toward treeing may be enhanced by exclusion of moisture, which is not always possible, and by good resin-to-conductor bonding

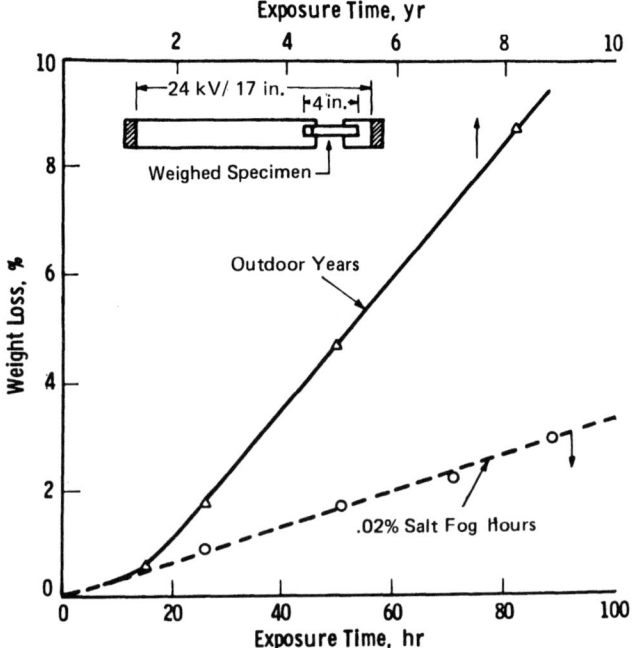

Figure 8.41 Correlation of weight loss during outdoor exposure and salt fog chamber exposure at 1.4 kV/in. (55 V/mm) surface stress with aluminum oxide trihydrate-filled cycloaliphatic epoxy. (From Ref. 56.)

and eliminating high concentrations of field—with smooth conductor surface—in the high-voltage stress regions. The insulation should, of course, be as void-free as possible. Good bonding could be established and maintained by resin-filler compositions whose coefficients of thermal expansion match that of the conductor. Table 8.6 lists coefficients of some insulators and it is noted that the cycloalphatic epoxy and silica formulation gives a good match with metal.

There are many ways to eliminate or minimize voids in which discharges that degrade the insulation can occur. Voids can develop within a solid insulator during resin cure, by gas entrapment, dissolution, or formation, and by mechanical fracture of the solid. Voids can also occur at a resin-filler or resin-conductor interface because of poor bonding or bond severance. As mentioned above, bond severance between metal and resin can be avoided by using a resin whose thermal coefficient of expansion matches that of the metal (Table 8.6). Fillers can be specially treated to bond well to a resin. Gas development in a resin can be minimized by recently advanced methods of vacuum-pressure impregnation (VPI) [58].

Table 8.6 Schedule of Dielectric Test Values for Outdoor Circuit Breakers and External Insulation

				Insulation withstand test voltages						
	Rated maximum voltage, kV, rms	Low frequency		Impulse test 1.2 × 50 μs wave				Switching impulse		Minimum creepage distance of external insulation to ground, in. (m)
		1-min dry, kV, Col. 2	10-s wet, kV, Col. 3	Full wave[a] withstand, kV, Crest Col. 4	Interrupter full wave, kV, crest Col. 5	Chopped wave, kV, crest[a]: Minimum time to sparkover		Withstand voltage terminal to ground with circuit breaker closed kV, crest Col. 8	Withstand voltage terminal to terminal on one phase with circuit breaker open, kV, crest Col. 9	
						2 μs withstand Col. 6	3 μs withstand Col. 7			
Line	Col. 1									Col. 10
1	15.5	50	45	110	—	142	126			9 (0.229)
2	25.8	60	50	150	—	194	172			15 (0.381)
2A	25.8[b]	60	50	125	—	—	—			15 (0.381)
3	38.0	80	75	200	—	258	230			22 (0.559)
3A	38.0[b]	80	75	150	—	—	—			22 (0.559)
4	48.3	105	95	250	—	322	288			28 (0.711)
5	72.5	160	140	350	—	452	402			42 (1.067)
6	121	260	230	550	412	710	632			70 (1.778)
7	145	310	275	650	488	838	748			84 (2.134)
8	169	365	315	750	552	968	862			93 (2.489)
9	242[c]	425	350	900	675	1160	1040			140 (3.556)
10	362[c]	555		1300	975	1680	1500	825	900	209 (5.309)
11	550[c]	860	Not required	1800	1350	2320	2070	1175	1300	318 (8.077)
12	800[c]	950		2050	1540	2640	2360	1425	1550	442 (11.227)

[a] 1.2 × 50 μs positive and negative wave. All impulse values are phase to phase and phase to ground and across the open contacts.
[b] These circuit breakers are intended for application on grounded wye distribution circuits equipped with surge arresters.
[c] These circuit breakers are intended for application only on systems effectively grounded, as defined in Neutral Grounding Devices, IEEE Publ. 32, 1972.

Source: ANSI C37.06-1979, Preferred rating structure for ac high-voltage circuit breakers rated on a symmetrical current basis.

Dielectric Properties of Circuit Breakers

To impart resistance to surface discharges, aluminum oxide trihydrate filler is very effective. Mechanical strength is increased by glass fiber filler. However, this lowers track resistance by making internal tracking possible. For both high mechanical strength and track resistance, the filler composition has to be carefully balanced.

High electrical stresses on the surface, which promote partial discharges or lower the flashover voltage, may be avoided or suppressed by attention to the conductor-insulator-air (or gas) interface. First, it may be possible to assure that there are no rough spots or sharp points in that region. An insulating coating may be affixed to the metal so that there is a solid in the high-stress regions. Finally, voltage-stress grading paint may be applied over the conductor and adjacent insulation, thus lowering the voltage stress at the conductor by distributing it more evenly over the thin coating. Such a coating consists of a varnish generally containing silicon carbide powder. It achieves the stress grading effect by its nonlinear resistivity, which decreases very steeply with increasing voltage stress by about the fifth power, so that the stress at the conductor is relatively low because of the stress-induced conductivity.

8.3.4 Vacuum

Vacuum Gap Breakdown

Gap Dependence If a voltage V is applied across a uniform field vacuum gap of separation d, then for nominally clean conditions the dependence of breakdown voltage on gap separation is as shown in Fig. 8.42. The voltage is proportional to the gap spacing for small gaps, but falls off with increasing spacing [59] above about 2 mm (or at voltages above 80 kV) according to

$$V = Kd^{\alpha} \tag{8.8}$$

where $3000 \leqslant K \leqslant 4500$ kV/mm and $0.4 \leqslant \alpha \leqslant 0.7$.

Small Gaps (d ⩽ 2 mm): The electric strength of short gaps is high (3000 to 8000 kV/mm) and is limited by electron emission currents from surface features such as protrusions, grain boundaries, and other preferentially emitting sites [60]. The electron emission follows the Fowler-Nordheim field emission theory and causes measurable currents at voltages just below the breakdown voltage, at which time emitter current densities of about 10^{10} A/m^2 are enough to cause thermal instability in the emitter by joule heating and breakdown in the vapor released into the gap.

With modern vacuum technology, residual gas initially in a gap is not a cause of low breakdown voltages, particularly for small gaps. For larger gaps with certain metals (e.g., Al, Cu, Ni, Nb) there is a low but finite gas pressure (e.g., 10^{-3} to 10^{-6} torr) at which the

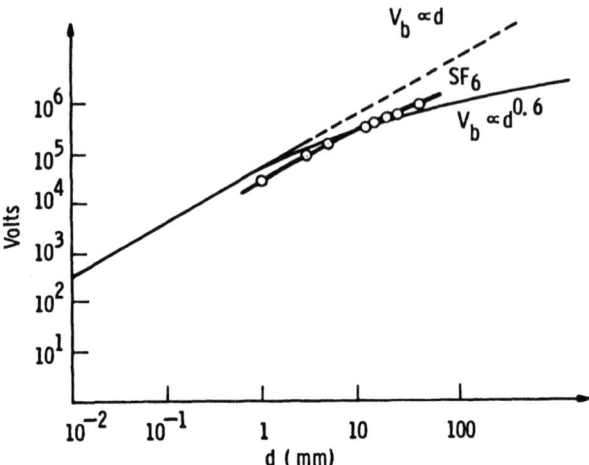

Figure 8.42 Vacuum breakdown voltages are directly proprotional to electrode spacing for gaps up to about 2 mm, and increase with (electrode spacing)$^{0.6}$ for larger gaps. The strength of SF_6 at 400 kPa is higher than for vacuum at spacings greater than about 10 mm. (From Ref. 59.)

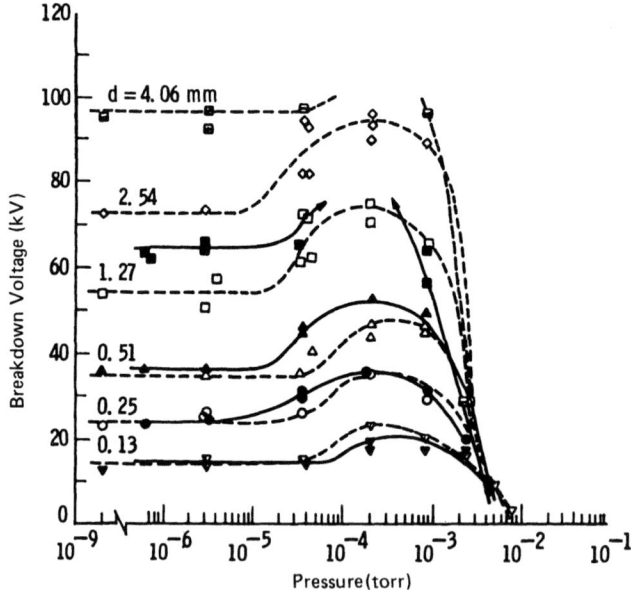

Figure 8.43 Breakdown voltages reach a maximum at gas pressures in the range 10^{-3} to 10^{-6} torr. Numerals indicate the gap spacing in mm. Niobium electrodes. Dashed line, dc; solid line, ac (peak). (From Ref. 61.)

breakdown strength can be up to 50% higher than for ultrahigh vacuum (Fig. 8.43). At similar pressures the emission current also reaches a minimum. The reduction in current at this pressure is large (e.g., by a factor of 6) (Fig. 8.44). The fall in emission may occur more slowly than the pressure changes, and is sometimes, but not always, reversible when the gas pressure is reduced again [61,62]. The effect is considered to be associated with the growth and decay of surface films and protrusions rather than a direct effect of gas pressure. In some applications not requiring high vacuum, this optimum pressure is sometimes obtained by coating the conducting walls of the vessel with an oil of appropriate vapor pressure. This approach is not possible in the arcing chamber of a vacuum circuit breaker, nor over a wide range of temperature.

Electric strength depends strongly on the metal of the electrodes. Figure 8.45 shows the electric strength decreasing in the sequence Ni, Nb, OFHC Cu, Al, with the respective strengths for a 1-mm gap relative to aluminum being in the ratio 2.2:1.7:1.27:1.0. Stainless steel and titanium alloys and possibly molybdenum are recognized good high-voltage electrode materials, while tungsten and copper are recognized bad high-voltage electrode materials [63]. An emitted electron beam can also heat a local region of the anode, creating an anode spot which

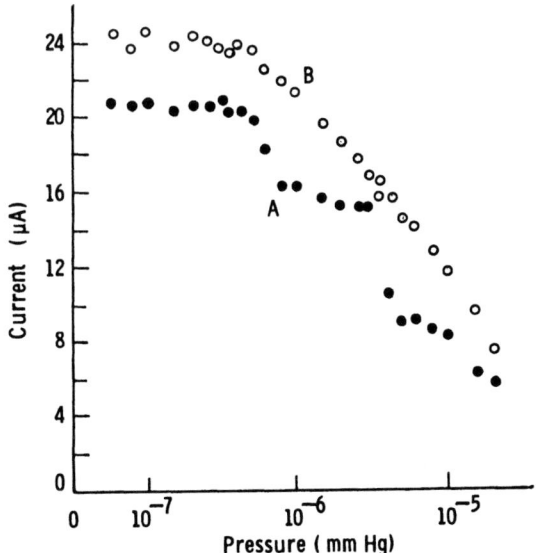

Figure 8.44 Emission currents fall with increasing pressure, corresponding to the increase in breakdown strength with gas pressure shown in Fig. 8.43. A, 16 kV dc, 0.5 mm; B, 34 kV ac rms, 1.5 mm. (From Ref. 62.)

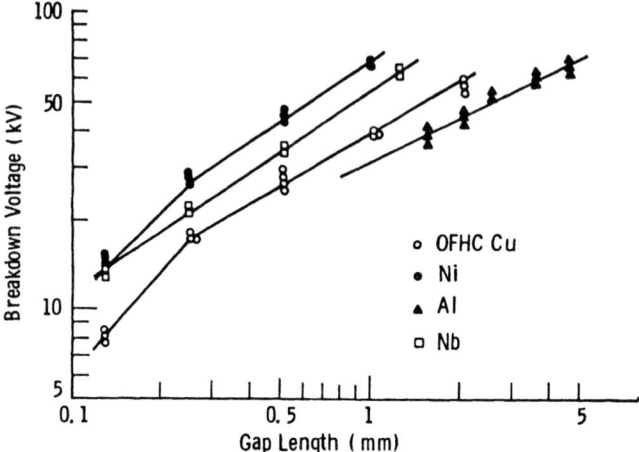

Figure 8.45 Dependence of electric strength on gap length and electrode material in vacuum. (From Ref. 61.)

can then release vapor to cause breakdown, particularly if there is a high-vapor-pressure material on the anode. Positive ions from the anode region can form a space charge, enhancing the field at the emitter and increasing the electron flux. X-rays released by such electron currents also produce secondary electrons which result in a distributed electron current and losses in the system. Conditioning by maintaining currents from local emitters leads to deactivation of the sites and an increase in the breakdown voltage. Sparking with only a limited availability of local energy (i.e., by use of limiting resistors and low-capacitance systems) can also raise the subsequent breakdown level. For small gaps, the use of very smooth polished electrodes reduces the number and activity of local emitter sites and thus raises the voltage withstand performance. The ac strength is usually equal to the dc strength or a little higher, the increase being associated with the limited time from buildup of prebreakdown currents in a quarter cycle of voltage and space-charge effects.

Large Gaps (d ⩾ 2 mm): Particle-initiated breakdown can occur in large gaps at stresses lower than for electron emission-initiated breakdown in small gaps. Particles present on an electrode surface of either polarity may be detached and driven across the electrode gap by electrostatic forces. The charge q on these particles is approximately proportional to the field, while the energy gained in crossing the gap is proportional to qV, or V^2, where V is the potential difference across the gap. Thus there is a strong absolute voltage dependence of the energy gained by such particles. Speeds up to 10 km/s have been measured for iron particles [63]. On impact at the electrode the energy tends to be dissipated in vaporizing the particle rather than the material of the impacted electrode, and breakdown occurs in the resulting vapor. It is also possible that the particle may make one or more transits where each transit ends with the particle bouncing on the electrode with a reversal in the polarity of its charge. The particle can then continue to be accelerated until it has enough energy to be vaporized.

Under impulse conditions, there may be insufficient time for even one transit of the electrode gap (e.g., 2 to 350 µs for a 2- to 10-mm gap at 20 to 200 kV), but the breakdown still follows a relationship $V = Kd^\alpha$, where $K \cong 4000$ kV/mm and $\alpha \cong 0.5$ [59]. Several mechanisms have been suggested involving one or more of the following processes: generation of vapor by a spark during detachment of a particle, heating and vaporization of a particle in flight by electrons and ions, and switching-on of nonmetallic electron emitters on the cathode.

The mechanisms discussed above show the need for ultraclean particle-free smooth-contoured mirror-finish conductors if high electric strength is to be achieved.

Electrode Geometry As with other insulating media, regions of electrodes with small radii of curvature result in high local stresses and consequent high breakdown probability. Quantitative estimation of the effect is very dependent on the manufacturing process and metals used, and some experimentation is usually necessary to obtain optimum results. For instance, the detachment of clumps of particles by the field will occur erratically with time and be dependent on their initial number, degree of adhesion, distribution, material, occurrence of mechanical and thermal shocks at the electrode, and so on. Regions of an electrode with a small radius of curvature, on the other hand, may be given a specially high degree of polishing and cleaning and may remain relatively free of breakdown despite high local stresses.

Breakdown Voltage as a Function of Electrode Area and Time The electric strength falls with increasing electrode area to an extent which depends on the density of emitters for small gaps and on the density of detachable clumps and particles for large gaps. Figure 8.46(a) shows results obtained by four authors for small gaps up to 1 mm and electrode areas in the range 10 to 3×10^3 mm^2, showing that the fall in strength is large, amounting to a factor of about 2 per decade increase in electrode area [59]. The reduction in strength with increasing time [Fig. 8.46(b)] is comparatively small, amounting to only about 6% per decade increase in time [60], so that there is not a probabilistic equivalence between time and area.

Breakdown Strength of Open Circuit Breaker Contacts The lowest breakdown voltages result from no-load opening of welded contacts and not from high-current interruption [64]. Opening of welded contacts (i.e., after closing on a high-current) in most cases leaves electrode

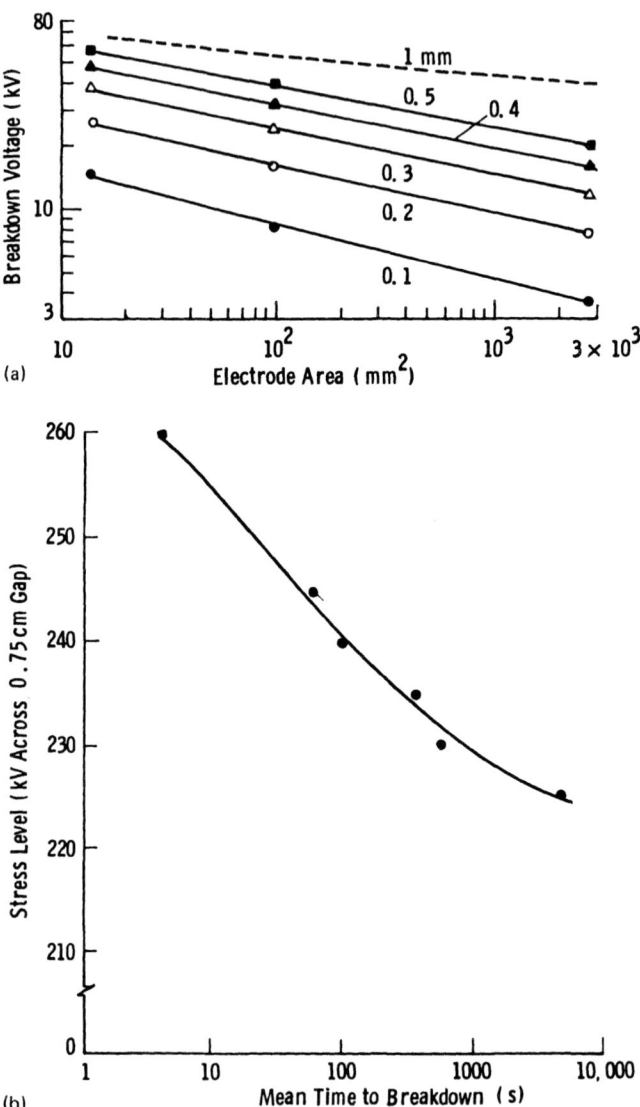

Figure 8.46 (a) Dependence of breakdown strength on electrode area for various gap lengths with stainless steel electrodes in vacuum. (From Ref. 59.) (b) Dependence of mean time to breakdown on electric stress in vacuum.

Dielectric Properties of Circuit Breakers

surfaces which emit electron currents before breakdown and for which a correlation exists between the current magnitude and breakdown. In addition, opening with a constricted arc present produces particles weakly attached to the electrodes and subsequent particle-initiated breakdown on reapplication of voltage. In practice, both mechanisms of breakdown occur independently and simultaneously. It follows that useful tests of vacuum interrupters require different procedures from those applied to fluid or gas breakers. The most meaningful results (i.e., the lowest breakdown voltages) require an adequate number of repetitions of a test in which the vacuum breaker is closed on a short circuit and then opened on no-load followed by a breakdown test. The essential point is that the breakdown strength depends on the preceding switching operations [64].

Flashover Along Insulator Surfaces in Vacuum

The surface flashover of solid insulators in vacuum is associated with positive surface charging of the insulator brought about by electron emission (usually from the cathode triple junction) coupled with secondary electron emission processes on the insulator surface [65]. With optimum design of the shape of the insulator (usually involving tapered shapes) the flashover strength of an insulator is comparable with that of the vacuum gap.

Electron Emission from Triple Junctions Electrons are emitted far more readily from a negative triple junction (i.e., a junction of a negative electrode, vacuum, and a dielectric) than by an electrode of the same shape. This is strikingly shown in Fig. 8.47(b), which shows the pattern of electrons arriving at a plane anode opposite a plane cathode containing a number of epoxy-filled holes [Fig. 8.47(a)]. The epoxy-cathode surface forms an uninterrupted plane. If the holes are not filled with epoxy, the electron pattern at the anode is random and does not reveal the pattern of holes in the cathode despite the high field at the edges of the holes [65]. This demonstrates the high electron emission from triple junctions. The electron beams impinging on the anode also cause ions of anode material to be emitted which travel to the cathode charging the epoxy positive. The high field between the positively charged epoxy and the edges of the holes in the negative electrode leads to the high electron emission into the gap. In practical designs the triple junction is often placed in a low-field region by recessing into the electrode [66], by making the insulator very long in comparison with the shortest vacuum gap (e.g., as in a typical vacuum interrupter), or by arranging for the insulator to acquire local negative charges which decrease the field at the triple junction [67]. Figure 8.48 shows typical breakdown voltages for a variety of insulator shapes, some of which are designed to acquire local negative charges.

Secondary Electron Generation, Charging, and Flashover of Insulator Surfaces Electrons emitted from a triple junction may impinge

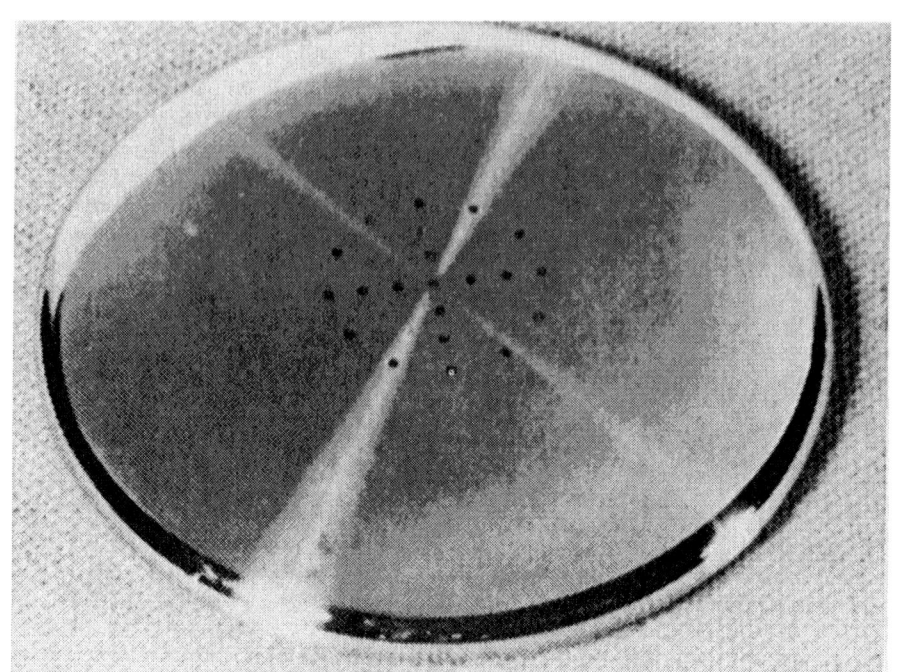

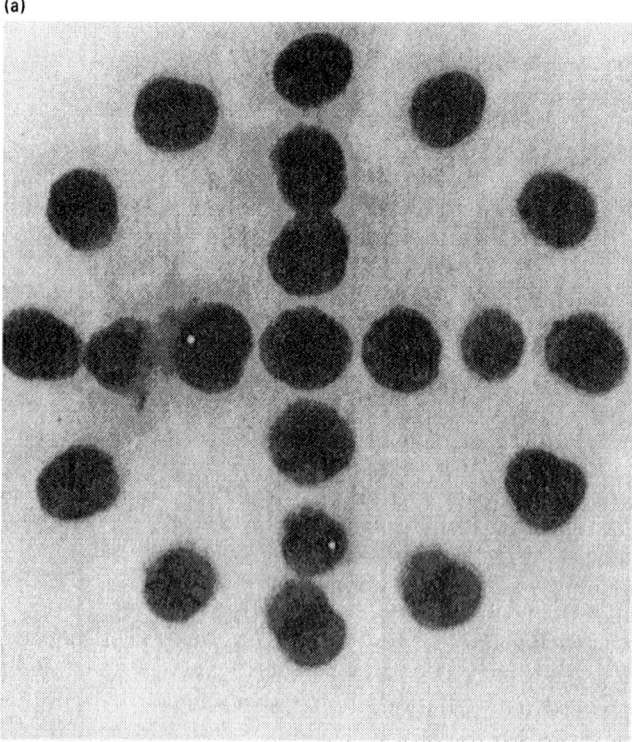

Dielectric Properties of Circuit Breakers

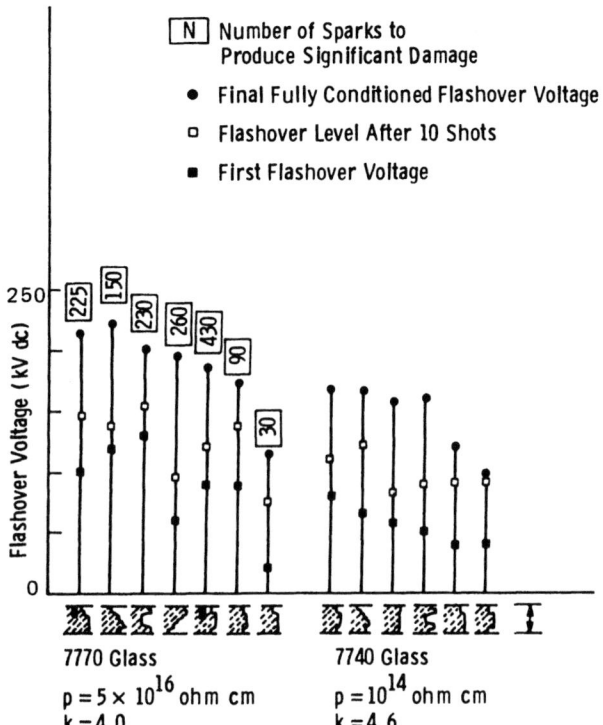

Figure 8.48 Dc flashover of insulators of various shapes in vacuum. (From Ref. 67.)

on an insulator surface and create secondary electrons [Fig. 8.49(a)] [63,68,69]. The original and secondaries then repeat the process creating still more electrons. The insulator surface is left positively charged, increasing the probability that further electrons emitted by the triple junction will hit the insulator rather than travel directly to the anode without multiplication. Eventually, an equilibrium situation may be set up with a constant positive charge on the insulator such that the electrons impact the insulator surface with an energy only just sufficient for the secondary electron emission coefficient to be greater than unity [Fig. 8.49(b)]. The impacting electrons deposit energy on the surface which desorbs gas, or generates gas by decom-

Figure 8.47 (a) Plane cathode with holes filled with epoxy resin in vacuum. (From O. Milton, IEEE Trans. Electr. Insul. EI-9: 68, © 1974, IEEE.) (b) Distribution of electrons arriving at a plane opposite the cathode shown in part (a). (From O. Milton, IEEE Trans. Electr. Insul. EI-9: 68, © 1974, IEEE.)

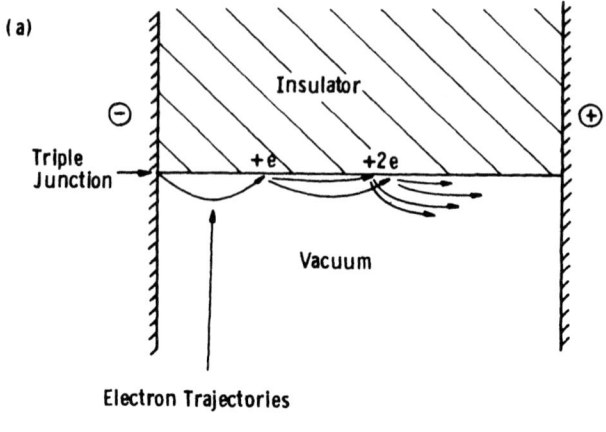

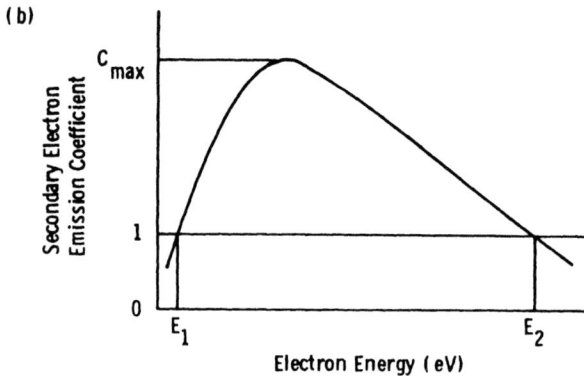

Figure 8.49 (a) Electrons emitted from a triple junction in vacuum gain energy from the field and produce secondaries from the insulator surface. (b) Dependence of the secondary electron emission coefficient on the energy of an incident electron.

position, allowing avalanching and breakdown in the gas. The triple-junction emission current is effectively space-charge limited, the space-charge being on the insulator.

Effect of Insulator Surface Angle Insulators in which the dielectric surface is inclined to the field lines have a higher breakdown voltage for short pulses if the electron trajectory is inclined away from the surface (positive values of θ in Fig. 8.50) than if the inclination of the trajectory is toward the insulator surface (negative values of θ in Fig. 8.50) [69]. This is a consequence of the process described in Fig. 8.49(a) being facilitated by negative values of θ and inhibited by positive values of θ. The effect of variations of θ is much smaller or

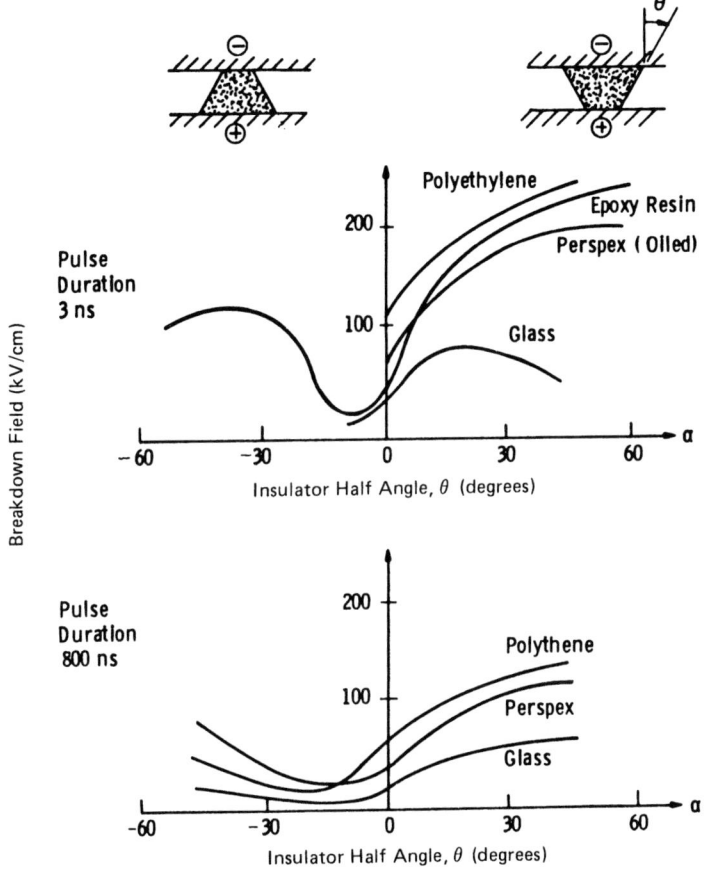

Figure 8.50 Variation of breakdown stress with half-angle across conical insulators in vacuum. (From C. H. deTourreil and K. S. Srivastava, IEEE Trans. Electr. Insul. *EI-8*:17, © 1973, IEEE.)

absent for long pulses and dc because the equilibrium situation described above is eventually set up whatever the sign of θ [69]. Figure 8.50 indicates a great variation of breakdown voltage for short pulses as θ increases from zero to small positive values.

Volt-Time Characteristics for Short Pulses (ns) The calculated space-charge-limited current for a triple junction is proportional to the component of electric field along the insulator surface, so that the maximum power dissipated at the insulator surface varies with the square of this field component. The experimentally measured proportionality between the time delay to breakdown and the inverse square of the

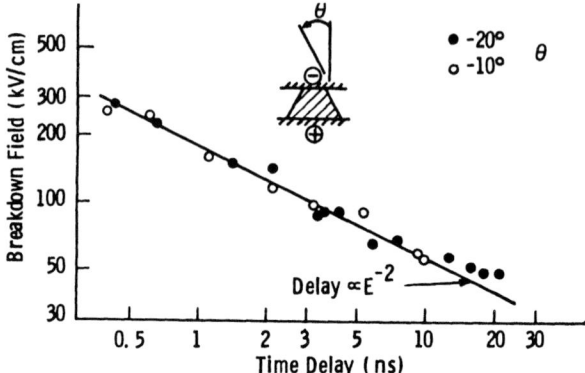

Figure 8.51 Variation of breakdown strength with time delay to breakdown for a conical insulator in vacuum. (From Ref. 68.)

applied electric field (Fig. 8.51), as well as a $(\cos \theta)^{-1}$ variation in breakdown field at constant delay, both imply that breakdown occurs after a fixed energy density has been deposited on the insulator surface [68]. Departures from the inverse square-law dependence on E occur for values of θ greatly different from zero, ranging from $E^{-1.7}$ for $\theta = -30°$ to $E^{-3.3}$ for $\theta = +60°$ [68].

Effect of Insulator Shape The highest breakdown voltages among the various shapes depicted in Fig. 8.48 are for insulators with conducting cathode inserts or with shapes designed to collect negative charge on the insulator and hence to reduce the field at the cathode triple junction.

Effect of Conditioning and Insulator Surface Coatings Coatings on insulators of materials having a low secondary electron emission coefficient inhibit the accumulation of positive charge on insulators and increase the flashover voltage. Thin films of materials such as Fe_2O_3 [70] and chromium oxide [71] increase the breakdown voltage. Carbon is unusual in that it has a secondary emission coefficient less than unity for all electron impact energies. It also increases the breakdown voltage [72]. Spark and voltage conditioning tend to deposit metallic and oxide coatings on insulators which often then have a higher breakdown strength both because of the decreased secondary emission and because the coatings tend to be resistive and hence disperse surface charges.

Effect of Surface Material and Texture Glasslike materials have been found to have higher breakdown voltages than porous ceramic materials [63]. It has also been observed that roughening the insulator surface near the cathode increases the flashover voltage, so that the former effect is apparently not due to the smoothness of glass surfaces [78]. Sharp dielectric edges reduce the breakdown strength [78].

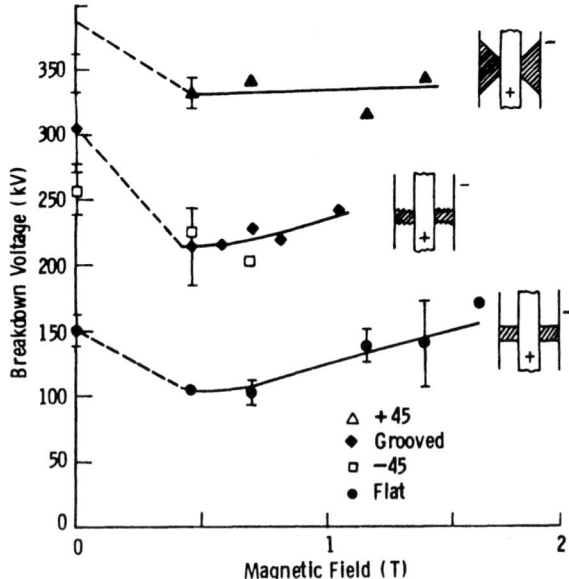

Figure 8.52 Breakdown voltage vs. magnetic field in vacuum. (From Ref. 73.)

Effect of Magnetic Fields Magnetic fields perpendicular to the electric field can reduce the breakdown voltage by about 30% for 45° grooved and flat insulators (Fig. 8.52). This is in contrast to the effect in a vacuum gap where "magnetic insulation" utilizes the effect of a magnetic field in *increasing* the breakdown strength [74]. The results shown in Fig. 8.52 indicate a minimum breakdown voltage for all three insulator shapes for a field of 0.5 T. At higher fields the strength is similar to that at 0.5 T for the 45° insulator, but increases with field for both the grooved and flat shapes. The 45° insulator has the highest strength of the three insulator types for all values of magnetic field.

8.4 BUSHINGS

Bushings are required to feed the high-voltage, high-current connection through to the circuit breaker contacts. The bushing design will vary depending on whether the circuit breaker is a dead-tank design, with the contacts inside a grounded metal tank, or whether it is a live-tank breaker, where the contacts can be inside a metal tank at high voltage or inside the bushing porcelain itself.

This section reviews the characteristics of the high-voltage, long-gap flashover in the air external to the bushing and then describes the

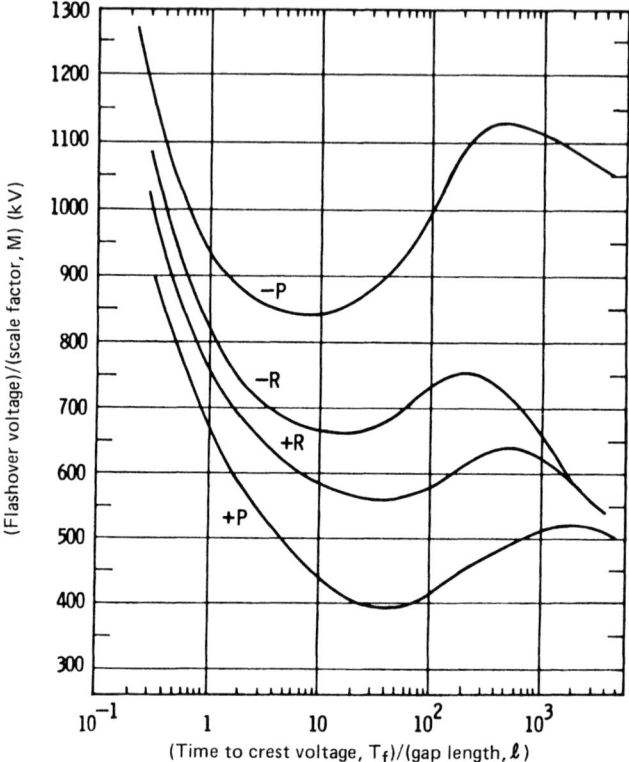

Figure 8.53 Scaled 50% flashover voltage in air vs. scaled time to crest for impulse voltages applied to rod-rod (R) and rod-plane (P) gaps. (From Y. Aihara, T. Harada, Y. Aoshima, and Y. Ito, IEEE Trans. Power Appar. Syst. *PAS-97*:342, © 1978, IEEE.)

design philosophy of several types of bushings, including condenser type oil/paper, cast epoxy, and gas bushings.

8.4.1 Bushing External Insulation Strength

Air Insulation

The withstand voltage level and the statistical spread of values for breakdown through the air gap around bushings are very similar to those for rod-rod gaps of the same length in air under both wet and dry conditions. Typical impulse results for both rod-rod and rod-plane geometries for gap lengths in the range 1 to 29 m are shown in Fig. 8.53 for both positive and negative polarity. This figure conveniently normalizes the data for lightning and switching impulses on a single set of curves. The time to crest (T_f) divided by the gap length in meters is plotted on the horizontal axis. The vertical voltage axis is scaled by an empirical factor $M = 2\ell^{0.4} - 4/(3 + \ell)$. Thus for a gap length of 1 m, both scale factors ℓ and M are equal to unity.

It may be seen that the lowest rod-rod curve (for positive polarity) is considerably higher (up to 50% higher) than the corresponding rod-plane curve for a gap of the same length, except for small values of T_f/ℓ, where there is little difference. Thus for a given length of gap (or bushing of a given length), there is a great advantage in using a rod-rod geometry rather than a rod-plane geometry, particularly for switching impulse voltages. A bushing may be made to behave like a rod-rod gap by mounting the bushing on a pedestal or on the top of equipment contained in a grounded enclosure. The positive withstand voltage level increases linearly with height of the ground rod or pedestal as shown in Fig. 8.54 until at infinite height the polarity effect disappears [75].

In estimating maximum permissible voltages for air breakdown, it is necessary to take into account the scatter in breakdown voltages in

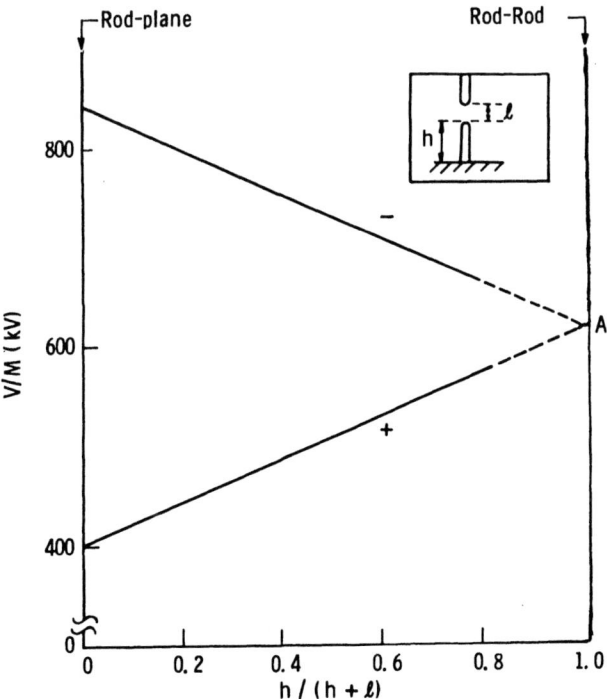

Figure 8.54 Dependence of scaled breakdown voltage in air on height of a rod-rod gap above ground. (From Ref. 75.)

addition to the mean values shown in Fig. 8.53. In general, insulation coordination procedures are based on the mean breakdown voltage minus three standard deviations. The scatter in breakdown voltages is generally 6 to 9%, being somewhat higher for long times to peak than for lightning impulse waveforms. Another factor to be aware of is that the breakdown path with switching impulses may take a path very much longer (by perhaps twice) than the shortest available air path. Lightning impulse breakdown paths tend to be only slightly longer than the shortest available path.

Breakdown tends to be on the front of the wave for long prospective times to peak and short gaps. The crossover from breakdown on the wavefront to breakdown on the tail of the wave depends on both the length of the gap and on the prospective time to peak, as shown in Fig. 8.55. The crossover occurs at ratios of (gap length)/(time to peak) in the range 2 to 3 cm/µs, the value tending to be smaller for long gaps [76]. The practical significance of such data is that there may be inadequate time for breakdown in chopped wave tests where the voltage may be applied for short times (e.g., 5 µs or less), and in front-of-wave tests (e.g., 1 µs or less). Also, breakdown over the surface of porcelain may be comparatively rapid, with little probability of late-time all-air breakdown.

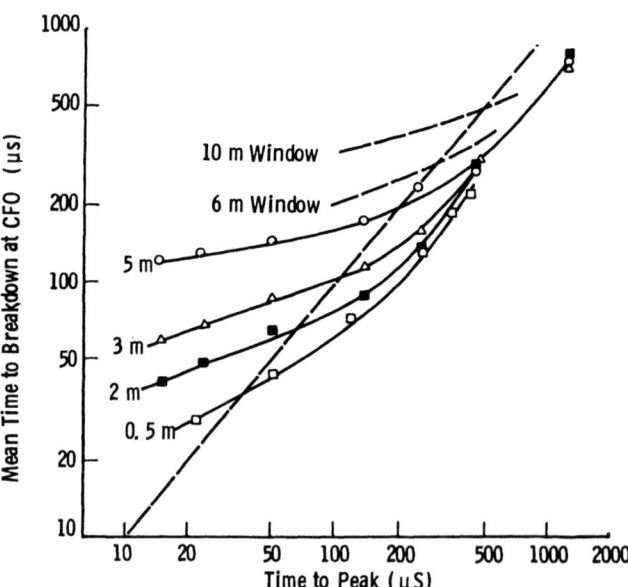

Figure 8.55 Dependence of mean time to breakdown on time to peak in air. (From Ref. 76.)

Dielectric Properties of Circuit Breakers

Surface Breakdown

For bushings exposed to the weather, the most critical design requirements are the switching and ac withstand characteristics under wet and/or contaminated conditions. Stresses due to impulse voltages are distributed capacitively across bushings and tend not to be affected by the presence of contamination. Wet uncontaminated insulators subjected to ac voltages tend to pass a conduction current through the moisture film and initially the field may be fairly uniform. Ac current causes heating, leading to the formation of "dry bands" across which most of the potential difference between the electrodes appears [77]. This can result in insulator damage within the "dry band" and flashover between the electrodes.

The use of sheds on insulators promotes the formation and maintenance of multiple dry bands in wet and contaminated conditions by keeping the underside of the sheds dry and free of contamination. Sheds are also beneficial where adequate total surface creepage length is important under atmospheric conditions when condensation is occurring on all exposed surfaces.

The reduction in positive switching impulse flashover voltage due to rain and contamination is rather unpredictable, and is dependent on the design and contamination level, but may be as large as 40% [78]. The flashover strength generally decreases nonlinearly with increasing density of contamination density and with precipitation rate.

8.4.2 Bushings Designs

General

This section reviews the dielectric design criteria for high-voltage bushings, primarily for application to circuit breakers. The U.S. standards for outdoor apparatus bushings are found in ANSI C76.1-1976 (IEEE Std. 21-1976) and ANSI C76.2-1977 (IEEE Std. 24-1977). The insulation withstand voltages required for outdoor circuit breaker terminals and the minimum creepage distances for external insulation to ground are given in Table 8.6.

Several designs of bushings are described. Bushings for dead-tank circuit breakers require full voltage insulation. In the past, condenser bushing designs have been employed in circuit breaker applications [79]. Modern condenser bushings can be classified into basically four types of construction: oil-impregnated paper, resin-bonded paper, epoxy-resin-impregnated paper, and full epoxy-resin castings. Noncondenser bushings fully insulated with SF_6 gas are used on both live-tank and dead-tank SF_6 circuit breakers; these are used to insulate series-connected interrupters and provide the line terminals of the breaker. The gas bushings are also used for entrance to gas-insulated substations and other gas-insulated equipment.

Oil-Impregnated Paper Condenser Bushings

Most dead-tank circuit breaker bushings rated above 350 kV (BIL) presently in service are condenser bushings, insulated by paper that is impregnated with well-processed mineral oil. These bushings are fabricated by winding paper sheets onto a cylindrical conductor and inserting cylindrical condensers made of metallic foils or conducting paper for stress grading. After oven drying (in some cases under vacuum), the paper condenser is vacuum impregnated with oil. The bushing is assembled by clamping the exterior (air end) insulator, the mounting flange assembly, and the interior (oil end) insulator (for oil breakers) together by the central conductor via a nest of precompressed springs located in the expansion chamber housing. After assembly, the precompressed springs are released, thereby providing the necessary loading force to hold the insulators in compression and achieving a vacuum-tight seal. The bushing is completed by vacuum filling with oil. Power factor, voltage withstand, mechanical cantilever, and thermal tests are performed as required by standards and manufacturer's practice (Chap. 3).

The design of oil-impregnated-paper condenser bushings is discussed briefly in the following section. The design approach presented is also applicable to the other types of condenser bushing construction such as resin-bonded paper, epoxy-resin-impregnated paper, and cast epoxy bushings to be discussed. Some extra-high-voltage (EHV) gas-insulated bushings also employ condenser stress grading using a limited number of nested metallic cylinders. In the case of gas bushings the condenser design is influenced by factors such as dielectric strength of the solid spacers used to locate and support the condenser cylinders, and the stress enhancement at the ends of the cylinders which are generally shielded by metallic rings. Therefore, the analysis of condenser bushings presented below is not in general applicable to gas condenser bushings.

Condenser Bushing Design

The basic design concept of a condenser bushing is to make the radial and axial electric field gradients as uniform as possible. This is accomplished by embedding metallic foils or metal-coated paper within the bushing insulation in order to grade capacitively the applied voltage from line potential to ground. Although there are a number of approaches to the design of a cylindrical condenser bushing, the most common practice is to divide the voltage into N equal steps by a series of N equal capacitors [80].

By properly sizing the lengths of adjacent conducting surfaces and employing as many condensers as is practical, the longitudinal (axial) stress over the condenser surface is made nearly uniform. The analytical formulation of a cylindrical condenser bushing based on equal capacitive grading is now described. The equations developed here

Dielectric Properties of Circuit Breakers

are applicable to any type of condenser bushing construction having essentially uniform insulation material between a large number of condenser layers.

Consider the cylindrical condenser depicted in Fig. 8.56. The condenser is made up of N metal foils (or conducting layers such as metal-coated paper) embedded in a roll of insulating paper. The condenser is wrapped around a conductor of radius r_0 at line potential U. The first foil has a radius r_1 and a length L_1 while the last foil, which is grounded, has a radius r_N and a length L_N.

The radial build of the condenser foils is designed so that the capacitance between any pair of adjacent foils is constant, and can be shown to be

$$r_n = r_0 \exp\left[\log \beta \left(\frac{2\alpha}{\alpha+1} - \frac{1}{N} - \frac{\alpha-1}{\alpha+1}\frac{n}{N}\right)\frac{n}{N-1}\right] \quad (8.9)$$

where $\alpha = L_1/L_N$ and $\beta = r_N/r_0$. The maximum electric stress in the $n + 1$ condenser, where $n = 0, 1, \ldots, N + 1$ is given by

$$E_{n+1} = E_1 \frac{r_0}{r_n} \Big/ \left(1 - \frac{\alpha-1}{\alpha}\frac{n}{N-1}\right) \quad (8.10)$$

where E_1 is the maximum stress between the conductor and the first foil given by

$$E_1 = \frac{U(\alpha+1)}{2r_0 \alpha \log \beta} \quad (8.11)$$

The ratio E_N/E_1, that is, the ratio of the maximum stress in the last condenser to that in the first, is given by

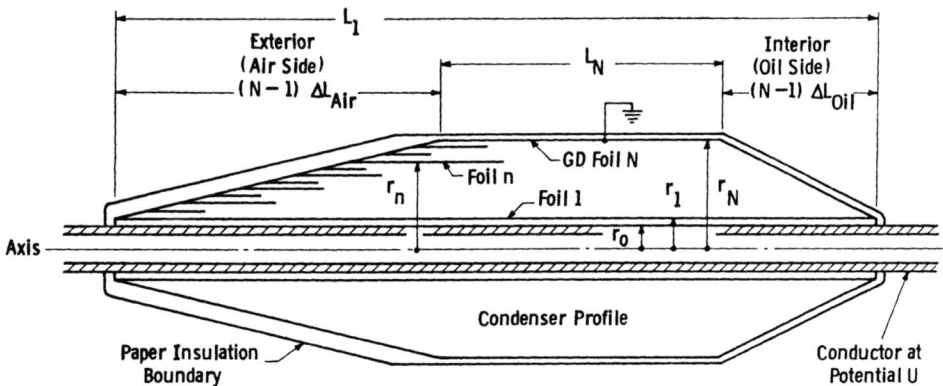

Figure 8.56 Condenser bushing sketch for analytic analysis.

$$\frac{E_N}{E_1} = \frac{\alpha}{\beta} \exp\left[\frac{2(\log \beta)}{N(\alpha + 1)}\right] = \frac{\alpha}{K\beta} \qquad (8.12)$$

where the factor $K = \exp[-2(\log \beta)/N(\alpha+1)]$ is very near unity for all practical condenser designs. The maximum stress occurs at the conductor surface when $K\beta > \alpha$ and is E_1 given by equation (8.11). When $K\beta < \alpha$ the maximum stress is in the ground condenser and is E_N given by

$$E_N = \frac{U(\alpha + 1)}{2Kr_N \log \beta} \qquad (8.13)$$

If $\alpha = K\beta$, the maximum stress occurs at both the conductor and ground end condenser, that is, $E_1 = E_N$.

In the case of condensers with a large number of foils the stress distribution in the condenser can be approximated by

$$\frac{E(\chi)}{E_1} \cong \exp\left[-\log \beta \left(\frac{2\alpha}{\alpha+1} - \frac{\alpha-1}{\alpha+1}\chi\right)\chi\right] \Big/ \left(1 - \frac{\alpha-1}{\alpha}\chi\right) \qquad (8.14)$$

where $0 \leq \chi \leq 1$ is the normalized distance from the conductor surface to the last condenser at this distance. Equation (8.14) is derived from equations (8.9) and (8.10) by replacing n/N with χ and setting $1/N = 0$. Figure 8.57 presents stress distribution curves based on equation (8.14) for four typical condenser profiles, that is, values of α and $K\beta$. Curve B in Fig. 8.57 is based on α and $K\beta$ values determined for a 161-kV insulation-class Westinghouse Type O condenser bushing. In this case the maximum stress occurs at the ground end condenser and is equal to $E_N = 1.035E_1$.

The total capacitance C_T of the condenser bushing based on the analysis above is

$$C_T = \frac{C}{N} = \frac{\pi\varepsilon\varepsilon_0(L_1 + L_N)}{\log(r_N/r_0)} = \frac{\pi\varepsilon\varepsilon_0 L_1(\alpha+1)}{\alpha \log \beta} \qquad (8.15)$$

where ε is the relative dielectric constant of the space between condenser foils and ε_0 is the permittivity of free space.

The potential difference between adjacent foils is U/N, and the charge on each condenser is UC_T, where U is the voltage applied to the conductor and the Nth foil is assumed ground. The average axial stresses along the condenser profile on the exterior (air) and interior (oil) sides are essentially uniform and are given approximately by

$$\frac{U}{[\Delta L_{air}^2 + (r_N - r_0)^2]^{1/2}} \quad \text{and} \quad \frac{U}{[\Delta L_{oil}^2 + (r_N - r_0)^2]^{1/2}} \qquad (8.16)$$

Dielectric Properties of Circuit Breakers

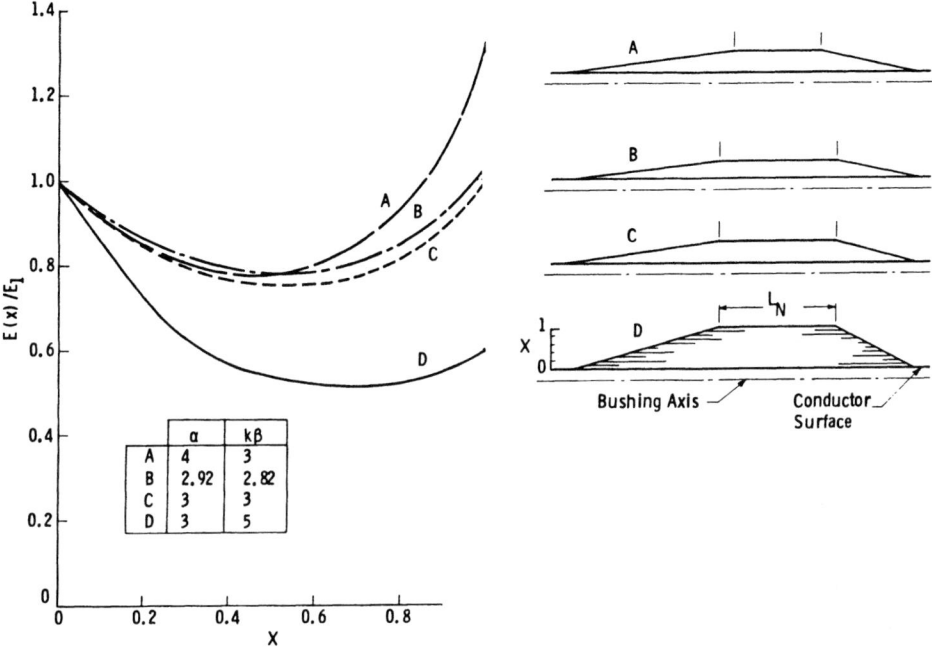

Figure 8.57 Theoretical electric stress distribution condenser bushings which have a large number of condenser foils for some α and $K\beta$ parameters.

respectively. In general, the parameters α and $K\beta$ usually lie in the range 2.5 to 4.5. A minimum bushing radius is obtained if $\alpha = K\beta = 3.15$.

Resin-Bonded Paper Bushings

In this type of bushing the major insulation is provided by paper sheet which has been coated with epoxy or phenolic resin. The resin-coated paper is wound continuously into a cylindrical mandrel under pressure and heat in order to bond the paper layers together. Condenser bushings are made by inserting metallic foil or conducting paper at appropriate intervals during the winding process to form cylindrical capacitors. The maximum design stress in this type of bushing is typically 1.4 kV/mm (35 kV/in.).

The radial build of resin-bonded paper bushings is limited because excessive dielectric heating in the resin can lead to thermal instability. As a consequence, the use of resin-coated paper in bushing construction is generally restricted to bushings rated below 550 kV (BIL).

Resin-bonded paper bushings are also made by nesting several resin-bonded paper tubes each of which contains a metallic condenser foil. The array of nested tubes makes up the condenser, which is then impregnated with mineral oil.

Epoxy-Resin-Impregnated Paper Bushings

Bushings are also made by winding with crepe paper tape and then impregnating under vacuum with epoxy resin. Capacitive grading is achieved by inserting metallic foils at predetermined radial intervals, usually about 50-mil (1.3-mm) spacings. Because the crepe paper insulation is impregnated with resin, rather than coated as in resin-bonded paper bushings, this type of bushing is essentially gas-tight and hence is applicable for high-pressure gas circuit breakers. However, since the impregnating epoxy must be extremely fluid, the physical properties that can be obtained are severely limited. In addition, the use of paper restricts the reliability of this type of bushing due to its susceptibility to moisture absorption.

Cast Epoxy Bushings

The development of epoxy resin systems which have thermal expansion coefficients near those of copper and aluminum have made possible the construction of cast epoxy condenser bushings [81,82]. The epoxy insulation is cast the entire length of the bushing conductor, encapsulating or embedding the metallic condenser structure. High mechanical stresses normally associated with differential thermal expansion between metals and the cured epoxy resin system are avoided by good thermal expansion matching. Such condenser bushing castings can be made with high corona inception voltage and a good dissipation factor. Filled-epoxy cast bushings, for example, have lower losses up to 1-min withstand levels than do oil-impregnated paper, epoxy-impregnated paper, or paper-based phenolic bushings.

The control of the thermal expansion of an epoxy resin system is accomplished by using low-expansion silica fillers. When the percent of fused silica filler is in the range 70 to 82%, the thermal expansion coefficient of a rigid bisphenol A, for example, can be controlled between 24 and 12×10^{-6} mm/mm°C over the temperature range 25° to 100°C. The thermal expansion coefficients for copper and aluminum are 18.0 and 22.0×10^{-6} mm/mm°C, respectively.

Epoxy resins are also used in a number of other applications in gas and oil breakers. Rigid and flexible cycloaliphatic epoxy resins, for example, have been used as replacements for porcelain lower arc shields in SF_6-gas-insulated circuit breaker bushings and as replacements for the lower porcelain in oil-paper condenser bushings. Cycloaliphatic epoxy weather casings can now be used in place of porcelain. Such epoxy weather casings have inherent advantages such as lighter weight,

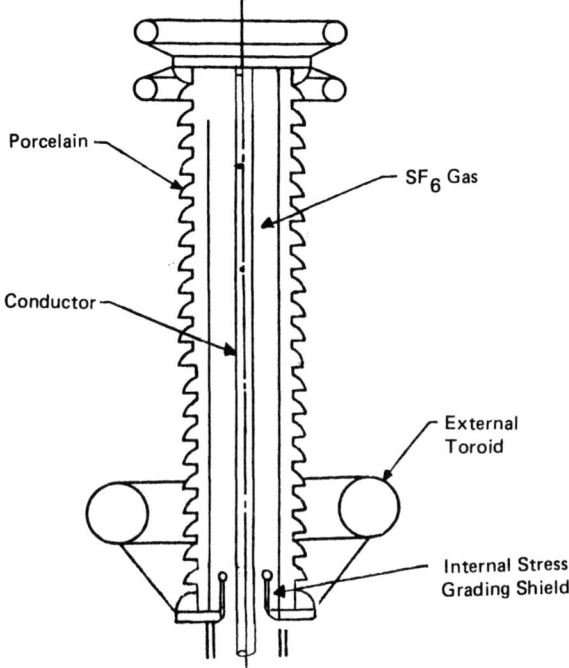

Figure 8.58 Typical gas bushing. (From Ref. 83.)

better dimensional stability, self-cleaning capability, field repairability, and design flexibility.

Gas-Insulated Bushings

SF_6-insulated equipment such as circuit breakers, compressed gas-insulated transmission lines, and test capacitors utilize the excellent SF_6 dielectric properties for the main internal high-voltage insulation in the bushing.

Figure 8.58 shows a typical design, the central high-voltage conductor, usually aluminum, fixed inside the porcelain weathershed with SF_6 typically at 0.45 MPa (50 psig). The voltage distribution along the weathershed is controlled by the height and diameter of the internal grounded throat shield and by the external high-voltage and ground toroids. The voltage distribution along the procelain is not uniform as it is for the condenser bushing. The 50% potential point may be typically at 35% of the length from the grounded bottom flange. Typically, the maximum design stress in the SF_6 gas for the full-wave basic impulse level (BIL) is 17 kV/mm (430 kV/in.). The internal SF_6

stresses can be estimated using the field utilization factors described in Sec. 8.2.2, but for the fields in the air at the toroids and at the porcelain air surface near the top of the throat shields, computer field analysis is required.

In some bushing designs the throat shield is not used, just the rounded metal flange; these are usually at the lower voltages and for example with bushings on the multibreak live tank circuit breakers (Chap. 10). Some designs may also have a cast epoxy or porcelain support cone on the conductor at the entrance to the circuit breaker.

Gas bushings generally are of greater length and larger base diameter than condenser bushings. However, they generally are of lower cost because of their simpler construction and have better thermal (i.e., heat transfer) characteristics.

REFERENCES

1. M. D. R. Beasley, J. H. Pickles, L. Baretta, M. Fanelli, G. Giusepetti, G. Gallet, J. P. Gregoire, and M. Morin, Proc. IEE (Lond.) *126*:126, 1979.
2. E. R. Hartill, J. E. McQueen, and P. M. Robson, Proc. IEE (Lond.) *104A*:17, 1957.
3. D. S. Ross and I. H. Qureshi, J. Sci. Instrum. *40*:513, 1963.
4. J. M. Mattingley and H. M. Ryan, Proc. IEE (Lond.) *118*:720, 1971.
5. H. Singer, H. Steinbigler, and P. Weiss, IEEE Trans. Power Appar. Syst. *PAS-93*:1660, 1974.
6. T. W. Dakin and A. I. Bennett, private communication.
7. J. M. Meek and J. D. Craggs, eds., *Electrical Breakdown of Gases*, Wiley, New York, 1979.
8. T. W. Dakin et al., Electra, No. *32*:61, 1974.
9. A. H. Cookson, Proc. IEE (Lond.) *128*:303, 1981.
10. J. G. Trump, U.S. Pat. 3,515,939, 1970.
11. S. Sangkasaad, Elektrotech. Z. Ausg. A *96*:515, 1975.
12. C. M. Cooke and R. Velazques, IEEE Trans. Power Appar. Syst. *PAS-96*:1491, 1977.
13. C. Masetti and B. Parmigiani, Proc. 3rd Int. Symp. High Voltage Eng., Milan, 1979, Paper 32.15.
14. A. Pedersen, IEEE Trans. Power Appar. Syst. *PAS-94*:1769, 1975.
15. A. H. Cookson and B. O. Pedersen, Proc. 3rd Int. Symp. High Voltage Eng., Milan, 1979, Paper 31.10.
16. J. Ozawa, F. Endo, S. Masuda, T. Isagai, and M. Hosokawa, IEEE Power Eng. Soc. Summer Meet., Anaheim, Calif., July 14-19, 1974, Paper C-74-464-4.
17. N. H. Malik and A. H. Qureshi, IEEE Trans. Electr. Insul. *EI-13*:135, 1978.

18. A. Pedersen, IEEE Trans. Power Appar. Syst. PAS-89:2043, 1970.
19. W. Zaengl and R. Baumgartner, Elektrotech. Z. Ausg. A 96:510, 1975.
20. I. M. Bortnik and C. M. Cooke, Sov. Phys.-Tech. Phys. 17: 1850, 1973.
21. I. M. Bortnik and C. M. Cooke, IEEE Trans. Power Appar. Syst. PAS-91:219, 1972.
22. W. Mosch and W. Hauschild, *High Voltage Insulation with Sulfur Hexafluoride*, VEB Verlag Technik, Berlin, 1979.
23. S. Berger, IEEE Trans. Power Appar. Syst. PAS-95:1073, 1976.
24. F. A. M. Rizk, C. Masetti, and R. P. Comsa, IEEE Trans. Power Appar. Syst. PAS-98:825, 1979.
25. R. E. Wootton et al., EPRI Final Report EL-1007, Palo Alto, Calif., 1979.
26. T. Nitta, Y. Fujiwara, F. Endo, and J. Ozawa, CIGRE, Paper 15-4, Paris, 1976.
27. S. A. Studniarz, G. A. Mullen, and T. W. Dakin, IEEE Trans. Power Appar. Syst. PAS-99:1130, 1980.
28. K. H. Weber and H. S. Endicott, AIEE Trans. 75(Pt. III):371, 1956.
29. K. H. Weber and H. S. Endicott, AIEE Trans. 75(Pt. III):393, 1957.
30. T. W. Dakin, S. A. Studniarz, and G. T. Hummert, 1971 Annu. Rep., Conf. Electr. Insul. Dielectr. Phenom., Washington, D.C., 1972, p. 411.
31. W. Hauschild, Elektrie 24:244, 1970.
32. Brown Boveri Rev. 54:368, 1967.
33. K. H. Weber and H. S. Endicott, AIEE Trans. 75:1091, 1957.
34. N. G. Trinh and C. Vincent, Canadian Electric Association Study 012T103, Prog. Rep. 4, August 29, 1980.
35. Y. Kawaguchi, H. Murata, and M. Ikeda, IEEE Trans. Power Appar. Syst. PAS-91:9, 1972.
36. J. K. Nelson, B. Salvage, and W. A. Sharply, Proc. IEE (Lond.) 118:388, 1971.
37. J. Binggeli, J. Froidevaux, and R. Kratzer, Int. Conf. Large Electr. Syst., CIGRE, Rep. 110, Paris, 1966.
38. M. Darveniza, Inst. Eng. (Aust.) Electr. Eng. Trans., September 1969.
39. Z. Krasucki, 1971 Annu. Rep., Conf. Electri. Insul. Dielectr. Phenom., Washington, D.C., 1972, p. 96.
40. K. N. Mathes and J. M. Atkins, Conf. Rec. 1978 IEEE Int. Symp. Electr. Insul., Philadelphia, p. 226.
41. K. Weschler and M. Riccitiello, AIEE Trans. 80:365, 1961.
42. A. M. Sletten and T. W. Dakin, IEEE Trans. Power Appar. Syst. PAS-83:457, 1964.

43. J. A. Kok and M. M. G. Corbey, Appl. Sci. Res. Sect. B 6: 197, 1956.
44. Z. Krasucki, Proc. R. Soc. (Lond.) A294:393, 1966.
45. T. W. Dakin, IEEE Trans. Electr. Insul. EI-9:121, 1974.
46. D. Goulsbra, H. M. Ryan, N. Shaw, and R. Wilkins, Proc. IEE Conf. Dielectr. Mater. Meas. Appl., IEE Publ. 67, 1970, p. 226.
47. F. Y. Tse, M. J. Mulcahy, and W. R. Bell, Proc. IEE Conf. Dielectr. Mater. Meas. Appl., IEE Publ. 67, 1970, p. 314.
48. O. Milton, Insulation 59, November 1967.
49. S. A. Studniarz and T. W. Dakin, Conf. Rec. 1982 IEEE Int. Symp. Electr. Insul., Philadelphia, p. 19.
50. D. Kind and L. Schiweck, 1969, Annu. Rep., Conf. Electr. Insul. Dielectr. Phenom., National Academy of Sciences, Washington, D.C., 1970, p. 128.
51. T. W. Dakin and S. A. Studniarz, Conf. Rec. 1978 IEEE Int. Symp. Electr. Insul., Philadelphia, p. 216.
52. G. W. Hewitt and T. W. Dakin, IEEE Trans. Power Appar. Syst. PAS-82:1033, 1963.
53. T. W. Dakin, H. M. Philofsky, and W. C. Divins, Trans. IEE 73:155, 1954.
54. L. Mandelcorn and T. W. Dakin, IEEE Trans. Power Appar. Syst. PAS-81:291, 1962.
55. G. A. Mullen and T. W. Dakin, Conf. Rec. 1982 IEEE Int. Symp. Electr. Insul., Philadelphia, p. 238.
56. T. W. Dakin and G. A. Mullen, IEEE Power Eng. Soc. Winter Power Meet., New York, February 1978, Paper A-78-075-4.
57. F. A. Yeoman and G. A. Gabriel, Proc. 14th Electr. Electron. Insul. Conf., Boston, 1979, p. 266.
58. L. W. Boss and R. J. Hillen, Proc. 10th Electr. Insul. Conf., Chicago, 1971, p. 90.
59. R. Hackam and S. K. Salman, J. Appl. Phys. 45:4384, 1974.
60. P. Bolin, F. Tse, W. Bell, and M. Mulcahy, Insulation/Circuits 59, September 1970.
61. R. Hackam and L. Altcheh, J. Appl. Phys. 46:627, 1975.
62. R. N. Allan and R. K. Bordoloi, J. Phys. D: Appl. Phys. 8:2170, 1975.
63. R. V. Latham, *High Voltage Vacuum Insulation: The Physical Basis*, Academic Press, New York, 1981.
64. K. Frölich and W. Widl, Proc. IEE (Lond.) 128:243, 1981.
65. O. Milton, IEEE Trans. Electr. Insul. EI-9:68, 1974.
66. R. C. Finke, Proc. 2nd Int. Symp. Insul. High Voltages Vacuum, 1966, p. 217.
67. J. P. Shannon, S. F. Philp, and J. G. Trump, J. Vacuum Sci. Technol. 2:234, 1965.

68. R. A. Anderson, 1975 Annu. Rep., Conf. Electr. Insul. Dielectr. Phenom., National Academy of Sciences, National Research Council, Washington, D.C., 1978, p. 475.
69. C. H. de Tourreil and K. S. Srivastava, IEEE Trans. Electr. Insul. *EI-8*:17, 1973.
70. A. Fryszman, T. Strzyz, and M. Wasinski, Bull. Acad. Pol. Sci. Ser. Sci. Tech. *8*:379, 1960.
71. J. D. Cross and T. S. Sudarshan, IEEE Trans. Electr. Insul. *EI-9*:146, 1974.
72. P. M. Gleichauf, J. Appl. Phys. *22*:766, 1951.
73. J. Golden and C. A. Kapetanakos, J. Appl. Phys. *48*:1756, 1977.
74. D. B. Seidel, C. W. Mendel, Jr., S. A. Slutz, and E. L. Neau, 3rd IEEE Int. Pulsed Power Conf., 1981, p. 241.
75. Y. Aihara, T. Harada, Y. Aoshima, and Y. Ito, IEEE Trans. Power Appar. Syst. *PAS-97*:342, 1978.
76. B. Jones and H. W. Whittington, Proc. 3rd Int. Conf. Gas Discharges, London, 1972, p. 32.
77. E. C. Salthouse, Proc. IEE (Lond.) *115*:1707, 1968.
78. *Transmission Line Reference Book 345 kV and Above*, Electric Power Research Institute, Palo Alto, Calif., 1982.
79. H. Barker, in *High Voltage Technology* (L. L. Alston, ed.), Oxford University Press, London, 1968, Chap. 14.
80. D. Legg, in *Power Circuit Breaker Theory and Design*, (C. H. Flurscheim, ed.), IEE Monograph Series 17, Peter Peregrinus, Stevenage, 1975, Chap. 12.
81. J. P. Burkhart and C. F. Hofmann, IEEE Winter Power Meet., New York, January 27-February 1, 1974, Paper C-74-064-2.
82. C. W. Upton and J. J. Dodds, IEEE Paper Eng. Soc. Winter Power Meet., New York, January 27-February 1, 1974, Paper C-74-178-0.
83. H. C. Doepken, Jr., Proc. 9th Electr. Insul. Conf., Chicago, 1969, p. 217.

9
SF₆ Breaker Research and Development

ROBERT E. FRIEDRICH[*]/Westinghouse Electric Corporation, Trafford, Pennsylvania

9.1 Introduction   354
9.2 Properties of SF_6 354
9.3 Early Research   355
9.4 Early SF_6 Applications   359
   9.4.1 Load-break switch   360
   9.4.2 Intermediate-interrupting-capacity circuit breaker   361
   9.4.3 High-power, high-voltage circuit breaker   362
   9.4.4 High-power, medium-voltage circuit breaker   365
   9.4.5 Extra-high-voltage circuit breaker   366
9.5 Recent Research   367
9.6 Materials Research   369
9.7 High-Power Single-Pressure Breakers   369
   9.7.1 "Super" puffer   370
   9.7.2 1200-kV breaker   370
9.8 Conclusions   370

References   373

---

[*]Currently Consultant, Pittsburgh, Pennsylvania

9.1 INTRODUCTION

In the early development of high-voltage electric power systems the inadequacy of air at atmospheric pressure as an arc quenching environment soon became evident. The substitution of insulating oil in circuit breakers was a brilliant success, so much so that the use of oil is still extremely widespread. However, in the period following World War II, the flammability, the carbonization by arcing, and to some extent the cost of oil led switch designers to search for other more nearly ideal arc quenching media. Compressed air was the choice of most, but Westinghouse engineers early recognized the unusual properties of the insulating gas sulfur hexafluoride (SF_6), and so pioneered in its application to switching equipment. This gas was found to have a unique combination of physical, chemical, and electrical properties which greatly favor its use in high-voltage circuit breakers. Although its dielectric properties were well known [1,2], it remained for research and development at Westinghouse to discover its exceptional arc quenching capabilities [3-8].

9.2 PROPERTIES OF SF_6

Physically, sulfur hexafluoride is a dense gas (five times the density of air, molecular weight 146) with a vapor pressure vs. temperature relation, shown in Fig. 9.1, such that it can be employed as a gas at considerably elevated pressures over a usefully wide temperature range. SF_6 is chemically inert, so it neither decomposes nor attacks

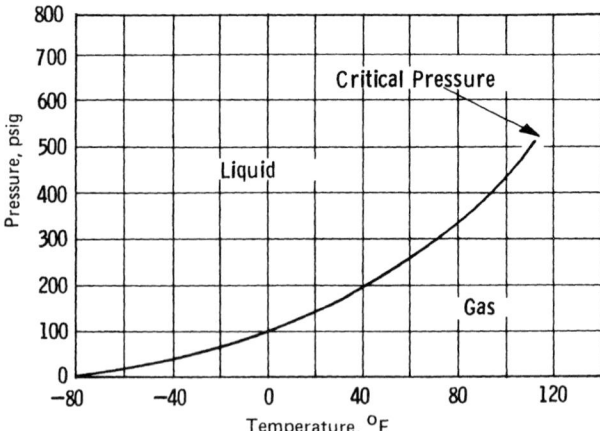

Figure 9.1 Pressure-temperature relationship for sulfur hexafluoride. [From W. M. Leeds, T. E. Browne, Jr., and A. P. Strom, The use of SF_6 for high power arc quenching AIEE Trans. 76(Pt. 2):906-909, © 1957, IEEE.]

ordinary materials of construction at temperatures below 150°C. It is dissociated within the high-temperature arc column, but the components recombine so that the net amount of decomposition products formed is actually very small. Metal vapors from the contacts tend to combine with active components of the arced gas to form soft insulating powders. Pure SF_6 does not hydrolyze and is nontoxic. Any chemically active impurities formed by arcing can be readily removed by inexpensive absorbents such as activated alumina.

Of particular importance for its use in circuit breakers are the electrical properties, which include voltage withstand (dielectric strength) and arc quenching capabilities. Figure 9.2 illustrates the comparable dielectric strengths of SF_6 and good insulating oil. At any given pressure the SF_6 strength is generally three to five times that of air.

Intensive research into the arc quenching properties of sulfur hexafluoride and the extremely superior performance over other media in this important property served to promote its application to various types of switching apparatus, beginning in the early 1950s. A short review of some of this early work in determining these properties is of interest.

9.3 EARLY RESEARCH

The first exploratory study of arc interruption in SF_6 was made by A. P. Strom and T. E. Browne, Jr., under the direction of H. J. Lingal [3], leading to the first U.S. patent on this use [4]. For the sake of simplicity this first study was made of the ac interrupting ability of plain-break arcs [Fig. 9.6(A)] in both air and sulfur hexafluoride, using the horizontal porcelain enclosure shown in Fig. 9.3.

To establish a basis for comparison, tests were made with this interrupter filled with air. The results of these tests at arc chamber pressures from 0 to 150 psig are plotted in Fig. 9.4. The results of similar tests with SF_6 are also plotted. The comparison with air is so extraordinary that it is difficult to show to the same current scale. Of considerable interest is the effect on interrupting performance of mixing air with SF_6. The results plotted in Fig. 9.4 show that the performance of a 50-50 mixture is roughly a geometric mean between that of air and of SF_6. At 30 psi, for example, the mixture is about 10 times as good as air but only one-tenth as good as pure SF_6.

It was found that the interrupting performance of the arc in SF_6 could be very greatly improved by only moderate rates of forced gas flow through the arc space. This is illustrated by the data of Fig. 9.5 for a simple puffer-type interrupter in which a piston containing the orifice was attached to the moving contact so that displaced gas was caused to flow through the orifice on the opening stroke [Fig. 9.6(B)]. The improvement thus obtained is shown by comparison

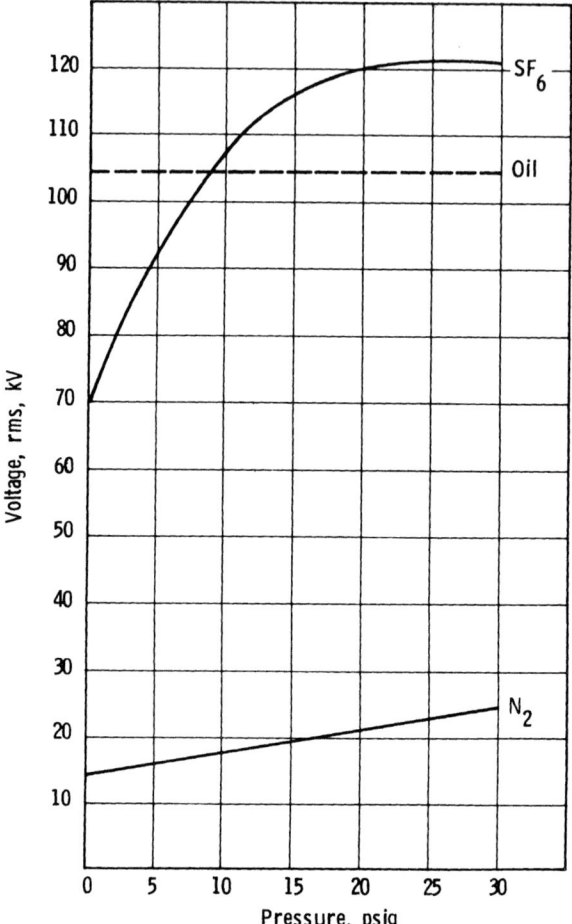

Figure 9.2 Sixty-hertz dielectric strength of SF_6 vs. N_2. (Courtesy of Westinghouse Electric Corporation, Trafford, Pa.)

with curve B of curve A in Fig. 9.5 for the 3-in. plain-break gap in SF_6. In certain cases the flow can be produced by using simple expansion chambers powered by arc heating [Fig. 9.6(C)]. Increased interrupting capability was obtained with the introduction of the two-pressure or blast-type interrupter [Fig. 9.6(D)], in which SF_6 gas is compressed to approximately 15 atm and by means of a blast valve is released into the arc when contacts are parted [6,9].

The basic requirement of an ac arc-interrupting medium is not primarily high dielectric strength but rather a high rate of recovery

SF$_6$ Breaker Research and Development

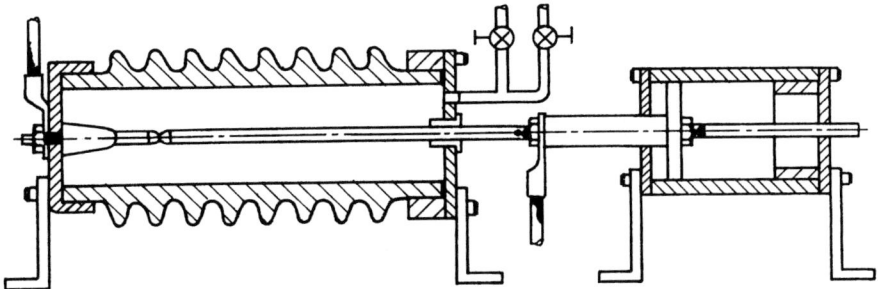

Figure 9.3 Schematic sketch of test apparatus. [From H. J. Lingal, A. P. Strom, and T. E. Browne, Jr., An investigation of the arc quenching behavior of sulfur hexafluoride, AIEE Trans. 72(Pt. III): 242-246, © 1953, IEEE.)

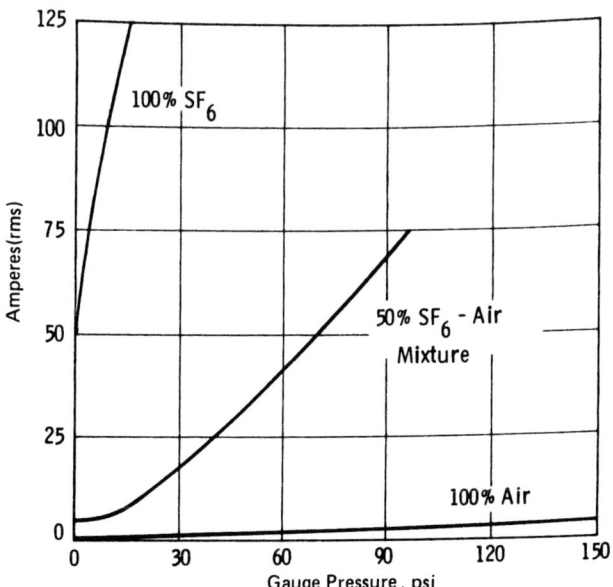

Figure 9.4 Interrupting performance of a 3-in. plain-break gap in SF$_6$ at 2300 V, compared with that of air. [From H. J. Lingal, A. P. Strom, and T. E. Browne, Jr., An investigation of the arc quenching behavior of sulfur hexafluoride, AIEE Trans. 72(Pt. III):242-246, © 1953, IEEE.]

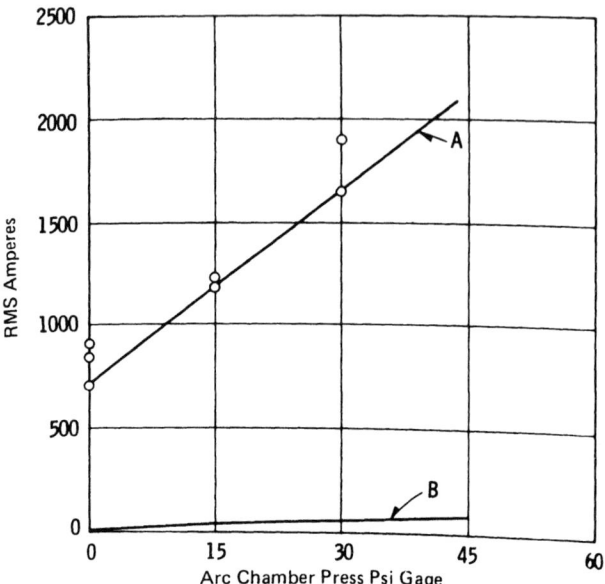

Figure 9.5 Zero-power-factor interrupting ability of a simple puffer-type circuit breaker with SF_6 gas at 13,800 V (curve A) compared with that of a 3-in. plain-break gap in SF_6 under similar conditions (curve B). [From H. J. Lingal, A. P. Strom, and T. E. Browne, Jr., An investigation of the arc quenching behavior of sulfur hexafluoride, AIEE Trans. 72(Pt. III):242-246, © 1953, IEEE.]

of dielectric strength. In important cases, this requirement can be alternately expressed as a high rate of loss of arc path conductance as the alternating current passes through zero. This rate of change in conductance is generally measured in terms of a "time constant" of the arc [10]. (See Chaps. 5 and 6.) This near-current-zero arc time constant is found to be exceptionally small when the gas medium is SF_6. Figure 9.7 shows comparative values of the time constant for a low-current plain break arc in SF_6, in air, and in a 50-50 mixture of the two gases [7]. Under blast conditions the comparison is less extreme, but it still accounts for about an order-of-magnitude improvement in ability of the SF_6 arc to withstand high initial rates of rise of recovery voltage [11].

For relatively low rates of rise of voltage, characterizing the interruption of normal load, line charging, and magnetizing currents, the ability of a nonconducting arc residue to build up dielectric strength at a high rate is limiting. Here the electronegative property of SF_6, causing rapid capture of any free electrons, aids both the rate

SF₆ Breaker Research and Development

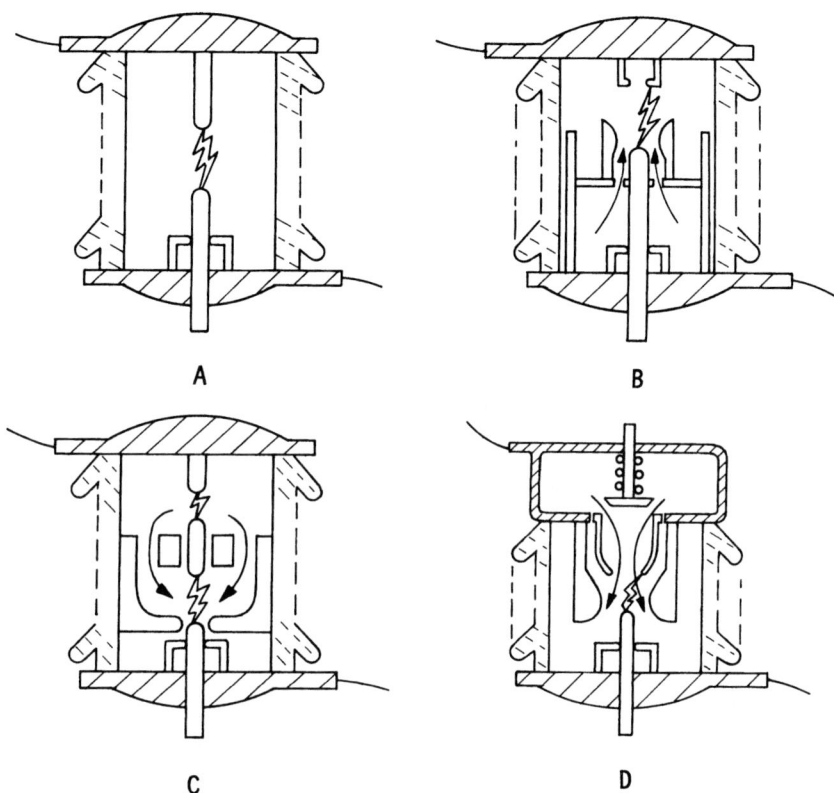

Figure 9.6 Single- and double-pressure SF$_6$ interrupter types: (A) Plain break, (B) puffer, (C) self-generated pressure, (D) blast (double pressure).

of recovery and the ultimate strength of this gas. Figure 9.8 is an example of estimated longer-time dielectric recovery rates in gas blasts of SF$_6$ and of air.

9.4 EARLY SF$_6$ APPLICATIONS

As research continued to provide these extremely encouraging fundamental data on SF$_6$, both dielectric and arc quenching, Westinghouse engineers built a number of different prototypes of switching equipment which led to the implementation of SF$_6$-filled apparatus on utility systems throughout the country, beginning in the early 1950s.

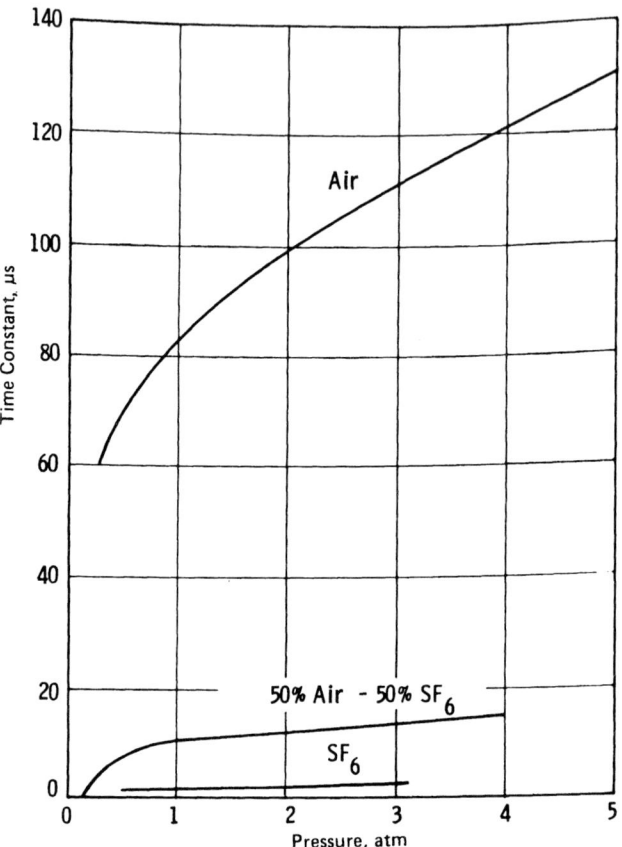

Figure 9.7 Time constant curves for low current arcs in air and SF_6 under plain-break conditions. (From Westinghouse Eng., March 1959.)

9.4.1 Load-Break Switch

The first application made in 1953 was to a high-voltage load-break switch with a sealed interrupter chamber containing SF_6 gas at 3 atm absolute pressure mounted on the jaw terminal of a conventional isolator switch [Fig. 9.9(A)]. Rapid arc extinction was obtained by the "puffer" principle [Fig. 9.9(B)] in which the relative motion of a piston and cylinder associated with the moving contact causes gas to flow close to the arc drawn in a Teflon nozzle which also moves with the contact. These load-break switches were designed for voltage ratings from 15 to 161 kV, opening load circuits with currents up to 600 A, at 50 to 100% power factor. Transformer magnetizing currents were safely interrupted without hazardous overvoltages due to current

SF₆ Breaker Research and Development

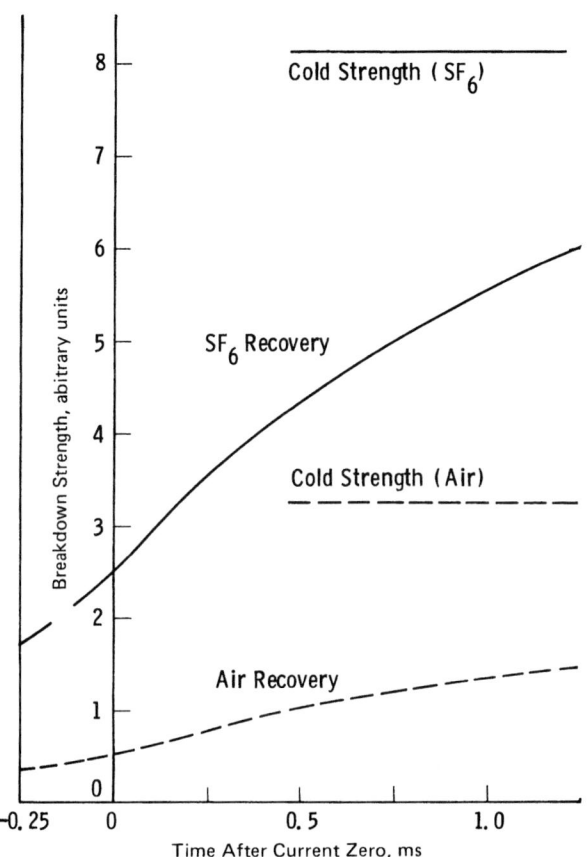

Figure 9.8 Estimated dielectric recovery by gas-blasted arc gap for SF$_6$ and air following arcing at 50 kA. (From Ref. 11.)

chopping because the gas flow velocity was low. Single banks of capacitors up to at least 5000 kVA were also switched satisfactorily.

9.4.2 Intermediate-Interrupting-Capacity Circuit Breaker

The first SF$_6$ power circuit breaker design, put into actual service in 1956, was a 115-kV 400-A breaker with 1000-MVA interrupting capacity shown in Fig. 9.10 [13,14]. It was based on the self-generation principle shown in Fig. 9.6(C). In the interrupter assembly there are a total of six breaks per phase, two for drawing pressure-generating arcs to produce gas flow, two for interrupting arcs within gas-blast nozzles, and two additional breaks to provide isolation in the open

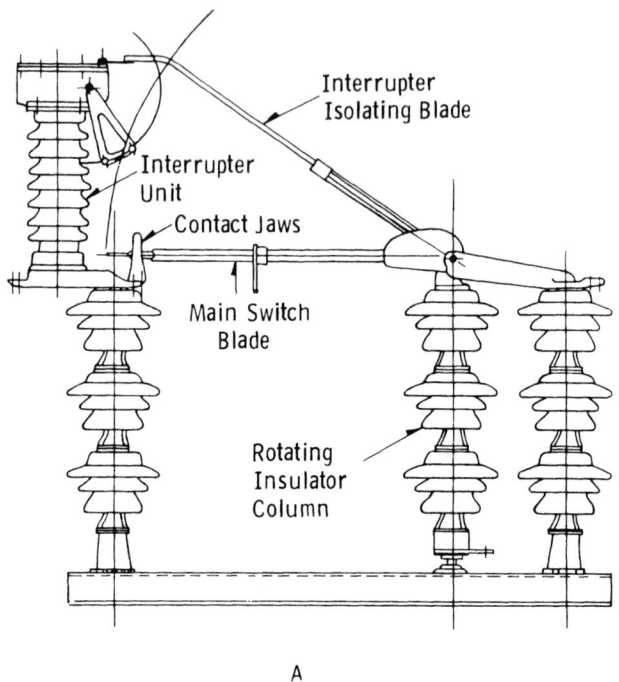

A

Figure 9.9 Outdoor load interrupter switch single-pole unit: (A) side view; (B) cross-sectional view of interrupter unit. (From Ref. 12.)

position and also to assist in interrupting charging currents. Operating experience with this type of circuit breaker demonstrated trouble-free performance, no appreciable loss of gas pressure, and ease of inspection and maintenance of the contact parts.

Another application of the same interrupting principle made in 1957 took the form of a 46-kV circuit breaker designed for multiple reclosing duty with an interrupting rating of 250 MVA [15]. Tests and field experience showed excellent performance not only when interrupting fault currents but also when interrupting transformer magnetizing currents and capacitance charging currents.

9.4.3 High-Power, High-Voltage Circuit Breaker

The expectation that sulfur hexafluoride might prove to be a new interrupting medium capable of competing successfully with oil and compressed air in the high-voltage field was borne out by the appearance in 1959 of fully developed breakers rated as high as 10,000 MVA and higher at 230 kV [6,8,16]. A survey of customer requirements in

SF$_6$ Breaker Research and Development

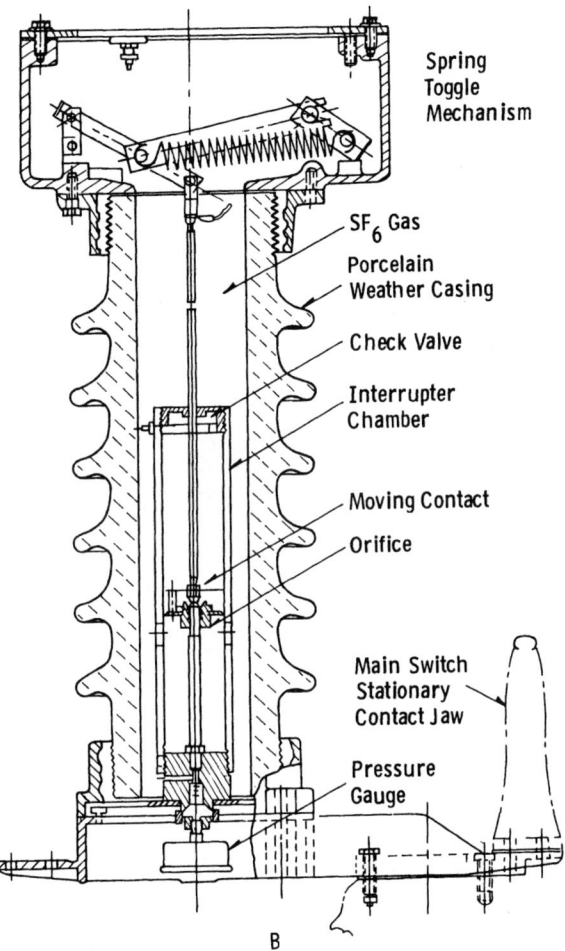

Figure 9.9 (Continued)

the United States and Canada for this class of equipment indicated a strong preference for the features associated with the bulk oil breaker, principally the safety of a strong steel tank solidly grounded, bushing mounted ring-type current transformers on both sides of the breakers, and contacts of all poles mechanically operated from one mechanism. Figure 9.11 shows an installation of three-pole 230-kV 15,000-MVA SF$_6$ breakers. These breakers were based on the two-pressure principle of Fig. 9.6(D). Gas is stored at 14.5 atm absolute and blasted through arcs drawn at three pairs of contacts in series per pole, exhausting into the main tanks held at 3 atm pressure. Reduced weight comparable

Figure 9.10 Field installation of 115-kV SF_6, 1000-MVA circuit breakers using the thermal expansion principle. [From C. F. Cromer and R. E. Friedrich, A new 115 kV, 1000 MVA gas-filled circuit breaker, AIEE Trans. 75(Pt. III):1352-1357, © 1956, IEEE.]

SF$_6$ Breaker Research and Development

Figure 9.11 Field installation of three-pole 230-kV 15,000-MVA SF$_6$ breakers. (From Ref. 17.)

to that of compressed air breakers is obtained, together with advantages gained by complete sealing, such as quiet operation and prevention of moisture penetration into the breaker assembly. Figure 9.12 shows a 161-kV single-tank breaker rated for 75 kA short-circuit interrupting capacity which utilized the two-pressure scheme.

9.4.4 High-Power, Medium-Voltage Circuit Breaker

In the early 1960s a line of puffer-type breakers was developed for voltage ratings of 34.5 to 69 kV and 25 to 30 kA interrupting capacity [18]. The design found popular appeal to the users but was not then economically competitive with the more conventional, lower-cost frame-mounted oil breaker. This breaker is shown in Figs. 10.1 and 10.2 of the next chapter.

Figure 9.12 161-kV single-tank SF_6 breaker rated for 75 kA interrupting capacity. (Courtesy of Westinghouse Electric Corporation, Trafford, Pa., and TVA, Knoxville, TN.)

9.4.5 Extra-High-Voltage Circuit Breaker

The extra-high-voltage line of Westinghouse Type SFA breakers for 362 to 800 kV which are presently in use was developed in the mid-1960s [19-22]. A 362-kV breaker shown in Fig. 9.13 is based on the two-pressure principle of Fig. 9.6(D). Over the years since its inception, the interrupting capability of this line has been increased greatly to currents in excess of 63 kA. At 362 kV each pole unit incorporates two breaks in one chamber mounted on a vertical porcelain column. This compact design permits mounting all three-pole units on a single frame and greatly simplifies transportation, erection, and maintenance.

SF$_6$ Breaker Research and Development

Figure 9.13 Extra-high-voltage Westinghouse Type SFA breaker for 362 kV. (From Proc. American Power Conf. 37, 1975.)

9.5 RECENT RESEARCH

To help guide the development of new and improved interrupting devices employing sulfur hexafluoride, a continuing program of research has been conducted into various aspects, including:

Arc-interrupting theory
Arc decomposition products
Dielectric properties
Effect and control of foreign particles
Seals: methods and materials
Leak detection methods

Of greatest interest for high-power circuit interruption is a study of arcs in rapidly flowing gas streams or "gas blasts." As an aid in understanding the effect of gas properties on the arc-extinguishing process in circuit breakers, a considerable amount of theory dealing with the dynamics of the arc near current zero can be called upon (see Chaps. 5 and 6). This theory deals with energy balance phenomena in the still somewhat conducting arc residue at and just after a current zero, before simple dielectric recovery begins. The use of mathematical techniques is made possible by adopting simplified models of the arc as a rapidly variable conductor definable by means of differential equations (Chap. 6). The approximate applicability of these equations has been verified by analysis of actual high-power laboratory tests of gas-blast circuit breakers using both air and SF_6. They have helped to clarify some of the main features of interruption in a gas blast.

More detailed analysis of interruption performance has been afforded by a broadened research effort shown in the flowchart of Fig. 9.14. The chart shows the logic needed to relate arc physics and fluid-flow theory to design procedure. In a circuit breaker arc, energy is transferred primarily by conduction, convection, and radiation. To understand the importance of these energy transfer mechanisms to circuit interruption, it was necessary to attempt to develop a theory for steady-state and dynamic nozzle arcs and arc-circuit interaction. Considerable effort was put into understanding energy balance phenomena before they could be applied to estimating interrupter performance under the severe short-line-fault condition. It is

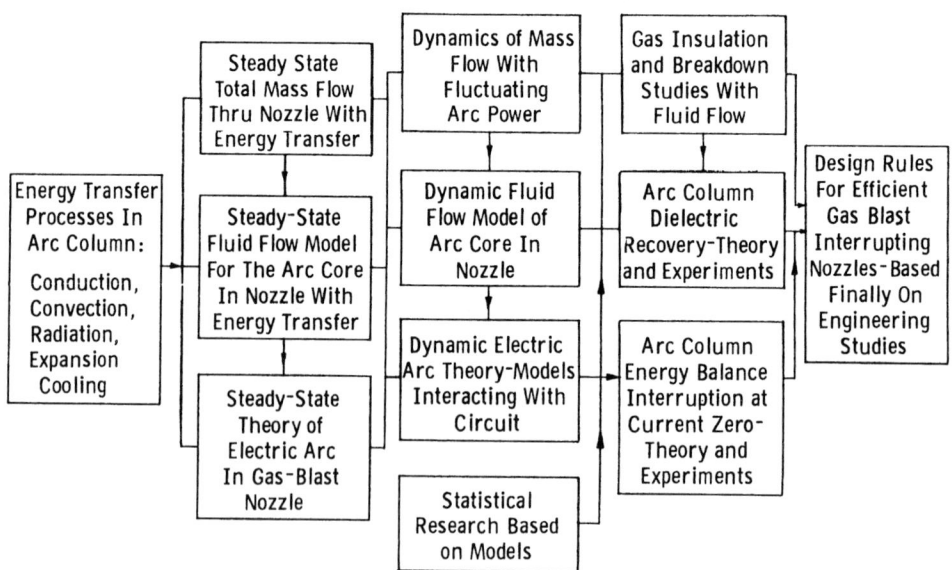

Figure 9.14 Diagram of interrelated arc interruption research. (Courtesy of Westinghouse Electric Corporation, Trafford, Pa.)

now possible to estimate arc model parameters from test data and to predict short-line-fault performance for a given interrupter [23-26]. This work is quite important to the circuit breaker designers in that they can minimize the number of expensive interrupting tests in the high-power laboratory and get more useful information from them. Such an analysis also permits putting into the mathematical model parameters established in a few tests and obtaining guidance for design modifications to achieve the required interrupting ability.

One method of increasing the current-interrupting rating of large circuit breakers, particularly for short-line faults, is to provide resistance or capacitance shunts to reduce the severity of the transient recovery voltage imposed on the breaker by the system. Because of the complexity added to the breaker in providing means for interrupting the resistor circuit following interruption, the capacitance shunt is the more desirable and economic solution. Figures 6.23 through 6.26 represent solutions to the mathematical model for capacitance-shunted short-line faults for a 242-kV system [26]. The designer thus has available an additional option for obtaining higher interrupting capabilities, the economics of which can be measured against other means, such as modifications to the interrupter or other breaker components.

9.6 MATERIALS RESEARCH

In addition to interrupting performance, the compatibility of construction materials with the SF_6 environment, especially in the presence of arcing, has received much study [27]. The unique suitability of polytetrafluoroethylene (Teflon) for the gas-blast nozzle was immediately established [4]. Further research and development resulted in antitracking coatings, the selection of suitable solid insulating materials, and the establishment of gas dryness standards (see Chap. 8). Suitable arcing contact materials were selected to assure durability of these contacts and to avoid harmful effects on insulation (see Chap. 15).

9.7 HIGH-POWER SINGLE-PRESSURE BREAKERS

This research, utilizing the logic chart of Fig. 9.14 and the computerized mathematical model of the arc to analyze test data, indicated the possibility of extending interrupting ratings of the puffer-type SF_6 breaker into the levels previously supplied only by the two-pressure approach. The model has been expanded (1) to simulate a double-flow nozzle configuration, (2) to provide better flow resistance calculation, and (3) to simulate nozzle arc and arc chamber clogging. The model was then used to investigate the effects of design variables on enthalpy clogging. In this theoretical study, changes were made in current, ambient pressure, nozzle area, arcing time, operating force, piston mass, and plenum volume. Results of these studies indicated that nozzle clogging is a function of pressure ratio and that current, gas pressure, piston velocity, and arcing time have a strong influence on interruption.

9.7.1 "Super" Puffer

Building on early breaker puffer data and the most recent research effort, the feasibility of interrupting more than 100 kA at transmission voltages became evident. Figure 9.15 is a photograph of a test model used in research for the Electric Power Research Institute to determine the parameters required to provide 100 kA plus interruption with the one puffer break.

9.7.2 1200-kV Breaker

The feasibility of utilizing the puffer principle is also being investigated for application to a 1200-kV breaker [25]. The research is sponsored by the Department of Energy (DOE). A computer model is being used to simulate a 1200-kV puffer breaker based on previous work which incorporates both the mechanical and arc theory considerations.

The logic chart of Fig. 9.16 illustrates how for the first time a determination of whether a breaker can be expected to work or fail is possible. The model permits inputting mechanical parameters associated with motion of control and operating valves as well as the main driving operator and linkages. These, combined with the position of the puffer cylinder, gas pressure, arc theory, and short-circuit current, provide the work-fail predictions. It is a powerful tool for designers and allows them to expedite appreciably the analysis of test results.

9.8 CONCLUSIONS

1. The original decision by the Westinghouse Electric Corporation to pioneer in the utilization of SF_6 was followed by most manufacturers.
2. User acceptance of SF_6 equipment has grown since its introduction as a load-break switch in the early 1950s until today it is the major medium being used in the United States and elsewhere.
3. Mathematical models based on research into the physics and fluid mechanics of arc quenching are continually being refined to the point where their application to the analysis of interrupting test data will aid optimization of designs.

SF$_6$ Breaker Research and Development

Figure 9.15 Test fixture for 100-kA puffer research project. (Courtesy of Westinghouse Electric Corporation, Trafford, Pa.)

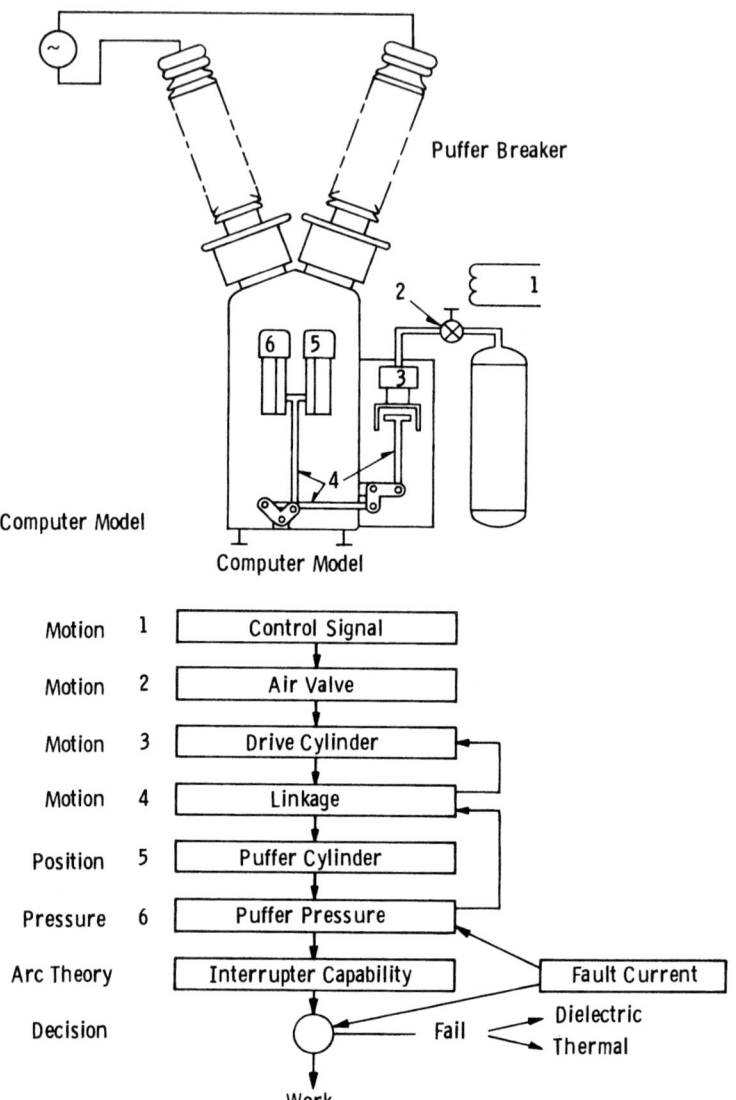

Figure 9.16 Application of computer simulation to 1200-kV power circuit breaker design. (Courtesy of Westinghouse Electric Corporation, Trafford, Pa.)

REFERENCES

1. F. S. Cooper, Gas dielectric media, U.S. Pat. 2,221,671, November 12, 1940.
2. G. C. Nonken, High-pressure gas as a dielectric, AIEE Trans. 60:1017-1020, December 1941.
3. H. J. Lingal, A. P. Strom, and T. E. Browne, Jr., An investigation of the arc quenching behavior of sulfur hexafluoride, AIEE Trans. 72(Pt. III):242-246, 1953.
4. H. J. Lingal, T. E. Browne, Jr., and A. P. Strom, Circuit interrupters, U.S. Pat. 2,757,261, July, 31, 1956.
5. T. E. Browne, Jr., and A. P. Strom, Interruption of capacitance charging currents in sulfur hexafluoride, AIEE Trans. 75(Pt. III):1357-1361, 1956.
6. W. M. Leeds, T. E. Browne, Jr., and A. P. Strom, The use of SF_6 for high power arc quenching, AIEE Trans. 76(Pt. III): 906-909, 1957.
7. K. H. Yoon and H. E. Spindle, A study of the dynamic response of arcs in various gases, AIEE Trans. 78(Pt. III):1634-1642, 1959.
8. T. E. Browne, Jr., and W. M. Leeds, A new medium for circuit interruption, CIGRE, Rep. 111, Paris, 1960.
9. R. E. Friedrich and G. Bates, The use of SF_6 in high voltage interruption, AIEE Conf., 1958, Paper 58-960.
10. T. E. Browne, Jr., An approach to mathematical analysis of a-c arc extinction in circuit breakers, Trans. AIEE Power Appar. Syst. PAS-40:1508-1517, February 1959.
11. W. M. Leeds, R. E. Friedrich, C. L. Wagner, and T. E. Browne, Jr., Application of switching surge, arc and gas flow studies to the design of SF_6 breakers, CIGRE, Rep. 70-1311, Paris, 1970.
12. H. J. Lingal and J. B. Owens, A new high voltage outdoor load interrupter switch, AIEE Trans. 72(Pt. III):293-297, 1953.
13. C. F. Cromer and R. E. Friedrich, A new 115 kV, 1000 MVA gas-filled circuit breaker, AIEE Trans. 75(Pt. III):1352-1357, 1956.
14. E. B. Henry, R. E. Friedrich, and F. L. Reese, Service experience and staged field tests on the 115 kV, 1,000,000 kVA gas filled power circuit breaker, AIEE Trans. 77(Pt. III):318-325, 1958.
15. R. N. Yeckley and R. H. Cunningham, A new 46 kV low capacity circuit breaker for multiple reclosing duty, AIEE Trans. 77(Pt. III):402-406, 1958.
16. R. E. Friedrich and R. N. Yeckley, A new concept in power circuit breaker design utilizing SF_6, AIEE Trans. 78(Pt. III): 695-706, 1959.

17. W. M. Leeds, R. E. Friedrich, and C. L. Wagner, EHV power circuit breakers using SF_6 gas, CIGRE, Rep. 117, Paris, 1966.
18. G. J. Easley and J. M. Telford, New design 34.5-69 kV intermediate capacity SF_6 circuit breaker, IEEE Trans. Power Appar. Syst. *83*:1172-1177, December 1964.
19. R. C. Van Sickle and R. N. Yeckley, A 500 kV circuit breaker using SF_6 gas, IEEE Trans. Power Appar Syst. *PAS-84*:892-901, December 1965.
20. R. N. Yeckley and C. F. Cromer, New SF_6 EHV circuit breaker for 550 kV and 765 kV, IEEE Trans. Power Appar. Syst. *PAS-89*:2019-2023, 1970.
21. W. M. Leeds, R. E. Friedrich, T. E. Browne, Jr., and C. L. Wagner, New goals for EHV and future UHV power circuit breakers, Westinghouse Eng., July 1970.
22. W. M. Leeds, SF_6 switchgear comes of age, Electr. Rev. (Lond.), August 14, 1970, pp. 225-227.
23. L. S. Frost, Dynamic arc analysis of short-line fault tests for accurate circuit breaker performance specification, IEEE Trans. Power Appar. Syst. *PAS-97*:478-484, March/April 1978.
24. T. E. Browne, Jr., Practical modeling of the circuit breaker arc as a short line fault interrupter, IEEE Trans. Power Appar. Syst. *PAS-97*:838-847, May/June 1978.
25. H. E. Spindle, T. F. Garrity, and C. L. Wagner, Develpment of 1200 kV circuit breaker for GIS system requirements and circuit breaker parameters, CIGRE, Rep. 13-07, Paris, 1980.
26. T. E. Browne, Jr., Simplified estimation of critical quantities for short line fault interruption, IEEE Trans. Plasma Sci. *PS-8*(4):400-405, 1980.
27. T. Ushio, I. Shimura, and S. Tominaga, Practical problems on SF_6 gas circuit breakers, IEEE Trans. Power Appar. Syst. *PAS-90*(5):2166-2174, 1971.

General References

Berg, D., and C. N. Works, Effect of space charge on electric breakdown of sulfur hexafluoride in non-uniform fields, AIEE Trans. 77(Pt. III):820-823, 1958.

Brado, J. J., and C. F. Sonnenberg, Insulation coordination qualities—500 kV SF_6 breakers, IEEE Trans. Power Appar. Syst. *PAS-84*:851-863, 1965.

Browne, T. E., Jr., B. W. Swanson, and L. S. Frost, Arcs in a flow field, Written notes for a short course on Plasma Arcs and Switching Phenomena conducted by the Current Zero Club, University of Wisconsin, Madison, July 15-17, 1971.

Colclaser, R. G., Jr., L. E. Berkebile, and D. E. Buettner, The effect of capacitors on the short-line fault component of transient recovery voltage, IEEE Trans. Power Appar. Syst. *PAS-90*: 1482-1491, 1971.

Cookson, A. H., O. Farish, and G. M. L. Sommerman, Effect of conducting particles on ac corona and breakdown in SF_6, IEEE Trans. Power Appar. Syst. *PAS-91*:1329-1338, July/August 1972.

Ellis, H. M., and R. E. Friedrich, Cable switching tests on 230 kV SF_6 circuit breaker, IEEE Paper 64CP118, 1964.

EPRI RP478 Project Final Report, Electric Power Research Institute, Palo Alto, Calif., 1982.

Friedrich, R. E., Circuit interruption with SF_6, Electr. South, March 1959.

Frink, R. E., SF_6 magnetic-puffer circuit breaker for 34.5 kV, 1500 MVA, IEEE Trans. Power Appar. Syst. *PAS-87*(3):742-749, 1968.

Frost, L. S. and R. W. Liebermann, Composition and transport properties of SF_6 and their use in a simplified enthalpy flow arc model, Proc. IEEE *59*:474-485, April 1971.

Hickam, W. M., and R. E. Fox, Electron attachment in sulfur hexafluoride using mono-energetic electrons, J. Chem. Phys. *25*:612, 1956.

Kane, R. E., and R. G. Colclaser, Jr., A 69 kV SF_6 common tank breaker rated 5000 MVA, IEEE Trans. Power Appar. Syst. *PAS-69*:1076-1082, December 1963.

Leeds, W. M., R. E. Friedrich, T. E. Browne, Jr., and C. L. Wagner, Breaker design spans many technologies, Electr. World, July 13, 1970, pp. 53-55.

Lowke, J. J., Characteristics of radiation dominated electric arcs, J. Appl. Phys. *41*:2588-2600, 1970.

Lowke, J. J., Electrical breakdown in a series of avalanches, Proc. Semin. Gas Breakdown, Tokyo, 1972, p. 41.

Lowke, J. J., A relaxation method of calculating properties of convection stabilized electric arcs, Bull. Am. Phys. Soc. *16*: 199, 1971.

Lowke, J. J., and E. R. Capriotti, Calculation of temperature profiles of high pressure electric arcs using the diffusion approximation for radiation transfer, J. Quant. Spectrosc. Radiat. Transfer *9*:207-236, 1969.

Lowke, J. J., and R. W. Liebermann, Predicted arc properties in sulfur hexafluoride, J. Appl. Phys. *42*:3532, 1971.

Perkins, J. F., and L. S. Frost, The behavior of gas blown SF_6 arcs in a nozzle, IEE Int. Conf. Gas Discharges, London, September 1970.

Perkins, J. F., and L. S. Frost, The enthalpy flow limitation of gas-blasted SF_6 arcs, Proc. IEE (Lond.), *118*(7):948-954, 1971.

Perkins, J. F., and L. S. Frost, Interruption properties of gas blasted air and SF$_6$ arcs, IEEE Paper 72CP530, 1972.

Perkins, J. F., and L. S. Frost, Magnetic pumping on dielectric recovery of blown SF$_6$ arcs, IEEE Trans. Power Appar. Syst. *PAS-91*:376-381, 1972.

Strom, A. P., Utilities study SF$_6$ as an interrupting medium, Electr. World, February 2, 1959.

Swanson, B. W., Nozzle arc interruption in supersonic flow, IEEE Trans. Power Appar. Syst. *PAS-77*:1697-1706, September/October 1977.

Swanson, B. W., A thermal analysis of short line fault interruption, IEEE Power Eng. Soc. Winter Power Meet., January/February 1974, Paper C-74-186-3.

Swanson, B. W., and R. M. Roidt, Boundary layer analysis of an SF$_6$ circuit breaker arc, IEEE Trans. Power Appar. Syst. *PAS-90*(3):1086-1093, 1971.

Swanson, B. W., and R. M. Roidt, Some numerical solutions of the boundary layer equation for an SF$_6$ arc, Proc. IEEE *59*:493-501, April 1971.

Swanson, B. W., and R. M. Roidt, Thermal analysis of an SF$_6$ circuit breaker arc, IEEE Trans. Power Appar. Syst. *PAS-91*:381-389, March/April 1972.

Swanson, B. W., R. M. Roidt, and T. E. Browne, Jr., Arc cooling and short line fault interruption, IEEE Trans. Power Appar. Syt. *PAS-90*:1094-1102, 1971.

Swanson, B. W., R. M. Roidt, and T. E. Browne, Jr., The effects of gas dynamics and properties of SF$_6$ and air on short line fault interruption, IEEE Trans. Power Appar. Syst. *PAS-89*:2033-2042, November/December 1970.

Swanson, B. W., R. M. Roidt, and T. E. Browne, Jr., A thermal arc model for short line fault interruption, Elektrotech. Z. Ausg. A *93*:375-380, July 7, 1972.

Whitehead, D. L., and H. E. Spindle, Pressurized SF$_6$ gas-insulated high voltage switchgear, IEEE Winter Power Meet., IEEE Conf. Paper CP31, Vol. 29, 1927, pp. 67-69.

Yeckley, R. N., R. E. Friedrich, and M. E. Thuot, EHV breaker rated for control of closing voltage switching surges to 1.5 per unit, IEEE Trans. Power Appar. Syst. *PAS-91*:399-403, 1972.

10
Single-Pressure SF$_6$ Circuit Breakers

BEN J. CALVINO / Westinghouse Electric Corporation, Trafford, Pennsylvania

10.1 Introduction 377
10.2 SF$_6$ Single-Pressure Interrupting Element 381
10.3 Development of Single-Pressure SF$_6$ Interrupters 383
    10.3.1 Pressure buildup characteristics 385
    10.3.2 Nozzle design concepts 393
10.4 Electrical Interruption Characteristics of a Puffer Circuit Breaker 397
    10.4.1 Operating characteristics of breaks in series 400
    10.4.2 Interruption of asymmetrical currents 402
10.5 Mechanical and Electrical Reliability 404
10.6 Industrial Circuit Breaker Configurations 408
    10.6.1 Live-tank designs 409
    10.6.2 Dead-tank designs 412
10.7 Conclusions 418
10.8 Additional Considerations for Sec. 10.3.1 419
    10.8.1 Blocked nozzle cold flow pressure rise 419
    10.8.2 Pressure rise with gas outflow but no arc 422
References 423

10.1 INTRODUCTION

In the early 1950s Westinghouse Electric Corporation engineers demonstrated the effectiveness of sulfur hexafluoride (SF$_6$) gas as an interrupting medium in circuit breakers [1,2]. Included in that research program were interrupters of the single-pressure type with gas flow produced either by arc heating or by a mechanical piston attached to the moving contact. Single-pressure interrupters using these methods were introduced in 1952 for load-break switches [3] and in 1956 for circuit breakers [4] of low breaking capacity. In 1957 it was revealed [5] that a piston "puffer" circuit breaker could be used for interruption of high short-circuit currents, which represented a breakthrough in circuit breaker technology.

However, full development of single-pressure puffer circuit breakers was delayed. The good service record obtained in the conventional circuit breaking techniques (i.e., bulk oil, small oil volume, compressed air, and two-pressure SF_6, also developed by Westinghouse Electric Corporation) did not allow its introduction.

A small oil volume circuit breaker could be considered a different packaging of a bulk oil circuit breaker and the double-pressure SF_6 circuit breaker represented an extension of compressed air design and was easily accepted and widely used. Single-pressure SF_6 circuit breakers had "no past" to claim and their development was not immediately welcomed. However, the inherent qualities of SF_6 as an arc extinguishing medium were known and the concept of using them more adequately was growing. Intensive development work on puffers was started in 1960 and a year later, in 1961, the first commercial SF_6 puffer-type circuit breakers were introduced in the United States [6] by Westinghouse. Figure 10.1 shows a section of that breaker and Fig.

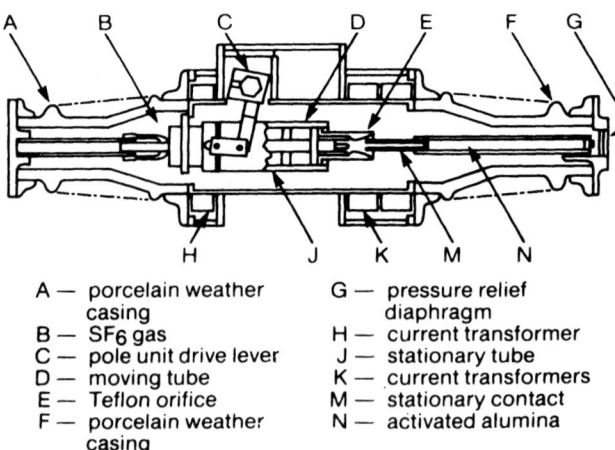

A — porcelain weather casing
B — SF_6 gas
C — pole unit drive lever
D — moving tube
E — Teflon orifice
F — porcelain weather casing
G — pressure relief diaphragm
H — current transformer
J — stationary tube
K — current transformers
M — stationary contact
N — activated alumina

Figure 10.1 Westinghouse 460SP3500 circuit breaker cross section. (Courtesy of Westinghouse Electric Corporation, Trafford, Pa.)

Single-Pressure SF$_6$ Circuit Breakers

10.2, a photograph taken in 1980, shows one of those circuit breakers in service 16 years after installation.

In 1965, Magrini SpA of Italy, a Westinghouse licensee at that time, developed and installed the first European EHV single-pressure SF$_6$ breaker [7,8]. It was the forerunner of the puffer circuit breakers covering the range of voltages from 123 to 1200 kV. Figure 10.3 shows one of the first puffer circuit breakers installed. From that date, it was possible to present circuit breakers having the simple construction of the more advanced minimum oil circuit breakers without using oil and the high interrupting capacity of the other techniques without their complexity.

The advantages of the single-pressure SF$_6$ technique of interruption are the following:

1. Oilless
2. No gas compressors
3. No heaters required above -40°F
4. Increased rated voltage per interrupting element as higher contact gaps can be attained
5. Increased rated breaking capacity per interrupting element
6. Decrease of wear by reducing arc energy
7. Suitability for high-speed reclosing without derating as the extinguishing pressure is not stored but recreated for each opening operation
8. Consistency of arcing times and electrical characteristics
9. Self-adjustment of the gas pressure needed to interrupt a given current and therefore no overvoltages when interrupting magnetizing and small inductive currents
10. Simplicity of design with small number of moving parts and sealing gaskets
11. Extension of electrical and dielectric capability which allows the conception of "modularity" for the whole range of extra-high-voltage (EHV) circuit breakers
12. No restrikes when interrupting capacitive loads
13. No external display when interrupting short-circuit currents
14. Reduced noise level
15. Short interrupting times
16. Practically no maintenance of conducting parts

Things that were thought to be disadvantages, such as the complexity of the interaction phenomena not being easily understood, the small gas volume used which could be polluted by arcing, the power needed to operate the piston of the puffer, and leakages without backup, have proved to be easily overcome by the designers. Today, puffer circuit breakers have achieved worldwide acceptance and all leading manufacturers have developed and installed them.

Figure 10.2 Westinghouse 460SP3500 circuit breaker in service. (Courtesy of Westinghouse Electric Corporation, Trafford, Pa.)

Single-Pressure SF$_6$ Circuit Breakers

Figure 10.3 Magrini SpA 245HMH30 circuit breaker installed. (Courtesy of Magrini-Galileo SpA, Bergamo, Italy.)

10.2 SF$_6$ SINGLE-PRESSURE INTERRUPTING ELEMENT

Two different types of interrupting elements have been developed: (1) the single-flow interrupting element and (2) the double-flow interrupting element. Other techniques are today under development, such as the self-pressure generating arc concept, the rotating arc column, and the magnetic coil booster. At this stage of development, their application in EHV is limited and they will not be examined here.

Figure 10.4 illustrates the section of a typical single-flow interrupter module Westinghouse Type LWE with two interrupting chambers, including:

1. Supporting insulators
2. Terminals
3. Operating box
4. Fixed contacts
5. Moving contacts and arcing tip

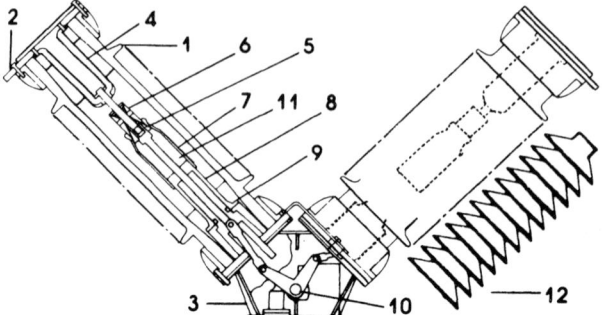

Figure 10.4 Westinghouse double-LWE interrupter module. (Courtesy of Westinghouse Electric Corporation, Trafford, Pa.)

6. Nozzles
7. Moving cylinders
8. Fixed pistons
9. Transfer contacts
10. Operating shaft
11. Compression chambers
12. Grading capacitors

Figure 10.5 shows the sequence of an opening operation, with the interrupting parts first in the closed position, then in the intermediate position, and finally, in the open position. During the completion of the stroke, the volume of gas inside the moving cylinder is compressed and escapes through the nozzle to extinguish the arc created between the contacts. It can be noted that the arc column and the movement of the piston create a pressure inside the compressing chamber which has to be overcome by the operating mechanism.

The inverse sequence is needed for a closing operation. The prestrike of the arc column will be much shorter but there will be a pressure buildup. It will therefore help the movement of closing and the operating mechanism will not need greater power to close under short-circuit conditions.

Figure 10.6 shows a typical double-flow interrupting element, Westinghouse Type SP, where the same elements as in Figure 10.4 have been identified and shows the sequence of an opening operation. In this case, the compressed volume of gas can escape from the path between the contacts through both of the hollow contacts to interrupt the arc and clear the ionized gas created.

Both concepts examined are good but, as in all technical matters, it is only the application that counts. Today, with both arrangements outstanding performance has been attained.

Single-Pressure SF$_6$ Circuit Breakers

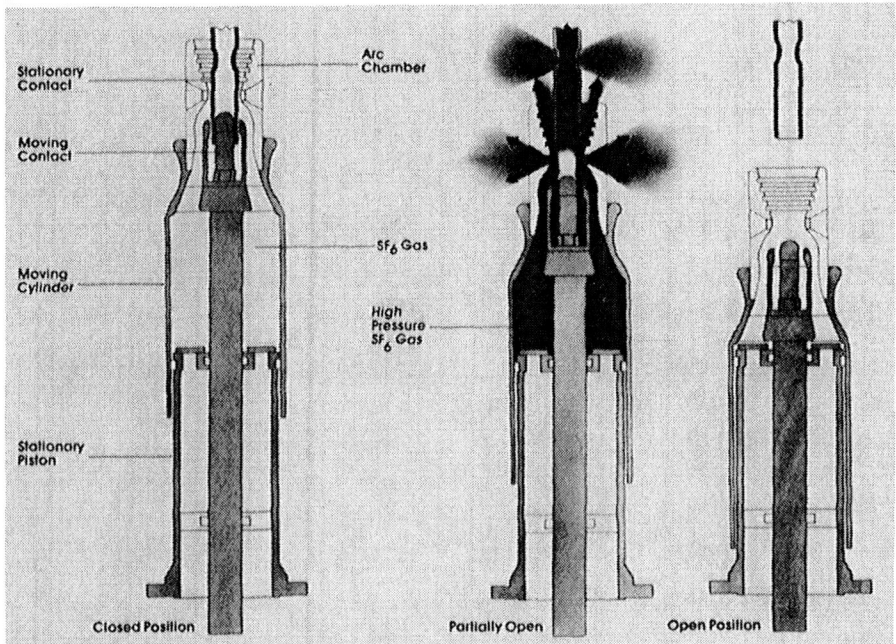

Figure 10.5 Sequence of an opening operation with a single-flow interruption. (Courtesy of Westinghouse Electric Corporation, Trafford, Pa.)

10.3 DEVELOPMENT OF SINGLE-PRESSURE SF$_6$ INTERRUPTERS

A simple solution of a problem is possible only when all its aspects are well known. There are no shortcuts or lucky breaks in simple technique. The principal difficulties in electrical circuit interruption have always been:

Arc energy
Dielectric recovery
Current chopping
Consistency of performance

They are present in the tasks a circuit breaker is called upon to achieve:

Interruption of all types of short circuits
Interruption of capacitive currents
Interruption of small inductive currents
Interruption in phase opposition

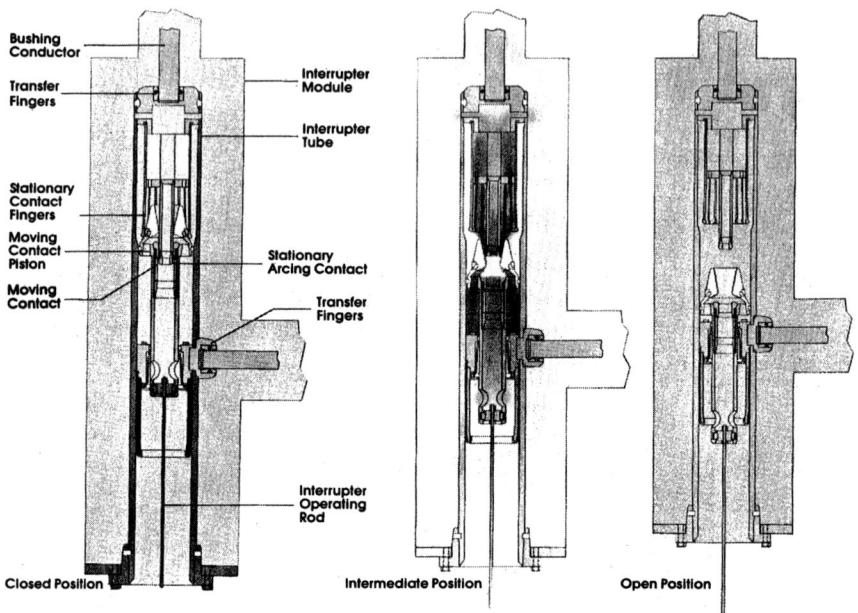

Figure 10.6 Sequence of an opening operation with a double-flow interrupter. (Courtesy of Westinghouse Electric Corporation, Trafford, Pa.)

Some of these functions, however, call for opposite interruption characteristics and the principal difficulty in developing a circuit breaker is in finding a design compromise which will fulfill all these requirements, with a common interrupting device, in an acceptable manner.

The analysis of some of the design parameters used in a single-pressure SF_6 interrupting element is important in the search for the best "compromise." One of them is the differential pressure Δp, the pressure difference between the upstream and downstream sides of a nozzle. Its variations as functions of piston diameter, length of stroke, rated SF_6 pressure, cold compressing ratio, contact speed, and overlapping dimension can be examined in a simplified way.

It is well known that to interrupt an ac arc in a conventional circuit breaker, the extinguishing medium has to be driven into the arc between the contacts. It is less well known that it is not the quantity of gas flow which is important but its rapidity of regeneration between the contacts. This characteristic is a function of the differential pressure

Single-Pressure SF$_6$ Circuit Breakers

value between the upstream and downstream ends of a nozzle at the instant of interruption, not of the volume of gas stored upstream.

10.3.1 Pressure Buildup Characteristics

The dimensions of the compression chamber, the length of the stroke, and the initial absolute gas pressure have an effect on the differential pressure and the size of the operating mechanism. To show the effects of design changes, the puffer dimensions of Figure 10.7 will be considered. They are:

- St = length of piston stroke
- P = puffer piston
- G_s = clogging section
- r = piston radius
- N = interrupting nozzle
- C = puffer cylinder
- P_n = rated pressure (absolute value)
- Tv = total volume of the compression chamber
- Sv = volume displaced by the piston
- Dv = dead volume inside the compression chamber
- N_s = minimum section of nozzle throat

Considering total clogging at the nozzle N at the minimum section N_s and a dead volume Dv equal to 1/10 of the displaced volume Sv, the following equations can be written:

$$\text{Displaced volume Sv} = \pi r^2 \text{St} \qquad (10.1)$$

$$\text{Total volume Tv} = \text{Sv} + \text{Dv} = \pi r^2 \text{St}(1 + 0.1) \qquad (10.2)$$

The compression ratio at the end of the stroke is

$$\frac{\text{Tv}}{\text{Dv}} = \frac{\text{Sv} + \text{Dv}}{\text{Dv}} = \frac{\pi r^2 \text{St}(1 + 0.1)}{\pi r^2 \text{St} \times 0.1} = \frac{11}{1} \qquad (10.3)$$

The differential pressure at the end of the stroke under the conditions examined would then be

$$\Delta P = 11 P_n - P_n = 10 P_n \qquad (10.4)$$

Differential pressure as a function of piston diameter

As developed in Sec. 10.8, a more general formula for the blocked nozzle pressure rise may be written as a function of the dimensionless piston radius ratio, r'/r, a dead volume ratio $\delta = \text{Dv}/\text{Sv}$, and the piston position expressed as the fraction f of the total stroke as

$$\Delta P_f = \left[\frac{(r'/r)^2 + \delta}{(r'/r)^2(1 - f) + \delta} - 1 \right] P_n \qquad (10.5)$$

Curves a and b in Fig. 10.8, plotted according to equation (10.5), show the effect of a 20% increase in piston and compression chamber diameter (44% increase in cross-sectional area), but with the dead volume unchanged. The end-of-stroke pressure rise is increased in the ratio 14.4/10 = 1.44, or by 44%. However, this gain cannot be considered. For the designer of a puffer circuit breaker the most important factor is the differential pressure necessary to interrupt a short-circuit current at minimum arcing time. For the circuit breaker considered, the minimum arcing time is reached at approximately half of the stroke. At this position the curves, or equation (10.5), show a gain in pressure of only 5.4% caused by this change [see equations (10.15) and (10.16)].

All conditions being equal, to increase by 44% the driving force necessary to operate the circuit breaker in order to increase by only

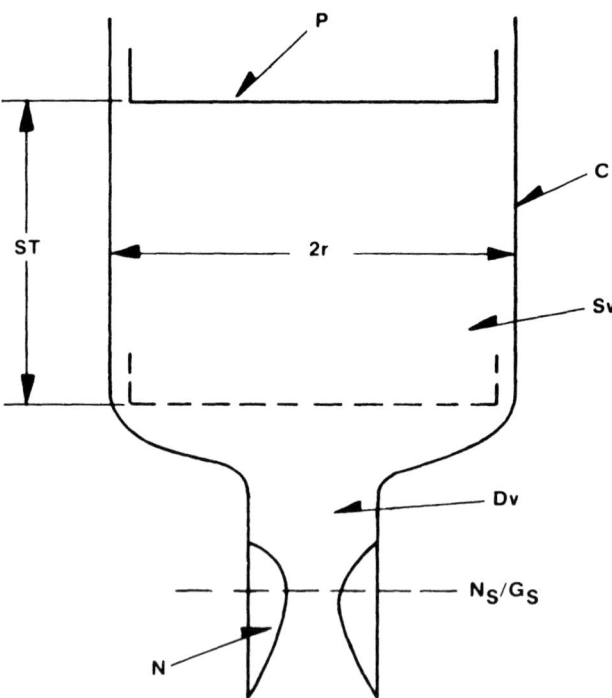

Figure 10.7 Typical puffer cylinder, piston, and nozzle.

Single-Pressure SF$_6$ Circuit Breakers

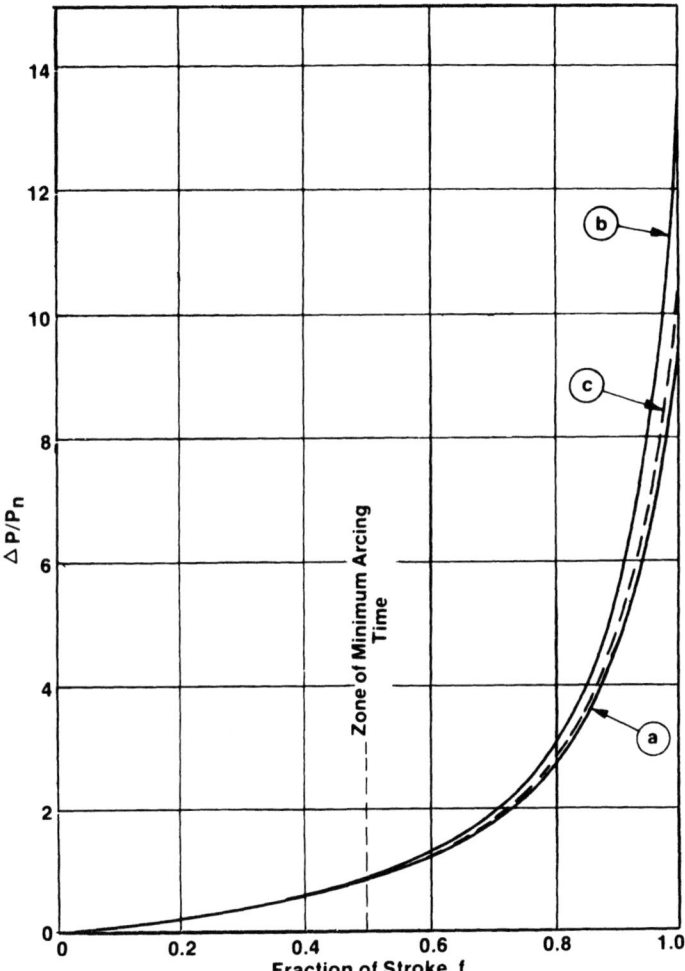

Figure 10.8 Differential pressure as function of piston position, diameter, and stroke.

about 5% the differential pressure at minimum arcing time, which is the most critical point, does not seem to be a promising approach.

Differential pressure as a function of stroke length

As affected by stroke length, the pressure rise expression from Sec. 10.8 is

$$\Delta P_f = \left[\frac{(St/St) + \delta}{(St'/St)(1-f) + \delta} - 1\right]P_n \qquad (10.6)$$

The dashed curve c in Fig. 10.8 shows the relative effect of a 10% increase in stroke, yielding a 10% increase in full stroke pressure. However, at minimum arcing time, half stroke, the increase is only 1.5%. Thus it can be seen that unless it is for voltage withstand between the contacts, increasing the stroke has a negligible effect on the interrupting capability of a breaker.

Differential pressure as a function of the rated pressure

The rated or fill pressure of an SF_6 puffer circuit breaker is generally determined by the lowest ambient temperature the circuit breaker will meet. A pressure of 5 bar (75 psig) is commonly used to be able to reach temperatures of approximately -40°F without danger of liquefaction.

Increasing this pressure by 20%, from 5 to 6 bar, increases in proportion the differential pressure if we consider only cold stream conditions. However, test results do not show such a gain. The half-stroke gain is theoretically only 0.833 atm and, under short-circuit conditions, the increase of pressure generated upstream by the arc column, as shown in Fig. 10.9, will render practically negligible the differential pressure gained by increasing the fill pressure. However, for other reasons such as dielectric withstand or dielectric restrikes at high-voltage values, an increase in rated pressure may be beneficial.

Effect of dead volume

The compression ratio $Tv/Dv = (Sv + Dv)/Dv$ at the end of the stroke can be increased by reducing the dead volume Dv for a given circuit breaker design. As assumed in equation (10.3), it is 11/1. If the Dv is reduced to half, the new full-stroke compression ratio will be

$$\frac{Sv + \tfrac{1}{2}Dv}{\tfrac{1}{2}Dv} = \frac{21}{1} \qquad (10.7)$$

but at 1/2 of the stroke, it will be only

$$\frac{Sv + \tfrac{1}{2}Dv}{\tfrac{1}{2}Sv + \tfrac{1}{2}Dv} = \frac{21}{11} = \frac{1.91}{1} \qquad (10.8)$$

and the differential pressure

$$\Delta P_{1/2} = (1.91 - 1)P_n = 0.91 P_n \qquad (10.9)$$

which represents a gain of only 9.1% in relation to the original value, $0.833 P_n$. The design modifications needed may not justify such a gain. On the other hand, if the dead volume Dv is doubled, the same calculations will give a differential pressure of $0.714 P_n$, representing a loss of 24%, which is excessive.

Single-Pressure SF$_6$ Circuit Breakers

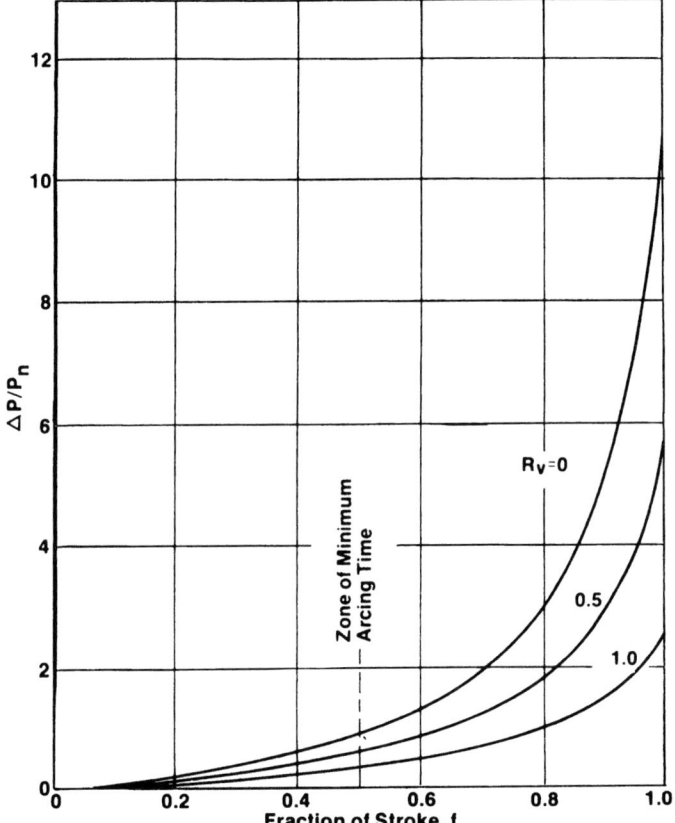

Figure 10.9 Effect of gas outflow and stroke speed on pressure rise.

Differential pressure as a function of stroke speed

With outflow blocked, the stroke speed does not change the compression ratios or differential pressures. As, at a certain speed, for a given design the minimum arcing time gives a minimum arc length for a successful interruption, the increase in stroke speed will reduce the minimum arcing time proportionally. However, the energy needed to accelerate the masses will be greatly increased and, if not absolutely necessary, does not seem worthwhile as a means to increase the short-circuit interrupting capacity of the circuit breaker.

At low currents the arc will not block the nozzle, so outflow of gas will reduce the pressure rise, and this reduction depends on the stroke speed. As suggested by T. E. Browne, Jr., the flow-reduced pressure rise can be calculated approximately by assuming a constant stroke velocity V and also continuous sonic gas outflow velocity Vs, as done by Yanabu et al. [10]: With additional inclusion of the dead volume ratio δ, the equation for pressure rise is shown in Sec. 10.8 to be

$$\Delta P = \left(\left\{ \exp\left[\left(1 - \frac{N_s V_s}{2\pi r^2 V}\right) \ln \frac{1 + \delta}{1 + \delta - f} \right] \right\}^\gamma - 1 \right) P_n \qquad (10.10)$$

Figure 10.9 is a plot of the relation for $\delta = 0.1$ and γ, the specific heats ratio of the gas, taken as 1.08 for SF_6. Three values of the volume outflow ratio, $Rv = (N_s V_s / \pi r^2 V)$; 0, 0.5 and 1.0 are illustrated. Since in actual designs this ratio is likely to be one or more, it can be seen that, with fixed stroke speed, the cold flow or light current pressure rise will be much smaller than that for the blocked nozzle condition. The slower the stroke, the greater will be this pressure reduction. However, the stroke speed is also determined by the requirement of satisfactory operation under line charging conditions without restrikes or to obtain the maximum interrupting time specified, and limits have to be determined.

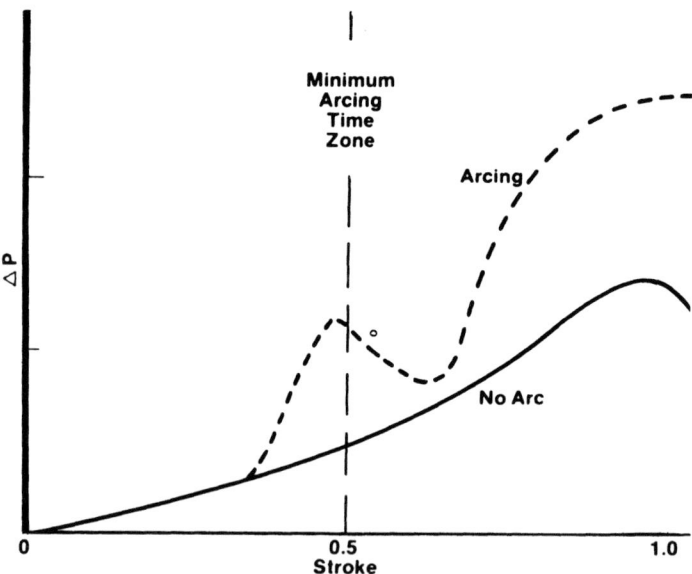

Figure 10.10 Differential pressure recorded during a 50-kA short-circuit test.

Single-Pressure SF$_6$ Circuit Breakers

Differential pressure as a function of contact overlap

When the overlap distance is increased, the differential pressure at contact parting is increased, but in practice at minimum arcing time under short circuit this increase is negligible. The main advantage of increasing the overlapping is an increase in contact parting speed, which may help the characteristics of the breaker in line charging conditions, but an increase of the overvoltage factor in small inductive current interruption may be expected due to higher current chopping, and a longer opening time has to be considered.

The six conditions examined above correspond to design parameters defined under cold conditions and with a theoretical total blocking of the nozzle. In real conditions, Fig. 10.10 shows that at half the stroke in the region of minimum arcing time, the upstream pressure is equal to nearly three times the adiabatic upstream pressure under cold conditions. Also, the real differential pressure will depend considerably on the pressure obtained downstream at the instant of interruption. This fact has led some designers again to consider self-blast interrupting chambers and development is in progress now, mainly in Japan [4,11].

As the deionization process has to be carried out between the instant of unclogging of the nozzle and the highest voltage stress appearing between the contacts, the time lapse available is of the order of 500 μs. In such a case, the quantity of clean gas stored has little importance.

The design of the nozzle and the exhaust path of the fixed contact (or contacts in a double flow) downstream regulate the differential pressure and the efficiency of the interrupting complex. In conclusion, in a single-flow puffer (for a double-flow interrupter the proportions may change slightly), a 10:1 ratio between the displaced volume and the dead volume is a reasonable value, considering the dimensions of the arcing contacts, the gas flow sections, and the power needed to operate the complete interrupter.

The contact parting speed is a function of the contact overlapping and should be determined for line-charging operations without restrikes. The stroke speed depends on the interrupting time chosen and the maximum arcing time expected. For this reason, the stroke speed cannot be increased indefinitely for a given design.

The increase in interrupter compressing chamber diameter, the stroke length, and the rated pressure seem to have little influence on the characteristics of an interrupter if the displaced volume and the nozzle throat have been adequately chosen. The breaking capacity efficiency factor depends mainly on nozzle design and upstream and downstream gas flow design features.

In the case of a double-flow puffer design, the upstream pressure generated by the arc may reach lower values than in a single-flow design, but downstream pressure may be lower and the resulting differential pressure may be higher. The concepts examined are, however, valid generally. The theoretical factors examined are not able to determine design completely. The complexity of arc interruption does not allow simple design formulas. Only test results can define a solution and its electrical characteristics. However, these considerations are useful, as the designer has to take into account a certain number of theoretical hypotheses to avoid loss of time and excessive testing.

Differential pressure at zero gauge pressure

The question of what happens in a single-pressure SF_6 circuit breaker when the pressure drops to atmospheric value is raised continually. In the case of an air-blast or double-pressure SF_6 circuit breaker, total loss of stored pressure is followed by total loss of breaking capacity and partial loss of insulation. The same consequences arise if an oil circuit breaker loses its oil.

In a single-pressure SF_6 circuit breaker, the results are different. At the end of the stroke, the differential pressure is given by equation (10.4). For the case considered, $P = 10 P_n$. At half the stroke in the region of minimum arcing time $P_{1/2} = 0.833 P_n$ [see equation (10.16)]. If the rated pressure of the circuit breaker considered was 5 atm, the differential pressure varies from 0 to 50 atm under ideal conditions. At atmospheric pressure, the differential pressure will vary from 0 to 10 atm, with the nozzle blocked. Figure 10.11 shows this variation for the considered case. However, taking into account the pressure generated by the arc column upstream, as shown in Fig. 10.9, a higher differential pressure will be obtained and a partial breaking capacity will still be reached. It is difficult to evaluate the percentage value of reduction, as it depends on the particular design of the inter-

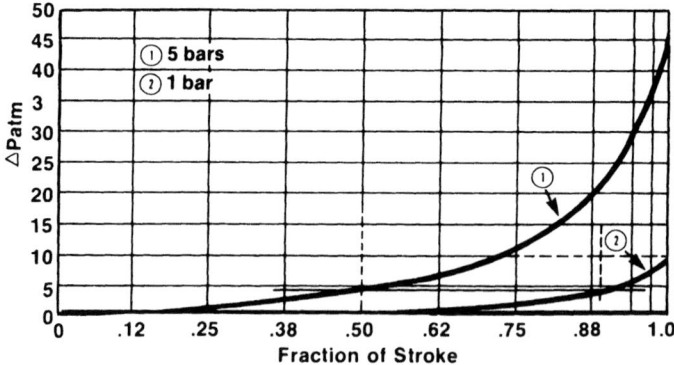

Figure 10.11 Differential pressure as function of fill pressure and stroke.

Single-Pressure SF$_6$ Circuit Breakers

rupter considered, but experimental results on a particular design have shown interruptions at 50% of the rated breaking capacity. There will be no problem in interrupting the normal load current; however, it must be noted that the insulation between contacts will be reduced and line dropping operations under such conditions are not recommended, due to the danger of restrikes.

10.3.2 Nozzle Design Concepts

Other important elements in the design of a single-pressure SF$_6$ circuit breaker are the shape of the nozzle and the dimension of the contacts. The interruption of a short-circuit current is characterized by an arc of high intensity appearing between the contacts with natural moments of easy extinction every half cycle. When interruption is obtained, a high recovery voltage appears between the contacts and has to be withstood. A typical oscillogram of such an interruption is shown in Fig. 10.12. Inside the nozzle, the arc plasma column generated changes its diameter as a function of the instantaneous current i_a. At its peak it can also have a temperature of the order of 20,000 K or higher.

The cold flow and adiabatic pressures inside the nozzle are greatly modified. Upstream, a stagnant region of hot gases is created. The pressure in the region of the contacts, due to the high temperature, increases and a negative flow may be observed which has to be compensated and swept away by the gas flow generated by the pressure obtained in the compression chamber. If this sweeping is not obtained, the hot ionized gas may break down under the transient recovery voltage stress and a definite interruption limit may be reached. For a given circuit breaker design, the upstream generated pressure is a function of stroke length. Under short-circuit conditions the effective upstream pressure generated depends on the stroke length at current zero. This upstream pressure will allow the sweeping of the arc path between the contacts and determine the minimum arcing time necessary for a successful interruption.

The factors dealing with the upstream pressure buildup have been examined and are important, but the amount of transient recovery voltage withstand will depend on the differential pressure ΔP obtained and therefore on the downstream flow and downstream pressure reached. The downstream flow depends on the diameter of the nozzle, the downstream divergence of the nozzle, and the size and shape of the contacts. The quicker the ionized gas portion is cleared, the higher the transient recovery voltage that can be withstood.

Two natural facts help the achievement desired. The arc column, when surrounded by a stagnant zone, is very conductive and so has a low arc voltage and consequently a relatively low arc energy, and within the plasma column, the plasma density is much lower than that of the surrounding gas. A much higher axial velocity is reached within the

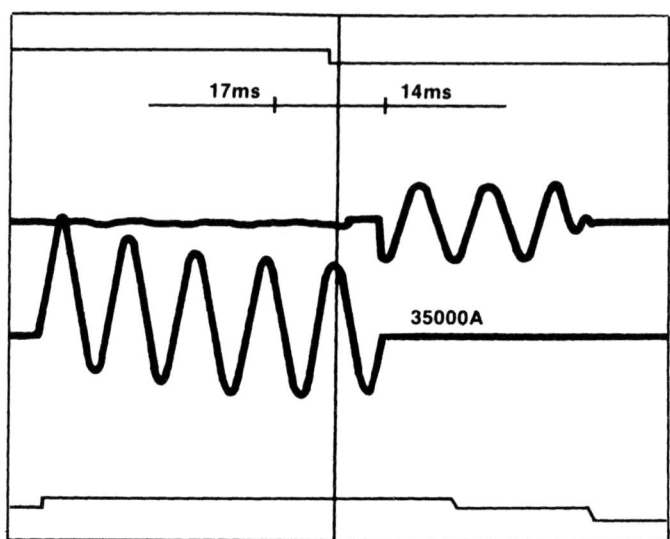

Figure 10.12 Typical oscillogram of a symmetrical short-circuit interruption

column and a continuous outflow of ionized particles is expelled naturally downstream. When the instantaneous current i_a reaches a value permitting the upstream gas to flow through the nozzle, hot gas will be involved at first, but a less conducting cooler and denser medium is eventually introduced between the contacts.

Scientific studies have determined the plasma column diameter [10, 12]. It is proportional to the square root of the peak value current multiplied by a factor k_p dependent on the surrounding pressure. In an industrial breaker, a column carrying a current of 40 kA can reach up to 25 mm in diameter, but at current zero, calculations and experimental data show a residual diameter of the arc column of the order of 1 mm. In this condition the temperature and thus the conductivity of the column drop very quickly, helping the process of interruption.

A typical recording of the arc voltage is shown in Fig. 10.13. Measurement of the parameters i_{ta}, ta, and Uc allows one to compare the severity of a circuit and different concepts of an interrupter. These parameters can be changed by modifying the nozzle geometry and the downstream flow. For a given design, they change as a function of the upstream pressure and are therefore a function of the arcing time.

Sometimes it is difficult to measure the parameters considered. To compare designs a practical method consists of establishing a circuit and

Single-Pressure SF_6 Circuit Breakers

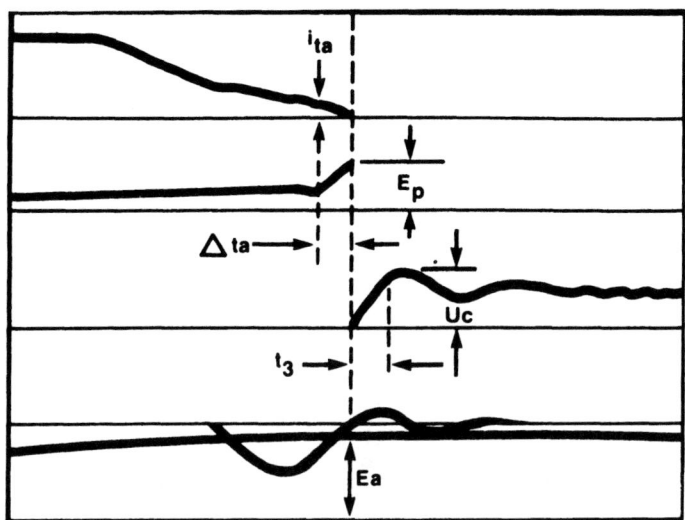

Figure 10.13 Typical recording of arc voltage.

determining the minimum arcing time of the chosen solutions. The shortest arcing time will show which interrupter is the most efficient. Considering a single-flow puffer interrupter, the design parameters that have an influence on the minimum arcing time are shown in Fig. 10.14. They can be determined by computer calculations. All the dimensions mentioned are interrelated and are functions of the maximum current to be interrupted. ϕ_N, for example, should be equal to approximately $0.8 \times I_{S/C}(kA)$ in mm. β_N also has to be determined as a function of the current to be interrupted, but a good value has been found to be about $0.3\phi_N$. Many experimental studies have been made to determine the angle α. In industrial breakers today, good experimental results have been obtained with a value of 15°. The diameter d_C should be as large as possible and L_C as short as possible to increase the downstream flow before interruption.

L_N, d_N, and L_H are more difficult to determine, but the total section of the d_H opening should be of the order of $(3/4)\pi\phi^2 N$ and $L_N \cong 2\phi$. γ_N should be around 2ϕ and L_H should not interfere with the minimum section of the nozzle. The same parameters should be considered in a double-flow concept. The inner diameter of the upstream hollow contact should be as large as possibly by design, as the tubular section is longer than that of the corresponding downstream contact (see Figs. 10.5 and 10.6).

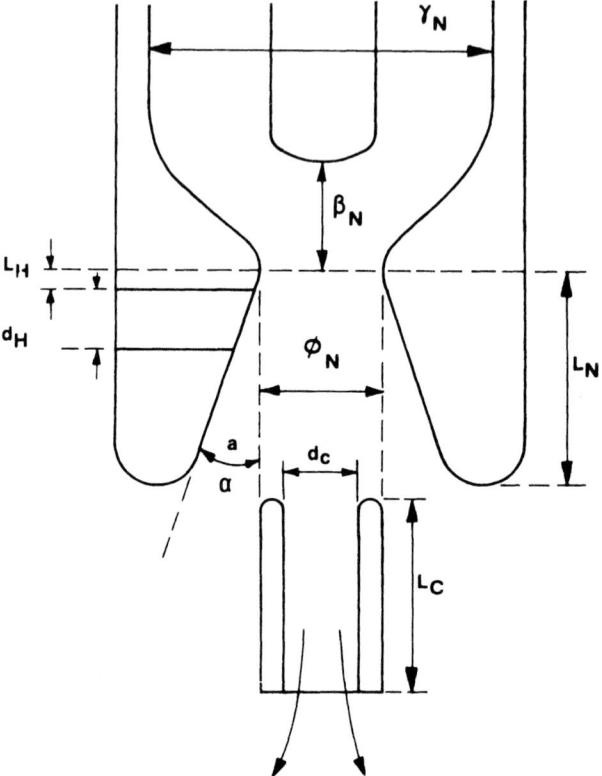

Figure 10.14 Typical single-flow interrupter nozzle.

The main concepts to follow are:

1. Choose ϕ as small as possible to avoid the "hiding effect," the tendency of the arc column to avoid the zones of maximum deionization or blast action [13].
2. Optimize plasma flow out of the nozzle before zero current.
3. Reduce ionization section upstream without reducing the upstream pressure to optimize a quick dielectric recovery after the current zero pause.
4. Increase downstream exhaust volume but control the position of the arc column to avoid the natural hiding effect.

The choice of the relations between all these factors determines the efficiency of the interrupter. The efficiency of an interrupter has to be judged around the minimum arcing time where generally the reigni-

Single-Pressure SF$_6$ Circuit Breakers

tion is of the thermal type) i.e., at a relatively low value of transient recovery voltage), but there is also a limit in peak transient recovery voltage which has to be determined. Generally, it is the maximum arcing time, which presents more risks of dielectric reignition; and the length of the stroke, the speed of the contacts, and the ionized gas control have to be examined to determine the limit of maximum voltage withstand.

10.4 ELECTRICAL INTERRUPTION CHARACTERISTICS OF A PUFFER CIRCUIT BREAKER

It has been noted during short-circuit tests that a particular characteristic of the arc voltage has been attributed to the functioning of the nozzle in a puffer circuit breaker. It can be seen as a sudden drop in arc voltage a few hundreds of microseconds after the zero pause, a steady portion at a relatively lower value, and a sudden rise a few hundreds of microseconds before the next zero pause. From there it reaches the extinction point.

Figure 10.15 shows the parameters considered necessary to compare test circuits and solutions. They are:

- t_a = arcing time
- U_a = stabilized arc voltage
- EP_a = extinction point arc voltage
- E_a = arc energy
- Δt_a = time between the instant of rapid change in arc voltage shape and current-zero pause at interruption
- i_{t_a} = instantaneous current at the instant of rapid change in arc voltage shape before the current-zero pause at interruption
- Δp_a = puffer piston differential pressure at the instant of interruption
- Δp_F = end-of-stroke differential pressure

The transient recovery voltage (TRV) and the initial transient recovery voltage (ITRV) determine the dielectric severity of the circuit. The peak voltage u_c defines the high dielectric reignition zone, but the minimum arcing time is frequently determined by the capability of the breaker to deal with the ITRV in the first or second microsecond after current zero. To compare results of different designs and the severity factor of a test circuit, the following parameters have been defined:

- $u_{c\,max}$ = maximum voltage the interrupting chamber for a definite short circuit is able to withstand after a time t_2
- t_2 = time after current zero to reach $u_{c\,max}$
- U_{i2} = instantaneous value of the inherent initial transient recovery voltage (ITRV) 2 µs after current zero
- U_{e2} = instantaneous value of the recorded initial transient recovery voltage (RITRV) 2 µs after current zero

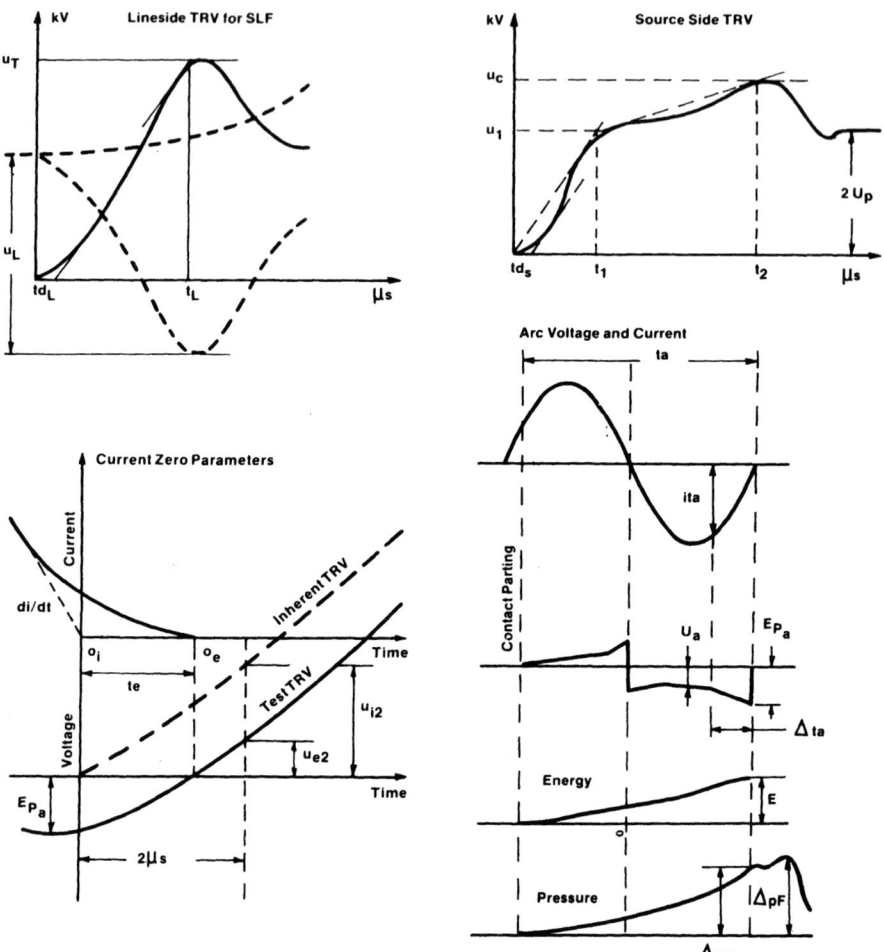

Figure 10.15 Definition of a test parameter.

The value of U_{e2} gives the measure of the breaker reaction when compared to U_{i2}. These values can be obtained only from CRO recordings having sweep speeds of 1 or 2 µs/cm and 1 or 2 kV/cm voltage sensitivity.

To be able to judge the efficiency of a solution as a first approximation without a great number of tests or testing circuits, an interrupting capability coefficient has been established. It is

$$\gamma_r = \Delta p_a \times \Delta t_a \times K$$

Single-Pressure SF$_6$ Circuit Breakers

where K is a proportionality coefficient. Figure 10.16 shows the values recorded for an industrial circuit breaker during short-line fault tests. They confirm the fact that the arc energy increases at a lower rate than the interrupting capability coefficient for 6/10 of the stroke and diminishes afterward until the end of the stroke. However, all short-circuit test results at maximum short-circuit capability have to be adjusted as a function of the interrupting chamber characteristics for capacitive current interruption without chopping, good out-of-phase interruption, and exceptional mechanical reliability. A best compromise in safety margins has to be determined. A circuit breaker in service has to deal more often with line charging operations, magnetizing current interruption, and generally short-circuit currents of 10, 30, and 60% of the breaking capacity of the circuit breaker. It is advisable to increase the margin of safety of these types of operations and to be sure of a great mechanical reliability under all ambient conditions and consistency of electrical operations without maintenance.

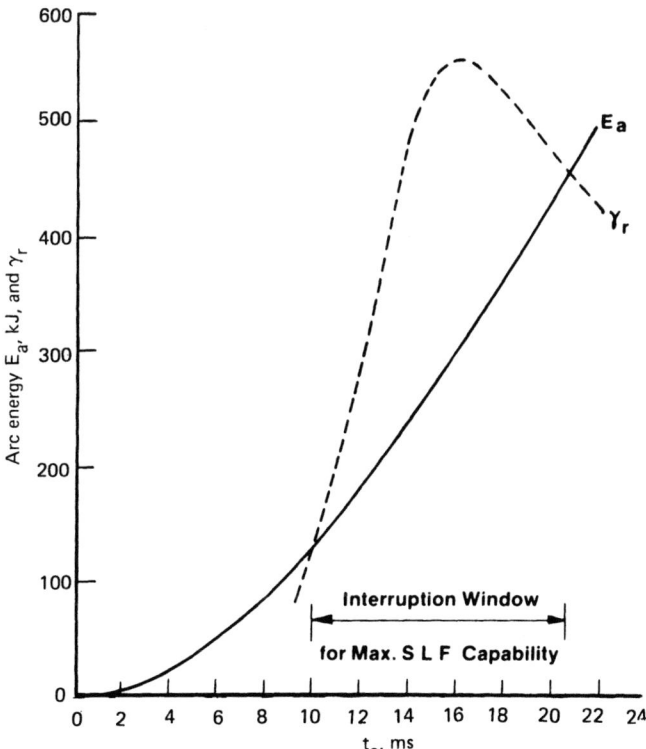

Figure 10.16 Interrupting capability curve as a function of arcing time recorded during short-line-fault tests.

10.4.1 Operating Characteristics of Breaks in Series

Even though a puffer SF_6 interrupting element is able to operate at higher voltages than any other known interrupter, it is necessary to connect in series several of these units for EHV circuit breakers. Capacitors connected in parallel with the breaks are necessary to ensure voltage division between the breaks during the buildup of TRV and when breaker is in the open position. If the high-power laboratory where the tests are made is able to apply the stresses requested on a complete pole of a circuit breaker, either in direct or synthetic tests, no problem is foreseen. However, this is generally not possible and unit test methods are used. The results of the unit tests are then extrapolated.

During field tests it was noticed that the interrupting time of certain types of circuit breakers was greater than expected and erratic, and with others was generally shorter than forecasted. The study of such phenomena led to what was called the interaction between breaks in series. The comparison of one break, two breaks, and four breaks in series has been checked in a laboratory [14]. The results are summarized in Table 10.1. They were obtained with a conventional SF_6 puffer circuit breaker, using the maximum current available the day of the tests and using a SLF test circuit. By deliberately changing the contact parting time spread up to 1.5 ms and the voltage division up to 56%, no noticeable changes in the interruption limit results were recorded. However, over the limits indicated, an increase in minimum arcing time was noticed.

Figure 10.17 shows the separate recording of arc voltage and TRV made on each break around minimum arcing time. This phenomenon disappears when the arcing time is greater. The interaction is therefore very effective in what is called the thermal reignition zone. At minimum arcing time the interrupting power of the interrupting element is at its lowest value and thermal reignition is most likely [15]. With two breaks connected in series, for example, the conditions for thermal reignition of the breaks do not occur at the same instant and therefore their interaction results in a lower post arc conductivity on one of the breaks. That break can withstand a higher value of initial transient recovery voltage (ITRV) and therefore reach its interrupting power earlier in the stroke than expected.

As a practical result, for an SF_6 puffer circuit breaker the extrapolation of the unit test results does not present any danger. The interruption characteristics of the breaker obtained during unit tests are not only maintained but are improved. Also, the interrupting time recorded during the unit tests can be reduced by 1 to 2 ms in predicting the interrupting time of a complete pole [14].

Single-Pressure SF$_6$ Circuit Breakers

Table 10.1 Comparison of Multibreak with Single Break SLF Interrupting Performance

I kA	Number of elements in series	Z Ω	RRRV, kV/μs	Static distribution maximum stress per element, %	Contact parting spread, ms	Minimum arcing time, ms
30	2	450	5.8	51	0	11.5
30	1	230	3	—	—	13
30	4	495	6.6	28	0	9.5
30	1	135	1.8	—	—	10.5

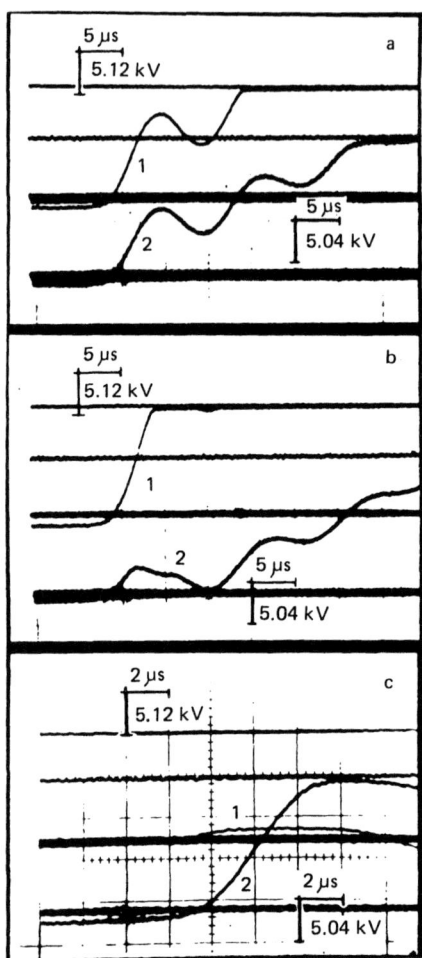

Figure 10.17 Typical behavior of a two-unit SF_6 single-pressure circuit breaker. (a) Dynamic voltage distribution between the two units with an arcing time higher than the minimum; (b) dynamic voltage distribution between the two units at the minimum arcing time; (c) dynamic voltage distribution between the two units at the minimum arcing time with a shift in synchronization of 3 msec. (From Ref. 14.)

10.4.2 Interruption of Asymmetrical Currents

In a puffer circuit breaker an asymmetrical current affects the interruption process in a more complex manner than with any other inter-

Single-Pressure SF$_6$ Circuit Breakers

rupter. Figure 10.18 shows the minimum and maximum arcing times for symmetrical short-circuit current in typical conditions. It can be established that

$$t_{amax} = (t_{amin} - \varepsilon) + 1/2 \text{ cycle}$$

For practical testing conditions ε is determined to equal 1 ms.

Figure 10.19 shows the minimum and maximum arcing times in typical asymmetrical conditions. The asymmetrical minimum arcing time is shorter than the symmetrical minimum arcing time, as the actual current flowing between the contacts is smaller. However, the asymmetrical maximum arcing time is longer than the symmetrical arcing time. It is

$$t_{amax\,asym} = (t_{amin\,asym} - \varepsilon) + t_{ml}$$

t_{ml} is the time between two consecutive zeros of a major loop which is greater than 1/2 cycle. In this case, the conditions at zero current are more favorable to achieve interruption. The di/dt is smaller, allowing a deionizing time Δt_a to be longer, the upstream pressure and consequently the differential pressure Δp are higher due to the longer arcing time, and the peak of the transient recovery voltage is lower [16]. However, the arc energy is greater. It is of the order of 1.5 times that for similar symmetrical conditions [14].

The probabilities of a thermal reignition are smaller, but due to the greater amount of ionized gas produced, the absence of a dielectric

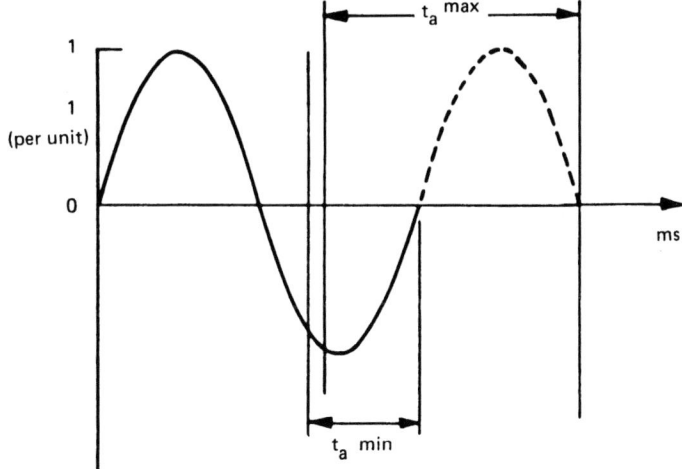

Figure 10.18 Minimum and maximum arcing time in symmetrical conditions.

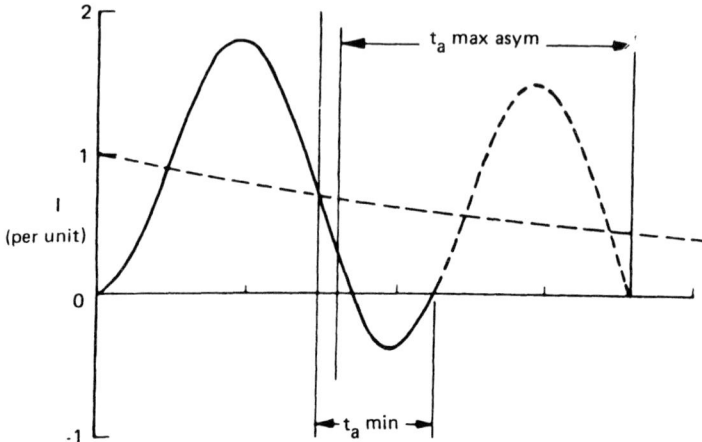

Figure 10.19 Minimum and maximum arcing time in asymmetrical conditions.

breakdown has to be checked by test even though theoretically the asymmetrical breaking capacity of a puffer circuit breaker is greater than with any other known apparatus.

10.5 MECHANICAL AND ELECTRICAL RELIABILITY

A power circuit breaker is one type of apparatus that nobody wants but everyone needs. It is therefore essential to design it with the concept that once installed, it will be forgotten. Statistics show that over 80% of in-service problems are due to mechanical failures. Mechanical reliability is a direct function of simplicity and single-pressure SF_6 circuit breakers present today a good compromise.

It is not realistic to consider the mechanical life of a power circuit breaker related to a fixed number of operations, whatever that number is. Except for special cases, an 800-kV circuit breaker protecting an EHV network should not be operated as much as a 145-kV transmission breaker. The number of mechanical operations should be related to the rated voltage. The formula

$$L_T = \frac{10^5}{\sqrt[3]{U_N}}$$

should cover the lifetime operations of 20 years of service (see Fig. 10.20).

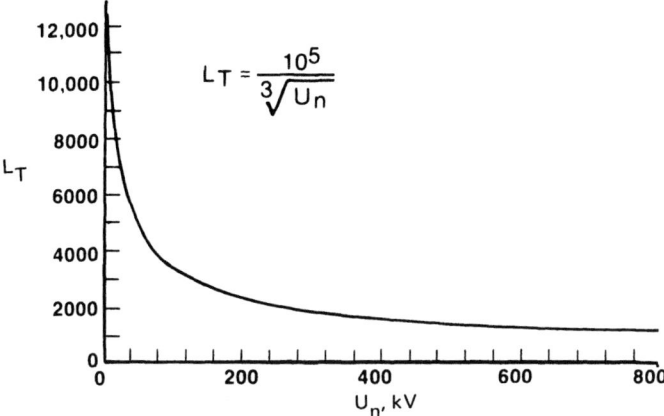

Figure 10.20 Number of mechanical operations without maintenance as a function of rated voltage.

A great number of mechanical operations during life tests cover the wear of the circuit breaker components, but it is not the only parameter to consider for reliability. The "reliability" factor in any ambient conditions and at any time is of primary importance. No standardized test procedure has been established so far and only design considerations and partial testing can give a feeling for reliability in service.

The number of moving parts, their weight, and their kinetic energy influence the stresses involved. Cold temperature operations, effect of corrosion, and a great number of monitoring devices are also of great concern. But simplicity of design remains the dominating factor.

In a single-pressure SF_6 circuit breaker, the conducting parts are of great simplicity. They can be compared to those of a disconnect switch. The only factor that may have a negative effect is the uprating of continuous current capability and the corresponding increase of moving masses. To design for higher continuous currents than needed may decrease the margin of mechanical safety.

EHV circuit breakers use a number of interrupting elements connected in series. In most cases, testing laboratory power available does not allow us to test a full pole of a circuit breaker. Unit testing is the procedure adopted, but to be able to extrapolate the unit test results, it is essential to have consistent synchronization between the interrupting elements connected in series [15]. The linkage connection between them and the operating mechanism should not allow variations in time. Solid linkage connections are considered one of the best solutions if, by reduced tolerances of manufacturing, linkages without adjustments are used.

Three types of operating mechanisms are used today:

1. Spring-operated mechanisms
2. Hydraulic operating mechanisms
3. Compressed air operating mechanisms

Figure 10.21 shows the spring-operated mechanism used by the Swiss firm Sprecher & Schuh. It was designed for the small oil volume circuit breaker and applied with small modifications to the single-pressure SF_6 circuit breaker. It has a good record of reliability, its noise level is very low, but it has a limitation in power and cannot be universally applied.

Figure 10.22 shows a hydraulic operating mechanism. It is used by a great number of European and Japanese power circuit breaker manufacturers. It is able to operate any type of circuit breaker. Its complexity and cost, however, limits its use to powerful circuit breakers needing an operating mechanism of high energy. It is noiseless and can be adapted to apparatus with mechanical linkages or hydraulic links.

Figure 10.23 shows the concept of a low-pressure compressed air mechanism developed by Westinghouse. This type of mechanism has the advantage of simplicity and does not have any latching device. It has two rest positions and is of great reliability. Its noise level can be limited to 100 dB. Being at ground level, it does not need special dried air. The number of operations available without the help of the compressors is a function only of the size of the air storage reservoir.

The electrical reliability of a circuit breaker is also an essential factor. The test results obtained in a laboratory show the soundness of the design concept but do not ensure the consistency of the results obtained in service. CIGRE study committees and IEC working groups are proposing solutions to the problem.

One of the proposals consists in doing the types of tests prescribed by the electrical standards in a predetermined sequence as per Fig. 10.24. These sequences stress the circuit breaker alternately electrically and dielectrically without maintenance between the tests. It does not call for supplementary power tests but it ensures the most frequent duties of a circuit breaker to be sure that the characteristics claimed are not modified by use. To test the circuit breaker dielectrically after the power tests is of great importance. The insulation coordination of the network must not be jeopardized by a used circuit breaker. The same applies to the current-carrying properties.

These new concepts of rating a circuit breaker allow reduction of the requested characteristics and reduce the margins between ratings and needs. They permit simplifying the designs to obtain reliability of operation and consistency of characteristics.

Single-Pressure SF$_6$ Circuit Breakers

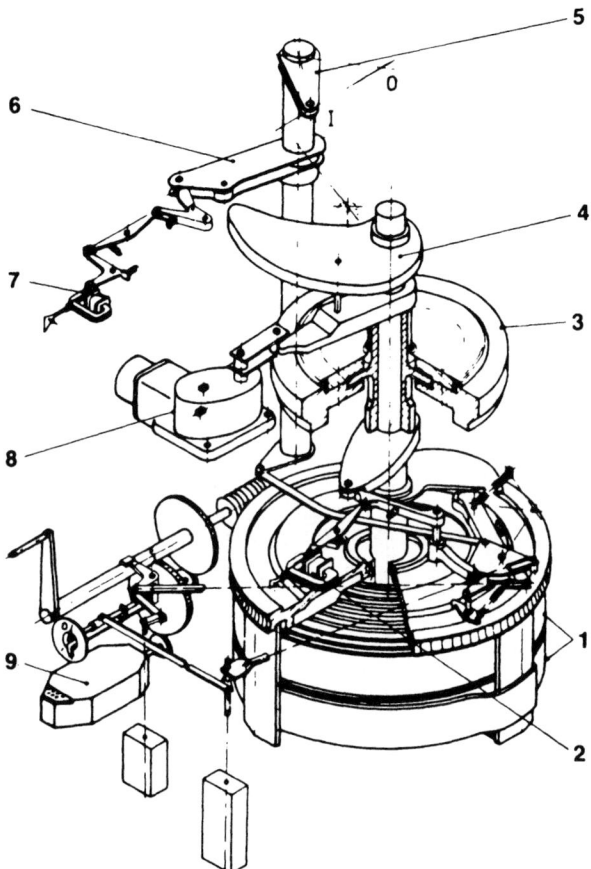

Basic Sketch of KFK 1-5 and FKF 1-6 Motor Wound
Spring Operating Mechanism

1 Closing spring   6 Roller lever
2 Closing coil   7 Trip coil
3 Flywheel   8 Dashpot for closing
4 Cam disk   9 Spring-winding motor
5 Operating lever

Figure 10.21 S & S spring-operated mechanism. (Courtesy of Sprecher and Schuh, Ltd., Aarau, Switzerland.)

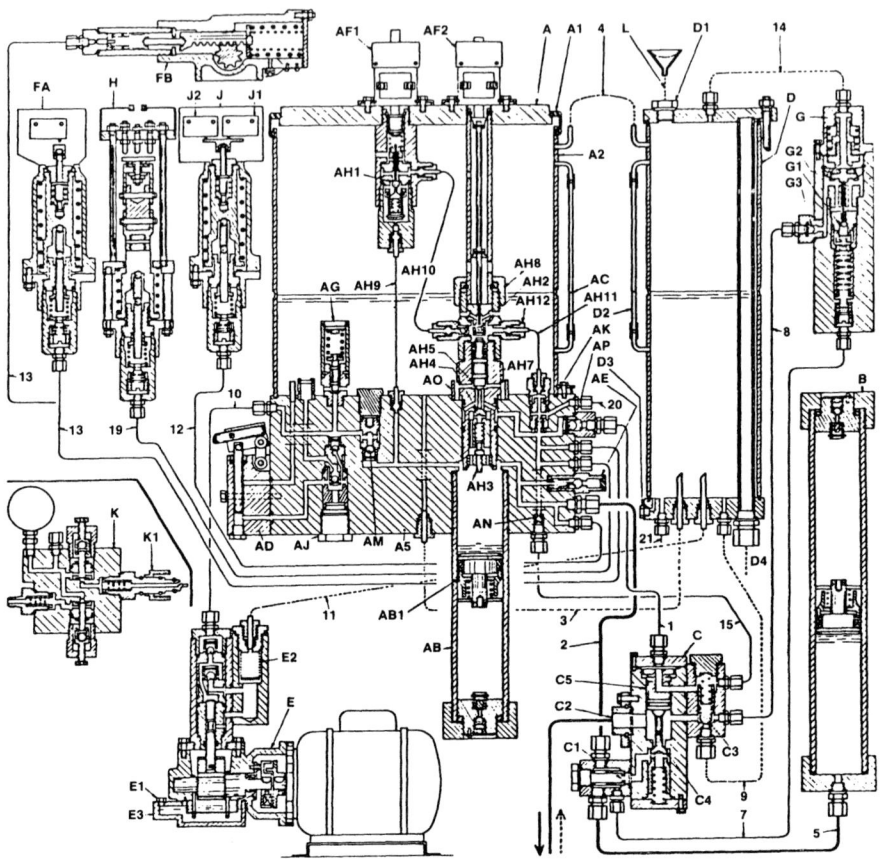

Figure 10.22 Cross-sectional drawing of a hydraulic operating mechanism (1960). (Courtesy of ETNA, Argenteuil, France.)

10.6 INDUSTRIAL CIRCUIT BREAKER CONFIGURATIONS

Since 1965, the use of single-pressure SF_6 circuit breakers for 72 to 800 kV has been rapidly expanding. Their breaking capacity has reached 63 kA on an industrial basis and prototypes of up to 100 kA breaking capacity have been tested successfully. All leading power circuit breaker manufacturers have developed single-pressure SF_6 circuit breakers. They can be divided into two main design categories.

Single-Pressure SF$_6$ Circuit Breakers

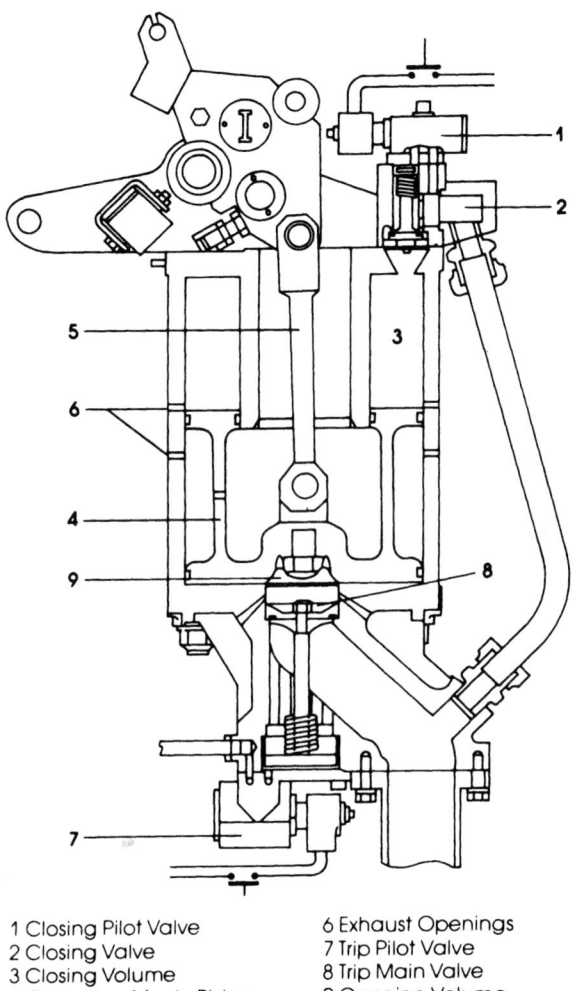

1 Closing Pilot Valve
2 Closing Valve
3 Closing Volume
4 Operating Mech. Piston
5 Connecting Rod
6 Exhaust Openings
7 Trip Pilot Valve
8 Trip Main Valve
9 Opening Volume

Figure 10.23 Westinghouse MWE250 compressed-air operating mechanism. (Courtesy of Westinghouse Electric Corporation, Trafford, Pa.)

10.6.1 Live-Tank Designs

The main characteristic of live-tank designs is having their insulation to ground constituted by porcelain. Figure 10.25 shows a typical 245-

			% of Rated S/C Current	
Type of Test	Duty Cycle	Type of Stress	Partial	Integrated
1. 100% Symmetrical bus fault	0-θ-CO-t-CO	Thermal	300	300
2. 100% Asymmetrical bus fault	0-θ-CO-t-CO	Thermal	300	600
3. 90% Short line fault	0-0-0	Thermal	270	870
4. Voltage drop measurement	Recording of the voltage drop between contacts to ascertain its continuous current capabilities.			

			% of Rated S/C Current	
Type of Test	Duty Cycle	Type of Stress	Partial	Integrated
1. 75% Short line fault	0-0-0	Thermal	225	225
2. 60% Short line fault 3. Line charging currents	0-0-0 12 x 0	Thermal & Dielectric Dielectric on	180	405
4. Magnetizing currents	12 x 0	a used	10	415
5. Inductive currents	12 x 0	interrupter		
6. Out of phase switching	0-0-0	Dielectric on a c.b. already stressed thermally and dielectrically	75	490
7. 60% Symmetrical bus fault	0-θ-CO-t-CO		180	670
8. 30% Symmetrical bus fault	0-θ-CO-t-CO		90	760
9. 10% Symmetrical bus fault	0-θ-CO-t-CO		30	790
10. B.I.L. withstand	75%	Dielectric		

Figure 10.24 Electrical endurance tables.

kV live-tank circuit breaker. These circuit breakers can be equipped, if needed, with closing resistors. Figure 10.26 shows a pole of a 550-kV circuit breaker equipped with closing resistors in a high-voltage laboratory.

The constant development of SF_6 puffer circuit breakers has permitted the increase of voltage per break up to 300 kV. Circuit breakers using one interrupting element per phase at 245 kV and two per phase for 420 kV have been installed. Figure 10.27 shows a 420-kV 40-kA 2500-A circuit breaker during mechanical life tests.

The live-tank circuit breaker design is generally appreciated for its reduced cost and modular construction, which reduces inventory for both manufacturer and user and permits component testing. They also can be shipped in relatively small crates and during erection the largest weight to be considered for lifting is never greater than 1500 lb.

Figure 10.25 245 LWE40 circuit breaker by WESA of Spain. (Courtesy of Westinghouse Spain, Córdoba, Spain.)

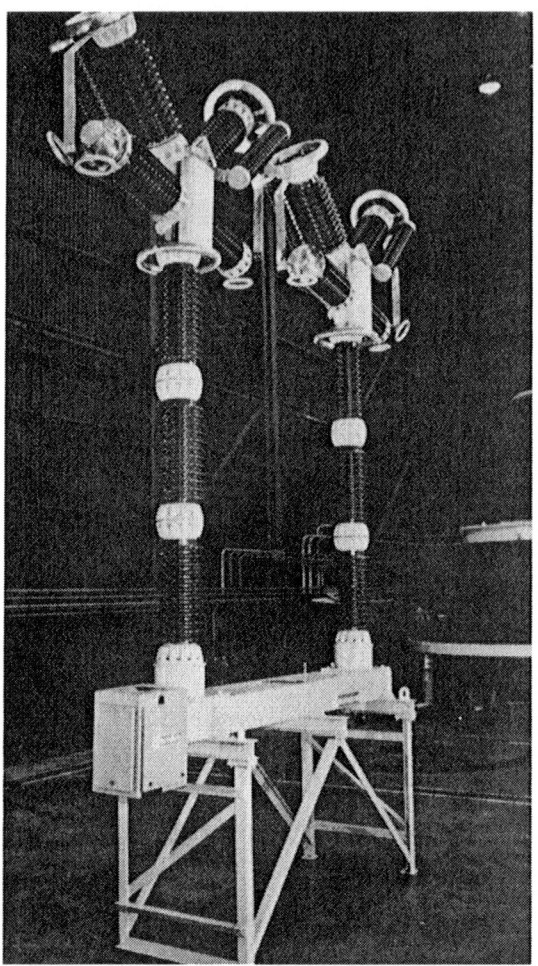

Figure 10.26 Westinghouse 550 LWER 40 circuit breaker during high-voltage tests. (Courtesy of Westinghouse Electric Corporation, Trafford, Pennsylvania.)

10.6.2 Dead-Tank Designs

SF_6 dead-tank circuit breakers have been designed following the concepts of bulk oil or GIS circuit breakers. Two examples of SF_6 dead-tank circuit breakers are shown in Fig. 10.28. They have been very successful in the United States and in countries where bulk oil circuit breakers were favored.

Single-Pressure SF$_6$ Circuit Breakers

Figure 10.27 420 LWES40 circuit breaker during mechanical tests. (Courtesy of Westinghouse Electric Corporation, Trafford, Pennsylvania.)

The main advantages of a dead-tank circuit breaker are:

1. Quick erection at the site because it is transported as a completed, ready-to-operate unit
2. Lower overall height, as shown by Fig. 10.29.
3. Reduced supporting structures; no current transformer pedestals
4. Integral current transformers of very simple design
5. Shorter and therefore lighter insulating operating rods

Figure 10.28 242SP63 dead-tank puffer circuit breaker and 242SF40 double pressure SF$_6$ circuit breaker installed. (Courtesy of Westinghouse Electric Corporation, Trafford, Pa.)

Single-Pressure SF$_6$ Circuit Breakers

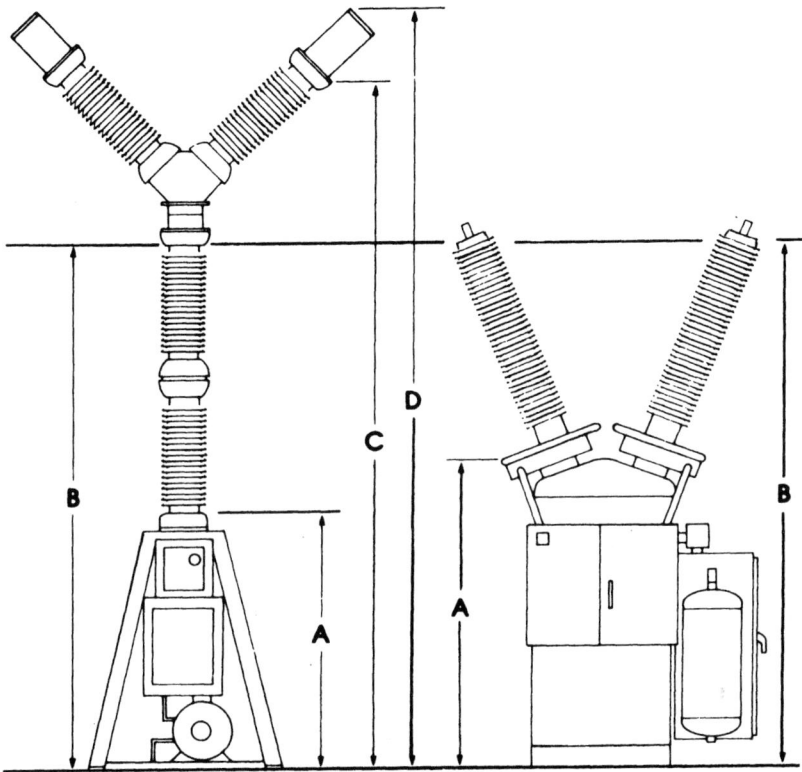

Figure 10.29 Comparison of overall dimensions between a live-tank circuit breaker and a dead-tank circuit breaker.

6. Insulation between terminals independent of ambient density or pollution
7. Better seismic withstand capability

Their main disadvantage is high cost due to linkage complexity and tank dimensions.

To maintain the advantages of dead-tank circuit breakers, minimizing their drawbacks, a new generation of single-pressure SF$_6$ dead-tank circuit breakers has been designed. Figure 10.30 shows the section of the interrupting module, and Figure 10.31, a three-phase assembly. This Westinghouse DWE design is based on a concept which allows keeping the interrupting components as near as possible to a live tank conventional circuit breaker, to reduce the grounded tank volume and to utilize integral toroidal current transformers.

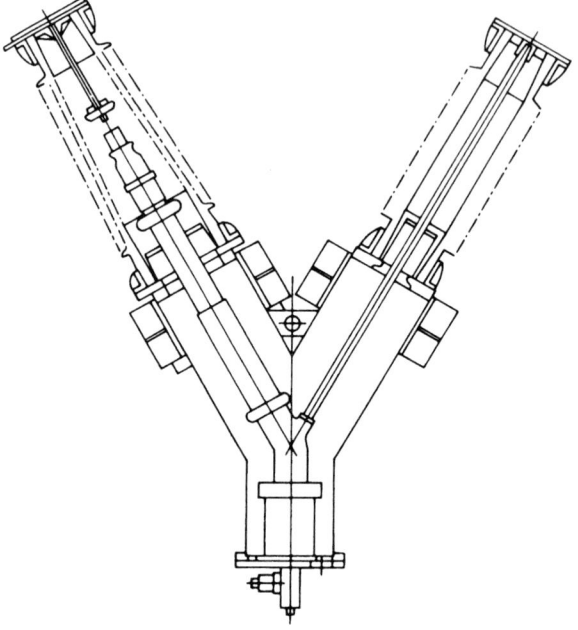

Figure 10.30 Cross section of a Westinghouse DWE-type circuit breaker pole.

   This concept enables the range from 121 to 245 kV to be covered with a single interrupting element, and its technical advantages allow room for future developments. The introduction of the interrupting chamber inside the porcelain instead of placing it inside the tank eliminates the insulating shields and gas coolers necessary to avoid flashover to ground during the interruption of heavy short-circuit currents. It separates the insulation to ground function from the interrupting components.
   The dimensions of the tank can be optimized to reduce tank dimensions as the conducting parts are static, regular, concentric, and of small diameter. The reduced dimensions of the tank permit standardizing its dimensions up to the 245-kV 1050-BIL class.
   The electric field between the contacts is practically identical to the one obtained in a live-tank design and very different from that of a conventional dead tank, as shown by Fig. 10.32. Similar results are obtained when the polarity of the live side is changed.
   One of the weaknesses of dead-tank circuit breakers or gas-insulated substation (GIS) apparatus is the solid insulation to ground. The

Single-Pressure SF$_6$ Circuit Breakers

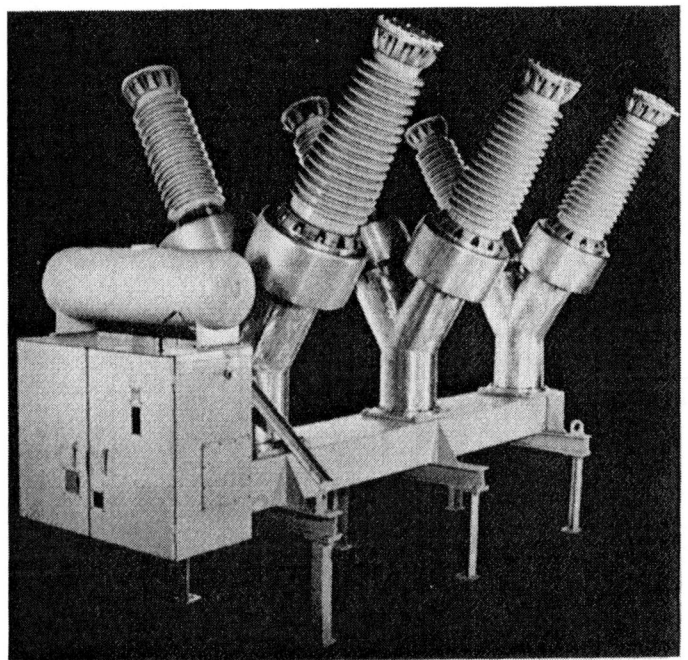

Figure 10.31 145 DWE40 circuit breaker. (Courtesy of Westinghouse Electric Corporation, Trafford, Pennsylvania.)

insulating parts have to be of reduced dimensions and may represent the weakest point of the insulation. A flashover along an insulator or an insulating operating rod even during dielectric tests may cause permanent damage if the insulation material has no self-restoring properties. In gas, a flashover has no negative effects. Gas has self-restoring insulation properties and as the arc energy is small during the dielectric tests, the conductors are not heavily scarred.

As can be seen in Fig. 10.30, the DWE breaker has only one support insulator and one insulated operating rod placed in a vertical plane, and its length does not change the tank diameter. The closed-position electric field, as shown in Fig. 10.33, is very regular and independent of the position of the insulating rod metallic ends. It is therefore relatively simple to reduce the dielectric stresses on the insulating parts and increase safety margins without jeopardizing the cost levels.

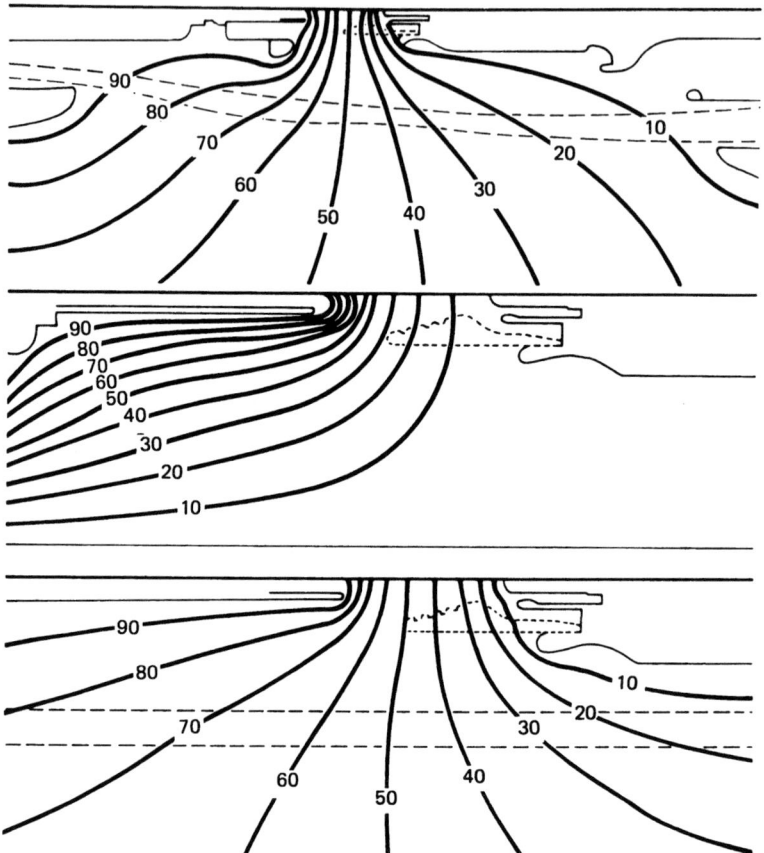

Figure 10.32 Field plot comparison of DWE dead-tank (top), conventional dead-tank (center), and live-tank (bottom) designs.

10.7 CONCLUSIONS

Single-pressure SF_6 circuit breakers have been developed all over the world. Today, they cover most of the needs of the electrical utilities up to 800 kV. Prototypes for the 1200-kV future grids are under development as shown by Fig. 10.34.

The success of the SF_6 puffer circuit breaker is due to the simplicity of its design and construction, which is one of the main components of reliability, and its consistency in characteristics after intensive use. These features render those circuit breakers practically maintenance free and only a technical breakthrough such as the development of static circuit breakers will be able to end their growth.

Single-Pressure SF₆ Circuit Breakers

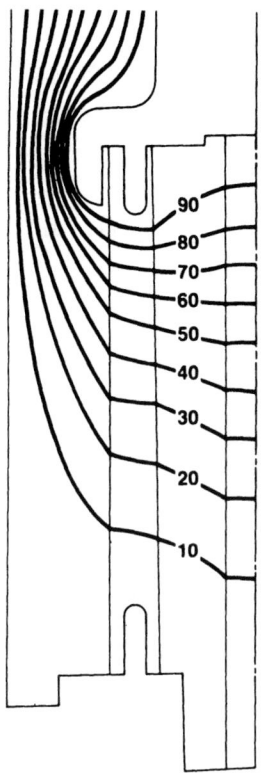

Figure 10.33 Field plot of the support insulator and operating rod of a DWE circuit breaker in the closed position.

10.8 ADDITIONAL CONSIDERATIONS FOR SEC. 10.3.1

10.8.1 Blocked Nozzle Cold Flow Pressure Rise

If, in equations (10.1) through (10.4), instead of the full stroke St, a partial stroke expressed as f × St is assumed, the compressed volume is not just Dv but includes the remaining cylinder volume $(1 - f)Sv$. Thus, equation (10.3) for the compression ratio now becomes

$$\frac{Sv + Dv}{(1 - f)Sv + Dv} = \frac{\pi r^2 St + Dv}{(1 - f)\pi r^2 St + Dv} = \frac{1 + \delta}{1 - f + \delta} \qquad (10.11)$$

where the dead volume ratio

$$\delta = \frac{Dv}{Sv} = \frac{Dv}{\pi r^2 St} \qquad (10.12)$$

Figure 10.34 1200-kV SF_6 puffer prototype circuit breaker. (Courtesy of Westinghouse Electric Corporation, Trafford, Pa.)

By inspection it can be seen that if the piston and cylinder radius is changed from r to r', and the dead volume Dv remains the same, equation (10.11) becomes

$$\frac{Sv + Dv}{(1-f)Sv + Dv} = \frac{(r'/r)^2 + \delta}{(r'/r)^2(1-f) + \delta} \qquad (10.13)$$

Single-Pressure SF$_6$ Circuit Breakers

As in equation (10.4), neglecting adiabatic heating of the gas, which is relatively small for SF$_6$, the blocked nozzle cold gas pressure rise is now

$$\Delta P'_f = \left[\frac{(r'/r)^2 + \delta}{(r'/r)^2 (1-f) + \delta} - 1 \right] P_n \tag{10.5}$$

With $\delta = 0.1$ and r increased by 20% ($r'/r = 1.2$), the full-stroke pressure rise is now

$$\Delta P'_1 = \left[\frac{(1.2)^2 + 0.1}{0.1} - 1 \right] P_n = 14.4 P_n \tag{10.14}$$

and at half stroke it is only

$$\Delta P'_{1/2} = \left[\frac{(1.2)^2 + 0.1}{(1/2)(1.2)^2 + 0.1} - 1 \right] P_n = 0.878 P_n \tag{10.15}$$

This compares with the original half-stroke pressure rise

$$\Delta P_{1/2} = \left(\frac{1 + 0.1}{1/2 + 0.1} - 1 \right) P_n = 0.833 P_n \tag{10.16}$$

an increase of only 5.4%.

If, instead of changing r, the stroke length St is changed to St', the new compression ratio becomes

$$\frac{Sv + Dv}{(1-f)Sv + Dv} = \frac{(St'/St) + \delta}{(St'/St)(1-f) + \delta} \tag{10.17}$$

and the new pressure rise is

$$\Delta P'_f = \left[\frac{(St'/St) + \delta}{(St'/St)(1-f) + \delta} - 1 \right] P_n \tag{10.6}$$

With a 10% increase in St and $\delta = 0.1$, the pressure rise at full stroke is

$$\Delta P'_1 = \left(\frac{1.1 + 0.1}{0.1} - 1 \right) P_n = 11 P_n \tag{10.18}$$

a 10% increase over the result of equation (10.3). At half stroke,

$$\Delta P'_{1/2} = \left[\frac{1.1 + 0.1}{(1/2)(1.1) + 0.1} - 1 \right] P_n = 0.846 P_n \tag{10.19}$$

which is only 1.5% larger than the half-stroke value by equation (10.16) for the original dimensions.

10.8.2 Pressure Rise With Gas Outflow But No Arc

Now, considering adiabatic compression, the cylinder pressure will be

$$P = \left(\frac{\rho}{\rho_0}\right)^\gamma P_n \tag{10.20}$$

where ρ/ρ_0 is the ratio of the gas density to its initial value and γ is the specific heat ratio C_p/C_v of the gas. The density

$$\rho = \frac{M}{\text{vol}} \tag{10.21}$$

where M is the mass of the gas in the cylinder and dead volume and vol is the volume of this gas. The rate of change of ρ is

$$\frac{d\rho}{dt} = \frac{1}{\text{vol}}\left[\frac{dM}{dt} - \rho\frac{d}{dt}(\text{vol})\right] \tag{10.22}$$

Here the compressed volume is

$$\text{vol} = \pi r^2 St(1-f) + Dv = \pi r^2 St(1+\delta-f) \tag{10.23}$$

and

$$\frac{d}{dt}(\text{vol}) = -\pi r^2 V \tag{10.24}$$

since the stroke velocity is

$$V = \frac{d}{dt}(fSt) \tag{10.25}$$

If the pressure is assumed to be high enough that the outflow velocity V_s is sonic and the constriction gas density is approximately $\rho/2$,

$$\frac{dM}{dt} = \frac{-\rho N_s V_s}{2} \tag{10.26}$$

Substituting equations (10.23), (10.24), and (10.26) in (10.22) and rearranging, we obtain the differential equation

$$\frac{d\rho}{\rho} = \frac{[(N_s V_s/2\pi r^2 V) - 1]dt}{St(1+\delta) - Vt} \tag{10.27}$$

Integrating and assuming that $\rho = \rho_0$ at $t = 0$,

$$\ln\frac{\rho}{\rho_0} = \left(1 - \frac{N_s V_s}{2\pi r^2 V}\right)\ln\frac{1+\delta}{1+\delta - Vt/St} \tag{10.28}$$

or

$$\frac{\rho}{\rho_0} = \exp\left[\left(1 - \frac{N_s V_s}{2\pi r^2 V}\right) \ln \frac{1+\delta}{1+\delta-f}\right] \quad (10.29)$$

Noting that

$$\Delta P = P - P_n \quad (10.30)$$

and substituting in equation (10.20) gives in final form

$$\Delta P = \left(\left\{\exp\left[\left(1 - \frac{N_s V_s}{2\pi r^2 V}\right) \ln \frac{1+\delta}{1+\delta-f}\right]\right\}^\gamma - 1\right) P_n \quad (10.31)$$

REFERENCES

1. H. J. Lingal, A. P. Strom, and T. E. Browne, Jr., An investigation of the arc-quenching behavior of sulfur hexafluoride, AIEE Trans. 72(Part III): 242-246, April 1953.
2. H. J. Lingal, T. E. Browne, Jr., and A. P. Strom, Circuit Interrupters, U. S. Pat. 2,757,261, July 31, 1956 (Application: July 19, 1951).
3. H. J. Lingal and J. B. Owens, A new high voltage outdoor load interrupter switch, AIEE Trans. 72(Part III):293-297, April 1953.
4. C. F. Cromer and R. E. Friedrich, A new 115 KV 1000 MVA gas-filled circuit breaker, AIEE Trans. 75(Part III):1352-1357, 1956.
5. W. M. Leeds, T. E. Browne, Jr., and A. P. Strom, The use of SF_6 for high-power arc quenching, AIEE Trans. 76:906-909, 1957.
6. G. J. Easley and J. M. Telford, New design 34.5-69 KV intermediate capacity circuit breaker, IEEE Power Appar. Syst. PAS-83:1172-1177, December 1964.
7. B. Calvino, La serie di interrutore di A. T. a SF_6 tipo MHM singola pressiore, L'Apparecchiature Elettrica, No. 24, April 3, 1968.
8. B. J. Calvino and P. Pezzi, Interruption characteristics of an SF_6 self-extinguishing circuit breaker, CIGRE, Rep. 13.04, Paris, 1968.
9. T. E. Browne, Jr., and A. P. Strom, Interruption of capacitance charging currents in sulfur hexafluoride, AIEE Trans. 75(Part III):1357-1361, 1956.
10. S. Yanabu, H. Mizoguchi, A. Kobayashi, Y. Ozaki, and Y. Murakami, Factors influencing the interrupting ability of SF_6 puffer breaker and development of 300 kV-50 kA one-break circuit breaker, IEEE Trans. Power Appar. Syst. PAS-101(6): 1511-1518, 1982.

11. T. Ushio, S. Tominaga, H. Kuwahara, T. Miyamoto, Y. Ueda, and H. Sasao, SLF interruption by a gas circuit breaker without puffer action, IEEE Trans. Power Appar. Syst. *PAS-100*(8): 3801-3810, 1981.
12. A. Kobayashi, S. Yanabu, S. Yamashita, and Y. Ozaki, Experimental investigation on arc phenomena in SF_6 puffer circuit breakers, IEEE Trans. Plasma Sci. *PS-8*(4):339-343, 1980.
13. B. J. Calvino, Electrical characteristics of an industrial SF_6 puffer nozzle, University of Sydney, H. V. Symp., May 1976.
14. B. J. Calvino, G. Mazza, and S. Rovelli, Study of current zero phenomena under various stress conditions of a single pressure SF_6 circuit breaker, CIGRE Rep. 1307, Paris, 1976.
15. B. J. Calvino, G. Mazza, B. Mazzoleni, and V. Villa, Some aspects of the stresses supported by H. V. circuit breakers clearing a short circuit, CIGRE, Rep. 1308, Paris, 1974.
16. V. di Marco and B. Mazzoleni, Effetto dell'asimmetria della corrente, Energia Elettrica Fasc. No. 7-8, Vol. 53, 1976.

11
Magnetic Air Circuit Breakers

THOMAS E. BROWNE, Jr.*/Consultant to Westinghouse Research and Development Center, Pittsburgh, Pennsylvania

JAMES D. FINLEY/Westinghouse Electric Corporation, East Pittsburgh, Pennsylvania

11.1 Introduction   425
11.2 Examples   426
    11.2.1 Lower-voltage breakers   426
    11.2.2 Medium-voltage breakers   427
11.3 Theory   432
    11.3.1 Magnetic field structure   432
    11.3.2 Short arcs with metal electrodes   433
    11.3.3 Effect of arc voltage   437
    11.3.4 Effect of arc space conductance after current zero   438
    11.3.5 Effect of insulating walls   439
References   452

11.1 INTRODUCTION

Circuit breakers of the magnetic air type, generally limited to circuits of 15 kV or less, are extensively used because of their relatively simple and inexpensive construction and their ease of inspection and maintenance. Their development has been gradual over the years from low-voltage "magnetic blowout" contactors to the breakers now available with 1000-MVA interrupting capability. This development has been guided by simultaneous theoretical and experimental studies to be described in Sec. 11.3, and each step has been checked by extensive laboratory testing.

    These breakers are of various forms of construction but depend upon arc quenching methods of only two types, the use of single "long" confined arcs or of multiple short arcs in series.

---

*Retired

11.2 EXAMPLES

11.2.1 Lower-Voltage Breakers

Figure 11.1 is a view of a modern Westinghouse Type DS breaker designed for circuits of up to 600 V with rated short-circuit currents up through 65 kA at 600 V or 85 kA at 208 or 240 V. With the aid of series-connected current-limiting fuses, breakers of this type (Westinghouse Type DSL) can be applied in circuits with prospective short-circuit currents as high as 200 kA.

Figure 11.2 is a view of an assembled arc chute and of samples of the individual plates and spacers in the chute. Note that the plates are of iron with generally V-shaped slots for confining and cooling the arc. These breakers have main and arcing contacts operated by energy stored in latched springs which are charged either manually or by electric motors. The main or current-carrying copper contacts are typically of the butt type with silver facings for the lower currents, and of the multiple-finger type for the higher continuous current ratings. Arcing contacts above the mains may be faced with silver-tungsten or copper-tungsten and are in the form of wedges between spring fingers. They are designed to withstand arcing for the longer time required to transfer the arc to the chute and quench it.

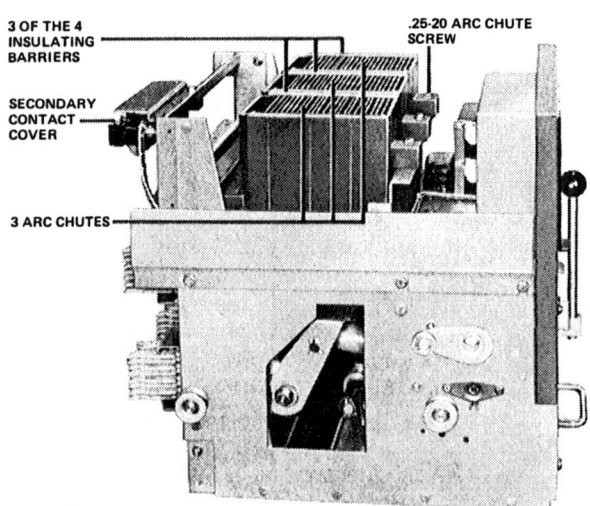

Figure 11.1 Low-voltage circuit breaker with barriers removed to show arc chutes. (Courtesy of Westinghouse Electric Corporation, Pittsburgh, Pa.)

Magnetic Air Circuit Breakers

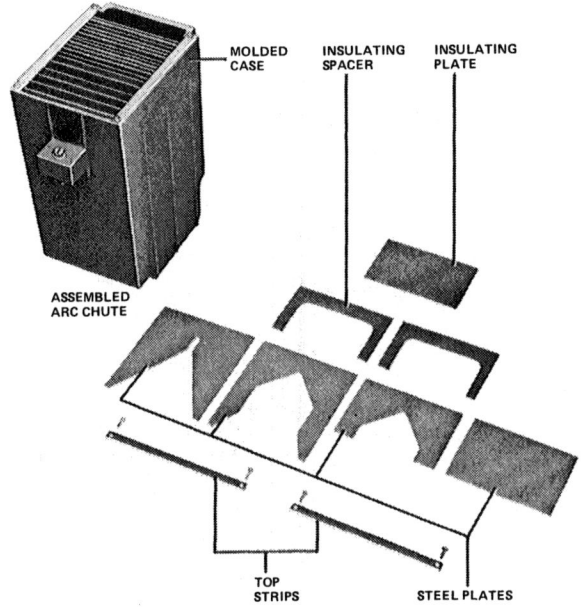

Figure 11.2 Arc chute with internal parts. (Courtesy of Westinghouse Electric Corporation, Pittsburgh, Pa.)

11.2.2 Medium-Voltage Breakers

The deion breaker

An early and novel breaker of this class, the Westinghouse deion or type U, employing the short-arc principle, resulted from an extensive physics-based development [1-5], first in the 1920s. Its construction and method of operation are illustrated by the section sketches, shown in Fig. 11.3. Figure 11.3(A) shows the usual main and arcing contacts with stationary horns to direct the arc into the quenching chute. Note that after arc transfer to the stationary horns the current flows through first a one-turn loop and then a two-turn loop, forming a coil to drive the arc upward into the multiple-plate extinguishing structure shown in Fig. 11.3(B). Here the initially single arc is driven into gradually narrowing slots in the copper plates and, further aided by additional magnetic flux from the surrounding steel plates, is transferred to short (1/16 in. or 1.6 mm) series gaps between the plates. When transferred, the arc is spun in a circular path at high speed by a radial magnetic field produced by the series-connected field coils

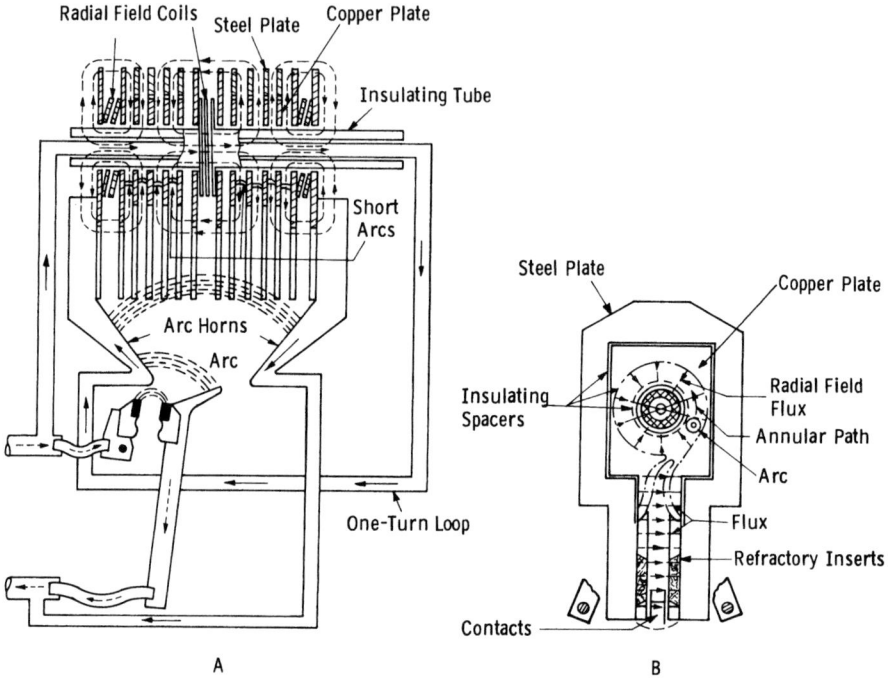

Figure 11.3 Section sketches of the Westinghouse Type U deion breaker. (Courtesy of Westinghouse Electric Corporation, Pittsburgh, Pa.)

shown. These coils are connected in opposition to each other so that their longitudinal field is suppressed and the radial field component is reinforced. Refractory inserts in the entry region protect surrounding organic insulation from ablation by the arc. Not shown is a surrounding insulating structure with metal foil inserts to balance electrostatically the voltage distribution among the series gaps when the recovery voltage appears across the plate stack. The number of plates and gaps is made proportional to the voltage rating of the breaker.

The first breaker of this type was placed in service in 1927 for starting duty of a 3300-hp 2300-V motor, and was still in operation without servicing in 1940. Voltage ratings were increased to 15 kV, and 34.5-kV service was attained by employing two interrupters in series per phase. In 1940 there were a total of more than 800 of these very durable breakers in trouble-free service with tested interrupting ratings up to 500 MVA at 15 kV.

However, manufacturing cost and limitation of current-interrupting ability to less than 30 kA caused this unique "air deion" breaker to be

Magnetic Air Circuit Breakers

superseded in later years by either compressed-air breakers or all-insulation-arc-chute magnetic breakers.

Insulating arc chute breakers

Figures 11.4 and 11.5 illustrate the Westinghouse version of the insulating chute breaker now designated as Type DHP. Figure 11.4 is a view of a 4.76-kV breaker with one arc chute tilted back to show the lower end of the stack of zircon ceramic plates within the chute. Figure 11.5, a schematic section view, illustrates the functioning of the breaker and shows the centrally located field coil in series with two sections of the ceramic plates. This is called the "H-magnet" design because when viewed from above the central magnet core and external pole pieces have the shape of the letter H. Earlier designs were U-shaped, with the core and series coil at one end. The H shape is more efficient magnetically and also improves electrostatic balance and,

Figure 11.4 Westinghouse Type DHP breaker with arc chute tilted up. (Courtesy of Westinghouse Electric Corporation, Pittsburgh, Pa.)

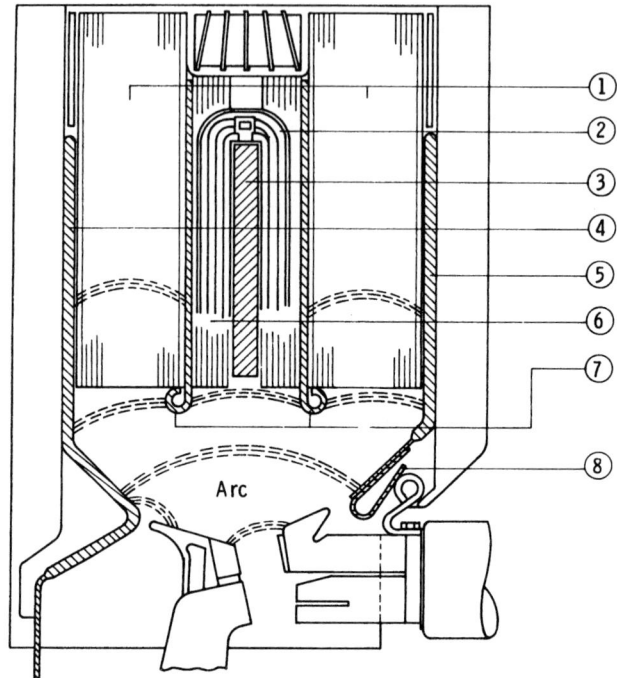

1	Main Interrupter Stacks	6	Transfer Stacks
2	Blowout Coils	7	Center Arc Horns
3	Blowout Magnet Core	8	Rear Arc Horn Disconnecting Contact
4	Front Arc Horn		
5	Rear Arc Horn		

Figure 11.5 Schematic drawing of contacts and arc chute shown in Fig. 11.4. (Courtesy of Westinghouse Electric Corporation, Pittsburgh, Pa.)

therefore, dielectric strength of the open breaker. The design also protects the field coil insulation from continuous electric stress when the breaker is open.

Figure 11.6 shows a characteristic shape of the ceramic plates within the chute. The gradually narrowing slots are offset from the center and alternate plates have opposite offsets, so that the arc is forced into a zigzag path and consequently lengthened and confined between paral-

Magnetic Air Circuit Breakers

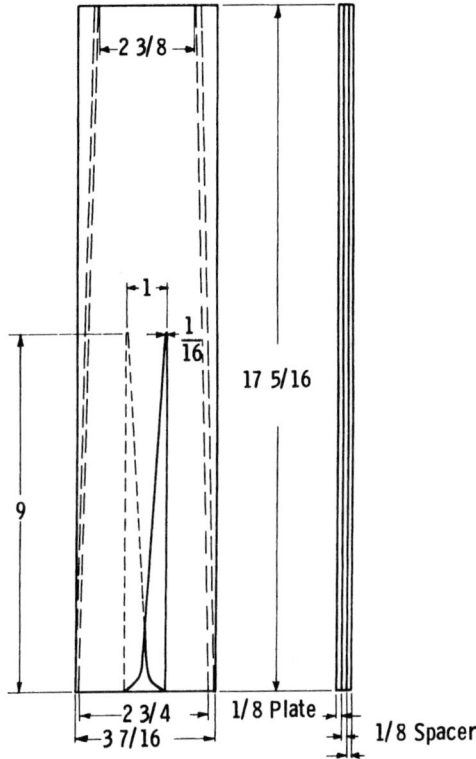

Figure 11.6 Sketch of sample arc chute ceramic plates as assembled. (From T. E. Browne, Jr., and A. P. Strom, A study of conduction phenomena near current zero for an a-c arc adjacent to refractory surfaces, AIEE Trans. 70:398-407, © 1951, IEEE.)

lel insulating walls 1/16 in. (1.6 mm) apart as it approaches and reaches the tops of the slots. At this final position the arc between the plates tends to bow upward somewhat and magnetically and thermally driven hot gases are exhausted up through the open spaces between the plates and thereby cooled before being discharged.

At high currents magnetic action rapidly drives the arc, sketched in successive positions in Fig. 11.5, from the opening contacts up into the chute, where transfer of the current to the series "blowout" field coil takes place. When interrupting very low currents, however, the magnetic drive is ineffective, so a puffer piston is normally employed to aid natural convection in quickly blowing the arc into the chute for such currents.

Insulating arc chute breakers are available for 2.3- to 15-kV ac service and have current-interrupting ratings as high as 48 kA. Although designed primarily for ac service, breakers of this type can also be designed for use in dc circuits.

11.3 THEORY

11.3.1 Magnetic Field Structure

In the low-voltage breakers (Westinghouse Type DS) the magnetic field to drive the arc into the chute is produced by the loop of the contacts and leads and accentuated by the iron plates, which tend to "draw" the arc into their V-shaped slots. These plates are separated and insulated from each other. Some manufacturers coat the plates with insulating enamel. In addition to the magnetic drive, convection caused by thermal buoyancy, relatively effective at lower currents, also urges the arc into the vented slots. At high currents the arc tends to transfer to the iron plates and so to be broken up into many short arcs in series. Insulating backup plates (Fig. 11.2) prevent these short arcs from being blown beyond the iron plates.

In the medium-voltage class, as already noted the Westinghouse Type U breaker has the leads arranged in one- or two-turn loops and also has iron outer plates to draw the arc upward. When transferred to the multiple short gaps between the copper plates, the arc sections are driven at supersonic speed, observed to reach 10 revolutions per half cycle, or 72,000 r/min, around circular paths by the radial field of the interspersed series coils. This high speed prevents melting or vaporizing of the copper plates and improves the dielectric recovery of these "cold cathode" arcs.

As already noted, the refractory insulating chute breakers are generally provided with series-excited U-shaped magnets to drive the arc, or with H-shaped magnets in the Westinghouse Type DHE breakers. In the latter a central set of reduced-size plates serves to transfer the current into the exciting coils (Fig. 11.5). A phase lag between the magnetic flux and the current is produced by eddy currents in the magnetic pole pieces and by closed resistive rings next to the coil turns. This phase shift prevents the magnetic flux from going to zero with the arc current, so that the driving force on the arc tends to fall near zero only as the first power of the current rather than as the square of the current. A marked increase in arc resistance attained at current zero, and hence in interrupting performance, has been shown to result from this. An advantage of the H-magnet construction is that the pole pieces are shorter, yielding a more nearly uniform field along the stack sections and smaller leakage of flux.

In an early development by Strom [6] the large component of leakage flux from the magnet pole pieces was largely suppressed by closed heavy copper turns around the pole pieces. This reduced the amount

Magnetic Air Circuit Breakers

of iron and copper required, especially for the U-shaped magnets, and also practically eliminated magnetic interaction between adjacent phases of the breaker. However, with the more modern H-shaped magnet design, this refinement has been found to be unnecessary.

It should be noted here that both theory and experience point to the existence of an optimum value of series-produced magnetic field strength for a given chute structure, so that increasing the field beyond this optimum is not advantageous [7].

In transverse magnetic fields employed in all these breakers, an arc behaves very much like a flexible conductor, so that the magnetic force on it can be calculated by the usual motor rule. This force causes the unconfined arc column to move rapidly through the air. It tends to push the cold surrounding air aside as a solid conductor would since its very high temperature and low gas density prevent all but a minor amount of the cold air from entering the column. However, if the arc is held stationary by barriers, such as the tops of the slots in ceramic plates, the magnetic force on that portion of the arc transverse to the flux has the effect of forcing some of the cold surrounding gas medium into and through the arc, thus tending to cool it. This cooling action raises the voltage gradient of the arc and speeds its loss of conductance near a current zero. In moving the arc into the plate structure and lengthening it, it has been found to be advantageous to obtain as nearly as possible a constant rate of increase of arc voltage with time and distance. Therefore, the shape of the plate slots is designed to accomplish this [8].

11.3.2 Short Arcs with Metal Electrodes

The improvement in per unit length ac interrupting effectiveness of short (1 to 3 mm) arcs between low-boiling-point electrodes was first reported in 1892 by Wurts [9], who also found the advantage of low-boiling-temperature electrodes (he called brass a "nonarcing" metal). These properties were first used by him and subsequently by others in the arc quenching gap structures of lightning arresters. As noted above, short arcs were also used in the arc-quenching structure of the Westinghouse Type U circuit breaker, where rapid arc motion appears to have achieved at high currents an effect similar to that of the low-temperature metal vapor from stationary arc terminals on brass. Both photographic and electrical evidence has shown that in low-voltage breakers with electrically isolated metal arc chute plates the arc can and does transfer to the plates at the top of the V slots, forming a series of short arcs. This short-arc principle is still being used in present designs of many low-voltage circuit breakers.

Figure 11.7, obtained in a study of post-current-zero dielectric recovery at low currents by short brass-electrode gaps [10], reveals the nature of the dielectric recovery by such gaps. In this case, the initial rise in breakdown voltage can be explained by increasing thickness,

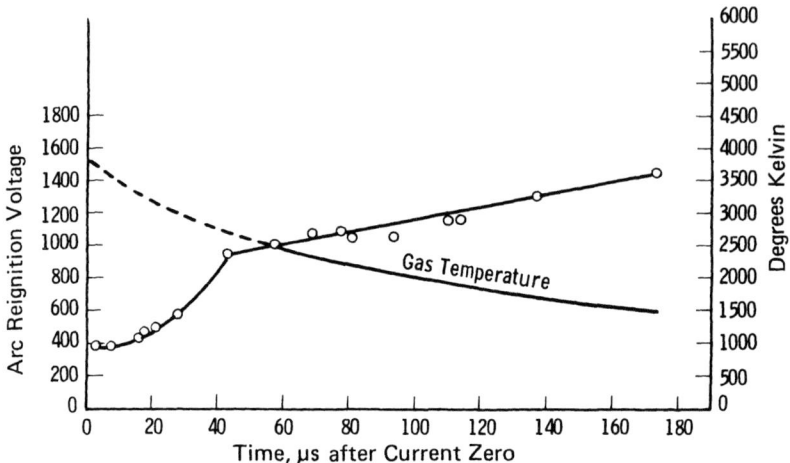

Figure 11.7 Postarc dielectric recovery and calculated gas temperature of a 1-mm gap between brass electrodes at atmospheric pressure. Arc current 50 A at 50 Hz. (From Ref. 10.)

for a given applied voltage, of a space-charge layer at the new cathode illustrated by Fig. A.10. This thickness increases with time as well as voltage as the ion density in the adjacent plasma diminishes, mainly by diffusion to its boundaries. At the sudden reduction in recovery rate, near 40 µs in this case, it is evident that the space-charge-layer thickness had reached the full gap length. After this, further recovery resulted only from increasing density of the gas in the then deionized constant-length gap as cooling continued, aided by the close spacing of the metal electrodes. The solid portion of the gas temperature curve was calculated on this basis to lie in the range near 2000 K in these experiments. The dashed extrapolation for shorter times to an initial value near the boiling temperature for brass was useful in analyzing conditions in the initial recovery period.

The effect of current magnitude is shown by Fig. 11.8, in which faster recovery for lower currents is evident. Also shown for rough comparison are curves for 1.6-mm gaps between brass and copper electrodes at the higher current levels of 300 and 1200 A. The latter curves were obtained in earlier studies associated with the Type U breaker development [3-5].

Figure 11.9 shows the effect on dielectric recovery during the first 380 µs of an increase in gap length between flat brass electrodes from 1/16 in. (1.6 mm) to 1/8 in. (3.2 mm). The dashed curves show the

Magnetic Air Circuit Breakers

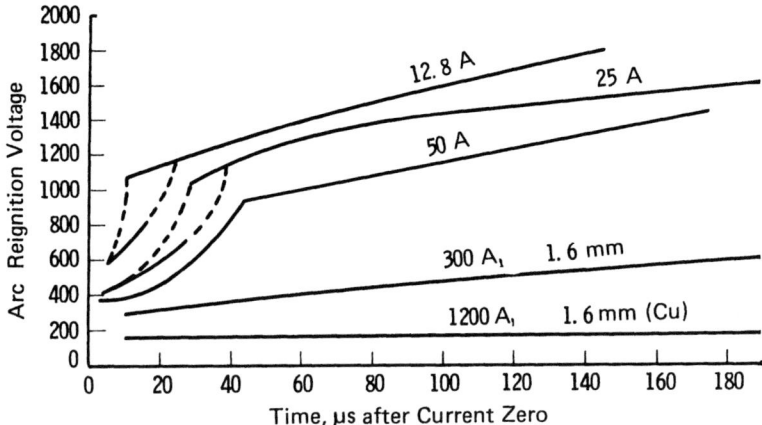

Figure 11.8 Postarc dielectric recovery of short (1 mm or 1.6 mm) brass- or copper-electrode atmospheric pressure arcs at various current levels. (From Ref. 10.)

critical recovery voltage waveforms controlled by resistance shunting for the 3.2-mm-gap case. Here, the dielectric recovery envelopes of the critical voltage curves reflect mainly the cooling of the deionized gap (later part of recovery shown in Figs. 11.7 and 11.8), which was

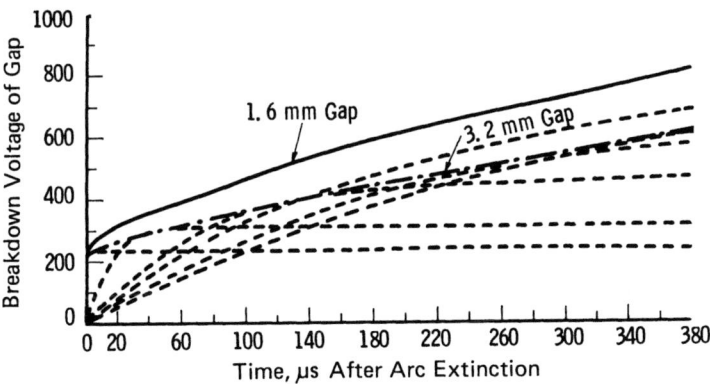

Figure 11.9 Postarc dielectric recovery envelopes of atmospheric pressure brass-electrode 300-A 60-Hz arcs with gap lengths of 1.6 and 3.2 mm. Critical resistance-controlled voltage rise curves shown dashed for 3.2-mm gap. (From Ref. 3.)

so much faster in the short gap that its breakdown voltage exceeded that of the longer gap throughout this time period.

In Fig. 11.10, the relative dielectric recovery rates are indicated by the initial slopes of critical volt-time recovery curves like those shown in Fig. 11.9 but with fixed voltage amplitude. Results with six different electrode materials are shown. The curve for mercury was obtained with amalgamated copper electrodes. With the lower-boiling-temperature materials mercury, zinc, and brass, the inverse effect of gap length shown in Fig. 11.9 is illustrated for gaps as long as 11 mm (0.43 in.), and also with copper and iron from 5 to 11 mm. Below 5-mm gap length, copper and iron, and especially the high-boiling-temperature tungsten, showed rather different behavior of the critical initial

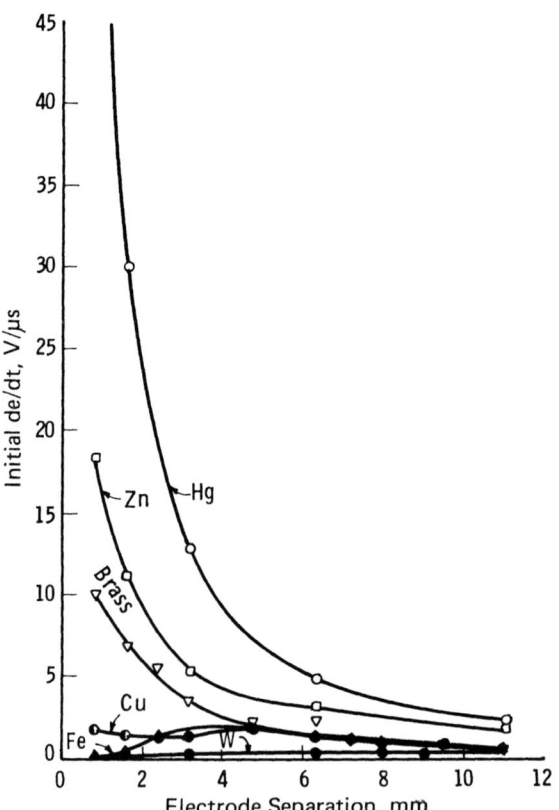

Figure 11.10 Critical initial recovery voltage rates for 300-A 60-Hz arcs in a 450-V circuit with various arc lengths and electrode materials. (From T. E. Browne, Jr., Extinction of short a.c. arcs, Trans. AIEE 50:1461-1473, © 1931, IEEE.)

Magnetic Air Circuit Breakers

de/dt. At least in the case of tungsten, the very low values suggest that thermionic emission of electrons from still hot arc terminal spots momentarily prevented dielectric recovery. With copper and iron, oxide layers at the terminals spots may also have been residual electron emitters, effective only at the shortest spacings.

Figure 11.11 shows that dielectric recovery for a copper-electrode gap 1.6 mm long but with the very rapidly moving "cold cathode" arc as used in the Type U deion breaker is very similar to that with the stationary arc of Fig. 11.9 between brass electrodes. However, these data were obtained at several thousand amperes instead of only 300 A circuit current. This shows that use of high-speed magnetic rotation to obtain "cold cathode" arcs made possible extension of the short-arc interruption principle to the power circuit breaker range of interrupted currents.

11.3.3 Effect of Arc Voltage

Direct-current interruption

In interrupting dc circuits the development of arc voltage by the breaker is essential because the circuit voltage must be opposed in order to bring the current to zero (see Chap. 2). Magnetic air circuit breakers which forcibly lengthen and also confine the arc are effective for this purpose. However, this special requirement of dc breakers

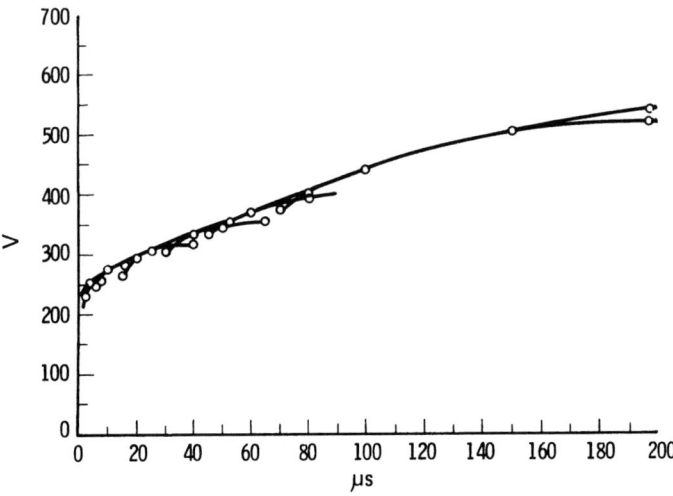

Figure 11.11 Postarc dielectric recovery envelope of a 1.6-mm arc of several thousand amperes rotated at high speed between copper plates at atmospheric pressure. (From Ref. 3.)

means that the magnetically stored energy in the circuit must be dissipated by the breaker as well as energy being fed in from the power source during the arcing period. Since many dc circuits have relatively high inductance, this energy dissipation requirement places a special burden on breakers used to interrupt direct current and so must be considered in their application. One means of limiting the switch energy is to develop the arc voltage as fast as possible, requiring special high-speed opening mechanisms. Some such mechanisms include trip devices that respond to the initial rate of rise of a short-circuit current in order to limit the current before it can reach its normal peak value, thus also limiting the stored energy to be adsorbed. In increasing numbers of cases, high-speed dc interruption is also required for short-circuit protection of energy-sensitive devices such as solid-state rectifiers or inverters.

In the case of breakers with metal plate arc chutes, although the arc is not lengthened, it can develop high voltage upon transfer to the plates because of the exceptionally high voltage gradient of short arcs between metal electrodes. This results from the effective cooling by thermal diffusion to the closely spaced plates.

Effect on alternating-current interruption

Especially when interrupting lower-voltage ac circuits, the breaker arc voltage, when it approaches the circuit voltage, will also serve to limit short-circuit current magnitudes (see Chap. 14). However, as shown by the current and voltage oscillograms of Fig. 11.12, an even more important effect of the arc voltage may be the phase advance of the final current zero moment to such a degree that the instantaneous circuit voltage at current zero is greatly reduced. In an extreme case this phase advance may bring the current zero *at* the voltage-zero moment, in which case the recovery voltage rate of rise would be only at the power frequency rate, very much lower than the oscillatory rate of recovery of normal low-power-factor ac circuits.

11.3.4 Effect of Arc Space Conductance After Current Zero

A characteristic of magnetic air breakers is that the arc resistance, or conductance, remains finite at and just after a current zero at which interruption occurs. The usual result is that the restored voltage oscillation is severely damped or even overdamped, as shown, for example, in test oscillograms plotted in Figs. 11.15 and 11.16. This damping is caused by the relatively large postarc current also shown.

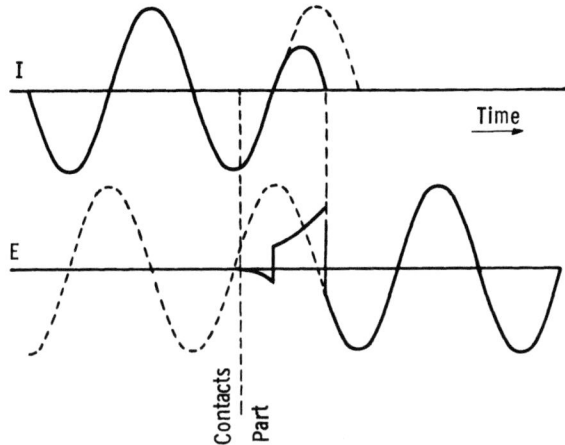

Figure 11.12 Arc current and voltage with typical interruption by high-arc-voltage type of breaker. Short-circuit current, if continuing, and prior generated voltage shown dashed. (From L. J. Linde and B. W. Wyman, The development, design and performance of magnetic-type power circuit breakers, AIEE Trans. 65: 387-393, © 1946, IEEE.)

11.3.5 Effect of Insulating Walls

Cooling by diffusion to walls

In medium-voltage breakers with insulating chutes the arc is driven into narrow spaces between barrier walls of refractory or semirefractory insulating material. The primary effect of this is to cool the arc column and thereby reduce the magnitude and increase the rate of fall of its residual conductance at current zero and thus its recovery voltage withstand ability. This diffusive cooling depends critically on close spacing of the walls, as illustrated by Fig. 11.13. Here in early experiments [1] the interrupted volts per unit length of an ac arc between soapstone walls was found to increase by a factor of nearly 20 when the slot width was decreased from 3/8 in. (9.5 mm) to 1/16 in. (1.6 mm). In later studies Frind [11] has shown this relation in terms of the current-zero "time constant" Θ of the arc conductance change (see Chap. 6). Here Θ was found by both theory and experiment to increase as the square of the slot width between ceramic plates. The energy balance limited interrupting ability is an inverse function of Θ.

Although arc interrupting performance continues to increase as wall spacing is reduced, practical considerations usually limit the narrowness of slots to about 1/16 in. in circuit breaker arc chutes.

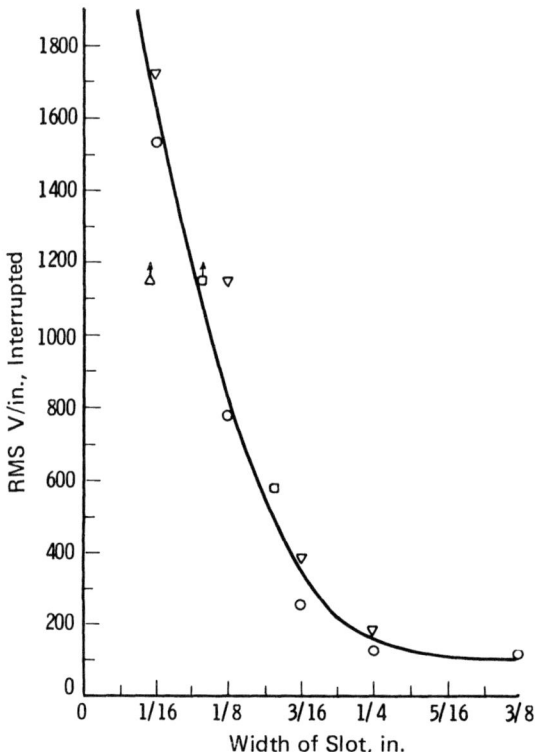

Figure 11.13 Recovery voltage gradient of arcs between soapstone walls vs. wall separation. (From J. Slepian, Extinction of an a-c arc, AIEE Trans. 47:1398-1408, © 1928, AIEE.)

Conductance of heated wall surfaces

Figure 11.14 shows the decay after current zero following heavy arcing of the temperature of a ceramic wall in studies by Frind and Hunziker [12]. In the temperature range near 3000 K ceramic materials have considerable electrical conductivity. The decay in this case may be seen to be quite slow, requiring 10 ms to drop from 3500 K to 2400 K. It should be observed, however, that electrical conductivity varies exponentially with temperature of the ceramic so that the "leakage" conductance dropped much faster than this. Analysis of test oscillograms by these experimenters did show rapid conductance decay, interpreted as first being due mainly to cooling of the arc plasma and later to be determined mainly by wall cooling.

Magnetic Air Circuit Breakers

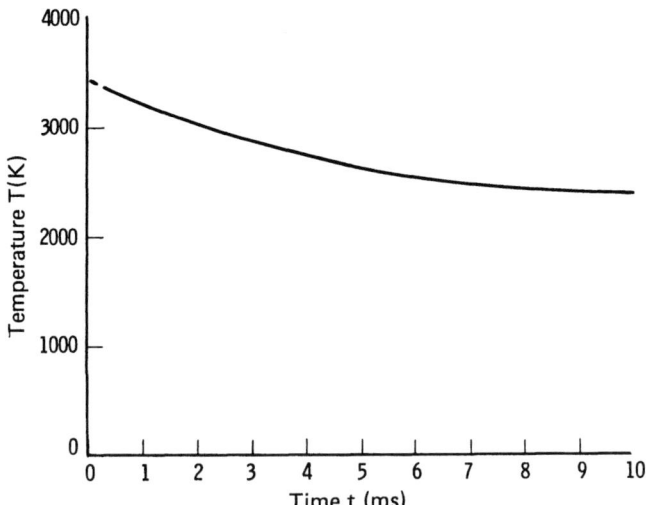

Figure 11.14 Ceramic wall cooling after heavy current arcing. (From G. Frind and R. Hunziker, A fundamental investigation of the interruption process in air-magnetic circuit breakers. Part. I. Measurements of arc motion, of wall temperature and of electrical wall conductivity, IEEE Trans. Power Appar. Syst. *Pas-90*: 360-365, © 1971, IEEE.)

Some earlier observations of near-current-zero arc conductance with model ceramic arc chutes by Browne and Strom [13] also help to reveal the separate influences of changing plasma and wall conductances on arc interruption. Figures 11.15 and 11.16 show the results of oscillographic measurements of currents and voltages and of the calculated conductances and power inputs to the arc space. Under the influence of the restored voltage and postarc current and power, the arc space conductance was found to change only slowly during the first 60 μs after current zero. In the test of Fig. 11.15, the 6250-A 1150-V 60-Hz circuit was interrupted by a chute with four zircon ceramic plates like those of Fig. 6. Here, there was a transverse magnetic field having a peak value of B_m = 0.015 T/kA produced by an air-core coil having short-circuited turns, causing the field to lag the current by 20°. In the test of Fig. 11.16 there was a failure to interrupt the same circuit when the magnetic flux was only slightly less, 0.0125 T/kA, and in phase with the current. Note that in this latter case the current-zero arc conductance was about 70% greater than in the Fig. 11.15 test, leading to a more rapid increase in postzero current and power input

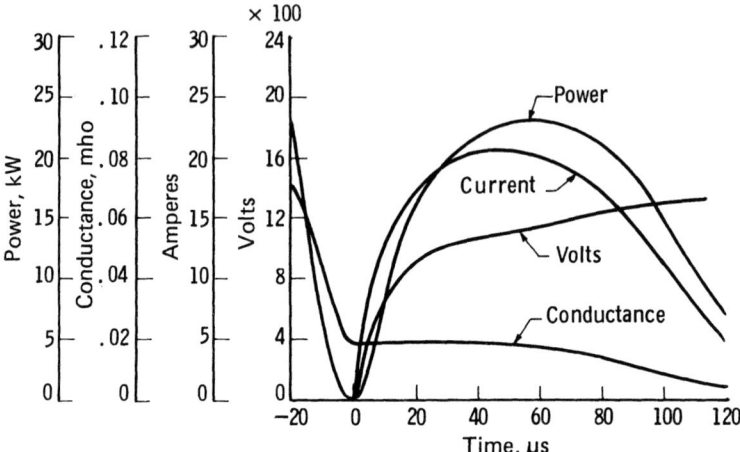

Figure 11.15 Observed near-current-zero quantities with near-critical interruption of a 6150-A 1150-V 60-Hz circuit by a model arc chute containing four zircon ceramic plates. Twenty-degree phase shift of magnetic field behind current. (From T. E. Browne, Jr., and A. P. Strom, A study of conduction phenomena near current zero for an a-c arc adjacent to refractory surfaces, AIEE Trans. 70:398-407, © 1951, IEEE.)

as the arc was reestablished. The arc conductance reached a minimum value of about 0.03 S near 10 µs after the zero and increased continuously thereafter. It may also be noticed, as stated above, that in these tests the recovery voltage was completely nonoscillatory and reduced by both the phase advance, evident in Fig. 11.15, and by the postzero current, as especially evident in Fig. 11.16, where the comparison is with the peak open-circuit voltage of 1625 V.

Since the arc conditions within the breaker chute are rather complex, attempts were made in both these early studies and also later to study the arc between parallel refractory plates under simpler conditions. Figure 11.17 illustrates one approach. Here, the arc between the carbon (graphite) electrodes represents the principal portion of the arc extending between the ends of the offset slots in the chute stack. Because this arc section is in a plane parallel with the applied magnetic field in the chute and so is not deflected by the field, no field was used in these model studies. The arc was started by a small fuse wire as shown in Fig. 11.17(A) and burned between the plate surfaces as indicated schematically in Fig. 11.17(B). Figure 11.17(C) shows the current "leakage" paths in the postzero recovery period, when the arc

Magnetic Air Circuit Breakers

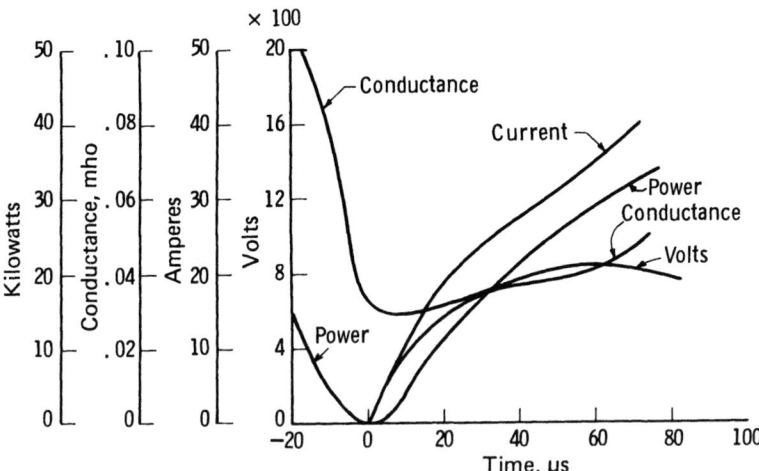

Figure 11.16 Observed quantities for failure to interrupt the circuit of Fig. 11.15 with a slightly smaller magnetic field in phase with current. (From T. E. Browne, Jr., and A. P. Strom, A study of conduction phenomena near current zero for an a-c arc adjacent to refractory surfaces, AIEE Trans. 70:398-407, © 1951, IEEE.)

plasma conductivity had become much smaller than that of the heated and more slowly cooling plate surface. The plate material was zircon ceramic, as used in all of the successive generations of Westinghouse Type DH breakers. In Fig. 11.17(C) situation, because of the difference in current paths from those of Fig. 11.17(B), the current was limited primarily by the plate surface resistance so long as the arc plasma retained a moderate amount of conductivity. Thus the arc space resistance appeared to be essentially that of the plasma before and near current zero, but as the plasma cooled more rapidly than the plate surfaces, there was a somewhat later time period when the applied voltage was mainly across the plate surface paths, which then served to limit the postzero current. Eventually, of course, in a successful interruption the plasma residue must lose its conductance completely and then be able to sustain the applied voltage.

In these parallel-plate model tests, an attempt was made [13] to analyze plots like those of Figs. 11.15 and 11.16 in terms of the approximate arc model differential equations discussed in Chap. 6. The equation due to Cassie, shown in many cases to represent arc behavior before a current zero, is

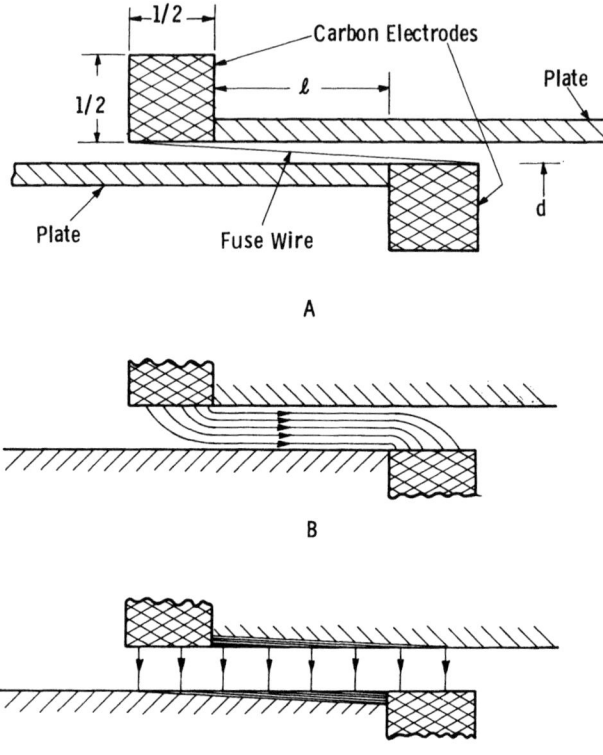

Figure 11.17 Arrangement for the parallel plate arc interruption test showing both gas and solid conducting paths. (A) Placement of plates and electrodes for test; (B) current flow through high conductivity gas; (C) current flow through low conductivity gas. (From T. E. Browne, Jr., and A. P. Strom, A study of conduction phenomena near current zero for an a.c. arc adjacent to refractory surfaces, AIEE Trans. 70:398-407, © 1951, IEEE.)

$$R \frac{d}{dt} \frac{1}{R} = \frac{1}{\Theta} \left(\frac{e^2}{E_0^2} - 1 \right) \quad (11.1)$$

where
    R = arc resistance
    Θ = arc time constant
    e = arc voltage or voltage gradient

E_0 = steady-state value of e

If both Θ and E_0 are assumed to be constant, it is evident that the left side of the equation should plot as a linear function of e^2. Figure 11.18 is such a plot for a parallel-plate test in a 192-V 300-A 60-Hz circuit with 3/16 in. (4.8 mm) plate separation and 1 1/8 in. (29 mm) path length between electrodes. Times with respect to the current zero moment are shown as the small numbers next to the plotted points. Most of the prezero (negative) time points may be seen to fall fairly close to the dashed straight line corresponding to the Cassie equation with Θ = 64 μs and E_0 = 117 V/in. (4.6 V/mm).

It has also been found that arc resistances in the postzero period tend to follow the Mayr differential equation

$$R \frac{d}{dt} \frac{1}{R} = \frac{1}{\Theta} \left(\frac{ei}{N_0} - 1 \right) \tag{11.2}$$

where i is the arc current and N_0 is the arc power loss. Again, if Θ and N_0 are constant, the left side of the equation is predicted to be a linear function of the arc power input ei. In Fig. 11.19 for the same test it may be seen that the postzero points from $t = 4$ through $t = 124$ μs did fit very closely the dashed straight line corresponding to the Mayr equation with parameters Θ = 167 μs and N_0 = 1250 W/in. (49 W/mm). It remained near this line for as long as 505 μs. Thus these arcs between refractory plates, like many other arcs, such as those in gas

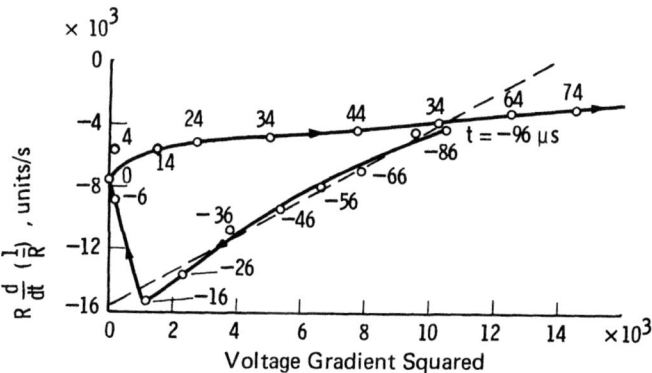

Figure 11.18 Per unit rate of change of conductance for parallel-plate interrupting failure test with 3/16-in. (4.3-mm) plate spacing plotted against the square of the arc voltage gradient. (From T. E. Browne, Jr., and A. P. Strom, A study of conduction phenomena near current zero for an a-c arc adjacent to refractory surfaces, AIEE Trans. 70: 398-407, © 1951, IEEE.)

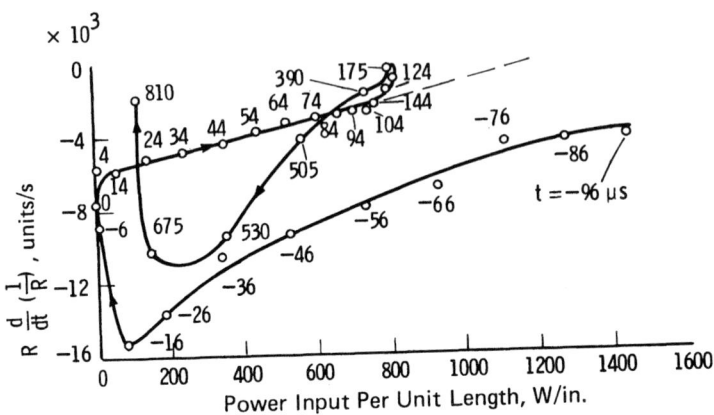

Figure 11.19 Plot of Fig. 11.18 data vs. arc power per unit length. (From T. E. Browne, Jr., and A. P. Strom, A study of conduction phenomena near current zero for an a-c arc adjacent to refractory surfaces, AIEE Trans. 70:398-407, © 1951, IEEE.)

blasts, were found to be amenable to modeling by either the Cassie or the Mayr arc model equations.

In a somewhat simpler test, not involving the varying series parallel combination of gas and surface conduction paths illustrated in Fig. 11.17, analysis of test results with the arc magnetically pressed against a refractory insulating surface, as shown in the insert of Fig. 11.20, throws some additional light on the observed arc behavior. It was found that the decay of observed total conductance, shown solid in Fig. 11.20, could be separated into two components, each following a Mayr equation like (11.2) but with rather different constant parameters. The dashed curve, departing from the solid curve before 80 µs and having a time constant value Θ_s of 200 µs, is believed to represent decay of the surface leakage path which dominated the total conductance in the later time period. The difference curve, shown dot-dashed with a time constant Θ_0 of only 31 µs, is believed to represent the more rapidly decaying arc plasma conductance, which also followed a Mayr-type equation. Thus it seems likely that the arc behavior illustrated in Figs. 11.18 and 11.19 also represented the actual behavior of two parallel conducting paths with time constants larger and smaller, respectively, than the apparent values for the combination. Before current zero and shortly after, the more rapidly changing arc plasma is the principal conducting element, as illustrated by Fig. 11.17(B). In a later period, possibly of duration 100 µs or more, the heated surface

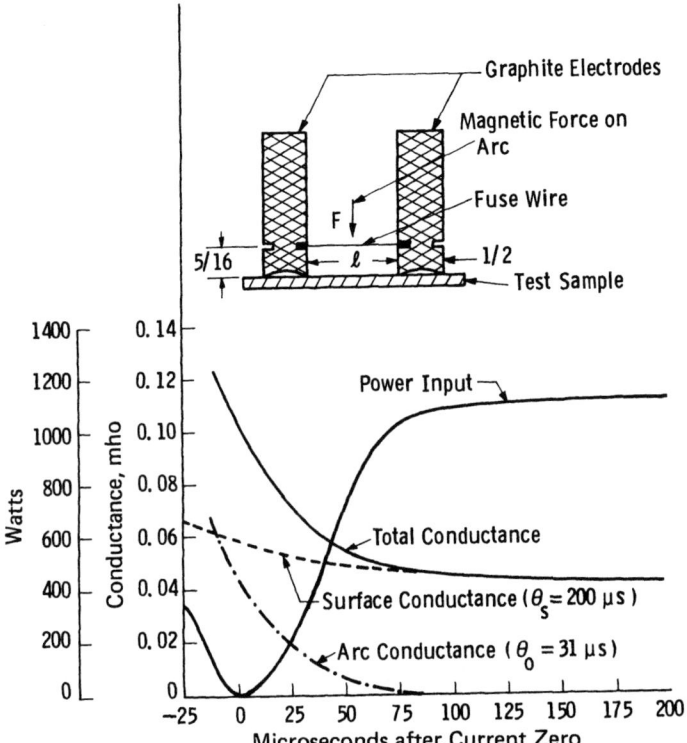

Figure 11.20 Analysis of apparent arc conductance decay in the plate screening test illustrated. (From T. E. Browne, Jr., and A. P. Strom, A study of conduction phenomena near current zero for an a-c arc adjacent to refractory surfaces, AIEE Trans. 70:398-407, © 1951, IEEE.)

path on the adjacent solid surfaces, decaying in conductance more slowly, tends to be the dominant conducting element, fed with current by the now small conductance residual plasma connected approximately as shown in Fig. 11.17(C).

Since the relatively large post-current-zero arc path conductance in these breakers tends to reduce both the restored voltage and its rate of rise, the behavior of the surface conductance of the arc chute plates in resisting the applied voltage at later times is very important. Some early approximate calculations of transient heating and cooling of zircon ceramic plates subjected to momentary heating of one surface help to illustrate the phenomena involved. The partial differential equation for transient heat conduction in one dimension is

$$\frac{dT}{dt} = \alpha \frac{\delta^2 T}{\delta x^2} \qquad (11.3)$$

where
T = absolute temperature
t = time
x = dimension into the plate
α = thermal diffusivity = $k/\rho c$
k = thermal conductivity of body
ρ = density of body
c = specific heat of body

Taking the cgs values appropriate for pure zircon ($ZrO_2 \cdot SiO_2$) as k = 0.0047 cal/s cm^2°C, ρ = 4.7 g/cm^3, and c = 0.132 cal/g, which gives α = 0.0076 cm^2/s, and assuming the surface of the plate, taken as a semi-infinite body, to be instantly raised from an initial temperature T_0 to a new elevated temperature T', equation (11.3) can be solved for the temperature penetration into the plate as a function of time. In Fig. 11.21 the solution is plotted as $\Delta T_x/\Delta T_s = (T - T_0)/(T' - T_0)$ vs. the dimension x into the plate for a series of values of the time expressed as fractions of a half cycle at 60 Hz. For small fractions of a half cycle the heat penetration into the plate is only a few mils and the temperature gradient—is very high, but the gradient and heat flow become much less at exposure times as long as one-half cycle. Figure 11.22

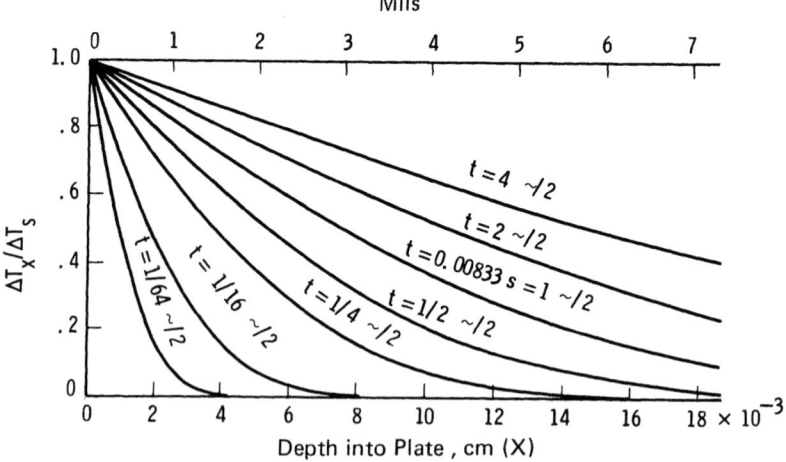

Figure 11.21 Transient heat penetration curves for zircon at various times after exposure of the surface to a sudden temperature rise. (Courtesy of Westinghouse Electric Corporation, Pittsburgh, Pa.)

Magnetic Air Circuit Breakers

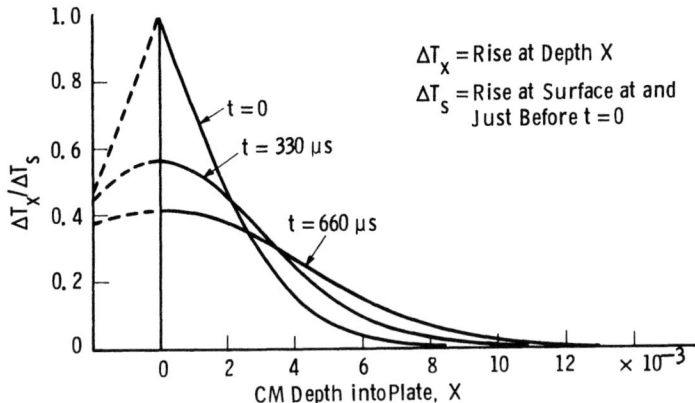

Figure 11.22 Transient heat flow into zircon plate at various times after cessation of exposure of the surface to elevated temperature for 1/16 half cycle (520 µs). (Courtesy of Westinghouse Electric Corporation, Pittsburgh, Pa.)

shows results of a calculation of continuing heat penetration and surface cooling for the case of sudden removal of the heat source after only 1/16 half cycle or 520 µs. This shows that after only 330 µs of continuing heat flow, the surface temperature would drop to about 56% of its initial value and the heat would have penetrated somewhat deeper into the plate. For longer times of exposure the rate of surface cooling would be much smaller. Of course, with a voltage gradient applied while the electrical conductivity is appreciable, the cooling would be slowed and even prevented in the extreme case, leading to failure to ininterrupt the circuit.

Although these simplified calculations are instructive, the problem is clearly much more complicated in the stressed breaker case. One complication, ignored in the calculations above, is the lack of constancy of the material properties k and c, and therefore α, over an extended temperature range, together with our inadequate knowledge of these properties at very high temperatures. Recently, Tslaf [14] has published results of much more complete calculations of heating and cooling of the plate under more representative conditions, including elevation of the plate material to the melting point and penetration of melting into the plate. His results for heat penetration are shown in Fig. 11.23 and for heating and cooling of the surface on a millisecond time scale in Fig. 11.24. In his plots, depth into the plate is given by the symbol Z and time during the cooling period is designated as τ. The time at

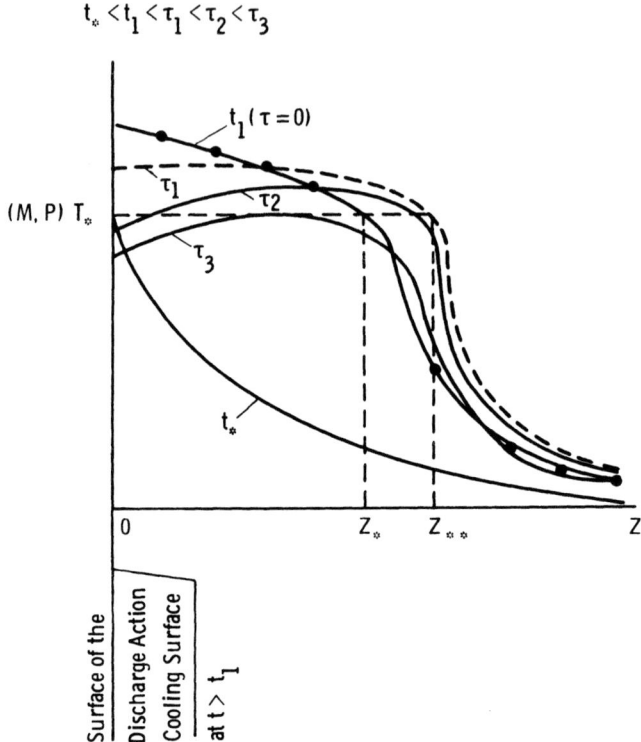

Figure 11.23 Calculated heating and cooling stages of an arc-heated ceramic plate raised above the fusion point: t_*, time at which fusion starts at the surface; t_1, time at which arc action ends; τ_1, time at which the advance of the fused zone into the insulation ends; τ_2 and τ_3, intermediate stages of cooling; Z_*, point of melt penetration at end of arcing; Z_{**}, maximum melt penetration. (From A. Tslaf, Calculation of the temperature in a conductive track of an arc on the surface of a magnet-blast chute, IEEE Trans. Plasma Sci. PS-8:455-460, © 1980, IEEE.)

which melting of the surface layer begins is t_*, when the temperature penetration curve marked t_* is similar to those in Fig. 11.21. After this time Fig. 11.23 shows that absorption of latent heat of melting drastically changes the temperature distribution within the plate, but Fig. 11.24 shows an actual increase in the rate of surface heating. Tslaf checked his calculation of surface temperature decay against the measurements by Frind and Hunziker [12] shown in Fig. 11.14 and found

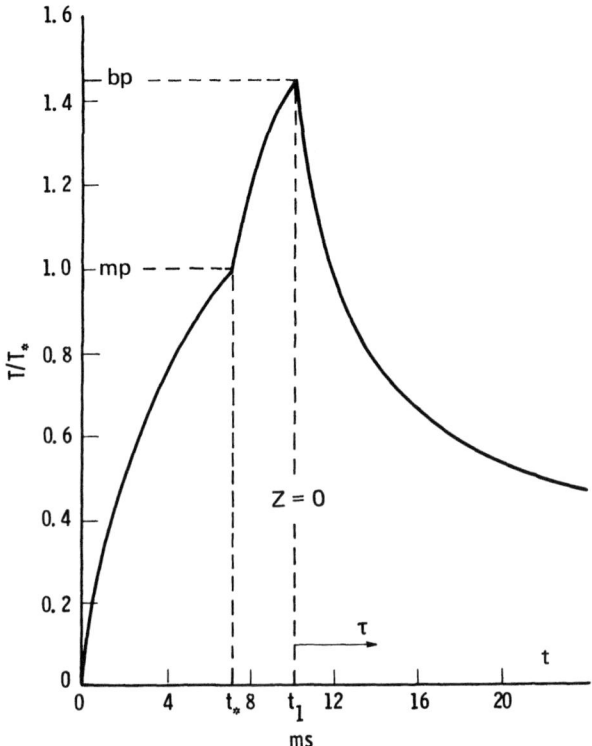

Figure 11.24 Heating and cooling of the active chute surface. Same data as in Fig. 11.23. (From A. Tslaf, Calculation of the temperature in a conductive track of an arc on the surface of a magnet-blast chute, IEEE Trans. Plasma Sci. PS-8:455-460, © 1980, IEEE.)

good agreement. The material assumed in Tslaf's treatment was corundum (Al_2O_3), which has a lower melting point (2320 K) than zircon (3173 K) but a higher thermal diffusivity.

Effect of wall material

In spite of the foregoing theoretical and experimental observations, the situation of the arc within the insulating arc chute is so complex that it is difficult to evaluate possible chute materials purely on a theoretical basis. Practical requirements may be enumerated as (1) mechanical strength and resistance to heat shock, (2) durability in the presence of high current arcing, (3) good insulating properties at moderately elevated temperature, and (4) ease of manufacture.

From the standpoint of arc-interrupting effectiveness, theory suggests that high heat capacity at temperatures below the melting point and high thermal diffusivity should be favorable characteristics. Examples of some possible materials are:

1. Zircon ceramic ($ZrO_2 \cdot SiO_2$)
2. Alumina (corundum—Al_2O_3)
3. Mullite ($3Al_2O_3 \cdot 2SiO_2$)
4. Thoria (ThO_2)
5. Beryllia (BeO)

In order to meet the thermal shock resistance criterion, any of these or other materials must be prepared as porcelain-type ceramics with a porosity level somewhat less than 10%. Hudis and Carroll [7] in comparative studies found a positive correlation between ability to withstand postarc voltage gradient and the thermal diffusivity of the material. A similar correlation was also found in the earlier studies by Browne and Strom [13], using the screening test illustrated in Fig. 11.20. Especially outstanding were results with beryllia, which has exceptionally high diffusivity values, but unfortunately this material also has high toxicity.

Another class of materials not mentioned above are those which release gases in the presence of arcing. One material which has been used, and which gave markedly superior results in initial screening tests, is asbestos lumber. However, the favorable behavior disappeared after either repeated arc exposure or baking, and the material exhibited low dielectric strength in arc chute tests. Another such material, described as a "phosphoric asbestos," has been used successfully in the form of a molding compound with which the zigzag lengthened arc path was achieved by interleaving vanes cast onto solid slab walls of the arc chute [15].

REFERENCES

1. J. Slepian, Extinction of an a-c arc, AIEE Trans. 47:1398-1408, 1928.
2. J. Slepian, Theory of the de-ion circuit breaker, AIEE Trans. 48:523-527, 1929.
3. T. E. Browne, Jr., and F. C. Todd, Extinction of short a.c. arcs between brass electrodes, Phys. Rev. 36(4):August 15, 726-731, 1930.
4. J. Slepian and A. P. Strom, Arcs in low-voltage a-c networks, AIEE Trans. 50:847, 1931.
5. T. E. Browne, Jr., Extinction of short a-c arcs, Trans. AIEE 50:1461-1473, 1931.

6. A. P. Strom, Suppression of leakage flux in magnetic air circuit breakers, AIEE Trans. III, 77(Part III):305-309, June 1958.
7. M. Hudis and J. T. Carroll, Modeling experiments applied to ceramic air-magnetic arc chutes, IEEE Trans. Power Appar. Syst. Pas-98:1522-1530, 1979.
8. R. E. Frink, A magnetic type circuit interrupter chute for an air breaker rated 50,000 kVA at 5 kV, Master's thesis, University of Pittsburgh, Pittsburgh, Pa., 1954.
9. A. J. Wurts, Lightning arresters, and the discovery of nonarcing metals, Trans. AIEE 9:102-138, 1892.
10. T. E. Browne, Jr., Dielectric recovery of short a-c arcs between low-boiling-point electrodes, Thesis, California Institute of Technology, Pasadena, Calif., 1936.
11. G. Frind, Time constants of flat arcs which are cooled by thermal conduction, IEEE Trans. Power Appar. Syst. PAS-84: 1125-1131, 1965.
12. G. Frind and R. Hunziker, A fundamental investigation of the interruption process in air-magnetic circuit breakers: Part I. Measurements of arc motion, of wall temperature and of electrical wall conductivity, IEEE Trans. Power Appar. Syst. PAS-90:360-365, 1971.
13. T. E. Browne, Jr., and A. P. Strom, A study of conduction phenomena near current zero for an a-c arc adjacent to refractory surfaces, AIEE Trans. 70:398-407, 1951.
14. A. Tslaf, Calculation of the temperature in a conductive track of an arc on the surface of a magnet-blast chute, IEEE Trans. Plasma Sci. PS-8(4):455-460, 1980.
15. E. W. Boehne and L. J. Linde, "Magne-blast" air circuit breaker for 5,000 volt service, AIEE Trans. 59:202-208, 1940.

12
Interruption in Vacuum

CLIVE W. KIMBLIN, PAUL G. SLADE, and ROY E. VOSHALL/Westinghouse Research and Development Center, Pittsburgh, Pennsylvania

12.1 Historical Background   456

12.2 Vacuum Interrupter Description   457
    12.2.1 Vacuum contactors   458
    12.2.2 Vacuum breakers   458

12.3 Electrode Phenomena in Vacuum   459
    12.3.1 Cathode spots   459
    12.3.2 Anode spots   461

12.4 Arcing and Interruption in Vacuum Interrupters During an Ac Wave   462
    12.4.1 Arc initiation   462
    12.4.2 High-current arc mode   464
    12.4.3 Current zero   470
    12.4.4 Dielectric recovery and voltage withstand   472

12.5 Range of Present-Day Applications   472
    12.5.1 Ac applications   472
    12.5.2 Frontiers to ac application   474
    12.5.3 Dc interruption using vacuum switches subjected to magnetic fields   479

References   480

The capabilities and advantages of the vacuum interrupter for use in medium-voltage switching circuits are widely recognized. Vacuum interrupters are applied in outdoor circuit breakers and distribution reclosers, metal-clad switchgear, tap changers, and contactors. Most present applications are confined to voltages less than or equal to 38 kV, although investigations at higher voltages have been reported.

In this chapter we review the development of the vacuum interrupter, and show how the present interrupter designs have benefited from extensive research and development efforts.

12.1 HISTORICAL BACKGROUND

The engineering concepts of power interruption in vacuum were demonstrated at the California Institute of Technology in the 1920s [1], and initial development work was continued [2] at the General Electric Co. in the 1930s. It was necessary, however, to await the development of ultrahigh-vacuum technology and gas-free metals before sustained performance could be achieved. In the 1950s, the Jennings Company [3] manufactured a line of high-voltage, low-current interrupting devices. The first commercial power vacuum interrupter was marketed in 1963 and resulted from a team effort [4] at General Electric in the United States. Also in the early 1960s, M. P. Reece [5] was conducting studies of power arcing in vacuum in England. Since 1963, many companies throughout the world have contributed to the improvement of vacuum interrupter devices.

The understanding of vacuum arc physics has assisted in improvements in vacuum interrupter technology. Tanberg performed pioneering experiments and observed [6] in 1929 that the low-pressure cathode spot moved in a retrograde direction when subjected to a transverse magnetic field. He also determined [7] that the vapor stream was partly ionized. Tanberg correctly determined [8] that the vapor streaming away from the cathode spot had a velocity of $\sim 1.5 \times 10^6$ cm/s. Berkey and Mason confirmed [9] Tanberg's velocity measurements, and also performed probe measurements which showed that the vapor was indeed ionized to a considerable degree. Furthermore, Slepian and Mason [10] commented that the distribution of density in the jet followed a cosine law. Many scientists and engineers have contributed to the physical understanding of vacuum arcs and interruption phenomena since those early experiments, and a recent compilation by H. C. Miller [11] lists about 2700 papers. Particularly pertinent to the early Tanberg observations [6-8] are the papers of Davis and Miller [12] and Plyutto et al. [13], who confirmed that the ionized vapor leaves the cathode region with an energy (in eV) higher than the arc voltage; of Kimblin [14] and Daalder [15], who confirmed that most of metal vapor is ionized; and the observation [16] of anode voltage drop formation when insufficient ions impinge on the anode. Other more recent, notable advances in the overall area of vacuum arc physics have been in the determination of the nonstationary aspect of cathode spot emission by Daalder [17] and Jüttner [18], investigation of the operating modes of high-current vacuum arcs [5,19-22], and exploration of dielectric recovery mechanisms [14,23-27]. An extensive treatment of these subjects appears in the book *Vacuum Arcs* edited by Lafferty [28], and recent

Interruption in Vacuum

advances in the understanding of arcing phenomena in vacuum are the subject of a paper by Kimblin [29].

12.2 VACUUM INTERRUPTER DESCRIPTION

The internal components of a typical vacuum interrupter are shown in Fig. 12.1. The ambient gas pressure within the evacuated envelope is $\sim 10^{-6}$ torr. Under normal circuit conditions the interrupter is closed and the contacts butt together. Arcing is established within the interrupter by withdrawing the bellows contact from the stationary contact. This arc burns in the metal vapor evaporated from local hot spots on the contact surfaces. The metal vapor continually leaves the intercontact region and recondenses on the contact surfaces and the surrounding metal vapor condensation shield. The latter is usually isolated from both contacts and serves to protect the glass or ceramic envelope from vapor deposition. At current zero, vapor production ceases and the original vacuum condition is rapidly approached. The

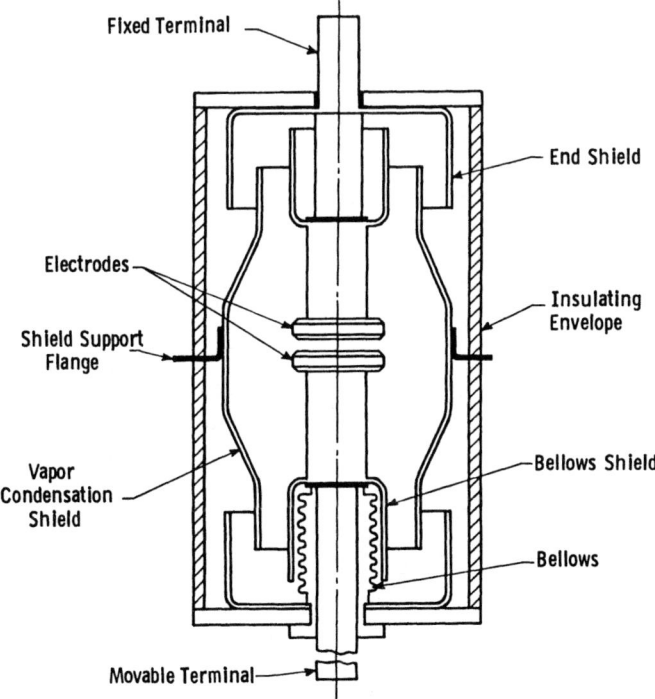

Figure 12.1 Vacuum interrupter schematic.

dielectric strength of the interrupter also increases, and the circuit is interrupted. With the contacts in the open position, the circuit voltage is withstood internally by the intercontact gap and externally by the insulating envelope.

In general, a major distinction can be drawn between vacuum interrupters designed for contactor applications, at circuit voltages of 1.2 to 7.2 kV, and those designed for fault current protection at circuit voltages of about 5 to 38 kV. This distinction arises from the different functions required of the vacuum devices.

12.2.1 Vacuum Contactors

Vacuum contactors are used to control and protect motors in relatively low voltage circuits. These devices must be capable of switching load currents during hundreds of thousands of operations, and must be capable of protecting the circuit against occasional faults, usually in coordination with a fuse. For contactors, the maximum interruption currents of about 5 to 15 kA rms are usually well below the fault currents associated with circuit-breaking devices. Furthermore, the continuous current of several hundred amperes is generally less than that for breakers.

For a low continuous current, the actuating mechanism does not have to provide a high intercontact force to reduce contact heating in the closed position. In fact, the mechanism can be much simpler than a breaker mechanism provided that the contacts have a low weld strength. This is one reason contact materials such as sintered alloys of tungsten and copper [30], silver tungsten carbide [31], and antimony within a molybdenum matrix [32] are used. These materials also provide adequate fault current interruption capability using plain disk contacts. Finally, the materials are selected to prevent [33,34] the creation of damaging overvoltages due to either arc instability prior to current zero or voltage escalation following current zero. Such phenomena are of lesser consequence in breakers where the protected apparatus usually has significantly higher insulation levels.

The low circuit voltages associated with many contactor applications can also lead to simplified designs for the metal vapor-condensation shield, which, in many contactor designs, is suspended from one of the metal end flanges.

12.2.2 Vacuum Breakers

Vacuum interrupters for circuit breakers and recloser duty must be capable of interrupting high fault currents, for example 11.5 kV/48 kA (15 kV/37 kA), and high circuit voltages of 5 to 38 kV. Switching of higher voltages means that the contact spacing must be larger than for the contactor application. This combination of high current and long contact spacings means that different arcing phenomena are encountered, with consequent changes in contact material, together with complex contact configurations for arc control. The contact materials most favored for vacuum interrupters used in breaker and recloser applications are copper-chrome [35] and copper-bismuth [36]. Copper chrome is a semirefractory material and the shape of the contact structure is maintained during high current arcing. The electrical conductivity remains high, and the material has a lower weld strength than pure copper. Copper-bismuth combines the high electrical conductivity of copper with the low weld strength [37,38] associated with the bismuth additive. Contact configurations are discussed in Sec. 12.4.2.

Many manufacturers suspend the center shield from the center of a cylindrical envelope as indicated in Fig. 12.1. Other manufacturers [30], however, make the center shield an integral part of the vacuum envelope, with the total envelope then comprising a central, large-diameter metal section, with smaller-diameter ceramic sections connected to each end of the shield.

12.3 ELECTRODE PHENOMENA IN VACUUM

During arcing, the overall arc gap is an excellent conductor, with an arc voltage of only several tens to several hundreds of volts for arc currents of tens of kiloamperes. At current zero, however, there is an abrupt reduction in overall gap conductance combined with a reversal of contact polarity, and tens of kilovolts are then required [23,24,39] to reignite the arc within microseconds of the natural ac current zero. This rapid change in conductance is intimately related to the cathode spot properties, and can be adversely affected by the presence of anode spots.

12.3.1 Cathode Spots

For arc currents to several thousand amperes, the vacuum arc can be characterized by a multiplicity of rapidly moving cathode spots (the number of spots being approximately proportional to the current and dependent on the contact material), a diffuse interelectrode plasma, and a diffuse collection of current at the anode; see Fig. 12.2. These cathode spots are essential to the maintenance of the discharge.

Most of the arc voltage drop occurs across a space-charge sheath at the cathode. The maximum current conducted by each of the multiple cathode spots varies with the electrode material, and is related [40,41] to the thermal parameter $T_B k^{1/2}$ of the cathode, where T_B is the normal boiling-point temperature and k is the low-temperature thermal conductivity of the cathode material. The current density at the cathode surface of an individual cathode spot is extremely high, and this region is therefore one of high power density. For arcs on copper electrodes, the cathode fall is ~ 18 V, the maximum current per cathode spot is

∿ 100 A, and the cathode spot current density at the surface based on crater sizes, is 10^8 A/cm^2 [18,42].

The phenomenon of multiple cathode spots results in distributed erosion at the cathode, and examination of the individual cathode spot craters [18,42,43] indicates a nonstationary emission process operating on a submicrosecond time scale [18,42]. Erosion rates have been determined for a wide variety of materials [42,44]. Of particular importance is the fact that the cathode spots are regions of intense ionization [44], with most of the material emitted from the surface spots moving away from the cathode region in ionized form. For current levels of several hundred amperes, the ion flux leaves the cathode spots with a spatial distribution approximating a cosine law [15,45-47]. For higher current levels, the ion flux more closely approximates an isotropic distribution [44,47]. Metal particles also stream away from the cathode spot regions, primarily in a direction parallel to the cathode surface [45,46]. For copper arcs, the erosion rate is about 10^{-4} g/C and, in common with most materials, the magnitude of the ion current *leaving* the cathode spot region is about 10% of the arc current [14,15,44]. The ions possess an energy (in eV) which exceeds the arc voltage, and can be considered to migrate away from a localized potential maximum

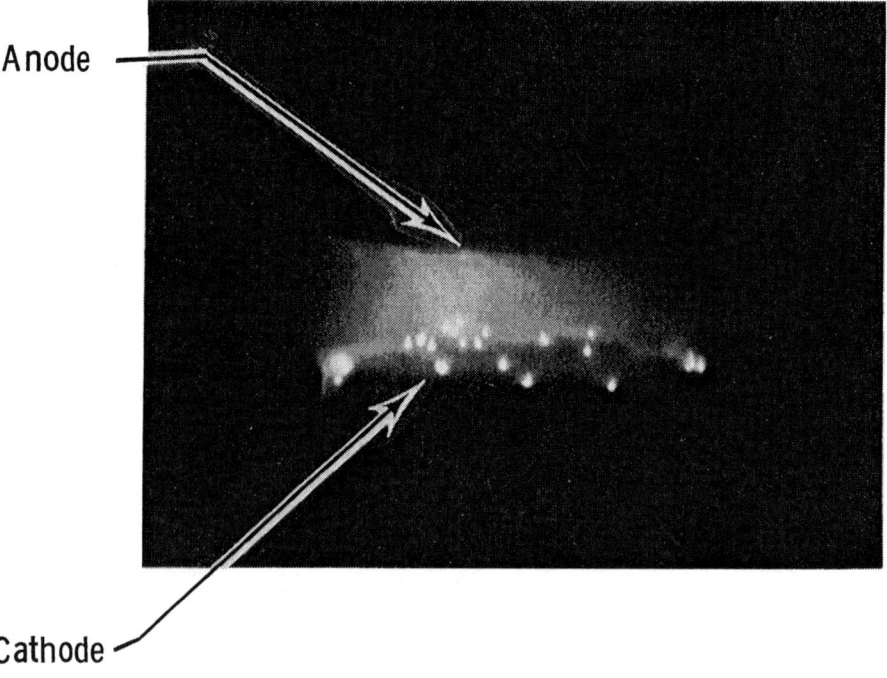

Figure 12.2 Photograph of a low-current arc.

[12,13,48,49] within the cathode fall. For copper the ions are multiply charged [12,13,49] and possess a mean energy $\sim$ 20 to 40 eV (about 10^6 cm/s) per unit charge, whereas the mean vapor energy is < 1 eV.

The primary cathode spot parameters that influence successful interruption at current zero are (1) the high velocity of the ionized metal vapor away from the cathode surface, which leads to a rapid decrease in the interelectrode plasma density, and (2) the absence of a cathode spot at the new cathode. Such a high power density spot on the former anode has to be created in order to conduct appreciable current in the next half cycle.

12.3.2 Anode Spots

During arcing, evaporation always occurs from the cathode. However, for certain arcing conditions, there may also be significant evaporation from the anode due to the formation of a single, grossly evaporating anode spot (see Fig. 12.3). This anode spot operates near the normal boiling temperature of the anode material, and can significantly increase the intercontact plasma density. Furthermore, since the erosion is not

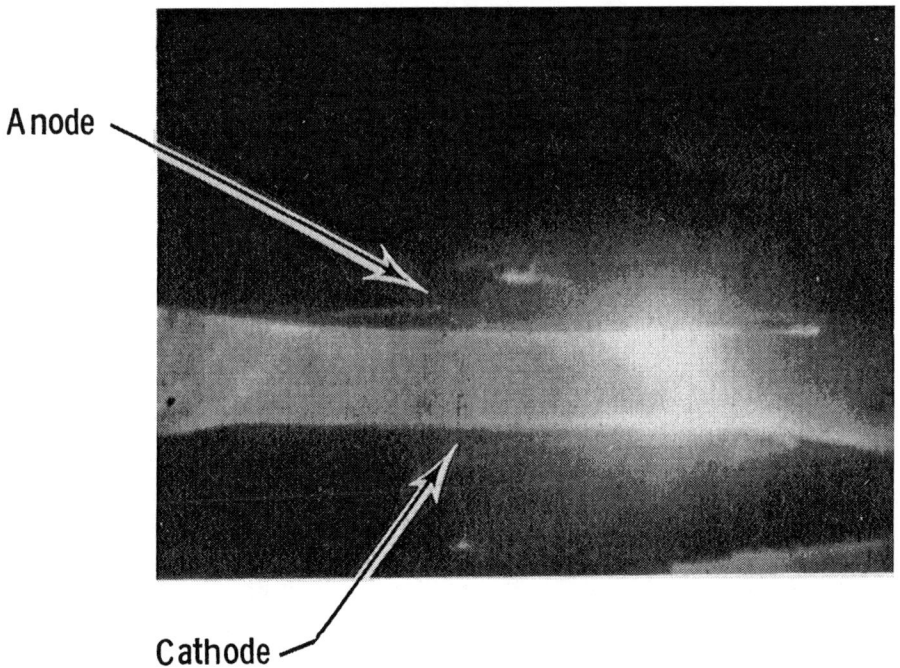

Figure 12.3 Photograph of a high-current vacuum arc with the anode spot visible.

distributed, anode spot formation can cause gross melting of the contact. Finally, the plasma jet from the anode spot can cause the cathode spots to bunch together [19,50], with resulting gross erosion on the cathode.

Anode spots can form from an initially diffuse vacuum arc. The probability of anode spot formation [16,19,20] increases with increasing contact separation, with increasing circuit current and arc duration, and with decreasing anode area. Furthermore, the probability of spot formation increases with a decrease in the anode thermal parameter $T_m(k\rho c)^{1/2}$, where T_m is the melting temperature of the anode material and k, ρ, and c are the thermal conductivity, density, and specific heat respectively. For small anode areas and long contact spacing, when most of the ions in the plasma stream from the cathode spots are no longer incident on the anode, the arc voltage increases due to formation of an anode sheath, and the anode surface is heated. This ion starvation concept [16,19] is strengthened by observations of anode phenomena in the presence of an axial magnetic field [21,51]. Here the cathode ions are confined [47,51,52] to the intercontact region, and spot formation is postponed to higher currents. Magnetic constriction [53,54] may also play an important role in triggering anode spot formation. Localized anode evaporation associated with the constricted arc formed after contact separation is discussed in Sec. 12.5.

The presence of stationary anode spots during the arcing half cycle can adversely affect dielectric recovery [55] due to (1) associated increases in the intercontact plasma and vapor densities, (2) continued evaporation from the localized hot spot following current zero, and (3) for the case of refractory materials such as carbon and tungsten, continued thermionic emission [39] of electrons following contact polarity reversal.

12.4 ARCING AND INTERRUPTION IN VACUUM INTERRUPTERS DURING AN AC WAVE

Figure 12.4 depicts the important arcing and interruption phenomena that occur during fault current interruption in vacuum [56]. These phenomena influence the design of the interrupter, and in particular the size, configuration, and material of the contacts [57]. In this section, arc initiation, the high current arc mode, current zero, and dielectric recovery voltage phenomena are discussed in sequence, with distinctions drawn, where necessary, between contactor and breaker applications.

12.4.1 Arc Initiation

When a vacuum switch is operated, an actuator separates the contacts a distance of 3 to 20 mm. This actuator must first provide sufficient

Interruption in Vacuum

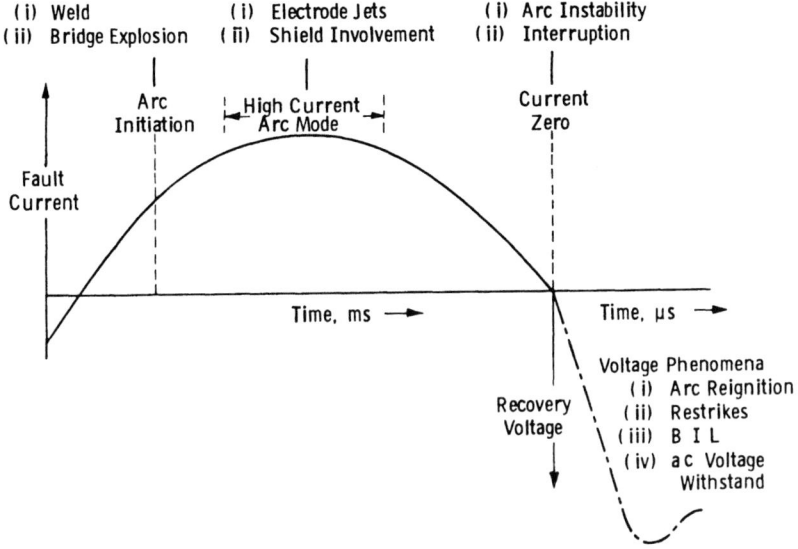

Figure 12.4 Ac arcing and interruption phenomena in vacuum.

force to break the intercontact weld, and contact materials are consequently selected [34,35,58] to ensure a relatively low weld strength.

The contacts are separated at random during the passage of ac current. At the last point of contact between the parting electrodes, a molten metal bridge [59] is formed. This bridge subsequently ruptures and a constricted arc is formed. At currents of a few thousand amperes, this constricted arc is short lived, and cathode spots migrate over the cathode surface from the previous bridge location. Essentially all ionization for this initial intercontact plasma occurs in these cathode spot regions, with recombination solely at the shield and anode. The phenomenon of spot splitting with increasing arc current results in a desirable distributed erosion at the cathode.

For vacuum contactors, with their limited contact stroke of only several millimeters and their limited fault current levels of usually less than 10 kA, during most of the arcing half cycle the vacuum arc can be characterized by a multiplicity of cathode spots, a diffuse interelectrode plasma, and a diffuse collection of current at the anode. There may, however, be limited anode spot activity around current crest.

With respect to vacuum breakers, if the current at the instant of contact separation approaches 10 kA, the bridge rupture can lead to formation of a single, high-vapor-pressure arc column which can remain constricted for 1 or 2 ms prior to changing the arc mode.

12.4.2 High-Current Arc Mode

The appearance of the high-current vacuum arc has been extensively studied by Heberlein and Gorman [22]. They showed that depending on the current level and contact spacing, the diffuse arc can form an anode spot and a columnar arc before going diffuse again close to current zero. It is possible, however, to form a constricted column (Fig. 12.4) from the initially confined arc which will stay constricted until just before current zero. At the highest currents, the electrode regions of this constricted column exhibit intense activity, with jets of material being ejected from the contact faces. In spite of this severe contact activity, even this arc mode can return to the diffuse mode just before current zero. Figure 12.5 shows an example of an arc appearance diagram developed by Heberlein and Gorman.

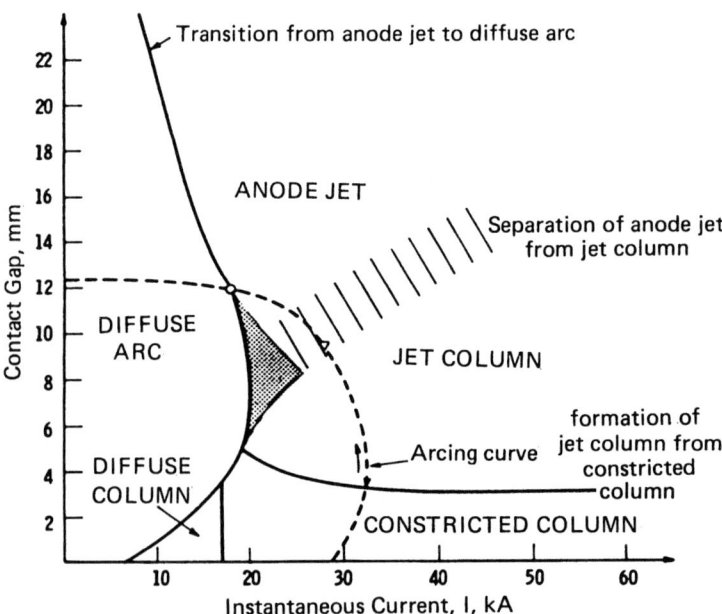

Figure 12.5 Physical arc appearance as a function of current and contact gap for one-half cycle of arcing, 50 to 60 Hz, electrode diameter 10 cm, $I_{sep} > 7$ kA. (From J. V. R. Heberlein and J. G. Gorman, The high current metal vapor arc column between separating electrodes, IEEE Trans. Plasma Sci. *PS-8*:283-289, © 1980, IEEE.)

Electric contacts that have a diffuse vacuum arc burning between them many milliseconds before current zero will have excellent dielectric recovery properties after current zero; this results from the properties of the cathode spots. First, the ions are ejected from the spot with very high velocities ($\sim 10^6$ cm/s), and this leads to a rapid decrease in plasma density within a few microseconds after current zero. Second, since the spots themselves are extremely small ($\sim$ 10 μm), they will cool very rapidly after current zero (microseconds or less), and will not continue to liberate metal vapor into the intercontact space. Finally, during the arcing period, small particles of contact material are ejected from the cathode spots tangentially to the contact face, and thus move away from the intercontact space in the shortest possible time. In contrast, if an anode spot or an uncontrolled columnar arc is present, the contact faces can be grossly heated and there can be liberal evaporation of metal vapor after current zero. It is therefore important to minimize the heating of the vacuum contacts during the current half cycle and to maximize the time during the half cycle for which the arc remains diffuse. Many electric contact designs have been proposed to accomplish this, but only two design types have found practical usage:

1. Contacts that cause the roots of the columnar arc to move rapidly over their surfaces
2. Contacts that force the high-current vacuum arc to remain diffuse

Contacts to enhance arc motion

At high currents, when the vacuum arc is confined, the arc moves in an amperian manner when interacting with a transverse magnetic field [60]. This has led to some very practical designs for vacuum interrupter contacts. Each of these designs accepts the occurrence of the constricted arc, but this arc is forced to move across the contact face through the interaction between the current flowing in the arc and a transverse magnetic field resulting from the current flowing in the contact. The earliest design for this type of electrode, the spiral contact, was patented by Schneider [61] in 1960; a typical design is shown in Fig. 12.6, with the mode of operation illustrated in Fig. 12.7(a). In the diffuse mode the cathode spots run over the cathode surface as if the electrode were a disk. When the arc constricts, the self-magnetic field generated by the current flowing in the spiral arms interacts with the current flowing in the arc, and forces the arc to move along the spirals. When the arc roots reach the end of the spiral arm, they are forced to jump the gap to the next spiral arm by the arc column continuing to move in the transverse direction. A typical photographic sequence of the arc motion is shown in Fig. 12.8.

Another contact design of this class is the contrate or cup contact [62,63] (see Fig. 12.9). It can be seen from Fig. 12.9 that the slanted

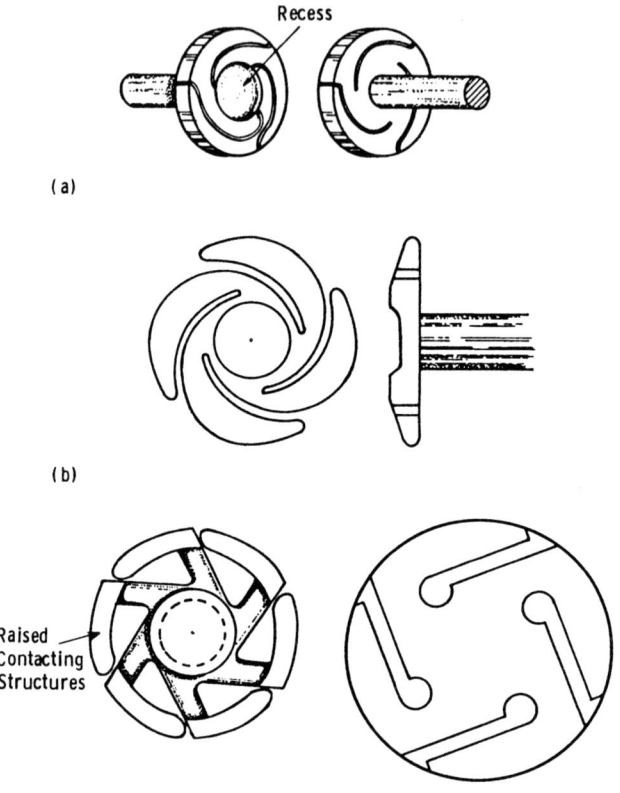

Figure 12.6 Example of a spiral contact structure for vacuum interrupters. (a) The original spiral contact structure proposed by Schneider; (b) an example of a spiral contact structure for vacuum interrupters; (c) variation of the spiral contact. (From P. G. Slade and M. F. Hoyaux, The effect of electrode material on the initial expansion of an arc in vacuum, IEEE Trans. Parts Hybrids Packag. *PHP-8*: 34-47, © 1972, IEEE.)

slots cut into the side of the cup provide a transverse component to the magnetic field which drives the arc around the rim [see Fig. 12.7(b)]. It has been found that the arc runs best in this design if the slots do not extend all the way to the rim [as shown in Figs. 12.7(b) and 12.9] or else with the slots pushed closed at the top.

Interruption in Vacuum

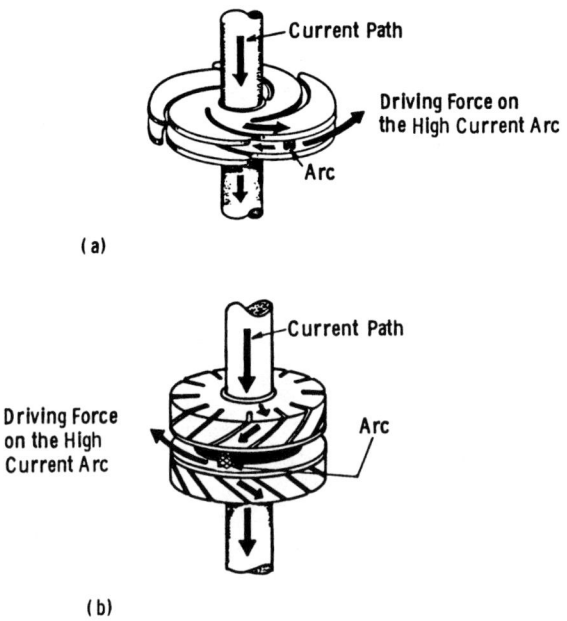

Figure 12.7 "Spiral" and "contrate" contact configurations which cause high-current vacuum arcs to rotate around the contact periphery: (a) "spiral" contact configuration; (b) "contrate" cup-shaped contact configuration.

Contacts that promote diffuse arcs

One method of maintaining a diffuse arc at high currents is to apply an axial magnetic field [52,64]. Heberlein and co-workers, and other investigators, have shown [51,52,64,65] that, for a sufficiently high axial field, the arc maintains the diffuse mode to much higher currents. After the rupture of the molten metal bridge, the confined arc forms, and this arc slowly expands into a diffuse arc which burns until current zero. The detailed expansion of the initial confined arc has not received comprehensive explanation. However, once the arc has gone diffuse, a large enough axial magnetic field allows the arc to remain diffuse. The electrons are confined by the magnetic field lines in the intercontact region and, because of the associated creation of radial electric fields, the ions are also confined to the intercontact region. An example of a 50-kA diffuse arc is shown in Fig. 12.10. During this high-current arcing the diffuse arc distributes the arc energy

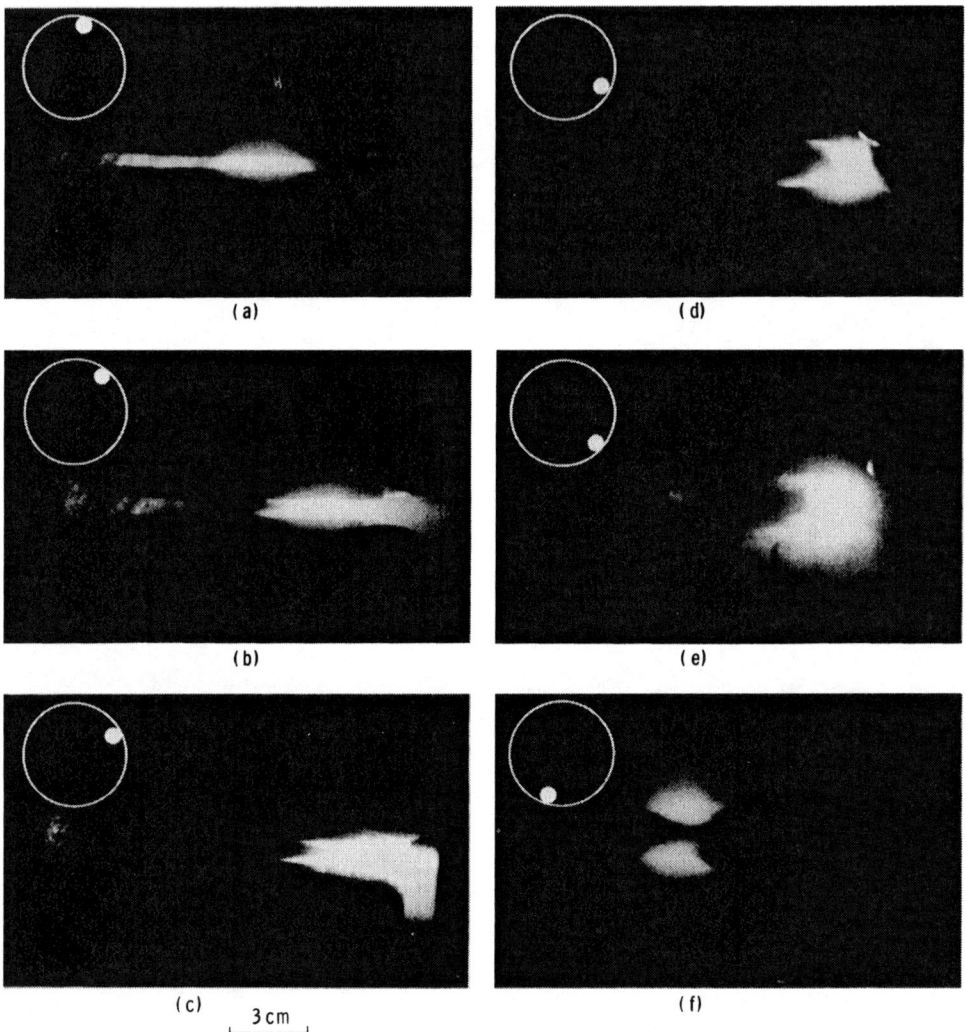

Figure 12.8 Sequence of arc motion on a spiral contact (1 ms between frames, 30 kA rms).

over the whole contact surface and thus prevents gross erosion of the contacts.

One method of applying an axial magnetic field is to wrap a coil, which carries the circuit current, around the outside of the vacuum interrupter. This arrangement has been used successfully for the repetitive switching of high-current dc circuits [51,66]. Although there

Interruption in Vacuum

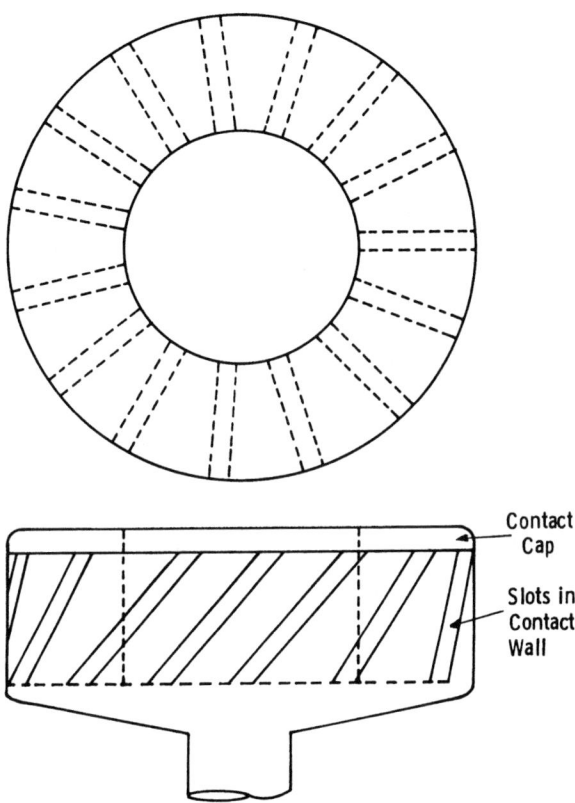

Figure 12.9 Example of a contrate or cup contact design.

have been some applications of external coils to ac switches, the method presents a number of disadvantages and limitations: (1) the resulting interrupter tends to be bulky and in three-phase circuits will affect the pole to pole spacing; (2) the coils must be insulated; (3) when the fault current occurs, it takes time for the magnetic field to penetrate the contact gap; and (4) even when the field penetrates the contact gap, eddy current effects in the vacuum interrupter structure can cause the magnetic field to be out of phase with the current.

A number of axial magnetic field electric contact designs have been proposed to overcome these disadvantages [67-69]. Figure 12.11 presents an example of Yanabu et al. [67]. Here a coil is placed behind a facing contact. The slots in the face of the contact are to prevent eddy currents. The major advantage of this class of contact structure is its ability to keep the arc diffuse during very high currents because

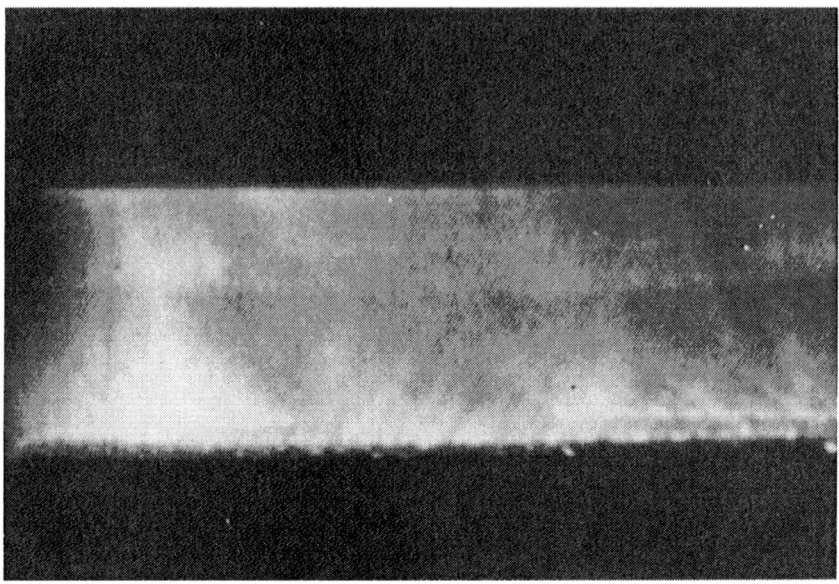

Figure 12.10 Photograph of a diffuse, 52-kA vacuum arc. Contact separation 25 mm, axial magnetic field ∿ 0.05 T.

the magnetic field increases as the current increases. This class of contact structure has been used to interrupt very high currents [67]. The major disadvantage is that the contact design increases the impedance between closed contacts and can cause the contacts to overheat when passing very high steady-state currents.

12.4.3 Current Zero

As the arc current decreases toward current zero, the arc plasma again becomes diffuse, with the electron current being emitted from a declining number of discrete cathode spots. As the current in the final cathode spot approaches the ac zero, the arc may become unstable and extinguish spontaneously. This "chop" phenomenon is intimately connected with instabilities in the cathode spot mechanism [70] and varies with contact material [71]. The current chop is undesirable [72,73] and can lead to high, steep-fronted surge voltages when switching low-current inductive circuits. However, for contactors there is considerable latitude in the selection of contact materials, and the chop phenomenon can be minimized [34].

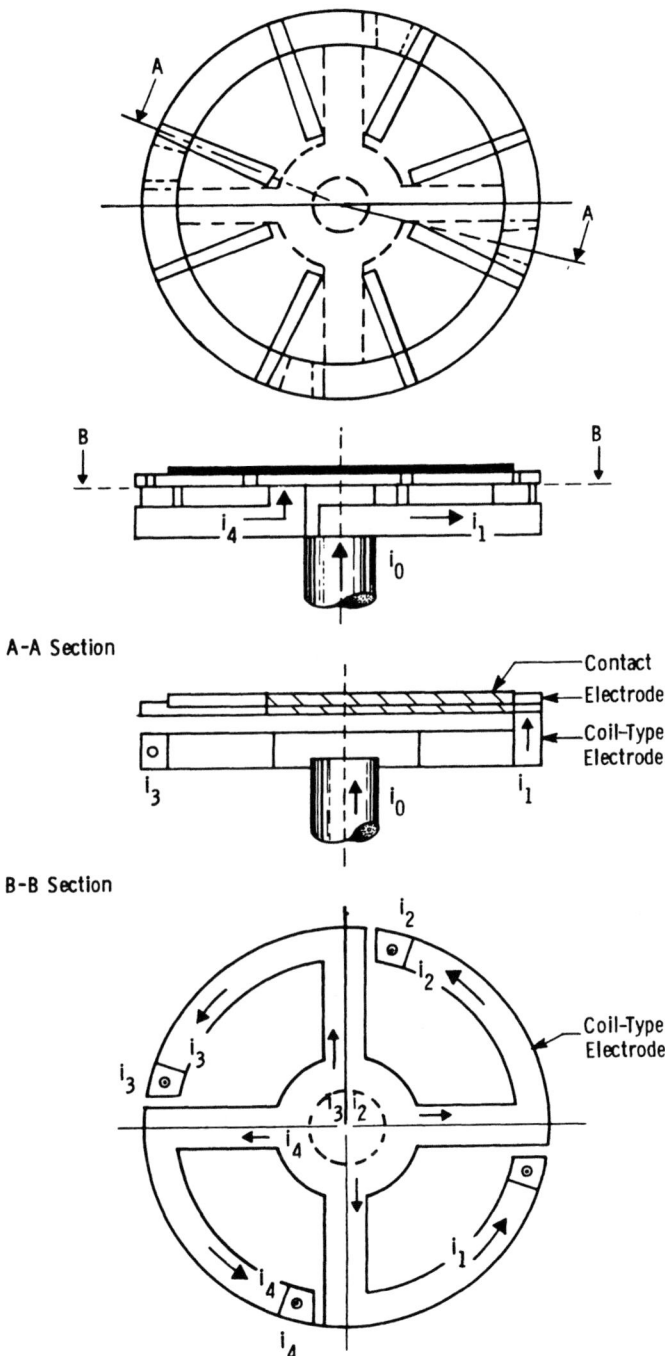

Figure 12.11 Example of an axial-field contact structure. (From Ref. 67.)

In some circuits, high-magnitude overvoltages and steep-front voltage surges may also result [72,74-76] from complex circuit interactions involving the ability of vacuum switches to interrupt high-frequency transient currents, following reignitions in the immediate post-current-zero period. For contactors, with their selected electrode materials, the experience of some manufacturers [34,77,78] indicates that overvoltage generation is not a problem. Other manufacturers suggest suppression for particular contactor applications. For breakers, many applications of vacuum switchgear require [76] no additional surge protection over that which would be applied as good engineering practice with other types of switchgear. As discussed in Chap. 13, applications can be evaluated by computer analysis [76] to determine, where necessary, appropriate means of surge suppression.

12.4.4 Dielectric Recovery and Voltage Withstand

At current zero, the vapor-producing cathode spots extinguish. The residual vapor and plasma within the intercontact region rapidly condense and recombine on both the shield and contact surfaces, and the original vacuum condition is rapidly approached. Analysis [25,79] of the recovery processes directly following current zero is complicated by the nonuniform distribution of the reapplied voltage in the recovering arc gap. In the presence of residual plasma, the circuit voltage is impressed across a narrow space-charge sheath at the new cathode [24,27]. One consequence of this sheath is that the vapor condensation shield, which is in good electrical contact with the intercontact plasma, initially assumes the potential of the new anode [14,24]. Then, as the plasma density decays within microseconds of current zero, the shield rapidly assumes midpotential.

The ac circuit is successfully interrupted if the instantaneous dielectric strength of the recovering intercontact gap always exceeds the circuit reapplied voltage. Full recovery can be attained [79] within microseconds or tens of microseconds of current zero. This ultimate breakdown voltage depends on both the spacing and geometry of the internal shields, and also on the electric field stress on the external envelope of the interrupter. Further, the ultimate breakdown voltage is critically dependent on the spacing of the contacts, the condition of the arced contact surfaces, and the magnitude and duration [80] of the recovery voltage.

12.5 RANGE OF PRESENT-DAY APPLICATIONS

12.5.1 Ac Applications

Vacuum interrupters have proved attractive for use in contactors, breakers, and reclosers because they are reliable, have a long life, and

Interruption in Vacuum 473

require minimal maintenance. The arcing process is also essentially noise-free, with the operating noise being restricted to the mechanism. The interrupters are also extremely compact, with typical vacuum interrupters having envelope diameters between 5 and 20 cm depending on the available circuit fault current, and with the stem diameter determined by the continuous current. An example of vacuum interrupter compactness appears in Fig. 12.12, where a 10-cm-diameter 500-MVA (15 kV/22 kA) interrupter is displayed alongside the arc chute of an air breaker of comparable rating. Of particular importance relative to the overall size of the contactor, breaker, or recloser is the fact that the maximum contact stroke need only be several millimeters to several centimeters. Thus, in commercial equipment, the contact

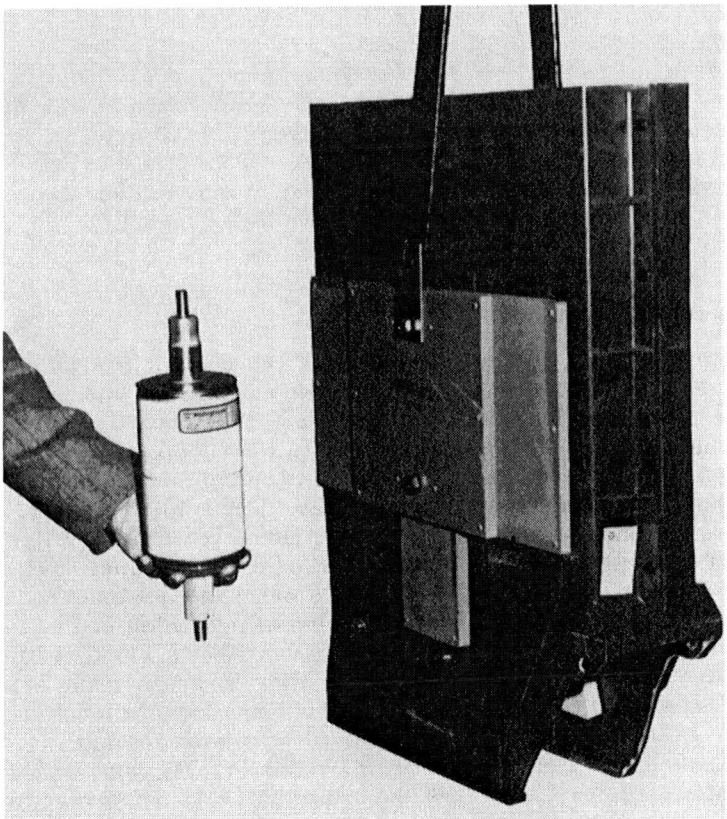

Figure 12.12 Size comparison between a Westinghouse 10-cm-diameter vacuum interrupter rated 500 MVA (15 kV/22 kA) and the arc chute of a Westinghouse air breaker of comparable rating.

stroke typically varies from 3 to 20 mm, with the longer strokes used for the higher circuit voltages. As a result, the actuating mechanism is also compact.

In the United States, vacuum interrupters are currently applied in 5- through 38-kV distribution breakers and reclosers. They are also receiving increasing application in 5- through 15-kV metal-clad switchgear rated for 250 through 1000 MVA. The continuous current for the applications described above varies from 300 to 3150 A. Vacuum interrupters in this voltage range are also used in transformer tap changer applications, where the long life and low-maintenance aspects of vacuum devices are particularly significant. Vacuum interrupters are also used individually for capacitor switching at 15 kV and below, and with two-in-series for capacitor switching at 38 kV [81]. Finally, vacuum contactors are receiving increasing application in the United States, particularly for mining applications, where the features of low maintenance and enclosed arc are most attractive.

In Japan, vacuum interrupters are widely applied in 3.6 through 7.2-kV contactors and motor starters, with some applications at circuit voltages as low as 1.2 kV. Metal-enclosed switchgear is also available over a voltage range of 3.6 through 36 kV. In Europe, vacuum interrupters are currently applied in contactors, in distribution apparatus, and in metal-clad switchgear for the voltage range 1.2 through 36 kV.

12.5.2 Frontiers to Ac Application

As a broad generalization, it can be stated that the present focus for commercial vacuum switching devices is on voltages of 38 kV and below, although there have been notable investigations [82-84] at higher voltages. The principal barrier to applying vacuum interrupters at transmission voltages of 72 kV and above is related to the high-voltage capability of a single gap between arced electrodes. For a 15-kV rating, the arced gap must sustain a recovery voltage rising to about 20 kV within 50 μs of current zero. Furthermore, the arced gap must have an ac withstand capability of 50 kV rms, and a basic impulse level of about 100 kV. All of these are possible for a contact spacing of 13 mm. At 72 kV, however, the recovery voltage can rise to 135 kV within 105 μs of current zero. In particular, the arced gap must have a 1-min ac withstand capability of 160 kV rms and a basic impulse level of about 350 kV. Individual long-gap devices, with several floating shields, have been made with these characteristics [82-84], and series-connected interrupters have also been investigated [85]. However, for the future, economic considerations favor SF_6 puffer technology for 72 kV and above (Chap. 10).

A further consideration relative to the application of vacuum interrupters at circuit voltages higher than 38 kV are the high-voltage re-

Interruption in Vacuum

quirements associated with capacitor bank switching. Such switching [80] differs from fault current interruption because, although the capacitor currents are usually less than the continuous current rating, the recovery voltages can be much higher and are usually unipolar. The generation of these higher voltages, which could cause voltage breakdowns across the interrupters, will now be discussed.

The neutral of many three-phase wye-configured banks is floating. Figure 12.13 shows a simplified diagram of such a bank connected to a line via a circuit breaker. The equivalent line inductance is L_ℓ. When the capacitor current is interrupted by the breaker, the recovery voltage on the first phase to clear achieves the highest value of 2.5 per unit due to the effects of charge remaining on the capacitor and the neutral voltage shift. The other two phases continue to conduct current, and clear 90° later. The recovery voltage across these two phases is 1.87 per unit. However, if these two phases arc for an extra half cycle of current, the recovery voltage on the first phase could increase to 3.0 per unit [86]. For vacuum breakers, an arcing time of this duration is extremely unlikely. Thus for vacuum breakers in

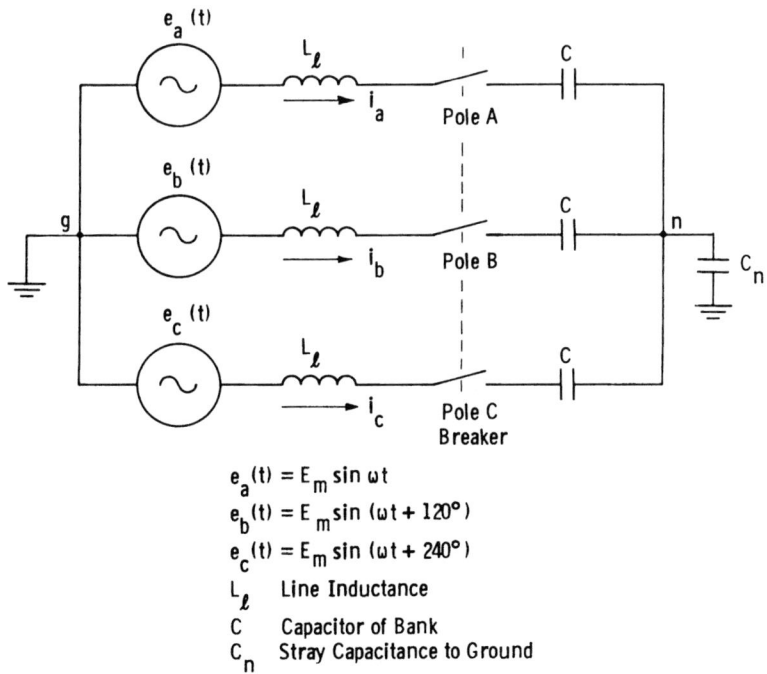

Figure 12.13 Capacitor switching; simplified schematic of source, breaker, and capacitor bank in a three-phase system.

25- and 34.5-kV circuits, where the maximum design voltages of the breaker are 27 and 38 kV, respectively, the recovery voltage across the first pole to clear is the highest, and this recovery voltage has values of 55 kV for a 27-kV circuit and 77.6 kV for a 38-kV circuit.

Figure 12.14 shows current interruption of the first phase to clear and the resulting recovery voltage across the interrupter in that phase. Note that it takes 180 electrical degrees or 8.3 ms for this voltage to reach peak value. Reignitions or breakdowns within 4.2 ms following current interruption are unlikely to occur in this first phase since the voltage has not reached peak value. However, delayed restrikes may subsequently occur. Here we define a delayed restrike as a breakdown which typically occurs several cycles or more after interruption. Such breakdowns result in high-frequency current flow and cause voltage swings on the capacitors and neutral point of the capacitor bank. This can lead to high overvoltages across the capacitor bank and system. Vacuum interrupters are normally applied for capacitor

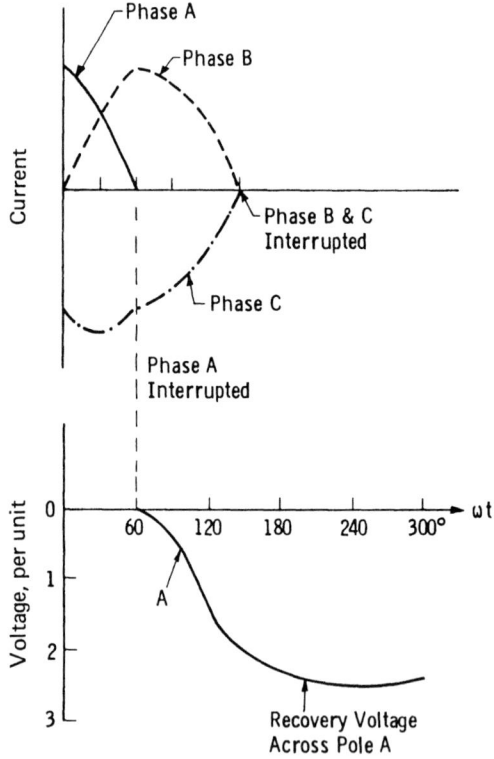

Figure 12.14 Currents in each phase and recovery voltage of first phase to clear.

Interruption in Vacuum

switching at a voltage level at which delayed restrikes will not occur.

Causes of delayed restrikes

Delayed restrikes usually occur 2 to 25 cycles after current interruption. During this time the contacts of the interrupter are in the fully opened position. As indicated in Fig. 12.14, during this period the voltage across the contacts has an ac 60- or 50-Hz component superimposed on a dc component. The mechanisms for this vacuum breakdown are very different from those which occur when an impulse type of wave is applied across the interrupter. Several vacuum breakdown mechanisms may cause delayed restrikes. These can involve the charging of macroparticles at the anode electrode or field emission sites at the cathode electrode. Two basic models of breakdown initiation have been proposed to explain the interaction of particles with contact surfaces:

1. Evaporation of a particle on impact with one of the contacts due to the change from particle kinetic energy to heat energy [87]
2. Evaporation of a slowly moving, positively charged, anode particle due to heating by field-emission current as the particle approaches the cathode [87]

Both particle heating mechanisms can produce sufficient vapor density to cause a local discharge that will break down the gap. Hence a delayed restrike occurs.

Since delayed restrikes can be triggered by the macroparticles between the contacts of a vacuum interrupter or by field-emission sites at the cathode [88], contact surface conditions, contact material, and contact configuration all influence the probability of delayed restrikes. In addition, vibration of the mechanism due to contact opening can trigger delayed restrikes. This vibration can cause macroparticles to be released from the contact with ejection into the gap. Thus the opening characteristics of the interrupter mechanism are important.

Effects of restrikes on breaker and system

Voltage escalation across the breaker and system can be caused by either a prestrike during the closing of the breaker, or by delayed restrikes following interruption of capacitor current. We will consider the effects of delayed restrikes on the breaker and system.

Table 12.1 shows calculated values of the voltage across each pole, from the load side of the breaker to ground, and from the neutral of the capacitor bank to ground, during the switch-off of a capacitor bank. For these calculations, the stray capacitance from the neutral of the capacitor bank to ground is assumed small compared to the capacitance in each phase of the bank. Also, the leakage current of the bank is neglected. After the first phase has cleared (assume phase A), the trapped voltage on the capacitor in phase A is 1.0 per

Table 12.1 Voltages Across the Breaker Poles (e.g., $V_{aa'}$) and Across the Capacitor Bank to Ground (e.g., $V_{a'g}$) as a Function of the Switching Operation [a]

Phase		First phase to clear	Second and third phases to clear	Peak voltage, $V_{aa'}$	Recovery after delayed breakdown, ϕA
A:	$V_{aa'}$	Clear	1.5	2.5	−2.5
	$V_{a'g}$	−1.0	−0.5	−0.5	−3.5
B:	$V_{bb'}$	0	Clear	−1.366	5
	$V_{b'g}$	0.5	−0.866	−0.866	−5.866
C:	$V_{cc'}$	0	Clear	0.366	5
	$V_{c'g}$	0.5	+0.866	+0.866	4.134
N:	V_{ng}	0	0.5	0.5	−4.5

[a]Voltages are per unit of the line-to-ground peak voltages; circuit resistance and damping are neglected.

unit, and the other two phases (B and C) continue arcing for an additional 4.1 ms. After this period the neutral-to-ground voltage of the capacitor bank is 0.5 per unit, and the voltages across the capacitors in phases B and C are −1.366 and 0.366 per unit, respectively. Subsequently, the maximum recovery voltage across phases A, B, and C are 2.5, 1.866, and 1.866 per unit. If a restrike occurs at the peak of the recovery voltage in phase A, a high-frequency current flows having a frequency f and magnitude I_{max} of

$$f = \frac{1}{2\pi(C_A L_1)^{1/2}} \quad (12.1)$$

and

$$I_{max} = \frac{V_{BD}}{(L_1/C_A)^{1/2}} \quad (12.2)$$

where V_{BD} is the breakdown voltage of phase A. If phase A experiences a breakdown at 2.5 per unit and the high-frequency current is interrupted at the first current zero, the voltage of the capacitor bank to ground would increase from 0.5 to −4.5 per unit, while the voltages across the capacitors would remain unchanged. This large shift in the neutral voltage produces voltages from the load side of the breaker to ground of 5.866 and 4.134 per unit. This voltage escalation could lead to a breakdown either across one of the poles of the breaker or from the load side of the breaker to ground.

12.5.3 Dc Interruption Using Vacuum Switches Subjected to Magnetic Fields

We have seen that vacuum interrupters are widely applied for switching ac circuits, with the arc gap presenting a low impedance to the circuit current until the natural current zero. The arc voltage is typically small in comparison to the circuit voltage, and at first sight this makes vacuum interrupters an unlikely candidate for high-voltage dc switching. However, vacuum interrupters can be applied in dc circuits by *creating* a current zero. This is achieved by either counterpulsing the arc current to zero, or by forcing the arc to become unstable through application of a rapidly rising transverse magnetic field.

Current counterpulse

In this technique, a dc arc of current I_0 is established in the vacuum interrupter by separating the current-carrying contacts. Then, at a contact spacing consistent with the required voltage withstand capabilities, the arc current is counterpulsed to zero [89] by connecting a precharged capacitor C in parallel with the arcing contacts. Following current zero, the arc gap rapidly recovers dielectric strength against the circuit-imposed recovery voltage, which has a rate of rise I_0/C. For dc interruption, this recovery voltage has the same polarity as that of the initial arc voltage. The counterpulse technique has been used for the interruption of high-voltage dc circuits [90,91] and for the repeated interruption of the ohmic heating coils of fusion machines [51,92,93]. For contact separation at the current levels of tens of kiloamperes associated with fusion machines, a columnar arc [22,51] develops from the initial bridge point. This columnar arc will go diffuse after several milliseconds and, for rapid dielectric recovery, it is important to retain this diffuse arc to the maximum contact stroke, at which time the current counterpulse is applied. To maintain a diffuse arc with the attendant properties of diffuse current collection at the anode, distributed cathode erosion, and rapid dielectric recovery, the vacuum arc can be subjected to a magnetic field parallel to the electrode axis. As discussed in Sec. 12.4.2, this field interacts [21,51, 52,64] with the discharge to confine the cathode spot ions to the intercontact region, and avoids anode spot activity. In the presence of this axial magnetic field, high-current arcs in the tens of kiloamperes range on copper contacts have the appearance of many highly collimated, parallel 100-A arcs.

Rapidly rising transverse magnetic field

For ac applications, the contacts are shaped to provide a magnetic field component transverse to the high-current arc. This drives the arc around the contact periphery, and the arc voltage remains low. If, however, a high-magnitude external magnetic field is rapidly applied to a diffuse vacuum arc, the arc voltage can rise to several kilovolts [94-96] within microseconds of field application. Here the magnetic field forces the plasma to bow away from the anode with consequent creation of an electron space-charge sheath in the anode region.

In the presence of a previously uncharged parallel capacitance across the vacuum arc, the rapid rise in arc voltage can commutate the current into this capacitance with consequent dc interruption. This concept has been used successfully [51,97] in developing a metallic return transfer breaker for the Pacific Intertie.

REFERENCES

1. R. W. Sorensen and H. E. Mendenhall, Vacuum switching experiments at California Institute of Technology, AIEE Trans. 45:1102-1105, 1926.
2. W. Kling, A new vacuum switch, Gen. Electr. Res. Rev. 38:515-516, 1935.
3. H. C. Ross, Vacuum switch properties for power switching applications, AIEE Trans. 77(Pt. III):104-117, April 1958.
4. J. D. Cobine, Research and development leading to the high-power vacuum interrupter: a historical review, AIEE Trans. 82: (Pt. III):201-217, April 1963.
5. M. P. Reece, The vacuum switch: Part I. Properties of the vacuum arc; Part 2: Extinction of an a.c. vacuum arc, Proc. IEE (Lond.) 110:793-811, April 1963.
6. R. Tanberg, Motion of an electric arc in a magnetic field under low gas pressure, Nature 124:371-372, September 1929.
7. R. Tanberg, A theory of the electric arc drawn under low gas pressure, Westinghouse Sci. Paper 408, December 1929 (unpublished).
8. R. Tanberg, On the cathode of an arc drawn in vacuum, Phys. Rev. 35:1080-1089, May 1930.
9. W. E. Berkey and R. C. Mason, Measurements on the vapor stream from the cathode of a vacuum arc, Phys. Rev. 38:943-947, September 1931.
10. J. Slepian and R. C. Mason, High velocity vapor jets at cathodes of vacuum arcs, Phys. Rev. 37:779, March 1931.
11. H. C. Miller, A bibliography and author index for electrical discharges in vacuum (1897-1982), General Electric Tech. Inf. Ser. No. GEPP-366c, March 1984.
12. W. D. Davis and H. C. Miller, Analysis of the electrode products emitted by dc arcs in a vacuum ambient, J. Appl. Phys. 40:2212-2221, April 1969.
13. A. A. Plyutto, V. N. Ryzhkov, and A. T. Kapin, High speed plasma streams in vacuum arcs, Sov. Phys.-JETP (New York) 20: 328-337, February 1965.

14. C. W. Kimblin, Vacuum arc ion currents and electrode phenomena, Proc. IEEE 59:546-555, April 1971.
15. J. E. Daalder, Erosion and the origin of charged and neutral species in vacuum arcs, J. Phys. D: Appl. Phys., 8:1647-1659, 1975.
16. C. W. Kimblin, Anode voltage drop and anode spot formation in dc vacuum arcs, J. Appl. Phys. 40:1744-1752, March 1969.
17. J. E. Daalder, Cathode spots and vacuum arcs, Physica, 104C: 91-106, 1981.
18. B. Jüttner, Formation time and heating mechanism of arc cathode craters in vacuum, J. Phys. D: Appl. Phys. 14:1265-1275, 1981.
19. G. R. Mitchell, High current vacuum arcs: Part I and II, Proc. IEE (London) 117:2315-2332, 1970.
20. J. A. Rich, L. E. Prescott, and J. D. Cobine, Anode phenomena in metal-vapor arcs at high currents, J. Appl. Phys. 42: 587-601, February 1971.
21. O. Morimiya, S. Sohma, T. Sugawara, and H. Mizutani, High current vacuum arcs stabilized by axial magnetic fields, IEEE Trans. Power Appar. Syst. PAS-92:1723-1730, 1973.
22. J. V. R. Heberlein and J. G. Gorman, The high current metal vapor arc column between separating electrodes, IEEE Trans. Plasma Sci. PS-8:283-289, 1980.
23. J. A. Rich and G. A. Farrall, Vacuum arc recovery phenomena, Proc. IEEE. 52:1293-1301, November 1964.
24. C. W. Kimblin, Dielectric recovery and shield currents in vacuum arc interrupters, IEEE Trans. Power Appar. Syst. PAS-90:1261-1270, May/June 1971.
25. R. E. Voshall, Current interruption capacity of vacuum switches, IEEE Trans. Power Appar. Syst. PAS-91:1219-1224, 1972.
26. G. J. Bauer and R. Holmes, Deionization of an interrupted vacuum arc, Proc. IEE (London)124:266-272, 1977.
27. S. E. Childs and A. N. Greenwood, A model for dc interruption in diffuse vacuum arcs, IEEE Trans. Plasma Sci. PS-8(4): 289-294, 1980.
28. J. M. Lafferty, ed. *Vacuum Arcs: Theory and Application*, Wiley, New York, 1980.
29. C. W. Kimblin, A review of arcing phenomena in vacuum and in the transition to atmospheric pressure arcs. IEEE Trans Plasma Sci. PS-10:322-330, December 1982.
30. H. Bettge, Design and construction of vacuum interrupters, Siemens Power Eng. J. (Special Issue on Vacuum Switchgear for Voltages of Up to 36 kV)3:4-8, 1981.
31. A. Nabae, M. Arii, O. Arakawa, and S. Sugiyama, Vacuum switch contact, U. S. Pat. 3,693,138, August 1972.

32. J. N. M. Legate and W. C. Brooks, Use of vacuum contactors in ac motor starters, J. Sci. Technol., 37:182-185, 1970.
33. F. Battiwala, H. Fink, M. Rimmrott, and W. Schultz, 3AF vacuum circuit-breakers and 3TL vacuum contactors in system operation, Siemens Power Eng. J. (Special Issue on Vacuum Switchgear of Up to 36 kV)3:43-50, 1981.
34. Present state-of-the-art switchgear, Brochure incorporating 13 papers circulated by the Toshiba Corporation, Japan, 1980. Also see Toshiba Rev. Int. Ed., No. 141:5-20, September/October 1982.
35. P. G. Slade, Contact materials for vacuum interrupters, IEEE Trans. Parts Hybrids Packag. PHP-10:43-47, March 1974.
36. T. H. Lee and J. D. Cobine, "Vacuum type circuit interrupter, U. S. Pat. 3,462,573, August 1969.
37. P. Barkan, T. H. Lee, J. M. Lafferty, and J. L. Talento, Development of contact materials for vacuum interrupter, IEEE Trans. Power Appar. Syst. PAS-90:350-369, January 1971.
38. P. G. Slade, An investigation into the factors contributing to welding of contact electrodes in high vacuum, IEEE Trans. Parts Mater. Packag. PMP-7:23-33, March 1971.
39. J. D. Cobine and G. A. Farrall, Recovery characteristics of vacuum arcs, IEEE Trans. Commun. Electron. 88:246-253, May 1963.
40. B. E. Djakov and R. Holmes, Cathode spot division in vacuum arcs with solid metal cathodes, J. Phys. D: Appl. Phys. 4: 504-509, April 1971.
41. L. P. Harris, Arc cathode phenomena, in *Vacuum Arcs: Theory and Application* (J. M. Lafferty, ed.), Wiley, New York, 1980.
42. J. E. Daalder, Cathode-erosion of metal vapor arcs in vacuum, Thesis, University of Eindhoven, The Netherlands, 1978.
43. D. R. Porto, C. W. Kimblin, and D. T. Tuma, Experimental observations of cathode spot surface phenomena in the transition from a vacuum metal vapor arc to a nitrogen arc, J. Appl. Phys. 53:4740-4749, July 1982.
44. C. W. Kimblin, Erosion and ionization in the cathode spot regions of vacuum arcs, J. Appl. Phys. 44:3074-3081, July 1973.
45. D. T. Tuma, C. L. Chen, and D. K. Davies, Erosion products from the cathode spot region of a copper vacuum arc, J. Appl. Phys. 49:3821-3821, 1978.
46. Z. Zalucki and J. Kutzner, Ion currents in the vacuum arc, Proc. 8th Int. Symp. Discharges Electr. Insul. Vacuum, Novosibirsk, USSR, August 1976, pp. 297-302.
47. J. V. R. Heberlein and D. R. Porto, The interaction of vacuum arc ion currents with axial magnetic fields, IEEE Trans. Plasma Sci. PS-11:152-159, September 1983.

48. L. P. Harris, A mathematical model for cathode spot operation, Proc. 8th Int. Symp. Discharges Electr. Insul. Vacuum, Albuquerque, N. M., September 1978, pp. F1-F18.
49. V. M. Lunev, V. G. Padalka, and V. M. Khoroshih, Plasma properties of a metal vacuum arc: Part 2, Sov. Phys.-Tech. Phys. *22*:858-861, 1977.
50. C. W. Kimblin, Anode phenomena in vacuum and atmospheric pressure arcs, IEEE Trans. Plasma Sci. *PS-2*:310-319, December 1974.
51. J. G. Gorman, C. W. Kimblin, R. E. Voshall, R. E. Wien, and P. G. Slade, The interaction of vacuum arcs with magnetic fields and applications, IEEE Trans. Power Appar. Syst. *PAS-102*:257-266, February 1983.
52. C. W. Kimblin and R. E. Voshall, Interruption ability of vacuum interrupters subjected to axial magnetic fields, Proc. IEE (Lond.) *119*:1754-1758, December 1972.
53. R. L. Boxman, Magnetic constriction effects in high-current vacuum arcs prior to the release of anode vapor, J. Appl. Phys. *48*:2338-2345, 1977.
54. D. Schuöcker, Improved model for anode spot formation in vacuum arcs, IEEE Trans. Plasma Sci. *PS-7*:209-216, 1979.
55. G. Frind, J. J. Carroll, and C. P. Goody, Recovery times of vacuum interrupters which have stationary anode spots, IEEE Trans. Power Appar. Syst. *PAS-101*:775-781, April 1982.
56. C. W. Kimblin, Arcing and interruption phenomena in ac vacuum switchgear and in dc switcher subjected to magnetic fields, IEEE Trans. Plasma Sci. *PS-11*:173-181, October 1983.
57. P. G. Slade, The vacuum interrupter contact, IEEE Trans. Components Hybrids and Manufacturing Tech., *CHMT-7*, March 1984.
58. J. M. Lafferty, P. Barkan, T. H. Lee, and J. L. Talento, Vacuum circuit interrupter contacts, U.S. Pat. 3,246,969, April 1966.
59. P. G. Slade and M. F. Hoyaux, The effect of electrode material on the initial expansion of an arc in vacuum, IEEE Trans. Parts Hybrids Packag. *PHP-8*:35-47, March 1972.
60. F. D. Althoff, Forschungsarbeiten auf dem Gebiet Elektrischer Vakuumschalter für grosse Ströme, Elektrotech. Z. Ausg. A. *92*(9):538-543, 1971.
61. H. N. Schneider, Contact structure for an electric circuit interrupter, U.S. Pat. 2,949,520, August 1960.
62. S. R. Smith, Contact structure for an electric circuit interrupter, U.S. Pat. 3,089,936, May 1963.
63. M. P. Reece, A review of the development of the vacuum interrupter, Philos. Trans. R. Soc. Lond, A, *275*:121-129, 1973.
64. S. Yanabu, S. Souma, T. Tamagawa, S. Tamashita, and T. Tsutsumi, Vacuum arc under an axial magnetic field and its in-

terrupting ability, Proc. IEE (Lond.) 126(4):313-320, April 1979.
65. J. V. R. Heberlein and J. G. Gorman, The columnar arc under the influence of an axial magnetic field, Proc. IEE Conf. Gaseous Discharges, Liverpool, England, September 1978, p. 281.
66. R. W. Warren, Vacuum interrupters used for the interruption of high dc currents, Proc. 7th Symp. Eng. Probl. Fusion Res. Knoxville, Tenn., October 1977, p. 1774.
67. S. Yanabu, E. Kaneko, H. Okumura, and T. Aiyoshi, Novel electrode structure of vacuum interrupter and its practical application, IEEE Trans. Power Appar. Syst. PAS-100:1966-1974, March/April 1981.
68. P. O. Wayland, J. G. Gorman, and R. E. Voshall, Vacuum type circuit interrupter with an improved contact with axial magnetic field, U.S. Pat. 4,260,864, April 1981.
69. Y. Kurosawa, Y. Kawakubo, H. Sugawara, and T. Takasuna, Behavior of vacuum arcs in transverse magnetic field and axial magnetic field, Proc. 10th Int. Symp. Discharges Electr. Insul. Vacuum, Columbia, S.C., October 1982, pp. 261-267.
70. G. Ecker, Theory of the vacuum arc; Part V: Chopping, General Electric Res. Rep. 74CRD018, January 1974.
71. G. A. Farrall, Current zero phenomena, in *Vacuum Arcs: Theory and Application* (J. M. Lafferty, ed.), Wiley, New York, 1980.
72. A. N. Greenwood, D. R. Kurtz, and J. C. Sofianek, A guide to the application of vacuum circuit breakers, IEEE Trans. Power Appar. Syst. PAS-90:1589-1597, July/August 1971.
73. M. Murano, T. Fujii, H. Nishikawa, S. Nishikawa, and M. Okawa, Current chopping phenomena of medium voltage circuit breakers, IEEE Trans. Power Appar. Syst. PAS-96(1):143-149, 1977.
74. Y. Murai, T. Nitta, T. Takami, and T. Itoh, Protection of motor from switching surge by vacuum switch, IEEE Trans. Power. Appar. Syst. PAS-93(5):1472-1477, 1974.
75. K. Yokokura, S. Masuda, H. Nishikawa, M. Ikawa, and H. Ohaski, Multiple restriking voltage effect in a vacuum circuit breaker on motor insulation, IEEE Trans. Power Appar. Syst. PAS-100:1940-1948, April 1981.
76. J. F. Perkins and D. Bhasavanich, Vacuum switchgear application study with reference to switching surge protection, IEEE Trans. Industry Applications, IA-19(5):879-888, Sept./Oct. 1983.
77. H. Fink, K. Müller, and E. Pflaum, Service experience with 3AF vacuum circuit-breakers and 3TL vacuum contactors, Siemens Power Eng. J. (Special Issue On Vacuum Switchgear of Up to 36 kV) 3:51-56, 1981.

78. T. Ishibashi, N. Ichihara, and T. Iida, Newly developed high voltage combination starters, Toshiba Rev. Int. Ed. No. 128: 22-28, July/August 1980.
79. G. A. Farrall, Recovery of dielectric strength after current interruption in vacuum, IEEE Trans. Plasma Sci. $PS-6(4)$:360-369, 1978.
80. R. E. Voshall and F. A. Holmes, An indirect test circuit for testing vacuum interrupters for capacitor switching duty, IEEE Trans. Power. Appar. Syst. $PAS-99(3)$:1276-1279, 1980.
81. R. A. Few and R. E. Voshall, 25 and 34.5 kV capacitor switching with vacuum circuit breakers, 45th Annu. Meet. Am. Power Conf., Chicago, April 1983.
82. E. Umeya, H. Hisatune, H. Yanagisawa, and T. Sano, A new high voltage interrupter, IEEE Power Eng. Soc. Winter Meet., New York, January 1975, Paper C-75-099-7.
83. R. E. Voshall, C. W. Kimblin, P. G. Slade, and J. G. Gorman, Experiments on vacuum interrupters in high voltage 72 kV circuits, IEEE Trans. Power. Appar. Syst. $PAS-99(2)$:658-666, 1980.
84. J. A. Rich, G. A. Farrall, I. Iman, and J. C. Sofianek, Development of a high power vacuum interrupter, EPRI Rep. EL-1895, Palo Alto, Calif., June 1981.
85. R. B. Shores and V. E. Phillips, High voltage vacuum circuit breaker, IEEE Trans. Power Appar. Syst. $PAS-94(6)$:1821-1830, 1975.
86. A. N. Greenwood, *Electrical Transients in Power Systems*, Wiley-Interscience, New York, 1971, pp. 99-106.
87. D. K. Davies, The initiation of electrical breakdown in vacuum— A review, J. Vacuum Sci. Technol. $10(1)$:115-121, 1973.
88. K. Fröhlich, and W. Widl, Determination of the microscopic field enhancement factor β of prestressed vacuum interrupter contacts, Proc. 14th Int. Conf. Phenom. Ionized Gases, J. Phys. (Paris) *40* (Suppl. to No. 7):407-408, July 1979.
89. A. N. Greenwood and T. H. Lee, Theory and application of the commutation principle for HVDC circuit breakers, IEEE Trans. Power Appar. Syst. $PAS-91$:1570-1574, July/August 1972.
90. S. Yanabu, T. Tamagawa, S. Irokawa, T. Horiuchi, and S. Tomimuro, Development of HVDC circuit breaker and its interrupting test, IEEE Trans. Power Appar. Syst. $PAS-101$:1958-1965, July 1982.
91. W. J. Premerlani, Forced commutation performance of vacuum switches for HVDC breaker application, IEEE Trans. Power Appar. Syst. $PAS-101$:2721-2727, August 1982.
92. S. Tamura, R. Shimada, Y. Kito, Y. Kanai, H. Koike, H. Ikeda, and S. Yanabu, Parallel interruption of high direct current by vacuum circuit breakers, IEEE Trans. Power Appar. Syst. $PAS-99$:1119-1129, May/June 1980.

93. R. Warren, M. Parsons, E. Honig, and J. Lindsay, "Tests of vacuum interrupters for the Tokamak Fusion Test Reactor," Los Alamos Rep. LA-7759-MS, Los Alamos, N. M., April 1979.
94. P. R. Emtage, C. W. Kimblin, J. G. Gorman, F. A. Holmes, J. V. R. Heberlein, R. E. Voshall, and P. G. Slade, Interaction between vacuum arcs and transverse magnetic fields with application to current limitation, IEEE Trans. Plasma Sci. *PS-8*: 314-319, December 1980.
95. R. Dollinger, D. R. Dettman, J. L. Lee, A. S. Gilmour, Jr., D. P. Malone, and P. R. Schwartz, Anode sheath growth and collapse in a hollow-anode vacuum arc, IEEE Trans. Plasma Sci. *PS-8*:302-307, December 1980.
96. P. D. Pedrow, L. M. Burrage, and J. L. Shohet, Performance of a vacuum arc-commutating switch for a fault-current limiter, IEEE Power Eng. Soc. Summer Meet., San Francisco, July 1982, Paper 82 SM 373-1.
97. A. L. Courts, J. J. Vithayathil, N. G. Hingorani, J. W. Porter, J. G. Gorman, and C. W. Kimblin, A new dc breaker used as metallic return transfer breaker, IEEE Trans. Power Appar. Syst. *PAS-101*(10):4112-4121, October 1982.

13
Vacuum Circuit Breaker Application and Surge Protection*

JOHN F. PERKINS/Westinghouse Electric Corporation, East Pittsburgh, Pennsylvania

13.1 Historical Perspective 488
13.2 Characteristics of Vacuum Circuit Breakers 488
    13.2.1 Switching surge phenomena 489
    13.2.2 Switching surge parameters and frequencies 490
13.3 Current Chopping 491
13.4 Multiple Reignitions and Voltage Escalation 494
13.5 Virtual Current Chopping 496
13.6 Circuits Most Subject to Surge 498
13.7 Determination of Switching Transient Voltages 498
13.8 Vacuum Interrupter Surge Program Input Parameters 505
    13.8.1 Data development 505
13.9 Validation of the Computer Program by Switching Tests 510
    13.9.1 Single-phase correlation 510
    13.9.2 Three-phase correlation: Unloaded transformer switching 512

---

*© 1983, IEEE. Based on material in the following IEEE Papers: J. F. Perkins, Evaluation of switching surge overvoltages on medium voltage power systems, IEEE Trans. Power Appar. Syst. *PAS-101*(6):1727–1734, 1982; and J. F. Perkins and D. Bhasavanich, Vacuum switchgear application study with reference to switching surge protection, (IEEE Trans. Industry Applications, Vol. IA-19, No. 5, Sept./Oct. 1983, pp. 879-888.)

13.10 Typical Switching Applications   514
    13.10.1 Unloaded transformer switching   514
    13.10.2 Switching motors under normal running conditions   517

13.11 Recommended Practice for System Protection   520
    13.11.1 Surge protection means   520
    13.11.2 Protection for specific systems   521

13.12 Summary   524

References   524

13.1 HISTORICAL PERSPECTIVE

The development of vacuum interrupters has been discussed in some detail in Chap. 12. There, a distinction has been drawn between vacuum contactors, designed for repetitive switching and motor-inrush current interruption, and vacuum breakers designed for higher-voltage, high-current fault interruption. Different contact materials are used for the two applications.

The present chapter focuses on vacuum circuit breaker applications and the measures that have to be taken, if any, to provide circuit protection against switching surges [1-7]. In the early years of vacuum technology, vacuum interrupters used in vacuum switches developed a reputation for generating overvoltages because of their high chop currents. Modern vacuum circuit breakers use vacuum interrupters having fast dielectric recovery rates and relatively low chop currents, so that they are much less prone to generating high overvoltages than were the early vacuum switches.

13.2 CHARACTERISTICS OF VACUUM CIRCUIT BREAKERS

Circuit breakers using different media for interrupting current may exhibit different characteristics with respect to all of the primary functional requirements of the breaker. Each type of circuit breaker has a unique set of characteristics which must be thoroughly understood before the breaker can be applied with safety and confidence.

All types of switchgear—magnetic air, vacuum, oil, SF_6—must meet certain basic requirements, including switching and interruption capability, BIL withstand capability, and mechanical and thermal specifications. A further requirement relates to limiting switching surges, and an important consequence of complying with the requirements is that no switching surge should cause insulation or arcing damage to any equipment on that system switched by the breaker. Because of the increasing use of vacuum switchgear in many applications, interest

Vacuum Circuit Breaker Application

has naturally arisen as to whether different considerations should apply with respect to the application of the newer switchgear.

Vacuum switchgear has some obvious advantages over the older equipment. Small size and reduced weight make for more efficient space utilization. The small weight of moving parts, short contact stroke, and minimal contact erosion permit fast interruption and prolonged service life of the operating mechanism and vacuum interrupters. Fast recovery of the dielectric strength of vacuum interrupters contributes to the excellent and reliable interruption performance of vacuum breakers. The hermetically sealed vacuum interrupters contain all arcing, so that vacuum breakers are relatively noise-free, and there is no explosion hazard. Vacuum breakers require less maintenance than do the established magnetic air breakers.

The excellent interruption and dielectric recovery characteristics of vacuum interrupters contribute to their current chopping and multiple reignition properties [8,9]. This chapter presents a proper understanding of such characteristics with a view to demonstrating what types of protection, if any, might be required to ensure control of switching surge phenomena so that safe and reliable application of vacuum metal-clad switchgear on medium-voltage power systems can be determined.

13.2.1 Switching Surge Phenomena

The switching surge phenomena usually associated with vacuum interruption include current chopping, multiple reignitions, voltage escalation, virtual current chopping, and prestriking.

Current chopping refers to the prospective overvoltage events that can result with certain types of inductive load due to the premature suppression of the power frequency current before normal current zero in the vacuum interrupter.

Multiple reignitions refers to the series of alternate reignitions and high-frequency (typically several hundred kilohertz) interruptions usually resulting in an increasing train of voltage peaks; this overall phenomenon is usually defined as voltage escalation [1,2].

If the high-frequency currents accompanying reignitions and voltage escalation in one phase couple into the other two phases, the process of virtual current chopping can occur [3]. Virtual current chopping involves the load current in the other two phases being forced to zero by the superimposed high-frequency reignition coupled current. It is important to appreciate that while current chopping and voltage escalation can occur in a single-phase circuit, virtual current chopping is specifically a three-phase phenomenon; the effects of normal current chopping, multiple reignition, and voltage escalation in one phase can generate surge overvoltages in the second and third phases.

Prestriking of the breaker in picking up a motor or transformer load is somewhat similar to the multiple-reignition events that occur on opening a breaker [6]. In both cases, a high-frequency current flow, governed by the circuit parameters. However, prestriking is less severe than multiple reignitions occurring during load rejection, first because the contact gap at the first prestrike is very small, and second, because the contact gap is rapidly decreasing, rather than increasing, with respect to time.

13.2.2 Switching Surge Parameters and Frequencies

The parameters that must be controlled to assure the safe application of vacuum circuit breakers are transient peak voltage magnitude and its rate of rise or frequency. Transient voltage magnitude must be limited so that breakdown in air or of solid insulation on system components can be avoided. Transient voltage rate of rise (frequency) must be limited to prevent the nonuniformly distributed voltage in coil windings of motors and transformers from overly stressing the turns in the first coil sections.

Transient voltage rate of rise is important because very fast-rising transients can cause the overvoltage to be nonuniformly distributed in motor or transformer windings. For example, a voltage transient with 0.2 µs rise time may result in 80 to 100% of a voltage surge appearing across the first coil of a multicoil motor winding; if there were six coils in a winding, the turns of the first coil could be stressed six times higher than if the transient were slow rising. Consequently, a surge whose magnitude is well below the insulation level of the machine could damage the interturn insulation of the first coil of the winding. An important factor to consider is that even if the nonuniform voltage distribution in a winding does not actually result in a failure of the interturn insulation, repetitive fast-rising transients can gradually degrade insulation to the point of possible failure over a long period of time. The multiple-reignition phenomenon can cause winding insulation to be subjected to fast transients more frequently with vacuum switchgear (several times per switching event) than with other types of switchgear. It is therefore important to determine those applications where fast transients could cause problems, and to take appropriate measures to control the voltage rate of rise in those applications.

In discussion of switching surge transients, there are three different frequencies which are used to describe the phenomena. The first of these is the source power frequency, which in North American systems is 60 Hz. The second frequency is the normal frequency load transient recovery voltage, normally in the range 500 Hz to 5 kHz. This normal recovery voltage frequency is governed by the effective inductance of the load and the effective capacitance from load terminal to ground. This capacitance may have three components: terminal bushing to ground capacitance, cable capacitance, and a surge capacitor, if one is provided at the load.

The third frequency is that due to high-frequency reignitions. Note that in all systems, irrespective of the type of breaker used, high-frequency currents are caused to flow whenever the breaker reignites or prestrikes. The value of this frequency is determined by the effective capacitance at the load and the effective inductance of the cable between breaker and load. When a reignition, restrike, or prestrike occurs in the circuit switching device (breaker), the collapse of voltage across the breaker injects a voltage surge into the cable and load system. This surge is reflected at the load terminal, returns to the source or breaker end of the system, and travels back and forth along the cable many times until attenuated by losses. The frequency of the current is related to the travel and return frequency of reflected surges propagated along the cable; the frequency is inversely proportional to cable length and modified by resistive and reflective attenuation losses. Typical values of high frequency vary in the range over 2 MHz for 50 ft of cable giving 0.2 μs rise time, to over 50 kHz for 2000 ft of cable with almost 10 μs rise time.

13.3 CURRENT CHOPPING

The process of current chopping is the premature suppression of power frequency circuit current before normal current zero due to instability of the low current arc in a vacuum interrupter [4,5]. Although the current in the vacuum interrupter can chop to zero almost instantaneously (fraction of a microsecond), the current in the load inductance cannot attain zero value instantaneously. Time is required for magnetic energy to be transferred from the inductance, and for the magnetic field associated with stored magnetic energy to collapse. When current chop occurs, energy stored in the effective load inductance is transferred to the available load-side capacitance to produce the so-called chop overvoltage, given by

$$I_c \sqrt{\frac{(1 - \gamma)L_b}{C_s}}$$

where I_c is the chop current level and $\sqrt{L_b/C_s}$ is the load surge impedance; γ represents circuit losses, especially iron losses, and is very significant in limiting chop overvoltage [10,11]. Long cable length or a surge capacitor or both reduce the surge impedance to limit chop overvoltage. Capacitance has the additional benefit of reducing both the transient normal recovery frequency (typically, 500 to 1000 Hz) and the number of reignitions following chopping.

When a current chop occurs from current level I_c, the voltage appearing across the load terminal to ground, E_c, is given by

$$E_c = I_c \sqrt{\frac{L_b}{C_s}} \sin \omega_0 t + E_0 \cos \omega_0 t$$

where L_b is the effective inductance, C_s the effective capacitance to ground at the load, E_0 the instantaneous 60-Hz ac voltage value across the load at the moment of current chop, and ω_0 the natural angular frequency of the load given by $1/\sqrt{L_b C_s}$ (Fig. 13.1). The peak value of transformer voltage is given by

$$E_c = \left(I^2 \frac{L_b}{C_s} + E_0^2 \right)^{1/2}$$

This magnitude of voltage is the theoretical maximum prospective overvoltage across the load to ground resulting from the chop, assuming no losses and that the voltage is not limited in magnitude due to a restrike in the breaker. In Fig. 13.1(C) and (D), the maximum prospective chop overvoltage is shown by the dashed curve, and the solid curve demonstrates how a restrike in the breaker limits the actual voltage appearing across the load.

Current chopping is an important application consideration when switching individual dry-type transformers, where the effective inductance is the large magnetizing inductance of the transformer, resulting in a high surge impedance. Because the magnitude of magnetizing current is very low, a vacuum circuit breaker frequently chops as soon as the contacts part. The chop overvoltage appearing across the very short gap is usually sufficient to cause reignition or restrike in the vacuum interrupter so that the voltage is limited by a succession of reignitions.

The other major factor that is important in limiting chop overvoltages is the circuit loss factor γ, described above. Iron losses and the proportion of energy stored in the magnetizing inductance which is unavailable to feed into the surge are the principal loss factors. In switching an unloaded transformer, the effective magnetizing inductance is inversely proportional to the percentage magnetizing current. If the magnetizing current were high (say 3%), this correponds to a relatively wide B-H curve in which $\int B \, dH$, the energy retrieved to feed into a surge, is relatively small (say 20%). A small magnetizing current (e.g., 1/2 to 1%) corresponds to a relatively narrow B-H curve in which the energy retrievable to feed into a surge is relatively larger (say, 50 to 60%). Where magnetizing current is less than the characteristic chop current of the vacuum interrupter, chop overvoltages are proportional to the square root of magnetizing current. In practice,

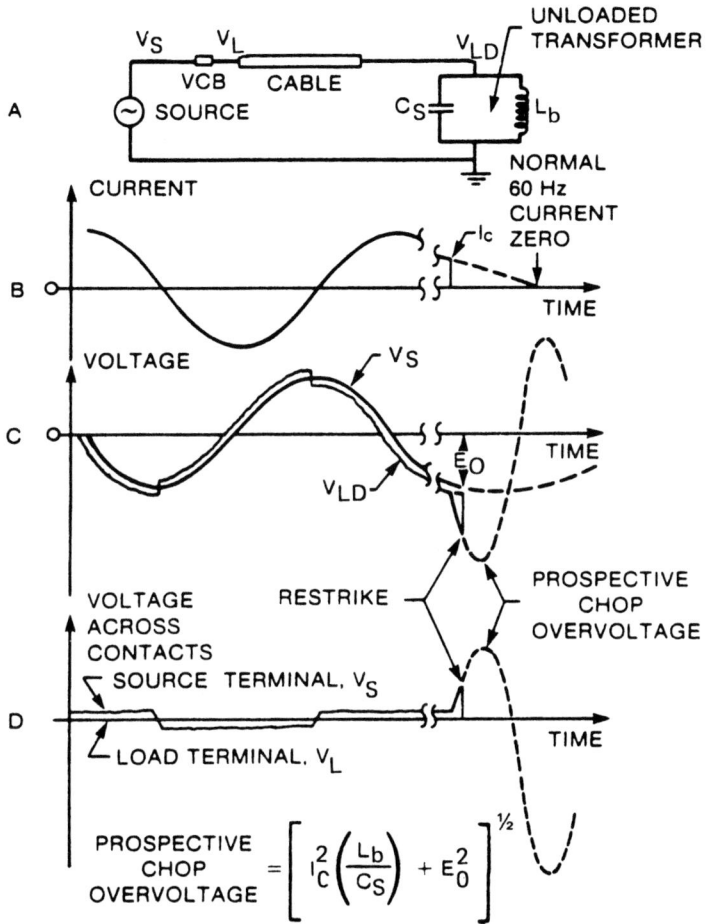

Figure 13.1 Mechanism of current chopping: (A) shows a single-phase circuit representation with a 60-Hz source, breaker (VCB), cable and unloaded transformer represented by parallel $L_b C_S$, where L_b is the effective inductance, and C_S is the effective capacitance to ground, including cable capacitance, bushing terminal-to-ground capacitance, and surge capacitance, if any; (B) shows the current through the breaker; (C) shows the voltage on the load terminal of the breaker, V_S, and the voltage on the load terminal, VLD; (D) shows the voltage across the contacts. (From J. F. Perkins, Evaluation of switching surge overvoltages on medium voltage power systems, IEEE Trans. Power Appar. Syst. *PAS-101*:1727–1734, © 1982, IEEE.)

transformers with lower magnetizing currents produce lower chop overvoltages, because reduction in magnetizing current predominates over reduced losses.

13.4 MULTIPLE REIGNITIONS AND VOLTAGE ESCALATION

The switching surge phenomenon of multiple reignitions [1,2] in a single-phase circuit is described with reference to Figure 13.2. The circuit in Fig. 13.2(A) shows a power frequency source, which might be a supply transformer, feeding an equivalent load (motor or transformer represented by C_s and L_b, through a vacuum circuit breaker (VCB) and cable; between the source and the vacuum circuit breaker is shown a parallel LC arrangement to depict the effective coupling for high-frequency currents between phase and ground for single-

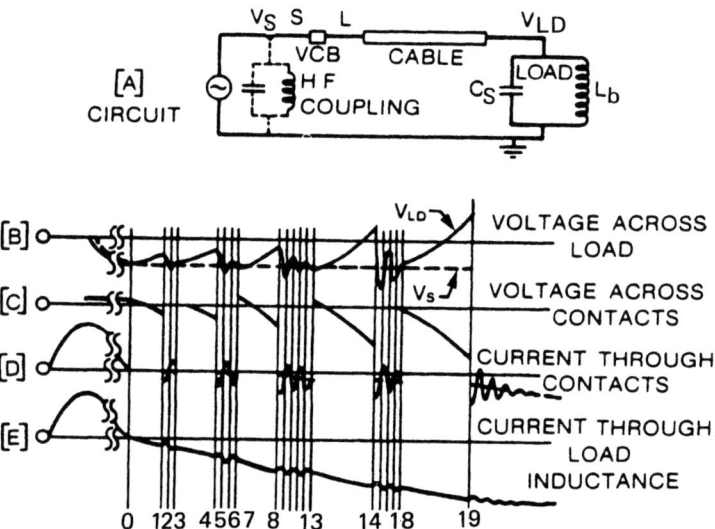

Figure 13.2 Mechanisms of multiple-reignition/voltage escalation switching surge phenomena: (A) is the circuit with a 60-Hz source feeding a breaker (VCB) and load via a cable. Load quantities C_s and L_b are as defined in the caption to Fig. 13.1. (B) shows the voltage across the load; (C) shows the voltage across the contacts; (D) shows the current through the contacts; and (E) shows the current through the load inductance. Note that the time scale in intervals 1-3, 4-7, 8-13, and 14-18 has been greatly expanded to demonstrate the very high frequency transient behavior there. (From J. F. Perkins, Evaluation of switching surge overvoltages on medium voltage power systems, IEEE Trans. Power Appar. Syst. *PAS-101*:1727–1734, © 1982, IEEE.)

Vacuum Circuit Breaker Application

phase, and between phase and phase for three-phase. The coupling path for high-frequency currents in a three-phase circuit may be via a power factor correction capacitor bank, or via mutual inductive coupling associated with cables on the source side of the breaker. For capacitive coupling, the coupling factor is close to unity (virtually zero impedance to high frequency), whereas the coupling factor for cables alone is generally smaller, typically 0.1 to 0.5.

Figure 13.2(D) shows the current through the contacts of the breaker, which are assumed to be parted in the time window before current zero which permits a reignition to occur, say up to 200 µs before normal 60-Hz current zero. If the 60-Hz current is interrupted at current zero, a normal frequency recovery transient voltage appears across the contacts between intervals 0 and 1 [Fig. 13.2(C)]; Fig. 13.2(B) demonstrates how this transient appears from load terminal to ground. At time 1 it is assumed that a reignition occurs, and this causes a high-frequency current to flow superimposed on the power frequency current [Fig. 13.2(D)]. The increasing current representing a continual accumulation of magnetic energy in the effective load inductance is shown in Figure 13.2(E). At time 3 it is assumed that the vacuum breaker interrupts the high-frequency current. When this occurs, the normal frequency transient recovery voltage reappears across the contacts between intervals 3 and 4. It is important to note that in the interval 1-4 between the first and second reignitions, the contact gap spacing has increased, so that a higher mean voltage is required to produce breakdown. Also, during the first reignition, additional magnetic energy from the power frequency source accumulated in the effective load inductance L_b, so that more energy is available to feed into the surge resulting from the second reignition. Both of these factors permit the second reignition to occur at a higher mean voltage than that for the first reignition [1,2].

This sequence of successive reignitions and high-frequency current interruptions can occur several times, as both the increasing amount of energy in the effective load inductance and the increasing contact spacing permit each successive reignition to occur at a higher mean voltage: hence the term "voltage escalation."

However, the statistical nature of breakdown and reignition in vacuum does not ensure that each successive reignition will occur at a higher voltage than the previous one. The normal frequency breakdown field of the vacuum interrupter, determined by several thousand measurements, is characterized by a log-normal statistical distribution. For one particular test vacuum interrupter, the median normal frequency breakdown field is 33 kV/mm. The log-normal statistical distribution means that 99% of all reignitions occur with a field less than 60 kV/mm, and 1% of all reignitions occur with a field less than about 15 kV/mm. These numbers give an extremely good picture of the statistical nature of reignition or breakdown in vacuum interrupters.

For circuits where chopping is likely to occur, test data indicate that current chopping has a smaller statistical scatter; 98% of current chops for the test interrupter occur in the range 2.5 to 5 A, with a mean of 3.6 A. The process of multiple reignitions occurs in a similar way to that described even if chopping rather than normal interruption of the 60-Hz current occurs. The only difference that a current chop makes is an increase in the amplitude of the initial normal frequency recovery transient voltage, which increases the probability of the first reignition, without which the whole multiple-reignition process cannot occur.

The multiple-reignition voltage escalation process can terminate in two ways. The first is for interruption of the high-frequency current to occur and for the contact gap to withstand the subsequent normal frequency recovery transient voltage; as time progresses and the contact gap spacing increases, the probability of this occurring increases with time. In a three-phase circuit, the high-frequency current must be interrupted in all three phases. The second way involves the high-frequency current continuing for a number of cycles until it is damped to the point where no further high-frequency current zeros of the superimposed high-frequency or power frequency currents occur; this is demonstrated following time step 19 in Fig. 13.2. When no further high-frequency current zeros occur, the 60-Hz power frequency current is reestablished. It is important to appreciate that multiple reignitions or voltage escalation can be either single-phase or three-phase phenomena.

13.5 VIRTUAL CURRENT CHOPPING

If a reignition in one phase (say phase A) causes a high-frequency current to flow which couples into the other two phases, virtual current chopping may occur [3]. The circuit paths for the high-frequency current are shown in Fig. 13.3. The high-frequency current in phase A, i_T, due to a reignition in phase A flows to ground via the terminal-to-ground capacitance at the load. If the three-phase system is balanced, i_T divides into two so that $i_T/2$ enters phase B and C via the respective terminal-to-ground capacitance. The high-frequency current ($i_T/2$) in phases B and C is shown to be capacitively coupled back into phase A on the source side of the breaker.

At the instant of reignition in phase A, which occurs some tens to hundreds of microseconds after the 60-Hz power frequency current zero, the power frequency current in phases B and C is approximately 0.87 × the crest value of the 60-Hz current. If the magnitude of the high-frequency current in phases B and C ($i_T/2$) is greater than 0.87 × the crest value of the 60-Hz current, the high-frequency current forces the 60-Hz current to zero; this forced current-zero phenomenon is virtual current chopping.

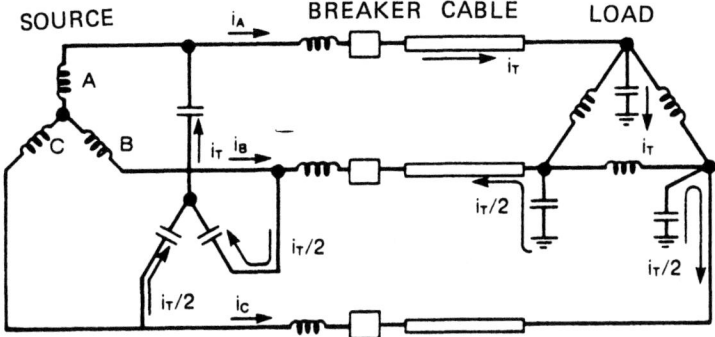

Figure 13.3 Circuit demonstrating how the high-frequency current resulting from a reignition, restrike, or prestrike in one phase couples into the other phases to produce the conditions for virtual current chopping. (From J. F. Perkins, Evaluation of switching surge overvoltages on medium voltage power systems, IEEE Trans. Power Appar. Syst. PAS-101:1727-1734, © 1982, IEEE.)

The high-frequency currents ($i_T/2$) in phases B and C are of the same polarity and equal in magnitude to each other, but are of opposite polarity to i_T in phase A. Because the normal 60-Hz phase relationships cause the 60-Hz currents in phases B and C to be of opposite polarities at the time of reignition in phase A, the forced current zeros are not time coincident. Also, the currents in phases B and C approach zero from different polarities, which means that the resulting transient voltages in phases B and C are of opposite polarities. Consequently, the surge overvoltage between phases B and C is twice the overvoltage from phase to ground on both phases, provided that the instantaneous 60-Hz voltages are significantly less than the virtual chop overvoltages.

Compared with normal current chopping, the effective level from which the load current is forced to zero (virtually chopped) is much higher, typically several hundred amperes instead of 3 or 4 A. However, the effective surge impedance of the load (several hundred ohms) is much lower than it would be in switching an unloaded transformer without protection (typically 10 to 30 kΩ). The overall effect is to make the phase-to-ground overvoltage comparable with or larger than the overvoltage due to a normal current chop, but the important difference is that the surge overvoltage between phases B and C is approximately twice the overvoltage from phase to ground on these phases.

13.6 CIRCUITS MOST SUBJECT TO SURGE

There are a group of electrical systems which are susceptible to switching surges with all types of circuit breakers. These circuits are characterized by having relatively highly inductive loads, high leakage reactance, and consequently, low power factors. This combination of characteristics means that the load surge impedance is also relatively high, which permits a high transient recovery voltage and increased probability of reignition if current chopping should occur. The systems having these characteristics include low-horsepower motors, unloaded dry-type transformers, and loads such as arc furnaces switched individually.

The equipment requiring most consideration in these circuits is that having low insulation levels. This includes all motors and dry-type transformers, whose basic impulse levels (BILs) are typically much lower than those of liquid-filled transformers.

If a transient condition should occur in any system, resulting in a prospective surge overvoltage in that system, the overvoltage will distribute around the circuit in accordance with how the circuit impedances are distributed. For example, if the surge impedance is highest at the load, the highest overvoltage will occur at the load. Conversely, if the surge impedance is highest at the source, the highest overvoltage will occur at the source.

13.7 DETERMINATION OF SWITCHING TRANSIENT VOLTAGES

During the past several years, a number of manufacturers have investigated switching transients by means of laboratory tests, transient system analysis studies, and instrumented tests performed on equipment installed in the field on actual industrial systems [12-15]. Frequently, major differences in system parameters have existed between different studies, and there has been no coherent approach to either reconciling parameter differences or to evaluating and correlating the data on a common basis. For instance, source- and load-side capacitance values have varied considerably, and there has been little consistency in the selection of cable lengths. In addition, it is known that switching operations performed with vacuum contactors, switches, or circuit breakers, each usually having different contact materials and configurations, can produce significant differences in overvoltage magnitude, wave shape, and duration, even when tests are performed under identical system conditions.

The optimum approach to developing a thorough understanding of switching surge behavior associated with medium-voltage vacuum circuit breakers involves the inclusion and analysis of all the parameters mentioned above. Such an understanding is necessary in order to determine the best methods of controlling switching surges over the

Vacuum Circuit Breaker Application

entire range of breaker applications. This type of encompassing approach is felt to be of more value to the spectrum of industrial users of medium-voltage metal-clad switchgear than, for instance, a limited field test program confined to a single, specific system.

To achieve the most efficient implementation of this approach, a computer program was used to model the interaction between vacuum breaker and electrical system with respect to switching operation transients [8,16]. The key inputs to the computer program included power system data (voltage, frequency, cable length and surge impedance, and number of source-side cables), load data (motor or transformer MVA capacity, number of poles of motor, load factor, power factor, transient recovery voltage amplitude factor or circuit damping), surge protection data (if surge protection was used), and vacuum interrupter or breaker data (see Fig. 13.4). To provide realistic data, it is recognized that the computer program input data must be accurate and it is important to ensure that the best data on the electrical system are used. Electrical systems, motor, and transformer specialists were consulted to ensure that correct data in the proper format was inputted to the computer program. The vacuum interrupter input data required several thousand tests to be performed to accumulate and analyze the data and convert them to a form suitable for inputting to the program. Since these data are so crucial if meaningful information is to be obtained from the computer study, a more detailed account of the acquisition and analysis of the vacuum interrupter data is presented in Sec. 13.8 [16].

The key output from the computer program is the cumulative appearance probability of peak transient voltages from load terminal to ground resulting from reignitions in the breaker. Because reignitions can occur at relatively low transient voltages when contact part occurs immediately before current zero, a given system can produce reignitions at voltages substantially below 1 per unit of crest line-to-neutral voltage. Depending on circuit or system parameters, damping, and power factor, prospective normal frequency transient recovery voltage peaks in the range 1 to 2 per unit typically occur.

The computer program uses a random number generator in four places in its Monte Carlo routine calculation procedure to ensure that the statistical nature of current chopping and voltage breakdown for a given interrupter is properly simulated. The first random number determines contact parting time relative to the 60-Hz current wave. The second determines the level of chop current from its statistical distribution, based on direct measurements. To determine whether reignition or high-frequency current interruption occurs the program again goes through a Monte Carlo routine, whereby the third and fourth random numbers are compared against analytically calculated normal frequency or high-frequency reignition voltages based on laboratory measurements.

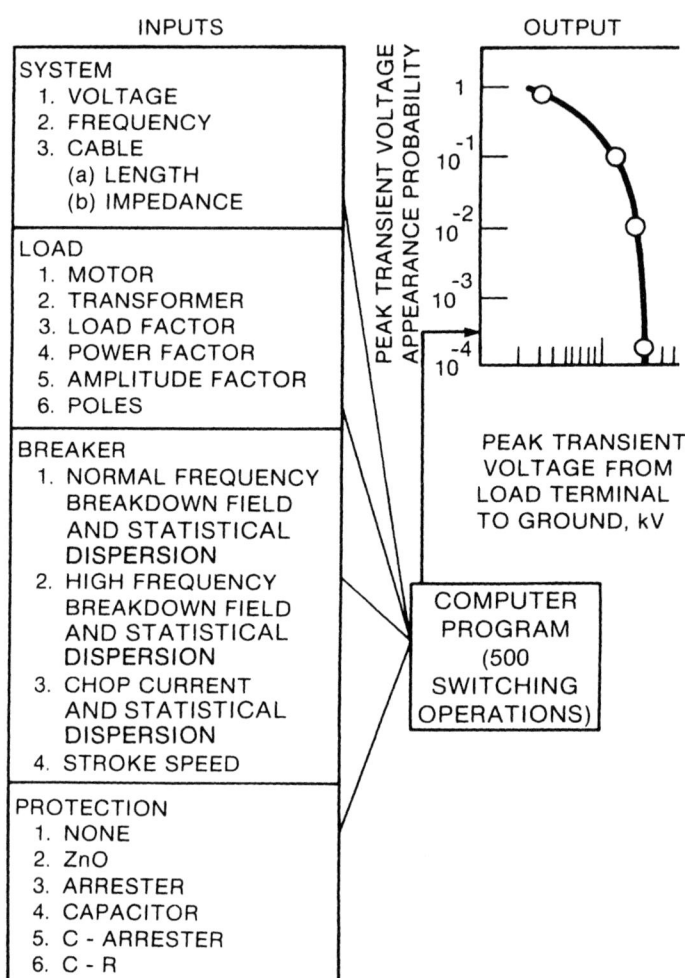

Figure 13.4 Surge analysis computer program, with main input parameters and graphical form of typical output showing the cumulative appearance probability of peak transient voltages from load terminal to ground. (From J. F. Perkins, Evaluation of switching surge overvoltages on medium voltage power systems, IEEE Trans. Power Appar. Syst. PAS-101:1727-1734, © 1982, IEEE.)

The procedure followed in the computer program during a switching operation is as follows. After the breaker contacts part randomly as described above, a level of current chop is selected (Fig. 13.5), and interruption occurs at or before the 60-Hz current zero, depending on

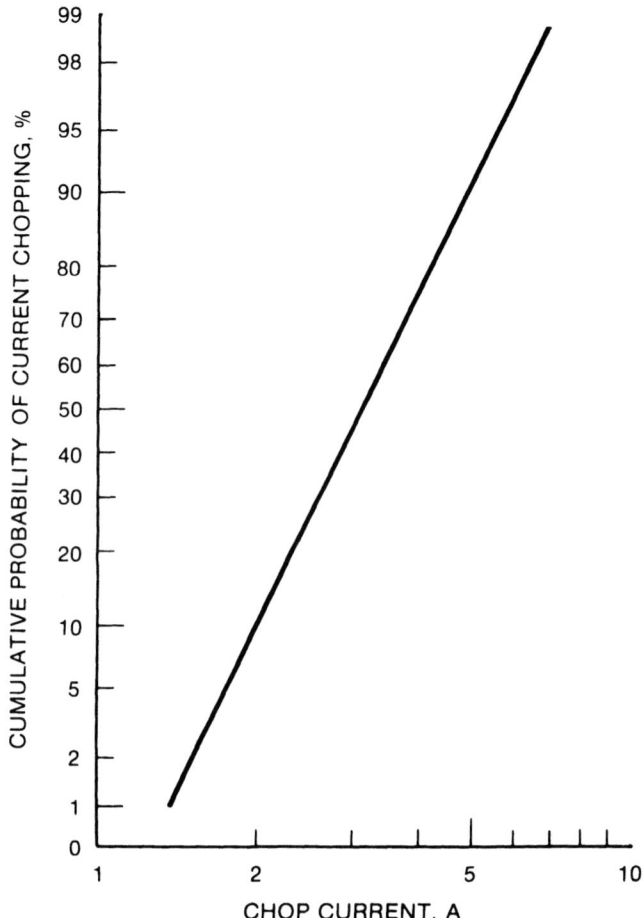

Figure 13.5 Cumulative probability of current chopping in a vacuum interrupter. (From J. F. Perkins and D. Bhasavanich, Vacuum switchgear application study with reference to switching surge protection, IEEE Trans. Industry Applications, IA-19(5), 879-888, © 1983, IEEE.)

whether or not chopping occurs; the collection of the data summarized in Fig. 13.5 is described in detail in Sec. 13.8. The TRV (transient recovery voltage) imposed by the circuit on the vacuum interrupter is computed, and an effective value of normal frequency reignition field is determined from a knowledge of the breaker travel curve. The probability that the TRV peak will cause breakdown (reignition) can be ascertained from Fig. 13.6, which shows a best-fit plot of cumulative

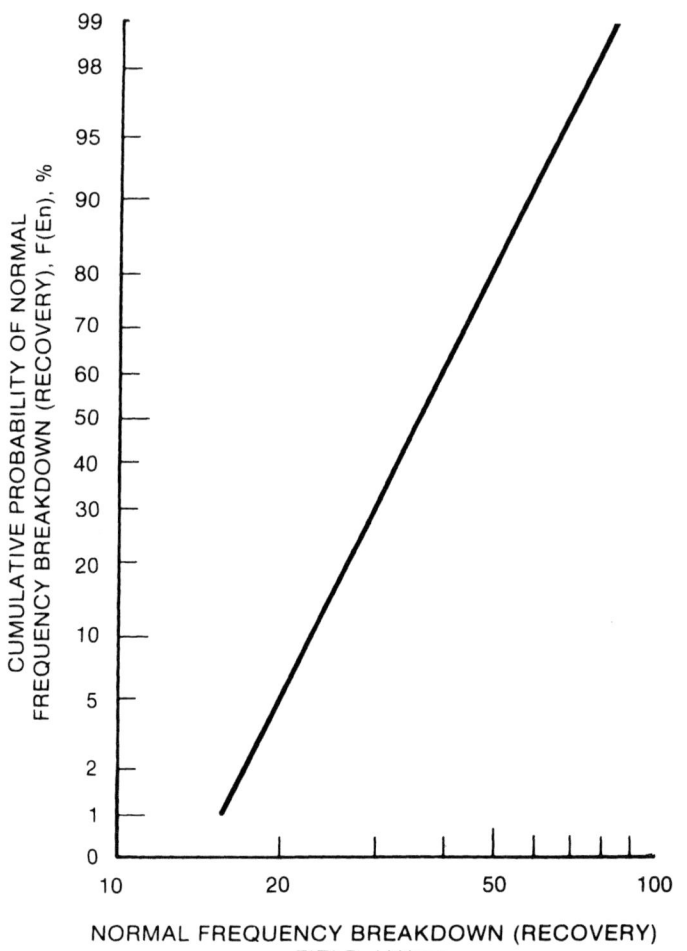

Figure 13.6 Cumulative probability of normal frequency breakdown (recovery) voltage in a vacuum interrupter. (From J. F. Perkins and D. Bhasavanich, Vacuum switchgear application study with reference to switching surge protection, IEEE IAS, Trans. Industry Applications, IA-19(5), 879-888, © 1983, IEEE.)

normal frequency breakdown probability $F(E_n)$. vs. average electric field, based on more than 1800 measurements. The function $F(E_n)$ is one of the vacuum interrupter inputs required by the computer program so that a determination of reignition can be made. The program goes through a Monte Carlo routine at this point, where a random num-

Vacuum Circuit Breaker Application

ber ν if generated and compared against the $F(E_n)$ function. If $F(E_n)$ is greater than ν, reignition occurs; if $F(E_n)$ is less than ν, there is no reignition and no voltage escalation surge.

If the first reignition occurs, a high-frequency (HF) current is caused to flow through the interrupter in a circuit loop involving the load- and source-side elements. The amplitude and frequency of this HF reignition current is computed from the inputted power system parameters. The HF interruption capability is presented in a form usable by the computer program in Fig. 13.7. This is a best-fit plot of cumu-

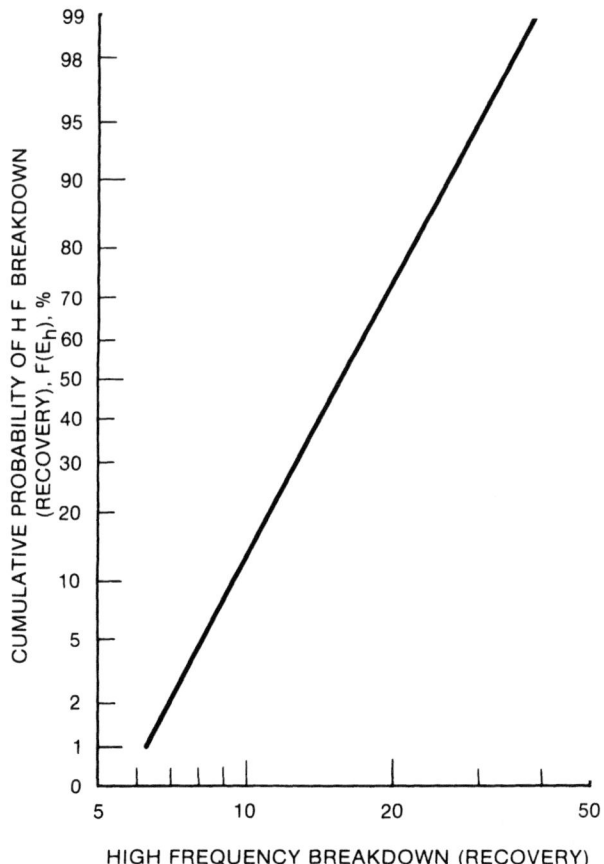

Figure 13.7 Cumulative probability of HF breakdown (recovery) field. (From J. F. Perkins and D. Bhasavanich, Vacuum switchgear application study with reference to switching surge protection, IEEE IAS, Trans. Industry Applications, IA-19(5), 879-888, © 1983, IEEE.)

lative HF breakdown probability $F(E_n)$ vs. average HF electric field, which is another vacuum interrupter input required by the computer program.

At the first HF current zero, it is assumed initially in the program that interruption occurs. The TRV produced by the HF circuit is computed, the contact gap is known from the travel curve, the average HF reignition field is determined, and the probability of breakdown (reignition) $F(E_n)$ occurring at the HF TRV peak is calculated. A second random number ν is generated, and reignition occurs if $F(E_n)$ is greater than ν, indicating that the initial assumption was incorrect. In this case, another half cycle of HF current flows, and the procedure is repeated at the next HF current zero.

If, however, $F(E_n)$ is less than ν, interruption of the HF current occurs, confirming the original assumption. At this instant, the instantaneous inductive current in the load-side circuit is greater than its initial value and the load circuit causes a second transient to swing to a higher prospective value than the peak of the first load-circuit (voltage escalation). Again, the contact gap is known, E_n and $F(E_n)$ are determined, and a new random number ν is generated to determine whether the second load-circuit transient causes reignition. This entire procedure is repeated for each load-circuit transient and each HF current zero until the load-circuit transient fails to cause a reignition, in which case the voltage escalation ceases.

For a given circuit and vacuum interrupter, a large number of switching operations as described above are simulated. The output of the computer program yields the switching surge severity in terms of the cumulative appearance probability of peak transient voltages appearing from load to ground.

When a reignition occurs, the high-frequency current in the first phase (say phase A) is assumed in the program to couple into the other phases B and C, so that virtual chopping can occur if the instantaneous value of the coupled HF current exceeds the instantaneous power frequency current.

The computer program was used to analyze a specific system by calculating switching surge activity during a statistically significant number of opening switching simulations. If the probability of surge activity was low, a large number of switching simulations were performed; if the surge activity probability was high, a small number of switching simulations were required. Normally, 500 switching simulations on a given system yielded a sufficient degree of surge activity to establish confidence that the output and predicted performance were accurate. If the output peak transient voltages were characterized by a normal distribution, a number of switching simulations producing about 3000 reignitions would establish 99% confidence that the computed overvoltages were within 5% of a standard deviation of the true overvoltage values.

Vacuum Circuit Breaker Application

To achieve a thorough understanding of switching surge phenomena and to determine safe application and proper surge protection for equipment fed by vacuum circuit breakers, the computer program was used to analyze a broad range of utility and industrial applications. Several hundred computer runs using more than 100 h of computer time on an IBM 370-155 computer were required to ensure that the most critical transformer and motor applications were encompassed in the study.

13.8 VACUUM INTERRUPTER SURGE PROGRAM INPUT PARAMETERS

The effectiveness of the computer program in determining transient voltages caused by switching vacuum circuit breakers is critically dependent on proper representation of the vacuum interrupters in the computer program. Since all the physical phenomena, such as current chopping, voltage breakdown, and interrupting performance, are known to be highly statistical in nature, the determination of the interrupter parameters must include the statistical dispersion of the distribution of parameter data as well as the center value of the distribution. The characteristics of the three vacuum interrupter parameters related to circuit stresses and interrupter properties required as input to the computer program are summarized in Table 13.1. These data apply to a specific interrupter, and it is important to appreciate that every type of interrupter must be characterized by its own numerical data. In addition, the breaker mechanism characteristic is needed, because surge behavior is critically dependent on the interaction between the vacuum interrupter parameters, mechanism characteristic, and the electrical system.

13.8.1 Data Development

Experimental techniques

Evaluation of switching surges was performed in a single-phase circuit (Fig. 13.8) designed to produce electrical stresses on the vacuum interrupter at levels experienced in practical field conditions. In a typical experiment, a voltage escalation surge was generated by the system and interrupter interaction, from which the normal and HF breakdown fields described in Table 13.1 could be directly measured. From the same switching event, the third parameter (chopping current) could be obtained by observing the interruption of the power frequency load current. A typical set of oscillograms from a switching experiment is shown in Fig. 13.9. To establish a statistical behavior of the parameters, a large number of switching experiments were performed, as described below.

Table 13.1 Characteristics and Measured Values of Vacuum Interrupter Surge Parameters with Data Fitted to Log-Normal Statistical Distribution

PARAMETER	CIRCUIT STRESS	INTERRUPTER PROPERTY	EXPERIMENTAL DATA
NORMAL FREQUENCY BREAKDOWN (RECOVERY) FIELD	TRV (TRANSIENT RECOVERY VOLTAGE) (0.5 TO 5 kHz)	DIELECTRIC STRENGTH	E_{n50} (MEDIAN) = 36.2 kV/mm β (DISPERSION) = 0.16 DATA SAMPLE = 1848
HIGH FREQUENCY BREAKDOWN (RECOVERY) FIELD	hf TRV AND hf REIGNITION CURRENT (50 kHz-1 MHz)	hf RECOVERY STRENGTH hf di/dt CURRENT INTERRUPTION CAPABILITY	E_{h50} (MEDIAN) = 15.9 kV/mm β (DISPERSION) = 0.17 DATA SAMPLE = 1534
CHOPPING CURRENT	POWER FREQUENCY 60 Hz CURRENT	ARC STABILITY PREMATURE ARC EXTINCTION BEFORE NORMAL 60 Hz CURRENT ZERO	I_{c50} (MEDIAN) = 3.1A β (DISPERSION) = 0.15 DATA SAMPLE = 170

Source: J. F. Perkins and D. Bhasavanich, Vacuum switchgear application study with reference to switching surge protection, IEEE Trans. Industry Applications, Vol. IA-19, No. 5, Sept./Oct. 1983, pp. 879-888, © 1983, IEEE.

Vacuum Circuit Breaker Application

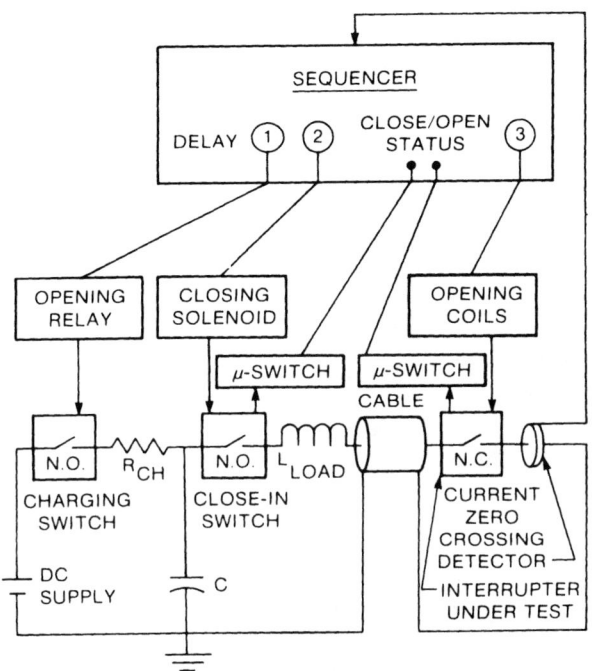

Figure 13.8 Schematic drawing of the experimental circuit showing the controller sequencer operating the charging, close, and vacuum interrupter switches in a precise sequence to generate a voltage escalation event. (From J. F. Perkins and D. Bhasavanich, Vacuum switchgear application study with reference to switching surge protection, IEEE Trans. Industry Applications, Vol. IA-19, No. 5, Sept./Oct. 1983, pp. 879-888, © 1983, IEEE.)

Data acquisition

To facilitate data development from a very large number of experiments, a semiautomated experiment controller was utilized. The controller repetitively and reliably produced voltage escalation in every switching operation. Acquisition and processing of the data were performed on a computer-based system which comprised digital oscilloscopes interfaced with a mainframe computer having custom waveform analysis programs. A schematic drawing of the experimental circuit and its controller is given in Fig. 13.8. A block diagram in Fig. 13.10 shows a layout of the data acquisition system, featuring an experiment controller and a computer-based management system. As an indication of the data-gathering capability of the system, the development of data for the first interrupter from 500 switching experiments, involving some 20,000 pieces of data, was accomplished within 1 week.

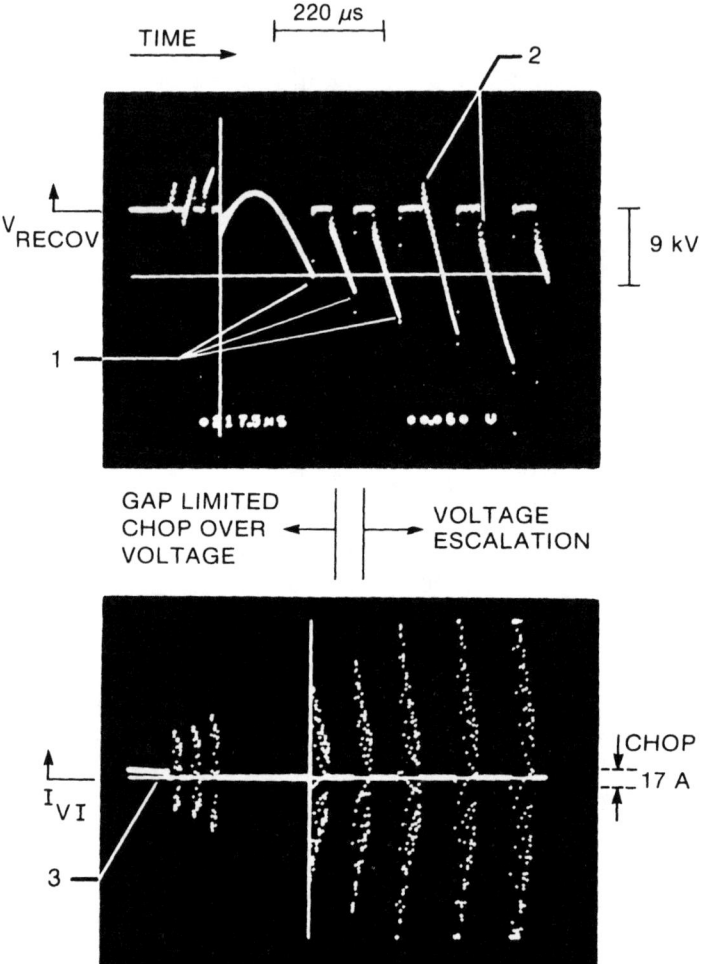

Figure 13.9 Sample oscillograms from a typical switching experiment in which a voltage escalation switching surge is produced. Three important surge parameters are identified: 1, breakdown voltages, upon normal frequency TRV; 2, recovery voltages, upon interruption of HF reignition current; 3, current chopping, upon interruption of power frequency load current. (From J. F. Perkins and D. Bhasavanich, Vacuum switchgear application study with reference to switching surge protection, IEEE Trans. Industry Applications, Vol. IA-19, No. 5, Sept./Oct. 1983, pp. 879-888, © 1983, IEEE.)

Vacuum Circuit Breaker Application

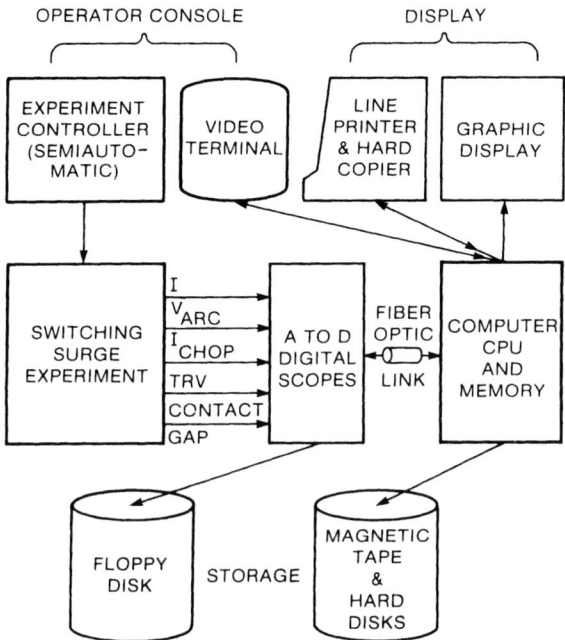

Figure 13.10 Schematic layout of a data acquisition system showing an experiment controller (upper left) which semiautomatically runs the switching surge experiment; and a computer-based data acquisition system (bottom right) comprising digital scopes as experimental data buffer and waveform analysis software programs to process and analyze the experimental data. (From J. F. Perkins and D. Bhasavanich, Vacuum switchgear application study with reference to switching surge protection, IEEE Trans. Industry Applications, Vol. IA-19, No. 5, Sept./Oct. 1983, pp. 879-888, © 1983, IEEE.)

Data processing and analysis

The digital data acquired from the experiment were transferred to the computer to catalog and record the values and order of occurrences of breakdown and recovery voltages. These values were then correlated with the instantaneous interrupter gap separation to permit calculation of the normal-frequency and high-frequency breakdown fields. Handling of the chop current data was much less complex. Since there was only one chop current value per switching operation, chop current values were read manually from the digital oscilloscope records without the aid of the computer. Once the three major parameters described in Table 13.1 were recorded for the entire test schedule, they were analyzed by the final software program for estimation of their statistical behavior. The results in terms of probability density and cumulative probability of occurrences were then plotted on log probability paper. The best-fit cumulative probability plots are presented in Fig. 13.5, 6, and 7 for the three required parameters, and a summary of the results is given in Table 13.1.

From a detailed study of the vacuum interrupter characteristics relative to its susceptibility for surge generation, such as the one outlined in this section, a reliable set of interrupter data is produced as input to the computer model. The sophistication of the computer model allows for a statistical representation of the vacuum interrupter, and the speed of the computer allows for a large number of contingency studies for breaker-circuit interaction to be performed in a short time.

13.9 VALIDATION OF THE COMPUTER PROGRAM BY SWITCHING TESTS

The comprehensive approach taken toward developing surge application recommendations for vacuum switchgear depends, in part, on the realistic calculation of transient peak voltages in the computer program as the result of switching operations. Before a high degree of confidence can be placed in such calculations, it is necessary to demonstrate that good correlation exists between computations and practical measurements of vacuum switching operations on an identical system to that on which the computations were made. The requisite high degree of correlation has been established in two ways: (1) on a single-phase laboratory circuit, and (2) on a three-phase unloaded transformer circuit in the field.

13.9.1 Single-Phase Correlation

The single-phase correlation was established on a circuit similar to that used to generate the data on normal-frequency breakdown field and high-frequency breakdown field for the vacuum interrupters (Fig. 13.8). The single-phase LC circuit was tuned to a nominal frequency of 60 Hz, achieved when the test current of approximately 330 A rms was obtained by discharging a 96-μF capacitor bank charged to 12 kV through the vacuum interrupter and a 76-mH air-core reactor fed by 750 ft of 350 MCM single-phase cable. This circuit was designed specifically to simulate multiple-reignition phenomena. The normal-frequency recovery transient had a frequency of about 2200 Hz, and the prospective TRV waveform without reignition could rise to about 26 kV in 110μs, as determined by the 76-mH inductance and effective capacitance to ground of 0.07 μF; the effective surge impedance at the load was 1040 Ω. The value of high frequency associated with reignitions was approximately 105 kHz, dictated primarily by the cable of

Vacuum Circuit Breaker Application

37-Ω surge impedance. The single-phase data were obtained automatically. The capacitor bank was first charged, the close-in switch was closed, and the vacuum breaker was opened to interrupt the simulated load current. The transient voltage peaks were stored in a digital data acquisition system, with a link to a minicomputer system for data analysis. The data captured over 6000 reignitions occurring during more than 450 opening operations of the vacuum breaker in the test circuit.

The parameters of the single-phase circuit were inputted to the computer program to determine the correlation between experimental and computed data. Since the single-phase circuit voltage relates to crest line-to-neutral voltage of a three-phase circuit, the voltage inputted is ($\sqrt{3}/\sqrt{2}$) × single-phase voltage. The other inputs required are the peak current, circuit power factor, the load surge impedance of 1040 Ω, and amplitude factors for the normal-frequency and HF recovery transients. The surge computer program simulated approximately 450 opening switching operations, so that the computed cumulative appearance probability of peak transient voltage from load terminal to ground could be compared with the measured experimental data described above. The results are shown in Fig. 13.11 and the correlation between computed and experimental data is shown to be good.

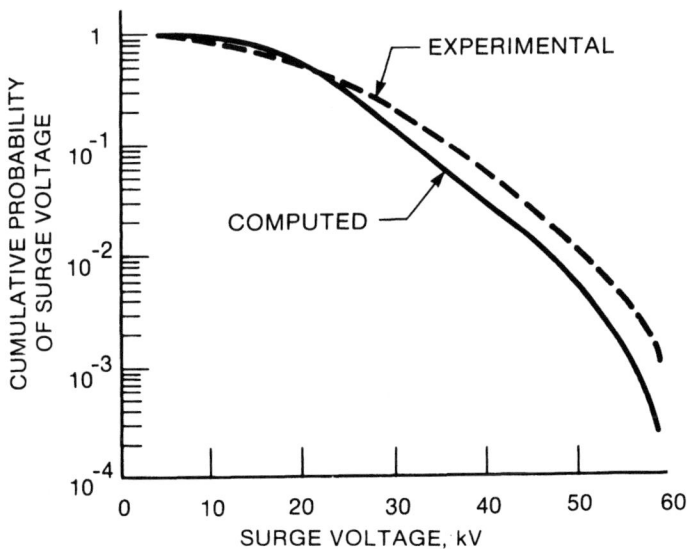

Figure 13.11 Single-phase verification of a surge analysis computer program. (From J. F. Perkins and D. Bhasavanich, Vacuum switchgear application study with reference to switching surge protection, IEEE Trans. Industry Applications, IA-19, (5), pp. 879-888, © 1983, IEEE.)

13.9.2 Three-Phase Correlation: Unloaded Transformer Switching

The three-phase correlation between measured and computed data was made relative to a system [16] where a transformer was switched unloaded. Some typical practical measurements involving a large number of opening operations, where the transformer was switched unloaded by a vacuum breaker, were reviewed and only the highest one-third of measured peak transformer voltages, taken as representing the highest transformer voltage per switching operation, were considered. The practical measurements were correlated with computed data for a 13.8-kV vacuum breaker switching a 2-MVA transformer fed by approximately 100 ft of cable. The loading factor in the computation was assumed to be 1% of full-load current, the power factor was assumed to be 20%, and the amplitude factors associated with the normal frequency and HF transient recovery voltages were assumed to be 1.3 and 1.5, respectively. A comparison of computed and practical peak transient measured voltages is presented in Fig. 13.12, and it can be seen that the correlation is very good.

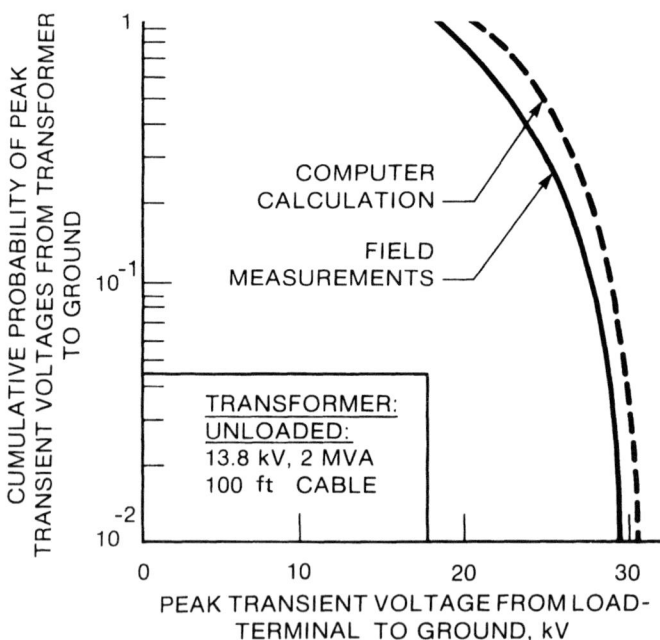

Figure 13.12 Validation of a surge analysis computer program switching an unloaded transformer. (From J. F. Perkins and D. Bhasavanich, Vacuum switchgear application study with reference to switching surge protection, IEEE Trans. Industry Applications, IA-19, (5), Sept./Oct. 1983, pp. 879-888, © 1983, IEEE.)

A more detailed comparison of measured and computed data relating to the switching of an unloaded transformer by a vacuum circuit-breaker is shown in Fig. 13.13. In this case, a random-wound dry-type transformer was set up in the laboratory fed by 100 ft of 350-MCM single-phase cable between breaker and transformer. One hundred switching operations in which the breaker closed in on no load, remained closed for 5 s to permit the magnetizing inrush current to decrease to the normal steady-state no-load current, and then opened to interrupt the current were performed. The laboratory data were then compared with computed data, again assuming a magnetizing current of 1%, power factor of 20%, and normal- and high-frequency transient recovery voltage amplitude factors of 1.3 and 1.5. Figure 13.13 shows that the measured transient voltages are lower than the computed values (21.1 kV measured, 24 kV maximum computed), so that generally the computed program results are somewhat conservative.

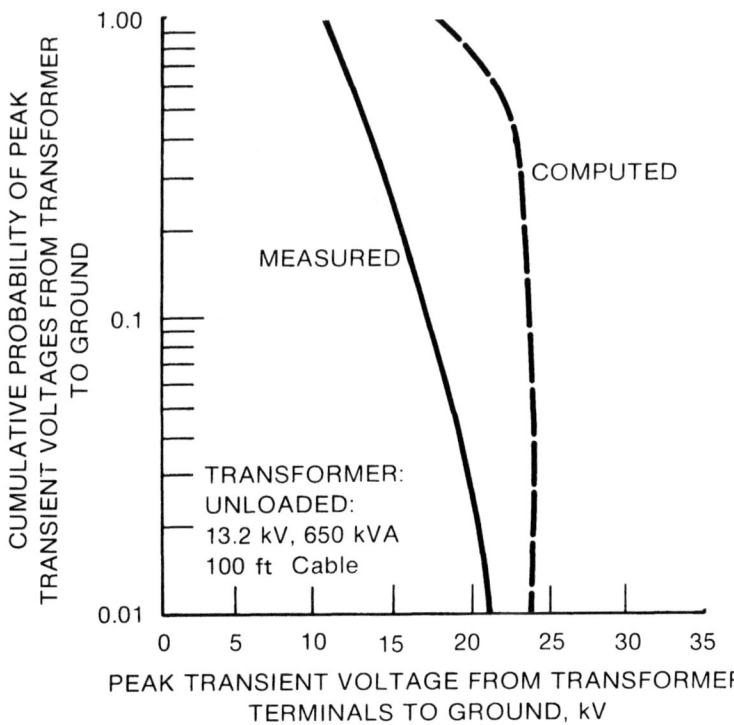

Figure 13.13 Comparison of laboratory measurements and computed data for switching an unloaded transformer. (From J. F. Perkins and D. Bhasavanich, Vacuum switchgear application study with reference to switching surge protection, IEEE Trans. Industry Applications, Vol. IA-19, No. 5, Sept./Oct. 1983, pp. 879-888, © 1983, IEEE.)

It is interesting to compare the computed and measured data of Fig. 13.12 with numbers derived from a single analytical expression in Sec. 13.3, which gives the chop overvoltage as $I_c\sqrt{(1-\gamma)L_b/C_s}$, where I_c is the chop current, L_b the transformer magnetizing inductance, C_s the effective capacitance to ground seen by the transformer, and γ is a circuit loss factor. With a magnetizing inductance of 25 H and an approximate capacitance to ground of 5000 pF, the load surge impedance is about 70 kΩ. Because the loss factor γ limits the amount of energy available to feed into a surge to about 40%, the effective surge impedance reduces to about 45 kΩ. Since the chop current level essentially corresponds to the magnetizing current level of somewhat more than 1 A, the maximum prospective chop voltage neglecting losses would be about 80 kV, and including losses, approximately 50 kV. Figure 13.12 shows that the maximum value of transient peak voltage from transformer terminal to ground is about 30 kV, with the computed values being somewhat higher than those from measurements. If it is assumed in this case that 60% of the transformer magnetic energy either remains in the transformer core or is dissipated as losses, the computed and measured results imply that around two-thirds to three-fourths of the remaining 40% is also dissipated [6]. The mechanism for this additional energy dissipation is reignition in the breaker. Furthermore, since the magnetizing current is less than the median chopping current level of the vacuum circuit breaker, chopping would frequently occur as soon as the contacts part. Since the gap is small, the chopping transient is usually sufficient to cause reignition in the breaker, so that transient peak voltage is limited by number of successive reignitions following the chop.

13.10 TYPICAL SWITCHING APPLICATIONS

13.10.1 Unloaded Transformer Switching

Some of the effects of switching unloaded transformers with vacuum circuit breakers have been mentioned in the discussion of current chopping in Sec. 13.3. An unloaded transformer is a highly inductive load (magnetizing inductance typically many henries) with a magnetizing current typically in the range 0.5 to 3% of rated full-load current. If the magnetizing current is less than the median characteristic chop current of the vacuum interruptor, and if contact part should occur at the magnetizing current peak, the resultant immediate current chop traps energy proportional to the magnetizing inductance and to the square of the magnetizing peak current in the transformer core. Typically, 20 to 60% of the energy remains in the transformer core or is dissipated as losses, and some of the remaining 40 to 80% is dissipated in reignitions in the breaker. Frequently, the proportion of the total magnetic energy trapped in the transformer core, which charges the

Vacuum Circuit Breaker Application

effective load capacitance and generates an overvoltage, is in the range of 10 to 40%. Taking into account the energy loss (or unavailability) factor γ, the chop phenomenon is governed by the energy equation

$$\frac{1}{2}(1 - \gamma)L_b I_c^2 = \frac{1}{2} C_s V_c^2$$

where the symbol definitions are consistent with those given in Sec. 13.3. It is apparent that crest chop voltage V_c is inversely proportional to the square root of the load-side capacitance C_s, which comprises three components: (1) transformer terminal-to-ground capacitance: (2) cable capacitance; and (3) any lumped surge capacitance.

The surge computer program was applied to the analysis of switching unloaded transformers. The program takes into account the energy losses due to the load, and the statistical vacuum breaker inputs permit the energy dissipation due to reignition to be accommodated. The program was used to compute how transformer terminal-to-ground overvoltages vary with MVA rating, cable length, and surge protection means when switched by a 15-kV-rated vacuum breaker. Figure 13.14 shows the results.

The general trend of the data is that overvoltages increase as transformer MVA rating increases, and this behavior reflects the fact that magnetizing current also increases with MVA, so that chopping from higher current levels is possible. Although transformer surge impedances decrease with MVA rating, the computed results indicate that the possibility of chopping from a higher current predominates to give

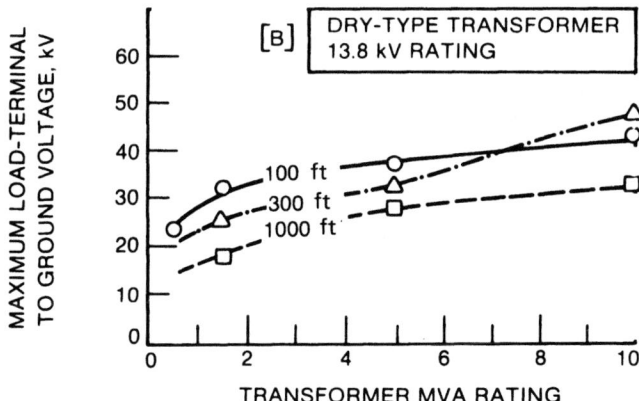

Figure 13.14 Effect of cable length on maximum load-terminal-to-ground surge voltage. (From J. F. Perkins, Evaluation of switching surge overvoltages on medium voltage power systems, IEEE Trans. Power Appar. Syst. *PAS-101*:1727-1734, © 1982, IEEE.)

higher overvoltages. For similar reasons, switching transient inrush magnetizing currents represent very severe duty, because inrush current may be significantly higher than the median chopping current level of the vacuum interrupters. In this situation, chopping may not occur instantaneously on contact parting, so that some appreciable gap spacing may be achieved before chopping occurs. This larger gap spacing would support a higher voltage without reignition than if chopping had occurred instantaneously. Consequently, the mechanism whereby reignitions in the breaker limit prospective overvoltages results in the limiting action occurring with respect to a voltage level that would be appreciably higher than for instantaneous chopping.

Another trend of the results is that overvoltages tend to increase as cable length between breaker and transformer decreases. There is, however, a second-order effect which shows a critical cable length giving maximum overvoltage for any given MVA rating. For transformer ratings below 2 MVA, this critical length is less than 50 ft, but the overvoltage here is only about 3% higher than for a 10-ft cable length and the effect may be ignored. However, at higher MVA values this effect is significant. For a 10-MVA transformer, the critical cable length is about 300 ft, and the overvoltage here may be 20% higher than at a reference length of say 10 ft. This effect is shown in Fig. 13.14, where the overvoltage at 300 ft is higher than at either 100 or 1000 ft. The reasons for this behavior are similar to those involved in motor switching, described below. For higher transformer MVA ratings, the computer program also demonstrates how a given cable length results in a lower overvoltage at the transformer terminals than is calculated analytically assuming that current chopping occurs at magnetizing current peak. The limitation on overvoltage due to reignitions in the breaker produces this effect.

The computer calculations correlate well with field measurements, and it is evident that second-order effects such as iron, circuit, and reignition losses are included in the analysis.

Figure 13.15 shows an example of how different surge suppression means are inputted to the computer program. With ZnO surge suppressors at the vacuum breaker, the overvoltages computed at lower MVA ratings are essentially the same as with no surge protection, because the conditions here result in overvoltages below the ZnO clamp level. At higher transformer MVA ratings, the damping action of the ZnO surge suppressors is evident, as the overvoltages are now limited below those occurring with no protection. The computed results with surge capacitors at the transformer terminals confirm the earlier discussion, which indicates that surge capacitors at the load reduce surge impedance and hence prospective overvoltage magnitude.

Vacuum Circuit Breaker Application

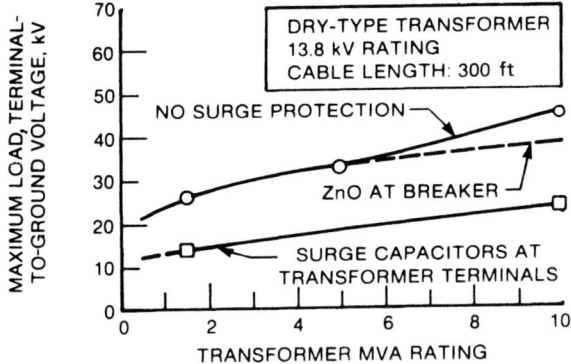

Figure 13.15 Comparison of the effects of surge capacitors at the transformer and ZnO surge suppressors at the breaker in reducing the maximum load-terminal-to-ground surge voltage. (From J. F. Perkins, Evaluation of switching surge overvoltages on medium voltage power systems, IEEE Trans. Power Appar. Syst. *PAS-101*:1727-1734, © 1982, IEEE.)

13.10.2 Switching Motors Under Normal Running Conditions

The surge analysis computer program has also been applied to ascertaining the overvoltage performance of vacuum circuit breakers switching motors under normal running conditions. Figure 13.16(a) shows computed overvoltages from motor terminals to ground with no protection and with surge capacitors applied at the motor terminals. From ratings of 5-MVA upward, overvoltages are generally within the insulation level for 0.2-μs fast-rise-time voltage transients V_2, defined by the IEEE Working Group on Impulse Voltage Standards for Rotating Machinery [17]. However, surge capacitors are very effective in reducing overvoltage values well below this V_2 level. Surge capacitors reduce the rise time of the normal frequency transient recovery voltage typically by a factor of 10, and this greatly reduces the probability of reignition in the vacuum circuit breaker. If reignition should occur, the resulting high-frequency current is more readily interrupted, but the high-frequency transient voltage accompanying high-frequency interruption has a much lower rise time (typically, about 10 μs), so that the resulting voltage is fairly uniformly distributed throughout the entire winding [9]. Figure 13.16(a) shows this significant reduction in surge overvoltage. Although not shown in Fig. 13.16(a), the computer analysis also demonstrates that C-R suppressors (capacitor-resistor in series line to ground) at the load are also effective in im-

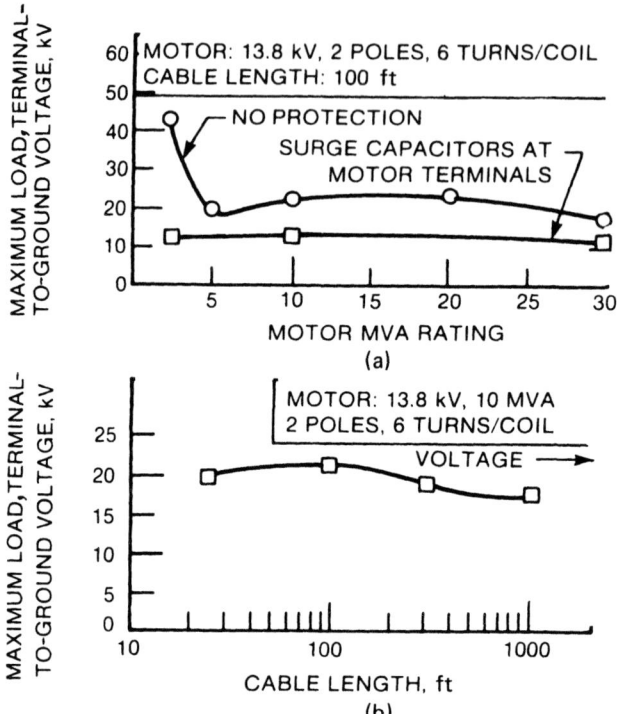

Figure 13.16 (a) Effect of surge capacitors at motor terminals in reducing the maximum surge voltage from load terminal to ground as a function of motor rating. (b) Variation of maximum load-terminal-to-ground surge voltage with cable length between breaker and motor. (From J. F. Perkins, Evaluation of switching surge overvoltages on medium voltage power systems, IEEE Trans. Power Appar. Syst. *PAS-101*:1727-1734, © 1982, IEEE.)

proving switching reliability of motors; although the R of a C-R suppressor is effective in damping high-frequency reignition currents, C-R suppressors are generally not as effective as capacitors alone in suppressing surge overvoltage effects, without higher values of C.

Figure 13.16(b) shows how maximum overvoltages from motor terminals to ground vary with cable length. In this particular case, the worst conditions for switching occur with 100 ft of cable between the breaker and motor. As cable length is increased to 1000 ft, the overvoltage decreases. The reasons for this behavior are as follows. Very short cable lengths create very high reignition frequencies, with many traveling-wave reflections producing losses and rapid damping of the high-frequency currents, with relatively less severe surges than with somewhat longer cables. Very long cables behave like lumped capacitance, reducing the surge impedance, frequency, and transient voltage rise time. It is an intermediate cable length which produces the most severe surge conditions; in Fig. 13.16(b), this length is 100 ft. Generally, the effective series inductance associated with a long cable would be expected to permit a higher voltage drop across the cable and higher overvoltage at the motor than at the breaker terminals than would a shorter cable. However, in practice, the effect of cable capacitance to ground in reducing surge impedance with very long cables (>1000 ft) predominates, and motor terminal-to-ground overvoltages are almost invariably lower than with short cables.

Figure 13.17 demonstrates how surge overvoltages from motor terminal to ground vary with motor MVA (horsepower rating) and with motor design parameters when opening switching operations were performed on a 15-kV test circuit breaker connected by 100 ft of cable. Most of the calculations were performed with a two-pole motor having six turns per coil, and this represents one of the most severe conditions as far as overvoltages are concerned. A six-pole motor with 10 turns per coil is probably closer to the center of gravity of 13.8-kV motor designs, and Fig. 13.17 indicates that overvoltages are lower in this case.

Surge activity during vacuum switching is directly proportional to motor surge impedance, which is higher for motors having fewer pole pairs. Also, the smaller number of turns per coil stresses each turn more severely in the event that very fast front transients cause

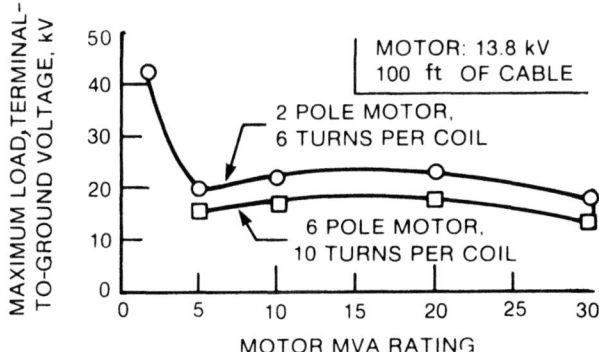

Figure 13.17 Effect of motor design parameters on overvoltages appearing at motor terminals to ground. (From J. F. Perkins, Evaluation of switching surge overvoltages on medium voltage power systems, IEEE Trans. Power Appar. Syst. *PAS-101*:1727-1734, © 1982, IEEE.)

most of the voltage to appear across the first coil [7,9]. It is for these reasons that overvoltages are higher for the two-pole motors with six turns per coil than for six-pole motors with 10 turns per coil.

13.11 RECOMMENDED PRACTICE FOR SYSTEM PROTECTION

13.11.1 Surge Protection Means

Surge protection devices may be divided into two categories: (1) those that protect against surge voltage magnitude, and (2) those that modify transient voltage rate of rise. The first category includes all types of surge arresters, from conventional gapped station class arresters with low protective levels, to intermediate and distribution class arresters having higher protective levels. The new technology component in this area is the ZnO surge arrester, which has much better voltage vs. current characteristics than does the SiC used in a conventional arrester. ZnO arresters may be gapless or they may have either series or parallel gaps. Several manufacturers throughout the world have developed ZnO surge suppressors specifically to be applied with vacuum switchgear. A ZnO surge suppressor is basically similar to a ZnO arrester. It is usually (but not always) gapless and is generally a lighter-duty unit having somewhat lower energy absorption capability than that of a ZnO arrester.

The second category of surge protectors includes surge capacitors, C-R suppressors, and series reactors. In accordance with industry practices and standards [18,19], surge capacitors should be located at the load and as close to the load terminals as possible. This minimizes series inductance effects which reduce the effectiveness of capacitors. The purpose of surge capacitors is to reduce the natural frequency of the load circuit, decrease the rate of rise of voltage transients governed by the load, and reduce the effective surge impedance of the load circuit. All of these factors are important in reducing the probability of degradation of solid insulation, which could result from the nonuniform distribution of fast transient voltages in machine windings. The rationale for these recommendations is in accordance with IEEE Standard 141 (4.9)-1976 [18].

C-R suppressors [20,21], which comprise resistors in series with surge capacitors to ground for damping HF reignition currents, and HF iron-core reactor chokes [7] for the same purpose, are not commonly used in the United States, but have been used extensively in Japan on 3.3-kV systems.

It is important to locate surge capacitors at the load and not at the breaker. If the capacitor is located at the load, the high-frequency current resulting from reignition is limited in amplitude by the surge impedance of the cable located between the capacitor and the breaker. This limitation of the amplitude of high-frequency current reduces the probability of virtual current chopping. In addition, location of

Vacuum Circuit Breaker Application

capacitors at the load minimizes series inductance effects. If, on the other hand, the capacitor is located at the breaker, reignition can cause the capacitor to discharge back through the breaker. In that case, the high-frequency current could have a higher frequency and amplitude than if the capacitor were located on the load side of the cable at the load [21].

13.11.2 Protection for Specific Systems

The surge computer program was used primarily to analyze motor and dry-type transformer circuits, since these have relatively low insulation levels and comprise the largest proportion of systems on which medium-voltage metal-clad switchgear is applied. Rather than presenting a detailed account of the study of switching a wide range of motors and transformer MVA ratings, several examples are presented to demonstrate how the computer program can be used to analyze specific systems.

First, consider the situation where a 13.8-kV 10-MVA unloaded transformer fed by 300 ft of cable is deenergized by switching a vacuum circuit breaker. Since the peak magnetizing current exceeds the median chop current level of the vacuum interrupters in the breaker, chop transient voltages from transformer terminals to ground can be expected to be higher than for a lower MVA-rated transformer. The maximum transient load-to-ground voltage resulting from 250 switching (opening) operations from computer analysis is 45 kV (4.0 per unit of crest line to neutral voltage). Although this is below the transformer BIL, the margin is such that some form of additional surge protection is desirable. The application of ZnO surge suppressors at the vacuum circuit breaker reduces maximum transient voltage to 3.4 per unit, a level which is higher than the ZnO clamp level due to the effect of inductance between the ZnO suppressors and the transformer terminals. ZnO surge arresters located at the transformer terminals would reduce the maximum transient voltage level significantly below 3.0 per unit. However, the application of surge capacitors at the transformer terminals reduces the maximum terminal to ground voltage to 2.0 per unit, basically because the wave-sloping effect of the capacitors reduces both the effective surge impedance of the load and the probability of restriking in the breaker. Note that the surge capacitors very significantly reduce the number of restrikes per phase during a typical switching operation. The results are summarized in Table 13.2.

If a smaller transformer, for example, a 1.5-MVA transformer fed by the same length of cable, is switched by a vacuum breaker, the maximum transient voltage is much lower (2.3 per unit). In this case, as in the case described in Sec. 13.9.2, the magnetizing current is less than the breaker median chop level, so that chopping occurs often at contact part, and transient peak voltage is limited by breaker reign-

Table 13.2 Summary of Computer Analysis on Typical Industrial Loads (13.8 kV)

LOAD	CABLE LENGTH FEET	SURGE PROTECTION	MAXIMUM TRANSIENT VOLTAGE (LOAD-GROUND)[a]	TYPICAL NUMBER OF RESTRIKE/PHASE
TRANSFORMER 10 MVA	300	NONE	4.0 pu	10
		ZnO AT BREAKER	3.4 pu	10
		CAPACITORS AT LOAD	2.0 pu	0.3
TRANSFORMER 1.5 MVA	300	NONE	2.3 pu	5
MOTOR 8,000HP NORMAL RUNNING	100	NONE	1.6 pu	0.4
		CAPACITORS AT LOAD	1.1 pu	0.1
MOTOR 8,000HP LOCKED ROTOR INRUSH	100	NONE	4.6 pu	1.3
		CAPACITORS AT LOAD	2.8 pu	0.3

Source: J. F. Perkins and D. Bhasavanich, Vacuum switchgear application study with reference to switching surge protection, IEEE IAS IA-19(5):879-888, © 1983, IEEE.

itions. More severe voltage transients can occur during the interruption of transformer magnetizing inrush current, and surge protection is usually recommended if this condition might be encountered.

The other examples illustrating the use of the surge computer program in evaluating systems relative to surge application considerations involve switching a large (13.8-kV 8000-hp) motor fed by 100 ft of cable; the motor in question is a six-pole machine having 10 turns per coil. Under normal running conditions, the maximum transient voltage surge from motor terminals to ground is 18 kV (1.6 per unit), and the HF transient of about 1 MHz imposes fast oscillations of about 0.3 μs rise time on the motor winding. Although 18 kV is significantly below the ground wall insulation level of 50.5 kV, the fast rise time of the HF transients can cause the surge voltage to be nonuniformly distributed in the machine winding, so that turn insulation in the first coil would be more highly stressed than in the other coils. When surge capacitors are connected at the motor terminals, the maximum transient voltage is reduced to 1.1 per unit, and the average rise time of the HF transient is increased, so that the lower-level surges are more uni-

uniformly distributed throughout the motor winding. Since the rise time of the normal frequency recovery transient is also significantly reduced, the number of restrikes per phase per switching operation is decreased by 75% (Table 13.2), so that the lower-level, more uniformly distributed surge voltages occur much less frequently. The result is that the stress imposed on the motor turn insulation is reduced to a harmless level.

Interruption of locked rotor inrush current, which can occur at startup due to improper relay settings, can cause higher surge voltages than in normal load switching. When this occurs, in the example above, it can cuase a maximum transient voltage of 4.6 per unit to appear from motor terminals to ground, which is approximately equal to the ground wall insulation level. Similar or higher values have been reported in the technical literature [22,23]. If 80% of this voltage were to appear across the first coil [7], there would be about 4000 V between turns, so that the turn insulation would be stressed at about 200 V/mil, which corresponds to the normal ac test level applied to turn insulation. Although turn insulation will typically withstand three to four times this level, it is desirable to reduce the stress to maximize life.

With surge capacitors at the motor terminals, the maximum transient voltage is reduced to 2.8 per unit. Since the rise time of the HF transient tends to be reduced, so that any surges would be more uniformly distributed in the winding, the voltage between turns and the stress on the turn insulation are both significantly reduced. Consequently, the stresses imposed on motor turn insulation are diminished to completely safe levels.

In applying surge capacitors, it is recognized that the small effective inductance in series with each capacitor tends to block a fast change in current, partially isolating the capacitor from the machine terminals and reducing the wave-sloping effectiveness [24]. However, the surge capacitors modify system transient behavior in several ways that offer real improvement relative to surge protection. Surge capacitors reduce the rate of rise of the normal frequency recovery transient so that the probability of breaker reignition is reduced. In addition, the effective surge impedance at the load is decreased. The combined result is to lower the magnitude of surge voltage appearing between motor terminals and ground. Therefore, the component of surge voltage, which is sufficiently fast rising to be nonuniformly distributed in the winding, has a significantly lower amplitude than if no capacitors were connected.

In the example above, the motor switched under normal running conditions with no protection could produce about 1500 V/turn (75 V/mil) on the most highly stressed turns. Although ideal capacitors might reduce these stresses by about 10 to 15%, the effective reduction produced by real capacitors would be somewhat less. Similarly, with locked rotor inrush interruption, approximately 4000 V/turn (200 V/mil) could be imposed on the most highly stressed turns, and these stresses would be considerably reduced by the presence of surge capacitors.

The mechanism whereby series inductance reduces the effectiveness of surge capacitors highlights the importance of locating capacitors as close as possible to the load, so that lead length and inductance are kept to the absolute minimum.

13.12 SUMMARY

This chapter has reviewed the most recent work concerning the application of vacuum circuit breakers and the interaction between the vacuum circuit breaker and the electrical system relating to switching surge protection. Switching surge application considerations have been studied by means of computer analyses, which have been validated by excellent correlation between computed and measured transient switching voltages on single-phase and three-phase systems. Several examples of vacuum circuit breakers switching a variety of motor and transformer loads under different conditions have been presented. Although vacuum circuit breakers have current chopping and high-frequency current interrupting characteristics which differ in degree relative to other types of breakers, standard application practices of utilizing surge capacitors and arresters located at the equipment to be protected prove effective for vacuum breakers and should be adhered to.

ACKNOWLEDGMENT

This chapter is based on copyrighted material in Refs. 8 and 16. The author wishes to acknowledge the assistance of D. Bhasavanich, his co-author of Ref. 16.

REFERENCES

1. M. Murano, T. Fujii, H. Nishikawa, S. Nishikawa, and M. Okawa, Voltage escalation in interrupting inductive current by vacuum switches, IEEE Trans. Power Appar. Syst. *PAS-93*:264–271, 1974.
2. T. Itoh, T. Murai, T. Ohkura, and T. Yakami, Voltage escalation in the switching of the motor control circuit by the vacuum contactor, IEEE Trans. Power Appar. Syst. *PAS-91*:1897–1903, 1972.
3. J. Panek and K. G. Fehrle, Over-voltage phenomena associated with virtual current chopping in three-phase circuits, IEEE Trans. Power Appar. Syst. *PAS-94*:1317–1325, 1975.
4. M. Murano, S. Yanabu, H. Ohashi, H. Ishizuka, and T. Okazaki, Current chopping phenomena of medium voltage circuit breakers, IEEE Trans. Power Appar. Syst. *PAS-96*:143–149, 1977.

5. F. A. Holmes, An empirical study of current chopping by vacuum arcs, IEEE Power Eng. Soc. Winter Meet., New York, January 1974, Paper C-74-088-1.
6. A. N. Greenwood, D. R. Kurtz, and J. C. Sofianek, A guide to the application of vacuum circuit breakers, IEEE Trans. Power Appar. Syst. PAS-90:1589-1597, 1971.
7. Y. Murai, T. Nitta, T. Takami, and T. Itoh, Protection of motor from switching surge by vacuum switch, IEEE Trans. Power Appar. Syst. PAS-93:1472-1477, 1974.
8. J. F. Perkins, Evaluation of switching surge overvoltages on medium voltage power systems, IEEE Trans. Power Appar. Syst. PAS-101(6):1727-1734, 1982.
9. K. Yokokura, S. Masuda, H. Nishikawa, M. Okawa, and H. Ohashi, Multiple restriking voltage effect in a vacuum circuit breaker on motor insulation, IEEE Trans. Power Appar. Syst. PAS-100(4), pp. 1940-1948, April 1981.
10. A. N. Greenwood, The effect of current chopping in circuit breakers on networks and transformers, Part II, Trans. AIEE 79(Part III):545-555, 1960.
11. E. J. Tuohy and J. Panek, Chopping of transformer magnetizing currents, Part I: Single-phase transformers, IEEE Trans. Power Apparatus Systems, PAS-97(1):261-268, 1978.
12. M. T. Wright and K. McLeay, Interturn starter voltage distribution due to fast transient switching of induction motors, IEEE IAS Petrol. Chem. Ind. Conf. Rec., September 1981, Paper PCI 81-14, pp. 145-150.
13. L. G. Ananian, K. W. Miller, G. H. Titus, and G. W. Walsh, Field testing of voltage transients associated with vacuum breaker no-load switching of a power transformer in an industrial plant, IEEE Trans. Industry Applications, IA-19(6):914-919, Nov./Dec. 1983.
14. W. M. C. van den Heuvel, Overvoltages after current chopping in a three-phase inductive circuit with isolated neutral, IEEE Trans. Power Appar. Syst. PAS-100(12):4795-4801, December 1981.
15. E. W. Buss, R. C. Dugan, and P. C. Lyons, Vacuum circuit breakers and dry-type transformers: special considerations for mining operations, IEEE IAS Annu. Meet. Conf. Rec., 1977, Paper 31-A, pp. 769-775.
16. J. F. Perkins and D. Bhasavanich, Vacuum switchgear application study with reference to switching surge protection, IEEE Trans. Industry Applications, IA-19(5):879-888, Sept./Oct., 1983.
17. IEEE Power Eng. Soc. Rot. Mach. Committee, Insul. Subcommittee Report, Impulse voltage strength of ac rotating machines, IEEE Power Eng. Soc. IEEE Trans. on PAS, Vol. PAS-100, pp. 4041-4053, Aug. 1981.

18. IEEE Standard 141-1976 (Red Book), Recommended practices for electric power distribution for industrial plants (4.9).
19. IEEE Standard 242-1975 (Buff Book), IEEE recommended practice for protection and coordination of industrial and commercial power systems (9.3.16).
20. R. E. Pretorius and A. J. Eriksson, A basic guide to RC surge suppression on Motors and Transformers, Trans. S. Afr. IEE, August 1980, pp. 201-209.
21. R. E. Pretorius, The suppression of internal overvoltage surges in industrial high voltage systems, Certificated Eng. (Pretoria, S. Afr.), 938-956, July 1981.
22. F. Battiwala, H. Fink, M. Rimmrolt, and W. Schultz, Vacuum switchgear for voltages up to 38 kV, Siemens Power Eng. J. 3: 43-50, 1981.
23. E. Slamecka, Interruption of small inductive currents, CIGRE, W. G. 13.02, Study Committee 13, Electra, No. 72, Paris, October 1980, pp. 73-103 and Electra. No. 75, March 1981, pp. 5-30.
24. D. W. Jackson, Surge protection of rotating machines, surge protection in power systems, IEEE Tutorial 79-EHO-144-PWR, 1979, pp. 90-111.

14
Molded-Case Low-Voltage Circuit Breakers

ANTHONY LEE and PAUL G. SLADE / Westinghouse Research and Development Center, Pittsburgh, Pennsylvania

14.1 Introduction   527

14.2 Breaker Functions   532
    14.2.1 Conduction function   532
    14.2.2 Interruption function   534

14.3 Breaker Components   535
    14.3.1 Molded case   535
    14.3.2 Trip system   537
    14.3.3 Operating mechanism   542
    14.3.4 Arc chamber   543

14.4 Interruption Process   549
    14.4.1 Arc column   549
    14.4.2 Electrode region   553
    14.4.3 Interruption   553

14.5 Trends in Low-Voltage Molded-Case Circuit Breakers   555
    14.5.1 Current-limiting circuit breakers   555
    14.5.2 High-interruption-capacity breakers   561

References   561

14.1 INTRODUCTION

In previous chapters circuit breakers for high- and medium-voltage applications have been discussed. The first ten chapters, for example, have concentrated on circuit breakers for and circuit interactions with high-voltage ($\geqslant 72$ kV) transmission circuits. In Chaps. 11 through 13 the emphasis has been on medium-voltage (5 kV $\leqslant$ V < 72 kV) distribution breakers. In this chapter we present a brief discussion of the low-

Figure 14.1 Photograph of an early Westinghouse molded-case circuit breaker type. (Courtesy of Westinghouse Electric Corporation, Beaver, Pa.)

Molded-Case Low-Voltage Circuit Breakers

voltage (<1000 V), molded-case circuit breaker (MCCB). According to the IEEE Standard 242 [1], a molded-case circuit breaker is "one which is assembled as an integral unit in a supporting and enclosing housing of insulating material." The MCCB is the one with which the general public is most familiar because it has found application in the switching and protection of many low-voltage distribution circuits.

MCCBs were first introduced after World War I in both Europe and the United States, but their development proceeded more rapidly in the United States [2]. Figure 14.1 shows an early model which has limited performance and operational capabilities, but which exhibits the major features of a modern MCCB. Similar to present-day MCCBs, the air arc provides the interruption function, and the mechanism, trip system, and arc chambers are enclosed in an insulating molded case. Figure 14.2 shows examples of a wide variety of MCCBs. There has been a continual evolution of the MCCB due to a number of factors, such as market demands, performance improvements, the development of new materials, and advances in technology such as solid-state and microprocessor trip and control systems. These developments have greatly increased the system flexibility of MCCBs and it is now possible to tailor an MCCB to satisfy a wide range of applications. Present-day MCCBs, for example, can be found in all low-voltage applications: in the residential electrical distribution panel, in the industrial power distribution center, and in the main power feed panels used in large buildings such as offices, hospitals, and shopping centers.

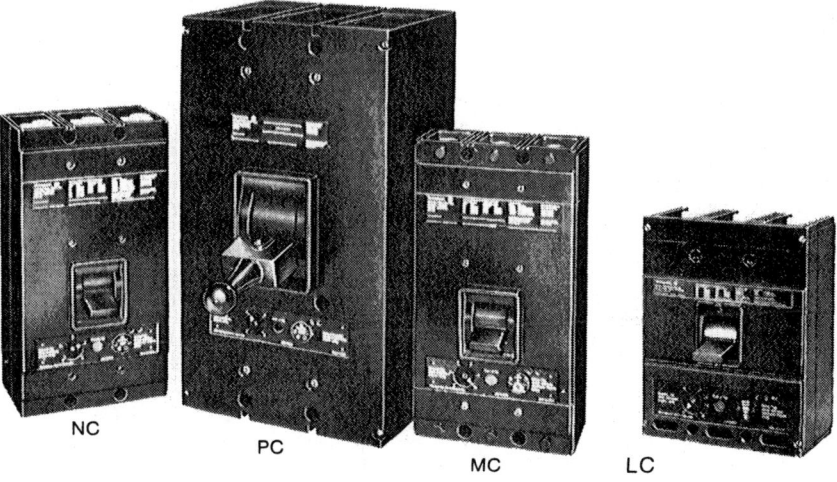

Figure 14.2 Examples of molded-case circuit breakers. (Courtesy of Westinghouse Electric Corporation, Beaver, Pa.)

A number of distribution system circuits are used in the United States [3]. The most common type of circuit, however, is the radial type. Figure 14.3 shows an example of such a distribution system. Although it is possible to use fuses and fused switches for some of the protection functions, the example shows MCCBs used exclusively. The principal advantages of a radial system are that they are (1) easy to operate, (2) very easy to protect, and (3) low in cost. The main disadvantage is that a fault near the source of the system can result in a major loss of power to the loads. For a well-planned and well-maintained system, however, the reliability of a radial system is very high.

The dominant U.S. standard for MCCBs is the Underwriters' Laboratories (UL) Standard 489 [4]. To obtain UL listing, a MCCB must

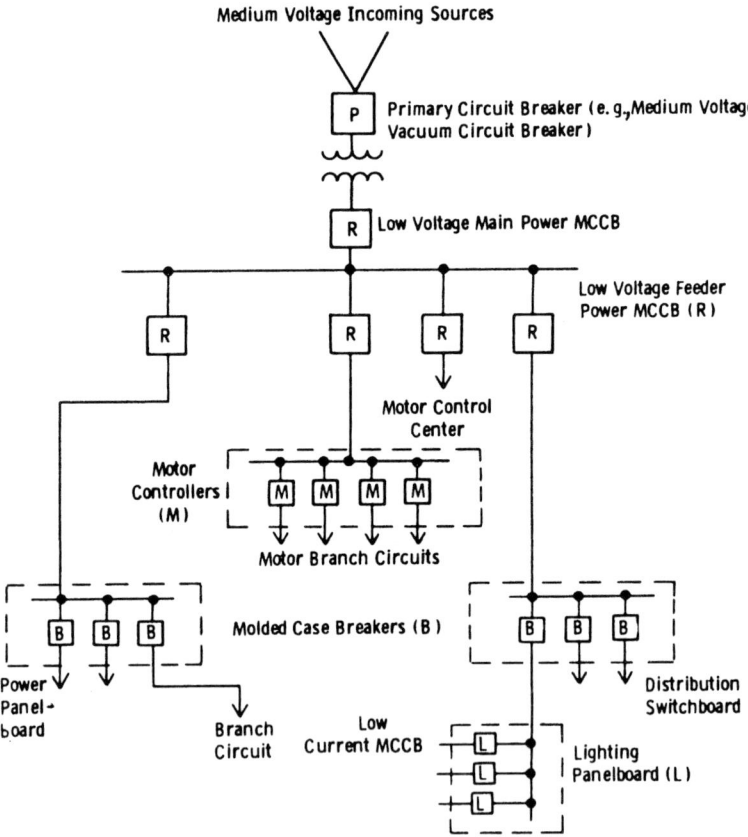

Figure 14.3 Radial distribution circuit system.

initially undergo a well-defined, rigorous series of test sequences. Furthermore, all MCCBs are also required by the UL to pass, on a regular basis, a series of follow-up tests similar to the initial test, to ensure the maintenance of the initial performance specifications.

The IEEE Industry Applications Society has also sponsored publications to guide in the selection of MCCBs for electrical system protection. These are IEEE Standards 141 and 242 [1,5] and ANSI Standard C37.13-1981 [6]. These publications serve as excellent references for system protection and coordination, fault calculation, equipment protection, procedures for maintenance, testing, and calibration, and preferred ratings.

Using the interruption capacity of an arc in air inside the molded case, the MCCB has been designed to interrupt currents with a range from a few amperes to a few hundred kiloamperes. For example, IEEE Standard 141 [4] gives "standard ratings" for 60 Hz (see Table 14.1). It contains the basic ratings for low-voltage alternating currents that are common in the United States. This list is by no means complete because other frame sizes are available (e.g., 25, 3000, 4000, and 5000 A [4]). There are other interruption capacity ratings as well (e.g., 14 kA). Further information should be obtained from manufacturers' bulletins. The frame size is characterized by the maximum continuous current that the breaker is designed to carry. In general, the higher the continuous current, the larger the physical size of the molded case.

It is interesting to note that, although the MCCB's primary function is to protect downstream circuit elements in the event of an overcurrent, most MCCBs spend their entire lives in a closed position with perhaps a few operations to switch normal load currents and in isolating electrical circuits. These functions are discussed in Sec. 14.2. In Sec. 14.3, the individual MCCB components are described: the molded case, the trip systems, the mechanism, and the arc chamber. Each component is important, but the MCCB must be able to interrupt the circuit. Section 14.4 presents a qualitative discussion of the interruption process and shows how the individual parts of the arc chamber

Table 14.1 Standard Ratings of MCCB As Recommended by IEEE Standard 141

Frame size, A	50,100,125,150,200,225,400,600,800, 1000,1200,1600,2000,2500
Voltage, V	120,120/240,240,277,277,480,480,600,
Interrupting capacity (rms symmetrical, kA)	7.5,10,13,18,22,25,30,35,42,50,65, 85,100,125,150,200

each play a role in interrupting the arc at an ac current zero. Finally, in Sec. 14.5 we discuss briefly the future trends in MCCBs by concentrating on current-limiting breakers and high-current power breakers.

14.2 BREAKER FUNCTIONS

14.2.1 Conduction Function

A typical MCCB in service will be conducting current when the system is energized. In this mode, the contacts are closed and a continuous

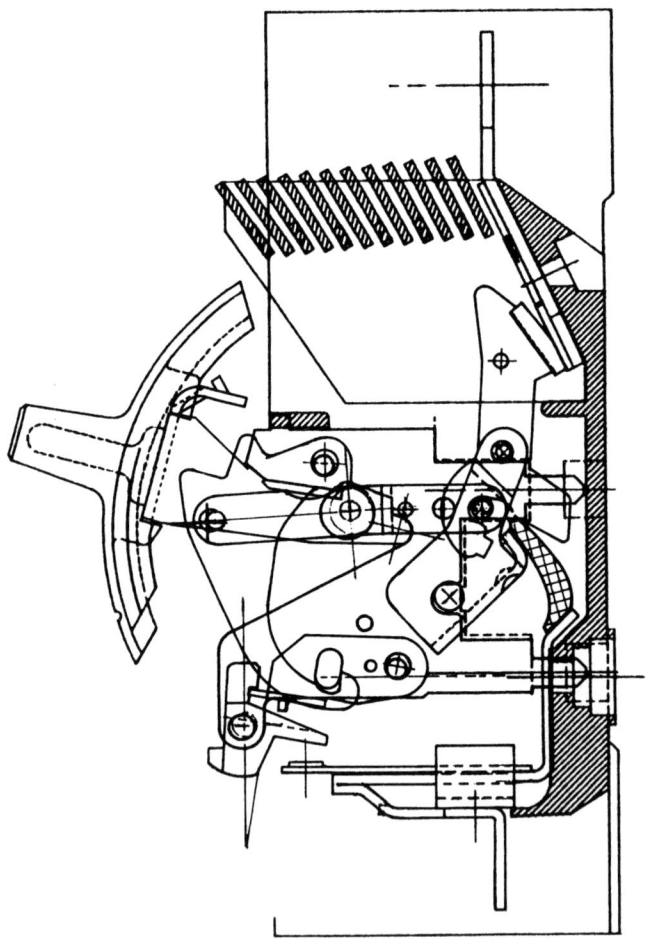

Figure 14.4 Typical MCCB showing its internal components. (Courtesy of Westinghouse Electric Corporation, Beaver, Pa.)

Molded-Case Low-Voltage Circuit Breakers

current goes through the breaker. This results in Joule heating of the current conducting components, such as the contacts, the trip units, and the flexible shunt connections. The breaker therefore has to dissipate effectively the heat that is generated within it by these components. Figure 14.4 shows the sketch of a typical MCCB, and a photograph of a cutaway view is shown in Fig. 14.5. The conducting components that change most with usage are the contacts (see Chap. 15). During each opening operation under load, an arc is always generated between the contacts, and this arc is maintained until extinction at a natural or forced current zero. During the arcing period, the electrodes are under high thermal loading, especially at the arc roots, where the temperature is at or above the metal's melting point. There can also be a high level of contact erosion at higher currents. If the contacts are Ag-W, for example, there can be depletion of Ag in the contact surface, and there can also be formation of nonconducting oxides [7]. The design of the mechanism takes this effect into account, thus eliminating the undesirable excess Joule heating and consequently higher breaker temperatures. UL standard tests for MCCB include temperature-rise measurements at the MCCB terminals while carrying rated current after the MCCB has interrupted an overload current (600% of rated current) 50 times.

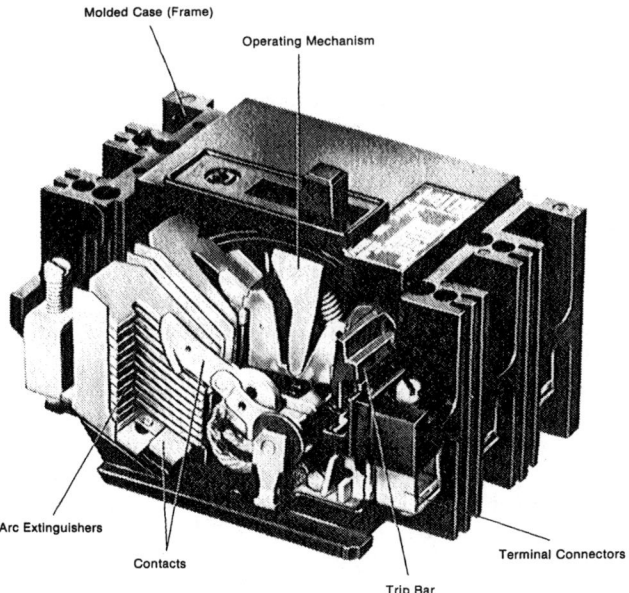

Figure 14.5 Cutaway view of a MCCB showing major components. (Courtesy of Westinghouse Electric Corporation, Beaver, Pa.)

14.2.2 Interruption Function

MCCBs are designed to switch a wide range of currents. They not only have to provide overcurrent and short-circuit current protection, but must switch load current and isolate sections of the low-voltage distribution system.

The isolation duty under no-load condition represents the easiest duty, because no arcing is involved. The breaker can be operated manually, with a motor operator, or by a remote tripping signal, and executes an opening operation. The final contact gap and clearances between the line-side contact structure and other components need only be sufficient to hold off the open-circuit voltage.

When the breaker operates with a load current flowing, an arc forms between the contacts inside the arcing chamber. The circuit is interrupted when the arc is extinguished, usually at the first current zero of the ac wave following arc initiation. The open gap must then withstand the line voltage; it is here that the power factor has an effect on the interruption. In highly inductive circuits, for example, current zero occurs near voltage crest, resulting in a high voltage stress across the contacts immediately after current zero. Fortunately, for normal load currents, the arc energy for a half cycle of arcing is low. For example, one half cycle of a 100-A arc at an arc voltage of 200 V dissipates a little over 150 J. This energy level is insignificant to the contact structure and the arc chamber, which are designed to handle arcs at full fault current of many tens of kiloamperes. Most MCCBs are designed to switch load currents for many thousands of operations.

The arc chamber of the MCCB is designed primarily to interrupt overcurrents in the circuit. To do this, there has to be a reduction in the number of operations that the breaker can withstand. This results from the higher arc energy that the arc chamber must absorb during each arcing period. It is not hard to see that, in general, arc chamber volume, contact size, and arc chute size have to increase with increased fault current rating of interrupting capacity, because both the arc erosion and degradation become more severe. For load or rated currents, the arc is small with respect to the chamber size, and the arcing duration is short. Erosion is usually limited to the contact region. At higher arc currents such as the level encountered in overload conditions, the arc is more intense. The arc energy loading not only involves the contact region but also the components near the contacts, such as the arc runners and the front part of the arc chute. At full or near-fault-current levels, the arc is extremely intense, occupying a large volume, and is driven into the arc chute region, which absorbs most of the arc energy.

The UL standard overload test is designed to ensure breaker performance at moderate overcurrent. For example, a breaker with a 100-A rating must interrupt a current of 600 A 50 times. After this

Molded-Case Low-Voltage Circuit Breakers

switching duty the MCCB is expected to carry the load current without overheating.

At the full fault-current level, the UL standard requires performing a number of interruptions, including both "open" and "close-open" operations. For example, a 480-V 100-A-rated breaker must interrupt 8660 A at each pole for an open and a close-open plus an open and a close-open on all three phases at 10,000 A. The breaker must then pass a 200% calibration test on its trip unit and a standard dielectric withstand test with the breaker open [8].

Fortunately, under fault-current conditions the arc voltage can reach values that are a good fraction of the line voltage. This arc impedance, together with the impedance of the circuit, give rise to a current-limiting effect, and hence MCCBs are never subjected to the full fault current. For example, two 20-A 100-V breakers in series have to be able to interrupt a potential 240-V 10-kA fault, but the current is usually limited to less than 2 kA. The current-limiting ability of MCCBs has recently resulted in a new class of MCCBs which have been designed specifically to limit current [9]. These breakers reduce both the peak let-through current level and the arcing time. The resulting lower level of energy dissipation has resulted in smaller arc chamber designs and more compact MCCBs. Furthermore, lower let-through current means better downstream protection.

The current-limiting MCCB is not the only major development in recent years. Continued research and development have also resulted in (1) better contact materials for both low contact resistance and low erosion; (2) solid-state trip units for greater flexibility and control, and (3) new materials and configurations for the arc chamber for higher reliability, interrupting duty, and withstand.

14.3 BREAKER COMPONENTS

14.3.1 Molded Case

A cutaway view of a typical MCCB, showing the different breaker components, is shown in Figure 14.5. The external parts are the case and terminals. The molded case provides both the insulation and the support structure for mounting all other components. The material is typically a reinforced polymer such as glass polyester which possesses both mechanical and dielectric strength. Typical construction consists of a case and a cover. The entire internal assembly is first mounted into the case and the cover is put on during the final assembly. Figure 14.6 shows such an assembly. For certain single-pole designs, the case can have a side cover instead of the top cover shown in Fig. 14.6.

The breaker case involves a multitude of design considerations. For example, it is important to consider not only the mechanical strength

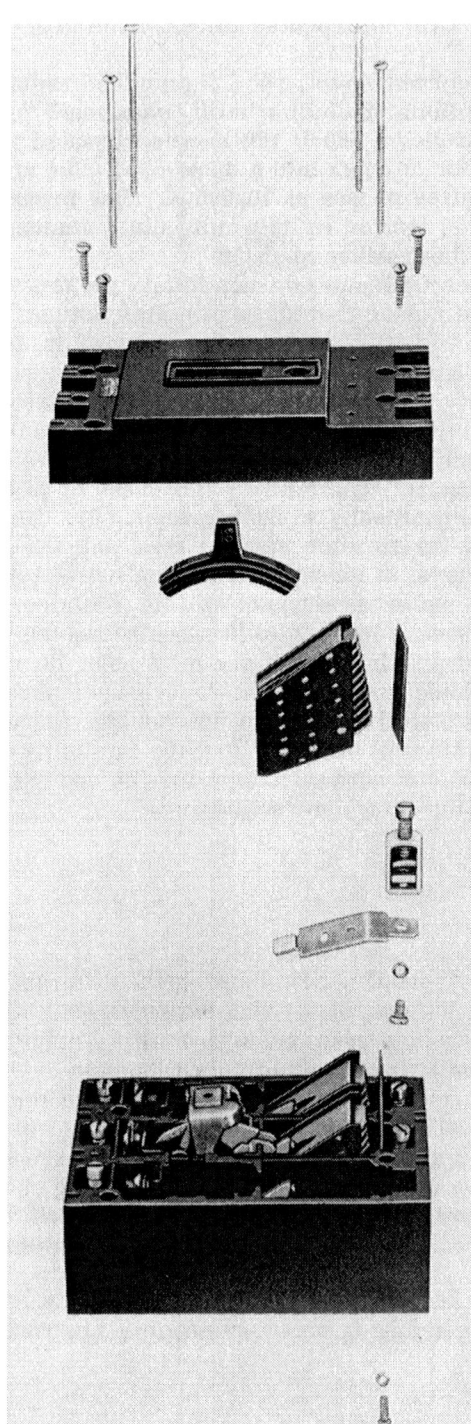

Figure 14.6 Final assembly of a MCCB. (Courtesy of Westinghouse Electric Corporation, Beaver, Pa.)

Molded-Case Low-Voltage Circuit Breakers

required to hold both the stationary and moving parts, but also the strength to withstand the magnetic force exerted through the current-carrying members, the thermal loading during short-circuit interruption, and the gas pressure generation during high current arcing. For low voltages, the dielectric strength of the case is usually more than adequate, although surface degradation may occur along the inside of the case which faces the arcing chamber. The dielectric strength can be increased by appropriate selection of case material or by application of an arc-resisting layer.

14.3.2 Trip System

A typical current-time characteristic of a MCCB is shown in Fig. 14.7. The trip system provides the function of activating the operating mechanism in the event of a prolonged overload or of a high fault current. The overload trip action is usually provided by heating a bimetal element. This can be done by direct heating, indirect heating, or by induction heating. An example of trip action is illustrated in Fig. 14.8. As the overload current flows through the heater adjacent to the bimetal, the bimetal bends as its temperature increases. A trip action

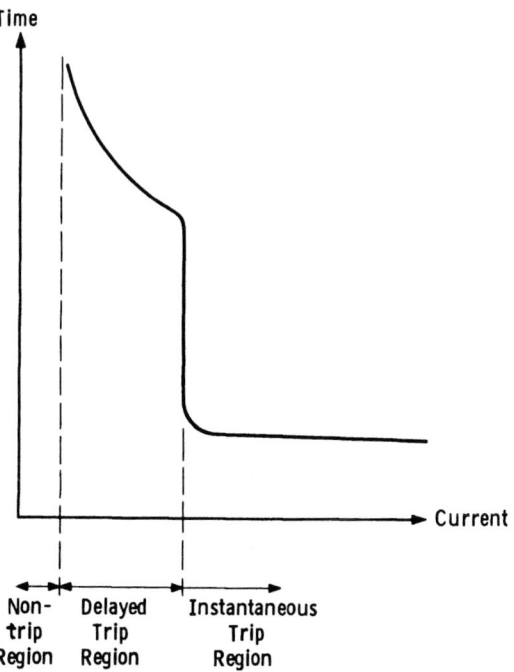

Figure 14.7 Circuit breaker time-current curve for a typical MCCB.

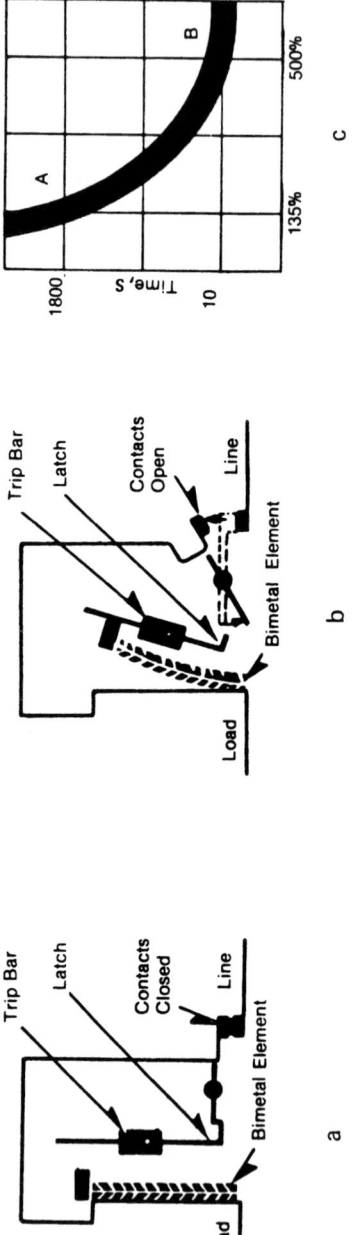

Figure 14.8 Thermal overload trip action: (a) normal; (b) overload situation with bimetal heating and bending; (c) typical time-current characteristics.

Molded-Case Low-Voltage Circuit Breakers

is accomplished when the bimetal deflects sufficiently to unlatch the operating mechanism.

Each rating and time-current characteristic requires a specific bimetal element. The bimetal design parameters that control the thermal trip action are (1) materials, (2) heater resistance, (3) thermal capacity, and (4) overall length. Because of the slow tripping feature, the desirable delay time is built into this type of element. A higher current will induce a faster temperature rise and a shorter trip time. For high short-circuit current levels, however, an instantaneous trip is required (as shown in Fig. 14.7). Here most MCCB designs utilize a magnetic action, as illustrated in Fig. 14.9. The plunger or a keeper that bridges the magnetic circuit is usually biased in placed by a spring so as to leave an air gap. As a fault current occurs, a force develops which closes the air gap. This action, illustrated in Fig. 14.9, has a threshold type of behavior: when the magnetic force overcomes the spring force, a trip occurs. Thus, a time-current characteristic similar to that shown in Fig. 14.9(c) is generated.

It is possible to design an adjustable magnetic trip merely by varying the air-gap length or the spring force. Point A in Fig. 14.9(c) can then be varied along the current axis. This feature allows one trip element design for a wide range of current ratings; even more flexibility can be achieved if it is used to give breakers trip settings which can be adjusted by the user.

These two types of trip elements can be combined, providing the type of time-current characteristic shown in Fig. 14.7. This is the most common type of trip system used in MCCBs and is known as the "thermal-magnetic" type.

With recent advances in semiconductor devices, a new type of trip unit has been developed [10]. As an illustration of this type of trip

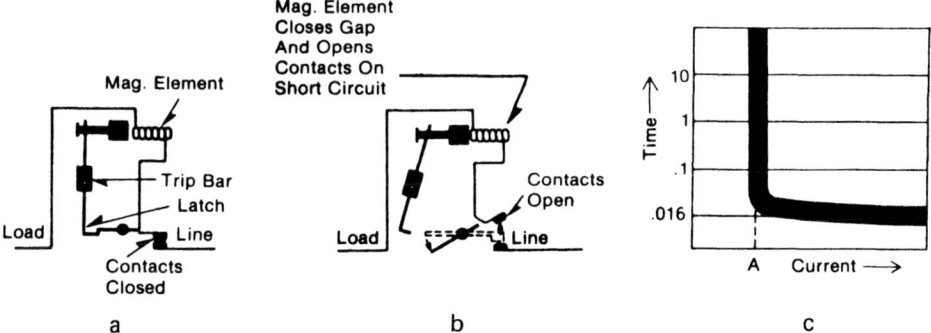

Figure 14.9 Magnetic trip action: (a) normal; (b) short-circuit situation with a magnetic element that pulls the plunger, releasing the latch; (c) typical time-current characteristic.

unit, the Westinghouse Seltronic (solid state) trip system will be discussed. There are four main parts in this system:

1. Current sensors
2. Solid-state circuit
3. Flux transfer device
4. Rating plug

The current sensors or transformers perform two functions: they provide an output which is proportional to the breakers's current, and they provide energy to operate the solid-state circuitry and to trip the breaker.

The solid-state circuit rectifies the output signal from each of the current sensors, senses the current level, and initiates a trip if the current is in excess of the rated value. For modest overloads, a relatively long delay is allowed before tripping occurs. This time delay has an inverse relationship to the current similar to that shown in Fig. 14.8(c). The standard time-delay curve shape can easily be altered to fit special applications. For high overloads (those above a level

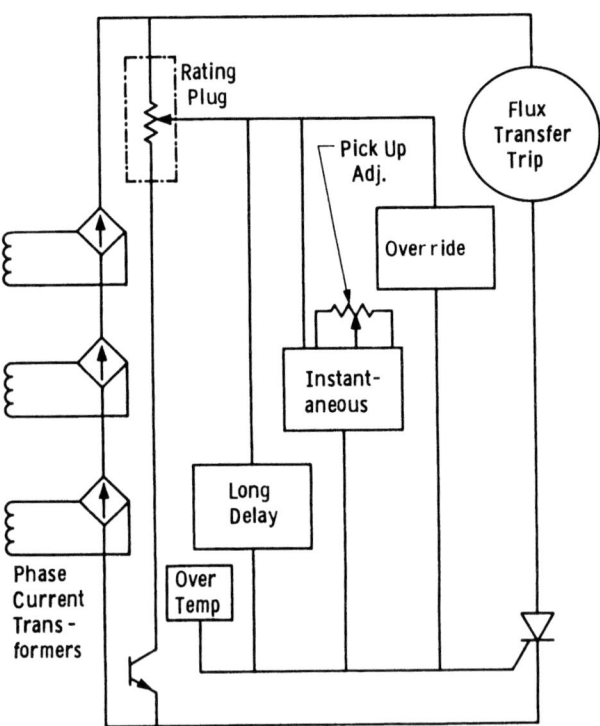

Figure 14.10 Simplified schematic of the Seltronic trip systems.

Molded-Case Low-Voltage Circuit Breakers

adjustable by the users) the breaker trips either instantaneously or with a short delay of a few cycles, depending on the application. To protect against short circuits, a separate override trip is provided which causes an instantaneous trip on high faults. The solid-state circuit also contains a temperature sensor which causes a trip if the ambient temperature inside the breaker exceeds a predetermined value. Figure 14.10 is a simplified block diagram of the solid-state trip unit.

The flux transfer trip is a small stored-energy trip device which can be triggered with a small pulse of power from the solid-state circuit.

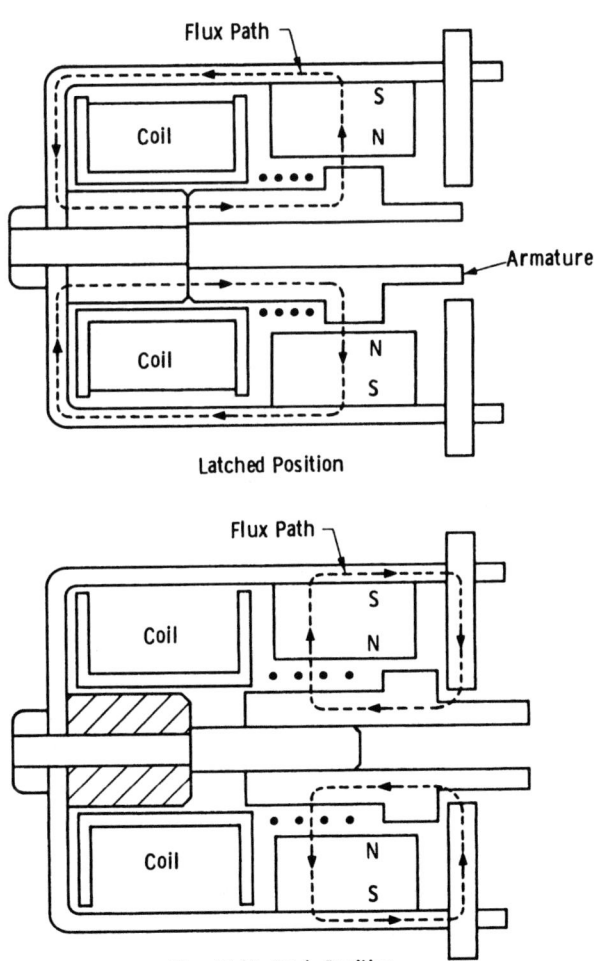

Figure 14.11 Sketch showing operation of the flux transfer device.

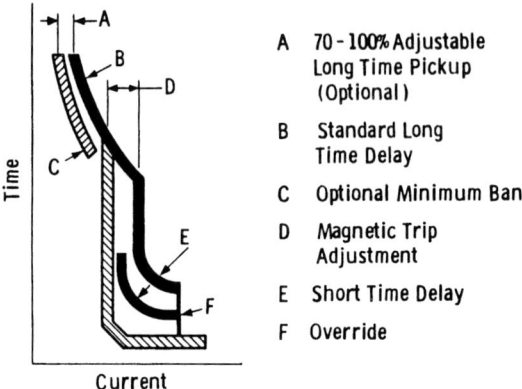

Figure 14.12 Time-current characteristics for a Westinghouse Seltronic trip unit showing some of the adjustable features.

It is a permanent magnet device which latches an armature that is spring loaded as the mechanism is reset. A 3 mJ pulse of energy is all that is required to shift the flux, thereby releasing the armature. The armature moves to unlatch a trip lever with about 5 lb of force, which in turn trips the mechanism's opening spring latch, releasing up to several hundred pounds of spring force, which will open the contact arms rapidly. Figure 14.11 ahows a cross-sectional view of this device in the latched and flux-shifted trip positions. The continuous current rating can be changed easily by changing a rating plug located on the breaker case cover. This is equivalent to changing a complete thermal magnetic trip unit. The adjustable features of such trip systems include continuous current, long delay, instantaneous pickup, short delay pickup, short delay time, ground fault pickup, and ground fault time. Solid-state logic provides a means to shape the time-current curve to take full advantage of the circuit breaker's capabilities and to provide maximum flexibility for the systems designer. Figure 14.12 shows a time-current curve with many of these features.

14.3.3 Operating Mechanism

The mechanism shown in Figs. 14.4 and 14.6 provides the means to execute "open" and "close" functions of the typical MCCB. The operating handle links the manual push or pull to the mechanism. The closing operation also charges a spring, which provides the force for the opening operation either through the handle or through the trip unit. A quick-make, quick-break operation is provided by a toggle-type mechanism. The contacts snap closed and open independently of the

Molded-Case Low-Voltage Circuit Breakers

handle and of the trip unit speed. Many variations of mechanism design have been developed, and an excellent survey of these can be found in the book by Dzierzbicki and Walczuk [11].

The snap action of the contacts is essential. During the opening operation, the arcing duration depends on contacts attaining a sufficient gap for successful arc extinction and dielectric recovery at an ac current zero. To reduce arc chamber degradation and to limit fault duration, this time should be as short as possible. A snap-action opening is therefore important. For closing, the contacts are made under full voltage conditions. As the contacts come together, the gap becomes increasingly small. Since the electric field will increase as the gap is reduced, a breakdown will occur, and a pre-strike arc will develop. The duration of this arc will depend on the time it takes to close the gap completely (i.e., the speed of the closing operation). If this time is too long, the arc roots can actually provide a sizable welding spot of molten metal on the contact faces. It is also important to design the mechanism to minimize the duration of contact bounce on closing and therefore to minimize the repetitive arcing that occurs as the contacts bounce open. Too much contact bounce can result in severe contact erosion and even in contact welding.

Another common design element is the "trip-free" feature. This is a safety feature that allows the trip action to open the contacts even if the handle is held in the "on" position. In the event of a fault, the breaker will still trip and interrupt the fault. This feature is important because an operator may desire to close the breaker, presupposing that the fault has been cleared. If, in fact, the fault has not been cleared, a trip will quickly occur even if the operator is executing the closing operation manually. The trip-free feature ensures proper protection of the electrical system and of the operator.

The most notable recent development in MCCB mechanisms is the stored energy type with multiple-operation capacity [12]. While maintaining the features of the earlier mechanical designs, the mechanism of the Westinghouse System Pow-R Breaker, for example [13], can execute a closing operationg in fewer than five cycles after the breaker is opened. The stored energy system can be charged either electrically or manually. The sequence of operation is expanded from the traditional close-open to charge-close-recharge-open-close-open. A photograph of a System Pow-R breaker is shown in Fig. 14.13. Note the mechanism status indications and the solid-state trip system selections.

14.3.4 Arc Chamber

The arc chamber is where the crucial function of circuit interruption occurs. In MCCBs, there are three major arc chamber components: (1) contacts; (2) arc runners; and (3) arc chute. Each component serves a specific function in the interruption process. The following discussion addresses each of these components and their functions.

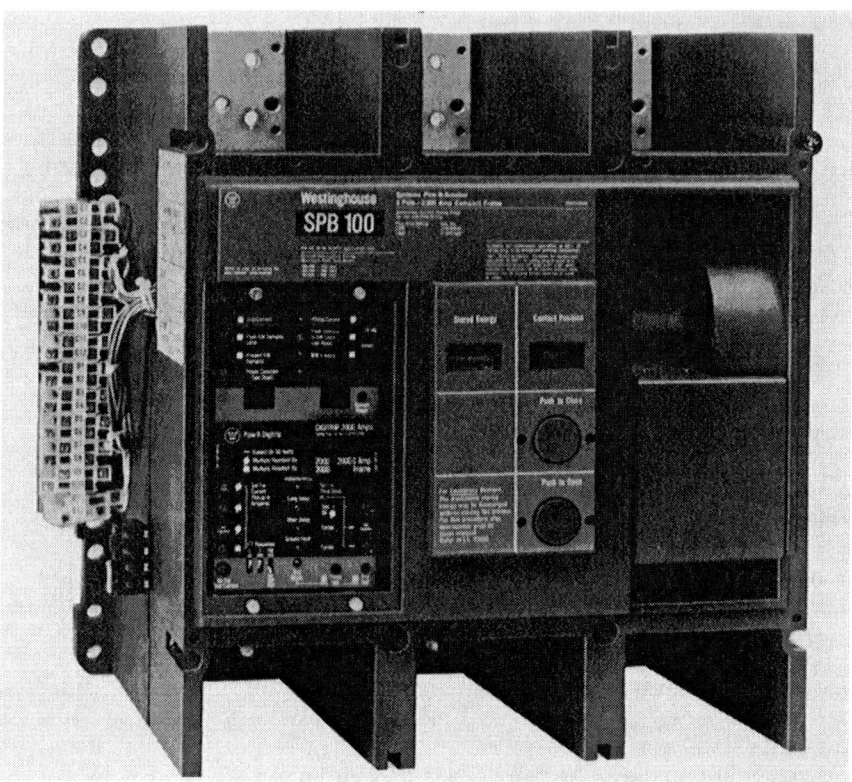

Figure 14.13 Photograph of a Westinghouse System Pow-R breaker showing the stored energy mechanism control and indication and the trip system adjustments. (Courtesy of Westinghouse Electric Corporation, Beaver, Pa.)

Contacts

The contacts are involved in both the conduction and interruption functions. In the closed position the mating interface must have a low resistance for carrying the continuous current without overheating. Upon opening under load or overcurrent conditions, an arc will form. The arc (see Chap. 4) is a high-conductivity plasma column with a temperature greater than 10×10^3 K. Here the contacts serve as the electrodes (anode and cathode) in direct contact with the arc column. In such a high-temperature environment, a certain portion of the contact material will vaporize, leading to contact erosion. Furthermore, being exposed to air, oxides can form, leading to an increased contact resistance. The ideal material would be one that has low contact resistance, high resistance to arc erosion, soft oxides such that the oxide layer is easily broken by the closing operation, and structural strength to withstand the impact of the mechanical operation. As discussed in Chap. 15, powder metallurgy has been successful in providing many useful contact materials. Those commonly found in MCCBs are Ag-W, Ag-Ni, Ag-C, Ag-metal oxide (e.g., CdO, SnO_2), Ag-Mo, and Ag-WC. The refractory part provides the high-temperature withstand and structural strength, while the Ag provides the high conductivity and low contact resistance.

For high-continuous-current-duty breakers such as those rated at 1000 A or higher, parallel contacts are used to separate the conduction function from the interruption function (see Chap. 15). For example, a set of main contacts parallel to the arcing contacts can be arranged such that the mains will make last and break first. Under these conditions, the main contacts will not experience excessive arc erosion. This also allows them to have higher Ag content in order to accommodate the high continuous current. On the other hand, the arcing contact that is relieved of the conduction function may now be designed strictly for resistance to arc erosion and circuit interruption.

In the opening operation the coordination between the arcing and the main contacts is critical. The current must first be commutated from the main contacts to the arcing contacts. This commutation is accomplished by the arc voltage generated between the mains as they separate. The time t_c it takes to complete the commutation is given by [14]

$$t_c = K \frac{L}{R}$$

where K depends on the current magnitude, the arc voltage, and the contact resistance of the arcing contacts, L is the loop inductance, and R is the contact resistance of the arcing contact. Figure 14.14 illustrates this commutation process. Figure 14.15 shows empirically determined commutation times at high currents. The commutation time must be short relative to the differential opening time between the two types of contacts. Furthermore, the commutation time should be kept short to avoid excessive erosion of the main contacts.

Arc Runners

The arc runners are extensions of the contact structures having the same electrical potential as the contacts, (see, e.g., Fig. 14.16). The functions of the arc runners are (1) to stretch the arc as it moves off the contacts onto the diverging portion of the runner, (2) to quickly channel the arc into the arc chute away from the contact region, and (3) to quickly establish the arc in a noncontact region to prolong the contact life. In other words, the arc runners serve as new electrodes for the arc.

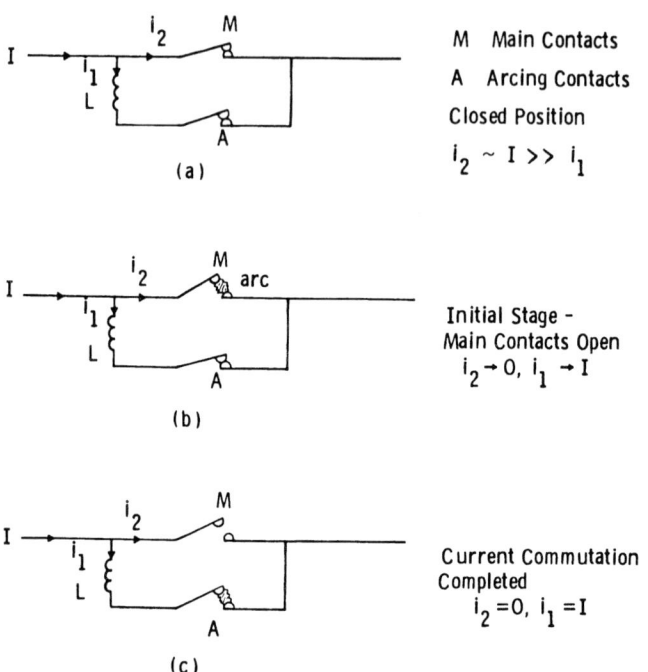

Figure 14.14 Sequence of operation showing the current cummutation process from the main contact to the arcing contact.

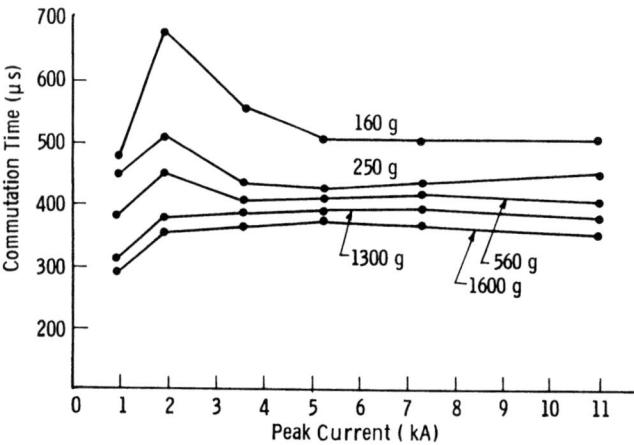

Figure 14.15 Measurements of current commutation time from the main to the arcing contact for different contact opening accelerations. Single-break Ag-Ni contacts were used.

Molded-Case Low-Voltage Circuit Breakers

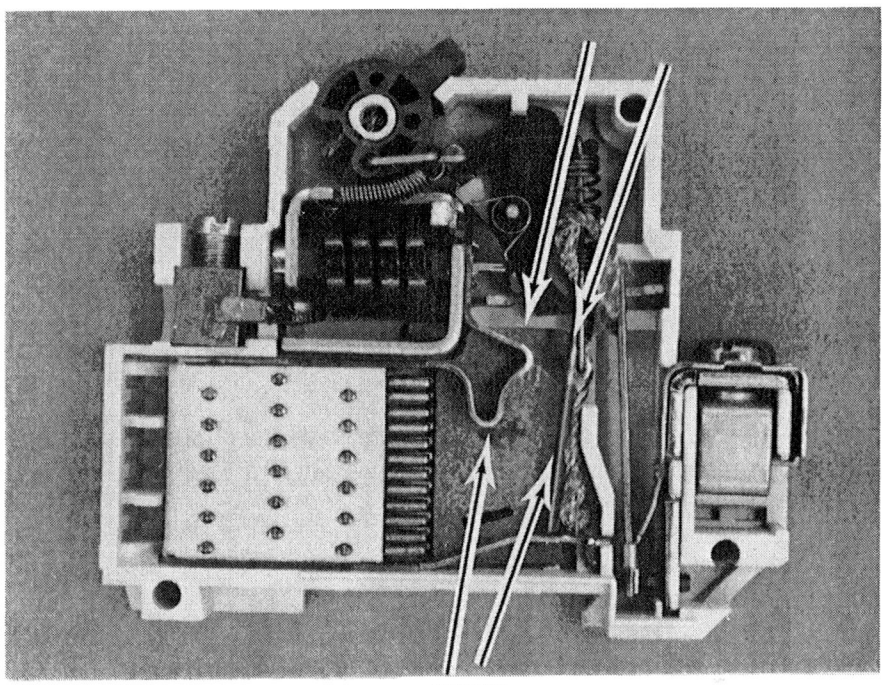

Figure 14.16 Photograph of a miniature current-limiting (CL) MCCB. (Courtesy of Westinghouse Electric Corporation, Beaver, Pa.)

Common material used to make the arc runners are copper and iron. The surfaces may be plated to reduce corrosion or to aid arc movement.

The arc runner shown in Fig. 14.16 is a structure from a miniature current-limiting MCCB. The self-generated $\bar{J} \times \bar{B}$ force causes the arc to migrate quickly along the arc runners. As it travels, its length increases and it is forced into the arc chutes, where it is cooled. Such a configuration requires the contact to open very rapidly and the arc to move along the runners when the contact gap is only a few millimeters. The arc voltage must reach a value of several hundred volts within 1 to 2 ms for effective current limitation. Figure 14.17 shows an example of the voltage and current waveforms of this design when interrupting a potential 10-kA fault current in a 240-V circuit. This type of breaker has a performance similar to that of a current-limiting fuse. The let-through current is a small fraction of the potential fault current, and the interruption or arcing time before extinction is shorter than a half cycle. Since this type of miniature MCCB has

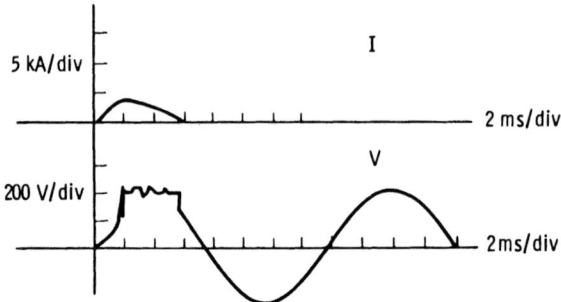

Figure 14.17 Arc voltage and current from an actual test of a CL-MCCB. (Courtesy of Westinghouse Electric Corporation, Beaver, Pa.)

received much attention in recent years [2], much effort has gone into optimization of arc runners for rapid arc motion [15-17].

Arc chute

Modern MCCBs use variations of the basic deion plate (first proposed by Slepian [18]) to form the arc chute. Usually, the arc chute is a stack of closely spaced steel plates supported by insulating material,

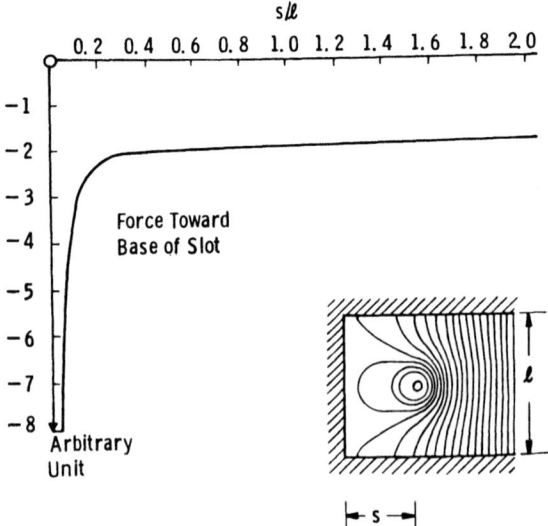

Figure 14.18 Calculated field line and force on a linear current inside an infinite slot. (From Ref. 19, Chap. 6.)

Molded-Case Low-Voltage Circuit Breakers

as shown in Fig. 14.5. Occasionally, other materials, such as copper or ceramic, are used instead of steel. Figure 14.18 illustrates why steel is commonly used. The arc is formed between the contacts, which are located close to the legs of the U in the deion plate. The magnetic field resulting from the current flowing in the arc interacts with the deion plate and causes the arc to be drawn into the slot [19]. For currents up to a few thousand amperes the arc is driven into the gaps between the plates, where it splits up into a series of arcs. Figure 14.19(a) shows the arc tracks on a set of deion plates made by such an arc splitting. Here the arc voltage will be approximately

$$V_{arc} = 25(n + 1) \quad \text{volts}$$

where n is the number of deion plates. The sum of the anode fall, the cathode fall, and column voltage drop is 25 V per gap.

At higher currents the arc is so large, filling the arc chamber volume, that it cannot form individual arcs between the plates. In this case the plates tend to cool the arc by thermal conduction and by ablation of the chute material. When Fig. 14.19(a) is compared to Fig. 14.19(b), the effect of the high current arc can be seen.

Figure 14.20 illustrates the effect of an arc chute made from insulating plates. In this case, the arc is stretched over the plates, increasing its length and surface area for cooling. The solid insulating surfaces have a high thermal capacity and hence are able to absorb the arc energy. The arc chute construction often has a diverging fan-shaped stack. This shape preferentially channels the gas flow out through the vents behind the chute. This relieves the pressure build-up in the arc chamber which results from arc heating and ablative gas generation.

The insulating support structure is commonly made from fiber material, polyester, melamine, or a large number of reinforced plastics. The selection is based on mechanical requirements, ability to withstand direct exposure to high current arcing, and high dielectric strength. Quite often, metal vapor and gases from the insulating material and from the case material influence the interruption ability of the arc.

14.4 INTERRUPTION PROCESS

14.4.1 Arc Column

A general description of the arc is given in Chap. 4. In this section we focus on the type of arc that appears in the MCCBs. The arcing medium is a complex mixture of air, vapor from the arc chamber walls, and metal vapors from the contacts and arc chute plates. In a device like an MCCB, the arc looks quite different from the idealized case shown in Chap. 4, where the arc column is cylindrical with well-defined electrodes. Figure 14.21(a) and (b) show photographs of a high current

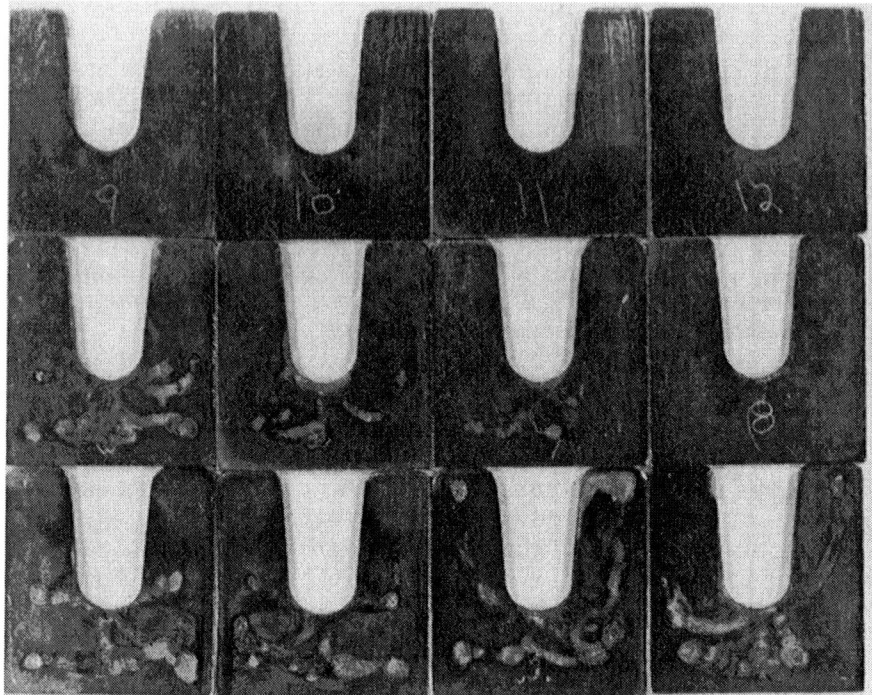
(a)

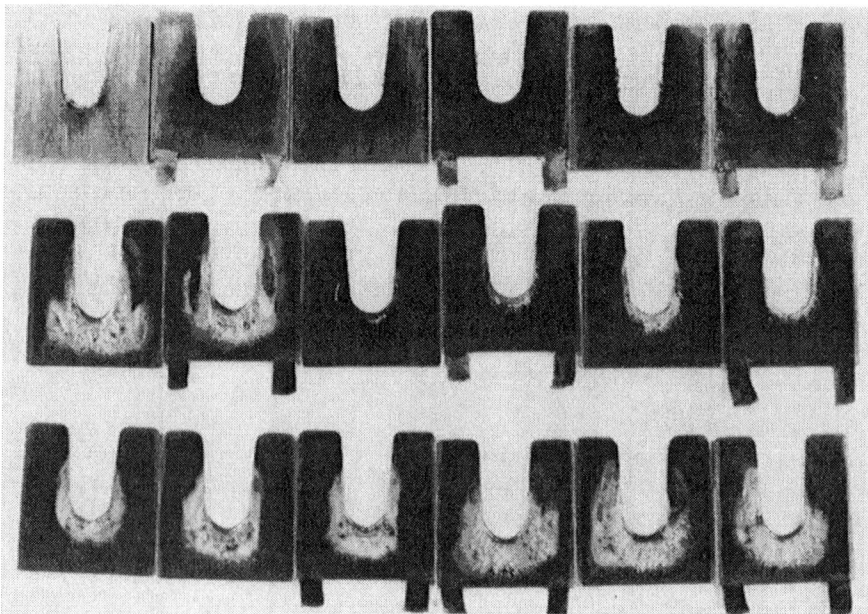
(b)

Figure 14.19 (a) Deion plate arc tracks, 1000 A; (b) deion plates, 20 kA.

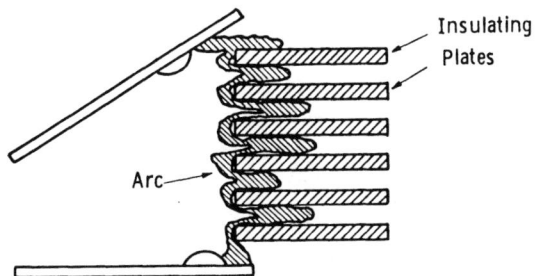

Figure 14.20 Arc stretching action of insulating plates.

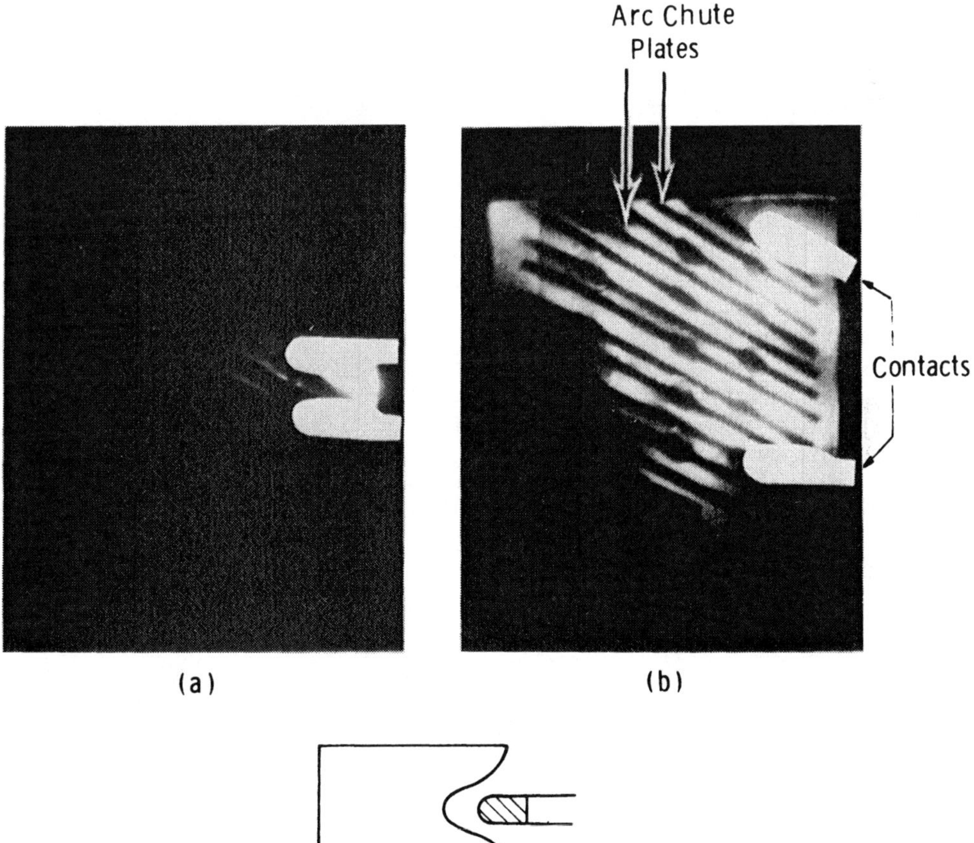

Figure 14.21 High-speed photograph of the arc (50 μs exposure time) in a Westinghouse current-limiting breaker (FCL): (a) 150 μs after contacts parting with an arc current of 13 kA: (b) 400 μs later with an arc current of 27 kA; (c) top view of the arc chute and electrodes.

arc in a practical MCCB. Figure 14.21(c) presents the top view of the position of the contacts relative to the arc chute plates.

At small gaps [Fig. 14.21(a)] the arc voltage is about 40 V. This arc is electrode dominated and the arc power loss results only from radiation, thermal conduction, and electrode effects. The next stage occurs when the electrode gap is increased further and the arc length is increased. Here the arc column experiences a $\bar{J} \times \bar{B}$ force, pushing it toward the arc chute plates. The arc is now free burning with a longer arc length and a higher arc voltage. In the third stage the arc is in contact with the arc chute plates, as shown in Fig. 14.21(b). The exposure time of 50 μs in this photograph is not short enough to resolve any rapid arc movement. There are, however, strong suggestions of series arcs between the plates as well as arc plasma filling the arc chamber volume. At this stage, where the arc is effectively cooled and confined, the arc voltage reaches 300 to 400 V. Figure 14.22 shows a typical arc voltage for a half cycle of high current arc. (N.B.: The contact gap is also increasing in time.) The arc voltage first appears 1.5 ms into the current loop; this corresponds to contact parting and arc initiation. For the next 2 ms, the arc is stretched and the arc voltage gradually increases. Finally, near current crest, the arc is in intimate contact with the arc chute. The arc voltage now reaches a peak of ~400 V. Note the fluctuation of the voltage signal. This results from the unstable arc motion and the periodic shorting of a number of series arcs by a long arc outside the arc chute plates. These effects can result in sudden drops in the arc voltage.

The temperature of these arcs is between 10×10^3 and 20×10^3 K [20,21] with higher arc temperatures at higher arc currents [21]. These high arc temperatures result in a conductance that is high enough to maintain the current flow. At a current zero, however, the gas in the arc chamber must be quickly cooled to a level suitable for withstanding the transient recovery voltage. If this is done successfully, the circuit current will be interrupted. Effective arc chamber cooling is, therefore, one of the most important MCCB design requirements.

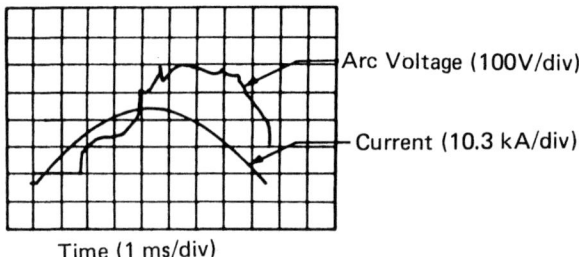

Figure 14.22 Typical arc-voltage waveform for a high-current MCCB arc.

Molded-Case Low-Voltage Circuit Breakers

When considering the arc chamber design it must be noted that the plasma enthalpy allows the arc column to store only a small fraction of the total arc energy. For example, at 1 atm pressure, a 2-cm-long 4-cm-diameter arc in air at 20,000 K has a specific enthalpy of 43.04 kcal/g and an energy content of 20.5 J. The total arc power for this arc when it is carrying 10 kA at an arc voltage of 200 V would be ~2 MW. Thus the dwell time of the energy in this arc, or its thermal time constant (cooling time), is only about 10 µs.

14.4.2 Electrode Region

To maintain high current conduction in the arc column, the cathode must be an efficient electron emitter. As discussed in Chap. 4, the cathode spots serve this important function. The transition from the solid contact to the surface cathode spot and through the high-temperature arc is only partially understood [22]. Qualitatively, the cathode region emits electrons, dissipates energy, and loses mass. Heat gains result from (1) particle bombardment, both ions and neutrals; (2) thermal conduction from the arc; and (3) radiation from the arc. This is balanced by vaporization of contact material and heat conduction from the surface to the bulk of the contact.

The anode's function is to collect electrons from the arc. Again, the physical processes that occur inside the transition region are rather complex [22,23]. At high current densities, the anode is heated by the electrons as they transport the current from the arc column. The cathode and anode can even become so hot that vigorous evaporating and violent jets of metal vapor form to help remove the heat.

The amount of metal vapor in the plasma column as a result of anode and cathode activities will influence the gap recovery and interruption. One recent study [24] showed that even at low currents the breakdown voltage is affected by small amounts of impurities in the contacts. The recovery voltage is a linear function of the effective ionization energy of the impurity metal vapor.

14.4.3 Interruption

Circuit interruption in an MCCB can be considered as a three-step process: (1) Rapid adjustment of the arc plasma within microseconds after current zero to withstand the immediate transient recovery voltage across the contact gap; (2) deionization of the arc plasma, and (3) the cooling stage, where the arc chamber recovers to its ultimate dielectric strength. The first stage is dominated by electrode sheath effects at the new cathode; the second stage is governed by recombination and diffusion processes in the plasma; and the third stage is controlled by cooling.

It has been experimentally demonstrated [18,25] that a short gap in air can recover to over 300 V in 40 µs (see Fig. 11.9). The accepted physical picture of this rapid recovery is that at the anode, an electron space-charge sheath exists before current zero. Immediately after arc extinction at current zero, reverse voltage appears across the contacts, and the former anode now becomes the new cathode. Because of their high mobility, the electrons are quickly driven from this space-charge region, leaving a layer of essentially nonconducting gas shielding the new cathode. Although this sheath region is initially very thin, it is a very important component of the recovery process in low-voltage MCCBs. The immediate recovery of the contact gap to ~300 V allows time for the longer-term effects of recombination and cooling to take place.

The second stage is the decay of the arc column by deionization and recombination. The rapid cooling effect of the arc chute plays an important role during this stage. For steel deion plates, where the arc has been driven into the plates and burns as separate short arcs, each short arc will recover immediately to ~300 V. The close proximity of the plates will then rapidly cool the plasma to a level where recombination takes place. At high currents where the arc does not split into separate arcs between the steel deion plates or at all currents when insulating plates are used, the arc is stretched and brought into intimate contact with the plate surfaces. The plates conduct energy away from the arc column by storing it and by evaporation of the deion plates themselves. At current zero, the plates again cool the plasma to a level where rapid recombination and deionization can take place. The cooling efficiency of the arc chute can actually be gauged during the arcing stage by observing the arc voltage. For example, a free-burning arc in the air at 20 kA has a field of about 25 V/cm [26]. Figure 14.22, on the other hand, shows the voltage of a 20-kA arc inside a MCCB's arc chamber. The total arc length is about 4 cm and the arc voltage is about 400 V, giving a field of 100 V/cm. This much higher field implies high-energy losses from the arc. These additional energy-loss mechanisms will still be present at current zero.

Once the recombination process is completed, the gas in the arc chamber must then be cooled to regain its complete dielectric strength. This third stage can take place many milliseconds after current zero. This process is much slower, as can be seem from Fig. 11.9. Arc reignition during this period is by a spark breakdown mechanism. This stage can be complicated by the presence of conducting deposits on the arc chamber walls, especially after a number of high current interruptions. Sometimes what appears to be a spark breakdown between the open contacts may actually have been started by a spark from one contact to a closer conducting deposit on the chamber wall. The current then passes through the conducting surface of the chamber wall to the other contact.

Molded-Case Low-Voltage Circuit Breakers

14.5 TRENDS IN LOW-VOLTAGE MOLDED-CASE CIRCUIT BREAKERS

14.5.1 Current-Limiting Circuit Breakers

For low-voltage dc applications the utilization of the arc voltage to artificially create a current zero has been well established [27]. To some extent MCCBs have also used this effect to limit the let-through current. Until recently, however, most of the current limiting in low-voltage ac circuits has been done by current-limiting fuses [3]. The 1970s saw the development of current-limiting CL-MCCBs to the point where they rivaled the protective ability of fuses [9,28,29]. In simple terms, the arc in a CL-MCCB serves the additional function of suddenly injecting a resistive element into the circuit to limit the fault current. This type of breaker will continue to be developed because of its ability to protect systems with large potential short-circuit currents to limit the energy let-through. The essential elements of CL-MCCB are:

Rapid contact motion after fault initiation
Rapid arc voltage development
High final arc voltage level
Fast gap dielectric recovery

The effectiveness of the CL-MCCB is given by both the peak let-through current I_p and the $\int I^2 \, dt$ value. The latter value represents the energy that the breaker will allow to pass to the downstream circuit.

Techniques for rapid contact motion

The traditional contact operation discussed in Sec. 14.3.3 is much too slow for current limiting, because it takes from 4 to 6 ms for the trip to operate and the mechanical linkage to open the contacts. For current limitation (see Fig. 14.23), the contact parting must take place ~1 ms after fault initiation. Two techniques have been widely applied to achieve this: (1) the solenoid kicker and (2) contact arm repulsion systems. Both use the electromagnetic force of the fault current itself. Figure 14.24 illustrates the operation of these two systems. In both cases the opening force is proportional to I^2. The mechanical linkage to the contact of the solenoid kicker system is direct, thus minimizing the total mass needed to be accelerated and minimizing the operating time. The force can be controlled by the number of turns on the solenoid. The repulsion system is most direct. Here the driving force depends on the separation of the contact arms. Thus, as the contacts open, the force is reduced. The initial opening force can be varied by changing the arm length.

Techniques for rapid arc voltage development

After the contacts open, the next stage is development of a high arc voltage. There are two common approaches. The first uses the mag-

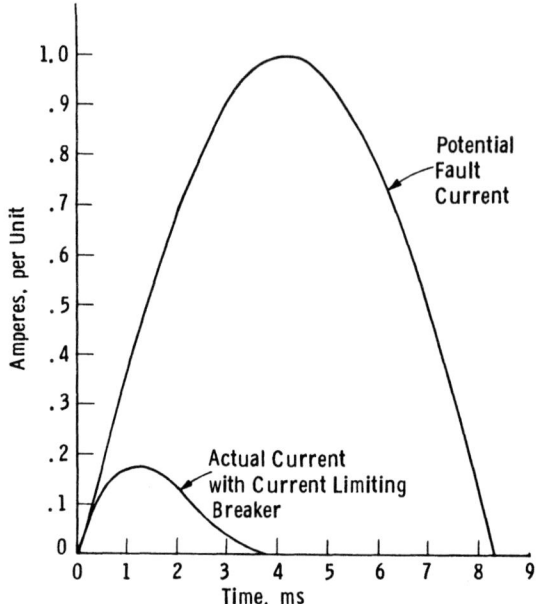

Figure 14.23 Current limitation by arc voltage. Note the reduced let-through current level and the shorter arcing time.

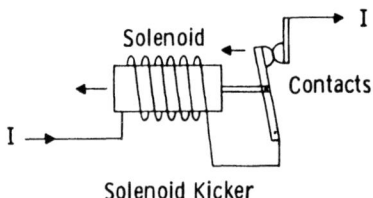

Solenoid Kicker

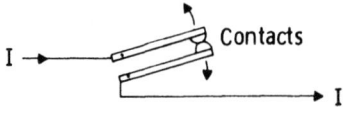

Contact Arm Repulsion

Figure 14.24 Rapid contact motion techniques.

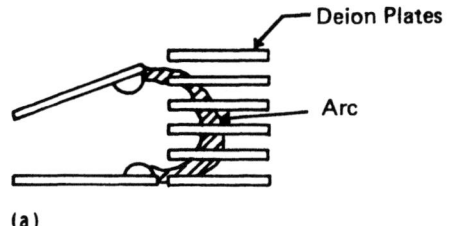

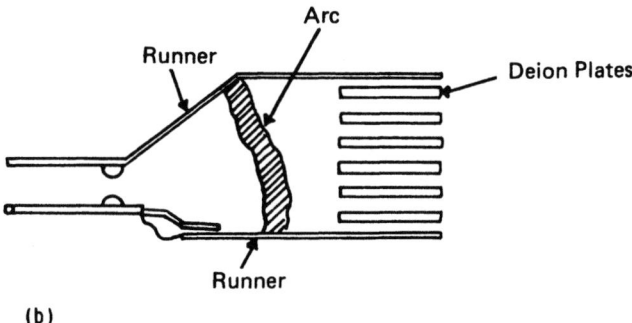

Figure 14.25 Rapid arc-voltage development schemes. (a) Magnetic blowout type; (b) arc runner type.

netic blowout effect common in the standard MCCB. In this case, however, the arc is stretched very rapidly, as shown in Fig. 14.25(a), increasing its length and thus the surface area for thermal conduction. Furthermore, the arc is split up and cooled by the deion plates. This approach requires the contact separation to be large enough so that the arc stays in the stretched state and does not break down and reignite between the contacts. This requires a very high contact opening speed. The arc roots stay close to the contacts, which can result in contact erosion.

The second way of obtaining a rapid increase in arc voltage is to use arc runners [see Fig. 14.25(b)]. Here the arc is driven off the separating contacts by the self-magnetic force of the current loop. The migration of the arc along the diverging arc runners will stretch the arc and cause the arc voltage to increase. As the arc is driven into the arc chute, it is further split up and cooled. For this approach to be effective (1) the arc must be forced to migrate off the contacts (this will not happen until the contact gap is greater than ~1 mm), (2) the arc must move quickly from the contact region in order to reduce contact erosion and to allow the contact gap time to regain its dielectric strength, and (3) the arc must then travel at a high speed along the runners (~100 m/s) and into the deion plates in order to develop a high arc voltage [30]. At present, the arc runner type of CL-MCCB has been effective for potential fault currents of $\leq$10 kA. The major problem is that at higher currents, the arc tends to stay on the contacts longer, causing excessive erosion. Furthermore, contact materials that resist arc erosion, such as Ag-W, are poor in promoting rapid arc motion even when strong magnetic driving forces are used.

An example of CL-MCCB

Figure 14.26 shows a Westinghouse Current Limit-R MCCB which uses the contact arc repulsion principle together with a "slot motor" to enhance the repulsion. The slot motor is made of U-shaped laminations of soft magnetic steel. The contact arms are placed inside the slot with the line conductor under the bottom part of the U. Figure 14.27 shows the influence on the magnetic repulsion as the field is modified by the slot motor. Here the contact arms are about halfway open.

For interruption of load and overload currents, this breaker operates as a conventional MCCB with the trip or handle operation moving the upper contact arm. To coordinate the repulsion and instantaneous trip, the trip unit is designed so that the trip current level is below the current level at which the repulsion force overcomes the contact force. This is defined as the "threshold current." Furthermore, to prevent reclosure a latch mechanism is designed to respond fast enough to catch the upper contact arm as it is forced to open under a current-limiting fault condition. This is a critical feature for successful cur-

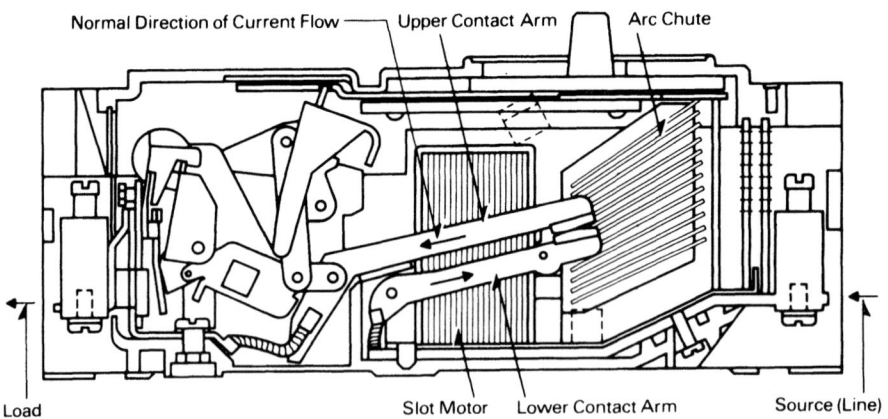

Figure 14.26 Westinghouse Current Limit-R MCCB. (Courtesy of Westinghouse Electric Corporation, Beaver, Pa.)

Molded-Case Low-Voltage Circuit Breakers

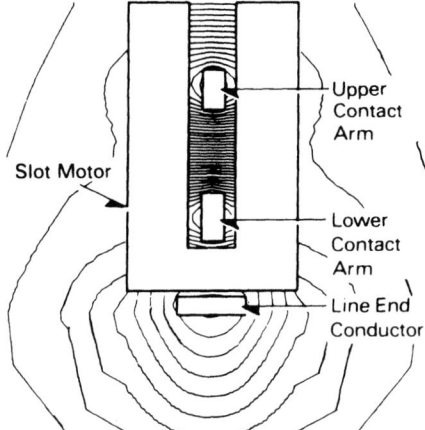

Magnetic Field With Slot Motor

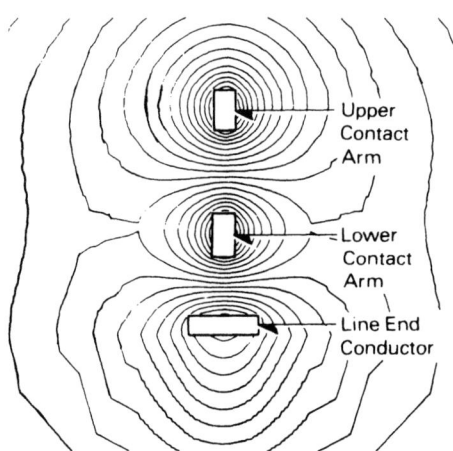

Magnetic Field Without Slot Motor

Figure 14.27 Magnetic field lines with and without the presence of the slot motor.

rent-limiting performance. As the current is driven to zero, the repulsion force disappears and the contact arms will tend to reclose if the mechanism does not hold them in the open position.

This slot motor concept has been used in current-limiting circuit breakers with continuous currents of 100, 250, and 400 A. These

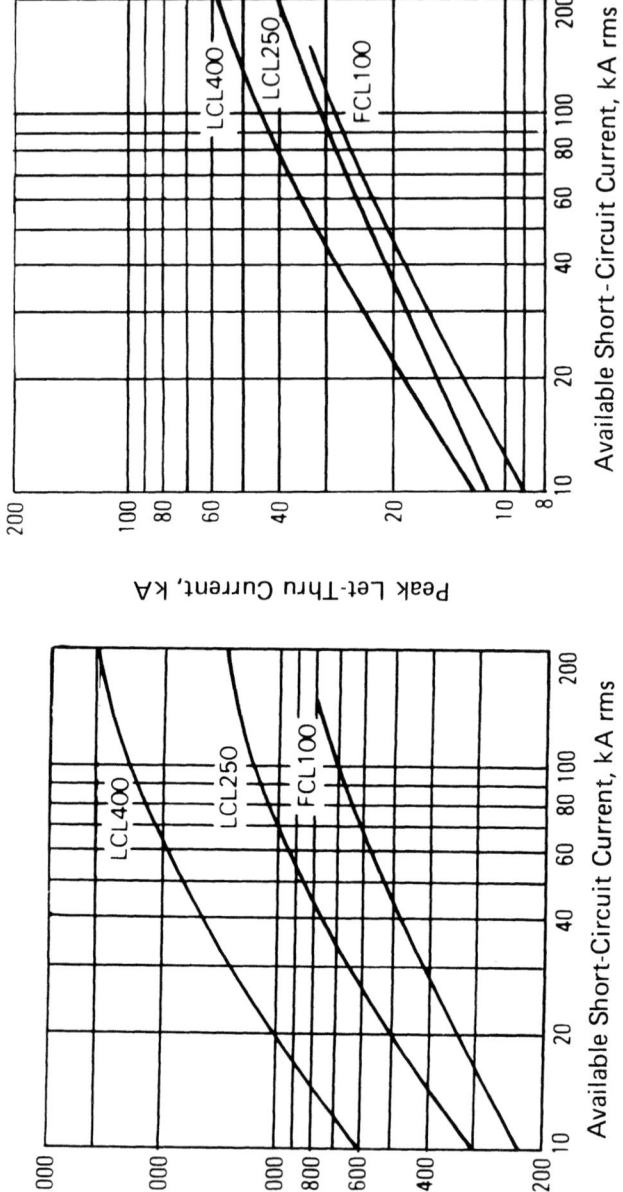

Figure 14.28 Peak let-through current and I^2t at 480 V ac of the Current Limit-R breakers with 100-, 250-, and 400-A frames.

Current Limit-R breakers have successfully interrupted potential fault currents up to 200 kA in a 480-V ac circuit, 100 kA in a 600-V ac circuit, and 25 kA in a 1000-V ac circuit.

The performance at 480 V ac is shown in Fig. 14.28. Note that the current-limiting action is in effect over the entire range of fault current. Because the repulsion force goes up roughly as I^2, the higher the fault level, the faster the contact opening. This, in turn, gives a shorter arcing time before the current zero is forced.

The economic advantage of this type of breaker comes in two ways. First, the amount of arc energy absorbed in the MCCB for a given interruption capacity is lower, resulting in a longer-lasting and more compact breaker. Second, the let-through level is lower such that downstream breakers can have lower interrupting capacity to take advantage of the lower $\int I^2 \, dt$.

14.5.2 High-Interruption-Capacity Breakers

The application need of present-day electrical systems has led to increasingly higher continuous current, higher interruption capacity (IC), and greater functional selectivity and coordination. The combined development and application of solid-state technology in trip and breaker control, of stored energy multioperation mechanisms, and of advanced arc chamber designs is expected to meet this need. In many respects, the performance of high-IC MCCBs rival those of low-voltage power circuit breakers [12]. As shown in Fig. 14.13, the Westinghouse SPB type of MCCB has been established as the industry standard for main and feeder breakers at the switchboard. No doubt, this is just the first step; future high-IC MCCBs will include:

New stored energy mechanisms
Higher interruption capacity
Even greater flexibility and adjustability in performance characteristics
More "intelligence" as solid-state devices improve and use integrated circuits and microprocessor-based controls

REFERENCES

1. IEEE Standard 242, IEEE recommended practice for protection and coordination of industrial and commercial power systems, 1975, Chap. 6.
2. H. W. Wolff, Integration of miniature circuit breakers into distribution networks, IEEE Rev. *117*:154, August 1970.
3. IEEE JH 2112-1, Protective fundamentals for low voltage electrical distribution systems in commercial buildings, D. S. Brereton, Ed., 1974.

4. UL-489, Molded-case circuit breakers and circuit-breaker enclosures, 1972.
5. IEEE Standard 141, IEEE recommended practice for electrical power distribution for industrial plants, 1976, Chap. 8.
6. ANSI/IEEE C37.13-1981, American National Standard for low-voltage ac power circuit breakers used in enclosures.
7. P. G. Slade, Effect of the electric arc and the ambient air on the contacts, J. Appl. Phys. 47:3438, 1976.
8. ANSI/IEEE C37.13-1981, American National Standard for low-voltage ac power circuit breakers used in enclosures, Tables 19.1, 21.1.
9. J. A. Wafer, Low voltage current limiting breakers using the slot motor principle, IEEE Ind. Appl. Meet., Cincinnati, October 1980.
10. J. A. Wafer, The impact of solid state technology on molded case circuit breakers, IEEE Petrol. Chem. Ind. Conf., San Diego, September 1979.
11. Dzierzbicki and Walczuk, Wy-Raczniki Organiczajace Pradu Przemiennego, Wydawnictwa Naukowo-Techniczne, Warsaw, 1976, Chap. 3.
12. R. O. D. Whitt, Power circuit breakers versus molded case— economics and system coordination, IEEE Rubber Plastic Conf., Akron, Ohio, April 1976.
13. Westinghouse system Pow-R breaker, Descriptive bulletin, Westinghouse LVBD, Beaver, Pa.
14. T. H. Lee, *Physics and Engineering of High Power Switching Devices*, MIT Press, Cambridge, Mass., 1975, Chap. 11.
15. A. Lee, Y. K. Chien, P. P. Koren, and P. G. Slade, High current arc movement in a narrow insulating channel, IEEE Trans. IEEE Trans. Comp. Hyb. and Manf. Tech., CHMT-5:51, 1982.
16. K. Poeffel, Influence of the copper electrode surface on initial arc movement, IEEE Trans. Plasma Sci. *PS-8*:443, 1980.
17. N. Behrens, Lichtbogenwandering zwischen divergierenden Laufschienen in Abhangigkeit der Gestallung des Anfweitungsbereiches, Elektrotech. Z., Ausq. A *101*:155, 1980.
18. J. Slepian, Theory of the deion circuit breaker, AIEE Trans. 48: 523, 1929.
19. B. Hague, *Principles of Electromagnetism Applied to Electrical Machines*, Dover, New York, 1962.
20. B. Bowman, Measurements of plasma velocity distribution in free-burning DC arcs up to 2160 A, J. Phys. D, 5:1422, 1972.
21. J. J. Lowke, Calculated properties of vertical arcs stabilized by natural convection, J. Appl. Phys. 50:147, 1979.
22. A. E. Guile, Arc-electrode phenomena, Proc. IEE (Lond.) *118*: 1131, September, 1971; T. H. Lee, *Physics and Engineering of High Power Switching Devices*, MIT Press, Cambridge, Mass., 1975, Chap. 5.

23. C. W. Kimblin, Anode phenomena in vacuum and atmospheric pressure arcs, IEEE Trans. Plasma Sci. *PS-2*:310, 1974.
24. M. Lindmayer, Effect of work function and ionization potential on the reignition voltage of arcs between Ag metal oxide contacts, *Proc. 26th Holm Conf. Electri. Contacts*, 1980, p. 185.
25. T. E. Browne, Jr., and F. C. Todd, Extinction of short a.c. arcs between brass electrodes, Phys. Rev. *36*:728, 1930.
26. D. C. Strachan, High current free burning graphite arcs, Int. Conf. Phen. Ionized Gases, 1975, p. 150.
27. R. Rüdenberg, *Elektrische Schaltvorgänge*, Julius Springer, Berlin, 1933, p. 245.
28. H. Arnhold, Heavy duty automatic circuit breaker, U.S. Pat. 4,001,743, 1977.
29. R. D. Ohlson, Performance and application of a new molded case non-fused current limiting circuit breaker, IEEE Petrol. Chem. Conf., Milwaukee, Wis., September, 1975, Paper PCIC 75-25.
30. W. Rieder, Interaction between magnet-blast arcs and contacts, *Proc. 28th Holm Conf. Electr. Contacts*, Chicago, 1982, p. 3.

15
Electric Contact Phenomena

PAUL G. SLADE/Westinghouse Research and Development Center, Pittsburgh, Pennsylvania

15.1 Introduction   565
15.2 Contact Fundamentals   566
    15.2.1 Making contact, contact area, and contact resistance   566
    15.2.2 Calculation of contact resistance   568
    15.2.3 Effect of oxidation   571
    15.2.4 Fretting   571
    15.2.5 Contact resistance and contact temperature   574
    15.2.6 Contact parting   577
    15.2.7 Oxidation of the contact surface following arcing   578
15.3 Considerations in Contact Design   582
    15.3.1 Holding force   582
    15.3.2 Magnetic blow-apart force   582
    15.3.3 Parallel contacts   589
15.4 Classes of Contact Materials   592
    15.4.1 Contact materials for air, oil, and SF_6 ambients   592
    15.4.2 Contact materials for vacuum interrupters   602

References   603

15.1 INTRODUCTION

The electric contact is an essential circuit breaker component. It is, however, a component often taken for granted until it does not work as it is supposed to; only then does the circuit breaker designer become aware of its interesting and varied world. When closed it permits the flow of current in an electric circuit. The current must flow through the contacts without overheating them; overheating could cause them to weld or could cause their surfaces to deteriorate through oxidation. In some applications the contacts may stay closed for many years; in others, frequent opening and closing may be normal. When contacts open under load they are subject to the very high temperature of the electric arc. Here they must resist excessive erosion and maintain their mechanical integrity. The contact surfaces must also maintain a reasonably high electrical conductivity so that, when they again close, current can flow without excessive heating. When the contacts are opening, the designer has to consider whether or not it is best for the arc roots to be forced off the contact faces. When current zero is reached in an ac circuit, the arc extinguishes and the gap between the electric contacts has to recover its dielectric properties very rapidly. In most low-voltage applications and in all vacuum interrupters the properties of the electric contacts are vital to successful dielectric recovery. For example, the contacts must not exhibit severe field distortion or stay hot enough to liberate large numbers of electrons. In some applications the contacts are required to close and latch on very high short-circuit currents. Here they not only have to withstand the mechanical forces involved, but also the circuit breaker mechanism should be able to break any contact weld that forms. Thus you can see that the electric contact participates in every phase of a circuit breaker's operation. No matter what the voltage range, current level, or ambient, the design of the electric contact is important.

In a power system the use of the electric contact is not confined to circuit breakers. It finds wide application in other current-connecting devices: joining bus bars, stab connectors to bus, no-load disconnect switches, isophase bus disconnects, and so on. These applications usually do not have the problem of arc erosion, but they must maintain their contact integrity for many years without failure.

In this chapter some contact theory and its application are discussed. Then, how this theory can be applied to contact design is described. Finally, there is a discussion of the classes of contact material and of the materials that should be used for given power applications.

15.2 CONTACT FUNDAMENTALS

15.2.1 Making Contact, Contact Area, and Contact Resistance

If the two cylinders shown in Fig. 15.1 are butted together and the resistance between points a and b is measured, the resistance will be found to be

    Total resistance a ↔ b

        = bulk resistance of the cylinders + contact resistance

or

$$R_T = R_B + R_C \tag{15.1}$$

Electric Contact Phenomena

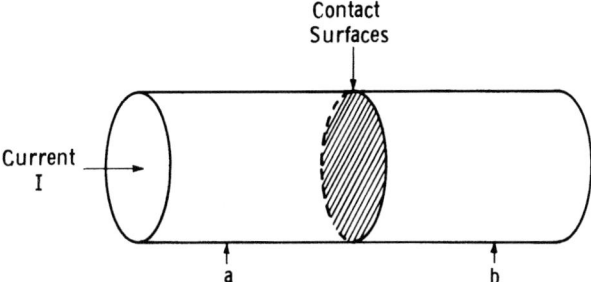

Figure 15.1 Electric contact.

The reason for this is that no matter how carefully the cylinders' faces are prepared, they will never be perfectly flat. Indeed, they will only make contact at a number of discrete points on these flat surfaces [1,2]. If a high-magnification picture were taken of a contact surface, it would be seen to contain a number of micropeaks and valleys. When two surfaces are brought together, they initially touch at two micropeaks. Even under light loads the pressures at these peaks can be very high [1] and the peaks deform elastically and plastically until the force on the contacts is fully supported. This is shown conceptually in Fig. 15.2. In this example any surface film has also been disrupted. This process can be represented by [1]

Contact closing force $\simeq$ material hardness X real microarea of contact

or

$$F = \xi \times H \times A_r \qquad (15.2)$$

where ξ is a constant less than 1 and typically in the range 0.1 to 0.3 [1]. It can be seen that equation (15.2) implies that the actual area of contact depends only on the force and the material properties of the contact material and not on the total area of the contact face. Figure 15.3 illustrates this for a 10-fold change in contact face area.

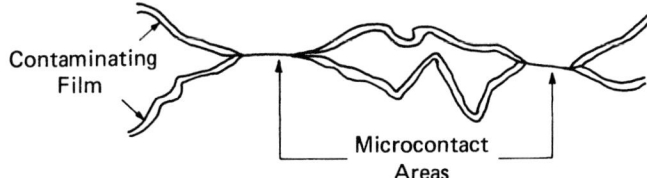

Figure 15.2 Plastic deformation and the real area of contact.

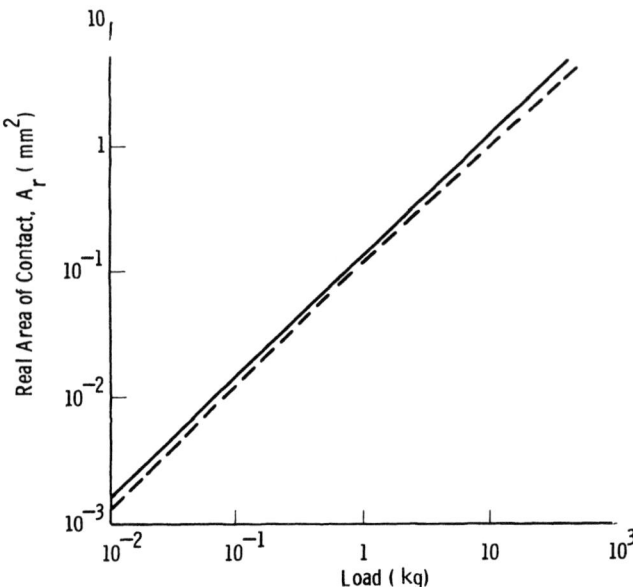

Figure 15.3 Example of the real area of contact vs. load for two contact sizes: (a) solid line, 10-cm^2 nominal area; (b) dashed line, 1-cm^2 nominal area. (From Ref. 3.)

15.2.2 Calculation of Contact Resistance

The Real Area of Contact a Small Disk of Radius a Consider as a first approximation Fig. 15.4, where a disk-shaped area A_r of radius a is achieved after the contacts have been forced together. The flow of current from one contact to the other would then be constrained to flow through this area. The constriction resistance is given by [4,5]

$$R_k = \frac{\rho}{2a} \qquad (15.3)$$

where ρ is the resistivity of the contact. Substituting from equation (15.2) and noting that $A_r = \pi a^2$,

$$R_k = \frac{\rho}{2} \sqrt{\frac{\pi \xi H}{F}} \qquad (15.4)$$

If R_F is the resistance of any film, then the total contact resistance R_C is given by

$$R_C = R_k + R_F \qquad (15.5)$$

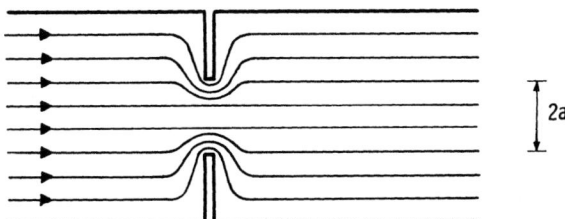

Figure 15.4 Schematic of the lines of current flow constricted to flow through the real area of contact.

Figure 15.5 presents data for Cu contacts showing how R_C varies with F and how closely equation (15.4) describes the data.

The current density at the surface of the contact disk at a distance r from the center is [15]

$$J = \frac{2V_0}{\pi \rho \sqrt{a^2 - r^2}} \quad \text{for } r < a \tag{15.6}$$

where V_0 is the potential drop from a to b in Fig. 15.1. At the center of the contact area, the current density has the value

$$J_{r=0} = \frac{2V_0}{\pi \rho a}$$

and it tends toward infinity as $r \to a$.

Equal Contact Spots, Each with Radius α Within a Circle of Radius β Holm showed [7] that for n such microspots,

$$R_k = \rho \left(\frac{1}{2n\alpha} + \frac{1}{2\beta} \right) \tag{15.7}$$

Random Array of n Spots, the Radius of the ith Spot Being α_i Greenwood [8] investigated this realistic contact model and showed that

$$R_k = \rho \left(\frac{1}{2\Sigma \alpha_i} + \frac{3\pi}{32n^2} \Sigma\Sigma \frac{1}{S_{ij}} \right) \tag{15.8}$$

where S_{ij} is the distance between spots. If we note that $\Sigma \alpha_i = n\bar{\alpha}$, where n is the total number of spots and $\bar{\alpha}$ is the average spot diameter, and we let

$$\bar{\beta} = \frac{3}{16n^2} \Sigma\Sigma \frac{1}{S_{ij}}$$

then

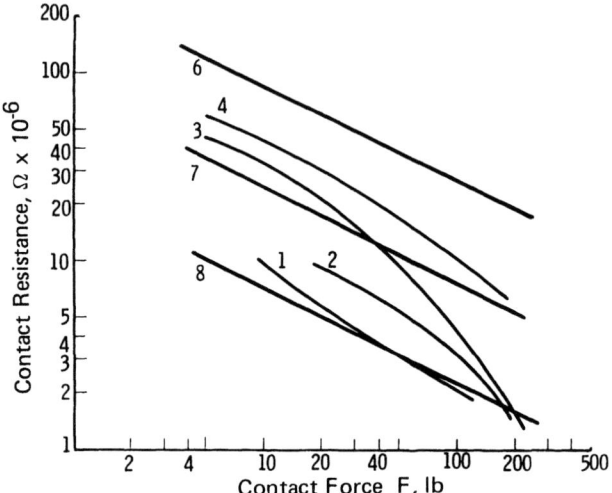

Figure 15.5 Contact resistance vs. force for Cu contacts. 1. Line contact 1 in. long, 3 in. radius, dry. 2. Flat contact 3 in. X 3 in. surface scraped flat on ground cast iron, dry. 3. Flat contact 1 in. X 1/4 in., glass finish, dry. 4. Flat contact 1/2 in. X 1/2 in., glass finish, dry. 6. Eq. (15.4) $\xi H = 175,000$. 7. Eq. $\xi H = 14,200$. 8. Eq. $\xi H = 1180$. (From Ref. 6.)

$$R_k = \rho \left(\frac{1}{2n\overline{\alpha}} + \frac{1}{2\overline{\beta}} \right)$$

that is, equation (15.8) reduces to the form of (15.7). Now if we define the radius of the contacting area to be

$$\overline{a} = \frac{n\overline{\alpha}\overline{\beta}}{n\overline{\alpha} + \overline{\beta}}$$

then

$$R_k = \frac{\rho}{2\overline{a}} \tag{15.9}$$

This is the form given by equation (15.3). When contact properties are calculated, we are usually using the value $\overline{a}$. This actually proves to be a very useful approximation. Figure 15.6 gives examples of different contact spot clusters and their effect on the size of $\overline{a}$ and $\overline{\beta}$.

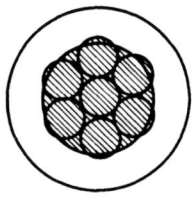 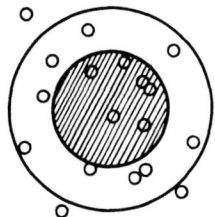

Figure 15.6 Contact clusters showing the equivalent single contact (shaded, radius $\bar{a}$ and Holm radius $\bar{\beta}$). (From Ref. 8.)

15.2.3 Effect of Oxidation

An insulating oxide film covering one of the micro contact spots in Fig. 15.6 will have only a small effect on the total contact resistance. It is important to ask, however, how many spots would have to stop conducting before R_C would begin to increase. This is an especially important question for electric contacts that remain together for long periods of time and are subjected to very slow oxidation of the contact spots (e.g., power connectors, bolted bus).

Figure 15.7 illustrates the process. The contacts are in the closed position. Even though the mating surfaces present a barrier to the influx of any ambient gas, the gas can penetrate the contact spot after a long period of time. Any gas that can react with the contact material to form an insulating layer is a potential problem gas (e.g., O_2, SO_2, H_2S). The contact spots will gradually reduce in size as the gas reacts with the contact surface. Williamson [9] studied this process and has shown that the effects of the loss of contact area show up only very slowly. Figure 15.8 gives calculations of the expected increase in R_C resulting from gaseous diffusion into the contact spots. In this example the diffusion of the oxide is assumed to be continuous, but as can be seen, the R_C hardly changes at all until just before failure, when a run-away situation occurs. Similar results were obtained by Lemelson [10] when he modeled the failure of high-current copper contacts under oil (see Fig. 15.9). This phenomenon occurs only if the oxides formed are insulating. It is possible that conducting or semiconducting films could form. In this case the R_C might even decrease if there were an increase in A_r.

15.2.4 Fretting

There is a form of accelerated atmospheric oxidation which occurs at the interfaces of contacting metals that undergo very slight cyclic motions relative to each other. This phenomenon was described many years ago [11], but its significance to electric contacts has been rec-

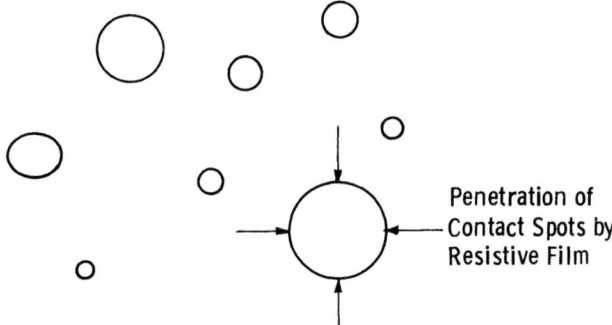

Figure 15.7 Diffusion of ambient gases to the contact spot.

ognized only recently [12]. The effect is illustrated in Fig. 15.10. When one contact moves relative to the other, part of the old contact area is exposed to the ambient. An insulating film can form upon it. If the contact moves back to its former position, it can break this film and push it to one side. This process will continue until sufficient insulator builds up and the actual regions of contact are modified. The rapid change in R_C shown in Figs. 15.8 and 15.9 will again be observed.

Much of the research on this phenomenon has been performed on low-current connections. Braunovic, for example, experimented with Al with many plating materials and showed that fretting occurs in two stages [13]. In the first stage there is a stable metallic contact between the two surfaces and R_C is unaffected. During this time the

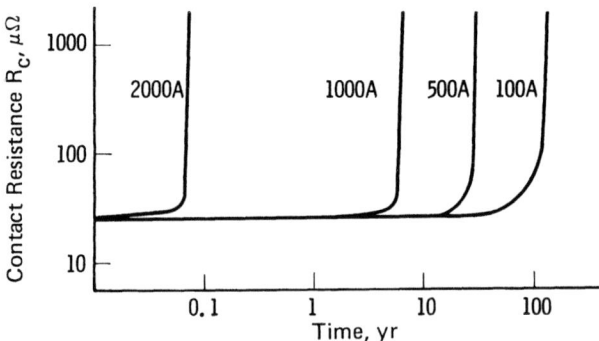

Figure 15.8 Increase in contact resistance as a function of time for a 20-$\mu\Omega$ connector normally used for 100 to 500 A, in an ambient temperature of 20°C. (From Ref. 9.)

Electric Contact Phenomena

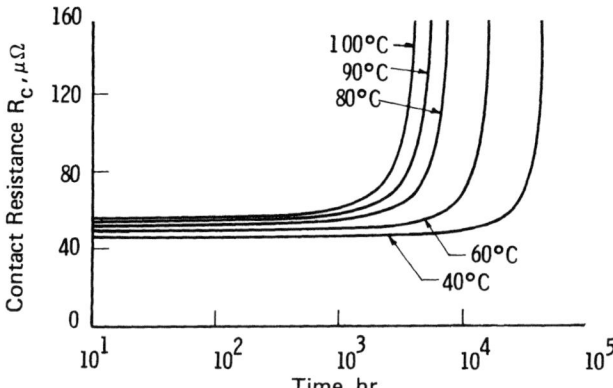

Figure 15.9 Increase in contact resistance as a function of time for a Cu contact under oil for a load current of 400 A and for different temperatures of the ambient oil. (From Ref. 10.)

oxide film is penetrated and mechanical seizure of the contact spots occurs. As the contacts move, these microscopic bridges shear. As a result, the first wear products formed will be metal particles. Although a good proportion of these will be oxidized, good metallic contact will be maintained. In the second stage, after long exposure to the fatigue-oxidation process, the metal layers become softened and the layers progressively separate. The oxide layer grows and metallic contact is lost. The increase of R_C during this stage is a strong function of current, contact force, and the plating material.

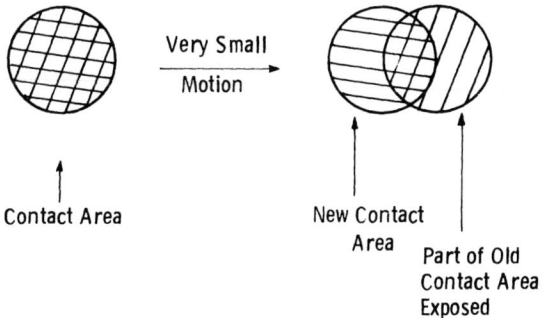

Figure 15.10 Process of fretting in electric constants.

At higher currents Johnson and Moberly [14] have shown that similar phenomena occur between Al bus and high-current stab contacs. They found that by plating Al bus with Ag, low R_C could be maintained even at high currents and under severe fretting conditions.

15.2.5 Contact Resistance and Contact Temperature

As R_C increases so does the power input to the contact junction, which results in a temperature increase at the contact junction. An increase in temperature can result in an increased rate of oxidation and an increase in R_C: the runaway condition illustrated in Figs. 15.8 and 15.9. In extreme cases the contact interfaces can even melt and severe contact failure can occur. It would therefore be useful to estimate the temperature of the contact junction without having to actually measure it.

The calculation of contact temperature

It is a very straightforward procedure to measure the voltage drop across a closed contact. If we take any conductor and assume that the lines of equipotential are the same as the line of equitemperature and that there is no heat loss by radiation, it can be shown that [15-17]

$$V^2 = 8 \int_{\theta_0}^{\theta_p} \rho \lambda d\theta \qquad (15.10)$$

where V is the voltage drop across the conductor, θ_0 the ambient temperature, θ_p the maximum temperature the conductor reaches, ρ the resistivity, and λ the thermal conductivity. Using the Wiedemann-Franz law, we have

$$\rho\lambda = L\theta$$

where L is the Lorenz constant given by [18]

$$L = \frac{\pi^2}{3}\left(\frac{k}{e}\right)^2$$

where k is Boltzmann's constant and e is the electronic charge, $L = 2.7 \times 10^{-13}$ esu/degree^2 or $L = 2.45 \times 10^{-8}$ W Ω/degree^2.

$$V = [4L(\theta_p^2 - \theta_0^2)]^{1/2} \qquad (15.11)$$

If $\theta_p \gg \theta_0$ this can be reduced to

$$\theta_p = \frac{V}{2\sqrt{L}} \qquad (15.12)$$

Electric Contact Phenomena

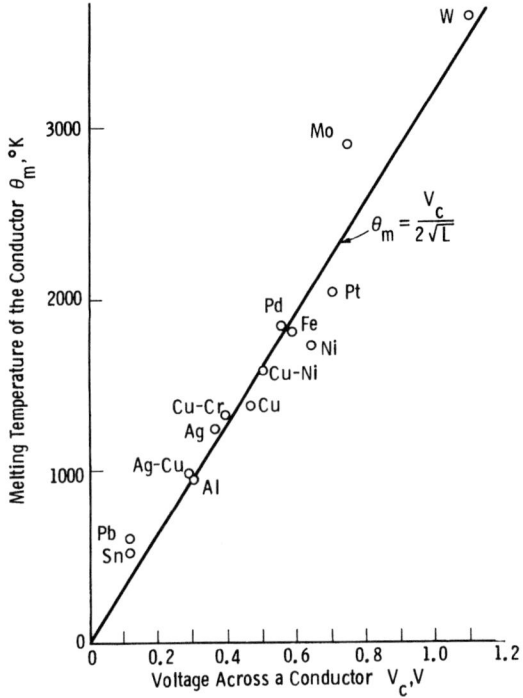

Figure 15.11 Comparison of the measured melting voltages with those calculated from Eq. (15.12).

Figure 15.11 presents a comparison of the measured voltage [16] across a conductor at its melting point with equation (15.12). It can be seen that it is possible to obtain a very good estimate of the contact temperature by just measuring the voltage across the contacts. It is interesting to note that very small voltage drops represent very high temperatures (e.g., if the voltage across a Cu contact is only 0.43 V, the contact faces will be molten).

Another important consideration is the softening temperature of metals. This temperature, usually from about 25 to 45% of the melting temperature, is where the metal softens and can deform plastically very easily. Equation (15.11) then allows us to define V_s as the softening voltage.

$$V_s = [4L(S\theta_m)^2 - \theta_0^2]^{1/2} \tag{15.13}$$

where S is a constant between 0.25 and 0.45 and θ_m is the melting temperature. Table 15.1 shows experimentally determined softening voltages.

Table 15.1 Comparison of Softening Temperatures and Softening Voltages

Metal	Melting temperature [19], K	Softening temperature [19], K	Softening temp. / Melting temp.	Measured softening voltage V_s	
				Holm [19]	Sato [20]
W	3653	1273	0.35	0.6	0.39
Mo	2883	1172	0.41	0.25	0.34
Ni	1726	793	0.46	0.22	0.21
Cu	1356	463	0.34	0.12	0.13
Au	1336	373	0.28	0.08	0.09
Ag	1233	453	0.37	0.09	0.1
Al	933	423	0.45	0.1	0.1

The melting current and softening current

It is now possible to investigate how much current can flow through a contact junction before the metal in the region of contact softens or melts. A softening current I_s and a melting current I_m can be defined as [20]

$$I_s = \frac{V_s}{R_C} \qquad I_m = \frac{V_m}{R_m} \qquad (15.14)$$

Assuming that there is no insulating film at the contact spots, R_C can be given by equation (15.4), in turn giving

$$I_s = \frac{2V_s}{\rho}\sqrt{\frac{F}{\pi \xi H}} \qquad (15.15)$$

and

$$I_m = \frac{2V_m}{\rho}\sqrt{\frac{F}{\pi \xi H}} \qquad (15.16)$$

that is, for a given material I_s and I_m should be proportional to $\sqrt{F}$. Figure 15.12 gives the experimental values of I_s as a function of F and it will be noted that $I_s \simeq K\sqrt{F}$. We would expect some sticking or minor welding of the contacts at these currents.

Because the contact spots soften, we would also expect A_r to increase, giving rise to lower contact resistance. This may in part explain why in Fig. 15.13 the I_m is not proportional to $\sqrt{F}$, but to an exponent less than 1/2.

Electric Contact Phenomena

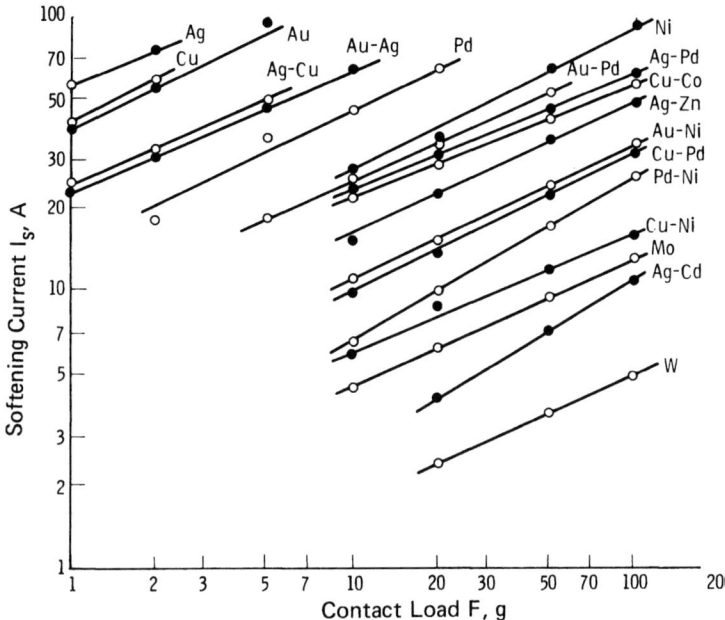

Figure 15.12 Softening current as a function of contact load. (From Ref. 20.)

15.2.6 Contact Parting

As the contacts begin to part the contact area begins to decrease. Eventually, the voltage across the contact reaches the melting voltage V_m (i.e., the R_C has increased until $V_m = R_C I$), and the metal in the contact region melts. At this time, a molten metal bridge forms between the contacts and is now stretched as the contacts continue to part. Figure 15.14 shows a molten metal bridge between steel electrodes. This bridge was formed at a very high current and is shown because it is a good example of a large stable molten metal bridge. There are a number of detailed studies of this phenomenon [21-23]. The bridge in Fig. 15.15 was frozen and one electrode was detached from it. Figure 15.15 shows that the bridge has a hollow center. In fact, we might have expected such a structure if the distribution of current through the contact spot [equation (15.6)] is considered. As the bridge stretches the voltage across it increases, and (as can be seen from Table 15.2) the voltage does not have to increase much before the boiling temperature of the metal is reached (these values assume that the Wiedmann-Franz law remains valid). In practice the molten bridge tends to become unstable before boiling occurs because of surface tension ef-

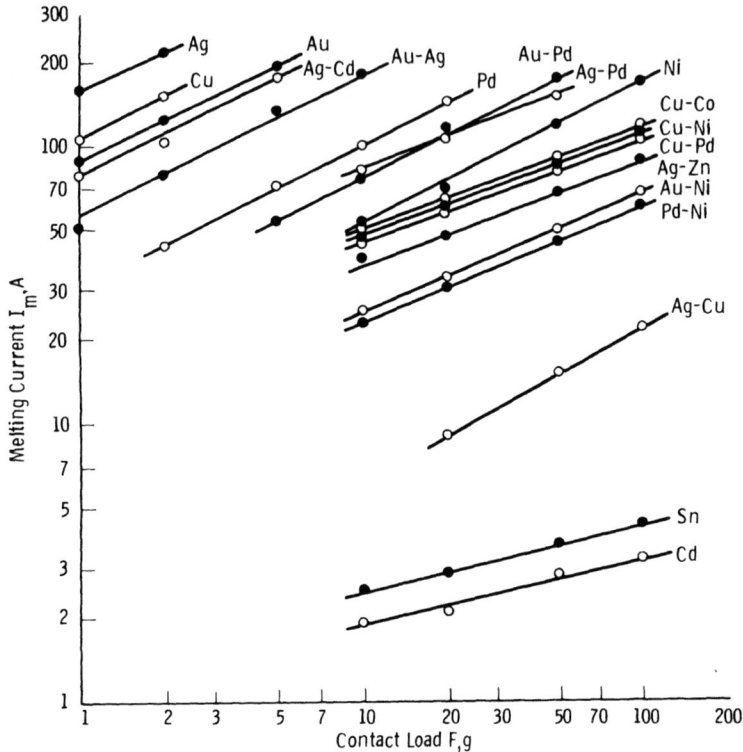

Figure 15.13 Melting current I_m as a function of contact load F. (From Ref. 20.)

fects. Also, under dynamic conditions [23], it has been shown that higher voltages are obtained than those given in Fig. 15.11 and Table 15.2; there is a lag between the time when the voltage reaches a given value and the moment when the temperature increases to its full value. Eventually, the molten bridge ruptures and an electric arc is formed. It is only after the arc has formed that the design of the arc chamber determines the interruption performance of the circuit breaker.

15.2.7 Oxidation of the Contact Surface Following Arcing

The only interrupting device whose contact surfaces are not oxidized after arcing is the vacuum interrupter. These devices thus make use of Cu-rich contacts very successfully. Other circuit breakers that operate in an oxygen-free environment (such as SF_6 or oil) also use Cu-rich contacts (usually Cu-W), but they usually operate with high con-

Contact Surface

Contact Surface   ⊢——⊣ 1 mm

Figure 15.14 Molten metal bridge between steel contacts.

tact forces and sometimes with a good wiping action. For circuit breakers that operate in air, Ag is the only contact material of reasonable cost that can resist oxidation and can be capable of modification for arcing currents of less than 1 A to greater than 100,000 A. At low currents pure Ag finds wide application. It is easy to disrupt any oxide or sulfide films that occur, and the contact maintains a low R_C over its entire lifetime. For medium currents (10 to 3000 A) the Ag is strengthened with CdO (see Sec. 15.4). This compound contact still maintains a low R_C and successfully resists arc erosion during switching. For the highest current arcing contacts, the compound contact of Ag and W is used. This contact has excellent resistance to arc erosion and high-current welding, but is very susceptible to oxidation. Figure 15.16 shows an example of the change in the surface character-

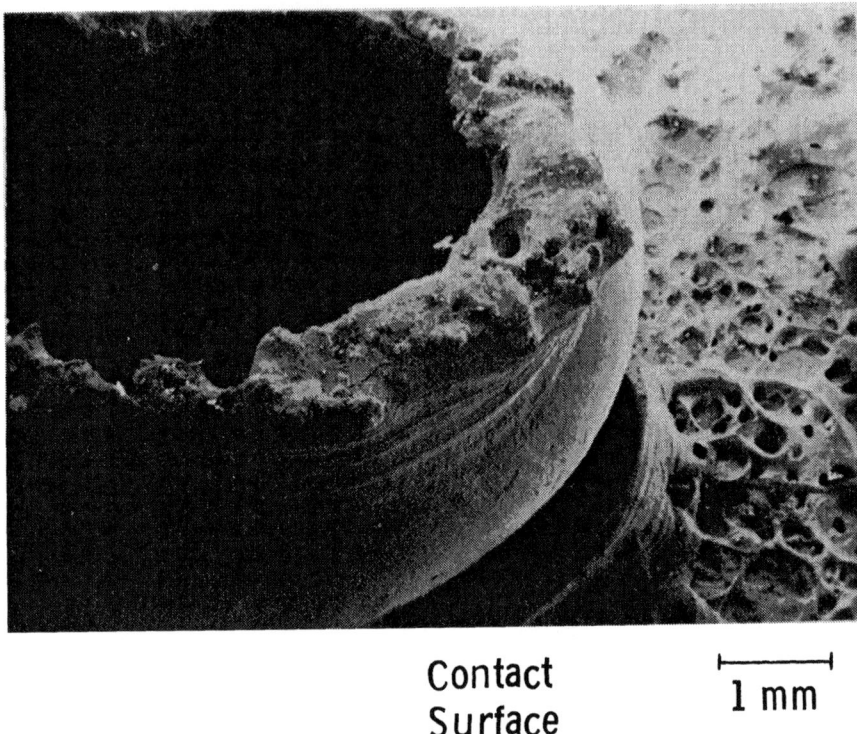

Contact Surface   ⊢——⊣ 1 mm

Figure 15.15 Frozen molten metal bridge (Fig. 15.14) separated from one contact showing the hollow center and the thin wall.

Table 15.2 Boiling Temperatures and Boiling Voltages for Some Common Contact Materials

Metal	Boiling temperature [19], K	Boiling voltage, V [Eq. (15.12)]
Ag	2460	0.77
Cu	2850	0.89
W	5800	1.82
Au	3090	0.97
Ni	3140	0.97
Sn	2780	0.87

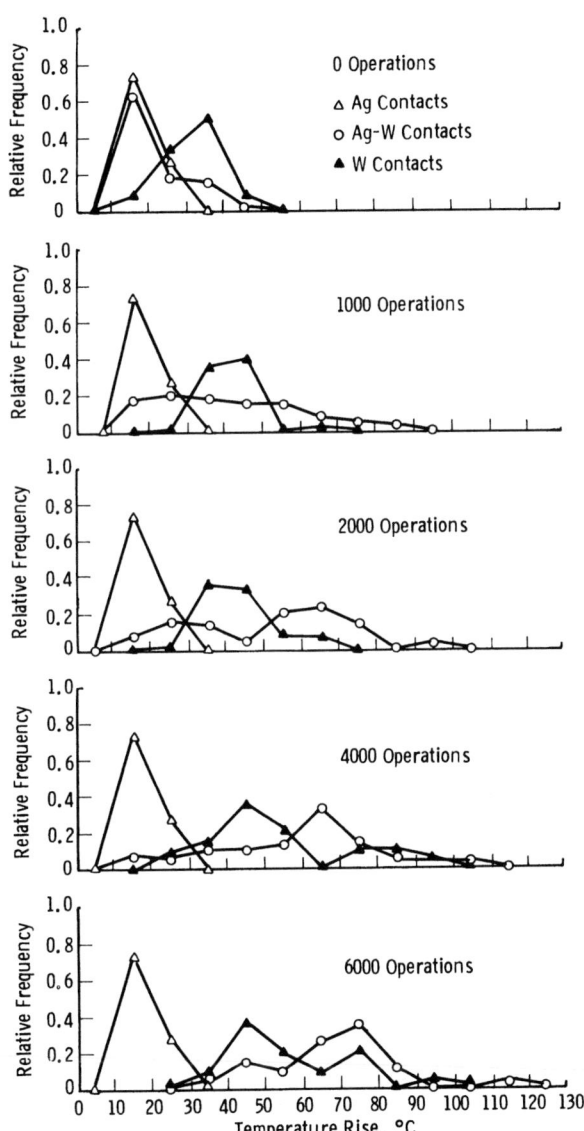

Figure 15.16 Change in the distribution of temperature rise values at the fixed contact (i.e., contact resistance) for Ag, Ag-W (35 wt % Ag), and W contacts as a function of the number of switching operations in a 20-A 110-V circuit. (From Ref. 24.)

istics of Ag-W contacts after switching 20 A in a low-voltage circuit. The temperature at the back of the stationary contact is measured (this temperature is directly related to R_C) while passing the steady-state current. It can be seen that R_C varies considerably. If Ag-W is used for a switching application, the designer should ensure that there is sufficient wiping action and a high enough contact force to disrupt any oxide films and thus to maintain a low R_C.

15.3 CONSIDERATIONS IN CONTACT DESIGN

15.3.1 Holding Force

Section 15.2 shows the importance of the force holding the contacts together. Sufficient force must be applied to maintain a low R_C in order to prevent softening or melting of the contact faces. This is especially important when the contacts are subjected to very high momentary short-circuit currents. Unfortunately, the current constriction at the point of contact also supplies a magnetic force that tends to blow the contacts apart.

15.3.2 Magnetic Blow-Apart Force

Figure 15.17 shows schematically the current flowing into the constricted contact area. It can be seen that a component of the current flows in opposite directions close to the contact surfaces. This current flow gives rise to a repulsion force which tries to force the contacts apart. The repulsion force P is given by [25,26]

$$P = \frac{\mu I^2}{4\pi} \ln \frac{r}{\bar{a}} \qquad (15.17)$$

where I is the instantaneous current, r the measured contact radius, and $\bar{a}$ the average constriction radius (see Sec. 15.2.2). Using equation (15.2) and noting that $\ln(r/\bar{a}) = (1/2) \ln (r/\bar{a})^2$,

$$P = \frac{\mu I^2}{8\pi} \ln \frac{\xi H A}{F} \qquad (15.18)$$

where A is the total area of the contact face.

Figures 15.18 and 15.19 give examples of how P depends on H and r. The difference between measured and calculated values of P are given in Table 15.3.

The flaw in using (15.17) is that as the blow-apart force on the contacts increases, the contact force F decreases. Barkan [27] has considered this problem. The force on the contacts

$$F = F_S + F_A - P \qquad (15.19)$$

Electric Contact Phenomena

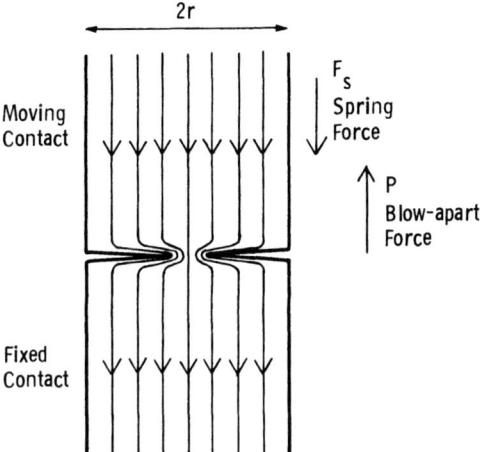

Figure 15.17 Current path in mating contacts for a single contact area A_r. Opposing currents in the contact surfaces tend to force the contacts apart.

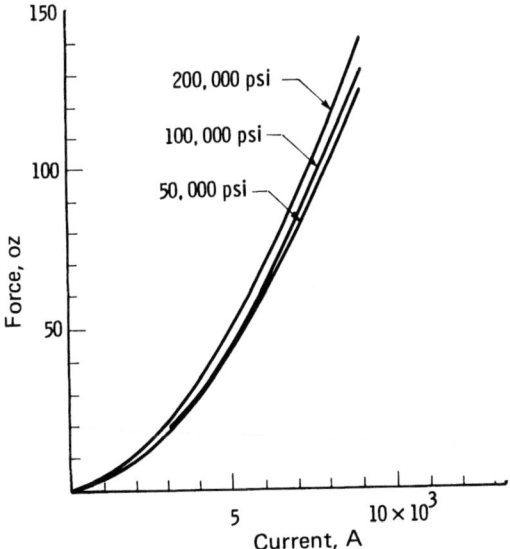

Figure 15.18 Calculated effect of contact material hardness H on separating forces for 1-in.-diameter contacts. (1) Higher forces are generated by harder contact material. (2) Soft contact material usually work hardens after use, especially at high closing forces, and therefore generates higher forces after use. (From Ref. 26.)

Table 15.3 Difference Between Measured and Calculated Contact Forces Required to Keep Contacts from Separating at a Given Current

Current, A	Force, oz		Percent difference
	Calculated	Measured	
5500	65	80	19
6000	70	72	2
5100	52	66	21
5700	65	66	2
5600	61	64	5
5800	44	45	3
4200	36	45	20
5000	49	61	20

Source: Ref. 26.

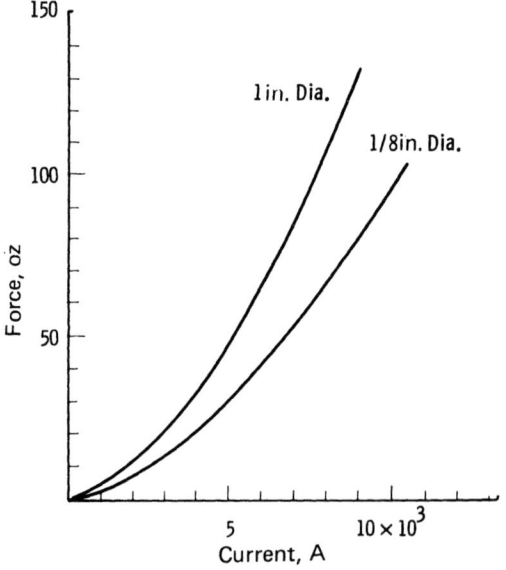

Figure 15.19 Calculated effect of measured contact diameter (2r) on generated contact forces: Larger diameter contacts generate higher forces. (From Ref. 26.)

Electric Contact Phenomena

where F_S represents the spring force holding the contacts together and F_A represents a force resulting from the current flow in the whole contact structure. This force is very often arranged to be a blow-on force that can counteract P and can usually be represented by

$$F_A = kI^2 \tag{15.20}$$

where k is a constant. Substituting for F in equation (15.18) from (15.19) yields

$$P = \frac{\mu I^2}{8\pi} \ln \frac{\xi HA}{F_S + F_A - P} \tag{15.21}$$

P appears in both sides of this equation and the equation cannot be solved explicitly for P. Barken put equation (15.21) into the form

$$e^{-(8\pi/\mu)(P/I^2)} = \frac{F_S + F_A - P}{\xi HA} \tag{15.22}$$

and showed that if the right-hand side = α and the left-hand side = β, a solution of (15.22) can be found when

$$\alpha = \beta$$

and

$$\frac{d\alpha}{dP} = \frac{d\beta}{dP}$$

The second expression gives

$$P = \frac{\mu I^2}{8\pi} \ln \frac{8\pi \xi HA}{\mu I^2} \tag{15.23}$$

Substituting for P in equation (15.23) into (15.22) gives the minimum spring force required:

$$F_S = \frac{\mu I^2}{8\pi} \left(1 + \ln \frac{8\pi \xi HA}{\mu I^2} - \frac{8\pi}{\mu I^2} F_A \right) \tag{15.24}$$

Butt contacts

Let us take the contacts shown in Fig. 15.17. Here $F_A = 0$, so equation (15.24) reduces to

$$F_S = \frac{\mu I^2}{8\pi} \left(1 + \ln \frac{8\pi \xi HA}{\mu I^2}\right) \tag{15.25}$$

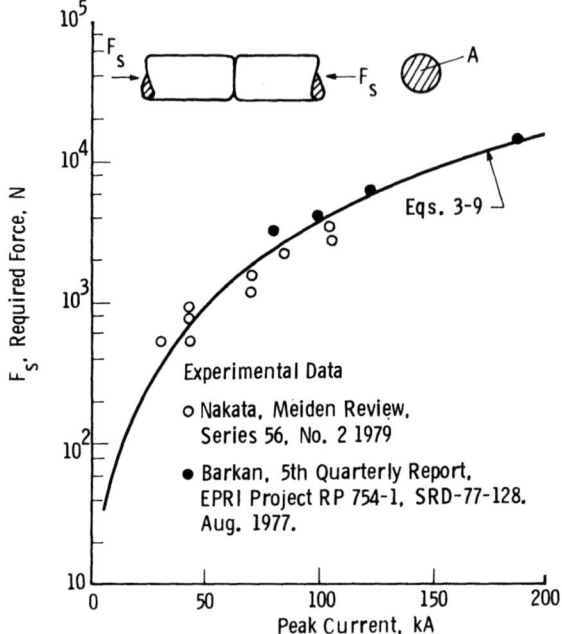

Figure 15.20 Spring force F_s which must be applied to butt contacts to maintain contact stability: comparison of equation (15.25) with experimental data. (From Ref. 27.)

Figure 15.20 shows data for the contact force F_S required to maintain stability for vacuum interrupter contacts. It can be seen that equation (15.25) correlates well with the experimental data.

Tulip contact structure

Barkan [27] made a number of simplifications in his analysis, but illustrated very well the excellence of this contact structure. Figure 15.21 illustrates the design. Assume that the total cross-sectional current-carrying areas of the fingers and the center rod are about equal (i.e., $D_0 \simeq \sqrt{2}\, D_i$), and that the cross-sectional area of each finger is

$$A \simeq \frac{\pi D_i^2}{4N}$$

Neglecting end effects, the average instantaneous circumferential magnetic field encircling the finger cluster is then

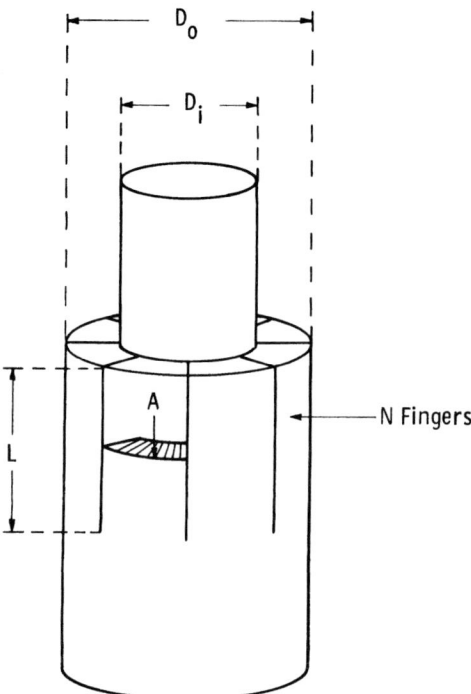

Figure 15.21 Tulip contact structure.

$$B_e = \frac{\mu I}{2\sqrt{2}\pi D_i}$$

If the magnetic force is uniformly distributed along the length of the finger, the instantaneous force excited at the contact point of each finger is approximately

$$F_A \simeq \frac{B_e IL}{2N} = \frac{\mu I^2 L}{4\sqrt{2}\pi N D_i}$$

Substituting in equation (15.24) and noting that each finger carries a current of I/N,

$$F_S = \frac{\mu}{8\pi}\left(\frac{I}{N}\right)^2 \left(1 + \ln\frac{2\pi^2 \xi D_i^2 HN}{\mu I^2} - \sqrt{2}\,\frac{NL}{D_i}\right) \qquad (15.26)$$

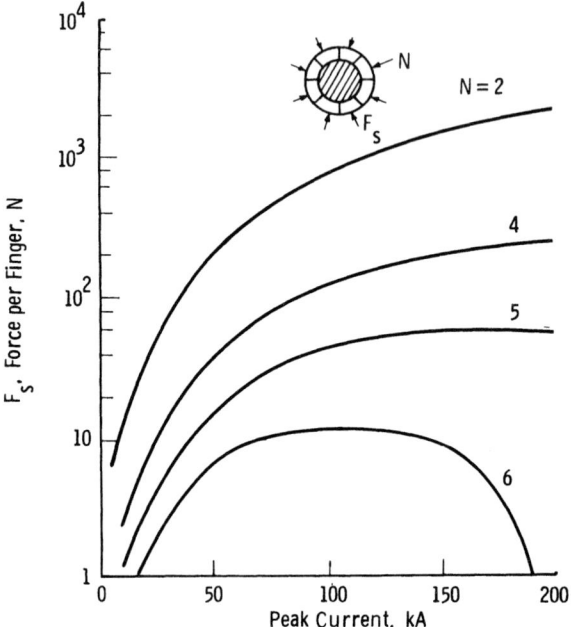

Figure 15.22 Solution to equation (15.26) showing the spring force F_S which must be applied to maintain contact stability for a tulip type contact where $L/D_i = 1.1$.

Barkan showed (see Fig. 15.22) the effect of the number of contacts on F_S. He claimed that with a Cu-based contact set with $L/D_i = 1.1$, $D_i = 0.038$ m, and $N = 8$ was stable for peak currents of up to 0.25×10^6 A with contact forces of only 80 N.

Knife blade

The force on two thin parallel conductors carrying currents I_1 and I_2 of length l and distance d apart has been shown by Hauge [28] to be

$$F'_A = \frac{2I_1 I_2}{d} l \sqrt{1 + \frac{d^2}{l^2}} - 2I_1 I_2 \qquad (15.27)$$

For contacts of finite cross section, Hague showed that the true force F_A is given by

$$F_A = K F'_A \qquad (15.28)$$

Electric Contact Phenomena

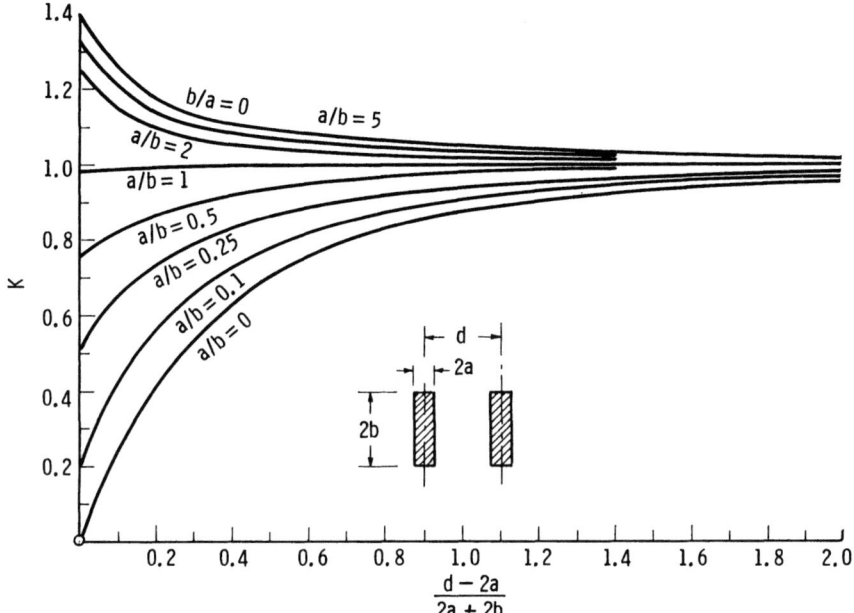

Figure 15.23 Correction factor K in equation (15.27) plotted as a function of the ratio of the air space between the contact structures to the sum of the sides of the cross section. (From Ref. 28.)

where the correction factor K is illustrated in Fig. 25.23 for parallel conductors.

Barkan took the contact structure shown in Fig. 15.24 and suggested the use of the very approximate relationship

$$F_A = \frac{\mu I^2 L}{8NW[(3t/2W) + 1]} \qquad (15.29)$$

Frost [29], using Fig. 15.23, showed that equation (15.29) is approximately true over only a limited range of W/t (see Fig. 15.25).

15.3.3 Parallel Contacts

In Sec. 15.2.2 it was shown that $R_k \propto F^{-1/2}$ [equation (15.4)], and in Sec. 15.2.5 there are limiting currents for a given F at which softening and melting occur. In order to overcome these problems at high currents, circuit breaker designers frequently use parallel contact structures similar to the example in Fig. 15.24. This type of structure can present a problem to the designer, because the current will not distri-

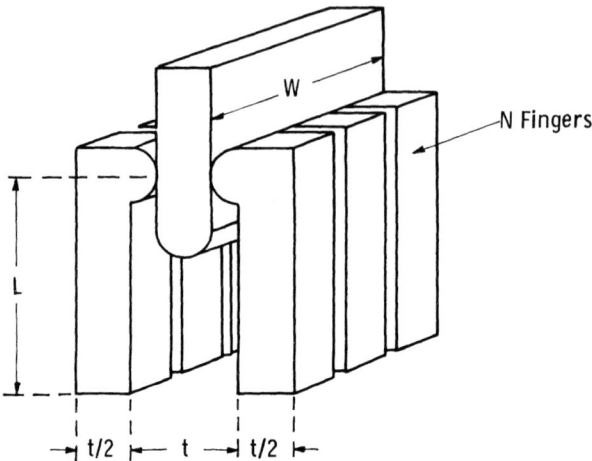

Figure 15.24 Knife-blade contact structure.

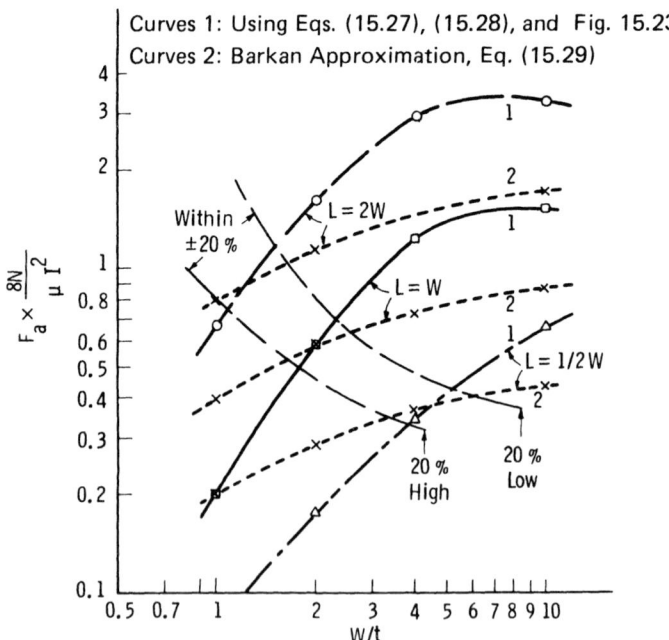

Figure 15.25 Magnetic attraction force on the knife-edge contacts in Fig. 15.24. A comparison of the values of F_a using equations (15.27) and (15.28) with (15.29).

Electric Contact Phenomena

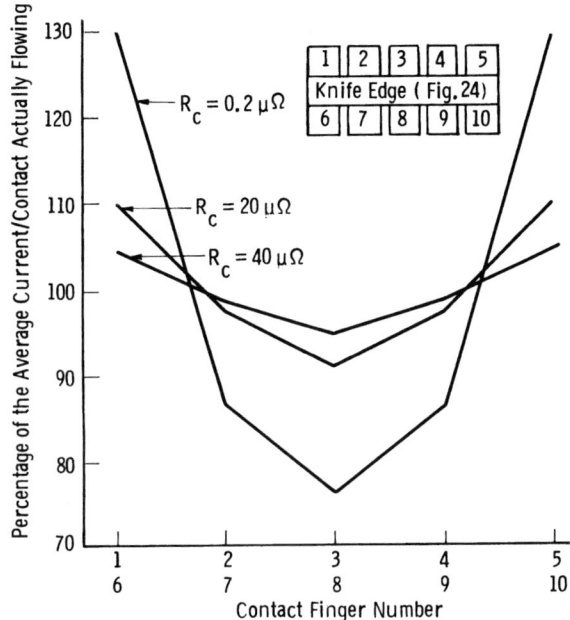

Figure 15.26 Current distribution in 10 contacts of a knife-edge structure. (From Ref. 30.)

bute itself equally among the contact arms. This results from the difference in the mutual inductance between fingers, which varies according to contact position. Yoshioka [30] made a careful analysis of a large number of linear contact structures. He considered the ith finger contact, and obtained

$$V = \sum_j L_{ij} \frac{dI_j}{dt} + (R_i + R_{Ci})I_i \qquad (15.30)$$

where

L_{ij} = mutual inductance between the ith and jth fingers
R_i = bulk resistance of the ith contact body
R_{Ci} = contact resistance of the ith contact
I_i = current in the ith contact
V = voltage drop across the ith contact

He subtracted the ith from the (i + 1)th equation and solved the resulting differential equations.

Figure 15.26 shows the current distribution of parallel contacts with 10 fingers and one blade. The very strong influence of R_C on the dis-

Table 15.4 Current Distribution in a Number of Finger Contact Arrangements

No.	Arrangement of Finger Contact	Dimension (mm) d	Dimension (mm) e	Dimension (mm) g	Parameters	Length of Finger Contact (cm)	Contact Resistance (μΩ)	Contact Material	Current Distribution Max. (%)	Current Distribution Min. (%)
1		10	-	3	Contact Resistance	10 10 10	0.2 20 40	Copper Copper Copper	131 114 106	74 92 96
					Material of Finger Contact	10 10 10	20 20 20	Copper Aluminum Brass	114 108 102	92 95 97.5
					Length of Finger Contact	10 20	20 20	Copper	114 119	92 86
2		-	-	3	Number and Arrangement of Finger Contact	10	20	Copper	106	96
3		-	-	3		10	20	Copper	125	87
4		10	30	3		10	20	Copper	132	88
5		10	-	3		10	20	Copper	146	77
6		0	-	0		10	20	Copper	133	71
7		20	-	10		10	20	Copper	127	76

a = 8 mm, b = 15 mm.

Source: Ref. 30.

tribution of current between the fingers should be noted. Table 15.4 gives Yoshika's data for many linear contact combinations.

In three-phase circuits, the distribution of current in contacts is also affected by the phase-to-phase interaction. Methods for solving three-phase current distribution problems are given by Silvester [31].

15.4 CLASSES OF CONTACT MATERIALS

15.4.1 Contact Materials for Air, Oil, and SF_6 Ambients

All electric contacts for power applications contain a high percentage of a good conductor such as Cu or Ag. Bus bars are usually made of Cu

Table 15.5 Contact Materials for Power Applications

Material	Contact production and connection	Contact resistance, oxidation and fretting	Performance under arcing (Switching, erosion recovery and welding)	General uses
Aluminum	The contacts are extruded from bar stock of Al or Al alloy. Connection is by bolting, crimping or welding. Great care must be taken when connecting to this material.	Once the oxide film has been broken on this soft material, low contact resistances can be achieved. The oxide is thin but tough and Al is often plated to prevent its formation. For unplated Al, fretting is a serious problem. Where fretting is likely to occur, it is wise to use Ag plated Al.	This material cannot be used where arcs may occur. Its low melting and boiling temperatures permit severe erosion. In oxidizing atmospheres the metal can even burn. The recovery is relatively poor and strong welds can form in nonoxidizing atmospheres.	In power engineering Al finds wide use in electric wiring and in electric bus. It is often Ag plated where contact has to be made to it. Bolted joints are made to pure Al if the interfaces are protected by special greases and the bolts are spring loaded.
Copper	The contacts can be made from the extruded bar stock, they can be stamped from plate, they can be forged and machined. The metal is quite ductile, but can be hardened by additions of Zn and Sn to form various bronzes. Contacts can be bolted, welded, or brazed. It is also possible to make a spring-loaded contact holder or terminal for other contact materials.	The material is harder than Al and when no oxide is present has a low contact resistance. This material also has high conductivity. The oxide can be removed easily and does not form very quickly at normal temperatures. In bolted contacts or stab contacts R_c remains low for long periods of time. In power circuits fretting is not a serious problem. It should be prevented,	Arc erosion is severe on this material. When it is subjected to an arc, the arc should not be allowed to remain stationary. The arc can be moved, for example, by using a $\bar{J} \times \bar{B}$ force. In oxidizing atmospheres the formation of oxides can cause problems for switching performance. It is usually used under oil and requires a circuit voltage > 50V and circuit	Finds wide usage in bus and in wiring. It forms an excellent stab or bolted joint. It is frequently plated with Ag for protection against oxide formation. It is a good opening contact in nonoxidizing atmospheres and under oil. It finds uses in switches for voltages > 50V and currents < 20A.

Table 15.5 (Continued)

Material	Contact production and connection	Contact resistance, oxidation and fretting	Performance under arcing (Switching, erosion recovery and welding)	General uses
Silver	Ag is a low melting point, ductile material. Contacts can be formed in many ways, e.g., extruded, stamped, forged, etc. The most common contact material is fine Ag (99.9%), but it is possible to obtain sterling Ag contacts (7.5% Cu) and coin Ag (10% Cu) contacts. In power circuits, and in oxidizing atmospheres, this material is very commonly used. It can be bolted, welded or brazed and forms excellent spring-loaded connections. Ag is often plated over Cu and Al to act as a terminal or a contact holder. It is also possible to produce a Ag inlay for contact applications.	This material has the highest conductivity and is relatively soft. Therefore it gives a very low contact resistance. With the addition of Cu it becomes harder and more force has to be used to maintain low values of R_c. Ag oxide is very volatile and decomposes under arcing. Oxide and sulfide films are easily ruptured by reasonable contact forces and/or wiping action. The hard Ag gives more of a problem with the formation of Cu oxides. Fretting is not a problem in power circuits. Ag should be used to plate other surfaces (e.g., Cu or Al) where fretting may occur.	however, in low current, low voltage circuits. It has excellent switching performance for currents up to 30A and for circuit voltages > 50V. The low boiling point, however, does result in severe erosion by the arc even at moderate currents (tens of amperes). As with Cu, the erosion can be minimized by using $\bar{J} \times \bar{B}$ forces to move the arc off Ag contact faces. Fortunately, arcs move very well on Ag. The recovery after current zero is good with Ag. Welding becomes a problem with make-break contacts for currents > 20A in 110V circuits.	currents < 20A. It forms strong welds and has good recovery characteristics. Commonly used as a plating material over Cu and Al for bus bar applications and for contact holder parts. It is frequently used for no-load switches for continuous currents of many thousands of amperes. It is also commonly used in make-break systems, in light duty, low contact force devices (e.g., light duty relays, switches and contactors). It has found use in miniature low voltage circuit breakers where rapid contact parting and rapid arc motion off the contacts prevents severe erosion and contact welding.

Silver-Graphite	Composition up to 5% by weight C. Higher graphite content contacts are used for sliding	Blocks are sintered from Ag and C powders and then extruded to size with the required contact profile. The contacts are then cut to size. The contacts can also be manufactured directly by pressing and sintering Ag and C powders. Before these contacts can be brazed it is necessary to have an Ag layer on the brazing surface. One method of achieving this is to "burn-off" the graphite from one face.	For the low percentages of C used in making this contact, it has very similar properties to Ag. It may be a little harder than Ag, but the contact resistance characteristics will be similar to those of Ag. The C has a small reducing effect which allows this contact to be used opposite materials more prone to oxidation, e.g. Ag-Ni. Fretting is not a problem. In fact the small amount of C will act as a lubricant.	The switching performance is excellent, but because of its low melting and boiling point, severe arc erosion results even at modest currents of a few tens of amperes. It also is made more susceptible to arc erosion because arc mobility on Ag-C is extremely poor. For small percentages of C recovery is very similar to that of Ag. The major advantage of this material is that it *does not weld* if at least 3% by weight C is used.	Increasingly used for low voltage circuit breaker applications, especially where arcing is not severe, e.g. on minimal arcing, load-carrying contacts, or where the arc is forced off the contacts rapidly. This contact is now frequently used opposite Ag-Ni contacts and has even been used opposite pure Cu contacts.
Silver-Cadmium Oxide		An extremely important contact material. It has found wide usage in low voltage applications and has been subjected to much R&D to improve its properties. The major manufacturing methods are: (a) Press and sinter Ag and CdO powders into contact shapes. Contact structures can be improved by using	This material is harder than Ag or Ag-C and its conductivity depends upon the percentage of CdO contained within the contact structure. The contact resistance is therefore higher than that of Ag. It usually maintains this contact resistance throughout its life. The thermal and electrical conductivities are usually more than adequate for most applications.	Excellent switching performance for currents up to 3 kA. The contact resistance does not increase as the contact structure is eroded. Significantly lower arc erosion than Ag and Ag-C contact materials in the range 50A to 3kA. Recent improvements in manufacturing techniques have made this material even more resistant to arc erosion. After	It is close to being the best contact material for less than 3 kA. It finds wide usage where many load switching operations are required, especially in low voltage circuits. It is used in switches, relays, low current circuit breakers and many forms of motor contactor (limited only by overload currents > 3 kA).

Table 15.5 (Continued)

Material	Contact production and connection	Contact resistance, oxidation and fretting	Performance under arcing (Switching, erosion recovery, and welding)	General uses
	fine powders. It is also possible to roll this material into strips and then cut it or stamp contact shapes. (b) Make contacts of Ag-Cd alloy and heat under high pressure oxygen. This results in the internal oxidation of the Cd to produce very fine CdO sites. (c) Internally oxidize Ag-Cd powders, press and sinter. The properties of this material can be improved by the addition of very small percentages of Li & Ge. The Ag-CdO material has limited workability, but it is possible to roll it into strips and to inlay it into Cu. A Ag rich layer is required before brazing is successful.	Oxidation is similar to Ag's, but any free Cd on the surface will form an oxide before Ag. Its fretting properties are similar to those of Ag and no problem is experienced in power applications.	arcing any free Cd in the contact surface forms CdO and helps prevent oxidation of the Ag. Heavy arcing, however, will cause a depletion of CdO from the region of the arc roots. Arcs move more easily on Ag-CdO than on Ag-C and Ag-W, but they still move faster on Ag, Cu and Ag-Ni. Arc mobility decreases as the CdO content increases. This material has the best recovery properties of all contact materials and hence is widely used for low voltage ($\leqslant 600V$) applications. It has excellent anti-welding properties for currents $\leqslant 3kA$. This property improves as the CdO contact increases, but is never as good as Ag-C.	

Silver-Metal Oxide

This class has been developed to replace Ag-CdO, examples:

Ag-SnO
Ag-SnO/InO
Ag-NiO
Ag-ZnO
Ag-SnO/BiO
Ag-BiO
Ag-SnO/WO$_2$

These contacts are manufactured with all the methods used for Ag-CdO i.e. (a) press-sinter-repress of powders and (b) internal oxidation of an alloy. The connecting and joining techniques again are similar to those used with Ag-CdO.

The contact resistance and oxidation properties are similar to those of Ag-CdO. Recent developments in Ag-SnO contacts have improved the potential problem posed by insulating layers of SnO forming on the contact surface. The SnO compositions may not perform as well as CdO contacts under fretting conditions.

The switching performance of the best of these materials is similar to that of Ag-CdO. Some of these materials resist arc erosion better than Ag-CdO does. None of the oxides sublime like CdO, but melt before they evaporate. This could pose a problem if insulating layers formed on the contact surface. The recovery properties are usually good. The welding properties are usually acceptable, but usually inferior to Ag-CdO and Ag-C.

These contacts are designed for use in all devices that presently use Ag-CdO contacts. It may also be possible to use some of these contact structures in low voltage circuit currents than is possible for Ag-CdO materials.

Silver-Nickel

The Ni content can range from 10% to 40% by weight.

Blocks of press-sinter-repress Ag and Ni powders are extruded and the contacts cut to shape. The contacts can be attached by brazing; a special Ag rich brazing surface is usually unnecessary.

This material is quite hard and its thermal and electrical conductivities decrease as the Ni content increases. Contact resistance values can be high for Ni content > 20%. Susceptible to oxidation; better than Ag-W but not as good as Ag-CdO. Its fretting performance is worse than that of Ag and Ag-CdO.

Its switching performance is inferior to Ag-CdO and Ag-C. Its resistance to arc erosion is as good as Ag-CdO at low currents. At higher currents it is usually superior. This material can be used where high-current short-circuit arcs are experienced. Arc mobility is good, but decreases as the Ni content increases. The recovery is

Ac and dc relays for currents < 10A, contactors and switches for currents < 25A and circuit breakers for currents up to 400A. It finds use for the main current carrying contacts in high current, low voltage circuit breakers when an arcing contact is provided to bear the brunt of the arc ero-

Table 15.5 (Continued)

Material	Contact production and connection	Contact resistance, oxidation and fretting	Performance under arcing (Switching, erosion recovery and welding)	General uses
		Its fretting performance is worse than that of Ag and Ag-CdO.	only slightly inferior to that of Ag-CdO. It generally does not weld for currents < 150A (therefore superior to Ag). It is not as weld resistance as Ag-CdO or Ag-C.	sion. Where welding is a potential problem, it can be mated against a Ag-C contact.
Silver-Refractory Material These materials can have a wide range of composition, but in Ag-W materials 35% and 50% Ag by weight	(a) For contacts where the refractory content is > 50% by volume, it is possible to infiltrate a sintered refractory matrix with Ag or Cu.	The resulting contacts are usually hard and long wearing. The contact resistance is usually high. These materials are very susceptible to oxidation (oxides, silver tungstates, silver molyb-	This class of material is not recommended for applications that require frequent load switching in air unless enough force is used to break through the oxide films. These materials resist	High voltage transmission circuit breakers use Cu-W contacts in SF_6 or under oil. Medium voltage distribution breakers also use CuW contacts in SF_6 and oil. For medium

are very common. Composition examples:

Ag-W, Cu-W
Ag-WC, Cu-WC
Ag-Mo, Cu-Mo
Ag-W-WC
W-Ni-Cu
Ag-W-C, Ag-WC-C

The Cu containing materials have similar properties to those containing Ag. They can only be used, however, in non-oxidizing atmospheres such as SF_6 or under oil.

(b) Press-sinter-repress of powders can be employed for all compositions.

dates). The tungstates and molybdates are troublesome, because they can lock up the surface Ag into an insulating compound. Contacts containing molybdenum should not be used where high humidities will be experienced. For contacts used in air, high forces plus wiping action should be employed to break through oxide layers. This class of material is not usually used where fretting would be a problem. This is fortunate, because these materials would be very susceptible to fretting.

arc erosion extremely well and are the preferred arcing contact materials for very high currents at all voltages. Arcs do not move well and therefore require strong magnetic fields, or high liquid or gas flows for arc motion. The recovery characteristics are worse than Ag-CdO. In most of the applications of these materials, however, the recovery of the contact regions is not as important as the recovery of the medium between the contacts. These materials have good resistance to welding and can be used for currents of > 100 kA if sufficient opening force is applied to break any welds that form.

voltage magnetic air breakers Ag-W or Ag-WC contacts are used. Ag-W, Ag-WC and occasionally Ag-Mo are used extensively in low voltage, molded case, circuit breakers.

599

Table 15.6 Contact Materials for Vacuum Interrupters

Pure Metals	Alloys	Refractory Materials plus a Good Conductor
Copper is the first metal that comes to mind, but it has the major disadvantage of forming very strong welds. This is a fault likely to occur with all pure metals, with the possible exception of the brittle refractory metals like W and the liquid metals like Hg and Ga. The disadvantage of W is that it is an efficient thermal emitter and consequently cannot interrupt very high currents. Nor can the liquid metal Hg be used in simple vacuum devices for very high powers. Gallium, on the other hand, does show some promise, but at this time no commercial vacuum interrupters using gallium are available.	There are a large number of alloys that have been proposed for use with the vacuum interrupter. The majority of these consist of Cu as the major constituent with other metals added to increase its weld resistance and/or mechanical strength and/or lower the current at which chopping occurs. It was the consideration of the chopping phenomenon that led to the development of the well-known Cu-Bi type of binary alloys (e.g., Cu-Sn, Cu-Pb, Cu-Sb, Cu-Zn). It has found, however, that a very high percentage of Bi is required ($>$ 5%) in order to obtain a desirably low chopping current. This amount of Bi makes the resulting contact material mechanically very weak, its voltage-withstand ability undesirably low and its erosion undesirably high. It has been found that if less than 1% Bi is added to Cu an effective contact material resulted which is able to withstand severe welding. The addition of a small percentage of Bi to Cu,	The most common type of material in this class is W-Cu or Mo-Cu and variations of it, e.g., W-Cu-Ti-Bi, W-Cu-Ti-Sn, W-Cu-Ti, W-In-Cu, W-Cu-Zr, W-Zr and of course W-Cu-Bi. In these materials the refractory W is usually a sintered matrix ($>$ 50% by volume) and infiltrated by the good conductor. The infiltrate is Cu or Cu alloy. The elements like Ti, Zr and In are used to aid the vacuum infiltration of the W and the elements Bi and Sn are there to aid the chopping and also the antiwelding capability. Borides and carbides have also been suggested for the refractory material. The major disadvantage of this whole class of materials is their inability to interrupt very high currents ($\lesssim$ 10 kA) in high voltage circuits. These materials do have three very good properties: 1) the erosion is low, 2) the chop current is usually low and 3) they are resistant to welding.

results in the Bi migrating to the grain boundaries of the Cu during solidification. This makes the resulting contact more brittle than pure Cu. The inherent defects of these materials, however, are their high erosion rate and mechanical weakness. This class of materials includes all alloys where the major constituent has a boiling point of less than 3500 K and where the minor constituent has a freezing temperature of less than that of the major constituent and little solubility in the solid state of the major constituent (e.g. Ag-Bi, Ag-Pb, Ag-Te, Cu-Te, Cu-Th, Al-Pb, Al-In, Al-Sn and Ni-Bi, Ni-Te). Some ternary systems have been proposed e.g. Cu-Al-Bi, Cu-Be-Bi, Cu-Ni-Bi, Cu-Ni-Te and Cu-Co-Bi.

The most commonly used material in this class is Cu-Bi. It is used in vacuum interrupters for circuit breaker applications. Silver is not used because it is more costly than Cu and offers no real advantage over Cu in a vacuum environment.

Cu-W contacts find use in low current, high voltage switches. Ag-WC [34] and Sb-Mo [35] have low chop characteristics, but have enough current interruption ability to find use in vacuum interrupters for high voltage contactors.

Non-Refractory Materials plus a Good Conductor

This class of materials retains some of the desirable properties of the refractory type, but has a very much improved current interrupting ability. The matrix or additive materials will have a melting point of greater than 1500°C and a boiling point of less than 3400°C and will be typically metals like Cr, Fe, Co and Ni.

The Cu-Cr contact has found wide usage in vacuum interrupters for medium voltage circuit breaker applications.

or Al, which may be plated with Ag. Other materials are added to Cu and Ag according to the particular application.* Table 15.5 presents the characteristics and properties of the contact materials used in power systems today[†] with the exception of vacuum interrupter applications, which are discussed in the following section.

15.4.2 Contact Materials for Vacuum Interrupters

Even though many new materials have been proposed for use in the vacuum interrupter, there are very few published data on these materials [32,33]. It is therefore necessary to rely on the extensive patent literature on this subject [33]. The vacuum environment does offer definite advantages to the contact material. For example, there is no ambient gas to contaminate the contact surfaces, and therefore materials that are quite unsuitable for applications in gases such as air or sulfur hexafluoride (SF_6) can be contemplated. Clean contact surfaces also help to maintain a steady contact resistance, something that is especially desirable when the contact is closed for long periods. The lack of ambient gas also allows for a high-voltage withstand ability across the open contacts, a fact that enables the manufacture of a relatively compact high-voltage switch (see Chaps. 11 and 12). The fact that the contacts are completely enclosed allows for the possible use of toxic materials, which could not be considered in an open system. Yet the vacuum environment does have some severe drawbacks. For example, the clean contact surfaces can result in severe contact welding. There is also the problem of contact erosion; this can, in turn, lead to a limit to the voltage withstand ability if the erosion is severe enough to deform the contact surfaces. Most of the proposed contact materials fit into four major classes shown in Table 15.6.

---

*"The Holm Conferences on Electric Contacts" are excellent source references to the latest contact materials research. From 1966-1984, the proceedings were published by the Illinois Institute of Technology and from 1985 have been published by the IEEE.

†Much information on the uses and properties of electric contacts can be found in the catalogs published by contact manufacturers (e.g., Gibson Electric, Stackpole Carbon, Texas Instruments, Engelhard, Art Wire-Doduco, Degussa, Advance, etc.).

REFERENCES

1. R. Holm and E. Holm, *Electric Contacts: Theory and Application*, Springer-Verlag, New York, 1967, Chaps. 1, 2.
2. J. B. P. Williamson and R. T. Hunt, The microtopography of surfaces, Proc. Oxford Conf. Metrol. Surfaces, April 1968 (Burndy Corp. Res. Rep. 59, January 26, 1968).
3. J. A. Greenwood and J. B. P. Williamson, Contact of nominally flat surfaces, Proc. R. Soc. Lond. *295A*: 300-319, 1966.
4. R. Holm and E. Holm, *Electric Contacts: Theory and Application*, Springer-Verlag, New York, 1967, Chaps. 2-9.
5. F. Llewellyn Jones, *The Physics of Electrical Contacts*, Clarendon Press, Oxford, 1957, pp. 11-17.
6. E. Shobert, Calculation of electric contacts under ideal conditions, ASTM Proc. *46*:1139, 1946.
7. R. Holm, Wiss. Veröff Siemens-Werk. 7(2):217-258, 1929.
8. J. A. Greenwood, Constriction resistance and the real area of contact, Brt. J. Appl. Phys. *17*:1621-1632, 1966.
9. J. B. P. Williamson, Deterioration processes in electrical connectors, Proc. 4th Int. Conf. Electr. Contact Phenom., Swansea, Wales, 1968.
10. K. Lemelson, The failure of closed heavy current contact pieces in insulating oil at high temperatures, Proc. 6th Int. Conf. Electric Contact Phenom., IIT, Chicago, June 1972, pp. 251-258.
11. G. A. Tomlinson, P. L. Thorpe, and H. J. Gough, An investigation of the fretting corrosion of closely fitting surfaces, Proc. Inst. Mech. Eng., *141*, No. 3, pp. 223-237, May 1939.
12. J. H. Whitley and E. M. Bock, Fretting corrosion in electric contacts, Proc. 20th Holm Semin. Electr. Contacts, IIT, Chicago, 1974, pp. 128-138.
13. M. Braunovic, Effect of fretting on the contact resistance of Al with difference contact materials, Proc. 9th Int. Conf. Electr. Contact Phenom./24th Holm Conf. Electr. Contacts, IIT, Chicago, Stepember 1978, pp. 81-86.
14. J. L. Johnson and L. E. Moberly, Separable electrical-power contacts involving aluminum bus bars, Proc. 21st Holm Conf. Electr. Contacts, IIT, Chicago, October 1975, pp. 53-59.
15. F. Kohlrausch, Über den stationären Temperaturzustand eines elektrisch geheizten Leiters, Ann. Phys. *1*:132, 1900.
16. R. Holm and E. Holm, *Electric Contacts: Theory and Application*, Springer-Verlag, New York, 1967, Chaps. 13-19.
17. F. Llewellyn Jones, *The Physics of Electrical Contacts*, Clarendon Press, Oxford, 1957, pp. 17-26.
18. A. H. Wilson, *The Theory of Metals*, 2nd ed., Cambridge University Press, Cambridge, 1953.

19. R. Holm and E. Holm, *Electric Contacts: Theory and Application*, Springer-Verlag, New York, 1967, pp. 89, 136, 161, 438.
20. M. Sato, Studies on the silver base electrical contact materials, Trans. Nat. Res. Inst. *18*(2): 19-37, 1976.
21. F. Llewellyn Jones, *The Physics of Electrical Contacts*, Clarendon Press, Oxford, 1957, Chaps. 2-5.
22. M. J. Price and F. Llewellyn Jones, The electrical contact: the properties and rupture of the microscopic molten metal bridge, Br. J. Appl. Phys. (J. Phys. D), Ser. 2, *2*:589-596, 1969.
23. P. P. Koren, M. D. Nahemow, and P. G. Slade, The molten metal bridge stage of opening electric contacts, IEEE Trans. Parts Hybrids Packag. *PHP-11*(1); March 1975.
24. P. G. Slade, Effect of the electric arc and ambient air on the contact resistance of silver, tungsten and silver-tungsten contacts, J. Appl. Phys. *47*(8):3438-3443, 1976.
25. R. Holm and E. Holm, *Electric Contacts: Theory and Application*, Springer-Verlag, New York, 1967, Chap. 11.
26. A. C. Snowdon, Studies of electrodynamic forces occurring at electrical contacts, AIEE Trans. *80*:24-28, March 1961.
27. P. Barkan, A new formulation of the electromagnetic repulsion phenomenon in electric contacts at very high currents, Proc. 11th Int. Conf. Electr. Contact Phenom. (VDE-Verlag GmbH, Berlin) June 1982, pp. 185-188.
28. B. Hague, *Principles of Electromagnetism Applied to Electrical Machines*, Dover, New York, 1962, pp. 336-340.
29. L. S. Frost, Westinghouse R&D Center, private communication.
30. Y. Yoshioka, Calculation of current distribution in heavy current contacts with many parallel finger contacts, Proc. 7th Int. Conf. Electr. Contact Phenom. (publ. SEER), Paris, June 1974, pp. 382-388.
31. P. Silvester, Skin effect in multiple and polyphase conductors, IEEE Trans. Power Appar. Syst. *PAS-88*(3):231-237, 1969.
32. P. Barkan, J. M. Lafferty, T. H. Lee, and J. L. Talento, Development of contact materials for vacuum interrupters, IEEE Trans. Power Appar. Syst. *PAS-90*:350, January/February 1971.
33. P. G. Slade, Contact materials for vacuum interrupters, IEEE Trans. Parts Hybrids Packag. *PHP-10*(1):43-47, 1974.
34. A. Nabae, M. Arii, O. Arakawa, and S. Sugiyama, Vacuum switch contact, U.S. Pat. 3,693,138, August 1972.
35. J. N. M. Legate and W. C. Brooks, Use of vacuum contactors in ac motor starters, J. Sci. Technol. *37*:182-185, 1970.

16
The Mechanical Operation of Power Circuit Breakers Rated Over 15 kV

ROSWELL C. VAN SICKLE*/Westinghouse Electric Corporation, Trafford, Pennsylvania

16.1 Circuit Breaker Functions Affecting Mechanical Operations 606

16.2 Improving Opening Performance 606
    16.2.1 Early circuit breakers 606
    16.2.2 Reductions in rated interrupting time 606
    16.2.3 Improvements in operating mechanisms 607

16.3 Operating Mechanisms 607
    16.3.1 Solenoid mechanisms 607
    16.3.2 Pneumatic mechanisms 608
    16.3.3 Hydraulic mechanisms 609
    16.3.4 Spring-driven mechanisms 609
    16.3.5 Mechanisms with other sources of power 609

16.4 Operating Linkage 610
    16.4.1 Skeleton linkage 610
    16.4.2 Functioning 610

16.5 Descriptions of Pneumatic Mechanisms 612
    16.5.1 MWE 250 pneumatic mechanism 612
    16.5.2 AA-7 pneumatic mechanism 615

16.6 Resistor Switching Mechanism 617

16.7 Conclusions 619

References 620

---

*Retired

16.1 CIRCUIT BREAKER FUNCTIONS AFFECTING MECHANICAL OPERATIONS

High-voltage power circuit breakers operate on command, usually an electrical signal from a protective relay or a manual control switch. A breaker may remain in the closed position for days or months carrying varying loads, even short-circuit currents, but must be ready to operate at any time if commanded to do so.

The ability of the breaker is described by a number of ratings. Among them, and directly related to the design of the moving parts, are: rated continuous current, rated short-circuit current, rated standard operating duty, rated interrupting time, rated reclosing time, rated load switching capability and life, rated control voltage, and rated fluid operating pressure. The rated short-circuit current establishes the highest currents that the breaker shall be required to close and latch against, to carry for short times, and to interrupt, respectively. Additional specifications may cover the synchronization of the various contacts in a multibreak breaker for both the closing and opening operations. The operating mechanism must be able to operate the breaker to meet these requirements over a range of operating conditions specified in standards [1]. In general, the demands on the mechanism increase with the rated voltage, rated short-circuit current, and with reductions in the rated interrupting time and rated reclosing time.

16.2 IMPROVING OPENING PERFORMANCE

16.2.1 Early Circuit Breakers

Much of the design of the moving parts of a circuit breaker depends on the required opening speed of the contacts. Over the years these requirements have increased greatly, although the basic opening function remains the same.

Paul Martyn Lincoln had been a superintendent at the Niagara Plant No. 2, which was put in service by the Niagara Falls Power Company in 1904. About 20 years later, as Director of the School of Electrical Engineering at Cornell University, he related to his students that in the plant, when one of the 2200-V 25-cycle air circuit breakers was to be opened because of an overload or short circuit, a loud bell alarm sounded, warning personnel to stand clear and to be prepared to wave a hat or cap, if necessary, to help blow out a persistent arc or arcs. Interrupting time?

16.2.2 Reductions in Rated Interrupting Time

In the early 1920s, 220-kV breakers for 60Hz systems had an interrupting time that was about 12 cycles, with 4 cycles to start the arc followed by about 8 cycles of arcing. As breaker performance was im-

Mechanical Operation of Power Circuit Breakers 607

proved, interrupting times were reduced in steps to 8 cycles, 5 cycles, 3 cycles (about 1936), and to 2 cycles in the 1960s. Breakers for 11-kV 25-Hz single-phase railway line service with one cycle of total short-circuit time were brought out in the 1920s [2]. More recently, one-cycle interrupting time at 60 Hz has been achieved with special high-voltage breakers [3].

16.2.3 Improvements in Operating Mechanisms

Improvements have been made not only in arc interruption techniques but also in each of the components controlled by the mechanism. This included reducing the time to release the latch and improving its ability to withstand the effects of long periods without operating, the time to relax the restraining forces in the mechanical linkage, the time to disengage the contacts and start the arcs (which sometimes involved shifting the current from main current-carrying contacts to arcing contacts before drawing the arc), and the time to establish arc extinguishing ability, which could extinguish the arc at the next current zero. It is desirable to have all of these events occur as soon as possible after the tripping signal is received. However, after the arc extinguishing ability is sufficient for the extinction of the arc, further acceleration is not desirable because increased contact speed results in increased arc length and increased arc energy. Reclosing ability was developed to improve system performance.

The newest Westinghouse mechanism, the DWE 250, eliminates the latches and uses compressed air to drive during both the closing and opening strokes [4]. A spring, external to the mechanism, biases it to both the closed and open positions.

16.3 OPERATING MECHANISMS

16.3.1 Solenoid Mechanisms

The changes and improvements in operating mechanisms have been in the source of the energy, the amount of energy input, the speed of closing, the speed of opening, the ability to perform reclosing duties, reliability, and simplicity.

In the early 1920s the operating mechanisms were usually driven by electric solenoids with levers that could be attached for manual maintenance operations. Accompanying stern warnings prohibited closing an energized breaker by hand. The problem of closing against a short circuit was recognized and provided for in applying mechanisms. "Trip-free" features were introduced to release the contacts from the operating mechanism and let the contacts open normally even though the closing effort was maintained or decreased slowly. This feature was important for power-closed breakers even though they were able to close against the short-circuit current. With solenoid mechanisms, the closing force decreased with the decay in flux after the control circuit was opened, but the rate was slow enough to make the opening sluggish.

The breakers required no external energy to open and were always ready to interrupt the circuits. Consequently, theey were provided with a "hand trip" so that they could be opened by manual command in an emergency when no control power was available.

Rapid reclosing was possible with mechanisms which, if the solenoid was not energized, opened without detaching the solenoid core and which tripped trip-free if the solenoid was energized. The non-trip-free opening eliminated the time required to reengage the latches, and the solenoid could be energized to reverse the opening movement of the contacts after interruption had been accomplished. This reduced reclosing time by several cycles.

Solenoid mechanisms of several sizes and outputs were used. In addition, the energy output of a solenoid mechanism could be fitted to the needs of different breakers by using coils which produced different magnetizing forces or by a coil having three sections, which could be used in several series-parallel combinations to produce several different magnetizing forces.

Larger solenoids produced more closing energy but were slower in operation and required more electrical power. The large storage batteries required for operating them, the heavy currents, and high-capacity conductors for carrying the currents imposed practical limits. Consequently, mechanisms taking electrical energy at a low rate and storing it in pneumatic reservoirs, hydraulic reservoirs, or springs to be used at a much higher rate were developed for the operation of power circuit breakers.

16.3.2 Pneumatic Mechanisms

Operating mechanisms driven by compressed air met the demand for more powerful and more versatile mechanisms [5]. The air fed through a relatively large, short pipe from an individual reservoir maintained a much more uniform force throughout the stroke. Electrically operated, three-way, fast-acting valves admitted compressed air to the cylinders and discharged air quickly when the closing stroke was completed.

The energy for the pneumatic mechanism is obtained through electrically driven air compressors which store the air for several operations in an air reservoir. The relatively low power demand is a big advantage over the high demand of the earlier solenoid mechanisms, which drew the power as used and operated at a low efficiency.

Several types of pneumatic operating mechanisms were developed and are still being applied on power circuit breakers. Two now being built by Westinghouse are described in Sec. 16.5.

Mechanical Operation of Power Circuit Breakers

16.3.3 Hydraulic Mechanisms

Hydraulic cylinders have also been used for many years in the operating mechanisms [6]. They, too, have low power demands and store in reservoirs the energy for several operations. They operate at much higher fluid pressures than do pneumatic cylinders, and consequently the cylinder diameter, fluid passages, and related fluid-handling parts are smaller. Leakage is much more important than in a pneumatic mechanism because the fluid is conserved and recycled. Consequently, the manufacturing tolerances for many of the parts are much smaller. They have been used for many years, particularly in Europe. The cross-sectional drawing of one used by many European and Japanese power circuit breaker manufacturers is shown in Fig. 10.22. It is noiseless and can be adapted to operate any type of breaker. Its complexity and cost, however, limit its use to powerful circuit breakers needing an operating mechanism of high energy output.

16.3.4 Spring-Driven Mechanisms

Spring-driven mechanisms have been used for many years. Their energy is stored in springs by motors acting through gearing, and consequently the power demands are low. They are held in the end positions by latches that can be tripped. The closing stroke stores energy in the springs, which will supply the energy for opening when the opening latch is released. The Sprecher & Schuh motor-wound spring operating mechanism [7], which has been in production, with improvements, for a number of years, was originally designed for their small oil-volume power circuit breaker. It operated a rotating vertical insulating shaft which carried pairs of contacts into and out of stationary contacts and interrupters. With small modifications it is now being used on single-pressure sulfur hexafluoride (SF_6) circuit breakers. The closing springs are located at the bases of the breaker columns. It has a good record of reliability and its noise level is very low, but it has a limitation in energy output and cannot be universally applied. It is shown in Fig. 10.21.

16.3.5 Mechanisms with Other Sources of Power

Operating mechanisms using power stored in other forms have been used or proposed. For example, small breakers were operated by mechanisms that were driven by the force of gravity acting on suspended weights. Mechanisms driven by expanding gases from explosive charges, in containers like shells for shotguns, were proposed many years ago and have recently been restudied. They have the advantages of small size because of the extremely high pressures and, if electrically fired, practically instantaneous development of maximum operating force. Electromagnetic repulsion with energy stored in capacitors has also been used to obtain very high accelerations [3].

16.4 OPERATING LINKAGE

16.4.1 Skeleton Linkage

The linkage connecting the operating mechanism to the contacts depends in part on how many contacts are to be operated by one mechanism. Most three-pole breakers have only one mechanism, but in the case of extra-high-voltage (EHV) and ultrahigh-voltage (UHV) breakers with large distances between poles and which may have multiple modules in a pole, one mechanism for each pole may be a better solution. Moreover, single-pole reclosing may be specified for these classes of breakers and necessitates the use of one mechanism per pole.

A suitable linkage can be laid out to tie the mechanism to the contacts. It may include rods, bell cranks, shafts through seals, and springs to provide biasing and accelerating forces. It must be stiff enough to operate the contacts simultaneously. Simplicity is essential.

16.4.2 Functioning

Mechanisms Without Latches

When the breaker and mechanism are held in the closed and open positions by a biasing spring, the linkage is normally under relatively little stress. The mechanism applies the force to open or close the breaker after the command signal is given and while the energy is being used. The relatively small closing force overcomes the biasing spring, closes the contacts, and resets the moving interrupting elements, consisting of gas-directing orifices and the puffer cyclinder. The opening force overcomes the biasing spring, opens the contacts, and drives the puffer, which forces the arc-extinguishing gas out of the cylinder and through the orifice. The opening force required is much greater than the closing force because of the energy needed to compress the gas. It is transmitted from the mechanism to the contacts and gas cylinder by members, most of which are in tension.

The pressure in the puffer is discussed in detail in Chap. 10 and is determined, together with the closely related gas chamber size, orifice, stroke, and speed, by calculation and development tests.

Mechanisms with Latches

The linkage in a breaker held in the closed position by a latch is under strain in the closed position until an opening signal is given and the latch releases the linkage. Immediately thereafter a moment exists in which the restraining force of the latch has disappeared and the contacts are still held in the closed position. The stresses pull on the released end of the linkage and it leaves the closed position. A wave of tension-release travels through the linkage to the contacts before they move. The time interval before the contacts start to move is about the time it takes the pressure wave to relieve the tension in the linkage.

Mechanical Operation of Power Circuit Breakers

The small amount of potential energy released in the linkage appears as kinetic energy which produces a small initial acceleration of the contacts.

With the restraining forces removed, the contacts are driven to the open position by accelerating and biasing springs. They are selected to move the contacts with speeds that will get the contacts to the critical points in their travel in the short times necessary for high-speed interruption as determined by tests on prototypes of the interrupter.

Forces to Operate The selection of the sizes and weights of parts and the driving forces are interdependent because increasing forces may necessitate increasing the strength and weight of parts. The problem is further complicated by the levers, which may change the mechanical advantage and relative speed of the parts as a function of the position in the opening stroke.

A first step in selecting the sizes of levers and connecting rods is taken by calculating the forces and springs necessary to drive the contacts and parts close to them to produce the travel-time and time-distance relationships shown necessary by interrupting tests on a prototype interrupter. The rest of the linkage is designed strong enough to transmit the forces for storing this energy plus the energy needed to accelerate the other members of the linkage to their respective speeds. This gives a preliminary design for which the opening time-distance relationship of the contacts can be calculated.

In complicated or long linkages, it is desirable to determine how much force is needed for each member or group of members and to locate the accelerating springs as close to them as possible so that all move as a coordinated assembly with as little force as possible having to be transmitted between members.

The operating condition desired for breakers held closed by latches and opened by springs is similar to that which exists when a chain consisting of three or four links suspending a weight is released. The weight and links fall freely, with gravity supplying each with the energy needed to give all the same speed at the same time.

Stiffness

The stiffness of the system is important for simultaneous operation of all contacts and for holding the correct engagement positions of all contacts in the closed position. This needs special emphasis. Experience has shown that breakers during development testing operated normally under no-load conditions, but when operated under short-circuit conditions, the contacts in one pole moved momentarily toward the closed position while simultaneously, the contacts in another pole moved toward the open position. The linkage had to be made stiffer.

Energy Supply Range

An operating mechanism must be able to perform its function over a range of voltage and fluid pressure. The voltage range assures operation of electrical elements such as magnetic trips and control valves. The fluid pressure range provides that the mechanism can operate the breaker a number of times without having the fluid replenished in the pressure reservoir and with the operating speed in a normal range for the breaker. The minimum operating pressure is stated on the mechanism nameplate and the volume of the reservoir provides for the required number of operations.

16.5 DESCRIPTIONS OF PNEUMATIC MECHANISMS

Many manufacturers of power circuit breakers have developed pneumatic operating mechanisms. The breakers they operate must meet approximately the same condidtions, so their performances must be similar. As examples of how the conditions can be met, two pneumatic operating mechanisms now being built and applied by the Westinghouse Electric Corporation on their high-voltage, high-power breakers will be described.

16.5.1 MWE 250 Pneumatic Mechanism [4]

One, the Westinghouse MWE 250, is relatively new and outstanding for its simplicity, as shown in Figs. 16.1 and 16.2, and also as Fig. 10.23. It was designed and developed to operate high-voltage, high-power, single-pressure, puffer-type SF_6 power circuit breakers. It has a double-acting piston which for tripping is driven upward by air acting on the bottom of the piston. The air is controlled by the trip coil, which activates the three-way pilot valve and large three-way opening valve. The air fed through the large air pipe from the nearby reservoir and through the large valves raises the pressure in the cylinder very rapidly and drives the piston upward. The linkage separates the main contacts of the Westinghouse 242,362 LWE and 550 LWER puffer circuit breakers 17 ms after the trip coil is energized and has started compression of the SF_6 gas and its flow through the interrupting area. Conditions capable of preventing the reignition of the arc are established in about one cycle. The arc is extinguished at the next current zero, so that the maximum interrupting time in a 60-Hz circuit is 2 cycles.

The contacts and the mechanism reach the end of their opening stroke in from 2.1 to 2.4 cycles. At this time the breaker operating rod, connected to the main lever about 75° from the piston rod, has gone over toggle and prevents any rebound of the contacts at the end of the opening stroke. The linkage has another over-center toggle driven by a biasing spring which acts to hold the contacts in the end

Mechanical Operation of Power Circuit Breakers

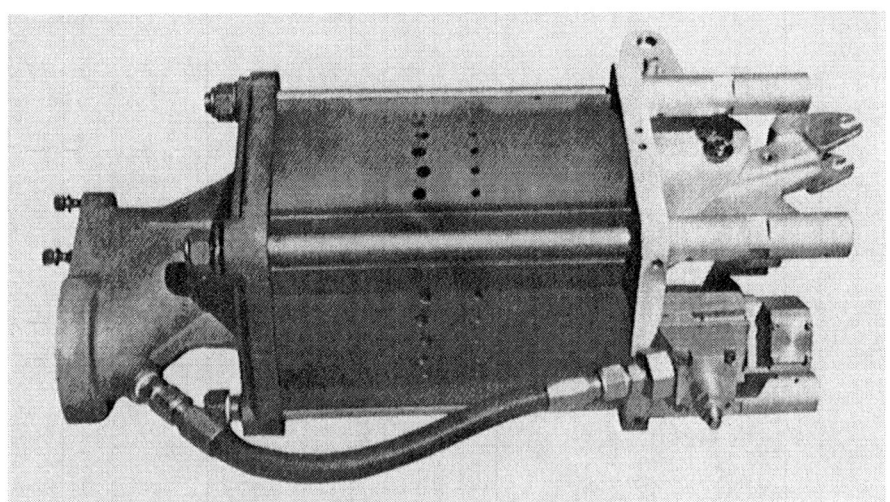

Figure 16.1 MWE pneumatic mechanism (on side; base to left). It operates SF_6 puffer power circuit breakers from 121 to 550 kV. It uses compressed air to drive both the closing and opening strokes, has no latches, and is kept in the end positions by small external biasing forces. (Courtesy of Westinghouse Electric Corporation, Trafford, Pa.)

positions. The piston uncovers exhaust ports in the cylinder wall at the end of the stroke, thereby venting the cylinder. The closing of the opening valve to shut off the air also vents the cylinder below the piston and keeps the pressure low while the piston later moves downward on the closing stroke.

The breaker is closed by energizing the closing coil, which opens the closing valve and admits air to the top of the cylinder. Its initial movement pulls the operating rod to the right and over center, breaking the toggle which had held the breaker open, and further movement closes the circuit through the contacts and positions the puffer, recharging the cylinder. The biasing spring, which offered a small force to keep the contacts in the open position, went over center during the closing stroke and now holds the contacts in the closed position. The area of the piston exposed to the air for closing is about 70% as large as the bottom of the piston because the forces needed to close the breaker are smaller than those required to open contacts and drive the puffer. An external hydraulic dashpot system damps the mechanical forces after operation. A hand jack can be coupled to the main lever for slow operation of the mechanism and breaker during maintenance work.

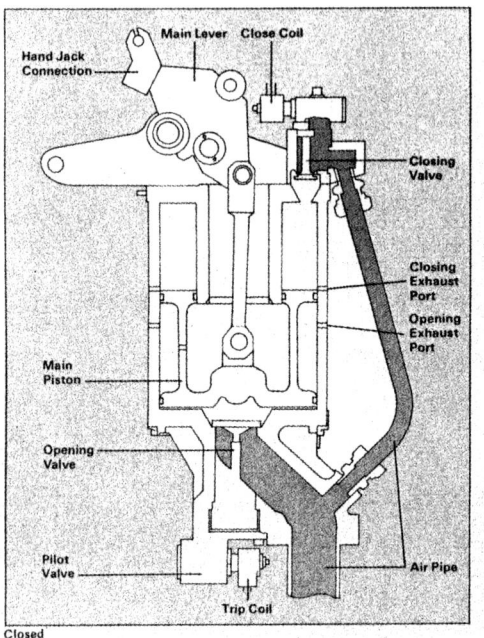
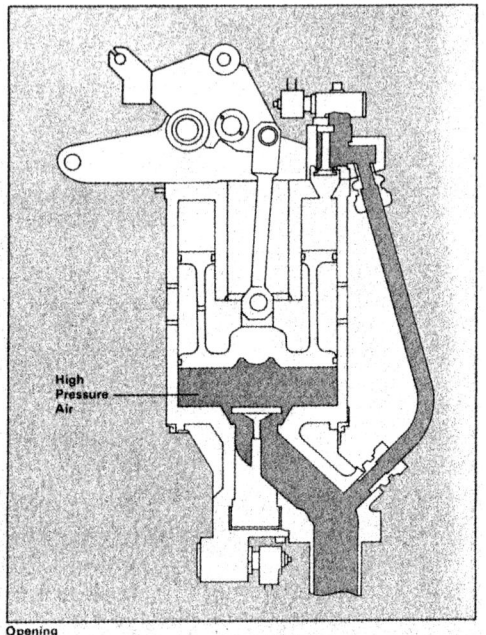
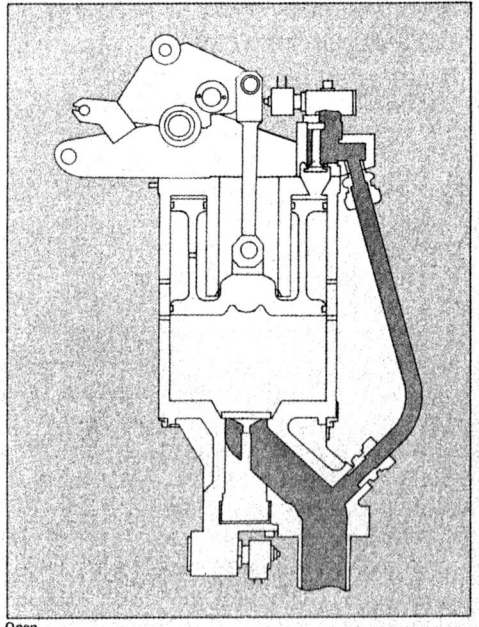
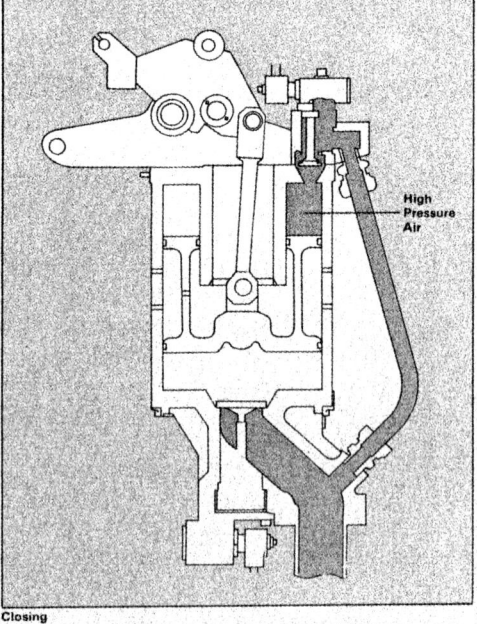

Figure 16.2 Four cross-sectional views of the MWE pneumatic mechanism. The one marked "closed" shows the piston at the bottom of the cylinder with compressed air in the air pipe and both the closing valve

Mechanical Operation of Power Circuit Breakers

16.5.2 AA-7 Pneumatic Mechanism

The other pneumatic mechanism, the Westinghouse AA-7, is used on single-pressure SF_6 puffer circuit breakers for voltages of 34.5 and 72.5 kV which require less energy to operate. It was developed as the first and smallest of a line of pneumatic mechanisms for operating oil power circuit breakers up to the 345-kV level and was applied on SF_6 breakers up to the 115-kV level [5]. This mechanism is shown in cross-sectional views for the open, closed, and trip-free positions in Fig. 16.3 and will be described using the nomenclature of the figure.

The end of the piston rod and the end of the breaker pull rod both have pins through them, which are provided with rollers. These rollers move between guide rails and constrain the pins to move in the same straight line. They are held in almost a fixed relation to each other by two links, designated "closing link" and "cam link," connected between them and held in a partially folded or toggle position by the intermediate link, held at its other end by the trip-free lever. This effectively connects the piston to the pull rod and causes the force of the compressed air on the piston to be transmitted almost unmodified to the pull rod during a closing stroke. At the end of the closing stroke a holding latch engages the cross-head on the piston rod and holds it in the closed position. When the piston reaches the closed position, the closing valve is deenergized, thereby both shutting off the air supply and opening a large discharge passage to vent quickly the air in the cylinder.

The breaker is opened by releasing the pull rod from the piston. This is accomplished by energizing the trip magnet to move the trigger, which relases the trip-free lever and thereby the intermediate link, which keeps the two folded links in their substantially constant relative positions. They open up as the breaker goes to its open position under the action of its accelerating springs. As the links open, the cam link pushes the holding latch out of engagement with the cross-head and the piston returns to the open position under the action of

---

and opening valves closed. The one marked "opening" shows high-pressure air being admitted by the opening valve, which has been opened by energizing the trip coil. The piston is being driven upward to open the circuit breaker. The one marked "open" shows the piston at the upper end of its stroke. The closing valve has closed, shutting off the high-pressure air supply. The air in the cylinder below the piston has been vented through holes in the side of the cylinder walls and the closed three-way air valve. The one marked "closing" shows high-pressure air being admitted above the piston through the closing valve. The initial downward movement of the piston rotated the main lever clockwise, so the vertical opening rod (not shown) has moved to the right and no longer acts over the center of the pivot for the main lever. (Courtesy of Westinghouse Electric Corporation, Trafford, PA.)

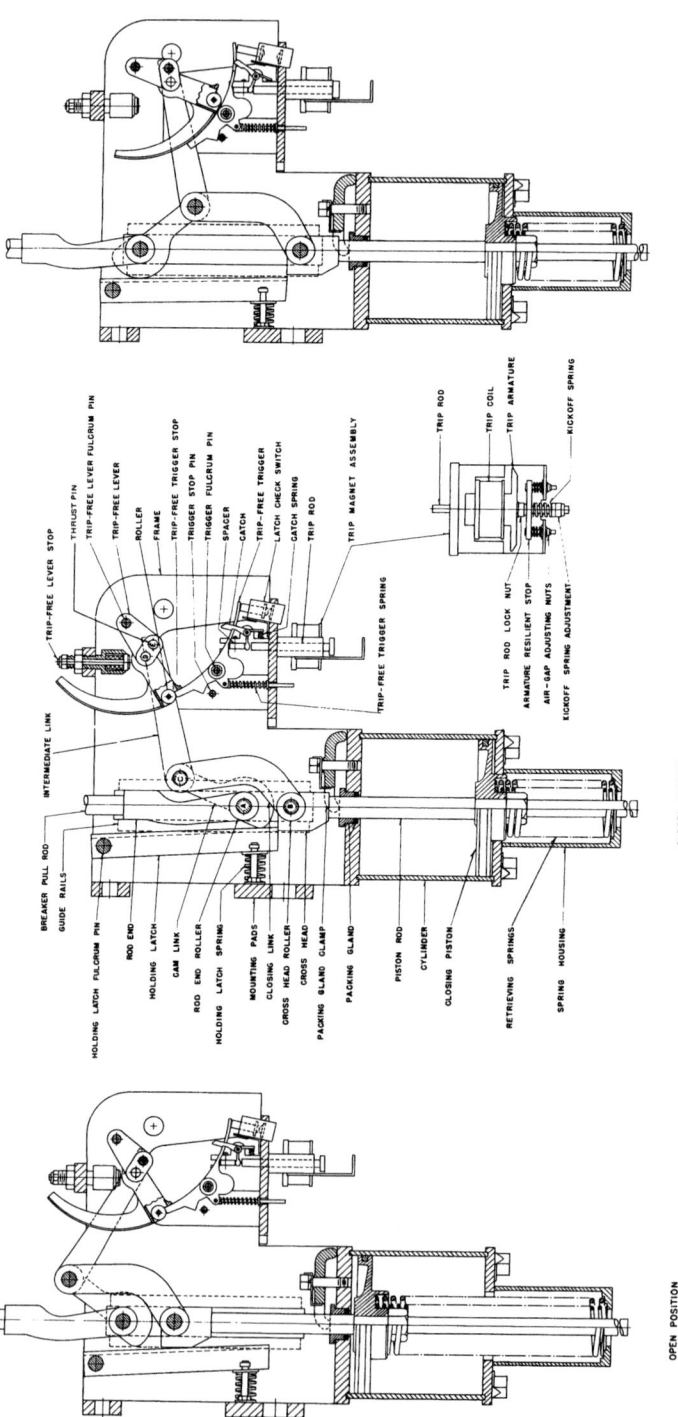

Figure 16.3 Cross-sectional views of the AA-7 pneumatic mechanism. The view on the left marked "open position" shows the piston at the top of the stroke with no load on the vertical pull rod. The view marked "closed position" shows the piston at the bottom of the cylinder with the cross head on the piston rod held down by the holding latch. The pull rod is in tension and is holding the breaker closed. The cylindrical projection on the lower side of the piston has entered the hole in the lower cylinder head and trapped air to cushion the end of the closing stroke. The view marked "trip-free position" shows how the mechanism is tripped to permit the breaker to open even though the piston were held in the closed position by air pressure. As soon as the force of the air on the piston is overcome by the retrieving springs, the piston is pushed upward and the open position is attained. (Courtesy of Westinghouse Electric Corporation, Trafford, PA.)

Mechanical Operation of Power Circuit Breakers 617

its retrieving spring. This restores the two toggle links to their folded position and thereby pulls the intermediate link back to its normal position. The trigger then engages the roller on the trip-free lever. In actual operation the piston starts to move to the open position before the breaker is fully open, even on a close-open operation, because of the extremely fast discharge of the compressed air from the cylinder. Consequently, the diagram labeled "trip-free position" in Fig. 16.2 shows an extreme condition of the levers which is provided for but never actually reached except during maintenance.

An air cushion is built in to absorb excess energy of the moving parts at the end of the closing stroke. Near the end of the closing stroke a cylindrical downward projection, on the underside of the piston and concentric with it, enters a hole in the lower cylinder head and traps air under the piston. This air is compressed by further movement of the piston and its pressure becomes very high at the end of the stroke and cushions the piston and attached parts as they come to a stop. The clearance between the cylindrical projection and the cylinder head is critical to provide cushioning without rebound.

A small catch, unlatched by the initial movement of the trip rod, prevents accidental release of the trigger by shock. The mechanism can be tripped by hand by pulling a ring on a rod projecting slightly form the protective housing and attached to a triangular lever pivoted on the trip-magnet frame. Pulling the ring pushes the trip rod upward and pushes the trigger out of engagement with the roller on the trip-free lever.

Hand closing during maintenance operations is by means of a screw jack which, when it is to be used, is attached to the lower end of the piston rod. It can hold the linkage and contacts in any desired position. To prevent accidental tripping of the mechanism during maintenance and while the hand-closing device is being used, a safety rod is pushed by hand through holes so located in the frame that the rod prevents a movement of the trip-free lever to release the pull rod from the piston.

16.6 RESISTOR SWITCHING MECHANISM

For higher-voltage service, a circuit breaker may be fitted with closing resistors to reduce voltage surges when closing on transmission lines [8]. In such a case, initial closure of the circuit occurs through the resistors about 10 ms before complete contact closure, which then short-circuits the resistors. On opening, these resistors are not normally in the circuit, so they must be disconnected shortly after full breaker closure.

Figure 16.4 shows in section the resistor switching mechanism [9] as used in a Westinghouse LWER breaker for 550-kV service, illustrated in Fig. 10.26. As shown, the porcelain tubes containing the resistors

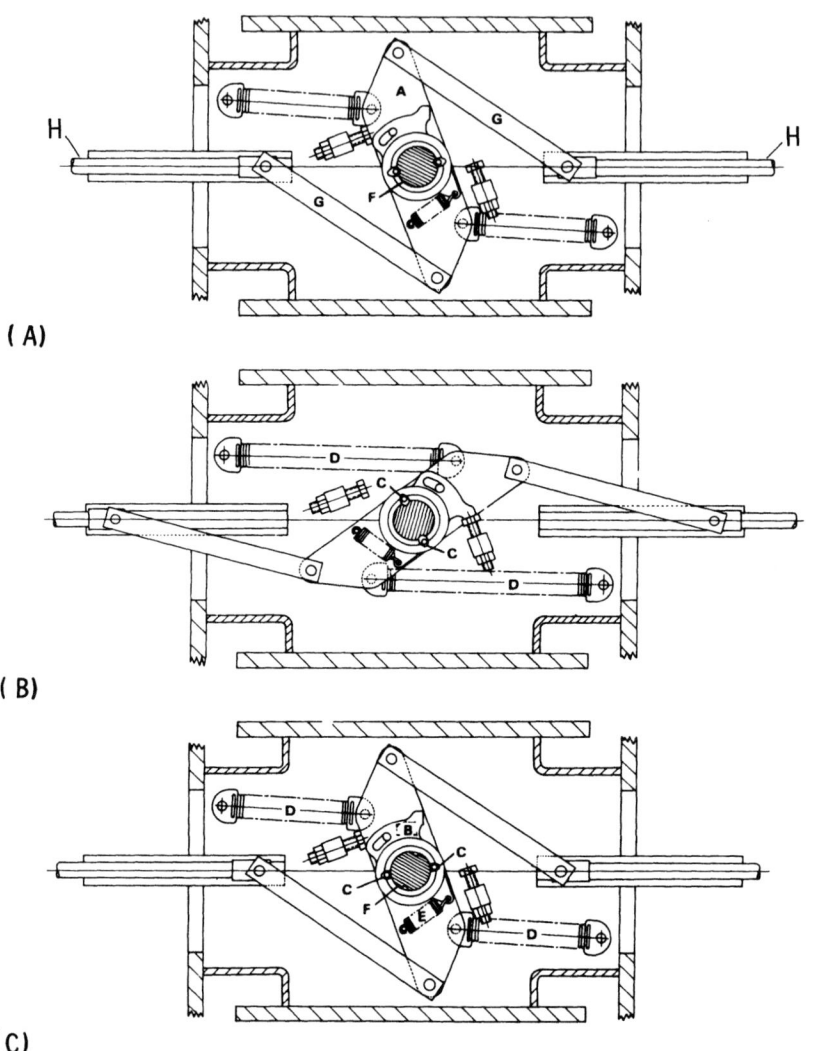

Figure 16.4 Cross-sectonal views of the resistor switching mechanism in three positions: (A) breaker opening, resistor open; (B) resistor contacts fully closed; (C) breaker closed, resistor open. (Courtesy of Westinghouse Electric Corp., Trafford, PA.)

and resistor switching contacts are mounted horizontally in pairs below the V arrangement of two interrupter modules supported by each vertical column. In Fig. 16.4, the central shaft F, shown crosshatched, is rotated through a shaft seal by an external lever system from the main breaker mechanism. Figure 16.4(A) shows the mechanism near

Mechanical Operation of Power Circuit Breakers

the end of an opening operation with the central shaft F rotated so that the rollers C can be moved radially inward to engage the central shaft and the resistor lever A. They are moved inward and held in engagement with the shaft F by the ring lever B, which, biased by its spring E, moves clockwise with respect to the resistor lever. After the ring lever locks the rollers in engagement with the shaft and resistor lever, the normal open position is reached and is maintained until a signal to close is received.

In Fig. 16.4(B) the resistor lever A, locked to the shaft by the rollers, has been rotated clockwise to project the contact rods H and complete the resistor circuit through contacts not shown. The mechanism and resistor lever have continued their motion even after the ring lever was stopped by its arm engaging a fixed stop. It has rotated counterclockwise with respect to the resistor lever and in the position shown permits the rollers to be pushed out of engagement with the slots in the shaft. The shaft continues to the closed position shown in part (C) and the resistor lever is free to return under the action of its biasing springs D to the open position shown in parts (A) and (C). However, inertia delays the opening of the resistor contacts until after the closing of the breaker contacts, but they open before the breaker contacts would open on a close-open operation.

The resistor switch does not operate during an opening operation of the breaker. The shaft rotates counterclockwise from the position shown in part (C) to the position shown in part (A), where the rollers can be pushed into it and locked by ring B rotating clockwise by action of its biasing spring E.

In brief, the resistor switch operates only on the closing stroke and inserts the resistor for about 10 ms before the main contacts of the breaker close. It then opens before the main contacts can open on a close-open operation.

16.7 CONCLUSIONS

The operating mechanism of a power circuit breaker performs an essential part in the operation of the breaker. It may stand for days or months without operating but must be ready at all times to respond to a signal to open or close. When the signal comes to the mechanism, it must respond instantly and faultlessly to operate the breaker to interrupt or to energize the circuit within specified times, and for circuit breakers using SF_6 gas, to produce the flow of gas needed to extinguish the arcs. In addition to simply opening or closing the breaker, the mechanism must be able to operate it in combinations of these functions when needed.

If the circuit breaker by closing a line completes a short circuit, the mechanism must be able to open the breaker to interrupt the circuit within its rated interrupting time. If the mechanism opens a breaker and clears a fault on a line, it must be able to reclose the breaker quickly in an attempt to restore service and limit disturbances on the system. Sometimes more than one reclosure is desired for persistent faults.

Operating mechanisms have been developed in parallel with the interrupting elements so that power circuit breakers have continued to meet the requirements of service as these breakers have grown with increases in size and complexity of electric power systems.

REFERENCES

1. ANSI/IEEE C37.04-1979, Standard rating structure for high voltage circuit breakers rated on a symmetrical current basis.
2. H. M. Wilcox, High-speed circuit breakers for railway electrification, AIEE Trans. 47:1285-1292, 1928.
3. T. Morita and T. Nonomura, Synchronous power circuit breaker for one cycle interruption: ten years field experience, IEEE Trans. Power Appar. Syst. *PAS-99*:January/February 1980, Abstract p. 16.
4. B. J. Calvino, Double flow compressed gas operating mechanism for high voltage circuit breaker, U.S. Pat. 4,205,208, May 27, 1980.
5. R. C. Van Sickle, W. T. Parker, and F. E. Florschutz, Pneumatic operating mechanisms for power circuit breakers, AIEE Trans. 71:725-734, 1952.
6. P. Barkan, Some considerations in the design of high-pressure, rapid response hydraulic actuators, Israel J. of Technol. 10(4): 285-293, 1972.
7. F. Bachofen, P. Steinegger, and R. Glauser, A novel solution for high-speed SF_6 puffer type power circuit-breakers of high rated capability with low operating mechanism energy, CIGRE, Rep. 13-07, Paris 1982.
8. W. M. Leeds, R. E. Friedrich, C. L. Wagner, and T. E. Browne, Jr., Application of switching surge, arc and gas flow studies to the design of SF_6 breakers, CIGRE Rep. 13-11 Paris, 1970.
9. B. J. Calvino, Circuit breaker closing resistor mechanism, U.S. and British patents applied for (U.K. Patent Application No. 2,117,975, Oct. 19, 1983).

General Reference

Barkan, P. The Mechanical phenomena of circuit breakers, in *Physics and Engineering of High Power Switching Devices* (T. H. Lee, ed.), MIT Press, Cambridge, Mass., 1975, pp. 461-536.

17
Interruption Testing

C. DONALD FAHRNKOPF* and DAVID VALLO/Westinghouse Electric Corporation, East Pittsburgh, Pennsylvania

17.1 Introduction 621
    17.1.1 Importance of interruption testing 621
    17.1.2 Interruption-testing requirements and procedures 622
    17.1.3 Interruption-testing facilities 623
17.2 Types of Interruption Tests 624
    17.2.1 High-current tests 624
    17.2.2 Low-current tests 626
17.3 Interruption-Testing Techniques 627
    17.3.1 Direct testing 627
    17.3.2 Unit testing 628
    17.3.3 Two-part testing 628
    17.3.4 Synthetic testing 629
    17.3.5 Instrumentation for interruption testing 637

References 639

17.1 INTRODUCTION

17.1.1 Importance of Interruption Testing

Testing is extremely important for circuit breakers, since they are the main protective devices to clear faults and isolate faulted sections. Failure to perform this duty when called upon can result in serious damage to equipment, loss of stability, and even collapse of the system. As protective devices, circuit breakers must be extremely reliable.

---

*Currently Consulting Engineer, Pittsburgh, Pennsylvania

In the design and development of circuit breakers, interruption testing is an essential procedure. As is well known, the sudden interruption of a high current arc subjects the breaker to extremely high mechanical, thermal, and electrical stresses which are difficult to calculate. Testing is therefore relied upon heavily as a design tool.

In a development testing program, results should convince the test engineer that the breaker will meet all of its ratings and will perform satisfactorily in service. Finally, a demonstration of performance within the framework of the applicable standards informs the user that the breaker will be adequate for the task when properly applied in the utility system.

17.1.2 Interruption-Testing Requirements and Procedures

Standards and guides for testing circuit breakers have been developed over the years from the accumulated experience of both manufacturers and users. Testing according to such standards provides that a circuit breaker will have at least those capabilities required by the standards. As a result, the nameplate rating of the breaker implies those capabilities.

Tests that will demonstrate the ratings under the American National Standards and methods for making tests are described in ANSI/IEEE C37.09-1979, Test Procedure for AC High-Voltage Circuit Breakers Rated on a Symmetrical Current Basis. This and other standards and guides [1-4] relate to required capabilities and procedures for interruption testing.

In addition to meeting American standards, circuit breaker manufacturers must be prepared to prove that their breakers meet International Electrotechnical Commission standards [5-7] if they intend to compete in a world market. Since IEC standards are based on a current frequency of 50 Hz, laboratories with only 60 Hz available must adjust parameters, as nearly as possible, to be equivalent to a 50-Hz test.

The difficult technical nature of the circuit-breaking problem calls for close control of the testing procedure. Breaker performance can be greatly influenced by the parameters of the circuit in which it operates. Often, breakers are sensitive to minor changes, some of which result from interaction of the breaker itself with the circuit. In the attempt to cover these situations, standards often become complicated. This occurs even though they undergo continual revision to keep up to date and to become more useful and readable. The guides [4,7] mentioned above are written to increase understanding and to offer guidance in the proper application of the standards. Together with this, experience from previous tests and experience of others should be fully utilized. In addition, there are special performance requirements or special applications requiring nonstandard tests which must be determined by agreement between the manufacturer and user.

Interruption Testing

17.1.3 Interruption-Testing Facilities

Early testing stations consisted simply of a transformer fed by the utility network to provide short-circuit current and voltage for an interruption test. Even today, some of the largest testing stations get their power from a cooperating utility, but they now have refined control systems and use highly versatile test procedures. However, most high-power laboratories for testing circuit breakers have their own short-circuit testing generators. In addition, they have multitapped transformers and banks of reactors, capacitors, and resistors for current and voltage control and for representing a wide range of interrupting and switching conditions.

Figure 17.1 shows a typical single-phase direct testing circuit with a local source of generation. A variable bank of air-core reactors provides current control with low loss and at the required low power factor. A closing switch initiates the current for the few cycles required for the circuit breaker to operate. Operation of the closing switch can be timed to provide a symmetrical current or an offset current with any degree of asymmetry.

This basic circuit has been the foundation of circuit breaker testing for many years. The generating station and associated equipment represent the major investment in a high-power testing laboratory. With the advent of techniques now called "synthetic testing" to increase the

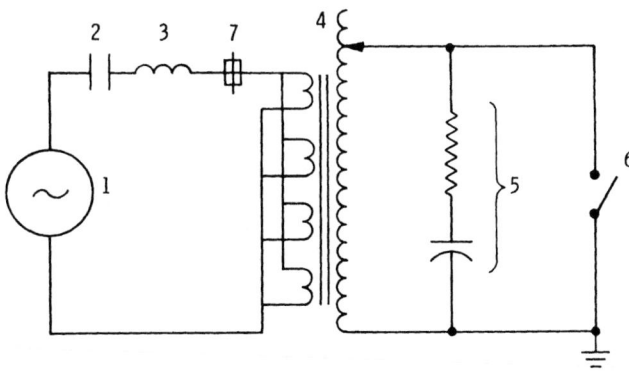

1 Test Generator
2 Closing Switch
3 Current Limiting Reactor
4 Testing Transformer
5 TRV Shaping Circuit
6 Test Breaker
7 Back-Up Breaker

Figure 17.1 Direct testing circuit.

capability of the testing stations, the power circuit remains the main current source in a dual or multiple-source system.

A worldwide view of interruption testing facilities, as reported by the Electric Power Research Institute (EPRI) in 1975 [8], shows 10 major high-power laboratories in North America and seven overseas. Descriptions of these laboratories include data on the physical plant, testing capabilities, and test equipment ratings. In Sec. II of the EPRI report a summary of the major capabilities of these laboratories is given in tables for direct testing, both single-phase and three-phase, and for synthetic testing.

17.2 TYPES OF INTERRUPTION TESTS

Of the many tests that circuit breakers must undergo, interruption of high currents representing faults on the station bus (bus faults) or faults a short distance out on a transmission line (short-line faults) are often the most difficult to accomplish and the most expensive to perform. Much of the theory of the nature of the arc and arc-circuit interaction are directed toward performing these interrupting tasks.

In addition, there are other interruption tests that must be demonstrated, including the ability to switch normal load current and to switch capacitor banks and long lines. Furthermore, circuit breakers are expected to be able to perform certain switching functions which generally do not require a demonstration test. These include the switching of reactors, unloaded transformers, and transformers with secondary faults.

17.2.1 High-Current Tests

Bus-fault tests

The basis of rating of a circuit breaker is its ability to interrupt short-circuit current on a three-phase ungrounded fault occurring on the bus near the terminals of the circuit breaker. A test representing this event is called a bus-fault test or a terminal-fault test. Bus-fault tests are characterized as high-current tests, although in addition to testing at the maximum short-circuit current rating, tests at lesser current levels are also required.

Three-phase bus-fault tests may be made for the smaller breaker ratings where sufficient three-phase power is available. For higher ratings it may not be economically feasible to provide adequate power and associated test facilities for a three-phase test.

The single-phase bus-fault test is designed to represent the worst condition during interruption of a three-phase ungrounded fault. This is the condition that immediately follows interruption of the first phase to clear, at which time the recovery voltage attempts to return to a

Interruption Testing

level 1.5 times the line-to-ground voltage. By taking this factor into account, a single-phase test of the proper severity can be set up.

The test procedure to establish that a circuit breaker can meet a certain rating is set forth in ANSI C37.09-1979 [1], which gives conditions for the test and methods of obtaining the most severe switching conditions. This specification tabulates duties for two sets of tests, one (Table 1) for testing a three-pole breaker on a three-phase circuit and the other (Table 2) for testing a single pole of a three-phase breaker on a single-phase circuit.

Short-line-fault tests

Although bus-fault tests establish and confirm the rating, a related capability required of a circuit breaker is the ability to interrupt single-phase line-to-ground faults on a transmission line. Of most interest are faults at short distances, a few spans to a few miles out on the line from the circuit breaker. These are called short-line faults, and although the short-circuit current is less than for a bus fault, the recovery voltage rate of rise on interruption is high because of the high surge impedance of the line.

This line-side recovery voltage appears as a sawtooth wave caused by reflections back and forth on the line: negative reflections from the shorted end at the fault and positive reflections from the open breaker at the source end. In addition, the opening contacts are subjected to a separate recovery voltage on the station-bus side of the breaker. The total recovery voltage across the interrupter contacts is then the sum of the bus-side and line-side recovery voltages. The steep initial rate of rise in the first few microseconds and up to the first sawtooth crest is the most difficult part of a short-line fault for the breaker to withstand. This initial rate of rise of recovery voltage may be 5 to 10 times that on the source side of the circuit breaker.

To perform a short-line-fault test, the testing facility must have a circuit to represent a transmission line of the desired surge impedance and capable of representing various distances to the faulted section. Short-line-fault tests are identified by the current to be interrupted rather than by the distance to the fault. Thus, a 70% SLF test would represent a fault at a distance out on the line such that the line impedance would reduce the fault current to 70% of rated short-circuit current. Of most concern are short-line faults in the range 60 to 90%. It is often known where in that range certain types of breakers are most sensitive, and that is where the test should be made. If the breaker's characteristics are not well known, a search should be made over the range.

17.2.2 Low-Current Tests

In completing a test program on a circuit breaker, certain low-current interrupting tests, usually referred to as switching tests, must be performed. Other low-current tests may or may not be performed, depending on the application of the breaker and the requirements of the user.

Line-charging current and capacitor-bank switching

Interrupting the current in an unloaded circuit such as a transmission line, an underground cable, or a capacitor bank is easy because the initial transient recovery voltage is low. However, it is so easy that the breaker may interrupt at a very early current zero with a short contact gap that may not hold voltage when the bus voltage reverses. A trapped charge in the capacitive circuit causes twice line-to-ground voltage across the breaker a half cycle after interruption. If a restrike occurs because of this, double voltage or even higher voltages can appear across the breaker. Multiple restrikes on successive cycles can reach four to five times normal voltage.

The test code according to ANSI standards [1-3] requires that for 50 random operations the voltage to ground shall not exceed 2.0 or 2.5 or 3.0 times normal, depending on the type of breaker and voltage rating.

Transformer exciting-current switching

In switching the magnetizing current of unloaded power transformers, some breakers may tend to chop the current (force current zero), resulting in transient overvoltages. Although circuit breakers are required to be able to switch unloaded power transformers, a test for this is usually made only to meet special requirements of a potential user. For this reason no ANSI standard or test code for this requirement has been developed.

Shunt-reactor switching

The switching of shunt reactors may be difficult because the high-frequency resonance effects produce transient recovery voltages similar to a short-line fault, although at relatively low current. As with the switching of transformers, this is considered a special requirement not covered at this time by an applicable standard.

Load-current switching

The ANSI test code [1] says that all types of breakers shall interrupt currents up to rated continuous current with normal 60-Hz recovery voltage and with a power factor of 80% lagging or less. Breakers rated 121 kV and up are required to be capable of 125 full-load current-

Interruption Testing

switching operations without maintenance. Demonstration tests are not required, as a rule, since the breaker should be able to interrupt these currents at high power factor with little difficulty. It is assumed that a breaker capable of meeting its short-circuit rating will have no trouble handling these switching operations. On the other hand, it may be a problem for the laboratory to supply the high power factor allowed by the standard. Also, there may be special applications where higher recovery voltages could occur. These are handled as special cases not yet covered by standards.

Out-of-phase switching

Although switching two parts of a system during out-of-phase conditions is not required of standard breakers, there are certain applications requiring a special breaker with an out-of-phase switching current rating. ANSI standards [1] provide a test code for those breakers, requiring essentially that the breaker interrupt up to 25% of its rated short-circuit current at twice normal voltage. This requirement is usually somewhat less severe than that for short-circuit fault current interruptions.

17.3 INTERRUPTION-TESTING TECHNIQUES

17.3.1 Direct Testing

A direct test implies a test circuit with a single power source having an MVA capability at least equal to the full power demand of the test. A direct test may be made with a single-phase circuit as illustrated in Fig. 17.1 or with a three-phase circuit. Most of the low-current switching tests described previously and many short-circuit interruption tests are direct tests.

Although direct testing may be preferred, where there is a choice, over more intricate techniques, there are limits, as noted earlier, beyond which it is not practical or economical to provide the power for direct testing. Considering present-day short-circuit ratings, these limits are quickly reached. For example, a direct three-phase test on a circuit breaker rated at 145 kV and 40 kA would require a three-phase source capable of supplying 10,000 MVA. However, for high-voltage breakers the test codes permit a single-phase test which reduces the power required to one-half, or 5000 MVA. Even so, there are few laboratories in the world that can make single-phase direct tests at 5000 MVA, and they soon find their capability inadequate for direct testing at slightly higher ratings.

Although limited in power capability, direct testing is widely applicable for lower-voltage switchgear ratings. Direct three-phase tests are required where there is a possibility of magnetic or thermal influence between poles, as in metal-clad or metal-enclosed breakers, where all poles are within a single enclosure. Both ANSI and IEC standards require that such breakers be tested as complete three-pole circuit breakers in a three-phase circuit. However, recent developments in three-phase enclosures for gas-insulated systems have led to higher voltage ratings for which only a few high-power testing stations can meet the three-phase direct testing requirements. To overcome this limitation, three-phase simulation methods have been proposed as described by Damstra and Kempen in a paper entitled "Synthetic Three-Phase Testing Methods for Metal-Enclosed Circuit-Breakers" [10].

Direct testing, where possible, has the advantage of providing a straightforward representation of the environment of the breaker in service, allowing little question as to the breaker being properly stressed. Disadvantages are in the size and extent of the facilities required and in the wear and tear on the facilities and test breaker, expecially the possibility of severe damage to the test breaker as a result of interruption failures under full power.

17.3.2 Unit Testing

A reduction in power required for testing high-voltage breakers can be realized if a breaker pole is made up of two or more interrupters in series. Testing one unit of a series requires full short-circuit current but at a reduced voltage proportional to the fraction of the pole being tested. However, in evaluating the overall performance from a unit test, the voltage distribution across the units must be considered. If equalizing resistors or capacitors are not installed, the test voltage for a unit test must be increased to compensate for possible variation in voltage division across individual units. This may result in the test being slightly more severe than necessary to assure that uneven voltage distribution can be handled.

Other conditions to consider in making a unit test are the force on the mechanism, the supply of arc-extinguishing medium, and its condition if more than one interrupter is in the same housing. Also, it makes a difference whether or not the breaker is a live-tank or dead-tank breaker (interrupter housing at ground potential).

Unit testing is less suitable for dead-tank breakers (1) since it is difficult to reach the intermediate point between interrupters, and (2) because of the unequal dielectric stresses on individual units because of the presence of the grounded tank.

17.3.3 Two-Part Testing

A method for obtaining useful test data when the testing station is clearly inadequate for a full-scale test is to make a two-part test. This is really two separate tests: one at full short-circuit current but at reduced voltage; the other at full voltage but at reduced current.

Interruption Testing

Although both tests may reach the power limit of the station, neither test may provide in full the desired conditions over the entire range. Still, a good indication of breaker capability may be obtained for some types of breakers. This method has worked fairly well in the past on bulk-oil breakers, as later confirmed by long years of service in the field. The two-part testing idea is used occasionally to prove a point, together with more sophisticated methods. For development work the idea of a less than full-scale test is often useful, economical, and efficient for studying a particular aspect of breaker performance.

17.3.4 Synthetic Testing

Methods to extend the capability of high-power testing facilities have been proposed since the early days of testing and are still being developed. One of the earliest circuits designed for "special tests" on circuit breakers was described in 1936 by Skeats [11]. This was followed in 1954 by a report, "The Weil Circuit for Testing High-Voltage Circuit-Breakers with very High Rupturing Capacities" [12]. Both are so-called "synthetic testing" methods which propose to subject the circuit breaker to realistic full-scale stresses without the necessity of providing full direct testing power from the station. Present methods are extensions and variations of these early circuit simulation schemes.

The basic idea in synthetic testing is the use of two power sources in a single test: one to supply full short-circuit current during current conduction, the other to supply full recovery voltage following current interruption. The basic circuit is shown in Fig. 17.2(a) and Fig. 17.2(b) shows the addition of components to make a practical operating circuit.

The circuit of Fig. 17.2 is a "parallel current-injection" circuit based on the original Weil circuit. It is the type most widely used in testing laboratories and the one covered primarily by the ANSI guide for synthetic fault testing [4]. However, the guide recognizes and describes other methods, such as "series current-injection" and "voltage injection," which may have some advantages for specific situations. Also, the IEC report on synthetic testing describes both current-injection and voltage-injection methods [7]. A typical circuit with test results is described in the paper, "Results from a Basic Current-Injection Synthetic Circuit" [13].

From the high-current circuit, which is usually a generator and transformer source at power frequency, the circuit breaker conducts full short-circuit current during the complete arcing period preceding interruption. Following current interruption the opening circuit breaker is subjected to full recovery voltage from the high-voltage source, which is usually a large precharged capacitor bank. The short-circuit current from the generator is supplied at considerably less than rated

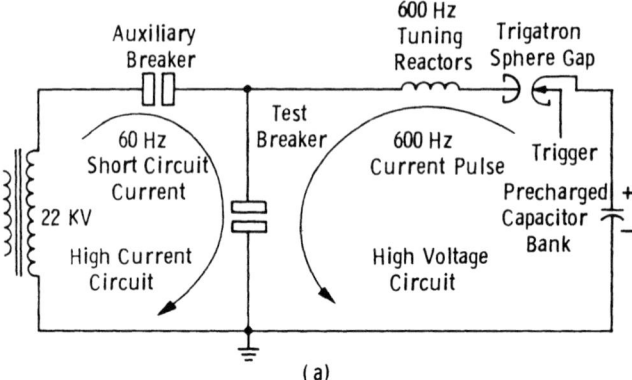

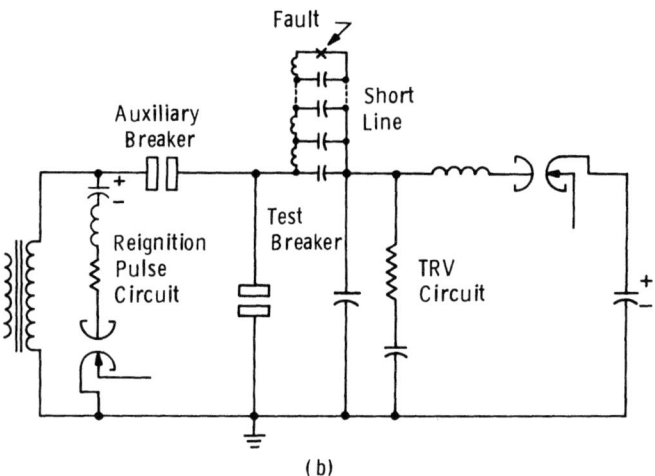

Figure 17.2 Current-injection synthetic testing circuit: (a) basic circuit; (b) typical operating synthetic circuit.

voltage, although the voltage must be several times the arc voltage. A power frequency crest voltage at least five times the arc voltage is considered acceptable. The high recovery voltage is supplied from the capacitor bank through a high-impedance RLC circuit with a relatively low power rating.

The application of two power sources in a single test is possible because of the time separation of the high-current phase and the high-voltage phase in the interruption process. Since the time separation occurs instantaneously at current zero, precise synchronization is required between the current and voltage sources. To relieve this

Interruption Testing

problem of close synchronization and to make the transition more smooth, the high-voltage source is applied in parallel with the high-current source at a point a few hundred microseconds ahead of that particular current zero where interruption is expected. For a brief time, then, both sources are supplying current through the test breaker. The current from the high-voltage source is the "injection current" of the parallel circuit. The firing of the trigatron in the voltage circuit initiates this high-frequency pulse of current through the test breaker. The timing is such that during this pulse the generator current reaches its normal zero and is isolated from the test breaker and the high-voltage source by the opening of the auxiliary breaker (sometimes called the isolation breaker) (Fig. 17.2). This leaves connected only the high-voltage source, which is still supplying the pulse of current through the test breaker until it is interrupted at its first current zero, provided that the interruption is successful.

The current-injection pulse is designed by selecting its amplitude and frequency so that its rate of change, di/dt, as it approaches zero is the same as that of the power current. For example, an injection

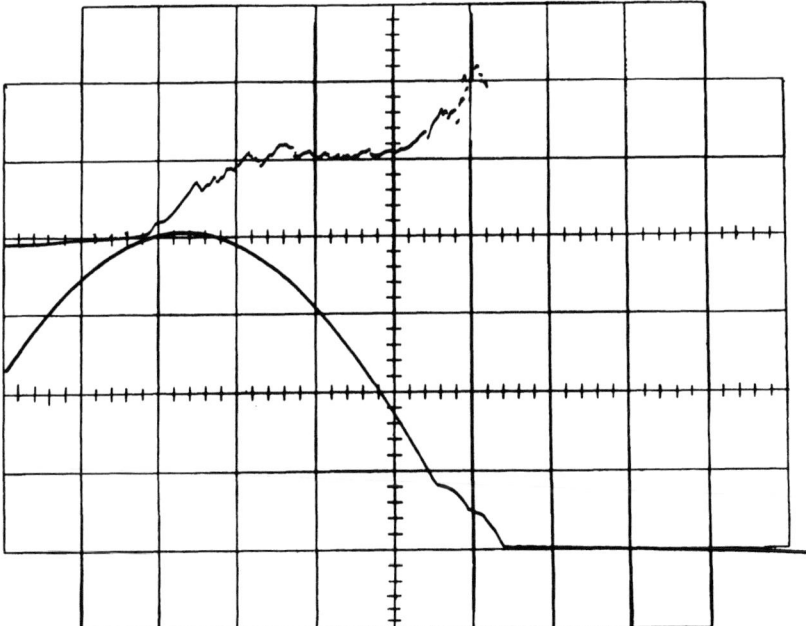

Figure 17.3 Record of successful interruption after short arcing time on a synthetic test. Upper trace: arc voltage across opening contacts of the test breaker. Lower trace: current successfully interrupted in the test breaker. Time scale: 1 ms per major division.

current of 600 Hz frequency having an amplitude equal to one-tenth of the 60-Hz current is a desirable selection for ANSI-based tests. Test codes allow a frequency range from 200 to 1000 Hz with the amplitude adjusted to give the proper di/dt. The overall effect is to delay current zero by a few hundred microseconds, as shown in Fig. 17.3 from an actual test. The slight node in the middle of the added current pulse, about 400 μs before current zero, indicates the point at which the isolation breaker interrupts the generator current at the 60-Hz current zero.

After the generator is isolated by the auxiliary breaker, the high-voltage source alone provides the remaining current, accompanied here by a significant increase in the arc voltage. This is the beginning of the "interaction period," which continues through the slightly delayed current zero and past any postarc conductivity. From this point on, the circuit breaker no longer interacts but becomes a passive element in the circuit as the contact gap widens and dielectric strength is established. The success or failure of the circuit breaker to interrupt depends very critically upon the realism with which the circuit is simulated in the interaction period around current zero (see Chap. 6).

The upper trace of Fig. 17.3 shows the arc voltage as it builds up following contact part. This test represents a short arcing time of slightly more than a quarter of a cycle. Following interruption, the

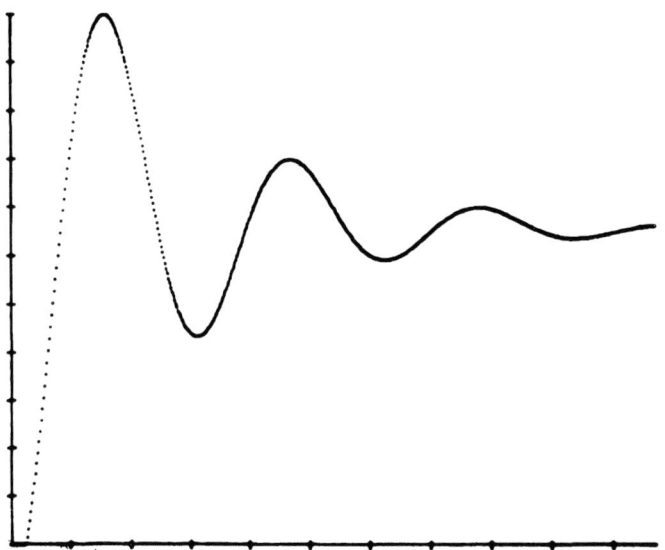

Figure 17.4 Digital record of transient recovery voltage following current interruption on a bus-fault test. Vertical scale: 30 kV per division. Horizontal scale: 30 μs per division.

Interruption Testing

arc voltage is replaced by the rapidly increasing recovery voltage. This recovery voltage, recorded on a different amplitude and time scale, is shown in Fig. 17.4.

Figure 17.3 shows a successful interruption. Had it not been successful, the high-frequency injection current would continue to arc through the breaker for one or more additional high-frequency current loops. Figure 17.5 illustrates a failure to interrupt, with high-frequency current continuing for the one additional (negative) current loop. It is obvious that this failure did not dissipate much energy. Had the breaker failed to interrupt on a full-power direct test, severe erosion of the breaker contacts might have curtailed the test. It is clear that the low power supplied by the high-voltage recovery circuit is a very desirable feature of synthetic testing. Although Fig. 17.5 represents a failure to interrupt at the desired current zero, it also represents a longer arcing time, of about 3/4 cycle. In addition, it shows a tiny injection pulse at the early current zero at the last loop of 60-Hz current. This pulse is injected by firing the reignition circuit of Fig. 17.2, which forces the 60-Hz current through zero, allowing the

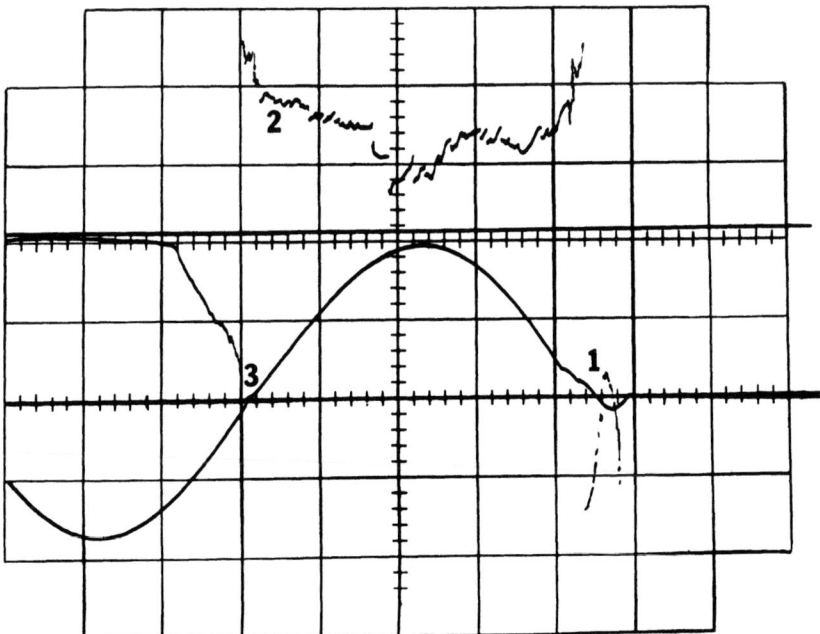

Figure 17.5 Total current and arc voltage showing (1) failure to interrupt; (2) arc voltage for long arcing time; (3) reignition pulse at early zero.

arc to continue through the final 60-Hz current loop. Without this reignition pulse the 60-Hz arcing current might go out at the early zero since it is being supplied at a comparatively low voltage.

The recovery voltage of Fig. 17.4 is from a bus fault test as described in Sec. 17.2.1. For short-line-fault tests a lumped section representing a transmission line, with a fault a short distance out on the line, is added to the circuit as shown in Fig. 17.2(b). The result is the addition of a steep sawtooth component to the initial part of the transient recovery voltage, as illustrated by Fig. 17.6 and described in Sec. 17.2.1.

Checking the validity of a synthetic test circuit

Since the standards do not define a specific circuit for synthetic testing but instead define requirements for fault current, transient recovery voltage (TRV), and time parameters, it is up to the test laboratory to demonstrate the credibility of the circuit for each test. This can be done by preliminary checks on the circuit to confirm that requirements will be met.

A typical set of requirements is given in Table 17.1 for making a 100% bus fault test for the same 145-kV 40-kA rating used in the previous example of direct testing (Sec. 17.3.1). The required TRV envelope and its characteristic features are defined in Fig. 17.7.

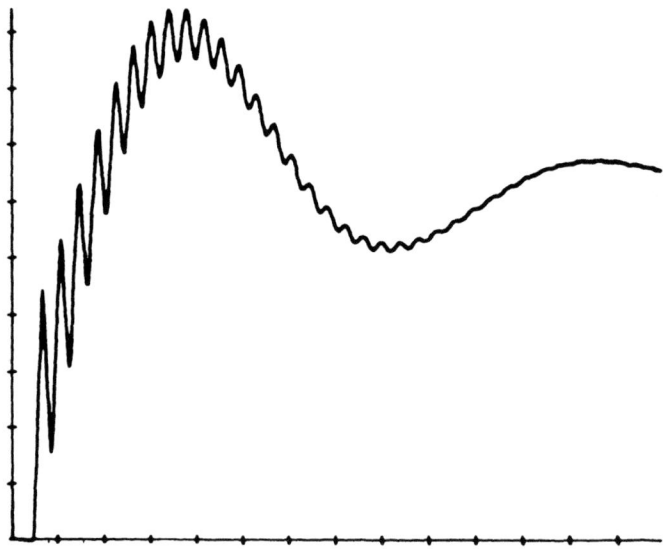

Figure 17.6 Short-line-fault TRV. Vertical scale: 20 kV per division. Horizontal scale: 100 µs per division.

Interruption Testing

Table 17.1 Requirements from Table 5 of ANSI 37.06-1979

V = 145 kV	Rated maximum voltage, kV, rms
I = 40 kA	Rated short-circuit current, kA, rms
E_2 = 225 kV	Crest value of TRV envelope
R = 1.8 kV/μs	TRV rate of rise (dv/dt)
T_2 = 310 μs	Time to crest of TRV
T_1 = 3.2 μs	Allowable time delay
E_1 = 178 kV	Asymptotic limit of (1-exponential) portion of TRV envelope

Source: Ref. 2.

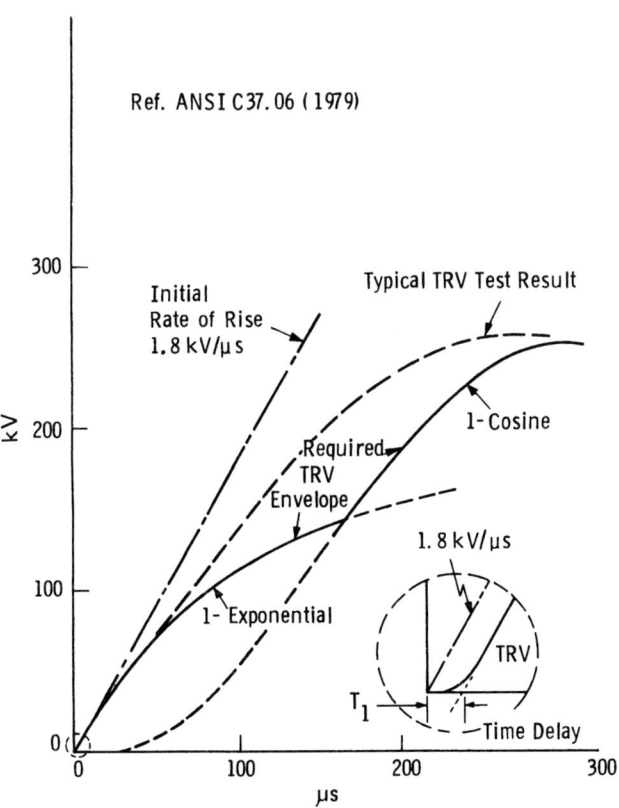

Figure 17.7 Transient recovery voltage envelope for 100% bus fault for 145-kV 40 kA rating.

In addition to the requirements listed in Table 17.1, synthetic testing requires that the high-frequency injected current should have the same value of di/dt at current zero as does the rated short-circuit current, which for this case is

$$\frac{di}{dt} = \omega I \sqrt{2} = 21.3 \text{ A}/\mu s$$

The source voltage to produce 40 kA is not specified, but it should be several times the expected arc voltage. For a source voltage of 22 kV, the power required for 40 kA would be 880 MVA. Compared to the 5000 MVA required in the previous direct test example (Sec. 17.3.1), this represents an increase in capability of more than 5:1 by the application of the synthetic testing method.

To confirm that the requirements above will be met by the circuit, preliminary checks can be made by firing the high-voltage circuit of Fig. 17.2 into the closed test breaker to check the injected current and by firing into an open circuit to check TRV characteristics. For a short-line-fault test the shorted line may be checked with a low-voltage current injector which produces a miniature circuit interruption and records the resultant sawtoothed, multireflected wave.

Limitations and advantages of synthetic testing

Certain tests present practical difficulties for synthetic testing. Some of these can be overcome by adding circuitry or by making additional tests. Various approaches for coping with these limitations are offered in Synthetic Testing Guide C37.081 [4].

1. *Duty cycles:* Synthetic circuits cannot easily comply with test duties that call for operating sequences of opening and closing operations separated by short time intervals such as 0 to 3 s CO. Other methods and allowances are discussed in the guide [4].

2. *Breakers with impedances in parallel:* In a synthetic test on a breaker with a resistive shunt impedance it may not be possible to attain peak recovery voltage as in a direct test because of the limited energy from the high-voltage source. Therefore, additional tests are necessary.

3. *Long-arcing-time tests:* For arcing times greater than one-half cycle, because of the relatively low voltage of the power source, some method is needed to maintain or reestablish conductivity at the current zero, or zeros, before the interrupting point. This can be done at one early zero by the use of a reignition circuit as described earlier and shown in Fig. 17.2, and with the test result illustrated in Fig. 17.5. For even longer arcing times additional reignition circuits or the multiple application of the voltage circuit may be required.

4. *Inherent limits:* Synthetic circuits unavoidably deviate a certain amount from the direct test circuits they represent. Because of this

it is necessary to define the limits of admissible deviation. Both ANSI and IEC Standards and guides for synthetic testing propose to define these limits. With these guidelines and proper control of the circuit, synthetic testing is generally accepted for testing ratings as high as 10 times the power rating of the station. Further work in defining the limits of admissibility from the theoretical point of view by the use of models is given by Slameka et al. [14].

The foregoing limitations notwithstanding, synthetic testing offers some overriding advantages:

1. *Amplification of testing power:* In addition to being the only practical way the highest ratings can be tested, synthetic methods can increase the testing capability of stations by as much as 10:1, and at a relatively low cost.

2. *Nondestructive testing:* Synthetic testing not only amplifies the testing capacity of a station, but it does so without the risk of destroying the breaker under test in the event of its failure to interrupt. This benefit is a result of the isolation provided by the auxiliary breaker, so that there is no power follow into the still arcing breaker.

The economic and technical advantage of saving the test breaker for continued and repetitive testing is obvious. Also, the wear and tear on the power-testing facility itself is much less than for full-power direct testing.

3. *A versatile tool for development testing:* In the development stages of circuit breaker design, synthetic testing techniques can be modified and applied to study particular aspects of circuit breaker performance. For example, synthetic tests may be made with greatly reduced power frequency current but with full injection current and recovery voltage. Used with caution for some repetitive tests, this practice can yield useful results with little erosion of interrupter contacts. This and other techniques in development testing are described by Gardner and Urwin [15].

17.3.5 Instrumentation for Interruption Testing

Results from an interruption test are essentially a set of measurements that represent the stresses applied to the breaker and the reaction of the breaker to those stresses. The ability to analyze the test results effectively depends on the quality of the measurements: the accuracy of transducer outputs, the integrity of transmission, and the resolution of the recording devices.

Traditionally, one of the biggest problems has been that of operating in a severe environment of high electromagnetic fields. Operation requires the recording of transients in an extremely short time frame when the transients themselves are the source of disturbance. Fortunately, the newer digital and fiber-optic techniques are being applied very successfully in coping with this problem. Optical isolators and transmission of signals by fiber-optic light pipes offer a great advantage in eliminating electromagnetic interference in high-power testing.

For any interruption test a number of records of current and voltage are needed and they must be recorded on several different time scales. Other time-varying parameters, such as the position of interrupter contacts, are necessary to describe the operation of the breaker. Pressure and temperature of the interrupting medium describe the environment within the interrupter. For these records a magnetic oscillograph provides a relatively slow but inclusive record over the complete time span of a test, which may last for several cycles of current. With a time resolution of about 0.2 ms, the magnetic oscillograph gives an overall record of current, voltage, and position of mechanical parts for adjusting the timing and travel of the breaker elements.

For faster records, the cathode-ray oscillograph (CRO) provides greatly increased resolution over a short time span. In the critical interaction period around current zero a resolution of 1 to 0.1 µs is needed and can be provided by a CRO capable of reproducing frequencies of 1 MHz or higher.

More recently, digital recorders are being applied which have sampling rates of one point in 50 ns or even faster. Digital recording techniques not only increase the available resolution, but provide other advantages. The overriding advantage is that once the data are in digital form they can be analyzed by computer, reconstituted in any form, and stored permanently for future recall. One attractive feature for the test engineer is that the digital data can be presented on a cathode ray screen in the form of analog waves, which the engineer is accustomed to seeing, and any portion can be expanded and analyzed. The value of any point can be read out, and any other parameter, such as rate of change of current or power in the arc, can be read out if programmed properly. Applications of digital techniques in high-power laboratory testing with a real-time computer are described by Vallo and Thout [16] and Corcoran [17].

Even before the advent of such highly efficient recording equipment, the accurate recording of currents and voltages depended on noninductive current-measuring shunts and high-frequency voltage dividers. Shunts should have extremely low mutual inductance for indicating a sharp cutoff of current. A clean indication of actual current zero needs to be defined relative to the initial transient recovery voltage (ITRV) in order to determine the ITRV characteristics from the true initial zero point. The di/dt of the current as it approaches zero should be recorded accurately since the validity of the test is dependent on having the correct di/dt.

Voltage dividers should be of the differential type to measure voltage directly across the interrupter, without including voltage drop in the ground lead or other leads. Also, dividers should be of high impedance to prevent damping of the recovery voltage.

Interruption Testing

Even the best current-measuring shunts and voltage dividers have some residual inductance. Also, when installed in position for testing, there may be mutual inductive couplings with nearby circuits. Errors from these sources can often be reduced or compensation made by methods described by Spindle et al. [18].

Although accuracy can be no better than that provided by the transducers themselves, recent improvements in instrumentation have made it possible to overcome many traditional problems. Optical isolators and fiber-optic transmission techniques go far in eliminating interference from the severe electromagnetic environment of power testing. High speed cathode-ray and digital recording devices offer resolution of the fastest transients, while computer processing rapidly provides all the worthwhile output possible from the raw data.

Further development of the circuits, the instrumentation, and the testing techniques now going on in many laboratories throughout the world will continue to advance the capability for interruption testing.

REFERENCES

1. ANSI/IEEE C37.09-1979 (Consolidated Edition), Test procedure for ac high-voltage circuit breakers rated on a symmetrical current basis.
2. ANSI C37.06-1979 (Consolidated Edition), Preferred ratings and related required capabilities for ac high-voltage circuit breakers rated on a symmetrical current basis.
3. ANSI/IEEE C37.04-1979 (Consolidated Edition), Rating structure for ac high-voltage circuit breakers rated on a symmetrical current basis.
4. ANSI/IEEE C37.081-1981, Guide for synthetic fault testing of ac high-voltage circuit breakers rated on a symmetrical current basis.
5. IEC 56-2(1971), Part 2: Rating, high-voltage alternating-current circuit-breakers.
6. IEC 56-4(1972), Part 4: Type tests and routine tests, high voltage alternating-current circuit-breakers.
7. IEC 427(1973), Report on synthetic testing of high voltage alternating current circuit-breakers.
8. Study of transmission and distribution laboratory facilities, Res. Project 327, prepared by Ebasco Services Inc. for Electric Power Research Institute, Palo Alto, Calif., August 1975.
9. L. E. Berkebile and R. Pfleiderer, Synthetic capacitor switching tests on a SF_6 circuit breaker, IEEE Paper A-79-541-4; abstract published in IEEE Trans. Power Appar. Syst. *PAS-99*(1), 1980.
10. G. C. Damstra and H. W. Kempen, Synthetic three-phase testing methods for metal-enclosed circuit-breakers, CIGRE, Rep. 13-03, Paris, 1980.

11. W. F. Skeats, Special tests on impulse circuit breakers, AIEE Trans. 55(6):710–717, 1936.
12. J. Biermanns, The Weil circuit for testing of high voltage circuit breakers with very high rupturing capacities, CIGRE Rep. 102, Paris, 1954.
13. C. D. Fahrnkopf, Results from a basic current-injection synthetic circuit, IEEE Trans. Power Appar. Syst. PAS-91(3):819–822, May/June 1972.
14. E. Slameka, W. Rieder, and W. T. Lugton, Synthetic testing: the present state, Proc. IEE (Lond.) 125(12), 1978.
15. G. E. Gardner and R. J. Urwin, Development testing techniques for airblast and SF6 switchgear, Proc. IEE (Lond.) 127(5), 1980.
16. D. Vallo and M. E. Thout, High power laboratory testing with a real-time computer: a new generation of testing methods, IEEE Trans. Power Appar. Syst. PAS-95:982–987, May/June 1976.
17. R. P. Corcoran, Applications of digital data acquisition and analysis in high power laboratory testing, IEEE Paper A-80-010-9; abstract published in IEEE Trans. Power Appar. Syst. PAS-99(4), 1980.
18. H. E. Spindle, T. E. Browne, Jr., and P. E. Martin, Synthetic testing techniques, IEEE Trans. Power Appar. Syst. PAS-91(3): 768–773, 1972.

APPENDIX
Basic Concepts of Gaseous Conduction

THOMAS E. BROWNE, Jr.*/Consultant to Westinghouse Research and Development Center, Pittsburgh, Pennsylvania

A.1 Nature of Gaseous Conductors: Kinetic Theory 642
    A.1.1 Importance, nature, and mechanism of gaseous conduction 642
    A.1.2 Nature of a gas 643
    A.1.3 Gas pressure, temperature, and energy 644
    A.1.4 Molecular velocities 647
    A.1.5 Distribution of molecular velocities 648
    A.1.6 Free paths of molecules and electrons 652
    A.1.7 Distribution of free paths 654
    A.1.8 Diffusion 655

A.2 Movement of Ions and Electrons 657
    A.2.1 Nature and function of ions 657
    A.2.2 Mobility in the electric field 658
    A.2.3 Deflection by magnetic field 663
    A.2.4 Space-charge effects 664
    A.2.5 Diffusion in an electric field: Ambipolar diffusion 667

A.3 Ionization and Deionization 669
    A.3.1 Atomic structure and ionizing processes 669
    A.3.2 Ionizing agents 671
    A.3.3 Deionization 676
    A.3.4 Summary of deionizing influences 680

A.4 Self-Maintained Discharges 681

References 683

---

*Retired

A.1 NATURE OF GASEOUS CONDUCTORS: KINETIC THEORY

A.1.1 Importance, Nature, and Mechanism of Gaseous Conduction

Interrupting the flow of an electric current by separating contacts almost always involves a momentary continuing flow of current through the intervening gas in the form of a spark or an arc. Since the circuit is not finally interrupted until this gaseous discharge ceases to conduct, or is extinguished, some knowledge about gaseous conduction phenomena is essential to an adequate understanding of circuit-breaking devices. Such knowledge is needed especially by those concerned with the design and testing of circuit breakers used to interrupt large fault or short-circuit currents, since the control and extinction of the arcs drawn in such breakers is a basic and sometimes formidable problem. In the case of high-voltage breakers, corona and spark-over are also important phenomena.

Unfortunately, gaseous conduction is governed by nothing so simple as Ohm's law for metals. Unlike metals, electrically conducting gases have resisitivities which may vary through enormously wide limits and which may change by many orders of magnitude in exceedingly short spaces of time. As a well-known example, air at atmospheric pressure is an excellent insulator, but when a voltage above the sparking value is suddenly applied to a sphere gap, a filament of air between the spheres reaches a conductivity approaching that of some metals in less than a millionth of a second. Moreover, this conduction is accompanied by an intense radiation of light and heat and by violent expansion of the gas, leading to a great reduction in gas density as well as to the characteristic sharp sound wave. The interaction of heat, expansion, actinic radiation, and electrical and magnetic effects in so common a phenomenon as the electric spark and in the steady-state arc that may follow it are so complex that our understanding of them is still incomplete. Nevertheless, much progress toward such understanding has been made, principally by exhaustive study of gaseous conduction under simpler and more easily controlled conditions.

It is well established that current is conducted in a gas by the motion of charged discrete particles moving under the influence of an electric field. Because of the nature of a gas, these charged particles move relatively freely through the empty spaces between the gas molecules subject to the usual laws of mechanics for freely falling bodies. These are principally

$$f = ma \tag{A.1}$$

force is equal to mass times acceleration, and the corollary laws,

$$l = \tfrac{1}{2}at^2 \tag{A.2}$$

distance moved from rest equals one-half the acceleration multiplied by the square of the time, and

Appendix

$$W = \frac{1}{2} mv^2 \qquad (A.3)$$

the kinetic energy of motion equals one-half the mass times the velocity squared. When the charged particles collide with gas molecules, the laws of conservation of energy and momentum also apply.

Current-carrying particles in a gas may have charges of either positive or negative polarity. Positively charged particles called positive ions are generally molecules or atoms of the gas which have lost one or more electrons of -1.6×10^{-19} C charge, leaving a particle with net positive charge and with essentially the same mass as that of the original neutral particle. Negatively charged particles may be molecules or atoms with an extra electron attached (i.e., negative ions) or of more importance in the usual conduction process, they may be free electrons with masses about 1/1837 that of a hydrogen atom, or 9.11×10^{-31} kg. The total current density in a gas due to the motion of charged particles of both signs (conducting gases usually contain both positive and negative carriers) may be expressed by the vector equation

$$\bar{j} = n_+ e \bar{v}_+ + n_-(-e)\bar{v}_- + n_e(-e)\bar{v}_e \qquad (A.4)$$

where n_+, n_-, and n_e are, respectively, the average densities of positive ions, negative ions, and electrons; $\bar{v}_+$, $\bar{v}_-$, and $\bar{v}_e$ are the corresponding average velocities; and e is the electronic charge. In an electric field, the direction of the average motion, or drift, of the negative carriers is, of course, opposite to that of the positive ions, so that the component currents are additive. The drift velocities depend on the electric field strength $\bar{E}$, but are not generally simply proportional to E. Both the $\bar{v}$'s and the n's are generally more or less complicated functions of both $\bar{E}$ and $\bar{j}$, and in non-steady-state cases, also of t, the time. It is this complexity that necessitates an understanding of the basic physical processes involved before quantitative calculations can be made with gaseous conductors. This same complexity also leads to a great many valuable uses for gaseous conductors depending on the possible wide variations in conducting properties and characteristics of such conductors.

A.1.2 Nature of a Gas

To understand the motions of ions in gases, it is first necessary to have a clear mental picture of the nature and structure of a gas. The kinetic picture, or theory, of a gas will be employed without proof since it is generally accepted. According to this theory, the gaseous state of matter is a state in which the elementary particles, or molecules, are separated so that they can and do move freely with negligible intermolecular forces except those due to collisions, either with other free molecules or with enclosing walls. The molecules are assumed to behave as small perfectly elastic spheres obeying the laws of newtonian mechanics. The gas molecules are known to be in ceaseless chaotic motion, continually rebounding from each other and from the enclosing walls. The mean kinetic energy of this motion is identified with the heat content of the gas. This molecular motion also explains the pressure of a gas upon enclosing surfaces and the tendency of a gas to expand indefinitely when released. It is further assumed in mathematical derivations that the gas particles occupy only a small part of the total gas volume, that any given volume contains a very large number of particles, and that the particles, when in equilibrium, have a purely statistical distribution of velocities and of free path lengths.

A.1.3 Gas Pressure, Temperature, and Energy

A rough quantitative relation between gas pressure and molecular velocity can be derived fairly easily. When looked at in a microscopic way, the actual force on a small surface exposed to a gas is not strictly continuous in time but is made up of very many brief discrete impulses resulting from random collisions of individual particles with the surface, as illustrated in Fig. A.1. The solid lines represent the force due to one particular nearby molecule and the dashed lines the force due to another. These impulses are randomly distributed in time and vary in magnitude with the random velocities with which the particles strike the wall. Under the stated assumption that any considered gas volume contains a very large number of particles, there is an averaging effect which results in an essentially steady total force on a surface of appreciable area. To find the total force on a surface of *unit* area, or the *pressure*, it is only necessary to determine the average rate of occurrence of molecular impacts with the unit area and the average contribution to the total force of each single impact.

The rate of occurrence of impacts, or the total number of impacts in unit time, can readily be computed if it be assumed for simplicity that the molecules all have velocities equal in magnitude and equally distributed between six possible directions: directly toward and directly away from the surface and similarly in both directions along the two other orthogonal axes parallel to the surface. Under these assumptions only one-sixth of the molecules in a given volume strike the surface. Those striking a unit surface in unit time will come from a volume having unit area of base and a height equal to the common molecular velocity magnitude. Molecules farther from the surface than this will fail to reach the surface in unit time while all those within the volume, at or less than this distance away and moving toward the surface, will reach and collide with the surface during unit time. Thus the number of molecular impacts per unit time on unit area is

$$\upsilon = \frac{1}{6} nv \tag{A.5}$$

Appendix

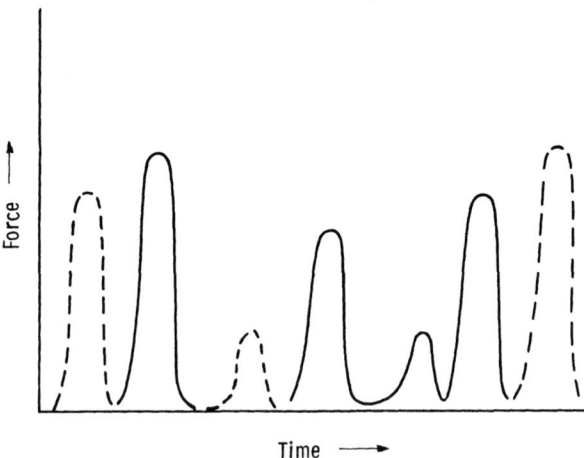

Figure A.1 Force due to individual molecules.

where n is the molecular density (average number of molecules per unit volume) and v is the molecular speed.

By a detailed analysis (omitted here for brevity), it can be shown that the pressure is simply the average *rate of change of momentum* due to the collisions between the moving molecules and the unit surface. Since perfect elasticity is assumed, each molecule rebounds with a momentum equal but opposite to its initial momentum, so each colliding molecule of mass m and velocity v experiences a net change in momentum of 2 mv. The total rate of change of momentum per unit area, or the pressure, is simply the product of the momentum change per molecule by the number of molecules striking unit area in unit time, or

$$p = 2mv \; \frac{1}{6} nv = \frac{1}{3} mnv^2 \tag{A.6}$$

This oversimplified procedure happens in this case to give exactly the same result as a rigorous derivation taking into account the actual random distribution of both speed and direction of the impacting molecules, that is, if v is the root-mean-square speed, to be designated hereafter by c.

It should be observed here that equation (A.5) is, however, not quite correct, but that a rigorous analysis considering random velocities gives for the important quantity, number of particles striking or crossing a unit surface in unit time,

$$\upsilon = \frac{1}{4} n\bar{c} \tag{A.7}$$

where $\bar{c}$ is the *average* molecular speed, to be defined later. If ρ is used for the average gas density,

$$\rho = nm \tag{A.8}$$

and equation (A.6) can be written

$$p = \frac{1}{3}\rho c^2 \tag{A.9}$$

which agrees with Boyle's law, pV = constant, for a perfect gas if it be assumed that c is constant when the temperature is constant and it is noticed that the specific volume V of anything is the reciprocal of its density.

The kinetic meaning of temperature may be deduced from other properties of a perfect gas. The Gay-Lussac or perfect gas law states that

$$pV = RT \tag{A.10}$$

where, if V is the volume of a kilogram-molecular weight of a gas and T is the absolute temperature, R is called the universal gas constant. Avogadro's hypothesis, based on observed gas behavior, states further that all gases at the same temperature and pressure have the same number of molecules per unit volume. Thus the mass of a given volume of gas is proportional to the molecular weight of the gas, or conversely, a kilogram-molecular weight, or mass, of any perfect gas will have the same volume under standard conditions. This kilogram-molar volume is 22.4 m^3 at 273 K and 1 atm and the fixed number of molecules it contains is 6.023×10^{26}, called the *Avogardo number*, to be designated by N_0. Under any nonstandard pressure and temperature conditions, the molecules per cubic meter, n, can be found by dividing N_0 by V, the kilogram-molecular volume under those conditions, or

$$n = \frac{N_0}{V} \tag{A.11}$$

Combining these last two equations gives

$$p = \frac{nRT}{N_0} = nkT \tag{A.12}$$

where the ratio $k = R/N_0$, applying to all gases, is known as *Boltzmann's gas constant*. Using the value of R in energy units, 8.31×10^3 J/K per kilogram mole, this constant has the value

$$k = \frac{8.31 \times 10^3}{6.023 \times 10^{26}} = 1.38 \times 10^{-23} \text{ J/K per molecule} \tag{A.13}$$

Combining equations (A.8), (A.9), and (A.12) leads to

Appendix

$$p = \frac{1}{3} mnc^2 = nkT \tag{A.14}$$

which shows on rearrangement that

$$\frac{1}{2} mc^2 = \frac{3}{2} kT \tag{A.15}$$

or the average molecular kinetic energy is proportional to the absolute temperature.

A.1.4 Molecular Velocities

Equation (A.15), rewritten in the form

$$c = \sqrt{\frac{3kT}{m}} \tag{A.16}$$

is useful for calculating the speed of gas molecules. For example, oxygen molecules, of mass

$$\frac{32}{6.023 \times 10^{26}} = 5.31 \times 10^{-26} \text{ kg}$$

have an rms speed at 273 K of

$$c = \sqrt{\frac{3 \times 1.38 \times 10^{-23} \times 273}{5.31 \times 10^{-26}}} = 461 \text{ m/s}$$

At any other temperature, the speed will vary with the square root of the absolute temperature, and for any other gas, the speed will vary inversely as the square root of the molecular weight. For mixtures of gases in thermal equilibrium (i.e., at the same temperature), equation (A.15) shows that

$$\tfrac{1}{2} m_1 c_1^2 = \tfrac{1}{2} m_2 c_2^2 \tag{A.17}$$

or

$$\frac{c_1}{c_2} = \sqrt{\frac{m_2}{m_1}}$$

a fact sometimes called the principle of *equipartition of energy*. The rms molecular velocity calculated by equation (A.16) is not far from the velocity of sound, the formula for which is

$$v_s = \sqrt{\frac{\gamma p}{\rho}} \tag{A.18}$$

Substituting values for p and ρ from equations (A.8) and (A.14) yields

$$v_s = \sqrt{\frac{\gamma nkT}{mn}} = \sqrt{\frac{\gamma kT}{m}} \qquad (A.19)$$

Comparison with equation (A.16) shows that

$$c = \sqrt{\frac{3}{\gamma}} \, v_s \qquad (A.20)$$

where γ, the specific heat ratio, c_p/c_v,* has the value 1.667 for monoatomic gases and about 1.4 for diatomic gases. Thus for diatomic gases like nitrogen and oxygen,

$$c = \sqrt{\frac{3}{1.4}} \, v_s = 1.46 v_s$$

or the rms molecular speed is some 46% greater than the speed of sound in the gas.

A.1.5 Distribution of Molecular Velocities

It can be shown by purely statistical methods that the stated assumptions about the nature of a gas lead to a definite velocity distribution. The random succession of elastic collisions may at one moment leave a particle almost stationary while at another moment the particle may be fortuitously driven to a speed far above the average, but most of the time its speed will fluctuate within a fairly narrow range not far from the average value. Figure A.2 shows such a distribution graphically, a total number of particles n being divided into groups of various sizes Δn having instantaneous speeds in equal successive ranges Δv from zero to infinity along the speed axis. The distribution can be shown more precisely by a smooth curve like that shown dashed, passing through the midpoints of the steps. This curve gives the step height, or Δn, as a function of the step position, or the speed v. Since the numerical value of Δn at any speed v is proportional to the total number of particles n and to the size of the speed interval Δv as well as to the relative height of the curve, it is convenient and customary to plot as ordinates of the distribution curve values of the ratio

$$\frac{\Delta n}{n \, \Delta v}$$

or in differential form (letting Δv and with it Δn approach zero)

$$\frac{dn}{n \, dv}$$

---

*c_p is the specific heat, the amount of heat required to change the temperature of a unit mass of a material by 1 degree, of a gas at constant pressure. c_v is the specific heat of a gas at constant volume.

Appendix

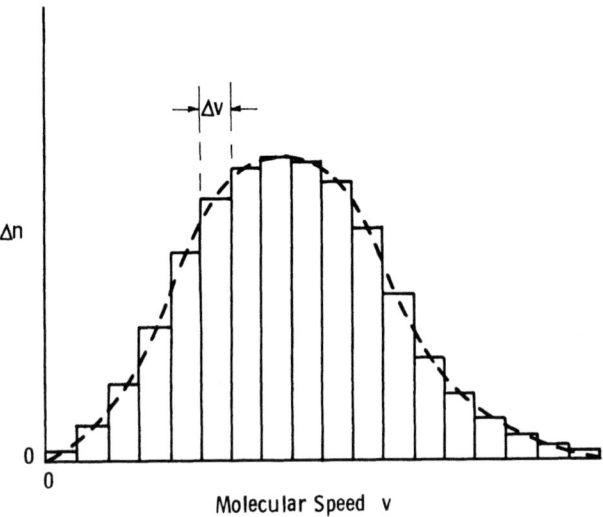

Figure A.2 Step-type graph showing distribution in speed of n gas molecules.

The statistical reasoning first carried out in somewhat different ways by Maxwell and Boltzmann leads to the definite functional relation

$$\frac{dn}{n\,dv} = f(v) = \frac{4v^2}{\sqrt{\pi}\,c_0^3}\,\varepsilon^{-v^2/c_0^2} \quad (A.21)$$

where ε is the natural logarithm base, 2.718, and c_0 is the value of v at which the distribution function has a maximum value and hence is the most probable molecular speed. In terms of gas temperature and molecular mass, it can be shown that

$$c_o = \sqrt{\frac{2kT}{m}} \quad (A.22)$$

and hence that

$$\frac{dn}{n\,dv} = 4\pi \left(\frac{m}{2\pi kT}\right)^{3/2} v^2 \varepsilon^{-mv^2/2kT} \quad (A.23)$$

The *average* molecular speed, defined by the integral relation

$$\bar{c} = \frac{1}{n} \int_{v=0}^{v=\infty} v\,dn \quad (A.24)$$

can be evaluated with the aid of the distribution function (A.23), obtaining

$$\bar{c} = \sqrt{\frac{8kT}{\pi m}} \tag{A.25}$$

By comparison with equation (A.22) it may be seen that

$$\bar{c} = \frac{2}{\sqrt{\pi}} c_0 = 1.129 c_0 \tag{A.26}$$

or the *average* speed is some 13% greater than the *most probable* speed. Similarly, by integration or by comparison between equations (A.22) and (A.16), it can be shown that

$$c = \sqrt{\frac{3}{2}} c_0 = 1.225 c_0 \tag{A.27}$$

or that the *root-mean-square* speed is 22.5% greater than the most probable speed.

Both the relative location of these values and the effect of temperature on the speed distribution in a given gas are illustrated by the example plotted in Fig. A.3, showing actual values of Δn for $n = 10^8$ molecules of argon at 300 K and at 900 K, with $\Delta v = 0.01$ m/s. This plot shows graphically just what is meant by mean or rms molecular speed in a gas at a given temperature. At the higher temperature the numbers of molecules in each speed interval at each corresponding part of the curve are reduced because the same total number are spread out over a larger range of speeds.

To express the distribution of molecular velocity *components*, for example along the x axis, a different function is required. This function, derivable in a manner similar to that for total speeds, is

$$\frac{dn}{n\, dv_x} = f(v_x) = \sqrt{\frac{m}{2\pi kT}}\; \varepsilon^{-mv_x^2/2kT} \tag{A.28}$$

and similarly for the y and z components of the total velocities. This can be expressed more simply in terms of the most probable total velocity magnitude, c_0, as follows:

$$p\left(\frac{v_x}{c_0}\right) = \frac{dn}{n}\, \frac{c_0}{dv_x} = \frac{1}{\sqrt{\pi}}\, \varepsilon^{-(v_x/c_0)^2} \tag{A.29}$$

This function, of dimensionless form, is plotted in Fig. A.4. It is necesarily symmetrical about the zero-velocity-component axis, since positive and negative components are equally likely, and unlike the total velocity distribution, it has its maximum (most probable value) at zero. As also shown in Fig. A.3, the number of particles having velocity

Appendix

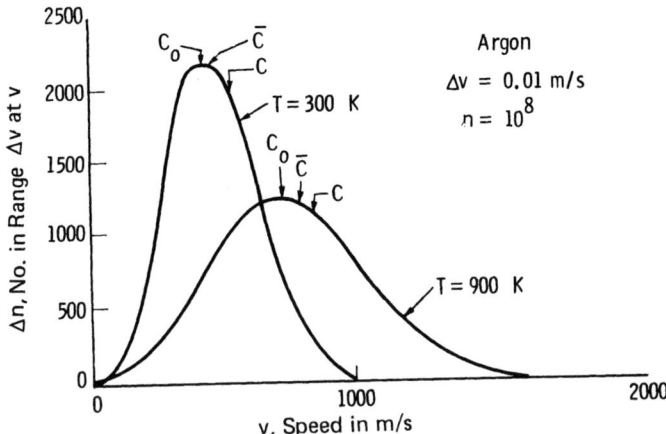

Figure A.3 Distribution of speeds for argon. (From Ref. 1.)

components or total velocities more than two or three times the most probable total speed are relatively very small.

These equations and curves describe what is generally known as either the *maxwellian* or *Maxwell-Boltzmann* velocity distribution. They are known to hold accurately for the molecules of a normal gas. They may also be nearly correct for the free electrons in a conducting gas, but conditions are often such that the maxwellian velocity distribution is considerably modified for electrons. The modification is especially drastic for the free electrons within a metal, a region outside the scope of this discussion.

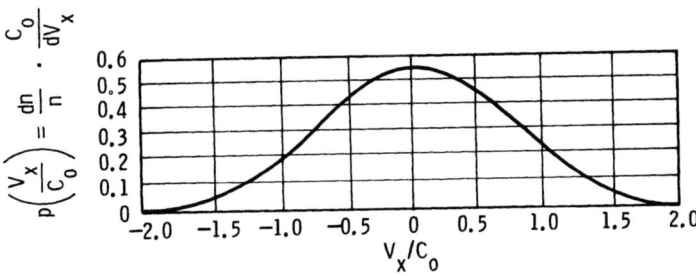

Figure A.4 Distribution curve for the x components of molecular velocities. (From Ref. 1.)

A.1.6 Free Paths of Molecules and Electrons

One of the most important quantities in the theory of gaseous conduction is the free distance through which particles, especially electrons or ions, may travel before being stopped or deflected by collision with a molecule of gas. Of course, the actual distance traveled between collisions in a gas will vary in a random manner, but the average distance, or *mean free path*, can be computed rather simply. The free path will obviously depend on the size of the moving particle and on both the size and mean density (number per unit volume) of the gas molecules. Two cases will be considered. The simplest is that of an electron, of relatively negligible dimensions, moving at such a velocity that the gas molecules may be considered as relatively stationary. The electron collides with any molecule whose center lies within one molecular radius of the electron path, or it collides with all molecules having their center within a circular cylinder about this path as an axis, of diameter equal to two molecular radii, or to the molecular diameter, to be called σ. The average number of collisions made in traveling unit distance, say 1 m, will then be equal simply to the average number of molecules within a cylinder of diameter σ and 1 m long; or collisions per meter =

$$\frac{\pi}{4} \sigma^2 n \qquad (A.30)$$

where n is the mean molecular density, or molecules per cubic meter. The mean free path between collisions, is, of course, only the reciprocal of this, or

$$\lambda_e = \frac{4}{\pi \sigma^2 n} \qquad (A.31)$$

where λ_e will be used to designate the electron free path as distinguished from λ, the molecular free path.

The molecular mean free path may be calculated in a similar manner, with two changes. In the first place, the region "swept out" by the moving molecule will be of diameter twice as great as before since the radius of interaction at collision is the sum of the radii of the two particles, or twice the radius of one. Second, the gas molecule is moving among other molecules of equal average speed, so its mean relative velocity of approach is increased by $\sqrt{2}$ due to motion across its path of the other molecules. Thus for a molecule moving among others of its own size and speed,

Collisions per meter = $\sqrt{2} \pi \sigma^2 n$

or

Appendix

$$\lambda = \frac{1}{\sqrt{2}\pi\sigma^2 n} \tag{A.32}$$

Comparison of the two formulas shows that

$$\lambda_e = 4\sqrt{2}\lambda \tag{A.33}$$

As an example of normal magnitudes, the mean free path of oxygen molecules ($\sigma = 3.6 \times 10^{-10}$ m) in oxygen gas at normal temperature and pressure

$$n = \frac{6.023 \times 10^{26}}{22.4} = 2.7 \times 10^{25} \text{ molecules/m}^3$$

is, according to formula (A.32),

$$\lambda = \frac{1}{\sqrt{2}\pi \times \overline{3.6}^2 \times 10^{-20} \times 2.7 \times 10^{25}}$$

$$= 6.4 \times 10^{-8} \text{ m}$$

This is nearly 200 times the molecular diameter or about 20 times the mean distance between oxygen molecules under standard conditions.

If n, the molecular concentration in equation (A.32), is replaced by the perfect gas law expression from equation (A.12),

$$n = \frac{p}{kT}$$

the resulting form,

$$\lambda = \frac{kT}{\sqrt{2}\pi\sigma^2 p} \tag{A.34}$$

shows the effect of gas temperature, pressure, and molecular diameter on the molecular mean free path. It should be noted particularly that the free path varies *directly* as the absolute temperature and *inversely* as the absolute pressure. High-vacuum conditions are said to exist in a vessel when the gas pressure, or density, is so low that the molecular mean free path is large compared with the principal dimensions of the vessel.

The mean free path, particularly of ions and electrons, is of fundamental importance in all calculations concerning conduction processes in a gas, but unfortunately, free path values are only imperfectly known. Because an atom, molecule, or ion is actually far more complex that the hard perfectly elastic sphere assumed in simple kinetic theory, the formulas above give only rough approximations to free path values actually observable in a number of indirect ways. The mean free paths of electrons are particularly indefinite, varying sometimes by several hundred percent as functions of electronic speed.

A.1.7 Distribution of Free Paths

As mentioned at the beginning of the preceding section, the free paths of particles moving in a gas are actually distributed at random between zero and values considerably greater than the mean. The distribution of these values about the mean can be determined by first considering the proportion of a given number of particles whose free paths will, on the average, *equal or exceed* a given distance, x. The free paths are plotted in Fig. A.5 as though they all started form the same plane and proceeded in the same direction, a fiction that does not alter the result to be obtained.

From such an arrangement, it can be seen that the number of particles n having free paths *at least as great as* x will equal the total number n_0 when x is zero and will continually decrease with increasing x as indicated by the dashed curve for the proportion n/n_0. The number of particles dn with free paths ending in any incremental distance dx will be proportional to n times dx, leading to the differential equation

$$\frac{dn}{dx} = -An \tag{A.35}$$

with the solution

$$n = n_0 \varepsilon^{-Ax} \tag{A.36}$$

where A, an arbitrary constant of proportionality, must be some function of λ, the mean free path, which is given by the defining equation

$$\lambda = \frac{1}{n_0} \int_0^{n_0} x(-dn) \tag{A.37}$$

Substituting the differential of the right-hand side of equation (A.36) for -dn in (A.37) leads to

$$\lambda = \int_0^\infty Ax\varepsilon^{-Ax} \, dx = \frac{1}{A} \tag{A.38}$$

Substituting in equation (A.36) gives the complete equation for n, the number of free paths reaching or exceeding x:

$$n = n_0 \varepsilon^{-x/\lambda} \tag{A.39}$$

The distribution function analogous to the one for molecular speeds is given by the differential form

$$f(x) = \frac{-dn}{n_0 \, dx} = \frac{1}{\lambda} \varepsilon^{-x/\lambda} \tag{A.40}$$

Appendix

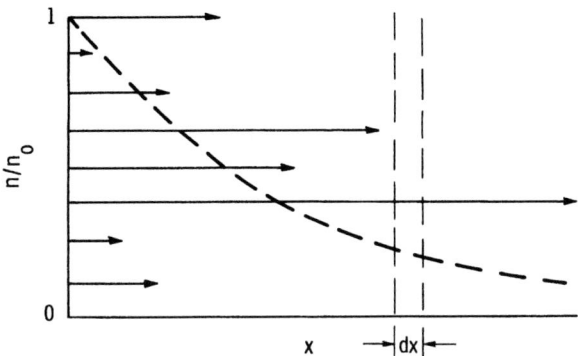

Figure A.5 Distribution of free paths.

This shows the somewhat surprising fact that the *most probable* free path length [at which f(x) is a maximum] is not close to the mean free path λ but is actually zero. Equation (A.39) shows also that 63% of all free paths are less than λ. The proportion of free paths exceeding λ by three times, for example, is only 5%, but these relatively few longer paths may be of considerable importance in many cases. This is especially true for electrons, which may gain energy in an electric field in proportion to their free paths in the field direction.

A.1.8 Diffusion

In addition to the expansion tendency, gases have the characteristic tendency to diffuse into each other as a result of the random motions of their molecules. Of particular interest is the diffusion of small quantities of gas, or impurities, into surrounding gas media, as for example the diffusion away of electrons and ions from conducting paths in gases.

Diffusion is a net transport of particles of a given kind across a surface resulting from a *gradient* in density or temperature of these particles perpendicular to the surface. The laws of diffusion of heat are identical with those for diffusion of impurities, but we will be concerned principally with the latter, the temperature gradient being tacitly neglected. From the definition above, a steady-state equation,

$$\eta_x = D\left(-\frac{dn}{dx}\right) \tag{A.41}$$

can be written in which η is the net diffusion in particles per square meter per second and D is called the diffusion coefficient. D can be evaluated by means of formula (A.7), which gives the rate at which gas particles cross a surface *in one direction*. Since the particles reach the surface after having made collisions on the average one mean free path away, the respective numbers arriving from opposite directions will be determined by the respective average densities at the previous collision points, or at surfaces plus and minus λ away from the surface in question, as illustrated by Fig. A.6. Thus the *net* flow from left to right across the plane at x is given by

$$\eta = \frac{1}{4}\bar{c}n_{x-\lambda} - \frac{1}{4}\bar{c}n_{x+\lambda}$$

$$= \frac{\bar{c}}{4}\left[n_x - \lambda\frac{dn}{dx} - \left(n_x + \lambda\frac{dn}{dx}\right)\right]$$

$$= \frac{\bar{c}}{4}\left(-2\lambda\frac{dn}{dx}\right) = -\frac{\lambda\bar{c}}{2}\frac{dn}{dx} \tag{A.42}$$

By comparison with equation (A.41),

$$D = \frac{\lambda\bar{c}}{2} \tag{A.43}$$

A more elaborate derivation, taking into account the distribution of the free paths in direction, yields the exact expression

$$D = \frac{\lambda\bar{c}}{3} \tag{A.44}$$

From the relations given previously for λ and $\bar{c}$, formulas (A.34) and (A.25), respectively, it may be seen that the diffusion coefficient should vary as

$$\frac{T^{3/2}}{p}$$

In addition to steady-state diffusion, *transient* diffusion is often important in gaseous conduction theory. The transient diffusion of an impurity in a gas is exactly analogous to the transient diffusion of heat, being governed by the readily derivable partial differential equation

$$\frac{\partial n}{\partial t} = D\left(\frac{\partial^2 n}{\partial x^2} + \frac{\partial^2 n}{\partial y^2} + \frac{\partial^2 n}{\partial z^2}\right) \tag{A.45}$$

In most applications, curves of density distribution along the important direction will be convex upward, so that the second derivatives are negative and hence diffusion results in a rate of *loss* of density.

A relation often useful in approximate transient diffusion calculations is that for the *mean square distance* of particles diffusing from a point, line, or plane of instantaneous origin. At any time t after the initial relase of the particles, the mean square distance to all the particles is given by the simple relations

Appendix

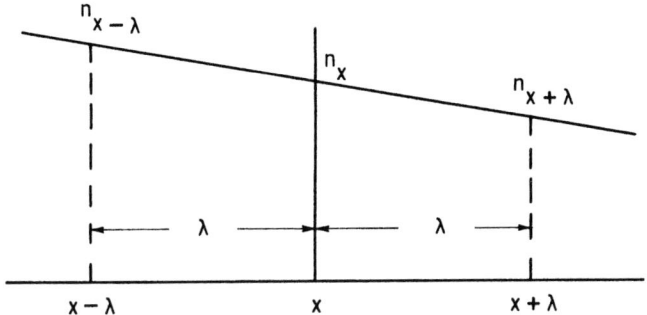

Figure A.6 Particle density plot in steady diffusion.

$$\overline{r^2} = \overline{r_0^2} + 6Dt \qquad (A.46)$$

for radial diffusion from a point source,

$$\overline{r^2} = \overline{r_0^2} + 4Dt \qquad (A.47)$$

for radial diffusion from a line source, and

$$\overline{x^2} = \overline{x_0^2} + 2Dt \qquad (A.48)$$

for diffusion from a uniform plane source. In many cases the initial values $\overline{r_0^2}$ or $\overline{x_0^2}$ may be neglected, leaving equations of extreme simplicity. These equations then state that the root-mean-square distance of diffusing particles from their source increases as the square root of the time.

A.2 MOVEMENT OF IONS AND ELECTRONS

A.2.1 Nature and Function of Ions

As already stated in Sec. A.1.1, the current carriers in a gas are particles of two principal types: *free electrons* and *ions*. They are formed from normally neutral atoms or molecules by processes which remove one or more of the bound orbital electrons of the neutral particles, yielding free electrons and *positive ions*, the latter retaining practically all of the original molecular mass. Most positive ions have lost only one electron and so have a net positive charge equal in magnitude to that of the negative electron, or 1.6×10^{-19} C. In addition, some gases called electronegative may form *negative* ions through attachment of extra electrons to neutral atoms or groups of atoms.

For an ion, the quantitative property of principal interest in gaseous conduction is the ratio of charge to mass, since this determines the rate of acceleration of the ion in an electric field. This ratio is radically greater for free electrons of only 9.11×10^{-31} kg mass than for atomic ions. A single-atom hydrogen ion has a mass which is 1823 times that of an electron, or 1.66×10^{-27} kg. Atomic ions of other gases are heavier in proportion to their relative atomic weights. Because of this great disparity in charge-to-mass ratio, electrons and ions generally play essentially different roles in gaseous conduction processes. It can be shown that the light and so comparatively mobile free electrons carry all but a very small fraction of the current in any ionized gas space, which, like the column of an arc, has appreciable conductivity in a practical sense. On the other hand, the principal function of the slower positive ions is neutralization of the "space charge" of the conduction electrons, or reducing the net charge per unit volume of the gas space near to zero. This charge neutralization is essential for the existence of high conductivity—current flow with only moderate voltage gradient. For example, high-vacuum electron tubes, employing only electrons as charge carriers, can carry only small currents at relatively high voltages because of the space-charge limitation. On the other hand, gas-filled tubes, Thyratrons, Ignitrons, and the like, carry relatively large currents at moderate voltages because positive ions produced in the gas are available for space-charge neutralization. In too great density, however, the positive ions may reduce electron mobility and so actually impede current flow.

A.2.2 Mobility in the Electric Field

The ratio of the mean drift velocity of a charge carrier through a gas in the direction of the electric field to the strength of the field is generally called the *mobility* of the carrier. In free flight through the empty space between the molecules of a gas, a carrier of mass m and charge e obeys the newtonian equation

$$m \frac{dv}{dt} = eE$$

or

$$\frac{dv}{dt} = \frac{e}{m} E \qquad (A.49)$$

when E is the electric field strength and v is the velocity of the particle. The charge-to-mass ratio, e/m, of the carrier is of basic importance in the theory of gaseous conduction. For an electron it has the value 1.76×10^{11} C/kg. This means that even in so weak a field as 1 V/m, the acceleration experienced by an electron is enormously greater than the acceleration due to gravity—nearly 20 billion times as great. Thus gravitational forces can almost always be neglected com-

Appendix

pared with the electric or magnetic forces on elementary charged particles.

For charged particles drifting through a gas under the influence of a relatively weak electric field—a field that does not appreciably raise the mean energy or temperature of the charged particles above that of the neutral gas molecules—an approximate expression for the mobility can be derived very simply. If the mean thermal velocity is $\bar{c}$ and the mean free path between collisions is λ, the mean time between collisions is

$$\bar{t} = \frac{\lambda}{\bar{c}} \tag{A.50}$$

If the mean velocity component in the direction of the field is presumed to be lost, or to return to zero at each collision, the velocity attained in the average case at the end of a free path is, from equation (A.49),

$$v = \frac{dv}{dt}\bar{t} = \frac{eE}{m}\bar{t}$$

with the mean value

$$\bar{v} = \frac{eE}{2m}\bar{t} \tag{A.51}$$

Combined with equation (A.50),

$$\bar{v} = \frac{eE\lambda}{2m\bar{c}} \tag{A.52}$$

From this, the mobility, designated as μ, is

$$\mu = \frac{\bar{v}}{E} = \frac{e\lambda}{2m\bar{c}} \tag{A.53}$$

More elaborate derivations, taking into account the distribution of free paths in length and direction and the persistence of attained velocity between collisions, give more exact values. A useful approximation to these is simply

$$\mu = \frac{e\lambda}{m\bar{c}} \tag{A.54}$$

with the units meters per second per volt per meter, or meters squared per second-volt. This is a "low-field" value for the mobility.

In most cases the complicating effect of the field on the mean velocity $\bar{c}$ of the charged particles must be taken into account. Because the charge carriers receive energy from the electric field, they tend to reach a mean equilibrium energy higher than the mean energy of the neutral gas molecules. In a somewhat more exact form,

$$\mu = \frac{e\lambda}{mc}\left(1 + \frac{mW_m}{MW_m}\right) \tag{A.55}$$

where m and W_m are, respectively, the mass and mean energy of the charge carriers, and similarly, M and W_M are the mass and mean energy of the neutral particles. The rms velocity c is that of the carriers, obtainable from the relation

$$W_m = \frac{1}{2} mc^2 \tag{A.56}$$

For the important case of free electrons moving through gas molecules with which they make perfectly elastic collisions, the equilibrium energy of the electrons is related to the field strength by the approximate relation

$$W_m = \frac{W_M}{2} + \sqrt{\left[\frac{W_M}{2}\right]^2 + \left[\frac{eE\lambda}{2.3}\sqrt{\frac{M}{m}}\right]^2} \tag{A.57}$$

Combining these equations, noting that for electrons

$$mW_m \ll MW_m \tag{A.58}$$

and slightly rounding off the constants, one obtains the relation

$$\mu_e = \frac{e\lambda}{\sqrt{2mW_M}} \left\{ \frac{2}{1 + [1 + \frac{M}{m}(eE\lambda/W_M)^2]^{1/2}} \right\}^{1/2} \tag{A.59}$$

At small enough values of the field strength E, the expression within the braces approaches unity and the electron mobility approaches the field-independent value given by equation (A.54). For relatively large values of E on the other hand, equation (A.59) reduces to

$$\mu_e \approx \sqrt{\frac{e\lambda}{\sqrt{mME}}} \tag{A.60}$$

in which the mobility varies inversely as the square root of the field strength. Values from equation (A.59) are plotted as the solid curve in Fig. A.7. The circled points represent experimental values of electron mobility in nitrogen determined by Wahlin [2]. They define a curve of the same general shape but about 100% higher than the computed curve. Both the small initial "constant" region and the more extensive field-dependent region of the relation are illustrated by both the theoretical and the experimental values. The dashed-line curve shows values given by the simpler equation (A.60), which is clearly adequate for field strengths of 1000 V/m or higher. In the case of hydrogen, Wahlin found values lying below rather than above the curve by equation (A.59), but again following a roughly parallel course. The discrepancies can be ascribed almost entirely to the uncertainty in the proper values of λ, the electronic mean free path. The free paths are somewhat indefinite because molecules are not actually hard,

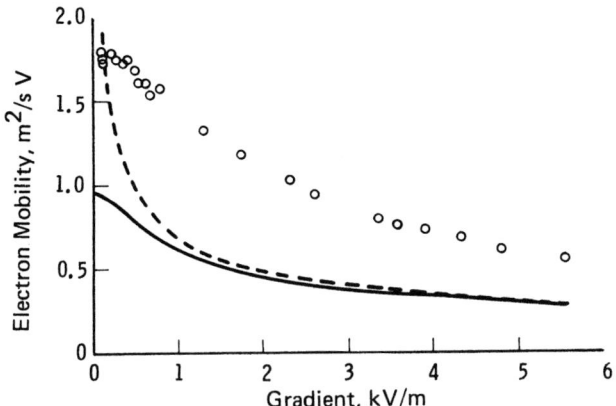

Figure A.7 Experimental values of electron mobility in nitrogen under standard conditions. Solid curve from theoretical equation (A.59); dashed curve from approximate equation (A.60) for high-field mobility. (From Ref. 2.)

smooth spheres as assumed in the calculations.

For very much higher ratios of field strength to mean free path, where the energy gained between collisions is enough for ionization or excitation of the molecules struck, the collisions will be mainly inelastic (with little or no "bounce") and a different type of calculation is required in estimating the electron mobility. Such a calculation leads to

$$\mu_{e_H} = \sqrt{\frac{2e\lambda}{\pi mE}} \quad (A.61)$$

which, because of the absence in the denominator of M, the molecular mass, gives much higher mobility values than equation (A.60). Thus it is true even though seemingly paradoxical that energy loss in inelastic collisions may serve to speed up the net motion of electrons in the direction of the applied electric field.

The mobility of molecular ions can be calculated by means of a formula like (A.59) but somewhat more complex. Discrepancies and uncertainties are still greater for ions, however, because of such complications as charge transfer between ions and molecules as well as uncertainty in path or effective cross-sectional area. In general, ion mobilities are much less than electron mobilities because of the very much greater mass of molecular or atomic ions. As indicated in the electron mobility formulas (A.59) and (A.61), the mobility tends to be reduced in proportion to the square root of the mass ratio.

In the discussion above it has been assumed for simplicity that the degree of ionization of the gas was so small that an electron or ion encountered for the most part only neutral atoms or molecules in moving through the gas. This situation is typical of relatively low-current discharges like glows or arcs of moderate current intensity in low-gas-pressure discharge tubes such as Thryatrons, Ignitrons, or fluorescent lamps, but does not hold true for high-current arcs in gases at atmospheric or higher pressures as found in circuit breakers interrupting currents of hundreds, thousands, or tens of thousands of amperes. In such high-current, high-pressure arcs, studies indicate that from 10 to 100% of the gas molecules are ionized. When the gas through which the free conduction electrons are moving contains a large proportion of *charged* particles, the mean free path of the electrons is drastically reduced. This is because the electric field surrounding the charged ions results in a much larger interaction distance between electrons and *ions* than between electrons and neutral particles. This effect is only partially offset by the higher temperatures of such arcs, which tend to reduce the number of particles per unit volume.

In computing the mean free path of an electron through a highly ionized gas, one must take into account collisions with both neutral particles with density n_0 and positive ions with density n_+. If the effective collision cross-sectional areas are expressed as A_0 and A_+, respectively, the average number of collisions made by an electron per meter of path length will be

$$\frac{1}{\lambda_e} = n_0 A_0 + n_+ A_+ \tag{A.62}$$

so

$$\lambda_e = \frac{1}{n_0 A_0 + n_+ A_+} \tag{A.63}$$

[similar to formulas (A.30) and (A.31)]. The molecular area A_0 is simply the kinetic theory cross section $(\pi/4)\sigma^2$, where σ is the molecular diameter. The effective collision cross section of the ions, however, must take into account deflections of the moving electrons by the charged ions at mean distances large compared with their normal collision radii. These effective distances depend also on the mean velocities of the particles, or on the gas temperature. Thus, at high ion densities, the electron mobilities are very considerably reduced by the electric fields surrounding the oppositely moving positive ions. This effect serves to reduce the electrical conductivity of highly ionized gases like arc columns by reducing the effective mean free paths to be used in electron mobility formulas like (A.54), (A.59), and (A.61). The reduction in mean free path and hence in mobility depends also on the degree of ionization, $n_+/(n_+ + n_0)$, as shown by formula (A.63).

Appendix

When the ionization approaches 100%, the electron mobility may be only 1% of what it would be in the absence of this electrical effect of the positive ions. In such a situation electron mobility depends only on ion density and temperature—not at all upon the identity of the ionized gas.

A.2.3 Deflection by Magnetic Field

A magnetic field acts only on moving charged particles in a gas, exerting a force on such particles proportional to their velocity multiplied by the component of field strength at right angles to their direction of motion. Unlike the electric field, a magnetic field cannot impart energy to moving charges but can only change their direction, because the force due to the field is always perpendicular to the motion. Under "vacuum" conditions (long free paths), a magnetic field may cause charged particles already possessing kinetic energy to move in circular or helical paths with radii determined by the equality between the centrifugal force mv^2/r and the electromagnetic force veB, where m and e are the mass and charge of the particle; v its velocity; B the magnetic field component perpendicular to the motion, or to the plane of motion in the circular case; and r is the radius of curvature of the path or radius of the circle. From this equality

$$r = \frac{mv}{eB} \tag{A.64}$$

and the angular velocity is given by the very simple relation

$$\omega = \frac{v}{r} = \frac{e}{m} B \quad \text{rad/s} \tag{A.65}$$

Where the orbital radius r is less than the mean free path length, the effect of a magnetic field is to inhibit appreciable progress of charged particles across lines of magnetic flux while permitting movement along the flux lines. Gas discharge tubes utilizing this effect have been built for making the shape of magnetic fields visible. Equation (A.65) is the basis for operation of the cyclotron. Equation (A.64) describes also the magnetic deflection of cathode rays in oscillograph or television tubes.

Where free paths between collisions are much shorter than the radii given by equation (A.64), the motion of charged particles through a gas is little affected, but through multiple collisions the electromagnetic force is immediately transmitted to the gas as a whole, just as it is in the case of electrons flowing in a metallic wire. Thus an arc column at atmospheric gas pressure is deflected in a transverse magnetic field like any flexible current carrying conductor, but distribution of current within the arc column is not appreciably changed.

A.2.4 Space-Charge Effects

In considering the motions of ions and electrons in electric fields, a frequent situation of special interest occurs when the field is due to or modified by the presence of space charge, that is, a net excess of charge density, either positive or negative. Such a field varies in accordance with Poisson's equation, which is, in the simplest case of plane symmetry,

$$\frac{\partial^2 V}{\partial x^2} = -\frac{\rho}{\varepsilon} \qquad (A.66)$$

where V is the electric potential, x a distance perpendicular to the plane of symmetry, ρ the net charge density at x, and ε the dielectric constant, having the value $10^{-9}/36\pi \cong 8.85 \times 10^{-12}$ F/m for empty space in the Giorgi system of units. In combination with the electric field equations

$$E = -\frac{\partial V}{\partial x} \qquad (A.67)$$

and

$$D = \varepsilon E \qquad (A.68)$$

equation (A.66) becomes

$$\frac{\partial D}{\partial x} = \rho \qquad (A.69)$$

where D is the electric displacement, or electric flux density. Equation (A.69) can be simply illustrated by Fig. A.8, in which the horizontal lines represent the flux lines of density D, increasing with the distance x as lines originate on elements of positive charge (ions) with mean charge density ρ.

Clearly, the introduction of ions between the charged parallel plates causes a considerable field distortion, as shown in Fig. A.9. With a given applied potential difference, there is a definite amount of positive charge per unit plate area, which will reduce the field strength at the positive plate to zero. With still higher ion density a zero-field region containing negative as well as positive ions may extend beyond the positive electrode and even fill most of the space. Such a region, containing nearly equal average densities of positive and negative charge, is called a "plasma." As illustrated in Fig. A.10, it may act like a positive electrode when it exists near a negative electrode. In this situation, the plasma boundary automatically adjusts itself so that the gradient at the boundary is zero. Ions are supplied to the "space-charge sheath" in front of the negative electrode only by diffusion across the field-free boundary. Positive ions crossing this boundary are mostly swept through the sheath by the field existing there, con-

Appendix

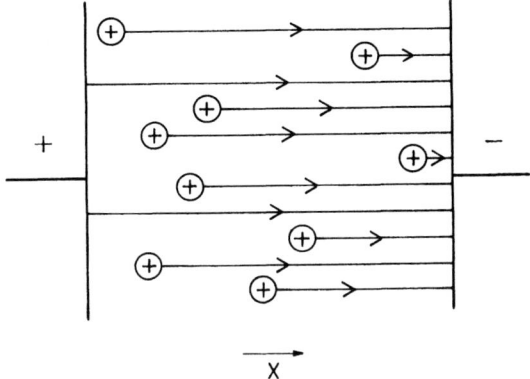

Figure A.8 Effect of space charge on electrical field strength.

stituting a current density, while negative particles (if electrons) are turned back into the plasma. The thickness x_0 of the sheath can be related to the diffusion current density i and the potential difference V between the plasma and electrode by solving the differential equation (A.66) in combination with the relation between average ion velocity and potential gradient E:

$$\bar{v} = \mu E \qquad (A.53)$$

where μ is the ion mobility discussed in Sec. A.2.2. This solution can be written in convenient form:

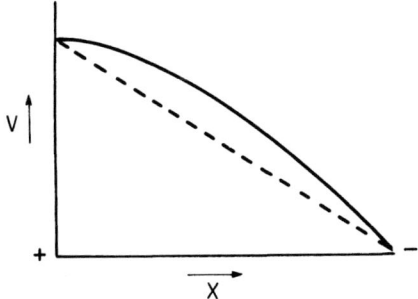

Figure A.9 Potential distribution between parallel plates. Solid curve with positive ions present; dashed line without ions.

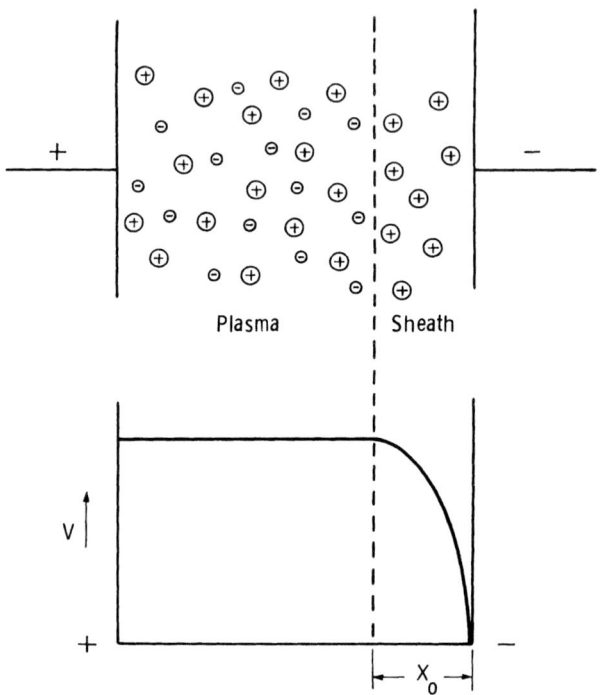

Figure A.10 Ion and potential distributions in a highly ionized gas near a negative electrode.

$$j = \frac{9\epsilon\mu V^2}{8x_0^3}$$

$$\approx \frac{10^{-11}\mu V^2}{x_0^3} \text{ A/m}^2 \tag{A.70}$$

This shows that for a given diffusion current and mobility (assumed constant), the automatically adjusted space-charge-sheath thickness x_0 is proportional to the two-thirds power of the potential difference V across the sheath and to the cube root of the diffusion current density. The latter can be computed with the help of an equation in Sec. A.1,

$$\upsilon = \frac{1}{4}n\bar{c} \tag{A.7}$$

for the rate at which particles cross a unit surface due to thermal motion only. If n_+ is the density at the sheath boundary of positive

Appendix

ions with charge e and mean thermal velocity $\bar{c}$, the "diffusion" current density is

$$j = \frac{1}{4} n_+ e \bar{c} \qquad (A.71)$$

and combination with equation (A.70) gives for the sheath thickness

$$x_0 = \left[\frac{9\varepsilon\mu}{2en_+\bar{c}}\right]^{1/3} V^{2/3} \text{ m} \qquad (A.72)$$

The distribution of potential gradient through this sheath may be obtained by differentiating V with respect to x in equation (A.70). It starts at zero at x = 0, the positive boundary, and reaches a maximum at the negative electrode, $x = x_0$, of $3V/2x_0$, 50% greater than the average gradient through the sheath.

Equation (A.70) may be called the "high-pressure" form of the space-charge equation. For the "high-vacuum" case, where ions pass through the sheath without colliding with gas molecules, solution of the equation gives the somewhat different form

$$j = \frac{4\varepsilon}{9} \sqrt{\frac{2e}{m}} \frac{V^{3/2}}{x_0^2} \qquad (A.73)$$

which applies to space-charge-limited electron currents in vacuum tubes or to ion sheaths smaller than a mean free path in thickness. In this case the maximum gradient at the negative electrode is only 4/3 of the average.

A.2.5 Diffusion in an Electric Field: Ambipolar Diffusion

In the preceding section a sharp division was tacitly assumed between the field-free "plasma" region where diffusion alone determines net charge flow and a space-charge region in which only the existing electric field acts on the ions present. In some cases both influences must be considered in combination. An example is the region next to an insulating wall or at an insulated electrode immersed in the "positive column," or main body, of a low-pressure glow discharge. If, as is commonly the case, the gas when ionized forms only positive atomic ions and free electrons, there will be a great disparity between the diffusion rates of the two types of charged particles primarily because of their greatly different masses. If thermal equilibrium exists, equation (A.17) shows that the rms velocities (and also the mean velocities) must vary inversely as the square roots of the particle masses. Even in helium gas this means that the electrons move on the average about 85 times as fast as the positive atomic ions or the neutral atoms. Because of the difference also in the mean free paths of electrons and ions [see equation (A.33)], the diffusion coefficients [$D = (1/3)\lambda \bar{c}$] will have the ratio

$$4\sqrt{2} \times 85 = 480$$

so that the electrons tend to diffuse out of the plasma immediately, leaving the positive ions behind. However, as soon as a charge density unbalance occurs Poisson's equation (A.66) or (A.69) shows that electric gradients automatically appear to check the charge separation. With zero net current flowing to the insulated surface, equilibrium exists when there is a retarding potential across a space-charge layer just sufficient to balance the rates of diffusion to equality by turning back almost all electrons before they reach the surface of the wall or electrode. This action has two principal results: (1) the insulated surface "floats" at some value of potential *negative* with respect to the adjacent plasma, and (2) the net rate of loss of ionization by diffusion from a plasma is governed by the diffusion rate of the *slower* particles, the positive ions. Because of interaction effects, this "ambipolar" diffusion may be shown to have a diffusion coefficient approximately twice that for the positive ions alone. The retarding potential difference may be calculated with the help of Boltzmann's relation, the diffusion formula (A.7), and the velocity formula (A.25), with the result

$$\Delta V = \frac{kT_e}{2e} \log_\varepsilon \left(\frac{T_e}{T_+} \frac{m_+}{m_e} \right) \text{ V} \tag{A.74}$$

when T_e is the electron temperature, which is generally higher than T_+, the temperature of the positive ions. m_+/m_e is the ratio of ion mass to electron mass. For atomic helium ions this mass ratio is about

$$4 \times 1820 = 7280$$

The thickness of the space charge layer may be estimated by means of equation (A.72) if the ion density n_+ in the plasma is known. The value ΔV represents the error in insulated "probe" measurements made, for example, to determine the potential of an ionized gas at a given point. The probe electrode will be more negative than the adjacent plasma by the amount ΔV, which may easily amount to 10 V or more. Fairly accurate measurements of plasma potential, and of electron temperature and ionization density as well, can be made by measuring the varying currents drawn by a probe connected to a source of varying potential. Equation (A.74) represents the zero-current point on a curve of probe current vs. probe potential.

Appendix

A.3 IONIZATION AND DEIONIZATION

A.3.1 Atomic Structure and Ionizing Processes

In visualizing the processes by which ions and free electrons are produced in a gas, and also how they may disappear, an approximate picture of the internal structure of an atom is helpful. Much progress has been made in the detailed understanding of this structure and even in unraveling some of the mysteries of its energy-packed heart, or nucleus.

For our purposes, however, it is sufficient to employ only the most elementary picture or "model" of the atom as devised about 1920 by Niels Bohr. This model was based on the prior discovery by Lord Rutherford that practically all of the mass of an atom is concentrated at its center in a very small positively charged nucleus surrounded by very much less dense negative charge. Bohr assumed that the negative charge consisted of electrons sufficient in number to balance the positive nuclear charge, revolving in orbits about the nucleus much as the planets revolve about the sun but held by electrical force rather than gravity. Bohr's key assumption underlying this model was that the electrons could move only in certain orbits for which their angular moments of momentum were restricted to integral multiples of a certain elementary amount,

$$\frac{h}{2\pi}$$

where h, known as *Planck's constant*, appeared in the relation

$$e = h\nu \tag{A.75}$$

expressing the energy e of a "quantum" of energy associated with electromagnetic radiation of frequency ν according to Max Planck's quantum theory of radiation. The unit of radiation carrying the energy quantum e is called a *photon*; so equation (A.75) gives the energy of a photon whose frequency is ν. Depending on the frequency, such a photon may be a unit of infrared or heat radiation, visible light, ultraviolet light, x-rays, or gamma rays. The importance of these photons is ionizing and deionizing processes will be explained later.

As an example, the simplest atom, that of hydrogen, is made up according to the Bohr theory, of two parts, a nucleus, consisting only of a single proton, and a single electron revolving around the nucleus as a satellite in a circular orbit. The proton possesses essentially the whole mass of the hydrogen atom, 1.67×10^{-27} kg, and a positive electric charge equal in magnitude to that of the electron, 1.6×10^{-19} C. Such an electromechanical system could readily be analyzed by the elementary laws of physics. For such an atom to exist as a stable entity, however, one additional postulate was necessary: that the very rapidly rotating electron would not radiate electromagnetic energy (as required by Maxwell's electromagnetic theory) so long as it remained in any one of the particular orbits having integral numbers of the elementary unit of angular momentum. Calculations show that such an atomic "solar system" has a total energy, sum of potential and kinetic energies, dependent on the radius of the electronic orbit. The lowest possible energy, actually a negative quantity according to classical definitions, is possessed by the system when the electron is revolving in its orbit of smallest possible radius where the integral number of momentum units is *one*. This minimum energy state is the *normal* state; the other possible states of larger energy associated with larger electron orbits are said to be excited states.

Generally, these excited states of an atom can exist only very briefly, usually for 10^{-8} s or less. After an excitation the electron drops spontaneously to a smaller, lower-energy orbit and the energy difference is radiated as a photon whose frequency depends on the energy according to equation (A.75). This radiation of discrete energy amounts or definite frequencies explains the "line" spectrum of radiation from a glowing excited gas. The process is generally reversible—photons can also be absorbed by a gas with the production of excited atoms or molecules. There are some exceptions to the normal brevity of excitation, the so-called *metastable* states, in which the energy is "locked up" until released by interaction with another molecule of the gas or a solid surface. Metastable atoms often play important parts in the ionization of gas mixtures or in releasing electrons from surfaces.

The *ionization* of an atom occurs as the extreme case of excitation, when one of the orbital electrons is moved so far from the nucleus that it is completely lost, leaving the atom with a net positive charge, in which state it is called a positive ion. Since most atoms have a number of orbital electrons in their structure, positive ions may have a charge equal to that of just one electron (single ionization) or of more than one electron (multiple ionization), but single ionization is by far the most likely and will be assumed unless otherwise stated. In the case of the hydrogen atom, the relatively simple two-body system of proton and electron can be completely solved and energies of excitation and ionization precisely calculated. The values, expressed in electron volts (energies associated with the change of one electronic charge from one potential to another through a certain number of volts potential difference), are 10.2 V for the first excitation potential and 13.6 V for the ionization potential. These represent the amounts of energy that must be internally absorbed by the hydrogen atom to raise it to its first or lowest state of excitation and to ionize it, respectively.

This atom model, then, helps to explain the principal important properties of gas atoms or molecules when the gas has electrical conductivity, or is in the process of gaining or losing conductivity. In the first place, ordinary collisions between atoms in a gas at normal temperatures are perfectly elastic because their mean kinetic energies

Appendix

are only a few hundredths of an electron volt, far below the threshold for internal absorption of energy to produce either excitation or ionization. In other than hydrogen atoms the minimum excitation potentials range mostly from about 1.5 to 20 V and the ionization potentials from 4 to 25 V. For a given atom, the ionization energy (or "potential") is approximately twice the first or minimum excitation energy.

A.3.2 Ionizing Agents

The ions and electrons needed to make a gas conducting may come from the gas itself or from containing walls or electrodes, and can be produced in a number of different ways.

Radiation It has already been mentioned that photons, units of radiated energy, can be absorbed in a gas. For ionization to be produced in this way the energy possessed by the photons, given by equation (A.75), must generally be equal to or exceed the minimum ionization energy of the gas molecules. The possible radiation frequencies for direct ionization may be obtained from the relation

$$h\nu \geqslant eV_i \tag{A.76}$$

where e is the electronic charge and V_i is the ionization potential. Generally speaking, only ultraviolet light or radiation of still shorter wavelength (higher frequency), x-rays or gamma-rays, possess enough energy per photon for direct ionization of gases. In gas mixtures it is possible for radiation coming from excited atoms of one gas to ionize atoms of a second gas if the second gas has an ionization energy below an excitation level of the first gas. For a penetrating radiation, the ionizing activity, rate of formation of ion-electron pairs per unit volume, is proportional to the product: radiation intensity times gas density. The activity is also a function of the radiation frequency, or wavelength, being a maximum when equation (A.76) is near equality and decreasing as

$$h\nu = \frac{hc}{\lambda} \text{ exceeds } eV_i$$

(c = velocity of light, λ = wavelength of radiation).

Of greater practical importance in most cases is the photoelectric effect at *surfaces*, the release of electrons from surfaces by incident photons. The required energy level for this action is given by an inequality of the same form as equation (A.76) but with the work function, usually designated by ϕ, replacing the ionizing potential V_i. Values of ϕ vary from 1.9 to 6.3 V, a range considerably lower than that of ionizing potentials for gases. Thus electrons may be introduced into a gas from a surface by radiation of much lower frequency or longer wavelength than that required to ionize gas atoms.

Instruments used to detect and measure radiation by means of its ionizing effect are Geiger-Müller counters, photoelectric cells, and the gold-leaf electroscope. Some low level of ionization is always present in a gas due to penetrating radiation coming from natural earth radioactivity and cosmic rays from outer space.

Chemical Action Ionization may occur as a result of chemical action. It is observed especially in flames, but may also be detected in air that has come into contact with moist phosphorus or has merely been bubbled through water. The ionizing action of water is believed to be an essential element in production of atmospheric electric effects, especially lightning.

Diffusion and Convection Under certain circumstances, diffusion and convection, ordinarily thought of as deionizing processes, may introduce ionization into a region. For example, ionization produced by a flashover arc to one wire of a transmission line may easily diffuse or be blown into the vicinity of an adjacent wire, causing one or more additional flashovers.

Excited Atoms An excited atom, particularly one in a semipermanent or metastable state, may possess enough stored energy to ionize on contact an atom of some more easily ionized gas, or such an atom may on contact eject an electron from a solid or liquid surface. This is analogous to the effect of radiation from excited atoms discussed above, but occurs as a result of low-speed collisions of excited atoms with gas molecules or with surfaces rather than by radiation. According to one theory, this ionizing process may be of primary importance in freeing electrons at metallic arc cathodes.

High-Speed Electrons Of most importance in electrical discharges is the ionizing activity of electrons which have been accelerated to sufficient kinetic energy by an electric field. The sufficient energy for the electron is, of course, eV_i, the minimum ionizing energy of a gas molecule or atom. As illustrated by Fig. A.11, the probability of ionization occurring at a given collision increases at first linearly with the excess amount of its kinetic energy over eV_i but reaches a maximum at several times eV_i and falls off for still greater energies. In Fig. A.11 ionizing activity is expressed as average number of ionizing collisions per centimeter of path length through gas at a given density, that existing at 1 mm Hg pressure and 0°C. At any other pressure and temperature both the total number of collisions per unit path length, $1/\lambda_e$, and the number of ionizing collisions per unit path length are proportional to the gas density. In Fig. A.11 the curve marked Hg^+ is for single ionization, removal of one planetary electron from a neutral mercury atom, and the curve marked Hg^{2+} is for double ionization, removal of two electrons. Ions with still alrger numbers of positive charges may be produced by high-energy electrons but with progressively smaller probabilities. The minimum energy for single ionization of mercury is

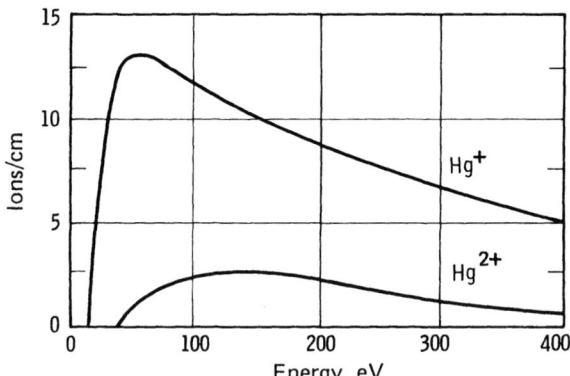

Figure A.11 Ionization rates for an electron in mercury vapor at 1 mm Hg pressure, 0°C. (From Ref. 3.)

10.4 eV, for double ionization 29.0 eV and for triple ionization 71.0 eV. Ionization by field-accelerated electrons is discussed in Chap. 8.

High-Speed Ions High-energy ions can also ionize neutral particles upon collision, but the probability of such ionization is negligibly small near the minimum ionizing energy. At very much higher energies, however, energies at which their *velocities* are comparable with those of effectively ionizing electrons, they may be even more effective than electrons. A well-known example is that of the alpha particles from radium (doubly charged positive helium ions with millions of electron volts energy) or ions accelerated to high energy by various types of "atom smashers." The probabilities of ionization by ions as a function of their kinetic energy follow curves very similar to those for electrons shown in Fig. A.11, but with tens of thousands of times the energy values for electrons. For example, alpha particles have a peak rate of ionization in air nearly 10 times that shown for electrons in mercury, but this peak occurs at an energy of 1.5 million electron volts. Ions may also release electrons from surfaces, a process that is of prime importance in glow discharges.

Thermal Ionization At high enough temperatures, such as those known to exist in electric arcs (6000 to 50,000 K), collision processes between neutral atoms and other atoms, ions, electrons, and photons, all in thermal equilibrium, can produce a total rate of ionization sufficient to balance the rate of loss of ions by recombination at some level of or degree of ionization which can be calculated by means of thermodynamic laws. The relation between the fraction f of atoms ionized and the pressure and temperature of the gas as derived by Saha [4] is

$$\frac{f^2}{1-f^2} P = 3.16(10^{-7})T^{5/2} \exp\left(\frac{-eV_i}{kT}\right) \quad (A.77)$$

where P is the pressure in atmospheres absolute, T the temperature in Kelvin, e the electronic charge, 1.6×10^{-19} C, V_i the ionization potential in volts, and k is Boltzmann's constant, 1.38×10^{-23} J/K. Figure A.12 shows f as a function of T at atmospheric pressure for the gases nitrogen, oxygen, hydrogen, mercury vapor, and copper vapor. The relations for silver and iron vapors (not shown) are very close to that for copper [5]. The Saha equation presumes thermal equilibrium, but in regions where it is applied, like the column of an electric arc, thermal capacities are so small compared to the rate of production and loss of heat that lack of equilibrium can last for only very short times.

Electron Emission In addition to processes within a gas, the introduction of charge carriers (in practical case electrons) from electrode surfaces is an essential feature of gaseous conduction. Principal proceses are:

1. *Photoelectric:* This has already been mentioned. It is important principally in spark breakdown, especially for short gaps with nearly uniform applied electric fields.

2. *Positive ion collisions:* Also mentioned earlier, this process is important in spark breakdown and in glow discharges. The rate of electron emission per ion increases linearly with the kinetic energy of the ions above 1000 eV and may exceed 20%. At low kinetic energies

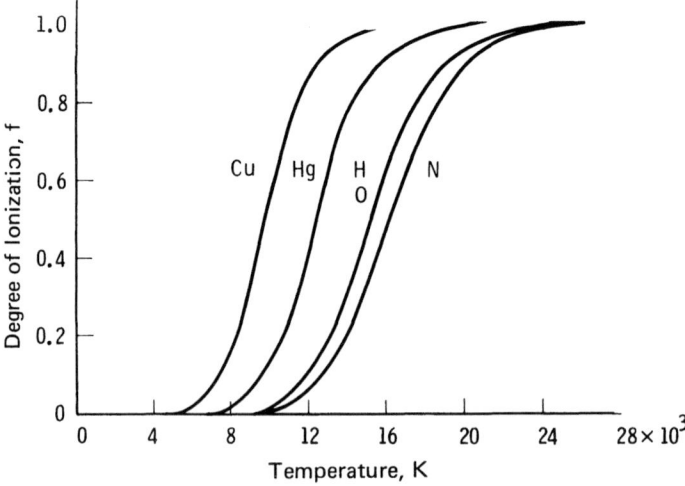

Figure A.12 Thermal ionization of some metal vapors and atomic gases at atmospheric pressure, from the Saha equation. (From Ref. 5.)

Appendix

the rate remains finite if the ionization energy of the ions exceeds twice the work function of the surface.

3. *Metastable atom collisions:* As mentioned in Sec. A.1, this process, important for glow discharges, may be appreciable when the excitation energy of the atom exceeds the surface work function.

4. *Thermionic:* When solid conductors are sufficiently heated, free electrons may escape if their kinetic energy normal to the surface is of the same order as the thermionic work function ϕ_0, close to the photoelectric work function. An equation for the emitted current density is

$$j = AT^2 \exp\left(\frac{-\phi_0 e}{kT}\right) \tag{A.78}$$

where T is the temperature, e the electronic charge, and k is Boltzmann's constant. The constant A is about 6×10^5 A/m^2 K^2 for most pure metals and ϕ_0 is 3 or 4 V. Since $\phi_0 e/k \cong 5 \times 10^4$ K, appreciable emission current densitites from pure metals require relatively high temperatures. As examples, at their melting points, calculated current densities are 4.6×10^6 A/m^2 for tungsten but only 8.4×10^{-2} A/m^2 for nickel. Thus only the most refractory metals or carbon are practical thermionic emitters in the pure state. Lower effective work functions characteristic of surfaces coated with alkali metals or their oxides permit operation of thermionic cathodes in vacuum at temperatures as low as 900 K.

5. *High electric fields:* Electrons can be pulled out of conducting surfaces by application of very high electric fields, generally of the order of 10^{10} V/m or more. The observed relation of current I to field strength E is of the form

$$I = aE^2 \exp\left(\frac{-D}{E}\right) \tag{A.79}$$

where a and D are constants for a particular case.

In the presence of very high fields and high temperatures in combination, these two effects aid each other. Useful approximate formulas [6] may be used in two cases:

a. For moderate field and high temperature,

$$j = j_T \left[\exp\left(\frac{e^{3/2} E^{1/2}}{kT}\right)\right] \frac{\pi d}{\sin \pi d} \tag{A.80}$$

where

j_T = thermionic expression (A.78)
e = electronic charge
d = $E^{3/4}/\pi kT$

At small values of d, the field-enhancing ratio $\pi d/\sin \pi d$ becomes unity.

b. For high field and moderate temperature,

$$j = \frac{j_0(\pi kT/c)}{\sin(\pi kT/c)} \tag{A.81}$$

where

j_0 = current density due to field alone, as equation (A.78)
c = $heE/6\pi m\phi^{1/2}$
h = Planck's constant
m = electronic mass
ϕ = surface work function

Such combinations may be important for maintaining the cathode of a metal vapor (vacuum environment) arc.

6. *Thermal explosion:* Recent studies of "vacuum" arcs suggest the possibility of a highly dynamic cathode process in which extreme momentary concentrations of power (current density times voltage gradient) explode away bits of the cathode surface material which are highly ionized and so make high electron current densities possible.

A.3.3 Deionization

Since the conductivity of a gas depends on the presence of electrons and ions, loss of conductivity by a gas and return to its normally insulating state can occur only as a result of disappearance of electrons and ions, or deionization. Thus deionization is fundamental to the arc quenching process in a circuit interrupter. Of particular importance in most cases is loss of the *free electrons,* which, because of their high mobility, carry most of the current and so account for most of the conductivity.

Electron-Ion Recombination Direct recombination between electrons and atomic positive ions is almost impossible as an isolated event because of the large energy difference eV_i between the ionized state and the normal neutral state of an atom. For this ionization energy to be lost by the electron-ion system it must reappear in other forms, such as radiation or kinetic energy of the resulting neutral particles. Loss by radiation alone is possible but relatively improbable, and gain of so large an amount of kinetic energy by the resulting single neutral atom is impossible because of the necessity also for conservation of momentum. Thus, although electrons are attracted toward neighboring positive ions, they are, in general, merely deflected in open curved orbits without recombining. In the case of *molecular* positive ions containing two or more atoms, the ionization energy *can* be carried away without creation of net momentum if the resulting particles have equal momenta but in opposite directions. In dense gases electronic recombination can also occur by three-body collisions, simultaneous collisions between electrons, positive ions, and neutral gas molecules, again yielding neutral particles moving in opposite directions.

Appendix

Attachment In electronegative gases such as oxygen and the halogens and most halogen-compound gases, electrons may become attached to neutral molecules to form negative ions. In some highly electronegative gases, of which sulfur hexafluoride (SF_6) is an outstanding example, low-velocity electrons have nearly 100% probability of becoming attached to molecules with which they collide. Even though the free electrons in a gas may have a nearly maxwellian velocity distribution corresponding to thousands of degrees temperature, there will always be a finite proportion (see Fig. A.3) with near-zero momentary velocities. A slight departure from maxwellian distribution because of inelastic collisions enhances the proportion of near-zero velocities. These facts, combined with the high rate $\bar{c}/\lambda_e$ at which electrons in thermal motion collide with the molecules of even moderately dense gases, show that free electrons may be expected to disappear extremely rapidly in a gas like SF_6. Although an ionized gas with no free electrons is still slightly conducting, it is able to support high voltage gradients with relatively negligible "leakage" currents and so may often appear to have returned almost completely to its normal insulating state.

Ion-Ion Recombination For complete deionization, of course, there must be complete neutralization by "recombination" of the negative ions with the positive ions, yielding neutral atoms or molecules. The rate of ion recombination depends on the rate at which encounters occur between positive and negative ions, which in turn depends on the product of their respective concentrations, or

$$-\frac{dn+}{dt} = -\frac{dn-}{dt} = \alpha_r n_+ n_- \tag{A.82}$$

where α_r is the coefficient of recombination. In most important cases positive and negative charge densities are nearly equal, so

$$\frac{dn}{dt} \cong -\alpha_r n^2 \tag{A.83}$$

where n represents charged particle density of either sign. When integrated, this equation leads to

$$\frac{1}{n} = \frac{1}{n_0} + \alpha_r t \tag{A.84}$$

or for $n \ll n_0$,

$$n \cong \frac{1}{\alpha_r t} \tag{A.85}$$

or transient ion densities vary inversely with time when determined by ionic recombination alone, following the production by some means of an initial relatively large density. Such a situation may exist in flames

resulting from momentary electric arcs. The coefficient of recombination α_r has been calculated by means of kinetic theory, considering both thermal motions and electrostatic attractions of ions for each other. One calculation by J. J. Thomson [7], neglecting ion mobilities, has been found to predict closely actual measured values at atmospheric pressure and below, while another calculation by Langevin [8], considering the effect of limited ion mobilities, gives correct values at several atmospheres pressure. Figure A.13 shows a curve of values obtained by Loeb [9] by combining these two theories so as to approximate experimental values over a wide range of pressures. At atmospheric pressure and below it may be seen that the coefficient increases rapidly with pressure while at several atmospheres and above it falls off with increasing pressure, as the ion mobilities also do. According to Thomson's theory, and in accord with experiment, increasing temperature rapidly reduces the rate of recombination.
At atmospheric pressure and fairly high temperatures the recombination coefficient varies approximately as the -7/2 power of the absolute temperature. Thus cooling an ionized gas has a strong *indirect* deionizing effect by speeding up ion recombination. Where electron attachment is a primary limiting process, cooling may have a still stronger effect by promoting electron attachment also.

In general, a third body of some sort is needed to bring about recombination between an electron and an atomic positive ion. In addition to other neutral atoms or molecules, dust particles or walls of containers may serve as the needed third bodies. Unfortunately, quanti-

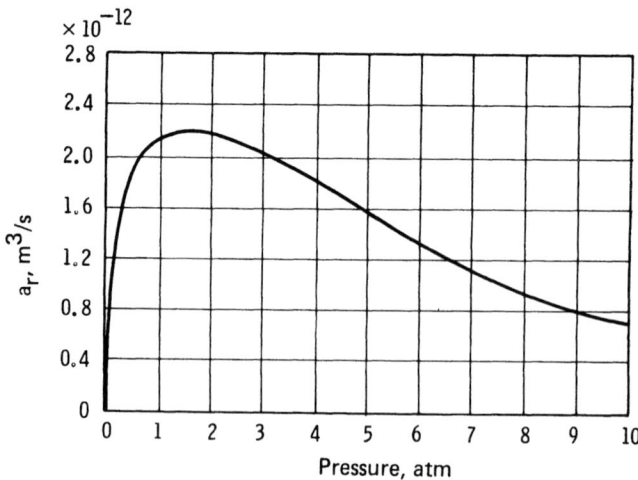

Figure A.13 Ion recombination coefficient vs. pressure in air. (From Ref. 9.)

Appendix

tative values of recombination coefficients are not generally available for accurate calculation of recombination rates in most practical cases, especially at the high temperatures existing in electric arcs and arc residues.

Diffusion In proximity to container walls or cooler gas regions where ions disappear rapidly by recombination, ion density gradients will generally exist. Across these gradients ions will be transferred by diffusion, following the mathematical laws for diffusion of a gas impurity or of heat. This may be the most important mechanism of ion loss from an ionized region in many cases, particularly at relatively low ion density levels where recombination may be negligible.

Diffusion loss varies in an inverse manner with the size of an enclosed space. For example, it will vary inversely with the diameter of a columnar gas region such as the heated and partly ionized path of a recently extinguished arc or corona streamer. In addition to the density gradient, diffusion loss is controlled by the diffusion coefficient, $D = \lambda \bar{c}/3$, which was shown in Sec. A.1 to vary inversely with the gas pressure (or density) and directly as the 3/2 power of the absolute temperature, or

$$D \sim \frac{T^{3/2}}{p}$$

Ambipolar Diffusion In an ionized gas containing nearly equal densities of electrons and positive ions, often called a "plasma," the diffusion coefficient for the electrons is always very much greater than that for the ions because of the greater free path λ_e and much greater mean velocity c of the lighter electrons. Thus bounding walls tend to receive more electrons than positive ions by diffusion and so become negatively charged. The effect of this charge is to establish a retarding electric field which holds back the electrons and slightly accelerates diffusion of the ions so that the resulting loss rates are equal. The net rate of loss of ionization by diffusion is therefore controlled by the diffusion rate of the heavier positive ions.

Electric Field Since charged particles of any kind tend to move in an electric field, ions or electrons are rapidly swept out of any space to which an electric field is applied. Thus the electric field may be both an ionizing and a deionizing agent. In the case of highly ionized plasmas, it was shown in Sec. A.2.4 that because of space-charge effects, the extent of an applied field may be limited to a small region near the negative electrode, and hence its deionizing effect may be correspondingly limited. With weak ionization, as in the initial stages of gas breakdown, the electric field may be effective in deionizing all of the space to which it is applied. The rate of sweeping away of the ions will depend, of course, on the product of the mobility (Sec. A.2.2) of the affected ions by the strength of the field.

A.3.4 Summary of Deionizing Influences

From the viewpoint of circuit interruption our principal interest is in promoting rapid loss of electrical conductance of gas spaces such as remnant arc paths or of flames from arcs. This means primarily promoting deionization of such regions.

Cooling When, as is most often the case, the region to be deionized is at a high temperature and initially highly ionized, thermal ionization according to equation (A.77) may be assumed and so deionization becomes synonymous with cooling. Even in the absence of thermal equilibrium or of the quasi-steady state to which equation (A.77) applies, cooling promotes deionization by slowing down ionizing processes and speeding up recombination. Furthermore, measures that aid cooling also generally aid the loss of ions by diffusion since diffusion of ions and diffusion of heat follow similar laws.

Constriction and Turbulence Such measures include constriction of the space (keeping distances to cooling and deionizing surfaces small) and enhancement of gas turbulence, which has the effect of speeding diffusion away of heat and ionization by a "stirring" action.

Compression and Expansion Other influences that may be used to aid deionization are pressure changes: either compression, which tends to increase attachment and recombination rates, or expansion, which increases diffusion rates and lowers gas temperatures and has been said to promote condensation of vapors on ions to immobilize them.

Gas Blast A much used technique for aiding deionization of arc paths in circuit breakers and power fuses is the application of "gas blasts" to the regions between the arcing contacts. At this point it may be well to mention what a gas flow or blast does and what it does not do. At the lowest level it is clear that the applied flow only augments natural convection, which is a very important normal mechanism of heat and ion loss in "plain-break" arcs. With intense blasts, those in which high-pressure gradients and sonic or supersonic velocities occur, the assumption is sometimes made that there is a radically different action: the physical cutting or shearing of the arc path with high-velocity separation of the arc plasma from a metallic terminal, or breaking-up of the plasma into separate parts. However, a detailed analysis of such a flow region always shows that no such cutting is actually possible in continuous fluid flow. Rather, there will generally exist some "dead water" regions which are shielded from the flow by contacts or other structural parts, and other regions where a combination of transverse and longitudinal fluid accelerations causes a stretching and thinning of the arc path (like the stretching of a rubber band) but where complete "pinching off" of the path cannot occur by simple fluid flow [10]. In addition to this stretching action, high rates of fluid acceleration in the regions of high density gradient in and near arc channels create in-

Appendix

tense turbulence which, as it builds up along the lines of flow, serves greatly to enhance normal diffusive loss processes.

Use of Favorable Gas Media Another practical aid to deionization is the employment of an atmosphere which by its nature is especially favorable for deionization. Various common gases show wide variation in the properties that determine deionization or ionization rates. Two outstanding examples, which are favorable in somewhat different ways, are hydrogen and sulfur hexafluoride. The use of these gases in circuit breakers is discussed further in Chaps. 8 through 10, but the reasons why they are favorable will be stated briefly here.

Hydrogen, which forms most of the arc atmosphere in oil circuit breakers, is outstanding among the simpler gases because of its low molecular weight and consequent high rate of diffusion, which promotes both simple cooling and diffusive loss of ions. In addition, it is a molecular gas—having two atoms per molecule—which further promotes cooling at and near arc temperatures. Ionizing processes leading to electric breakdown of this gas are also hindered by its molecular nature in comparison, for example, with the also light but monatomic gas, helium.

The most favorable practical gas medium so far discovered for aiding deionization of arcs in circuit breakers is the highly electronegative compound sulfur hexafluoride (SF_6). In spite of its relatively high density, which retards diffusive processes and diminishes convection and blast velocities, the molecular properties of this gas and of its decomposition products result in a highly favorable current and energy distribution in low-current arcs in the gas. As a result of this distribution, the arc column conductance follows instantaneous arc current values very closely (the arc has a relatively small "time constant," as discussed elsewhere), so that conditions are especially favorable for the quenching of an ac arc in SF_6 at or near a normal current zero. One part of this arc-quenching effectiveness is believed to reside in the outstanding electronegative, or electron trapping, property of the gas, which speeds up the disappearance of free electrons and hence of gas conductivity during the later stages of dielectric recovery when high values of voltage gradient must be withstood by the still hot arc residue.

A.4 SELF-MAINTAINED DISCHARGES

A detailed discussion of the "discharge" processes is beyond the scope of this appendix, but a general property is the existence of a near balance between combinations of ionizing and deionizing influences. Usual forms may be very briefly described as follows.

Glow A discharge form in which ionization is almost completely by electron or ion collision. In the body or column of the glow, electrons accelerated by the electric field are the chief ionizing agents, with deionization at low gas pressures mainly by lateral ambipolar diffusion and recombination at walls or other boundaries. At the cathode, an essential feature, electron emission is caused principally by positive ion bombardment of the surface. Glow discharges operate normally at relatively low current densities and gas temperatures. Their minimum maintenance voltages are a few hundred, concentrated near the cathode.

Arc (see Chap. 4) Here the electron emission process at the cathode is more efficient than in the case of the glow, requiring only a few tens of volts potential fall in a very thin region next to the cathode surface. With high-boiling-temperature cathode materials such as carbon or tungsten, thermionic emission can supply most of the electrons. With low-boiling metals such as copper, the emission process is not thoroughly understood; but since extreme values of current density and voltage gradient are required, thermal and high-field phenomena must be involved.

The column of the arc may in some cases be identical with that of the glow, but it generally takes a high-temperature and high-current-density form consisting of a plasma maintained by thermal ionization. Loss of ionization is primarily by radiative and diffusive loss of heat, aided by diffusion of ions and excited atoms carrying potential as well as thermal energy.

Spark A spark is a transitory phenomenon present during the transition of an electrically stressed but essentially insulating gas to a conducting state, which is usually that of an arc. It results from a runaway increase of very low-level pre-breakdown conduction processes dominated by ionization by collision. In the pre-breakdown phenomena, ionization by radiation may also be important, and deionization is mainly by the electric field and often by electron attachment, leading to enhanced recombination. Electric field distortion by space-charge formation may be an important part of the breakdown process, particularly in long gaps (lightning is a spark discharge across very long gaps). Spark breakdown related to circuit breakers is dealt with in Chap. 8.

Corona This is a partial breakdown, usually from sharp points or wires, in which the initially very nonuniform electric stress near a conductor is reduced rather than increased by space-charge formation, so that complete bridging of a gap by a conducting region is prevented. It usually precedes spark breakdown in very non-uniform-field gaps.

References A brief summary of electric discharges, based on early theory, is given by Ref. 11, and an up-to-date coverage of arcs may be found in Ref. 12 as well as Chap. 4.

Appendix

REFERENCES

1. F. A. Maxfield and R. R. Benedict, *Theory of Gaseous Conduction and Electronics*, McGraw-Hill, New York, 1941.
2. H. B. Wahlin, The motion of electrons in nitrogen, Phys. Rev. *231:*169-177, 1924.
3. W. Bleakney, Probability and critical potentials for the formation of multiply charged ions in Hg vapor by electron impact, Phys. Rev. *35:*139-148, 1930.
4. M. N. Saha, Ionization in the solar chromosphere, Philos. Mag. *40:* 472, 1920.
5. Knoll, Ollendorff, and Rompe, *Gasentladungs-Tabellen*, Springer-Verlag, Berlin, 1935, Fig. 98.
6. E. L. Murphy and R. H. Good, Jr., Thermionic emission, field emission, and the transition region, Phys. Rev. *102:*1464-1473, June 15, 1956.
7. J. J. Thomson and G. P. Thomson, *Conduction of Electricity Through Gases*, 3rd ed., Vol. 1, Cambridge University Press, Cambridge, pp. 20-84.
8. P. Langevin, Ann. Chim. Phys. *28:*287, 433, 1903.
9. L. B. Loeb, Theory of recombination of ions over an extended pressure range, Phys. Rev. *41:*1110-1111, 1937.
10. J. Slepian, Displacement and diffusion in fluid-flow arc extinction, AIEE Trans. *60:*162-167, 1941.
11. J. Slepian and R. C. Mason, Electrical discharges in gases-III, Electr. Eng., April 1934, pp. 511-518.
12. G. R. Jones and M. T. C. Fang, The physics of high-power arcs, in *Reports on Progress in Physics 1980*, Vol. 43, The Institute of Physics, London, pp. 1415-1465.

GENERAL REFERENCES

J. Slepian, *Conduction of Electricity in Gases*, Westinghouse Technical Night School Press, Westinghouse Electric Corporation, Pittsburgh, Pa., 1933.

J. D. Cobine, *Gaseous Conductors*, McGraw-Hill, New York, 1941.

A. von Engel and M. Steenbeck, *Elektrische Gasentladungen*, Vol. 1, Julius Springer, Berlin, 1934.

A. B. Cambel, *Plasma Physics and Magnetofluid mechanics*, McGraw-Hill, New York, 1963, Chaps. 5-7.

Index

A

Ablation, 166-168
  by the arc, 428
Adiabatic
  compression, 422
  heating of a gas, 421
Air
  blast, 2, 235, 246, 248
  breakers, 153
  compressed, 2, 4-6
Anode spot, vacuum arc, 461-465, 479
Arc
  air, 529
  ac voltage, Cassie model, 213
  chamber, 534
  chute, 426-431, 548, 549
  circuit interaction, 187-239
  cold-cathode, 432
  column, conductivity, 27
  conductance and power loss, 212
    after current zero, 442-447
  cooling, by diffusion to walls, 439
  discharge, definition of, 682
  effect on puffer pressure rise, 390
  energy, 534

[Arc]
  total, 553
  vs. time in puffer, 399
  gap, dielectric recovery in air and SF_6, 361
  horns, 430
  instability, 21
  interruption
    criterion, 21
    study of in SF_6, 355-359
  model, 158
    application to short-line fault, 224-228
  Cassie, 164, 173, 189, 190
  channel, 162, 165
  at high current, 159
  Mayr, 174, 188
  parameters
    Cassie voltage, 233
    determination of, 233
    time constant, 233
  Prince's wedge, 173
  relation to current rates of change, 223
  Swanson's, 179-181
  transient, at low current, 169
  modeling, 136

[Arc]
  nature of, 135-154
  peak recovery voltages, 216, 217
  physical description of, 136
    anode, 142
    cathode, 140
    column, 139, 143
    plasma conservation equations, 145-150
  quenching
    by gas flow, 4
    by hydrogen, 3
    methods, 425
  runners, 543, 545, 557
  residual conductance at current zero, 439, 441
  short, with metal electrodes, 433
    dielectric recovery by, 434-437
  stretching action of insulating plates, 551
  shunted by
    capacitance, 199-205, 217, 232, 233
    combined R and C, 202-203, 206
  temperature
    as function of axial distance, 179
    distribution after current zero, 26
    vs. radius, 26
  thermal reignition
    time constant, 153
    zone of, 400
  "time constant" of, 18, 439
    in air and SF_6, 360
  turbulent picture of, 177
  vacuum
    anode spot, 461-465, 479
    breakers, 458, 472-474
    cathode spot, 456, 459, 460, 463, 465, 470, 472
      erosion rates of, 460-462
      ion flux from, 460, 461

[Arc]
    465, 467
    multiplicity of, 459
    retrograde motion of, 456
  columnar, 464, 465, 479
    modes of, 464, 470, 479
    voltage, rapid rise by magnetic field, 480
  contact materials, 458-460, 462, 470, 472, 477
  contactors, 458, 463, 470, 472
  current chop, 470, 489, 492, 493, 501
    virtual, 489, 496, 497
  current zero, 459, 470, 474, 479
  delayed restrikes of, 476, 477
  dielectric recovery of, 462, 472, 474
  diffuse interelectrode plasma, 463-465, 467, 469, 470, 479, 480
  high current, 464, 465, 467, 479
  in interrupters during an ac wave, 462-472
  initiation, 462, 463
  interrupter, 455-480
  interruption in, 455-480
  magnetic constriction of, 462-465, 471
  prestriking of, 490
Arc chute
  metal plates, 433
  plate surface conductance 447
  zircon ceramic plate, 429, 447
Arcing
  contacts, 426, 564
    oxidation of following, 578
  maximum time of, 403
  medium, 549
  minimum time of, 395, 403
Arrester, surge, zinc oxide (ZnO), 520

Index

Asbestos lumber, 452
Atom
  excited states of, 670
  metastable, 670
Atomic structure, 669
Avagadro's hypothesis, 646
  number, 646

B

Barkan, Philip, 582, 586, 588
Basic impulse level (BIL), withstand capability, 488
Berkey, William E., 456
Bohr, Niels, 659
  theory, 669
Boltzmann's, 649
  equation, 259
  gas constant, 646, 674
  relation, 668
Boric acid fuse, 3
Boyle's law, 646
Brass electrode gaps, dielectric recovery by, 433
Braunovic, M., 572
Breakdown
  high frequency probability of, 503
  normal frequency, cumulative probability of, 502
  statistical, 495, 505
    log-normal, 495
    nature in vacuum interrupter, 495
Breaks in series, operating characteristics of, 400
Bridge, molten metal, 463, 577
Browne, Thomas E., Jr., 1, 187, 355, 425, 441, 452, 641
Bushings, 2, 3, 337
  designs, 341
    cast epoxy, 346
    condenser, 342
    gas insulated(SF_6), 347
    oil-impregnated paper, 342
    resin-bonded paper, 345

[Bushings]
  external insulation strength of, 338

C

California Institute of Technology, 456
Calvino, Ben J., 377
Capacitance
  channel, 244
  shunt, 187, 192, 198, 199, 202, 205-208, 217, 224, 226, 228, 230, 232, 234
Capacitor, surge, 491
  effect of, 517
Carrol, J. T., 452
Cassie, A. Morris, 159, 228, 443
  arc model, 163, 164, 167, 173, 193, 204, 209, 210-213, 220, 236, 237
  arc voltage, 236
  criterion, 228, 229
  equation, 174, 178, 187, 189-191, 193, 194, 209, 445
Cassie-Mayr-Cassie model, 187, 191, 224
Cathode spot, 456, 459, 460, 463, 465, 470, 472
Channel
  core, 251
  model, 163, 165
Charge-to-mass ratio, 658
Circuit
  capacitive, 35, 192
  distribution system, 530
  effective impedance, 228
  electrical, 5
  electric power, 1
  inductive, 35, 217
  interaction of arc with, 6, 187-238
  rate of rise of recovery voltage (RRRV), 2, 6, 197, 198, 209, 220, 221, 225, 234
  recovery, 6, 199

[Circuit]
  reignition, 633
  resistive, 34
  short-line fault, 187, 192, 193, 199, 206, 218, 224-230, 234-236
  synthetic test, 243
  Weil, for testing, 629
Circuit breaker
  air-blast, 235, 275
  air-break, 3
  air deion, 3
  arc, 4
  arc voltage, 438
    effect of on dc interruption, 437
  configurations, 408
  current-limiting, 547, 555
    peak let-through current, 547, 555-557
    slot motor for, 558, 559
  dead-tank, 4, 123, 413, 628
  direct-current, 4
  early, 606
  extra-high-voltage (EHV), 365, 379
  independent-pole operation (IPO), 125
  functions
    circuit interruption, low-voltage, 534, 535
    conduction, 532
    operation, 534
  gas-blast, 144, 218, 224
  gas-insulated-substation (GIS), 412
  high-power
    high-voltage, 362
    medium-voltage, 365
  insulating-arc-chute, 429
  insulating oil in, 354
  interrupting medium in, 377
  live-tank, 3, 4, 123, 410
  low-voltage, 433, 527-560
    frame sizes, 531
    molded case (MCCB), 527-560

[Circuit breaker]
  high-interruption capacity, 543, 544, 561
  magnetic-air, 425-452
  mechanical operation of, 605-619
    factors affecting, 606
  mechanism, 542, 605-619
    stored energy, 543, 561
    toggle-type, 542
    trip-free, 543
  minimum-oil, 379
  molded-case, 144, 527-560
  oil, 2, 3
  open, capacitive resonance, 129, 130
  performance, 195
  piston "puffers," 377-422
  pole disagreement
    circuit, 127
    relay, 128
  puffer, 5, 187, 224, 235, 236, 365, 369, 377-422
  single-pressure SF_6, 4, 201, 369, 377-422
  single-step resistor for, 59
  standards, 40, 205
  sulfur-hexafluoride (SF_6), 5, 197, 201, 210, 218, 227, 249, 250, 353-372, 377-422
  technology, 378
  tests, 209, 234, 621-638
  trip system for MCCB, 537
    bimetal, 537
    magnetic, 539
      time-current characteristic, 539
    thermal-magnetic, 539
      flux transfer, 541
      semiconducting devices, 539
    solid-state, 540, 561
  1200-kV, 370, 372
  two-pressure, 4
  type-U deion, 437
  vacuum, 5, 153, 455-479,

Index

[Circuit breaker]
488-523
Clogging, 158, 165, 167
  total at the nozzle, 385
Coatings, antitracking, 369
Colclaser, R. Gerald, Jr., 11
Compression chamber
  diameter, 386
  dimensions of, 385
  ratio, 419
    effect of dead volume, 388
Computer
  program, 499, 500, 512
    input parameters for, 505
    Monte Carlo routine, 499
    validation by tests, 510
Condensation shield, metal vapor, 457-459, 472, 474
Conducting parts, maintenance of, 379
Conduction, thermal, 152
Conductivity
  electrical, 151, 188-190
  thermal, 151
Correlation, of computer and experiment
  single-phase, 510
  three-phase, 512
Conservation
  of charge, 12
  of energy, 643
  equations, 151
  of momentum, 643
Contactors, vacuum, 458, 463, 470, 472
Contacts
  area of, 566, 568
  boiling voltage, 577, 580
  breaker, 2
  bus-bars, 566
  butt, 585
  current density at, 569
  current distribution in parallel, 545, 589
  disconnect switch, 566
  force
    blow-apart, 582

[Contacts]
    minimum spring force required, 582
    holding, 582
  fretting, 571
  knife-blade, 588
  material classes for, 592
    air, 592
    aluminum, 572, 594
    copper, 594
    -bismuth, 601
    -chromium, 602
    low-voltage breakers, 544
      arcing, 545
      main, 545
      rapid motion, 555
    molybdates, 600
    oil, 592
    silver, 595
      -cadmium-oxide, 545, 596
      -graphite, 545, 596
      -metal-oxide, 545, 598
      power-applications, for, 572, 594
      -refractory, 545, 579, 599, 601
      vacuum interrupters, for, 601
  melting
    current, 576
    voltage, 575
  molten-metal-bridge, 577
  oxidation, 571, 578
  parting speed, 391
  phenomena, 565
  potential drop, 569
  resistance, 5, 566, 568
    constriction, 568
  softening
    current, 576
    voltage, 575, 576
  temperature, 574
    calculation, 574
  tulip structure for, 586
  voltage-temperature relation, 574
  vacuum interrupter, 586, 593, 601

[Contacts]
  welds, 5
Control, of switching surge, 57
Cookson, Alan H., 275
Cooling, by diffusion to walls, 439
Corcoran, R. P., 638
Corona, 642
Cowley, M. D., 176
Criterion
  for arc interruption, 21
  Cassie, 228
Current, 211, 224
  arc, 190, 195-197, 200, 215, 218, 219, 224
  asymmetrical, 16
  bus-fault, 227, 230, 232
  capacitance, 219
  chopping by vacuum interrupter, 470, 489, 491-493, 508, 509
    cumulative probability of, 501
    median level of, 521
    virtual, 489, 496
  closing and latching, 82
  commutation, 545
  continuous rating, 62
  critical
    for interruption, 226-231, 233-235
    line, 231
  fault level, 535
  high, 190
  high-frequency transient, 472
  impressed, 187, 209
  initial, 15
  injection, 25, 631
    high-frequency, 636
  inrush, 116-120
  interrupted, 223, 224, 228
  interrupting
    limit, 224
    ratings, 65
  large, 189
  let-through, 547, 556, 560
    i^2 dt value, 555, 560

[Current]
  limiting, 23, 192, 217
    effect, 535
  line, 231, 232
  magnetizing, 492
  magnitude, 236
  momentary rating, 70, 82
  phase advance of zero, 438
  post-arc, 153, 197, 213-215, 218, 220
  pre-zero, 245
  ramp, 210, 213
  rate of change of, 187, 197, 218, 223
  reignition, 36
  residual, 243
  short-circuit, 193, 200, 227, 278, 642
    rated, 75
  short-line-fault, 194, 226, 229
  short-time load capability, 64
  sinusoidal, 210, 214
  symmetrical method of rating, 74
  threshold for contact repulsion, 558
  total, method of rating, 67
  transformers, 2
    ground current, 125
    overlapping protection of, 124
  zero, 4, 152, 153, 190-193, 210, 212, 213, 218-221, 228, 243
    by counterpulse, 479
    vacuum arc, 459, 470, 474 479
Current transformers, ring-type, 363

D

Daalder, J. E., 456
Damstra, G. C., 628
Data acquisition, 507
Davis, W. D., 456
Deion

Index

[Deion.]
 arc chutes, 153
 grids, 2
 plate, 548, 549, 554
Deionization, 6
 by attachment, 677
 by electron-ion recombination, 554, 676
 influences, 680
 by ion-ion recombination, 677
 process, 391
Demixing, of particle species, 257, 261
Design
 concepts, nozzle, 393
 dead-tank, 412
 live-tank, 409
 simplicity of, 379
Dielectric
 breakdown, 159, 171, 172, 191, 242, 257
 breakdown voltage, 252
  ac, 300
  front-of-cycle, 300
  impulse, 300
 failure, 31, 191
 properties, 359, 367
  of circuit breakers, 275
 recovery, 6, 7, 243, 244, 462, 472, 474
  pause, 244
  rated, 488
 reignitions, 397
 strength, 246, 249, 355
  of arc space, 6
  of gas, 4
Diffusion, 655, 679
 ambipolar, 667, 679
 coefficient, 655, 656, 668
 theory, 262
 transient, 656
Discharge, self-maintained, 681
 arc, 682
 corona, 682
 glow, 681
Displacement theory, 262
Double-flow, puffer design, 391

Duty factors, reclosing, 83
Dzierzbicki, S., 171

E

Eddy-current, effects on vacuum interrupter, 469
El-Akkari, F. R., 178
Electron
 diffusion, 142
 free, 643, 647, 658
 mass of, 658
 mobiliy of, 660-662
Elenbass-Heller equation, 151, 152
Eliason, B., 171
Emission
 electron, 141
 field, 141
 thermionic, 141
 thermionic and field (TF), 141
Energy
 balance, 7, 29, 141, 143, 368
 distribution of, 259
  nonmaxwellian electron, 257, 258
 loss mechanism, 241
 principle of equipartition of, 647
 transport, 252
Equation of state for a perfect gas, 149
Equilibrium
 chemical, 269
 chemical non-, 257, 259-261
 local thermodynamic (LTE), 136, 158, 159, 257, 258, 267-270
Erosion rates, of contacts, 460-462

F

Fahrnkopf, C. Donald, 621
Fang, M. T. C., 176
Faults
 line-to-ground, single-phase, 625

[Faults]
  low-current, 626
  short-line, 625
  switching, 626
Field
  distribution, 276
  plotting techniques, 276
    computer analysis, 276
    conducting paper analog, 276
    electrolytic tank, 276
    sketching, 276
  utilization factor, 277
Finley, James D., 425
Flashover
  along insulator surfaces in vacuum, 331
  mechanism of breakdown, 309
    tracking, 320
  of oil-solid surfaces, 309
  voltage, of wet surfaces, 320
Flux linkages, 12
Foreign particles, effect and control of, 367
Free paths
  distribution of, 654
  of electrons, 652
  mean, 652
  of molecules, 652
  most probable length, 655
Frequency
  high, for reignitions, 491
  normal load transient, 490
  power, 490
Friedrich, Robert E., 353
Frind, Gerhard, 176, 177, 439, 440, 450
Frost, Leslie S., 187, 267, 589
Fuses
  expulsion, 3
  power, 3

G

Gap, plain break in SF_6, 356
Gas
  Boltzmann's constant, 646

[Gas]
  compressed, breakdown, 278
  density, 646
  dryness standards, 369
  effect of, blast, 250
  electronegative property of, 681
  energy, 644
  mantel, 249
  perfect, law, 653
  pressure, 246, 644
  temperature, 644
Gaseous
  conduction, 642
  discharge, 642
  state, 643
Gay-Lussac, perfect gas law, 646
General Electric Company, 456
Glow, model, 158, 159, 169, 172

H

Hague, B., 588
Heat
  flux potential, 152
  -ing, Joule, 533
  penetration into plate, 448
  specific, ratio of, 648
Heberlein, Joachim V. R., 135, 467
Hertz, Walter, 181
Hermann, W., 166, 178
History, 1, 7
  of vacuum interrupter development, 456
Hochrainer, August, arc model, 187, 190
Hoyaux, Max F., 171
Hudis, M., 452
Hunziker, R., 440, 450
Hydrogen, 250
  arc atmosphere, 681
  atom, 643
    mass of, 669
  ion, 658
    mass of, 658

Index

[Hydrogen]
　product of arc in oil, 297

I

Impedance, surge, 491
Insulation, 3
　characteristics of circuit
　　breakers, 275
　coordination, 275
　flashover, 293
　liquids, 297
　　functions and requirements,
　　　297
　media in circuit breakers, 275
　properties of materials, 278
　puncture, 294
　short-time and impulse
　　strength of, 312
　solids, 310
　　epoxy
　　　aluminum-oxide-trihydrate
　　　　filler for, 322, 325
　　　breakdown voltage, effects
　　　　of mechanical stress, 315
　　　cast, voltage endurance
　　　　of, 316
　　　quartz-filled, 313
　　switching impulse levels, 61
　　thermal runaway breakdown
　　　of, 315
　　treeing breakdown of, 312
Interrupter
　puffer type, 355
　single-pressure SF_6, 377-422
　　development of, 383
　single-pressure type, 378
　two-pressure, 356
Interrupting
　capability coefficient, 398
　double-flow element, 381
　medium, 377
　single-flow element, 381
Interruption
　ac circuit, 6
　of all types of short circuits,
　　383

[Interruption]
　of asymmetrical currents, 402
　of capacitive currents, 362,
　　383
　chambers, self-blast, 391
　circuit, 1
　of dc circuit, 6
　dc, high-speed, 438
　of fault currents, 362
　of line charging currents, 358
　of magnetizing currents, 358,
　　362, 399
　of normal load currents, 358
　in phase opposition, 383
　power factor effect on, 534
　process, low-voltage, 549, 555
　types of tests, 624
　under short-line-fault condi-
　　tions, 368
　by vacuum switch, dc, 479
Ion, 657
　flux from cathode spot, 460,
　　461, 465, 467
　hydrogen 658
　-izing agents, 671
　nature and function of, 657
　negative, 643, 657
　positive, 643, 657
Ionization
　by chemical action, 672
　electron emission, 674
　　by high electric fields, 675
　　photoelectric, 674
　　thermionic, 675
　by excited atoms, 672
　by high-speed electrons, 672
　by high-speed ions, 673
　by metastable atoms, 675
　by positive ion collision, 674
　by radiation, 671
Ionizing processes, 669
Isentropic, 165, 167

J

Jennings Manufacturing Com-
　pany, 456

Jüttner, B., 456

K

Kelman, Joseph N., 2, 8
Kempen, H. W., 628
Kimblin, Clive W., 135, 455-457
Kinetic gas theory, 642, 643

L

Laboratories, high-power, 623
Lafferty, J. M., 456
Langevin, P.
  ion mobility calculation, 678
Laplace
  equation, 276
  transform, 13, 15, 32, 33
Leaks, detection methods, 367
Lee, Anthony, 135, 527
Liebermann, Richard W., 267
Lightning surge arrester, 5
Lincoln, Paul Martyn, 606
Line, transmission
  capacitance of, 199, 225, 233
  charging conditions, 391
  inductance, 199, 225, 229, 235
  length, 199, 205, 225-230, 232-235
    critical, 228
  short—fault, 187, 192, 193, 199, 203, 205, 206, 218, 224-230, 234-236
  surge impedance, 199, 224, 229, 625
Lingal, Harry J., 355
Linkage, operating
  determining required forces, 611
  location of springs, 611
  stiffness, 611
  weights and forces interdependent, 611
Liquefaction, danger, of, 388
Loeb, combined ion recombination theory, 678
Lowke, John J., 157, 178

Ludwig, Howard C., 178

M

Maecker, Heinz
  criterion, 258
  effect, 141, 144
Magnet
  "H"-construction, 432
  H-design, 429
  series "blowout" field coil for, 431
  U-shaped, 429
Magnetic
  blowout, 3
  constriction, 462-465, 471
  field, 2, 3, 4, 549
    axial, 462, 467-471, 479
    deflection by, 663
    $\bar{J} \times \bar{B}$ force, 547, 552, 555
    pumping, 249
    structure, 432
    transverse, 479
  flux phase shift, 432
  interaction between phases, 433
  pressure, 141, 143
Magrini SpA of Italy, 379
Maintenance, 5, 6, 399
  of conducting parts, 379
Mandelcorn, Lyon, 275
Mantel, gas, 251, 266
Mason, R. C., 456
Mass
  action laws, 144
  conservation of, 144
Material, ceramic, 440
Maxwell, Clerk, 649
  -Boltzmann velocity distribution, 651
  -ian distribution of energy, 242, 258, 263
  's electromagnetic theory, 669
Mayr, Otto, 159, 179, 202
  arc model, 188, 209, 238
    differential equation, 445, 446
    modified type equation,, 258

Index

Mechanics, newtonian laws of, 644
Mechanism, operating, 406
　circuit breaker, 6
　electromagnetic-repulsion, 609
　explosive-charge driven, 609
　functions, 619, 620
　gravity-driven, 609
　hand-closing, 607, 613, 617
　hand-tripping, 607, 613, 617
　high-speed opening, 438
　hydraulic, 6, 609
　improvements in, 607
　pneumatic, 608, 612, 615
　position, held by toggle and bias spring, 615
　range of operating voltages, 612
　rapid-reclosing, 608
　requirements by industry standards, 606
　size of, 385
　solenoid-driven, 607
　Sprecher and Schuh, 609
　spring-driven, 609
　trip-free feature, 607
　with latches, 610
　without latches, 610
Miller, H. C., 456
Mineral oil
　composition of, 298
　gas generation by arcs in, 299
　properties of, 298
　technical grade, electric strength of, 299
　　effect of area, 302
　　effect of carbon contamination, 303
　　effect of gap length, 302
　　effect of moisture, 303
　　effect of stressed volume, 302
Mobility, 659
　in the electric field, 658
　electron, 661

[Mobility]
　units, 659
Modularity, conception of, 379
Molecular
　diameter, 653
　kinetic energy, 647
　speed, 646
　velocities, 647
　　components, 650
　　distribution of, 648
　　distribution function, 649, 650

N

Newtonian equation, 658
Nitrogen, 179
　arc in, 26, 162
Noise level, reduced, 379
Norton equivalent circuit, 32, 33
Nozzle
　air-blast, 2
　clogging, 247
　cylindrical, 248
　convergence, 247
　design concepts, 393
　divergence angle, 248
　double, 250, 253
　entrance, 257
　exit, 268
　geometry, 242, 247, 270
　interrupting, 4
　laval, 263
　length, 247
　pressure, 268
　throat, 247, 257

O

Ohm's law, 145, 149, 151
Oil
　arc-quenching, 2
　blasts, 4
　circuit breakers, 2, 3, 7
　mineral, 2
　tank-type breakers, 2
Operations, mechanical, 405

P

Particle "traps", 281
Paschen
   curve, 252
   's law, failure of, 278
Perkins, John F., 487
Photoelectric effect, 671
Photon, 669
Piston
   diameter, 384, 385
   length of stroke of, 384
   puffer, 437
Plank's constant, 669
Plasma, 664, 667
   charge neutrality, 149
   column diameter, 394
   conservation equations, 145-148
   diffuse interelectrode, 463-467, 469, 470, 479, 480
   equations of state, 145
   radiation transport in, 145, 148
Plyutto, A. A., 456
Poisson equation, 276, 664, 668
Postarc current, Mayr model of, 215
Postarc dielectric recovery, 241-271
Power, electric
   high-voltage, 354
   systems, 1
Prandtl mixing length, 261, 269
Pressure
   differential, 385, 386
      as a function of contact overlap, 391
      as a function of rated pressure, 388
      as a function of stroke length, 387
      as a function of stroke speed, 389
      at zero gauge pressure, 392

[Pressure]
   distribution, 242
   fill, 388
   gas, 230
      difference, 209, 384
   rise
      cold-flow, 419
      with gas outflow, 422
Prestriking of breaker, 490
Prince, D. C., 159, 173
Probability of working, 209, 211
Protection of open breaker, 54
Puffer, "super," 370

R

Radiation transport, 142, 143, 145, 271
   equation, 148
   power density, 152
Radiative power emission loss, 152
Ragaller, Klaus, 176-178
Rated
   interrupting time, 84
   on a symmetrical current basis, 205
Ratings, standard of for MCCB's, 531
Reclosing, high speed, 579
Recording, digital techniques, 638
Recombination, 170, 181
   coefficient of, 678
   electrons, 242
   heavy particles, 242
Recovery
   free, 244
   rate of rise of—voltage (RRRV), 241
Reece, M. P., 456
Reignitions
   multiple, 489, 494, 496
   of arc, 472, 476, 478
Reliability
   electrical, 404
   mechanical, 399, 404

Index

Research
  and development at Westinghouse Electric Corp.,
    1, 4, 354
  on interrelated arc interruption, 368
  on materials, 369
  program, 367, 377
Resistance
  of channel, 244
  shunt, 187, 192, 197, 198, 205, 206, 208, 214, 218, 224, 228
Resistor, closing, 410
  switching mechanism for Westinghouse breaker, 617
Restriking, delayed, 476-478
Retrograde motion of an arc in a magnetic field, 456
Reynolds number, 262, 265
Rieder, Werner, 180
Roach, J. Franklin, 275
Roidt, R. Michael, 178
Rutherford, Lord, 669

S

Saha equation, 144, 149, 188
Scaling, 231
Schade, E., 171
Sealing, gaskets, 379
Seals, methods and materials, 367
Seismic withstand capability, 415
Self-absorption, 162, 163, 165
Self-generation principle, 361
Shunt
  capacitor, 4, 187, 198, 199, 202, 203, 205-208, 217, 224, 226, 228, 230, 232, 235
  -ed arc, 193
  -ing circuit elements, 219
  -ing, combined capacitance and resistance, 187, 206

[Shunt]
  -ing conductance, 194, 196, 228
  -ing ratio, 194
  resistance, 2, 187, 192, 197, 198, 205, 208, 214, 218, 221, 224, 369
Skeats, Wilfred, F., 173, 629
Slade, Paul G., 455, 527, 565
Slameka, E., 637
Slepian, Joseph, 6, 8, 9, 159, 169, 171, 173, 176, 456, 548
Solid-state devices, 7
Sound, velocity of, 647
Space charge, 252, 256
  effects, 664
  equation, 667
  neutralization of, 658
  sheath, 664, 666
Spark
  breakdown mechanism of, 554
  definition of, 682
Spark gap
  breakdown
    area effect, 285
    effect of electrode geometry, 282
    effect of electrode surface finish, 287
    effect of insulating coatings, 288
    effect of particles, 288
      Westinghouse "tri-trap" design, 292
  conditioning effect, 280
  stress conditioning, 281
Sparkover, 642
Spindle, Harvey E., 639
Sprecher and Schuh, spring-operated mechanism, 406, 407
  motor-wound, 609
Stagnation point, 257
Standards
  American National Standards Institute (ANSI)
    C37 Power Switchgear, 41,

[Standards]
  205, 622
  C37.13-1981, 531
  International Electrotechnical
    Commission (IEC), 42,
    622
  Institute of Electrical and
    Electronics Engineers
    (IEEE), 141, 242, 529,
    531
Steel-tank breaker, 363
Strom, Albert P., 355, 432, 441,
  452
Sulfur hexafluoride (SF$_6$), 157,
  175, 179, 181, 681
  arc in, 26
  -blast, 248, 250, 256
  coaxial geometry breakdown
    in, 290
  early applications, 359
  electronegative property of,
    358
  -gas-blast, 253
  gas-blast-breaker, 3, 197,
    210, 227, 230, 235, 245,
    249, 252
  gaseous insulation, 277
  plasma, 260, 261
  properties of, 354
  puffer breakers, 2, 5, 157,
    612
  single-pressure breaker, 7,
    201, 377-423
  uniform-field breakdown in,
    286
Suppressor, surge, C-R, 520
Swanson, Bruce W., 177-180
Switch
  load-break, 3, 4, 360, 378
  vacuum, 4, 479
Switchgear
  metal-clad, 474
  types of, 488
Switching
  of cable circuits, 120
  capacitance, 105, 475-479
  interruption currents, 106, 107

[Switching]
  inrush currents, 106, 116
  opening resistors, 114, 115
  recovery voltage, 115
  transient overvoltages, 113
  transient recovery voltage,
    106, 107
  equipment, 354
  of motors, 517, 518
  pulse-power, 136
  of unloaded transformer, 514
Synchronous operation
  closing, 5
  opening, 5
  switching, 7

T

Tanberg, Ragner, 456
Tarocinski, Z., 171
Techniques, mathematical, 368
Teflon (polytetrafluorethylene)
  nozzle, 360, 369
Temperature, arc, 188-190
  decay, 257
    rate of, 259
  distribution, 136, 242
    radial, 18
  peak, 139
  profile, 266
Tensor notation, 148
Test
  bus-(or terminal) fault, 624
    single-phase, 624
    three-phase, 624
  field, 400
  high-current, 624
  line-charging current, 626
  load-current-switching, 626
  short-line-fault, 625
    circuit, 400
  shunt-reactor switching, 626
  switching
    capacitor bank, 626
    line-charging-current, 626
    load-current, 626

Index

[Test]
    out-of-phase, 627
    shunt-reactor, 626
    transformer-exciting-current, 626
    unit methods, 400
Testing
    accelerated, for breakdown, 317
    circuit, direct, 623, 627
    development program, 622
    facilities, interruption, 623
    generators, 623
    interruption, 621, 622
    instrumentation for, 637
    nonstandard, 622
    standards, 622
    stations, 623
    synthetic, 623, 629, 630
        Weil circuit for, 629
    two-part, 628
    unit, 628
Theory
    of arc-circuit interaction, 187-239, 368
    of dynamic nozzle arcs, 368
    of fluid flow, 368
    steady-state, 368
Thermal
    breakdown, 242
    conduction, 152
    conductivity, 462
    inertia of arc, 188
    ionization, 139, 188
    radial–conduction, 188
    region for arc interruption, 29, 191
    reignition, 191, 242
        zone, 400
    time constant, 153
Thermionic emission, 141
Thermodynamic, local–equilibrium (LTE), 136, 144, 158, 159
Thompson
    ion mobility calculation, 678
    theory, 678
Thuot, Michael E., 638

Time
    constant, 358
        thermal, 553
    interruption, reduction in rated, 606
    maximum arcing, 397, 403
    minimum arcing, 386, 395, 400, 403
    to release latch, etc., 607
Transient recovery voltage (TRV), 29, 242, 263
Tslaf, A., 449, 450, 451
Tuma, David T., 178, 179, 241
Turbulence, 159, 176, 180, 261
    effect of, 252
    free-shear, 257, 262
Turbulent
    eddies, 252
    energy transport, 267, 269
    flow, 252, 263, 270
    mixing, 246
    momentum transport, 269
    region, 257
    shear flow, 262, 269
    transport, 261
Two-pressure principle, 363

U

Ultraviolet, 162
Urbanek, Joseph, 180, 191

V

Vacuum
    arc-quenching ambient, 2
    arcs, 141
    breaker, 5
    circuit breaker, 7, 488
        current chopping by, 489
        voltage escalation by, 489
    gap breakdown, 325
        dependence on
            electrode geometry, 329
            gap area and time, 329, 330
        gap length, 325
    effect of

[Vacuum]
    electron emission, 324, 331
    insulating surface angle, 334
    magnetic field, 337
    surface material and texture, 336
    volt-time characteristic for short pulses (μs), 335
 high, conditions, 653
 interrupter, 4, 453
    ac applications for, 472-474
    dc applications for, 479
 interruption in, 455-480
 switch, 4
Vallo, David, 621, 638
Van Sickle, Roswell, C., 605
Voltage
 arc, 4, 17-24, 197, 200
    Cassie-arc-model, 193, 204, 212, 213, 220, 233
 basic impulse level (BIL), 48
 breakdown
    calculation of, 295
    time to, 290
    waveform effects on level, 290
 bus-side, 192, 205
 chopped-wave withstand level, 48
 corona
    onset, 282
    stabilizing effect of, 282
 escalation, 478
 full-wave withstand level, 48
 impressed, 209, 215
 lightning surge, 48
 low-frequency-test, 47
 maximum withstand, 397
 over--
    chop, 492
    maximum prospective, 492
    power-frequency, 47
    switching-surge, 55
 ramp, 216, 217, 220, 223, 225, 229
 rated

[Voltage]
    maximum, 46
    range factor K, 46
 recovery
    capacitor switching, 478
    ratio of rise (RRRV), 208, 214, 218, 221, 358
    critical, 214, 222, 225
 slope, 208
 transient, 626, 632, 634
    envelope, 635
    recovery (TRV), 85, 198, 199, 203, 205-207, 214, 217, 220, 223, 397, 400
    capacitor switching, 478, 625
    initial (ITRV), 104, 192, 234, 397, 625
    exponential-cosine(ex-cos) wave, 87
    four-parameter method, 88
    peak, 202, 206, 215, 216, 224, 490
    rate of rise, 490
    short-line-fault, 101
    single-phase circuits, 85
    standard envelope, 87
    surge, 470, 472, 475-477
    switching, 498
    two-parameter method, 88
    transformer-terminal-fault, 91
 withstand, 472, 474
Volume
 compressed, 419
 outflow ratio, 390
Voshall, Roy E., 455

W

Wagner, Charles L., 39
Wahlin, H. B., 660
Water, quenching medium for arc, 3
Westinghouse
 breaker mechanism

Index

[Westinghouse]
  DWE 250, 607
  MWE 250, 612
  242, 362 LWE and 590 LWER, 612
  pneumatic AA-7, 615
  resistor switching on LWER for 550 kV service, 617
  deion, Type U breaker, 427, 432, 433
  developments by, 378
  engineers, 354, 359, 377

[Westinghouse]
  licensee, 378
  research and development, 354, 367, 368
Wiedemann-Franz law, 574
Wootton, Roy E., 275
Wurts, A. J., 433

Y

Yos, J., 267
Yoshioka, Y., 591, 592

ISBN 0-8247-7177-X